CAMBRIDGE LIBRARY COLLECTION
Books of enduring scholarly value

Technology

The focus of this series is engineering, broadly construed. It covers technological innovation from a range of periods and cultures, but centres on the technological achievements of the industrial era in the West, particularly in the nineteenth century, as understood by their contemporaries. Infrastructure is one major focus, covering the building of railways and canals, bridges and tunnels, land drainage, the laying of submarine cables, and the construction of docks and lighthouses. Other key topics include developments in industrial and manufacturing fields such as mining technology, the production of iron and steel, the use of steam power, and chemical processes such as photography and textile dyes.

Principles of the Manufacture of Iron and Steel

Sir Isaac Lowthian Bell (1816–1904) was a leading metallurgist and industrialist who served as president of both the Iron and Steel Institute and the Institution of Mechanical Engineers. He combined business skills with scientific expertise and was recognised as a world authority on blast furnace technology. His major works reveal both technical know-how and commercial awareness, and show that he was conscious of the threat to Britain's early lead in industrialisation from foreign competition. He supported free trade, and believed that British industry needed a firm scientific base in order to maintain its global position. Although his posthumous reputation has been eclipsed by that of his contemporaries, he was highly respected in his lifetime, receiving a baronetcy in 1885 for his contribution to industry. This book was first published in 1884 and deals with the economics of iron production in Britain and abroad as well as the processes themselves.

Cambridge University Press has long been a pioneer in the reissuing of out-of-print titles from its own backlist, producing digital reprints of books that are still sought after by scholars and students but could not be reprinted economically using traditional technology. The Cambridge Library Collection extends this activity to a wider range of books which are still of importance to researchers and professionals, either for the source material they contain, or as landmarks in the history of their academic discipline.

Drawing from the world-renowned collections in the Cambridge University Library, and guided by the advice of experts in each subject area, Cambridge University Press is using state-of-the-art scanning machines in its own Printing House to capture the content of each book selected for inclusion. The files are processed to give a consistently clear, crisp image, and the books finished to the high quality standard for which the Press is recognised around the world. The latest print-on-demand technology ensures that the books will remain available indefinitely, and that orders for single or multiple copies can quickly be supplied.

The Cambridge Library Collection will bring back to life books of enduring scholarly value (including out-of-copyright works originally issued by other publishers) across a wide range of disciplines in the humanities and social sciences and in science and technology.

Principles of the Manufacture of Iron and Steel

With Some Notes on the Economic Conditions of their Production

ISAAC LOWTHIAN BELL

CAMBRIDGE UNIVERSITY PRESS

Cambridge, New York, Melbourne, Madrid, Cape Town, Singapore,
São Paolo, Delhi, Dubai, Tokyo, Mexico City

Published in the United States of America by Cambridge University Press, New York

www.cambridge.org
Information on this title: www.cambridge.org/9781108026949

© in this compilation Cambridge University Press 2011

This edition first published 1884
This digitally printed version 2011

ISBN 978-1-108-02694-9 Paperback

This book reproduces the text of the original edition. The content and language reflect
the beliefs, practices and terminology of their time, and have not been updated.

Cambridge University Press wishes to make clear that the book, unless originally published
by Cambridge, is not being republished by, in association or collaboration with, or
with the endorsement or approval of, the original publisher or its successors in title.

The original edition of this book contains a number of colour plates, which cannot
be printed cost-effectively in the current state of technology. The colour scans
will, however, be incorporated in the online version of this reissue, and in printed
copies when this becomes feasible while maintaining affordable prices.

Additional resources for this publication at www.cambridge.org/9781108026949

Principles of the Manufacture of Iron and Steel.

Principles

of

The Manufacture

of

Iron and Steel,

with some notes on the

Economic Conditions of their Production.

by

I. LOWTHIAN BELL, F.R.S.,

HON. MEMB. OF THE AMERICAN PHILOSOPHICAL SOCIETY, &c., &c., &c.

LONDON:
GEORGE ROUTLEDGE & SONS, BROADWAY, LUDGATE HILL.
(NEW YORK: 9, LAFAYETTE PLACE.)
E. & F. N. SPON, CHARING CROSS.

1884.

[*All rights reserved.*]

TO
HIS GRACE THE DUKE OF DEVONSHIRE,

K.G., F.R.S., &c., &c., &c.,

First President of the Iron and Steel Institute.

DEAR DUKE OF DEVONSHIRE,

Most of the labours, of which a partial record is to be found in this volume, originated with the foundation of the Iron and Steel Institute.

This circumstance renders it natural that I should desire to connect the name of the Iron and Steel Institute with a work, undertaken in the hope of its being useful to the members of that body.

I feel convinced, that I cannot do this in a manner more acceptable to the Institute at large, than by inscribing on this page the name of your Grace, to whom, as its first President, it owes so much.

I remain, dear Duke of Devonshire,

Very faithfully and respectfully yours,

I. LOWTHIAN BELL.

Rounton Grange, Northallerton,
   15th Sept., 1884.

PREFACE.

THE circumstances which led me to undertake the present work are described in the beginning of the introductory section. Not only have many of my leisure hours been occupied in searching for and arranging information contained in many volumes of notes, but time which ought perhaps to have been otherwise employed has been devoted to the duty I had promised to perform. Nevertheless more than four years have elapsed since I was invited to place this result of my labours in the hands of the Iron Trade Association and of the members of the Iron and Steel Institute. This apparent delay has been chiefly due to a wish to extend my enquiries on some of the questions treated of in the papers I had published in the Transactions of the Institute. This observation is particularly applicable to the use of charcoal and of raw coal in the blast furnace; and in furtherance of these objects, not only had additional experiments to be undertaken, but various furnaces in this as well as in foreign countries had to be visited and examined.

The concluding section, in which it is attempted to compare the iron-producing powers of different nations in their economical aspects, has involved a good deal of additional work. In so fluctuating a trade as that of iron the cost of labour is constantly changing. My enquiries on this head extend over twenty years; and as it was desirable to obtain as much information of a contemporaneous date as possible, much time, having this object in view, was spent in correspondence with manufacturers in Europe as well as in America. In treating this branch of the subject, the value as well as the cost of the raw materials employed had to be considered. To explain this, an occasional repetition of what may be regarded as purely scientific matter will be met with; but I preferred this course, in order to make my meaning plain, and to avoid trouble in referring to previous sections. To some extent perhaps, the same observation as that just made is more or less applicable to the sections more exclusively scientific. This arose from a wish to insert my latest experiments, some of which were continued after the first portion of the work had been printed.

By many the subject of the quantity of fuel consumed in the blast furnace may be considered as having been treated at sufficient length in Section VII. This portion of the enquiry into the smelting process is the only one which demands much attention, so far as economical production is concerned. I have therefore, for reasons stated at page 234 and for others connected with the use of charcoal, extended my observations on fuel in Section X., which may be read by those who feel sufficiently interested in questions there dealt with.

It will no doubt occur to some who may care to examine with the requisite attention what has been said on comparative costs, to take exception to some of my conclusions. Such friendly critics will no doubt remember the differences which are to be found in the experience of individual mining and manufacturing establishments. My object at the time of my enquiries was one of a purely private nature made without any intention of future publication. They were undertaken for my own guidance, and the opinions I have formed, correct or otherwise, are faithfully recorded in the pages of the present work.

It will be readily understood how largely I am indebted to my fellow workers in the great field of industrial enterprise which I have attempted to explore, for information respecting their practice and experience. I only regret that the confidential character of many of the communications I have received, prevents my cordially thanking, by name, those who have so materially assisted me in the course of my investigations. Other sources of information will be found acknowledged in the proper places.

Under ordinary circumstances I might have dispensed with any allusion to the aid rendered by experimental research; but this, in my own case, has been done in so careful and conscientious a way by Mr. Rocholl in the Clarence laboratory, that I cannot refrain from acknowledging his zeal in this respect, as well as recognizing the thought he bestowed on the examination of the numerous calculations rendered necessary by the nature of the subject.

PREFACE.

I would also express my obligations to Mr. Walter R. Browne, the late Secretary of the Institution of Mechanical Engineers, for his help as the sheets were passing through the press. My son-in-law, Mr. Walter Johnson, has also afforded valuable assistance in examining many of the statistical figures they contain.

I. LOWTHIAN BELL.

Rounton Grange, Northallerton,
15th Sept., 1884.

CONTENTS.

SECTION I.

INTRODUCTORY 1

Origin of work, 1. State of iron trade, 2. Foreign competition, 5.

SECTION II.

HISTORICAL 4

Metallic iron as found in native condition, 8. Disappearance of iron in ancient monuments, 8. Presence of iron in its ores easily detected, 9. Earliest modes of manufacture, 9. Gradual development of the blast furnace, 11. Application of fossil coal to smelting, 13. Introduction of puddling process, and the use of grooved rolls, 14. Increase in make and changes in prices of pig iron, 15. Introduction of hot blast, 16. Extended application of iron, 17. Discovery of the Bessemer process, 19. Siemens' furnace and Price's, 21. Increase of dimensions of blast furnaces in Cleveland, 22. Utilisation of escaping gases from blast furnaces, 23. Fire brick stoves for heating air, 24. Improvements in rolling mills, 25. Nasmyth hammer, 26. Past, present and future of the manufacture of iron and steel, 27.

SECTION III.

DIRECT PROCESSES FOR MAKING MALLEABLE IRON 30

Catalan furnace, 30. Loss of heat and of metal by use of Catalan furnace, 31. Partial removal of phosphorus in direct processes, 32. Clay's direct process, 33. Chenot's process, 33. Direct process of T. S. Blair, 34. Dupuy's direct process, 37. Siemens' direct process, 37. Ritter v. Tunner on direct process, 39.

CONTENTS.

SECTION IV.

PRELIMINARY TREATMENT OF MATERIALS FOR THE BLAST FURNACE ... 46

Manufacture of coke, 46. Mode of estimating heat afforded by combustion of fuel, 47. Waste in coking process, 49. Utilisation of waste heat from coke-ovens, 50. Recovery of the products of distillation in the coking process, 52. Open hearths for coking coal, 53. Conversion of wood into charcoal, 53. Calcination of ironstone, 55. Calcination of limestone, 57.

SECTION V.

THE BLAST FURNACE 61

Manufacture of cast iron, 61. Economical action of, 61. Chemical action of, 62. Combustion of carbon in, compared with that in Catalan fire, 63. Reduction of oxide of iron in blast furnace, 65. Oxidation of metallic iron by carbonic acid, 65. Superiority of blast furnace compared with direct process as a means of reducing oxide of iron, 66. Combustion of carbon in upper region of furnace to be avoided, 67. Extent of loss by such combustion of carbon, 68. Combustion of carbon at the tuyeres, 69. Quantity of heat evolved in reducing iron by means of carbon and by means of oxide of carbon, 69. Influence of temperature in reducing iron by carbon and in carbonic oxide and of oxidising metallic iron by carbonic acid, 71. Influence of size of furnace in the economical reduction of oxide of iron, 72. Composition of escaping gases and loss from heat therein from furnaces of different dimensions, 73. Practical limit in enlargement of blast furnaces, 74. Absorption of heat in the reduction of oxide of iron, 76. Generation of heat in reducing zone, 76. Position of equilibrium in reducing zone, 78.

SECTION VI.

ON THE USE AND THEORY OF THE HOT BLAST 80

Introduction of heat at the hearth by heating the blast, 80. Neilson's discovery of hot blast, 81. Dufrenoy's examination of its use, 81. Difference of effects by heating the air, 82. Difference in susceptibility of reduction in oxides of iron differently prepared, 83. Same difference perceptible in different ores, 84. Mode of action of, 84. Analogy between heating the blast and enlarging the size of the furnace, 85. Point to which it is desirable to increase temperature of the blast, 86. Extent to which fuel burnt in the furnace can be substituted by heat in the blast, 91. Saving of coke by raising temperature of air from 1,000° F. to 1,700° F., utility of hot blast in small furnaces, 92.

CONTENTS.

SECTION VII.

ON THE QUANTITY AND QUALITY OF THE FUEL REQUIRED IN THE BLAST FURNACE USING AIR OF DIFFERENT TEMPERATURES 94

Different quantities of fuel required for different kinds of ores, 94. Quantity of heat required to smelt Cleveland ironstone, 95. Production of the necessary heat, 96. Large and small furnace workings compared, 97. Point of saturation by oxygen of reducing gas in the blast furnace, 98. Effect of irregular charging and of irregularities in quality of the coke, 100. Sulphur in coke, 103. Oxygen in coke, Parry's experiment, 105. Effect of using inferior coal or coke, 106. Fire-brick stoves, 107. Metal stoves at the Clarence works, 110. Disappearance of carbonic acid in upper zones of furnaces, 111. Action of large and small furnaces compared, 112. Excessive consumption of coke, 113. Effect of increased capacity of furnace in smelting Bilbao ore, 113. Analysis of gases from furnaces using Bilbao ore, 114. Different quantities of coke required for ores of different richness, 116. Limit of quantity of carbon in form of carbonic acid found in escaping gases, 117. Value of superheated air in smelting poor ore, 118. Use of raw coal in blast furnaces, 120. Composition of splint coal of Airdrie, 120. Increase of deoxidising power of gases when using raw coal, 121. Furnace of Will. Ferrie, 122. Raw coal in North Staffordshire, 122. Value of shape in the blast furnace, 124. Value of height in a blast furnace, 124. Apportionment of coke for different work performed in the blast furnace, 126. Use of anthracite in smelting iron, 126. Composition of anthracites and bituminous coals compared, 127. Use of charcoal in the blast furnace, 127. Heat required in smelting Styrian and Carinthian ores, 128. Charcoal used in Pennsylvania Pine furnace, 129. Use of charcoal in Sweden, 130. Use of torrefied wood in blast furnace, 135. Growth of timber, 135. Produce of charcoal from wood, 135. Charcoal and coal compared, 136. Lignite and peat in the blast furnace, 136. Raw coal blown in at the tuyères, 137. Use of petroleum in the furnace, 137. Loss of heat in application of fuel, 138. Economy effected by Siemens' furnace, 139. Heat rendered available in the blast furnace, 140.

SECTION VIII.

ON THE SOLID PRODUCTS OF THE BLAST FURNACE 147

Cast iron and its composition, 147. Hot and cold blast iron, 148. Temperatures of furnaces driven with hot and cold air, 149. Composition of hot and cold blast iron, 150. Strengths of mixtures of hot and cold blast iron, 150. Composition of pig iron before and after reaching the tuyeres of blast furnace, 154. Kish, 155. Carbon in pig iron, 157. Graphitic and combined carbon, 158. Conversion of combined into graphitic carbon, 158. Conversion of graphitic into combined carbon, 159. Cementation process, 160. Transference of carbon from cast to wrought iron, 160. Case of absence of carbon in iron in the blast furnace, 161. Silicon in pig iron, 161. Glazed iron, 162. Sulphur in pig iron, 163. Phosphorus, 165. Manganese, 165. Other metals in pig iron, 167. Slag or cinder from the blast furnace, 168. Furnace fume, 174.

SECTION IX.

CHEMICAL CHANGES AS THEY TAKE PLACE IN THE BLAST FURNACE ... 176

Order of changes ascertained by composition of gas in the furnace, 176. Temperatures of escaping gases, 178. Change in content of carbonic acid effected by charges of coke, 179. Cooling effects of different materials in charging, 179. Increase of temperature of gases on ceasing to charge, 181. Temperature at which ore commences to lose oxygen, 182. Rate at which reduction goes on at different temperatures, 183. Effect of rapid currents of reducing gas, 183. Carbonic acid decomposed by metallic iron, 184. Action of mixtures of carbonic oxide and carbonic acid at different temperatures, 185. Five reactions of carbonic oxide on oxide of iron, 186. Carbon deposition from carbonic oxide, 187. Caron and Schinz on carbon deposition, 189. Carbon deposition affected by quality of oxide of iron, 189. Presence of carbonic acid reduces amount of carbon deposited, 190. Temperature at which reduction of Cleveland ore commences, 191. No necessary correspondence between rate of reduction and of carbon deposition, 192. Presence of metallic iron not indispensable for carbon deposition, 193. Dissociation of carbonic oxide by other metals and their oxides, 193. Power of carbon to split up carbonic acid, 194. Effect of dissociation of carbonic acid, 195. Power of soft coke to split up carbonic acid, 196. Power of solid carbon to reduce ore, 197. Dissociation of carbonic acid of the limestone used, 198. Generation of carbonic oxide on the hearth of the blast furnace, 199. Carbonic acid in the gases of the Styrian furnaces, 201. Change of composition of gases as they ascend through materials filling the furnace, 201. Disadvantages of a low furnace, 202. Composition and temperature of gases from furnaces of 80 and 48 feet in height, 203. Irregularities in composition of gases taken at different levels of the furnace, 212. Potassium and sodium cyanides in gases, 216. Contribution of dissociation of carbonic oxide to

CONTENTS. xiii

observed irregularities in composition of gases, 219. Value of deposited carbon in the process, 222. Behaviour of cyanides in the furnace, 222. Composition of fume at different heights in the blast furnace, 224. Behaviour of zinc in the blast furnace, 226. Ammonia in the blast furnace, 228. Lithia in the furnace, 229. Silica, lime, alumina and magnesia vaporized, 229.

SECTION X.
ON THE EQUIVALENTS OF HEAT EVOLVED BY THE FUEL IN THE BLAST FURNACE 234

Value of the fuel dependent on fixed carbon it contains, 234. Heating power of fuel, how modified, 235. By quantity and nature of foreign matter, 235. By temperature and condition of air, 238. By state of oxidation, 238. By quantity of heat which escapes in gases, 240. Effect of imperfectly calcined ironstone, 240. Intricacy of heat estimates, 241. Equivalents of heat and of coke to each function in the furnace, 241. Causes which affect oxidation of carbon, 247. Heat carried off in the escaping gases, 248. Value of successive additions to the temperature of the blast, 249. Furnace gases applied to heating the blast, 251. Quantity of heat evolved by coke when burnt with air at a given temperature, 251. Use of superheated air at the Ormesby furnaces, 252. Effect of superheated air on furnaces under 25,000 cubic feet, 255. Use of superheated air in the manufacture of Bessemer iron, 256. In furnaces of sufficient capacity oxidation of carbon and loss of heat in gases are independent of temperatures of the blast, 260. Heat contributed by blast at different temperatures and by carbon in different stages of oxidation, 260. Calculations showing equivalents of coke required per ton of iron according to temperature of blast and state of oxidation of carbon, 262. Waste of fuel in a furnace of insufficient size, 263. Limit to which air can be heated, 264. Further observations on saving of fuel by successive increments of temperature of the blast, 265. Superheated blast supposed cause of reducing quantity of carbonic acid in gases, 267. Influence of shape of the blast furnace in economizing fuel, 268. Possible errors in stating the quantity of coke consumed per ton of iron, 269. Mode of checking this by analyzing escaping gases, 269. Heat equivalent of charcoal, 274. Heat equivalent of coke and charcoal compared, 282. Moderate temperature of blast used at charcoal furnaces, 283. Experiments on charcoal furnaces by M. Frederici, 284. Reduction of ores used in Styria by carbonic oxide, 285. Effect of hard coke, soft coke and of charcoal on carbonic acid, 287. Reducing power of carbonic oxide on mixtures of Styrian ore and charcoal and Cleveland ore and coke, 287. Similar experiments by M. Åkerman in Sweden, 288. Ratio of carbonic acid and carbonic oxide in charcoal furnaces, 289. Portion of ore always escaping reduction, 291. Quantity of carbonic acid in gases of charcoal furnaces less than that in Cleveland furnaces per ton of iron made, 291. Difference of composition of gases at different heights of the furnace between coke and charcoal furnaces, 292. Increase of temperature at different depths in charcoal furnaces, 295. Zone of reduction in charcoal furnaces different from that in Cleveland furnaces, 296. Supposed effect of using superheated air in charcoal furnaces, 299. American experience with superheated air in charcoal furnaces, 300.

SECTION XI.

ON HYDROGEN AND CERTAIN HYDROGEN COMPOUNDS IN THE BLAST FURNACE 305

Hydrogen produced by water, 305. Quantity of hydrogen in the gases of a coke furnace, 306. Quantity of hydrogen in gases at different depths of the furnace, 308. Value of hydrogen as a reducing agent doubtful, 309. Experiments to ascertain reducing power of pure hydrogen, 310. Mixtures of hydrogen and carbonic oxide as reducing agents, 310. Effect of hydrogen on carbonic acid, 312. Condensation of tar and ammonia from furnace gases where raw coal is used, 314. Use of raw coal in smelting iron, 315. Composition of escaping gases when using coal, 315. Heat absorbed and evolved in raw coal furnace, 319. Effect of a mixture of vapour of water and hydrogen on metallic iron at high temperatures, 321. Deficiency of carbonic acid in gases of furnaces using raw coal, 322. Use of raw coal in smelting Bilbao ore, 323. Raw coal in smelting Oolitic ores of Midland Counties, 324. Raw coal and coke compared, 325. Ammonia from furnaces using coke, 326. Ammonia from furnaces using raw coal, 327. Collection of ammonia, tar and oil from coke ovens, 327. Proposed injection of hydrogen at tuyeres, 329. Proposal to inject water-gas at tuyeres, 329. Experiment of use of coke and water-gas in smelting iron, 340. Use of petroleum in smelting iron, 342.

SECTION XII.

ON THE PRODUCTION OF MALLEABLE IRON FROM PIG IRON IN LOW HEARTHS 343

Use of low hearth as an instrument of oxidation, 344. High quality of iron produced by its means, 345. Slag from Lancashire hearth, 346. Amount of work and expense attending its use, 347. Quantity of iron yearly manufactured in hearths, 348. Extracts from books of Messrs. Knights of Kidderminster, showing costs, &c., in years beginning 1727, 348.

CONTENTS.

SECTION XIII.

ON THE REFINING AND PUDDLING FURNACE 351

Cort's invention of the puddling furnace, 351. Puddling on sand bottom, 351. S. B. Rogers' improved furnace, 351. Continued use of refinery to oxidise the silicon, &c., 352. Amount of oxygen required, compared with air used, 353. Change effected in pig iron by the process of refining, 354. Slags produced in manufacture of malleable iron returned to the blast furnace, 356. Proportion of phosphorus and sulphur which accumulates in slags, 357. Gradual change of composition of iron during process of refining, 359. Change of composition in puddling refined metal, 360. Composition of refinery cinders, 361. Use of unrefined pig iron in the puddling furnace, 362. Impurities of pig removed by oxide of iron, 362. Loss of weight and coal used in puddling pig iron, 363. Quantity of the metalloids left in iron obtained from puddling Cleveland pig iron, 365. Mechanical puddling furnaces, 365. Quality of iron produced by mechanical puddling, 366. Presence of slag in iron made in puddling furnace, 368. Relative cost of puddling by hand and by mechanical means, 372. Redshortness in iron made in mechanical puddling furnaces, 373. Use of mechanical furnaces for making iron by direct process, 374. Use of chemicals in puddling, 376. Prospect of ingot iron and steel displacing the puddling furnace, 376.

SECTION XIV.

ON MORE RECENT METHODS OF SEPARATING THE SUBSTANCES TAKEN UP BY IRON DURING ITS PASSAGE THROUGH THE BLAST FURNACES 381

Bessemer's invention, 381. Phosphorus, an obstacle to its success, 383. Mushet's discovery of value of Spiegel iron to correct red-shortness, 383. Henderson makes ferro manganese, 383. Rapid advance of Bessemer steel industry, 384. Its economy compared with the manufacture of iron, 385. Superiority of strength as compared with iron, 385. Progress of the manufacture of steel, 386. Consequent advance in value of hematite pig iron, 387. Diagrams showing rates of separations of metalloids from pig iron by different processes, 388. Conditions which influence their separation, 389. Slags from Bessemer converter, 393 M. Grüner's explanation of silica in slags as the cause of phosphorus being retained by iron in the converter, 394. Removal of phosphorus by oxide of iron, 395. Influence of temperature in dephosphorising, 396. Influence of temperature in removal of other metalloids from pig iron, 400. Removal of carbon, silicon, and phosphorus in different processes, 402. Experiments at Essen to dephosphorise pig iron, 404.

Mr. Snelus' trials of lining converter with lime, 406. Basic process, 406. Waste of iron in basic process, 410. Absorption of oxygen by iron and steel in manufacture, 413. Composition of rails made by acid and basic processes, 414. Addition of phosphorus to pig intended for basic treatment, 416. Composition of pig iron and basic-made steel, 417. Estimates of heat evolved during basic blows, 418. Heat evolved in an acid blow, 421. Expense of basic process, 425. Quality of steel by basic process, 425. J. M. Heath's mode of making steel, 426. Siemens-Martin's steel process, 426. Composition of iron rails, 428. "Ore process" for making steel, 430. Increase in make of Bessemer and open hearth steels, 432. Use of steam in open hearth process, 432. Pernot furnace for steel making, 433. Quality of Bessemer and open hearth steels, 434. Other methods of making steel, 435. Steel by cementation process, 436. Wootz, 436. German steel, 436. Puddled steel, 437. Soaking pits of Mr. Gjers, 440.

SECTION XV.

STATISTICAL 441

International exhibitions, 441. Progress of iron trade in Great Britain, 443. Increase of iron trade in some other countries, 444. Supremacy of British iron trade threatened, 446. Effect of protective duties levied in foreign countries, 446. Alteration in cost of labour in Great Britain, 447. Effect of duties on British exports of iron, 447. Statistics of coal production in Great Britain, Germany, France, Belgium, and in the United States, 448. Statistics of iron ore production, 449. Quality of ore required for acid process in making steel, 450. Ore for basic process, 451. Ore obtained in five above named countries, 451. Imports and exports of ore, 452. Increase of Bessemer steel made, 453. Different kinds of ore produced in Great Britain, 454. British ores raised and pig iron produced therefrom, 456. Effect of increased make of steel on manufacture of malleable iron, 459. Sources of limestone required in smelting iron, 461. Effect of abolition of duties by foreign governments, 461. Large investments of capital in iron manufacture, 463. Iron trade in Germany, 463. Belgian iron trade, 464. French iron trade, 466. Russian iron trade, 467. Iron trade in United States, 467. Consumption of iron per head by different nations, 470. Shipbuilding as a source of consumption of iron, 471. Power of United States in producing and exporting iron, 472. Export of iron from Germany, 474.

SECTION XVI.

BRITISH LABOUR COMPARED WITH THAT OF THE CONTINENT OF EUROPE ... 475

Cost of iron in Great Britain how made up, 475. Agricultural labour in Europe, 475. Cost of food in Europe, 477. British imports of articles of food, 480. Increase of price of provisions on continent of Europe, 480. Imports of food from United States, 480. Equalisation of prices in articles of food in Great Britain and abroad, 480. Dr. Young on costs of provisions in Europe, 482. Mode of life of foreign workmen, 484. Cost of maintaining families on continent of Europe, 484. Cost of living of English families, 486. John Bright on Free Trade, 490. Drinking habits in Great Britain and elsewhere, 490. Savings' banks in Great Britain, 494. Efficiency and cost of British and foreign labour, 495. Agricultural labour in France and Great Britain, 496. Effect of low wages on health and efficiency of labourers, 497. Wages of mechanics in Great Britain, 497. Cost of English and continental labour compared, 499. Labour in iron ship-building, 501. Increase of cost of labour in English chemical works, 504. Cost of labour in getting coal in Great Britain and on continent of Europe, 505. Restriction of output in Great Britain, 513. Earnings of miners in Great Britain and on continent, 514. Wages of ironstone miners in Cleveland, &c., 517. Same in other countries of Europe, 518. Labour at blast furnaces in England, 520. Same in other countries of Europe, 521. Labour in malleable iron works in in Great Britain, 525. Same in other countries of Europe, 527. British and continental rates compared, 529. Wages in Bessemer steel works compared with those in the continent, 535.

SECTION XVII.

ON LABOUR IN THE UNITED STATES OF AMERICA 539

Connection between wages and cost of necessaries of life, 539. Rent of land in United States, 539. Cost of growing wheat, 540. Cost of agricultural produce in Great Britain, 542. Agricultural wages in United States, 544. Cost of provisions, 546. Cost of living, 547. Increase of prices of provisions in United States, 549. Prices in Great Britain, 551. Price of labour in Southern States, 554. Labour in Anthracite and other coal districts, 555. Labour in ore mines, 559. Mechanics' wages in U.S.A. and Great Britain compared, 559. Same at blast furnaces, 562. Same in malleable iron works, 566. Same in steel works, 571. Mr. J. D. Weeks on British and American wages, 574.

xviii   CONTENTS.

SECTION XVIII.

PAGE,

CHIEF IRON-PRODUCING COUNTRIES COMPARED   578

Competition of continent of Europe with Great Britain, 578. Exports from Germany, Great Britain, Belgium, France, and United States. 579. Export of rails from Westphalia and Germany, 580. Import duties of German Zoll Verein, 581. Manufacturing skill in foreign nations, 582. Prices of rails by foreign and English firms, 583. Relative costs of minerals in the different countries, 584. Effect of improvement in the manufacture of iron of different competing nations, 586. Labour in Great Britain compared with that of the continent, 587. Prospect of further improvements in the manufacture of iron, 587. Railway dues in the different countries, 595. Royalty dues in the different countries, 602. Change in iron trade of United States, 604. Amount of royalties on minerals paid in France, Belgium, and Germany, 609. Coal of Great Britain, 610; coal used in Great Britain for smelting iron, 622; coal of Germany, 623; of France, 626; of Belgium, 629; and of United States, 632. Iron ores of Great Britain, 646; of German Zoll Verein, 652. Oolitic ironstone of Western Germany, Luxemburg, and Eastern France, 653. Ironstone of remainder of France, 656; of Belgium, 659; and of the United States, 660. Limestone in different countries, 681. Comparative cost of minerals required for producing one ton of pig iron in Great Britain, 683; in the German Zoll Verein, 685; in France, 687; in Belgium, 688; and in the United States, 689. Power of United States to compete in the exportation of iron, 695. Increased cost of production in the United States, 695. *Iron Age* newspaper on the manufacture of iron in America, 698. Southern States of America as a seat of the iron trade, 700. Future sources of economy in production in Great Britain, 701. Competition of Western Germany, 702; of Eastern France, 702; and of Belgium, 703. Competition in the manufacture of steel between Great Britain and the continent of Europe, 704. Effect of Basic process in this competition, 704. Extent of exports of iron from Great Britain, 707. The future foreign trade of Great Britain in iron, 708. Present prospects of the iron trade of the world, 711.

CORRECTIONS AND ADDITIONS   ...   .   *See end of Book.*

LIST OF PLATES, &c.

	PAGE
Plate 1, showing temperatures at which carbon and carbonate oxide begin to act on peroxide of iron, and those at which carbonic acid begins to act on iron and carbon	71
Plate 2, showing temperatures of zones of two furnaces of different heights	72
Plate 3, showing curve of cooling effect on furnace gases by increased furnace capacity...	75
Plate 4, showing differences in temperature of the zone of reduction ...	112
Plate 5, showing imperfect action in a furnace of large capacity, but deficient in height	124
Plate 6, showing temperatures of zones in furnaces of different heights...	203
Plate 7, showing temperatures of different zones in charcoal furnace of Styria	295
Plate 8, diagrams showing the rate at which certain metalloids are expelled from molten iron in different processes	388
Plate 9, diagram showing increase of powers of coal production in France, value of coal, and rise in men's wages...	511
Plate 10, diagram showing price of Scotch pig iron, and earnings of colliers in Scotland and in Westphalia	517

SECTION I.

INTRODUCTORY.

AFTER the completion of my duties as a Juror at the French International Exhibition of 1878, I was honoured by a request from the Board of Management of the British Iron Trade Association, to prepare a report on the present condition of the Manufacture of Iron and Steel, as illustrated by the objects displayed in the different buildings in the Champs de Mars, at Paris.

I had, however, previously proposed to myself a more extensive enquiry than that which would be covered by a mere examination of the products of the Iron Works of France and of other nations, as exhibited upon the occasion referred to.

I had published at intervals, in the "Transactions of the Iron and Steel Institute," certain investigations into points connected with the action of the blast furnace, in respect to which some explanation was, it appeared to me, still wanting. The experiments bearing on this question were communicated to that body, as the enquiry advanced, and were described in the language of the laboratory. It was not only my wish to present the general results I had arrived at in a more consecutive and less unattractive form than that necessarily adopted under such circumstances, but to correct any opinion therein stated, which further observation had shewn to require modification.

On communicating my ideas to the President and his Associates at the Board of the Iron Trade Association, a ready willingness was expressed to see my labours extended in any way likely to add to the utility of the object we had in view.

The depression in the iron trade, not of Great Britain only, but throughout the world, was the cause of great uncertainty and uneasiness in the minds of those engaged in its prosecution, and it was my wish, for reasons immediately connected with the mode of manufacture, to embrace in the scope of my labours some considerations of a more commercial character, than those usually found in scientific or technological works.

Considerable additions to the producing powers of other nations were no doubt partly the cause of the distress in question; but, under ordinary circumstances, a mere temporary excess of production brings with it a remedy more or less speedy in its nature. All additions to manufacturing appliances are suspended, establishments in the least favourable localities are brought to a stand, and this process continues until the growing demand, consequent upon wider civilization and increasing commerce, calls for increased activity on the part of the producer.

The iron trade was passing through such a crisis at the period in question; and the apprehension entertained by those, whose capital was embarked in certain mines of iron ore and in malleable iron works, was not that a demand for the metal would never revive at all, but that it would revive in a form incapable of being supplied from the produce of their pits, or dealt with by their present machinery.

As will be anticipated, this is one of the results of the inventions of Bessemer and Siemens, by which steel made by the so-called pneumatic, or by the open hearth process, has already largely taken the place of iron. So far, indeed, as concerns rails, in this country at least, it may be said that the puddling furnace bids fair to be ultimately superseded by the converter; unless at any time the demand for rails shall be found to rise in excess of our means of supplying them in steel, whether from a want of suitable ore or from a temporary deficiency in proper steel-making establishments.

This recent addition to our national industry requires, however, the use of a raw material hitherto regarded as incapable of being furnished by those ores, which, from their abundance and cheapness of extraction, have placed Great Britain in her present high position as an iron making nation. In consequence of this supposed unsuitability of our clay ironstones for Bessemer pig iron, smelting establishments and some rail mills have been constructed in the neighbourhood of the hematite mines of Cumberland and Lancashire, which almost alone, at least in this kingdom, produce ore sufficiently free from phosphorus to meet the requirements of the Bessemer or Siemens-Martin processes, as generally practised. It soon became apparent that the West of England ores would be inadequate to meet the additional demand made on them by all the new steel rail mills, and hence our supplies had to be supplemented by large importations from foreign countries.

The iron trade of Great Britain, with ample mining resources for supplying its blast furnaces, and with mills of sufficient capacity to meet all ordinary demands, has thus found itself constrained to import iron ore from abroad, and to construct new works to replace those which this sudden change of circumstances has silenced.

Is there any chance of a reversal of these conditions, forming, as they do, a great aggravation to the difficulties which threaten almost to overwhelm a considerable portion of the iron trade of this country? So far as a return to the puddling furnace is concerned, this seemed at one time hardly possible; for, incredible as it would have been at one time regarded, hematite ore can, in the opinion of some competent authorities, be brought oversea from a distance of 1000 miles, landed close to mines furnishing the cheapest made pig iron in Great Britain, and converted into steel rails, at a lower cost than the native ironstone of Cleveland can furnish similar rails in iron.

Speaking from individual conviction, I am not disposed to take a desponding view of our purely national iron trade. The Bessemer converter, as hitherto employed, retains almost all the original phosphorus in the product. From different quarters well-grounded pretensions are now set up, which promise to modify the action of the pneumatic process, so as to secure the elimination of the phosphorus in the more impure varieties of pig. Failing this, the experience of some time past has incontestably proved, that, by means of the open hearth, rails can be manufactured of good quality, and yet containing a percentage of the metalloid referred to, which hitherto has been regarded as inadmissible in the Bessemer converter. This modification of the steel process would, of itself, permit a certain portion of British iron being employed, at all events in the manufacture of rails; but in addition to this it has been demonstrated as physically possible that pig iron containing 1·75 per cent. of phosphorus can be freed from this element sufficiently well to afford steel of excellent quality, using the Siemens furnace instead of the Bessemer process in the subsequent stages of the process.

Can our own cheap ores, when burdened with the expense attending the differences of treatment referred to, compete with the purer mineral of the West of England and elsewhere? This is a subject which is largely occupying the minds of the ironmasters of this country. The experience of twenty years has reduced the cost of making Bessemer ingots to a mere fraction of its former amount. Giving it again as an individual opinion, I think that enough has been done to render it highly probable that, at no very distant day, any additional expense incurred in adapting the usual make of British pig iron to steel-making purposes, will be more than met by the lower cost of the mineral obtained from British soil.

A large amount of work, in the direction of thus adapting our native ores, more or less rich in phosphorus, to the purposes of steel-making, has been done since the year of the French Exhibition. The delay, therefore, arising from the extended scope of my labours, and augmented by other claims on my time, has had the advantage of enabling me to examine the progress and consider the future of this interesting question.

Since the above was written the so-called *Basic process*, as applied to the pneumatic treatment of pig iron containing an excess of phosphorus, has met with considerable success. In Germany particularly, where the difference between the cost of the class of crude metal hitherto exclusively used in the Bessemer converter and that of the commoner descriptions is greater than in Great Britain, a very large quantity of steel has been made from pig containing upwards of 2 per cent. of phosphorus. It is true that some English manufacturers have expressed doubts whether the relative cost of the two varieties of pig iron in Great Britain will permit any general application of this dephosphorizing system in this country. A most important step in advance, however, has unquestionably been made—it has been demonstrated that phosphorus can be readily separated from iron in the Bessemer converter. With this knowledge I have no doubt that steel will ultimately be made, if it has not been made already, on the East coast of England, and from Cleveland stone, in all probability more cheaply than it can be produced from ore imported from Spain, and more cheaply also than it can be supplied from works on the West coast to the banks of the Tees.

The chemical conditions attending the so-called Direct process of obtaining malleable iron from the ores, are such as to prevent the absorption by the metal of the greater part of the phosphorus which the ores contain. This has led some to regard as possible a return to a mode of manufacture more or less identical with that practised in prehistoric times. Enough indeed has been done and said in connection with this opinion to render it desirable, at the proper time, to consider the grounds upon which this revival of the most ancient method of producing the metal is founded. Again, by means of some very important improvements in the puddling furnace, hopes have been, and perhaps are still, entertained of securing for what is now understood and described as malleable iron, a much more extensive sphere of usefulness than some are disposed to imagine it is destined to occupy in the future.

It is, of course, at all times difficult to lay down any limits as to what may or may not be capable of achievement. There seems but one course to pursue, under circumstances requiring such serious and immediate consideration by those interested in the subject. This course consists in a careful and systematic study of the different processes, either proposed or in actual use, with a view to ascertain what margin there is for any material change in those processes themselves, and hence for any marked advance in point of economy or of efficiency.

A prominent feature in the competition which the British iron manufacturers have to encounter, is the great extension of iron works now taking place in foreign countries. Until a comparatively recent period, Great Britain enjoyed many advantages from the possession and advanced state of development of her large fields of coal, as well as of the iron mines already mentioned. To a considerable extent her position in this respect has been altered by the opening out of important carboniferous deposits in different parts of the world. If, in addition to this source of competition, circumstances compel British manufacturers to be dependent, to any great extent, on Spain for their supply of ore, our iron trade cannot fail to be greatly affected by such a change.

Again, the very improvements that have taken place in the production of the metal have sensibly improved the relative position of other nations, less favourably circumstanced than Great Britain in the matter of mineral fuel. To manufacture an iron rail in Scotland fifty years ago, 11 or 12 tons of coal would have been consumed. To-day less than one-half of this quantity suffices for the work, and one-fourth of it is enough to produce a rail of steel. Thus, instead of having to multiply any difference in the cost of fuel by twelve, which would represent our former position, the same sum has only to be multiplied by three in order to ascertain the extent of our present advantage.

Within the recollection of many, almost the whole world was more or less dependent upon Great Britain for its supply of iron of ordinary quality. So recently as 1871, the United States of America virtually relieved us, in one shape or another, of nearly one-fifth of our make of pig iron. This important market was almost closed against us from 1875 up to the end of the year 1878. An unexpected demand for iron and steel in the autumn of 1879 found the United States, as it was alleged, unprepared to meet its own requirements; in consequence large importations were made from Great Britain, and these to some extent still continue. So long, however, as the present prohibitive tariff remains in force, it is only a want of capital and labour which will prevent North America, save under exceptional circumstances, from supplying all its own requirements.

As to our relations with the Continent of Europe, we have not only to meet France, Germany and Belgium in the open market, but, to the alarm of many—as I think, an unnecessary alarm—German and Belgian iron is to a moderate extent imported into this country.

In the hope of adding to, or correcting when necessary, the impressions derived, upon former occasions, from an examination of continental works, I once more, after the close of the Paris Exhibition, revisited some of the chief seats of the manufacture of iron in Europe. In like manner, so recently as 1874, and again in 1876, I endeavoured to make

SECTION I.—INTRODUCTORY.

myself acquainted with the natural advantages possessed by most of the coal-fields and principal iron ore deposits of the United States. From the information thus obtained, an attempt will be made to compare the resources possessed by different localities in the old and new worlds as iron-making centres. This object would scarcely be complete, unless the cost and efficiency of the labour at the disposal of the mine owner and manufacturer received some consideration. This last-named subject may involve questions of a character not so easily dealt with as the natural resources of the country itself. Such questions will probably occupy hereafter—if, indeed, they do not do so already—a position of considerable importance with regard to the interests of Great Britain as a manufacturing nation. In the matter of cost, viewed in the abstract, the Continent of Europe has undoubtedly the command of cheaper labour than is to be found in Great Britain. We shall be better able, however, to compare the relative economy of the two, when other elements, besides that of mere price, are brought into the account.

SECTION II.

HISTORICAL.

It is certainly not essential to the main object of this work that the early history of the manufacture of iron should occupy any place in its pages. Nevertheless, a brief consideration of the various steps, by which the smelting of this metal has arrived at its present position, affords an opportunity for comparing the progress of recent years with that of preceding ages. Hence a section, historical rather than technical, may not be without its use as well as its interest.

Except in the form of those well-known meteoric masses, in which iron forms the chief ingredient, this metal, I imagine, was, until a comparatively recent date, unknown in its native state. Some years ago, Professor Nordenskjöld had heard of its existence in Greenland, and in 1870 he was fortunate enough to discover its source on certain islands in Davis Straits. It is there found as small detached masses embedded in Dolerite, a circumstance which accounts for its preservation in the metallic form. It is a remarkable fact that this native iron is usually associated with 2 or 3 per cent. of nickel and cobalt, in which particular it resembles the meteoric iron already referred to. The subject has been ably examined by my friend Dr. Lawrence Smith, of Louisville, Kentucky,[1] who supposes that the heat of the fused basalt, and the presence of carbon as found associated with the metal, had effected the reduction of the original oxide of iron, frequently found in this igneous rock.

The absence of iron, in its metallic form, in many of the monuments of very remote antiquity, has led to the belief that its existence was unknown to the earlier inhabitants of our globe. In the numerous instances when objects anciently used for warlike or for domestic purposes have been discovered, they have generally been fashioned out of stone, or in more recent times out of bronze.

[1] *Vide* "Annales de Chimie and de Physique," 5^e serie, t. xvi., 1879.

SECTION II.—HISTORICAL.

In many instances, unquestionably, the absence of this metal may be set down to its perishable nature, prone as it is, under atmospheric influences, to rapid oxidation. Iron implements, protected from these influences, have been found in some recent explorations, and have proved that mankind was acquainted with the existence and value of iron at a much earlier period of the world's history than was at one time believed.

The minerals containing iron afford, it is true, but little indication of their metallic nature; but this is so easily rendered conspicuous by mere contact with burning wood, at very moderate temperatures, that accident alone must, where the ore is plentiful, have led to a knowledge of the valuable nature of its composition. Metallic iron being thus so easily produced, little ingenuity or skill would be needed to forge it into a variety of useful articles, greatly prized even amongst the least civilized communities.

The facility with which a portion of the iron is reduced, and separated from the earthy substances with which it is usually associated in nature, would doubtless lead to its production, even among savage nations, between which no intercourse existed. Colonel Grant kindly made a sketch for me of a primitive forge which he had seen in operation in the interior of Africa, upon the occasion of his discovery of the sources of the Nile, in company with Captain Speke. In this drawing, two natives, each working a pair of single-acting bellows, are seen urging the combustion of a small heap of charcoal situate midway between the two. The ore, in small pieces, is added from time to time, with fresh supplies of fuel; and this is continued until a mass of iron of the required dimensions is obtained. As may be imagined, a process carried on by persons destitute of all mechanical contrivances, save those of the rudest kind, could make no progress; and Colonel Grant supposes that the produce of the two men did not exceed a dozen pounds per day. On the other hand, a furnace differing but little from the rude hearth just alluded to, in the hands of more skilful mechanicians, can turn out as much as 300 pounds in the twelve hours.

In some parts of Europe this simple mode of making iron, performed in what is known as the Catalan Forge, is still practised, as it has been for some hundreds of years; and in the United States of America as much as 30,000 tons are annually obtained by a process which is almost identical with that witnessed by Grant and Speke in Central Africa.

Whatever may have been the precise nature of the process by which in ancient times iron was produced, it is clear that the human race was early possessed of materials capable of forming tools of great hardness and power of endurance. It has been suggested that some forgotten method of tempering bronze, an alloy in common use, supplied the means of cutting and even carving the hardest kind of stone. It seems, however, equally probable that people capable of designing and constructing the monuments of Assyria and Egypt should have noticed that the process of making iron, in the rude hearth already spoken of, may be so conducted as to produce that partially carburized form of the metal known as steel. Indeed, it appears as if in later times the very same description of furnace was employed indifferently to obtain wrought iron, steel, and pig iron.

It would, at any rate, be hazardous in the extreme to ground any settled opinion upon the state of metallurgic art in ancient, or even in prehistoric times, by the state of our own more recent knowledge; for there is evidence which points to the belief that in some lands the art of making iron was, at a very early period, in a far more advanced condition than it had attained so late as the reigns of Elizabeth or Charles I. in this country. As an example of this, there is at Delhi an architectural column of very great antiquity, formed out of malleable iron; it is above 16 inches in diameter, and is calculated to weigh upwards of 17 tons. Its existence proves that subsequent to the date of its manufacture, a great decadence must have taken place in the art of working the metal; for I am not aware that there remains any later record of the existence in the Indian peninsula of the means of forging, in so perfect a manner, so large a mass of iron.

The hearth and Catalan fire—as the latter was called from its use in that province—blown by means of a fall of water known as the *trombe*, or by bellows, were the only means employed for ages in furnishing the world with the iron it required. We are, singularly enough, left to conjecture, not only when, but in what manner the next step in its

SECTION II.—HISTORICAL.

manufacture was brought about. This want of precise knowledge is the more remarkable, seeing that the change, consisting in the production of cast iron, has probably been effected within the last three centuries, and that it altered entirely the character of the operation in which it virtually constituted the first really great improvement.

Contemporaneously with the use of the Catalan hearth, at all events in later times, the Stückofen was worked in Germany, and the Osmund furnace was employed in Sweden, both being driven with compressed air. A mere addition to the height constituted all the essential difference between these two forms of furnace and their primitive predecessor. This change of construction would probably be suggested by an expected economy in fuel, or an increased production of metal, or by the hope of combining both these objects. It seems doubtful, however, whether the weight of malleable iron, or of "blooms," obtained by means of either description of furnace ever exceeded two or three tons per week.

The German Stückofen was usually about 10 feet high, but this was increased sometimes to as much as 16 feet, with a diameter at its widest part of 5 feet. The breast was built up temporarily with brickwork, which was removed when the bloom of wrought iron was ready for the hammer. It cannot be doubted that in such a form of apparatus cast metal would be occasionally formed.

This last-mentioned circumstance in all probability paved the way to the construction of the Blauofen, an invention for which we are likewise indebted to the German people. In it the height was sometimes as much as 25 feet, by probably 6 feet at the widest part. With such dimensions as these there would be no difficulty whatever, by proper treatment of the ores, in combining the metal with sufficient carbon to obtain a constant supply of cast iron; and there seems no doubt that by means of the Blauofen this was attained in actual practice. At the same time mere capacity will not of itself suffice for the production of what is now known as pig iron. If the ore and fuel were employed in such proportions that complete reduction of the oxide was rendered impossible, a substance containing little or no carbon would be the product, in whatever form of furnace the operation was conducted. Accordingly we find the Blauofen at one time used as a means of obtaining cast iron; an addition being then made to the quantity of charcoal consumed. On other occasions the same furnace was employed, like the Stückofen, for the production of malleable iron or of steel.

Iron-foundries are, it is true, spoken of by early writers as existing above 300 years ago. From the context, however, it seems clear that the term, notwithstanding its obvious derivation, was not applied, as it is in our own day, to the running of the metal in a state of fusion, into moulds, but was employed to designate establishments for forging or working up malleable iron, merely softened by heat obtained in the hearths already described.

In the work of Agricola, bearing the date of 1556, there is no allusion to the Blast-furnace; the only arrangement described for obtaining iron being a kind of Catalan hearth, in which the product, as we have already seen, is malleable and not cast iron. About 1618, as we learn from Dud Dudley's *Metallum Martis*, this unfortunate pioneer in the manufacture of iron was engaged in his attempts to substitute pit coal for charcoal; and, as he distinctly mentions mottled and grey iron, there is no doubt that, anterior to his time, the blast furnace was in existence. The writings of these two authors, Agricola and Dudley, appear, therefore, to fix the date of its introduction at somewhere about the beginning of the sixteenth century.

Seeing that the date and precise manner of the introduction of the blast furnace itself are so uncertain, it is needless to consider, with any minuteness, the circumstances which led to the making of cast iron the first stage in the manufacture of the malleable metal. Without the faintest glimmer of scientific knowledge to guide them as to the fundamental difference between the two forms of iron, it is nevertheless easy to comprehend that the forge owners of the period would be led to apply the same treatment to the crude metal, which had been found successful with the ore. This much is quite certain, that a hearth, not differing greatly in construction from the Catalan fire, remained, until near the close of the last century, the only means of producing malleable metal from pig iron: indeed, for the finer kinds of bars, the process is still practised in Sweden, Russia and elsewhere, one modification being known as the Lancashire hearth.

SECTION II.—HISTORICAL.

The product of the first blast furnaces, so long at least as they were of small dimensions, would be white cast iron: and this, being useless for the founder, would doubtless all be converted into the malleable form of the metal, and hammered in the way still pursued in the small forges of the countries above mentioned.

Down to the end of the sixteenth century, the only combustible employed in iron works was that afforded by the forests of the day; but the demands upon the resources of the latter became so large that regard for other requirements compelled legislative enactments to be passed in England regulating the felling of timber. Such a state of things naturally led to the consideration of the fitness of mineral coal as a substitute for charcoal; and towards the middle of the seventeenth century we find Dudley engaged in perfecting plans for "making of iron, and melting, extracting and refining all minerals and metals with pit coal, sea coal, peat and turf."

This early adventurer in the iron trade had many difficulties to contend with. Floods swept away his works, unscrupulous competitors destroyed his machinery, and dishonest partners completed his ruin. It would not, however, appear that very marked success had attended his efforts, if it be true that seven tons was the outside weekly product of his furnaces; being only about half that usually obtained from the charcoal furnaces of his time.

It is very probable that mechanical science, as it existed in the seventeenth century, although capable of contriving means of blowing a charcoal furnace, was unable readily to devise appliances adapted to forcing the air through pit coal, which was used just as it came from the mine. The effect of heat in charring coal, so as to convert it into coke, must have been known at that time; for in Plot's "History of Staffordshire" we learn that "Cokes for melting and refining of iron cannot be brought to do, though attempted by the most cunning and curious artists."

For something like half a century the idea, which had proved the ruin of Dud Dudley, slept. It was then revived by Abraham Darby, a name deservedly held in high esteem among the promoters of industrial science. Not only was the invention of his predecessor brought to a successful issue by his perseverance, but he did much at the begining of the last century to spread the use of cast iron for foundry purposes. It was about the year 1733 that, in spite of the failures of "the most cunning and curious artists," this distinguished man succeeded in using charred coal or coke in his iron furnaces. Notwithstanding the importance of the discovery, circumstances were such that the annual make of pig iron gradually fell off; so that, in 1740, of 300 furnaces which had been blowing in the middle of the previous century, only 59 were in blast. Their production in that year was only 17,350 tons, and this very small yield, due probably to interruptions in their work, left 30,000 tons to be imported to meet the wants of the country. The march of improvement seems to have been but languid, for, with the aid of the steam engine and the substitution of iron cylinders for leathern bellows, the average weekly make in 1788 of the 85 furnaces in the kingdom was scarcely $15\frac{1}{2}$ tons per furnace. Of the total annual quantity produced—viz., 68,300 tons—about four-fifths was smelted with coke, and the remainder with charcoal.

Such were the insignificant proportions of a trade, now so vast, when, about the year 1784, another inventor, Henry Cort, rendered as important service to the manufacture of malleable iron as Abraham Darby had afforded to the smelting of the ore. To his ingenuity we owe the grooved rolls and the puddling process of our iron works; and it is no exaggeration to say that to these two inventions we are mainly indebted for those changes, which separate so sharply the last 40 or 50 years from the world's previous social history. With no better implement than the hearth and the forge hammer, as they are used to this day in Sweden, it would have been practically impossible to produce a railway bar or a ship plate; and who can set a limit to the aid these two articles have afforded to the unprecedented strides which have been made during the period referred to?

A more recent invention may possibly complete the revolution in the manufacture of iron, of which the beginning has just been described, by superseding the puddling furnace in its turn; but the facts as stated will remain unaltered. Indeed, admitting that steel or ingot iron will ultimately occupy the place of puddled iron, there seems but little likelihood of a rail of either material being obtained without the assistance of a rolling mill, for which we owe so much to the inventor of puddling.

SECTION II.—HISTORICAL.

The absence of all continous records of our industry prevents our doing more than estimating, in a very rude way, the statistical progress made in the manufacture of iron by the assistance of the puddling furnace of Cort. Within seven years of its introduction, however, evidence was given before the Commissioners of the Navy that 50,000 tons of pig iron were annually converted by means of the process he had devised. If this statement be correct, it would appear that its value was quickly acknowledged by those interested in its application; although some refused to reward the inventor, who, from circumstances for which he was not personally responsible, was allowed to die without reaping any benefit from the undoubted service he had rendered to mankind.

Some thirty-two years after the first introduction of the puddling furnace—viz., in 1816—another unrequited discoverer, Samuel Baldwyn Rogers, greatly added to its value by substituting, for the sand bottom of Cort, one of iron, protected by a coating of oxide of iron. The very great importance of this change will be more conveniently indicated when the process itself is described; suffice it now to say that it added greatly to the durability of the furnace, as well as to the quality of the product.

The introduction of successive minor improvements in the development of the blast furnace, the invention of the puddling process, and the employment of mineral fuel instead of charcoal, both in smelting the ore and converting the pig into malleable metal, were followed by marked economy in both operations. An extended use of cast and wrought iron was the natural result of reductions in price. It will be remembered that in the year 1788 the weekly make of the blast furnaces then in existence averaged $15\frac{1}{2}$ tons; and although I have been unable to obtain any exact figures as to the make, it is highly improbable that the total annual production at that period exceeded, if it even reached, 70,000 tons. In 1796 the annual make per furnace in Great Britain is given at 1,032 tons, or just within 20 tons per week. In the event of the 121 furnaces, mentioned as having been in existence at this period, being in blast, the make would have amounted to 125,000 tons for the year.

As nearly as can be made out by reference to Scrivenor and other authorities, the following figures shew the weekly average make of the British blast furnaces, when in blast, at the above mentioned dates:—

    1788.—15½ tons, using ordinary clay ironstone.
    1796.—20 do.   do.
    1806.—21 do.   do.
    1827.—35[1] do.   do.

The subjoined values of English pig iron, for each fourth year, are taken from Tooke's History of Prices:—

	£	s.	d.		£	s.	d.		
1782.	6	0	0	to	7	10	0	per ton.	
1786.	3	0	0	,,	6	10	0	,,	⎫ No reason seems to be assigned for these
1790.	3	0	0	,,	7	0	0	,,	⎭ extreme variations of prices.
1794.	5	0	0	,,	8	0	0	,,	
1798.	5	0	0	,,	8	0	0	,,	
1802.	5	10	0	,,	9	0	0	,,	
1806.	7	0	0	,,	9	0	0	,,	⎫ The quotations of £7 to £9 remain
1810.	7	0	0	,,	9	0	0	,,	⎪ unchanged for 15 years—viz., from
1814.	7	0	0	,,	9	0	0	,,	⎪ the beginning of 1804 to the end of
1818.	7	0	0	,,	9	0	0	,,	⎭ 1818.
1822.	6	0	0	,,	7	0	0	,,	
1826.	6	10	0	,,	10	0	0	,,	⎱ The highest quotation occurs in the year 1825, viz., £12 per ton.

It has thus been made manifest what a wonderful change the forty years ending 1828 had effected in the iron trade of this kingdom. During this period, the weekly make of a furnace was more than doubled; and the total annual production rose from 70,000 tons to 700,000 tons. We shall subsequently see how, during the succeeding forty years, the 35 tons run weekly from a furnace came to exceed 400, and how the yearly make of 700,000 tons approached very closely to 7 millions.[2]

In 1828, under the condition of things just described, it occurred to the mind of James Beaumont Neilson that heating the air before it

---

[1] This refers to such furnaces only as were using ordinary clayband ironstone; the make, when the mineral was the rich black band of Scotland, being very much higher.

[2] Since this was written, the returns for 1881 give the quantity of pig iron made in that year as having reached 8,352,623 tons.

reached the fuel might be useful. It would, in my opinion, be difficult to justify the reasonableness of this expectation, with all the light which the previous and subsequent labours of physicists have thrown on the subjects of heat and combustion. The idea, however, possessed a certain attraction, and it was speedily put to the test of experiment. The result was that, within four years of its introduction, the furnaces at the Clyde works, in Scotland, were running more than double their former make, without requiring more fuel for the larger than for the smaller quantity. In other words, the consumption of fuel per ton of iron, when burnt with air heated to 600 deg. F. (315 deg. C.), was less than one half of what it was when using the blast at atmospheric temperatures.

The utmost surprise was excited among scientific men by these results, the examination of which will be best deferred until the action of the blast furnace, chemically and otherwise, forms the topic of consideration. I will then endeavour to explain the precise mode of action of the hot blast, the introduction of which was, at the time, of such immense value as to constitute undoubtedly one of the most marked epochs in the advancement of the iron trade.

At the period to which we have brought this history, the comparative cheapness of iron had considerably extended its application not only to those objects to which it had been long adapted, but to certain new inventions, many of which, but for its assistance, would have remained unthought of. Chain cables for mooring ships had taken the place of hempen ropes; pipes of iron, instead of wood, conveyed the water required in our towns; and before long all great centres of population were illuminated by gas, distilled in iron retorts, and led in metal piping to the different points of consumption.

For the economical transport of great weights of any material, water communication was, at the time at which we have arrived, deemed indispensable. Where natural facilities did not permit this, artificial means were resorted to; and from many of our more considerable manufacturing towns the produce was carried away, towards the sea coast or elsewhere, on canals of excellent design and capacity. At the same time, however, in some of our colliery districts (probably from the almost insuperable difficulty of cutting canals), the coal was conveyed from the mines to the ship on railways—an invention which has been in use for this purpose for something like 250 years. For about 120 years after their first introduction the sole material used in the construction of these primitive railways was wood; and it says something for the enterprise of the early coal owners that, at a period (about 1740) when the make of pig in the United Kingdom was only 17,340 tons, one of their body suggested and tried rails of cast iron.

Owing to various reasons—the high price of the metal probably underlying them all—it was 40 years later, or about 1780, when cast iron rails came into frequent use. But, long after this date, wood continued to be often employed; and I remember, so late as about 1840, seeing the old beech-wood rails in use on a colliery railway in the County of Durham.

The economy effected in the manufacture of malleable iron by the puddling furnace began, in due course, to bear fruit; and exactly 20 years after Cort's patent—viz., in 1804—we come upon the use of wrought iron rails, made in the shape of square bars, and only two feet long. These were found unsuitable; and it was not until 1820, when Birkenshaw succeeded in rolling bars of a convenient section, that malleable iron commenced to be the material employed, at first sparingly and afterwards generally, for railway purposes.

This humble form of transport, so long employed in the North of England, was destined to prove the nursery not only for railway communication itself, but for those steam engines without which it would have been useless to contemplate the building of tracks of iron upon anything approaching their present scale.

After driving the stage-wagon from the high-road, iron gradually took the place of wood as a material for naval constructions, in the propulsion of which the wind plays but a secondary part: the furnace and the forge again providing, in the steam engine, the means of locomotion.

About a quarter of a century after the pretty general adoption of Neilson's great discovery, it was becoming apparent that a metal, hitherto considered as typical of strength and endurance, was unequal to the heavy demands made upon it by the ever-growing magnitude of our railway traffic. From this apprehension we were relieved in 1855

SECTION II.—HISTORICAL.

by a further important invention, viz., that of the Bessemer process, made by the eminent man whose name it bears.

In the present section of this work, it will not be inappropriate to recall the various steps which may be regarded as having been the forerunners of this discovery—a discovery which will probably for some time, if not permanently, remain as a culminating point in the progress of the manufacture of iron.

The blast furnace furnished the old forge master with crude iron, so that in his low and expensively conducted hearth he had now to deal with a raw material containing nearly 95 per cent. of metal, instead of an ore often of only half this richness. When Cort proposed to dispense with the costly mode of treatment just mentioned, considerable difficulty was at first encountered from the corrosive action on the furnace, due to the silicon, which is always found in greater or less quantity in all pig iron. It was, in consequence, found more economical to submit the crude metal to the partial action of the old hearth, or refinery as it came to be called, so as to expel a considerable portion of this objectionable constituent. Since the earlier days of Cort's furnace, the process of refining has been generally superseded by the device of saturating the silica, produced by the oxidation of the silicon, with oxide of iron. This takes place in the operation of puddling; and it is only exceptionally that a refinery is now found among the appliances of a modern forge.

Those who have watched with attention the action of refining, will remember that towards its close a violent ebullition and evolution of flame manifest themselves, indicative of a great elevation of temperature. This is partly due to combustion of some of the carbon, but mainly to the oxidation of the silicon contained in the pig. If the operation be continued long enough, these two substances, which confer fusibility on the iron, are gradually reduced in quantity; and the temperature evolved by their action, even when assisted by burning coke, being insufficient to maintain the altered metal in a fluid state, it assumes a semi-pasty condition. In Sir Henry Bessemer's ingeniously contrived apparatus, the circumstances connected with the evolution of heat are greatly modified, by pouring a vast volume of air very quickly through the molten iron. The combustion of the carbon and silicon becomes so rapid, that the temperature of the mass rises to a pitch of intensity sufficient to maintain even malleable iron in a perfectly liquid state.

It was soon found that the plan thus proposed by Sir Henry Bessemer for burning off these foreign substances, the presence of which constitutes the difference between pig and malleable iron, was beset with practical difficulties; which at one time even threatened to arrest the progress of a discovery that has already wrought such changes in the manufacture of the metal we are considering. For reasons referred to in the previous section, it was speedily demonstrated that the ordinary run of British iron was entirely unsuited for treatment in the converter; and for some time even the use of those brands, which did not contain above one thousandth part of their weight of phosphorus, appeared to be hopeless. Recourse was had at length, on the suggestion of Mr. R. F. Mushet, to the known influence of manganese on the quality of steel; and this removed all impediment to the manufacture by the new method, from metal containing not more than the above mentioned proportion of the hurtful ingredient in question. So complete has been the success which has attended this modification of Sir Henry Bessemer's great discovery, that no one, who has studied the subject, will be surprised if, by it or other means having the same object in view, nearly all the old landmarks and dogmas relating to the manufacture of iron, although founded on long experience, are ultimately swept away.

The violent heat excited in the Bessemer converter depends, as we have seen, on the oxidation or combustion of the silicon and carbon contained in the original pig. This fact, therefore, necessarily limits the use of this process to iron containing a sufficient quantity of these two elements, or of some suitable substitute, the burning of which serves to melt the steel. If a system of producing steel is to be pursued, in which, from the nature of the materials employed, it is impossible to generate the heat in the manner of the Bessemer converter, then the use of ordinary fuel becomes unavoidable. In the reverberatory furnace, as it is commonly constructed, there would be very great difficulty in maintaining a steady temperature of the proper intensity. Each time fresh coal is thrown on to the fire, the vaporizing of the hydro-carbons, and the admission of a large volume of air, greatly tend to cool the contents of the hearth. Besides this inconvenience, there is an immense loss due to the arrival of the products of combustion at the chimney long before they have parted with more than a fractional portion of their heat, although they are already cooled to a temperature below that to which it would be advantageous to expose the substance under treatment. These difficulties are admirably met in the "regenerative furnace" of Messrs. Siemens. The fuel is employed in a gaseous condition, and the heat that would otherwise escape is imparted to the incoming air and gas; by which means the interior of the structure and its contents are raised to a temperature capable of maintaining any quantity of steel, or even of wrought iron, in a state of perfect fusion.

Exception has been taken to the extent of the service rendered by these gentlemen to metallurgic science, founded on the fact that the idea upon which their furnace is constructed had been described by a previous inventor. Whatever may have been the merits of the latter in connection with the subject, certain it is that Messrs. Siemens, with or without the knowledge of what had been previously done—it matters little—have introduced to the world an invention, invaluable as a means of commanding a most intense temperature for purposes never contemplated, so far as I know, by any former discoverer.

In respect to the immense loss of heat which, thirty or forty years ago, was uniformly permitted to take place from all puddling or mill reheating furnaces, it may be here parenthetically observed, that this serious waste is now avoided to a considerable extent by using the escaping flames for the purpose of driving the steam engines employed in the works.

To some extent the principle of the regenerative furnace has been introduced into the so-called retort furnace of Mr. Price. In this the heat otherwise lost is made to surround an upright retort, where the coal parts with a considerable portion of its gaseous constituents before it reaches the fire-place. The volatile portion passes through the furnace in the usual way, while the fixed carbon, more or less in a state of bright incandescence, descends to the grate bars, the cooling effect already spoken of in connection with reverberatory furnaces being thus very greatly diminished. The air required for combustion enters the fire-place in a state of compression, and is also previously heated by the waste heat of the furnace itself. Speaking from personal observation, it may be added that the furnace of Mr. Price has shewn itself quite capable of commanding a temperature sufficiently intense to melt steel with great facility.

During the thirty-five years, or thereabouts, which followed the application of heated air to the smelting of iron, little or no improvement was effected in the construction of the necessary appliances. Ironmasters, amazed in many instances at the unexpected results of Neilson's invention, were probably satisfied that its then unexplained economy could not be further improved upon. It had, it is true, been demonstrated by philosophical research that the calorific power of the fuel was far from being exhausted by the work it had performed in the blast furnace. As far as concerns Great Britain, however, it must be admitted that in those days the voice of true science was rarely heard in our ironworks, and still more rarely was it listened to.

Changes in the dimensions of the furnaces in use had, no doubt, been attempted; but the additional capacity was either insignificant, or the mechanical appliances were incommensurate with any great alteration in the general arrangement of the plant. Hence it happened that the several furnaces put down soon after 1851 in the new district now known as the Cleveland were, after due enquiry, erected on the old lines—*i.e.* they varied from 45 to 50 feet in height, with boshes having a diameter of from 14 to 16 feet. At that period, it may be said that the smelter on the Tees was satisfied, if he produced a ton of grey foundry iron with 35 cwts. of coke; while in Scotland, the other chief seat of the pig-iron trade, something like 50 cwts. of raw coal was the usual weight of fuel employed in the furnace. In both cases no fault was found with the furnace manager if the blast was hot enough to melt lead readily—indicating a temperature of about 650° F. (343° C.)

In this rapid sketch of the progress of the iron trade, an Englishman may be permitted to observe with satisfaction that—with the exception, perhaps, of the gradual growth of the high from the low furnace, which is attributed, on no very sufficient grounds, to Germany

SECTION II.—HISTORICAL.

—the different strides towards improvement have been exclusively of British parentage. The next, however, and a very important one it proved to be, is due to a Frenchman, M. Fabre Dufaur. He was the first to utilise the vast volume of flame which flashed from the throats of our older blast furnaces, lighting up the sky and the country for miles round the great centres of their operations. This French idea was first put into general practice in this country, for the purpose of raising steam and heating the blast, among the iron works of South Wales, where the apparatus connected with its use were simplified and improved by Mr. Parry at Ebbw Vale. The new furnaces, near Middlesbrough, were the next where it was applied, and so successfully that the direct saving in coals achieved by its means has amounted to not far short of a million and a-half tons per annum. This is, moreover, not the whole of the economy it has effected. The higher and more regular temperature maintained in the blast is the cause of a notable saving of coke in the furnace; and, in the matter of labour, it has been found that in an establishment of twelve furnaces it saves the work of thirty men or more, who were formerly engaged in coaling the fires of the blowing engine and hot air stoves.

No marked economy having attended a moderate increase of the dimensions of blast furnaces, the new works at Middlesbrough were constructed similar to those which had obtained favour by many years' experience elsewhere—*i.e.* they did not exceed a capacity of 5,000 to 6,000 cubic feet. Messrs. Whitwell and Company departed from the practice hitherto observed by adopting a height of 60 feet, which, however, was not more than what had been already tried in Wales. A trifling reduction in the fuel consumed appeared to attend this change in form, but not more than that which had been already obtained in the 48 feet furnaces at Port Clarence, by the use of air heated to nearly 1,000° F. instead of 700° or 800°.

The late Mr. John Vaughan, in the year 1864, erected a furnace of 75 feet, which was done chiefly, if not entirely, in the hope of increasing the weekly make; for this gentleman had not given any attention to the subject of the actual waste of heat in smelting iron. The experiment, however, was not only eminently successful in the direction in which he expected, but there was a most important saving in the consumption of fuel. Speedily other furnaces were erected somewhat higher—80 feet instead of 75—and of double the internal capacity of Mr. Vaughan's, with such beneficial results that before long all the small furnaces on the banks of the Tees were demolished, in order to make way for the colossal structures now universally employed in that district.

Hitherto the limit to which the blast had been heated was that imposed by the power of the iron pipes to resist the action of the fire. This impediment was removed by Mr. E. A. Cowper's proposal to adopt the regenerative principle of Messrs. Siemens, in which brick-work was raised to a high temperature, and the heat thus stored up was then conveyed into the furnace by passing the air over the hot surface of the bricks.

The ironmasters had thus placed at their disposal both larger furnaces and hotter blast—two very valuable modifications of the appliances they had been in the habit of using. Some among them were sanguine enough to believe that the margin of economy to be effected by one or both of these changes was very much larger than subsequent experience warranted. This, however, is a question that may be conveniently deferred until the action of the blast-furnace itself comes to be described.

It will be useful at this place to glance at the effect produced by the various improvements in iron smelting referred to. For this purpose I have extracted from different sources an approximate statement of the weekly make of a blast furnace in 1835, 1845, 1855, and 1865; together with the quantity of coal, reckoned in its raw state, required to make a ton of iron.

	Weekly make. Tons.	Coal per ton of iron. Cwts.
1835.—Height 40 to 50 ft., capacity 5,000 cub. ft., blast cold	70	120
1845.— Ditto, ditto, blast at 650° F.	120	85[1]
1855.—Height 40 to 50 ft., capacity 5,000 cub. ft., blast at 800° F., and using the escaping gases for steam and hot air	220	62
1865.—Height 80 ft., capacity 20,000 cub. ft., blast at 1,000° F.	450 to 550	40

---

[1] The saving consequent upon the introduction of heated air was very irregular in its amount. In South Wales it was stated to be something under one ton of coal per ton of iron, whereas no less than five tons was claimed in Scotland as having been economised by its use. The causes of this extraordinary discrepancy will be explained when the theory of the action of the hot blast is discussed.

SECTION II.—HISTORICAL.

In all these examples the mineral under treatment is supposed to be ordinary clay ironstone, yielding about 42 per cent. of iron in its calcined state. When mineral of a richer description, and more quickly reduced, is employed, such as the magnetic ore of the United States, as much as 1,200 tons has been run in one week from furnaces containing only about 10,000 cubic feet.

The last line of figures, in the table just given, contains a remarkable contrast with the particulars given by some[1] French engineers of a works they visited in Wales about forty-five years ago. Then, in order to produce weekly almost exactly half the make of iron from one of the above furnaces, no less than five furnaces were required. These were worked by 309 men, women, and children; 257 of the number being men.

It would be inconsistent with the object of the present work to attempt to do more than name, as occasion requires, the various furnaces or mechanical contrivances employed in the different processes referred to. A complete description of the whole of these is to be found in the admirable English work by Dr. Percy, as well as in those of many foreign writers, in France, Belgium, Germany, and Sweden.

In 1796, or a few years earlier, pig iron had fallen sufficiently in price, and the art of the founder had made sufficient progress, to permit the construction of a bridge, at Sunderland, with large segments of cast iron. It was at the time considered a work of sufficient risk to render appropriate the affixing to the structure the motto—*Nil desperandum auspice Deo;* and every traveller to the North of England considered himself bound to visit what then was regarded as a most daring example of metallic engineering.

For a long time antecedent to Cort's use of grooved rollers, plates or sheets of iron had been spread out between plain cast iron cylinders. The last named invention is ascribed by Coxe, in his "Tour in Monmouthshire," to John Hanbury, whose family, in the early part of the eighteenth century, were lessees of coal and iron under the Earl of Abergavenny, in South Wales. The power of Hanbury's plate mills and those in use for above a hundred years after his time was such

---

[1] "Voyage Metallurgique en Angleterre," by MM. Dufrénoy, Elie de Beaumont, Coste, and Perdonnet.

as to cause a piece of 400 lbs. to be held as one of unusual dimensions, and to be paid for accordingly. What a contrast this affords to the practice of the present time, in which are to be seen furnaces to heat, mechanical arrangements to move, and rolls to receive, masses of iron weighing nearly 40 tons, to be converted into armour plates for ships of war!

In the early days of public railway construction, the only experience possessed by engineers to determine the strength required for their materials, was that obtained with the slow speed and light weights of the colliery lines, or wagon-ways, as they were styled, in the North of England. A rail weighing 50 lbs. per lineal yard was considered to afford ample margin for a considerable increase in velocity and tonnage; and these our rolling mills were not generally required to deliver in lengths exceeding 15 feet, or weighing, therefore, more than 250 lbs. when finished. Latterly, machinery has been constructed to roll 90 feet of finished rail, in one piece weighing 2,460 pounds; and this with not above half the men required to produce the old rail of 250 lbs. Such a mill, driven by the reversing engines designed by Mr. Ramsbottom (formerly of Crewe), has been known to turn out above 3,600 tons of rails in a single week.

The machinery employed in hammering iron, which either from its shape or size could not be fashioned in the rolling mill, was formerly of the most insignificant dimensions; and this necessarily limited the weight of the work which could be undertaken. No hammer exceeded a few tons in weight, and of this only a very small portion was effective in giving the blow, the height of which never exceeded 3 or 4 feet. The peculiarity of construction of the Nasmyth hammer commands a blow due in some cases to the fall of a mass of 80 tons through a perpendicular height of 16 feet. Moreover the effective power of the impact is not limited to that produced by the mere falling weight, as in the old forges, but may be supplemented by steam pressure in the cylinder which works the hammer.

The application of cast steel, particularly in France, to articles of very large dimensions has necessitated the employment of apparatus of corresponding capacity. At Creusot, in 1878, the members of the Iron and Steel Institute were permitted to witness the casting of armour

SECTION II.—HISTORICAL.

plates of 25 or 30 tons, which were afterwards consolidated under the crushing blows of the 80 ton hammer. At St. Chamond, near St. Etienne, are to be found two ladles, each capable of holding more than 25 tons of fluid steel; and there the needful power of resistance is sought to be conferred on armour plating by annealing the pieces in a cistern containing 100 tons of oil.

Perhaps a still more impressive illustration of the vast power of the appliances employed in the manufacture of iron is furnished in forging the coils for ordnance purposes, as first suggested and first practised by Sir William Armstrong. It is, indeed, a marvellous sight to behold a mass of metal, sometimes weighing as much as 50 tons, drawn from a cavern heated to dazzling brilliancy, and then to watch the ponderous steam hammer effect a weld, many yards in length, so perfectly as to be able to resist the tremendous strain which modern artillery has to endure.

Thus it may be said that, about the commencement of the present century, 20 or 30 tons of metal per week, as dribbled from our blast furnaces, was handed over to the Lancashire fire and its accompanying hammer of puny dimensions. By their united action something like a ton of wrought iron bars was obtained every 24 hours; which bars could only be sold at prices that forbade their use except in very limited quantities. From this primitive state of the process we pass to the blast furnace, driven with heated air, and furnishing 20 puddling furnaces with a weekly supply of something like 250 tons of metal apiece. In these 20 furnaces the laborious exertions of 80 men were required to agitate the molten iron, so as to expose a succession of fresh surfaces to that chemical action, by which it was purged from the foreign substances acquired during the smelting of the ore. The heat of the puddling furnace, being far below that required to fuse the iron, when separated from its associated carbon and other bodies, all the workman could do was to soften his product, and then, by hard manual labour, cause it to stick piece by piece together in the furnace itself.

In the puddling furnace something like an hour and a half was needed for dealing with 4 to 5 cwts. of pig; and two good men did well if they laid 22 cwts. of puddled bars on the bank, for a day's work. In place of this laborious operation, 7 or 8 tons of metal are now brought direct in one charge from the blast furnace, to the converting vessel; and the same air which burns off the impurities, by its passage through the iron, dispenses with the use of any other motive power, than that furnished by the blowing engine. In 15 or 20 minutes the crude metal is malleable iron; and the change is wrought with such rapidity, that the heat evolved by the combustion of the impurities is intense enough to maintain the contents of the converter in a perfectly fluid condition.

Chemically, as stated above, the product is malleable iron; mechanically, it possesses certain points of dissimilarity. Upon our ability to modify these latter will depend the realization of those prophecies, which predict the complete supersession of iron by what is generally, but often somewhat improperly, denominated steel.

Leaving the past and looking to the future history of this national industry, the question uppermost in the minds of our iron masters is, undoubtedly, the application of ordinary British pig iron to the manufacture of ingot metal or steel. The Transactions of the Iron and Steel Institute are replete with proofs of the attention which this problem has received for some years past at the hands of its members and others.

By some it may be considered premature to declare that the elimination of the phosphorus, which constitutes the barrier to the use of pig iron obtained from our clay ironstones, bids fair to become a generally accepted process in our steel works. The laws, however, which govern the removal of this hurtful ingredient are sufficiently well understood, and enough has actually been done to justify the assertion that it is a subject which has progressed far beyond the region of experimental research. In illustration of the complete success which has attended the application of the Basic process, in the removal of phosphorus during the process of conversion by the Bessemer system, the average composition of 47 rails, made of the best hematite iron from the works of three makers, is contrasted with that of 20 rails made at the Eston works, from pig containing probably 1·75 per cent. of phosphorus. Both lots were taken indiscriminately as received by the North Eastern Railway, and analysed in the laboratory of that Company.

SECTION II.—HISTORICAL.

	Carbon. per cent.	Silicon. per cent.	Sulphur. per cent.	Phosphorus. per cent.	Manganese. per cent.
47 rails, from Hematite pig...	·434	·065	·091	·053	·915
20 do. Cleveland pig...	·451	·094	·095	·053	1·020

With such results before us as those recorded in the above figures, it is clear that the only question which requires consideration in connection with the removal of phosphorus from iron is the cost at which it can be accomplished. In other words, can the cheap pig of Cleveland, when burthened with the expense attending the Basic treatment, compete with the dearer metal of Lancashire and Cumberland, which latter can be converted into steel by a somewhat simpler manipulation? This is the question to which those interested in metallurgical science in Great Britain still await a definite answer.

The effect, in an economic point of view, produced by the various improvements briefly reviewed in the present Section, is little short of marvellous. During the 40 years already alluded to as ending in 1826, pig iron was rarely quoted below £5, while the average over the entire period may be taken at £7 per ton. Recently we have seen the same article sold at 32s., and steel rails of the highest quality have been obtained for a considerably less sum than the lowest average price of pig iron during any year of the 40 years above referred to. This great change in the cost of production is, to some extent, to be ascribed to the discoveries of such ores as the Black Band, and those found in Cleveland and analogous districts. The chief source of economy is, however, unquestionably due to the improvements introduced into the manufacture itself.

SECTION III.

ON THE DIRECT PROCESSES FOR MAKING MALLEABLE IRON.

MALLEABLE iron and steel are obtained, as has been described, by the circuitous method of first making pig iron, and then removing the foreign matter which the metal has taken up in the furnace from the materials used in smelting. Any process, therefore, having for its object the production of malleable iron from the ore at one operation, is distinguished by the use of the word "Direct."

No serious attempt has been made to revive in this country the obsolete and almost forgotten Catalan furnace—much less its more humble predecessor, the low hearth of Asia and of Africa. It is desirable nevertheless to consider briefly the conditions of this primitive mode of procedure, in order that we may more correctly appreciate the ground upon which it is now sought to resuscitate a process, now so long abandoned in almost every iron making community in the world.

I have had no opportunity of inspecting such a furnace as that previously referred to, and described to me by Colonel Grant; but, according to Dr. Percy,[1] in the operation as pursued in Asia, the structure costs under ten shillings, and the furnacemen are satisfied with three half-pence per diem. Even at this miserable rate of wages, the cost of labour on the raw mass or ball is probably much more than that at which the same ore could be converted into a steel rail ready for use, by men earning on an average 4s. or 5s. per day. In producing the bloom, six times its weight of charcoal is used, and half the iron contained in the ore is lost. In reheating the crude lump for forging into a bar, one half its weight is wasted, and again there is a considerable expenditure of charcoal. We are probably within the mark in accepting

[1] Work on Iron and Steel.

SECTION III.—DIRECT PROCESSES.

as a fact, that ten times as much fuel and three times as much ore are required, for every unit of metal produced, as is consumed in the blast furnace and Bessemer process together.

Some Catalan furnaces, which I had an opportunity of examining in North Carolina, were near 3 feet from back to front and 2 feet from side to side, by 18 inches or 2 feet in depth. They were blown by a *trombe*—a very simple form of apparatus, in which the current of air is produced by water falling through a square upright box of wood, the blast being conveyed to the hearth through stems of trees bored for the purpose. Into the furnace are thrown charcoal and ore, the latter in small fragments. The hot embers, and the masonry heated by the previous charge, quickly cause combustion to pervade the mass, when the blast is turned on.

In this direct mode of dealing with the ore, as in the blast furnace proper, there are two distinct stages through which the mineral has to pass: viz., first, the expulsion of the oxygen with which the iron is associated, and secondly, the raising of the reduced metal to a welding heat in the bloomary, or to the fusing point in the blast furnace. Again, so far as economy of combustible is concerned, it is essential that the intensely heated gases, generated at the tuyeres, should communicate as much of their heat as possible to the materials, on their way downwards in the furnace. Failing this, the loss by the gases escaping into the atmosphere at a high temperature is exceedingly great.

It will be most suitable to defer considering, with the proper degree of minuteness, the questions raised by the conditions just enumerated, until we are describing the action of the blast furnace. It will be convenient, however, to mention at the present moment that, as the office of reduction is to withdraw oxygen from the ore, it is of importance that the gases near the tuyeres should be as free from this gas as possible, otherwise there would be a risk of reoxidizing the metal. This is the case in the blast furnace, because reduction is performed in the upper part of the furnace during the gradual and slow passage of the ore downwards; but in the Catalan or any similar furnace, much of the mineral is deoxidized at or near the tuyeres. Hence arises the necessity of burning additional quantities of carbon, to give sufficient reducing energy to the gases at this point. Even with this precaution, complete reduction of all the iron is impracticable, which is the cause of so great a proportion of the metal being carried off as slag in low structures such as those last mentioned. It will hereafter be seen how admirably a well-appointed blast furnace discharges the threefold duty of intercepting heat, expelling the oxygen from the ore, and raising the product to the required temperature; while in the Catalan furnace these operations, from the nature of the apparatus, are all of them most inefficiently performed.

So much for the imperfections of the direct process—a word now for its advantages. In most iron ores phosphorus is found existing, either as a phosphate of iron or phosphate of lime. In the blast furnace the deoxidizing power is so intense, that both these compounds lose their oxygen, and are converted into phosphide of iron or phosphide of calcium, as the case may be. If the latter, phosphide of iron is still the final product, since the calcium is displaced by the molten iron in the lower zone of the furnace. The phosphide of iron so generated is dissolved by the metal, and, if in excessive quantity, gives rise to great inconvenience in the forge: and when exceeding 1 part in 1,000, it utterly unfits the pig for steel-making.

On the other hand we have the ancient, and to a great extent obsolete, method of direct procedure, which during the last 30 years has engaged more attention than its actual merits, as an economical process of making iron, perhaps deserve. To its apparent simplicity has to be added the desideratum of giving a product comparatively free from phosphorus; for, owing to the partially oxidizing tendency of the operation, the phosphates are chiefly left in the slag. Now that steel is the object which occupies the special attention of those engaged in the manufacture of iron, this direct process is deserving of some further examination, than that which would be its due apart from this consideration.

There are two plans for accomplishing the reduction of the ores of iron, without the operation of smelting as it is commonly understood. In the first the mineral is heated in contact with charcoal in closed vessels; in the other a mixture of ore and carbonaceous matter is heated on the floor of a furnace of the reverberatory principle, either stationary or revolving.

The actual quantity of carbon required for the reduction of the particular compound of iron and oxygen usually found in our ores as they reach the smelter—viz.: the peroxide—is in reality inconsiderable. Heat, applied to a mixture of the fuel and ore, converts the carbon into carbonic oxide, or into a mixture of carbonic oxide and carbonic acid; the oxygen of course being derived from the ore. Supposing charcoal to be pure carbon, which however is not the case, and such carbon to escape entirely as carbonic oxide, then 6·428 cwts. of this fuel would suffice to reduce 20 cwts. of iron to its metallic state, from a condition of peroxide. It will, however, hereafter be shewn that carbon can carry away from iron more oxygen than that required to form carbonic oxide; and we may therefore assume that $6\frac{1}{2}$ cwts. of common charcoal, or thereabouts, would suffice for the mere reduction of the ore.

The moderate temperature required for the operation, and the small weight of carbon necessary for the chemical part of the process, doubtless offer great inducements to pursue the enquiry, both from the point of view of cost and of simplicity. How far expectations in either direction have been verified by actual experience, we will now proceed to consider.

I will pass over the attempts of Mr. Clay, who laboured long and assiduously in the cause, and who, after very many extensive trials made at the Walker Works in the year 1846, failed entirely to convince me, or himself, that it was not cheaper for the forge owner to buy pig iron at 60s. per ton than to treat Lancashire ore in a reverberatory furnace by the direct process, although that ore cost less than one-third of this money.

In 1855 M. Chenot received the highest honour the Jurors of the Paris Exhibition of that year could bestow, for his direct process, declared by a high French authority to be the "greatest metallurgical discovery of the age." In spite of this recommendation there is, I believe, only one work in the world where it is in operation; or at least was so in the year 1872, when I visited the locality in Spain. Even in this case, its continuance seemed dependent upon the possession of a quantity of charcoal screenings, which would otherwise have been wasted. By the courtesy of the proprietors, Messrs. Y. Barra & Co., I was permitted to examine the process, as it was being carried on at that time.

The ore, in small pieces previously calcined, is placed along with charcoal in an upright retort of brick, about 33 feet high, its horizontal dimensions being about 4 ft. 9 in. by 1 ft. 4 in. Heat from a coal fire is applied externally, causing a wasteful expenditure of fuel. In three or four days reduction of the ore is complete, and the charge is then withdrawn and cooled in a vessel closed so as to exclude the air, which would otherwise reoxidize the porous metal, or "sponge" as it is termed. This immunity from reoxidation cannot, of course, be secured when the product has to be reheated as a preliminary to being drawn into bars. The waste of metal, by the fire acting on so large a surface of iron, is very great: 30 to 40 per cent. being the usual loss from this cause.

Fifty years ago, when five tons and more of coal were frequently used to make a ton of pig iron, the Chenot system, when dealing with *rich* and *cheap* ores, might have had some pretensions to hold a place, in competition with the blast furnace and the puddling process. Against the present more perfect mode of smelting iron, where the ton of pig metal is obtained with 40 cwts. of coal, the struggle, in my opinion, would be hopeless; and still less has this direct plan a chance of holding its own against the Bessemer mode of treatment, in which the waste gases from the blast furnace, with proper machinery, suffice to expel, in the converter, the impurities absorbed during the smelting of the ore.

In support of this opinion, I have taken the cost of making rolled Bessemer steel as unity, and against this have estimated approximately, from data in my possession, the cost of bar iron produced by the Chenot process:—

	Cost of Bessemer Rolled Steel.	Cost of Chenot Rolled Bars.
Fuel (charcoal) ...	100	(part charcoal) 230
Loss of Metallic Iron	100	... 350

Notwithstanding the overwhelming disadvantages with which the direct process is thus burthened, even after some years of experience, the idea of avoiding the circuitous treatment by the blast furnace, and the subsequent operations, has not been suffered to be forgotten.

Mr. T. S. Blair, of the Glenwood works, near Pittsburg, U.S.A., considered that the large waste of combustible, in heating the exterior of Chenot's retorts, was due to the difficulty the heat had in penetrating their contents. Accordingly, he sought to remedy the evil by applying fire to an annular space, which contained the mixture of ore and charcoal. By means of a central tube in the retort, the materials, having now a thickness of only $4\frac{1}{2}$ inches, were heated much more readily than when they occupied a space of $15\frac{1}{2}$ inches. In some cases the apparatus had a diameter of 4 feet, and a height of about 45 feet. Heat was applied only to the upper 8 or 10 feet; the remainder of the retort being used for allowing the product to cool before coming in contact with the air.

The weight of fuel consumed, per ton of sponge made, was stated to be about 8 cwts. of charcoal and 27 cwts. of coal—in all, 35 cwts. This is not so much as is consumed for making Bessemer pig metal; but, as nearly one-fourth of the quantity is charcoal, the cost will, generally speaking, be something in excess of that in the blast furnace, where only mineral coal is employed.

As much as two tons of sponge were obtained per day in such an apparatus, and at a very low cost for labour, according to the figures I received; but nevertheless, I am mistaken if it was not higher than that paid in England at the smelting furnace and Bessemer converter put together.

The inducement to revive this modification of the Chenot process was to use the product for making steel in the method known as the Siemens-Martin, which, as is well understood, consists in melting malleable and pig iron together. The sponge was thrown into the melted pig metal, when, according to Mr. Blair, the supernatant slag protected the porous iron from the oxidizing action of the fire. It was my misfortune not to agree with Mr. Blair, either as to the extent to which this protection was accomplished, or in his estimate of the advantages possessed by his direct process as a partial substitute for the blast furnace.

My reasons have been given above for regarding the sponge, in the matter of fuel and labour, as decidedly more expensive than an equal weight of pig iron. Now, in the so-called "Ore process" for obtaining steel—to be hereafter spoken of—to 100 parts of melted pig iron, as much ore as contains 10 to 15 parts of iron is added; and the result is that the carbon and silicon in the pig reduce (by a much cheaper "direct process" than any hitherto practised) as much iron, as gives the maker about the same quantity of steel that he uses of pig metal.

SECTION III.—DIRECT PROCESSES.

The actual loss of metallic iron in the operation just described is very small—not exceeding 6 per cent. of the entire quantity delivered to the melter in the form of ore; and on this no expense for smelting charges, &c., has been incurred.

The ore in use for the Blair process, on the occasion of my visit at Glenwood, was from the Iron Mountain of Missouri, and was said to contain 66 per cent. of iron. If this was correct, the sponge should have yielded 90 per cent. of iron—supposing all the metal it contained to be reduced, as was supposed to be the case. But the average composition, as given to me, consisted of the following ingredients, in the proportions annexed:—

Pig Iron.........	360 parts,	Metallic Iron, taken at	94	per cent...	338	parts.
Scrap Steel......	60 do.	Do. do.	90	do. ...	54	do.
Spiegel..........	110 do.	Do. do.	80	do. ...	88	do.
Sponge..........	470 do.	Do. do.	90	do. ...	423	do.
Total of charge	1,000 parts.	Total of Metallic Iron...............			903	parts.
Steel received...	768 do.	Do. do. at 99 per cent			760	do.

[1] Shewing a loss of 15·83 per cent. of Iron, or 143 parts.

When considerations of a commercial nature are applied to these figures, the result holds out little hope of the process occupying any important position as a means of making either iron or steel. It has already been shewn that the labour is higher, and the fuel quite as high, in producing a ton of sponge as in smelting a ton of pig iron. As there is practically no loss of metal in the blast furnace, it may safely be admitted that the cost of sponge will be *at least* as great as that of pig metal. We have thus, in the ore process, say 1,000 tons of pig iron added to 250 tons of ore, and producing 1,000 tons of steel; and in the Blair process, 1,000 tons of material costing fully more per ton than the pig, but only giving 760 tons of steel. Few manufacturers would, under such circumstances, hesitate in assigning a difference of 15/- per ton of steel in favour of the use of pig and ore

---

[1] It is only right to mention that Mr. Morrison Foster, the partner of Mr. Blair, does not agree with me in the opinion I formed of the direct process, as carried on at Glenwood. In a letter addressed to the American manufacturer, owing to some differences in the figures we made use of, he makes the loss in actual metal 14·62 per cent. His numbers are no doubt entitled to be preferred to mine, but the discrepancy is not such as to alter the views expressed, which I regret are not those of Mr. Foster.

over that of pig and iron sponge. Nor is this all; for it must not be forgotten that the metal made by the direct process constitutes only 47 per cent. of the whole charge, and that to this the whole of the extra cost is to be debited; so that, in reality, the steel obtained from the sponge has been produced at a cost of above 30s. per ton over that which the other ingredients are capable of affording, by the alternative mode of treatment.

A further loss falls on the steel made from the iron sponge, due to the smaller output of the open hearth in which it is melted, viz. 760 tons, as against probably 1,000 tons in the ore process. It is, therefore, not improbable that the difference of 30s. given above, may be increased to 33s.; and this would, it is submitted, amply justify the opinion already expressed respecting the future of Mr. Blair's process.

More recently Mr. Dupuy has communicated to the Franklin Institute of Philadelphia a mode of protecting the sponge, when made, by enclosing the ore and carbonaceous matter in cases of sheet iron. When reduction is completed the whole is brought to a welding heat, and in that state is drawn out into a bar. According to his own estimate the cases will cost 23s. 5d. per ton of iron obtained; which cannot fail to prove a serious obstacle to the introduction of the system recommended by this gentleman.

In all the trials hitherto spoken of in connection with the direct process, the purer ores of the hematite class alone constituted the subjects of experiment. These were purposely selected in order not to encumber the apparatus unnecessarily with inert matter, and thereby add proportionately to the expense of obtaining the actual product. But such ores were also at the same time free from phosphorus; whereas the elimination of this substance, had phosphoric ores being used, would have been at all events a partial set off against the costliness of the process itself. Dr. Siemens, Past President of the Iron and Steel Institute, impressed with the advantage of using the regenerative principle in combination with a rotating furnace, has not confined himself to the use of the purer oxides of iron, but has operated for a considerable time on the earthy ores of the oolitic measures in the Midland counties, containing a large quantity of phosphorus. A mixture of the crushed mineral and coal is introduced into the furnace, which is gradually heated up to a full red heat, until the ore is supposed to enter into a state of fusion. By this mode of treatment it was believed that reduction would be effected by the carbon suspended in the liquid mass, which would neutralize in a great measure, if not entirely, the oxidizing character of the carbonic acid and vapour of water, produced by the complete combustion of the coal.

A very instructive and highly interesting series of papers is to be found in the Transactions of the Iron and Steel Institute and of the Chemical Society, giving in ample detail the results obtained by Dr. Siemens in the course of his experiments. These results are such as to have inspired their author with great confidence in the future reserved for the process, on which he has bestowed so much thought and perseverance. In the course of the papers referred to, stress is laid on the fact that, whereas in the blast furnace the carbon chiefly escapes as carbonic oxide, in his regenerative apparatus the whole of it is practically converted into carbonic acid. Undoubtedly the heat obtained from the combustion of every unit of fuel is, on this account, much greater in the latter than in the former case. Owing to a variety of causes, however, even in this improved form of furnace not much more, if any more, than 25 per cent. of the heat evolved in the fire place is made available. On the other hand, it will be hereafter seen that in the blast furnace, with a less perfect character of combustion, the fuel actually performs a much larger comparative duty than that just mentioned.

I am quite prepared, from personal observation, to admit that a very considerable portion of the ore is ultimately fused by Dr. Siemens; and that possibly reduction under the favourable conditions claimed may, to some extent, take place. I am nevertheless assured, from my own experiments, that it will be extremely difficult to heat up a mixture of carbon and oxide of iron towards fusion, without the operation being accompanied by the reduction, previous to fusion, of a very large quantity of the metal contained in the ore. The moment this reduced iron meets the highly oxidizing atmosphere of the furnace—which must, in a rotating furnace, happen more or less—waste will begin.

Undoubtedly the useful product of this direct process contains only a fractional part of the phosphorus found in the ore; whereas the blast furnace, as has been mentioned, gives us metal containing practically the whole of this deleterious ingredient, as originally associated with the minerals employed. Its absence, however, from the iron obtained in the rotating furnace is entirely due to the oxidizing nature of the vast quantity of iron which either escapes reduction, or, being reduced, passes again as oxide into the cinder. Were it possible, therefore, to carry on the operation without this loss, it is highly probable that the malleable iron obtained would be found contaminated with phosphorus.

The view just expressed, as to the nature of the action, appears to coincide with the personal observations of my friend, P. Ritter von Tunner,[1] who calculated, from the particulars given him of 100 heats with a mixture of calcined Northampton ore and calcined Black band operated on, that 27·6 per cent. of the metallic iron was lost in the slag.[2]

The product obtained possessed an advantage over that of the Chenot process, inasmuch as it was in the form of a solid hammered bloom, containing 99·71 per cent. of iron, and free from the great liability to waste in the subsequent stages of the manufacture. It only contained ·074 per cent. of phosphorus, although made from materials which would probably give a pig iron having 1 or 1·25 per cent. of this metalloid, out of which ·2 or ·3 per cent. would remain in the bar iron when puddled in the ordinary way. We have here a material which can be used either for steel making or for the production of bar iron; but it is chiefly in connection with the former that Dr. Siemens attaches a value to the process.

Let us examine the subject under both these heads, regarding Dr. Siemens' product as a substitute, wholly or partially, for iron in the form of pig which, with the addition of ore, is capable of affording steel. The economy of the blast furnace, in the matter of fuel, has been already alluded to incidentally, where one ton of iron was stated to require only two tons of coal. A word now as to the labour. The very nature of the smelting process secures the exercise of an economy, in this respect, for which it would be difficult to find a parallel in the entire range of manufacturing operations. As an illustration, let us take a pair of furnaces making 1,000 tons of iron per week, which is by no means an uncommon rate of production. The three classes of raw

---

[1] "Das Eisenhüttenwesen der Vereinigten Staaten."

[2] Although Professor Tunner, in his Preface, mentions that the *centner* (cwt.) made use of is one of 112 lbs., his calculation of the loss is on one of 100 lbs. The waste of iron, therefore, in reality amounts to 35·3 per cent. instead of 27·6.

SECTION III.—DIRECT PROCESSES.

material consumed in the Cleveland works for this quantity will weigh together about 4,750 tons, the whole of which can be so dealt with as to be brought in train loads to two or three points in close proximity to the furnaces. From the depôts they slide by gravitation into the wagons or barrows, from which they are shot into the tunnel heads. In like manner the pig iron and the refuse are so concentrated, in point of space, that effective mechanical means can be largely employed as a substitute for manual labour. All this is entirely changed, if anything approaching the weight referred to above is to be treated in a range of furnaces, each of which occupies four and a half hours in the reduction of 20 cwts. of ore, according to the weights given by Professor Tunner. A large area of ground would necessarily be occupied; and the materials having to be distributed in small quantities at various points over its entire surface, such labour-saving contrivances as are usefully employed at the blast furnace become impracticable.

In my opinion, the advocates of these direct processes underrate the importance of the duty performed by the blast furnace, in connection with such ores as those operated on by Dr. Siemens at Towcester, while they magnify its cost. In the case of an ore containing 40 to 42 per cent. of iron, there will be found almost exactly the same weight of earthy matter as there is of metal. These earths have a very powerful affinity for oxide of iron, and therefore seriously interfere with the process of reduction at high temperatures. Besides this serious inconvenience, we are thus, in the direct process, crowding up the space of an expensively worked piece of apparatus with inert matter, equal in weight and more than equal in bulk to the iron itself. Under such circumstances as those just related, it is needless to say how important it is to rid the process of this foreign matter as speedily as possible.

It will be more convenient to wait till we are dealing with the action of the blast furnace itself, before we show how very small a quantity of fuel is there required to effect the fusion of these earths. Suffice it now to say that the entire cost—for fuel, wages and flux—of freeing the metal from its earthy admixture can be shewn not to exceed 5s., on the ton of metal obtained from an ore containing 40 to 42 per cent. of iron.

Professor Tunner gives no less than 36s. 8d. (20 gulden) per ton as the cost for labour at the Towcester works, carried on by Dr. Sie-

SECTION III.—DIRECT PROCESSES.

mens. This seems excessive, and may be partly due to the fact that the process at that establishment had scarcely passed—if, indeed, it had passed—the mere experimental stage. This item cannot however fail to be high, for in its nature the process closely resembles mechanical puddling, in which the wages per ton of product are at least four times as great as those incurred at the blast furnace. This difference will unquestionably be further increased by the circumstance that, while pig iron loses but little in the process of puddling, the ore in the rotary furnaces, with this direct process, only yielded about 33 per cent. of its weight in blooms.

In the matter of fuel, according to the authority already quoted, no less than 4 tons were being consumed per ton of hammered iron obtained; but for the purpose of comparing its cost with that of pig iron we will accept the weight named by Mr. Rose, the Manager of the works, to Professor Tunner, viz., 3 tons. From the data just described, we may, for our present object, accept the following as factors:—

	Pig Iron. Per Cent.	Hammered Blooms. Per Cent.
Waste of Metallic Iron in the Ore ...	Practically nil.	35
Labour	taken as 100	at least 600
Fuel	do. 100	150

The question the steel maker has to consider is the extent, if any, of the superiority of malleable iron so obtained over pig iron, for the future stages of his operation. Speaking from such information as I possess, it would not appear that a furnace melting a mixture of pig and wrought iron gives a much greater weight of produce than one working pig and ore. If this be so, the actual expense of melting either mixture will be about the same; and we are left simply to compare the cost of the one set of ingredients with that of the other.

Let it be assumed, where the charge consists of blooms and pig iron, that it is made up of 80 per cent. of the former and 20 of the latter, while in the Siemens Landore mode of treatment 80 per cent. of pig iron has added to it 20 to 25 per cent. of ore.

Under such circumstances, and having regard to the great waste of iron and extra cost incurred for labour and fuel, in producing the blooms by the direct process, I cannot see how the materials employed for each ton of steel can cost less by this process, than by that where hematite pig iron (at 60s.) and ore are used, as at Landore.

SECTION III.—DIRECT PROCESSES.

No doubt the presence of foreign subtances, such as carbon and silicon, in the pig iron, means an expenditure of fuel in the blast furnace; but this is not all loss, for these substances serve to reduce so much iron from the ore added, that the actual waste of metal is not one-third of that incurred in the direct process, if both are calculated upon the ore delivered at the very commencement in each case. Of course it may be alleged that in the direct process we have begun with materials containing an amount of phosphorus inadmissible for steel making, and that we have succeeded in separating this substance by avoiding the use of the blast furnace.

Recent experiments however have demonstrated the possibility of washing out almost all the phosphorus contained in cast metal, by means of oxide of iron, so expeditiously and easily, that the cost of doing it would, probably, be attended with much less expense than that connected with the direct mode of reducing the ore.

Practically, the inference to be drawn from what has been advanced in the present Section is, that the removal of the carbon and silicon, absorbed during the ordinary smelting of the ore, is done at so small a cost, as to render it more economical to employ pig iron, than to use the direct process in order to obtain a product free from both of these elements. Now that the Basic treatment has been found so efficacious in separating the phosphorus, it would seem as if the only apparent argument in favour of dispensing with the use of the blast furnace no longer exists.

If the final object in view be not steel but malleable iron, a process which gives us the article we seek in one operation, instead of two, has a most attractive sound; but it is by no means certain that the advantages are not rather apparent than real. Adopting the same classification as before, I am sure that the actual waste of metal accompanying the use of the blast and puddling furnaces will not, including the "fettling" employed by the puddler, amount to that spoken of in connection with the direct process. On the other hand, it seems pretty clear that the payments for labour and fuel will amount to less for the double mode of treatment than they do for the single one.

It is true that in the one case we have as our first result pig iron containing nearly all the phosphorus found in the minerals employed, while in the other we have malleable iron containing less than ·1 per

SECTION III.—DIRECT PROCESSES.

cent. of phosphorus. This same pig, however, containing as much as 1·5 to 1·75 of the objectionable metalloid, can have it separated in the rotatory puddling furnace so completely, as to show an equality in this respect with the analyses mentioned by Professor Tunner.

Having mentioned so high an authority as the learned Professor, I am bound to add that he views the future of these direct processes more hopefully than I have been able to do.

In the work already quoted, Professor Tunner alludes to the fact of the actual production of iron by these processes, 40,000 tons being now the yearly make in the low hearths situate near Lake Champlain, and in other places. I scarcely gather from the context that even here this ancient plan is regarded by him as being able to compete commercially with our present modes of making malleable iron. Without, however, being capable of this, it is easy to understand that an old industry may continue to exist, and even to prosper, in times of high prices and high import duties, notwithstanding that more economical methods may be carried on in its immediate vicinity. This is particularly true of a manufacture requiring such simple appliances as are used on Lake Champlain, and where the product is applied to purposes for which more phosphoric iron is unfit.

When we come to the matters of cost and waste of iron, I do not apprehend that the figures given by my friend differ in any degree from my own. The ore employed on Lake Champlain is valued at 15s. per ton. It contains 70 per cent. of metallic iron; but only 57 per cent. of rough slabs is obtained at the end of the process, showing a loss of nearly 20 per cent., or above double that involved in puddling pig metal.

The particulars of cost of these rough slabs are given as follows:—

	£	s.	d.
35 cwts. of ore at 15s.	1	6	3
13 bushels, say 46·5 cwts., charcoal	4	6	9
Labour	1	10	0
General charges	0	12	6
Total	£7	15	6

It may be thought, looking at the wasteful mode of its application, that the quantity of fuel consumed is unexpectedly small, as compared with the combined processes of smelting and puddling. This arises from the fact that there is little or no earthy matter to fuse, in treating so rich an ore; and also that the temperature required in the low hearth is not so intense as that excited in the blast furnace. As a matter of final cost, however, there are few ironmaking localities where ore valued at 15s. per ton, and containing 70 per cent. of iron, could not be converted into pig iron, and have its phosphorus reduced to ·075 per cent., in a rotating furnace, for very much less money than the sum just named. The expectation of substituting mild steel, or ingot iron, for malleable iron made in the ordinary way, has diverted attention from further improvement in the puddling furnace. Nevertheless in that modification of the revolving furnace with which Mr. Samuel Danks' name is associated, a considerable step in advance has recently been made. At Creusôt two such furnaces are at work, in which the shell is kept cool by a current of water; from each of these 8 to 10 tons of puddled blooms are produced every 12 hours, being double the make usually obtained from the furnaces designed by Mr. Danks. The natural effect of this increased output is a considerable diminution in the working expenses, thus rendering it still more difficult for the direct process to compete with the modern art of iron-making.

Tunner states that the Jersey or Champlain system, as this is often called, was the first method practised in America for obtaining the metal; it may, therefore, be concluded that long experience does not permit much hope of its cost being materially lessened. Such is not the case with regard to the other two direct processes more particularly referred to in these pages; and in predicting their future, we are left to apply our general knowledge of the natural properties of the materials dealt with, coupled with such experience as we possess in comparing the expense of a new system with those already in use.

In my short account of the Blair process, given on a former occasion,[1] it was stated that the porous sponge, when thrown into the steel furnace, would float on the melted pig iron, and there, to some extent, would be re-oxidized. Mr. Blair at the time dissented from that view, and asserted that the sponge, light as it is, nevertheless sinks; and Professor Tunner himself gives it as his opinion that it is incorrect to suppose that the sponge must be oxidized before it is dissolved by the

[1] "Transactions of Iron and Steel Institute, 1875."

bath of metal. From this it is inferred, that sooner or later, we may expect to see Mr. Blair's plan in successful practice—an opinion which, even when backed by the great experience and authority of the learned Professor, I do not see my way to agree with. In support of the incorrectness of my own views, Professor Tunner mentions the fact that the ore, when thrown into pig iron on the Landore system, sinks through the 2 or 3 inches of *slag;* and that the sponge at Mr. Blair's works, instead of being porous, was compressed "until it resembled pure iron"—it is persumed in density. Of course it cannot be meant that, because lumps of ore fall through melted slag, sponge, even when compressed, will sink through liquid iron; nor is this very important, because the process of compression, being costly, is only applied to those portions of the sponge which, from their fine state of division, would be difficult to deal with in the open hearth furnace. As a fact, only about 40 per cent. of the sponge is so manipulated, the remainder being used as it comes from the retorts.

I possess no means of resuming my former discussion with Mr. Blair on this head by reference to direct experiment; and it is, perhaps, only due to Professor Tunner to observe that operations at Glenwood were suspended upon the occasion of his visit. It may be that I am mistaken in my views as to the actual way in which oxidation occurs; but that loss from this cause really does take place seems proved by the figures obtained from Mr. Blair himself. The knowledge of how this loss is occasioned is only important, so far as it might lead to the adoption of some means for its suppression.

Notwithstanding the reasons here given against the probability of a general adoption of the Direct process, it is still largely practised, as has been stated, in the United States. Its continuance, however, in my opinion, is not justified by any superiority in respect to economy, as compared with those processes which are dependent upon the blast furnace as a starting point. The Catalan works, generally speaking, are located where it would be inexpedient to incur the large outlay involved in the erection of modern ironworks. High prices, fostered by protective duties, may suffice, in ordinary cases, to afford fair returns to those who still pursue this primitive mode of working; and, in the event of the market value of the produce falling below prime cost, the extinguishing of a few Catalan fires or bloomaries is not a serious matter.

SECTION IV.

ON THE PRELIMINARY TREATMENT OF THE MATERIALS INTENDED FOR USE IN THE BLAST FURNACE.

IF the conclusions arrived at at the close of the last Section can be maintained, we must continue, for the present at all events, to look to the blast furnace as the starting point of those operations which have to furnish the world with iron and steel. This must at least be the order of things, until some considerable improvement takes place in that more direct process which has been suggested as an alternative one.

It would conflict with the intense temperature required in the hearth of a blast furnace, if the materials, when they reached that point, retained any portion of those volatile constituents found with them in their natural state. Hence everything capable of being expelled by heat should be evaporated in the upper part of the structure. This however implies a redundancy of heat in that region, and a sufficiency of time to accomplish this change in composition. In the larger furnaces, as recently introduced, this redundancy of heat does not exist; and in the older furnaces the necessary time, owing to their limited capacity, was not at the disposal of the smelter. Under such circumstances some preliminary treatment, as indicated by the heading of this Section, becomes necessary.

PRELIMINARY TREATMENT OF THE FUEL.

There is but one variety of fuel which in its natural state approaches the conditions required for smelting iron, viz. anthracite coal. This mineral in many cases, and particularly in the United States, contains a mere trace of volatile matter; and in consequence it reaches the hottest part of the furnace without any apparent change in its physical properties.

There are nevertheless many kinds of coal which, although very rich in volatile ingredients, are capable of being used in their raw state

SECTION IV.—PRELIMINARY TREATMENT OF MATERIALS. 47

in smelting iron. In such cases, however, a certain additional amount of fuel is required, in order to provide the necessary heat for expelling the tar and hydro-carbons. On the other hand there are other varieties of coal, often less rich in volatile matter than those referred to, which are totally useless in the blast furnace. These are known as coking coals, and, when exposed to heat, enter into a state of semi-fusion, which is regarded as interfering with the entry and upward passage of the blast, as well as with the uninterrupted descent of the contents of the furnace. Such coal is usually coked, before being delivered to the iron smelter —an operation which, as it is usually conducted in this country, is attended with a considerable waste of the fixed carbon, and necessarily also with considerable expense for labour, and for the necessary buildings and plant. These two sources of cost are saved when the coal is used raw; but it must be borne in mind, that the value of the fuel in the blast furnace is represented solely by the quantity of coke it is capable of affording. This is the only portion which affords any useful heat in the operation of iron smelting: for the inflammable gas demands, as has been already stated, the combustion of a certain quantity of the fixed carbon to secure its expulsion.

For the present we will confine our attention exclusively to the process of coking, without reference to the subsequent use of the coke itself; and here it may be remarked that this country has little reason to be satisfied with the rude and unscientific manner in which this operation in most cases is still conducted. In fact, until very recently, no kind of progress had been made in the process for the last 50 years. In proof of this I propose to describe the practice of the County of Durham, in which something like 6 millions of tons of coal are annually converted into coke, and where, if in any locality, some advance towards perfection might be expected.

Before proceeding with this demonstration, it may be well briefly to give the figures upon which the calculations as to heat, frequently hereafter to be referred to, are based. These figures are all made dependent upon the number of thermometric degrees by which a given quantity of the combustible can raise the temperature of a certain weight of water; or, which amounts to the same thing, on the quantity of water which can be raised one degree in temperature by a given amount of the combustible. The use of the decimal system is so much

48   SECTION IV.—PRELIMINARY TREATMENT OF MATERIALS.

more convenient than any other, that scientific writers in this country now generally adopt the centigrade thermometer, and the decimal weights used in France. On this system of notation a kilogramme of water raised one centigrade degree is considered as equivalent to one *calorie* or *heat unit;* the kilogramme of matter burnt being also regarded as a unit of fuel. The heating power of different substances varies greatly; thus a kilogramme of sulphur gives out, during combustion in atmospheric air, only 2,320 calories or units, while the same weight of hydrogen gas affords 34,000 units. Again, the combustion of the same body is accompanied by the evolution of very different quantities of heat, according to the amount of oxygen with which it may be made to combine in different cases. As an example, one unit of carbon can take up either one or two equivalents of oxygen; but it gives out 2,400 heat units when burnt with one equivalent, forming carbonic oxide, and as much as 8,000 units when burnt with two equivalents, forming carbonic acid. It is also essential, for the proper understanding of such calculations as those in question, to bear in mind that when a compound body is split up into its component parts, exactly the same quantity of heat is absorbed by the act of decomposition as was evolved by that of combination. Thus, if a unit of hydrogen in the form of water has to be separated from its combined oxygen, the dissociation is accompanied by an absorption of the 34,000 units mentioned above; or again, if carbonic acid containing a unit of carbon is reduced to the condition of carbonic oxide, the absorption of heat, or cooling effect, is equal to (8,000 − 2,400) or 5,600 calories.

Occasions will arise during the course of these investigations, which will require us to understand the quantity of heat, *i.e.* the number of heat units, in a given quantity of heated matter. As the amount of heat required to raise the temperature of equal weights of different substances is by no means the same, it becomes necessary to make an allowance for this variation in what may be called their capacity for heat. This is done by taking water as a standard, and assuming as the unit of measurement what is designated the *specific heat* of water, in other words the quantity of heat required to heat a unit of water by 1 deg. C. say from 0 deg. C. If then W represents weight, t temperature (centigrade), and S H specific heat, taking that of water as unity, we have the equation $W \times t \times S H = x$,

SECTION IV.—PRELIMINARY TREATMENT OF MATERIALS.

to give the number of calories or heat units contained in a given weight of any known body at a known temperature. Let us apply this formula first to 10 kilogrammes of water, and then to the same weight of a substance having one-fourth the specific heat of water, each body heated to 100 deg. C.

For water we have $10 \times 100 \text{ deg.} \times 1 = 1,000$ heat units.
For the other body $10 \times 100 \text{ deg.} \times \cdot 25 = 250$,,

It is almost superfluous to add that the *quantity* of heat evolved by combustion has no necessary relation to its *intensity*—the latter being dependent on the rapidity of the combination, and upon other circumstances on which it would be foreign to the objects of this work to dwell.

A wide range of information, of the character just alluded to, has been placed at the service of industry by scientific investigation. This of itself leaves the Durham coke manufacturer almost, if not entirely, without excuse for the rude manner in which he has conducted, and in many instances continues to conduct, his business. It has been urged that to obtain a product of the quality sought for by the ironsmelter some sacrifice was unavoidable. The answer to this is that for the last twenty-five years or more the operation has been carried on in Belgium and France, so as to avoid much of the needless waste incurred in the North of England, and without the change, it is alleged, being accompanied by any of those disadvantages, the apprehension of which would seem to have impeded progress in this country.

We will proceed in the first place to ascertain the extent and nature of the waste, thus condemned as unnecessary, accompanying the method pursued in this, the chief coking district of Great Britain.

The coal of the County of Durham may be regarded as containing 30 per cent. of volatile substances; or, in other words, as being capable of furnishing about 70 per cent. of its weight of coke. As a matter of fact scarcely 60 per cent. is the average yield, implying that one-seventh, or nearly 15 per cent., of the desired product is lost. In addition to this source of waste, imperceptible perhaps to the casual observer, there is another which forces itself upon the attention of the most indifferent spectator. I allude, of course, to the vast loss of heat which accompanies the immense volumes of smoke and flame which issue from a Durham coke-work, blackening and desolating the country around it.

The following figures exhibit an approximate statement of the actual quantity of heat which is evolved, but from which no use is derived, in the coking of 100 kilogrammes of coal in the old so-called bee-hive oven :—

	Calories.
Heat evolved by the combustion of the 30 kilogr. of hydrocarbons, affording 7,400 calories per kilogr. ... =	222,000
Heat evolved by the combustion of 10 kilogr. of solid carbon, giving 8,000 calories per kilogramme ... ... =	80,000
	302,000
From which has to be deducted the heat required for expelling the 30 kilgr., reckoned as representing 2,000 calories per unit of volatile matter	60,000
Giving as net loss	242,000

The heat emitted by the combustion of different specimens of raw coal varies in amount, but for our present purpose we will take it at 8,500 calories per unit, or 850,000 for the 100 kilogr. From this statement we are justified in the assertion that something like 28 per cent. of the entire heating power of the fuel is lost in coking—a loss which, on the six million tons annually coked in the County of Durham, is equivalent to about $1\frac{3}{4}$ million tons of coal.

It is not to be expected that anything like the whole of this immense loss can be usefully intercepted; for in almost every operation where heat is required great waste is incurred by radiation, by loss at the chimneys, and from analogous causes. The vast area covered by coke ovens, and the extent of their heat-emitting surface, adds greatly to the unavoidable loss which accompanies their use.

Greatly as the temperature of the products of combustion from coking must fall, before they reach points where they can be usefully applied, it remains sufficiently intense to burn bricks; this having been effected, at a considerable distance from the ovens, with complete success. The purpose, however, to which this waste heat most readily

SECTION IV.—PRELIMINARY TREATMENT OF MATERIALS.

lends itself is the raising of steam for the engines required at the mine. At one establishment, where nearly the whole output is coked, it is calculated that 8 per cent. thereof would formerly have been burnt to ventilate the mine and draw the coal and water—duties now entirely performed by the ovens themselves.

The manner in which the waste gases are applied for obtaining steam is simple. A boiler is placed on a main flue, into which as many ovens as experience has shown to be required for its use, discharge their surplus heat. At the colliery in question twenty-five ovens, having each a diameter of 11 feet and producing 7 tons of coke per week, are attached to a boiler 60 feet long and 5 feet in diameter. The quantity of water evaporated averages about 24 cwts. per hour per boiler, a margin of power being required, owing to the somewhat intermittent nature of the intensity of the heat given off during the distillation of the coal.

A collateral advantage attending the system is the *entire* absence of smoke; but to secure this the flues must have ample capacity. The sulphur thrown into the atmosphere from the coke ovens in the County of Durham, in the form of sulphurous acid, has been estimated at 45,000 tons per annum, which accounts for the immense destruction of vegetation which takes place, when the gas leaves the oven within ten feet of the surface of the land. By the use of chimneys of sufficient height, this poisonous gas can be, and now often is, so diluted before it reaches the earth that its effects, as it is believed, cease to be visible.

A few words now as regards the serious loss involved in the destruction of the 10 kilogr. of coke, for each 100 kilogr. of coal employed. It is only proper to observe that the process of coking has been, and is still, somewhat extensively conducted in Durham by means of "flued ovens," in which the burning gases, after they leave the dome, are made to heat the charge of coal externally. This greatly accelerates the process, and by so much shortens the time during which the coke is wasting. But uniform testimony, in which I am disposed to concur, goes to prove that the coke thus obtained is not as good for blast furnace work as that burnt in the ordinary way. It is less dense and less silvery in appearance, and in consequence, for reasons best explained when dealing with the smelting process itself, is incapable of performing its full measure of duty in the furnace. Thus the increased yield in the oven is considered as being neutralised, by the coke having less value when it reaches the ironworks, than has the produce of the ovens constructed without such flues, and with no external heating of the contents.

Some of our most experienced authorities, who have given the subject much attention, were led to infer that for obtaining hard burnt coke the heat afforded by the combustion of the gases did not suffice, and that the high temperature required must be continued, after the process of distillation was completed. This, according to their opinion, can only be secured by the burning of a portion of the coke itself.

A careful examination of the ovens abroad, and analyses of the gases taken at various stages of the process in the collieries of my own firm, led me to adopt an opposite conclusion; and by a mere change in the way in which the air is admitted, this serious source of waste is in some measure removed.

By the two alterations just described, viz. that of utilizing the heat for steam purposes and a reduction in the waste of coke, it has been estimated that, of the 242,000 calories calculated as disappearing, something like 100,000 may be saved; viz. 56,000 by avoiding a part of the waste in the coke itself, and 44,000 calories in the generation of steam.

Among the more elaborate systems of coke making, that known as the Knab process, from the name of the inventor, is deserving of notice. In every case where air is allowed access to the gases immediately on leaving the coal, they are speedily converted into watery vapour and carbonic acid—the hydrogen of the hydro-carbons giving rise to the formation of water, and the carbon to that of carbonic acid. If, instead of this mode of procedure, the distillation is conducted in closed vessels, and the volatile matter is passed through proper condensers, the same products are obtained which accompany the manufacture of illuminating gas, viz., ammoniacal compounds and coal-tar, the latter yielding creosote, asphalte and aniline dyes. The inflammable gas, after being freed from these less volatile substances, is brought back to the oven, where by its combustion in external flues the coal is converted into coke.

SECTION IV.—PRELIMINARY TREATMENT OF MATERIALS. 53

In two localities, viz. at South Brancepeth Colliery belonging to my firm, and at Wigan, this system has had an extensive trial in this country; and the result was final abandonment in both cases, mainly owing to the difficulty of upholding the buildings. In France also, the land of its first introduction, there are not above two or three establishments where it is in operation. Nevertheless, it is difficult not to believe that the idea is a very valuable one; for M. Carvés, who has had several years experience in working it, has recently, at his own expense, erected an establishment at a cost of about £20,000, at some iron furnaces near St. Etienne. At the end of a certain term of years he will hand the ovens, etc., over to the furnace owners free of charge, his remuneration consisting exclusively of a *share* of the profits reaped in the meantime from the sale of the tar and ammoniacal salts.[1]

In South Wales and Staffordshire, coal is often coked in the open air, by piling blocks of coal round a low central chimney. This however, is a plan which cannot be recommended for adoption, the waste of coal, especially in high winds, being enormous.

PRELIMINARY TREATMENT OF WOOD.

When the forest, instead of the mine, is the source of the fuel, previous charring is almost invariably practised. The exception is where the wood is highly dried; but in this state it is a small proportion only which can be employed in iron smelting. The volatile matter, consisting of gas and a large quantity of water, which is capable of expulsion from wood by heat, is so great, that about 25 per cent. of the weight of timber employed represents a common yield in charcoal.

In the woods of the United States, a square acre of forest affords about 20 to 40 cords of timber, each measuring 128 cubic feet as it is stacked. This forest, if allowed to grow again, would not be fit for cutting for thirty or forty years. Very frequently, however, the lands used in the Western World for obtaining charcoal remain cleared for agricultural purposes. The actual weight of charcoal afforded by an acre of ground varies from 7 to 14 tons. This is evidently too small a quantity to

---

[1] I understand that Messrs. Pease are at present erecting a coking work, in the County of Durham, at which they propose collecting the products of distillation in the manner described.

permit of any permanent erections for charring, even if the wood-lands were intended to be exclusively devoted to supplying the ironworks with fuel. The reason is obvious. The conveyance of 100 tons of timber, giving at the outside 25 tons of charcoal, would entail so much expense, owing to the distances apart of the trees to be felled, that any particular building would speedily have to be abandoned from unsuitability of position. The consequence is that the wood is almost all charred in heaps formed on the ground; and from the cause already mentioned, a blast furnace so speedily consumes the trees in its immediate vicinity, that the expense of conveying the charcoal to the works becomes an important item in the cost of the fuel.

It is true that near Marquette, on the shores of Lake Superior, the charcoal is made in large close kilns; but this is done from necessity and not from choice, and is owing to the length and severity of the winters, any saving in labour or gain in produce being more than counterbalanced by the additional charge incurred on the carriage of the volatile constituents of the wood.

From numerous enquiries made during visits to the American forests, I have estimated the cost of a ton of charcoal to be as follows:—

	s.	d.
Price paid for timber as it stands on ground	0	8½
Cutting timber	5	3½
Charcoal burners and loading	18	0
Total cost loaded into wagons, at kilns	£1 4	0

The price of transport varies according to distance; sometimes it is a mere trifle, but 5s. or 6s. per ton is a common charge for this item, and occasionally the carriage comes to as much money as the cost of the charcoal itself.

In Austria the cost of a ton of charcoal delivered at the furnaces was given me as running from 30s. to 56s. per ton. In Sweden the iron master who owns the forest lands obtains his fuel for about 20s.; but inclusive of the value of the wood 30s. to 35s. will represent more nearly the price of a ton of charcoal in that country.

The price paid for charcoal, under the conditions of production just described, will vary greatly according to the state of the iron trade. With low selling rates of his produce, the smelter who is dependent on charcoal brought from long distances, and often over very bad roads, will curtail his make of pig; and so discontinue the use of fuel, one-half of the cost of which may be incurred in transporting it from the forest to the furnace.

These prices of charcoal have been quoted with a view to show under what disadvantages, in point of cost, iron is manufactured, when using this description of fuel. In the coal-fields of Great Britain, coke is usually delivered at the furnaces at about 12s. per ton, or at about one-half the sum paid for charcoal in America. Practically the manufacture of charcoal iron in the United Kingdom may be regarded as a thing of the past. Messrs. Harrison Ainslie and Co. still smelt about 1,800 tons of this article per annum, which is chiefly used for making malleable iron castings, that is, castings which are rendered malleable by a prolonged heating in contact with an oxidizing substance like peroxide of manganese. I am informed that at their furnaces charcoal costs 45s. to 75s. per ton. For use in tin works charcoal, down to a recent period, commanded 40s. to 50s. per ton, but here again Bessemer steel bars bid fair to drive iron out of the market, so that charcoal fires are fast disappearing.

PRELIMINARY TREATMENT OF THE ORES OF IRON.

The ores of iron, as they are delivered to the blast furnace, may be divided under three heads :—

1. —Simple oxides, viz. magnetic oxide ($Fe_3 O_4$) and peroxide ($Fe_2 O_3$), which contain no combined water, but often a considerable quantity of included moisture, as in the case of ores from the mines of Lancashire.
2. —Peroxide of iron containing combined water, as in brown hematite.
3. —Ores in which the metal exists as a carbonate of the protoxide ($Fe\, O\, CO_2$).

The first two are occasionally found nearly pure; in most cases however they contain from 10 to 15, per cent. or more of earthy matter. The carbonate, when occurring as spathose ore, is also sometimes almost pure; but in the usual run of clay ironstone, in which the metal is likewise combined with carbonic acid, the foreign matter exists in such quantity as to reduce the yield to from 28 per cent. to 40 per cent. Occasionally clay stone contains even more than 40 per cent. of metal; but 30 per cent. represents the ordinary produce of this variety of the mineral, which is by far the most plentiful of all.

Practically it may be said, that for all ores in which the iron exists in the form of carbonate, recourse is had to calcination, before their delivery to the blast furnace. The change effected by the operation is the expulsion of combined water and of moisture, while the metal is peroxidized, and a portion of any sulphur present is driven off.

In some cases, as in Sweden and elsewhere, the ores of the oxide class are occasionally calcined before they are submitted to the action of the blast furnace. The mineral is by this treatment more readily reduced, and to some extent the sulphur it may contain is removed.[1] As a rule, however, the oxides are smelted as they are raised from the mine, unless it has been necessary to remove foreign matter by any of the usual modes of washing when the same is capable of being so separated.

Clay ironstone is frequently calcined in the open air, in heaps or clamps as they are called. This is a very wasteful method both in fuel and labour; but in the case of Black Band clay stone the quantity of coal associated with the ore is so great, that it would be difficult to calcine it alone in a kiln. This is owing to the liability of the mineral to run together, by the high heat evolved during the process of calcination.

The apparatus used for the operation in question has been greatly improved, in late years, in the district of Cleveland. The kilns are as lofty as the blast furnaces formerly were, by which means the fuel consumed is reduced to about $3\frac{1}{4}$ per cent. of the mineral employed; and they are so constructed that they are self-discharging into the furnace barrows. The credit of these improvements is due to Mr. John Borrie of Middlesbrough, and they seem so complete as to leave little room for further amelioration.

It has been proposed to use a portion of the ironstone uncalcined; and making allowance for the loss of coal which is incurred, even in the well constructed kilns just described, it may be granted that the *quantity*

---

[1] By experiment I ascertained that previous calcination greatly facilitated the reduction of an ore by means of carbonic oxide. (See "Chem. Phen. of Iron Smelting.")

of heat escaping from the latest constructed blast furnaces is sufficient to expel the water and carbonic acid from the whole of the ironstone used. The *intensity* of the temperature of the gases is however on the whole much below that required for the operation; and hence the advantage of the change is very questionable. Besides this, there is a further objection, connected with the subsequent use of the escaping gases. By weight the combustible portion of this gaseous fuel only amounts to 15 per cent. of the whole; and any material addition to the uninflammable ingredients might, and we know does, tend seriously to diminish the combustibility of the useful portion of the gases. The actual labour in calcining is too insignificant to demand consideration; and the quantity of coal consumed in the operation is so small, as not to render it advisable to affect in saving it, the burning of the gases which have to serve for driving the machinery, and heating the blast for the furnaces. It may also be added that the calcining of the ironstone, with the excess of free air which passes through the kilns, effects a more perfect separation of the sulphur than can be accomplished in the blast furnace, where at the proper period for expelling this noxious element there is no free oxygen to be found.

PRELIMINARY TREATMENT OF LIMESTONE.

In the great majority of cases, the purer the iron smelter receives his limestone, the better it suits the purpose he has in view. It generally happens that the silica and alumina of clay ironstones are found in such proportions, that the mere addition of pure lime suffices as a flux for them. The presence therefore of either of these substances in the limestone itself merely adds to the inert matter to be fused. Fortunately for successful work in the furnace fusible slags can be produced, in which the constituents coexist in a vast variety of proportions, the extent of which will be most conveniently explained in a future Section.

Many of the great geological deposits of limestone consist of carbonate of lime almost pure; thus chalk, when dried, will contain as much as 99 per cent. of this substance, and the mountain limestone 95 to 97 per cent., while the limestone of the Silurian age is often nearly as pure as chalk. The impurities are a trace of oxide of iron, alumina, and silica; and occasionally magnesia, existing from a mere trace up to 10 or 20 per cent. or more. In the formation known as the magnesian limestone, the last mentioned earth is almost invariably present in greater or less quantity; sometimes nearly half the weight of the stone consisting of carbonate of magnesia.

As a mere question of fusibility, the presence of magnesia is the reverse of a disadvantage; because not only do certain mixtures of silica, lime and magnesia, even without any alumina, form compounds which are easily melted, but the presence of a fourth ingredient, or even of more, frequently promotes fusion in the remainder. Generally speaking however, magnesia can be dispensed with; because the earths of our clay ironstones are sufficiently easily melted without its aid, and it is at all times better to use lime, on account of the sulphur which is almost invariably found both with the iron and with the fuel. Lime has at high temperatures a certain affinity for this element, whereas magnesia possesses little or no action on it. To free iron as far as possible from so objectionable an ingredient as sulphur, the lime should be in excess; hence the substitution of a flux inert as regards this metalloid, which is the case with magnesia, is not desirable. At the same time it often happens that economy of cost, generally determined by transport charges, leaves no alternative but to employ limestone containing a considerable percentage of the earth in question. Thus in the United States the lime in the flux is often associated with so much carbonate of magnesia, that in smelting the magnetic ores of New Jersey, yielding 48 per cent. of iron, as much as 25 to 26 cwts. of flux is required per ton of metal produced. The carbon wasted by the reaction of the carbonic acid, entering the furnace as carbonate of magnesia, and the fusion of the needless amount of cinder, will probably require the presence of an addition of 5 to $7\frac{1}{2}$ per cent. to the coal, as was the case at the works at which I obtained my information on this head.

Among certain iron works, using furnaces of 4,000 to 6,000 cubic feet capacity, it has been the fashion to calcine the limestone before charging it; and inasmuch as the presence of carbonic acid in such cases is, for the reasons assigned, undesirable, such preliminary treatment of the flux may on certain occasions prove useful. So far, however, as it is possible to compare the performances of furnaces, engaged in an operation where there are so many disturbing causes, I

SECTION IV.—PRELIMINARY TREATMENT OF MATERIALS. 59

have never been able to satisfy myself that the full measure of economy was effected, due to the difference between employing carbonate of lime raw, and in its calcined condition.

The objection to the use of raw limestone in the furnace is mainly one involving consumption of fuel, which may be summarised as follows:—

	Calories.
The expulsion of carbonic acid from the quantity of flux used per 20 units of iron in Cleveland, say from 11 units of carbonate of lime, gives $11 \times 370 =$	4,070
1·32 units of carbon in the carbonate of lime would dissolve a like quantity of solid carbon, and would prevent its combustion at the hearth, where, burnt with hot air, each unit would yield 3,000 calories; now $1\cdot32 \times 3,000 =$	3,960
Making a total theoretical gain of ...	8,030

This number divided by 3,000, the number of calories afforded by fuel in the hearth, gives an estimated saving of 2·67 units of fuel on the 20 of iron produced.

Of course from this saving, namely 2·67 cwts. of coke per ton of iron, must be deducted the value of a much larger weight of coal, practically used in the kilns for calcining the limestone. Inasmuch however as the refuse coal used for the latter operation costs, in the case alluded to, only one-third of the price of coke, there remained a margin of economy; although in reality the saving was not more than, if indeed it reached, one-half of the estimated weight mentioned above.

This difference between the real and calculated saving seems to me to be due partly to the fact that burnt lime, even as it is used by builders, often contains a notable quantity of carbonic acid. In the case of that employed at the blast furnaces, the quantity of carbonic acid unexpelled by the act of calcination must have been considerable, looking at the relative quantity of the two which was used. Thus, when raw limestone was employed, 12·8 cwts. sufficed per ton of iron, equal therefore to 7·17 cwts. of actual lime. Or again, 14 cwts. of limestone sufficed in furnaces only 48 feet high, which would be equivalent to 7·84 cwts. of pure lime. To maintain the cinder however of a similar composition as regards lime, when the calcined lime was used, the weights were 10 and 11 cwts. respectively. It may therefore be inferred that there was still present (10—7·17) or 2·83 cwts. and (11—7·84) or 3·16 cwts. of carbonic acid respectively. Now as the full equivalent of carbonic acid was only 5·63 and 6·16 cwts., for the two quantities of limestone, it looks as if the actual quantity of this volatile constituent had only been removed by calcination to the extent of about one-half.

Another cause of the want of economy of fuel is probably due to the behaviour of carbonic acid as regards lime. Under certain conditions, as is well known, a high temperature dissociates the two substances, while at ordinary temperatures they combine with marked readiness. It has to be observed, however, that a heat up to that of redness greatly adds to the rapidity with which lime absorbs carbonic acid. In such furnaces as those to which the present remarks apply, the calcined lime would commence to absorb carbonic acid immediately after being charged; and it would continue so to act until it reached that zone of the furnace where elevation of temperature and other conditions would put a stop to further absorption. Such carbonic acid, however, as was thus taken up by the lime, would act on the carbon, and give rise to the same expenditure of heat for its subsequent expulsion, as if it had existed naturally in the flux.

When the larger furnaces were erected in Cleveland, the same practice as regards the limestone was applied to them; but the results were not such as to justify its continuance, for there was scarcely if any perceptible economy of fuel. The cause of this was probably due to the prolonged period of exposure to the more moderate temperature, extending as this does over a much wider space in a lofty than in a low furnace. This change would be accompanied by a more complete return of the calcined lime to its natural state of carbonate of lime.

No doubt the reabsorption of carbonic acid by lime would be accompanied by a rise of temperature; but this probably takes place so near the point where the gases leave the furnace that they have not sufficient opportunity of imparting the heat thus required to the materials filling its upper region.

The only case I am acquainted with, when caustic lime is employed in a modern large furnace, is that where a portion of the ore (clay ironstone) is used in a raw state. Such treatment therefore only means expelling carbonic acid from limestone instead of from ironstone.

SECTION V.

THE BLAST FURNACE.

THE difficulty any form of the direct process labours under, in taking the place of the blast furnace, meets us from almost every point of view in which the subject may be regarded.

For the founder the blast furnace is indispensable; for so far as our experience goes, it constitutes the only available means by which he can be supplied with cast iron. No doubt, for malleable iron purposes, the blast furnace, as has been admitted, is a circuitous mode of procedure, for in it the metal takes up substances which have to be got rid of by a subsequent process or processes. This disadvantage, as has been already explained, is more than counterbalanced by the very economical application of labour permitted by the character of the operation, and by the much higher duty performed by the fuel, when compared with that obtained elsewhere, whenever a high temperature is required. Along with these two claims for a position of superiority over the direct process, is the practical absence of all waste of iron in the operation of smelting. Doubtless against this last named ground of preference has to be set the loss of metal incurred during the subsequent process of separation from the contracted impurities. After allowing for this, however, a considerable balance remains, even in the item of waste, as against the production of malleable iron direct from the ore, and in favour of the use of pig iron.

For a proper appreciation of the nature and efficiency of the process, as performed by the blast furnace, a somewhat careful examination of its work, chemical and otherwise, is indispensable. Before entering into this question, it may be well to refer again to the circumstances which render it easy for the pig iron maker to economise in human labour. One of the most important of these is the compactness of his plant: for, including moderate but sufficient room for stocks of materials, a blast furnace, with its calcining kilns and other accompaniments, stands on a space not exceeding half an acre of ground. Within this small area will be annually received, and most economically handled in the manner already spoken of, about 120,000 tons of minerals; and from it upwards of 60,000 tons of iron and slag will be conveyed with equal expedition and cheapness. The consequence of such an arrangement is that, in some of our best establishments, for each day's work of the men employed, nearly two tons of pig iron are obtained. It would be difficult, looking at the quantities to be handled, to find an instance where, with so small a number of men, so large an amount of work is performed.

We may now proceed to the consideration of those agencies, by the exercise of which the blast furnace performs its duty in the efficient manner claimed for it. In dealing with this question, it is not proposed to enter, in the present Section, into all the details of the chemistry of the smelting process. The leading phenomena only of the theory will now be considered, reserving any further explanation to be discussed under separate Sections.

If a piece of carbon, such as charcoal, is heated to redness, and continues to burn, surrounded on all sides by air, it will be so without any visible flame. Each molecule of the combustible is converted at once into carbonic acid, with the evolution of the largest amount of heat its combustion can afford, viz.: about 8,000 centigrade units per unit of carbon.

If on the other hand several pieces of charcoal are placed in contact with each other, and the air required for their combustion is made to pass upwards through the mass, a blue flame will appear at the top. This is owing to a deficient supply of atmospheric oxygen causing the previous formation of carbonic oxide (CO), which is afterwards burnt to the state of carbonic acid (CO_2), the moment it meets the external air. This generation of the lower oxide of carbon may take place in two ways. Each equivalent of atmospheric oxygen may meet with one of carbon and unite with it, thus forming carbonic oxide at once; or else the carbon may first be burnt to the condition of carbonic acid, and then this, the highest oxide of carbon, may come in contact with a fresh equivalent of carbon in a *highly heated state,* and dissolve it, two equivalents of carbonic oxide being the result ($CO_2 \times C = 2CO$). In

SECTION V.—THE BLAST FURNACE.

a calorific point of view, this action is precisely the same as if the two units of carbon had been burnt directly to the state of carbonic oxide.

In all low fires, such as the Catalan hearth and refinery, the real nature of the combustion varies with circumstances, such as the depth of the fire, and the temperature of the fuel, whether it be charcoal or coke. As a rule we may take it that near the tuyeres of these shallow furnaces there is a certain amount of carbonic acid present, only part of which is reduced higher up to the state of carbonic oxide. In this way a portion of the carbon arrives at the surface of the fire in the form of carbonic acid, and there it mingles with that resulting from the combustion of any carbonic oxide which may reach this point as such.

It thus happens that in such arrangements as those last mentioned, the whole of the carbon, in the one way or the other, passes up the chimney as carbonic acid, having evolved by its oxidation the greatest amount of heat it is capable of affording. It is clear, however, that when the object of such a fire as the Catalan or refinery is the imparting of heat to some body immersed in the fuel, it cannot be otherwise than an exceedingly wasteful operation. That portion of the carbon which is burnt below the surface of the fuel, passes too rapidly upwards to have time to impart more than a mere fraction of its heat to the matter exposed to its influence. On the other hand, such combustion as takes place on the surface of the mass, is scarcely in contact with the body to be heated, and exercises little or no useful effect.

Let us now compare the nature of the combustion as it is effected in the blast furnace, and the application of the resulting heat, with that just described. For this purpose, we will suppose that the fuel is burnt in a furnace having a height of 80 feet. The result of several analyses[1] satisfied me that almost all traces of carbonic acid disappear within a foot or two of the level of the tuyeres: we may therefore infer that, in the absence of any subsequent change, the whole of the carbon burnt at the hearth would be given off at the throat as carbonic oxide. Imagine such a furnace filled with coke, along with a neutral substance such as slag, not liable to any chemical change. Fire is communicated below, and the blast applied. Combustion rapidly sets in, and the gases, as they arrive at the top, soon become sensibly

[1] Chem. Phen. of Iron Smelting, pp. 8 and 9.

warmer. Their temperature will continue to rise until at the rate at which the furnace is driven, the refrigerating influence of the cold materials, as they enter, establishes a position of heat equilibrium; and the mean temperature of the gases will then remain stationary. It is easy to note the time when this occurs, and to observe the exact quantity of coke which has been burnt between this epoch and that at which the blast was laid on. The number of heat units evolved by burning this weight of coke is easily computed, and along with it the weight of gases which has been generated by its combustion. The mean temperature of these gases having been noted, we can ascertain with tolerable nicety the quantity of heat they are carrying away with them.[1] The difference between the two sets of figures represents the quantity of heat intercepted by the incoming materials. The amount of this difference I ascertained upon two occasions, when blowing in a furnace,[2] and found it to be such that, for every calorie originally evolved in the hearth by the direct combustion of the fuel, 2·33 calories were brought back thither by the materials descending from the upper region of the furnaces. From these figures we have the following statement of the heat development in the hearth, per unit of carbon consumed therein :—

	Heat Calories.
One unit of carbon burnt at the hearth to carbonic oxide gives	2,400
Heat imparted to the gases by the combustion of preceding units of carbon, which heat being intercepted by the descending materials, is returned to the hearth, in the ratio given above, viz. 2·33 to 1, and gives	5,592
Together	7,992

Practically therefore the combustion of a unit of carbon burnt to carbonic oxide in a blast furnace of 80 feet gives nearly as good an effective result, although it evolves only 2,400 calories, as the same quantity of carbon burnt to carbonic acid in a low fire although in the latter case 8,000 calories per unit of carbon are generated. There is however this marked difference between the two examples, that whereas

---

[1] Certain disturbing influences must be taken into the account, but they need not be considered at the present moment.

[2] Chem. Phen. of Iron Smelting, p. 293.

the 7,992 heat units referred to in the case of the blast furnace are almost all usefully employed, a very large proportion of the 8,000 evolved in the low hearth escapes into the air unutilized. In the low fire, as experience tells us, there is an enormous waste of heat, which is indeed visible in the flame and incandescence at the surface of the fuel. On the other hand, in a blast furnace of 80 feet, the materials are, it is true, red hot, for more than 50 feet above the hearth; but the upper surface of the materials, instead of being red hot, exhibit little or no signs of incandescence, proving a comparative freedom from waste due to this cause.

Hitherto, in considering the behaviour of the blast furnace, regard has only been paid to its power of raising the temperature of any substance exposed to its action. But the duty of the blast furnace is of a two-fold kind: it has of course to fuse every non-volatilized substance which enters it, but, before this is done, it has also a chemical function to discharge, viz. the reduction of the ore. This duty is also admirably performed, as will be perceived on a short consideration of its nature, effected as it is by means of carbonic oxide gas. Although an oxide of iron is easily reduced by solid heated carbon, it is greatly to be preferred, for reasons which will be explained in detail, that the deoxidation should be performed by carbonic oxide; and therefore by carbon which in this case, by its combustion at the tuyeres, has already rendered valuable service in melting the slag and iron.

In order to enable the reducing gas (CO) to perform its office, an elevation of temperature is necessary. For the removal of the first part at least of the oxygen in the ore, a very moderate heat suffices—420° F. (215° C.) being enough for the purpose; but at anything below a low red heat reduction goes on slowly. The effect of the action upon the gas itself, is its conversion from carbonic oxide (CO) into carbonic acid (CO_2)—a change effected, as will be seen from the symbols, by the absorption by the carbon of an additional equivalent of oxygen, derived of course from the ore, $CO + O = CO_2$.

It is a fact well known to chemists that reduced iron in its spongy state, when heated in the presence of oxygen, unites readily with this element; indeed so strong is the affinity between the two substances that, under certain conditions, iron in this form is capable of splitting up the very carbonic acid produced from its own previous reduction by means of carbonic oxide. The conditions for this reversed action are dependent on differences of heat; the power of carbonic acid to oxidize iron increasing, at certain temperatures, more rapidly than does the reducing power of carbonic oxide. Hence, when certain mixtures of the two gases are passed over oxide of iron at certain temperatures, reduction is suspended. As an example it was found impossible to obtain metallic iron when,

1·5 vols. of carb. acid were added to 1 vol. of carb. oxide, the heat being a low red.
·47   do.   do.   1   do.   do.   full red.
·11   do.   do.   1   do.   do.   approaching white.

Further trials were made with the two gases in equal volumes, the mixture being passed over spongy iron, and over peroxide of iron at a white heat. The metal was oxidized to the condition of protoxide (FeO) and the peroxide (Fe_2O_3) was found to have lost exactly as much of its oxygen as brought it down to the same state of combination (FeO) as the other had attained, when it ceased to absorb more of the gas.

With a natural law such as that demonstrated by the results of these experiments, it is easy to imagine what must be the fate of ore thrown into a low fire. It speedily reaches a point where the temperature, and the proportion of carbonic acid to carbonic oxide, are such as to render reduction of the *whole* of the oxide of iron physically impossible. Or, even admitting the gases at the point in question to possess a deoxidizing power, under other conditions, sufficient for complete reduction, that power is so diminished in intensity, and therefore in rapidity of action, by the presence of intensely heated carbonic acid, that fusion of the earths and slagging of the iron oxide become inevitable.

By a parity of reasoning, it is difficult to see how the direct process, as carried on in a reverberatory furnace, with an intensely heated atmosphere containing a large quantity of carbonic acid, can be entirely satisfactory, so far as an approach even to complete reduction is concerned. Against the advantage of a saturated oxidation of the fuel, more or less complete in its character, has to be placed the loss of heat which takes place at the chimney of an ordinary furnace, or in the producers employed for generating the gas used in Siemens' furnace.

and also in the latter case, of the heat which escapes interception in the regenerators. The sum of these two sources of waste is such as to reduce, we are told, the actual work performed in a common reverberatory furnace to only 10 per cent., and in the regenerative furnace to less than 25 per cent., of the full heating power of the fuel employed. It must however be borne in mind, as a circumstance in their favour, that these descriptions of furnace employ raw coal and not coke; and further that the duty required to reduce iron, and raise it to the welding heat, is considerably less than that required to fuse pig iron mixed with nearly one and a half times its weight of slag. This makes it difficult to compare the two operations, with any close approximation to accuracy, in the matter of consumption of fuel; but there is no doubt that the effect of the whole of the carbon being burnt, in the reverberatory furnace, to the state of carbonic acid—the point upon which its superiority in the evolution of heat depends—must, for the reasons just assigned, materially interfere with the reduction of the metal.

In the blast furnace, the fuel no doubt is imperfectly burnt, that is it is far from being entirely converted into carbonic acid, but with such accompanying advantages as to be, in other respects, equal in beneficial effect to the most complete form of combustion. Nor is this all; for there remains, after the work of the blast furnace has been performed a reserve of useful heat in the waste gases, the discussion of the value of which will be most conveniently postponed to a future period. Again, the complete reduction of the metal by a simple enlargement of the capacity of the furnace, without any further cost or trouble, is invaluable. In such a furnace a slow and economical reduction of the ore is effected, at a temperature where the oxidizing power of the carbonic acid is effectually kept in check, by the opposing, or reducing force of the lower oxide of carbon.

Having now attempted to describe in general terms the two salient points which confer marked excellence of results on the blast furnace, we may proceed to consider more closely its mode of action. In the pursuit of this enquiry it will be assumed that the final operation of fusion, in the smelting of iron, is accomplished by the heat generated in burning the carbon in the hearth to the condition of carbonic oxide. It therefore follows that anything tending to consume carbon in the upper region of the furnace must be avoided, because it means that there will remain a smaller quantity for combustion where the greatest heat is needed, viz. at the tuyeres. It is not a difficult matter to ascertain pretty correctly, to what extent the carbon in a blast furnace is carried off before it reaches the hearth. For this purpose we must ascertain the quantity of atmospheric oxygen entering the furnace at the tuyeres, obtained by determining the weight of atmospheric nitrogen in the gases; we must also learn, so far as the somewhat rough nature of the operation permits, the quantity of carbon entering the furnace, say for each 20 units of iron made. From these two sets of figures the carbon oxidized at the tuyeres is estimated. The following calculation embodies the figures necessary for obtaining the information just referred to. The data are those furnished by a furnace 80 feet high, smelting Cleveland ironstone, with a consumption of 23·5 units of coke and 12·8 units of limestone per 20 units of pig.

The analysis of the escaping gases gave the following results, by weight:[1]—

			Carbon.	Oxygen.	Nitrogen.
Carbonic oxide (CO)	28·2	=	12·08	16·12	—
Carbonic acid (CO_2)	14·9	=	4·06	10·84	—
Nitrogen	56·9	=	—	—	56·90
	100·0		16·14	26·96	56·90

The carbon in the gases, reckoned on 20 units of iron, is obtained as follows:—

	Carbon.
Coke used 23·5 units, less ash and water 2·35 =	21·15
Limestone used 12·8 units, containing of carbon	1·53
	22·68
Less carbon entering into combination with the iron	·60
Carbon carried off in the gases, per 20 units of iron produced	22·08

Now the proportion of nitrogen to carbon in the gases has been shewn to be as 56·90 to 16·14. Hence the proportion to obtain the nitrogen per 20 units of iron, is:—

C 16·14 : N 56·90 :: C 22·08 : N 77·84

Thus it has been shewn that upon the occasion referred to, the 22·08 units of carbon found in the gases were associated with 77·84

---

[1] Hydrogen and marsh gas are omitted, as not seriously affecting the result.

SECTION II.—THE BLAST FURNACE.

units of nitrogen. When however the gases at the tuyeres came to be analyzed, the carbon present with the same quantity of nitrogen (77·84 units) only amounted to 19·35 units, shewing a disappearance of 2·73 units of carbon.

The proportions of carbonic oxide and carbonic acid to nitrogen having also been ascertained in the analysis given above, we have for every 20 units of iron,

Of carbonic oxide—N 56·90 : CO 28·2 :: N 77·84 : CO 38·57
and of carbonic acid—N 56·90 : CO_2 14·9 :: N 77·84 : CO_2 20·38

Returning to the carbon burnt at the tuyeres, it is necessary to ascertain in the first instance the full quantity of carbonic acid capable of being generated by the reduction of the 20 units of iron; because any result short of this implies direct loss, inasmuch as the absence of carbonic acid means, either that carbon has been the reducing agent, or that carbonic acid, if formed, has been afterwards reduced to the condition of carbonic oxide; the loss being the same in both instances.

Let us see how the quantity of heat evolved by a unit of carbon is affected by these three different ways of acting on oxide of iron.

Heat Units.

If the carbon unit reaches the tuyeres it is burnt there to the condition of carbonic oxide—giving off therefore... 2,400
If the same unit of carbon, in the form of carbonic oxide (CO), reduces oxide of iron, it becomes carbonic acid (CO_2), and, if it escapes as such, it will have generated a further amount of heat, viz. 5,600

Together 8,000

If on the other hand carbon be the reducing agent, and the gas escape as carbonic oxide, the heat evolved is ... 2,400
Or, if carbonic oxide be the reducing agent, and the resulting carbonic acid (CO_2) be afterwards reduced to the condition of carbonic oxide (CO), then we have the same result; for the total heat evolved by the combustion of carbon to form carbonic acid is ... 8,000
Less absorbed by subsequent reduction of carbonic acid to carbonic oxide ... 5,600
───── = as before 2,400

The Pig iron of Cleveland is associated with carbon, silicon, etc. to such an extent as to leave only about 18·60 units out of the 20 units for pure iron.

SECTION V.—THE BLAST FURNACE.

	Units.
The oxygen combined with 18·60 units of iron, as peroxide of iron, will, by oxidizing carbonic oxide, the reducing agent, form carbonic acid amounting to	21·92
The limestone employed, 12·8 units per 20 units of pig iron, contains of carbonic acid	5·63
A peculiar reaction, to be hereafter described as carbon impregnation, probably gives rise to the formation of carbonic acid to the extent, per 20 units of pig, of ...	2·20
Total carbonic acid generated per 20 units of pig iron...	29·75
But the actual quantity of carbonic acid (CO_2) in the escaping gases, per 20 units of iron, was only	20·38
Shewing a disappearance of carbonic acid amounting to...	9·37

Now the quantity of carbon required to reduce 9·37 units of carbonic acid to carbonic oxide is 2·55 units; hence, instead of finding 22·08 units of carbon in the gases at the tuyeres, we ought only to have (22·08 — 2·55), or 19·53 units.

As a fact there were really found, as has already been shewn, 19·35 units per 20 of iron, the samples for analysis in both instances being taken as nearly as could be at the same time. These two numbers are perhaps as near an approximation as the nature of the case would lead us to expect.

The means of diminishing this solution, as it were, of carbon by carbonic acid, and the examination of the laws which govern the relations between mixtures of the two oxides of carbon (CO and CO_2) with iron ore, are questions demanding the most careful attention of the smelter. Had it happened that solid carbon acted on oxide of iron by removing its oxygen at a lower temperature than that at which carbonic oxide performs the same office, it is probable that we should have had the reduction of the ore chiefly effected by the former; and consequently should have been compelled to content ourselves with the conversion of all the fuel into carbonic oxide, and an evolution of 2,400 heat units per unit of carbon; whereas in reality a portion passes into the state of carbonic acid, and affords as we have seen 8,000 units per unit of carbon. Precisely the same effect would have been caused, had carbonic acid dissolved carbon at a lower temperature than that, at which carbonic oxide is able to withdraw oxygen from an oxide of iron. Here also the sole product of the combustion of the coke in the blast furnace would have been carbonic oxide.

Plate 1.

DIAGRAM SHEWING TEMPERATURES AT WHICH
CARBON (C) AND CARBONIC OXIDE (CO)
BEGIN TO ACT ON PER-OXIDE OF IRON (Fe_2O_3)
AND THOSE AT WHICH
CARBONIC ACID (CO_2) BEGINS TO ACT ON
IRON AND CARBON

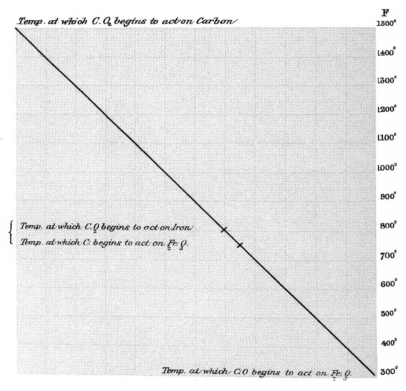

SECTION V.—THE BLAST FURNACE.

It is further obvious that the nearer the temperatures, at which the two classes of reaction just mentioned take place, the greater must be the difficulty of securing the greatest practicable weight of carbon raised to its higher state of oxidation.

Having the importance of the last named object in view—the determination of the different degrees of heat, at which the various changes are liable to occur in the blast furnace—forms an indispensable object of research for the furnace manager.

Fortunately, as will be seen immediately, the range of temperature, which embraces the series of reactions in question, is very considerable, running over 1,100° F. (611° C.), and extending from about 400° F. (204° C.) to 1,500° F. (815° C.)

The following is a general summary of the result of a long series of experiments undertaken with the view to ascertain the laws affecting the action of carbon and carbonic oxide on oxide of iron, and of carbonic acid on carbon and on metallic iron. At the same time it must be understood that the numbers given are liable to considerable fluctuations, dependent on the nature of the materials employed. Thus some varieties of ore are more susceptible of reduction by carbonic oxide than others, and are therefore more speedily reduced. There are differences also in the readiness with which carbonic acid acts on the carbon in different kinds of coke. This want of uniformity is important, because upon it, as will be subsequently shewn, practical questions connected with the use of the blast furnace are greatly dependent.

It was found from the experiments referred to :—

1st.—That the action of carbonic oxide on oxide of iron may be considered to begin at a temperature of about ... ... 400° F. (204° C.)

2nd.—That the action of carbon on oxide of iron begins at a temperature of about ... 750° F. (399° C.)

3rd.—That the action of carbonic acid on metallic iron begins at about ... ... 800° F. (426° C.)

4th.—That the action of carbonic acid on carbon, as it exists in hard coke, begins at about 1,500° (815° C.)

A diagram, Plate I., embracing the figures just given is inserted, in order to shew at one view the gradually increasing temperatures at which the four reactions just described take effect.

Let us suppose a furnace of such dimensions that the gases escape at something above 1,500° F. (815° C.), or any temperature at which carbonic acid quickly dissolves carbon ($CO_2 + C = 2CO$). In such a case, if carbonic oxide did reduce any of the ore, the resulting carbonic acid would be subsequently converted, in the manner already described, to carbonic oxide. This, as we have seen, would in all respects be the same in calorific effect, as if the carbon itself had been the reducing agent. If on the other hand the temperature of the reducing zone were 800° F. (426° C.), then any carbonic acid in the mixed gases would have a tendency partially to re-oxidize the metallic iron, or prevent the reduction of the ore; and would thus leave carbon to perform the task of reduction, which it commences to do at 750° F. (399° C.) From what has preceded it follows that the most economical conditions for working would be secured, if the temperature of the entire reducing zone did not exceed that point where the reduction of the ore would be exclusively effected by means of carbonic oxide, without the stability of the resulting carbonic acid being affected either by metallic iron or by carbon.

Such lines of absolute demarcation, as might be inferred from the four reactions just described, do not however exist in reality. A portion of the oxide of iron would in every case be reduced, without the resulting carbonic acid having time to return, under the influence of the carbon, to the condition of carbonic oxide; or the temperature might be such that a position of equilibrium would be reached, when half, a quarter, or any other proportion of the carbonic acid would be regularly acted upon by the highly heated carbon, and then further change in this respect would end.

This will be understood by a reference to the drawing, Fig. 1, Plate II., which represents an ideal section of a 48 feet furnace, drawn to a scale from actual observation as to the conditions of temperature. These it is attempted to exhibit by the mode of colouring; the intense heat at the tuyeres is shewn white, and passes gradually into dull red, and from this to black, the last representing the absence of any marked approach to incandescence. In such a furnace, when ore is shot in at the top, it is speedily heated to a point where deoxidation sets in; but, long before this is completed, the partially reduced mineral has reached first the level where the second, then where the third,

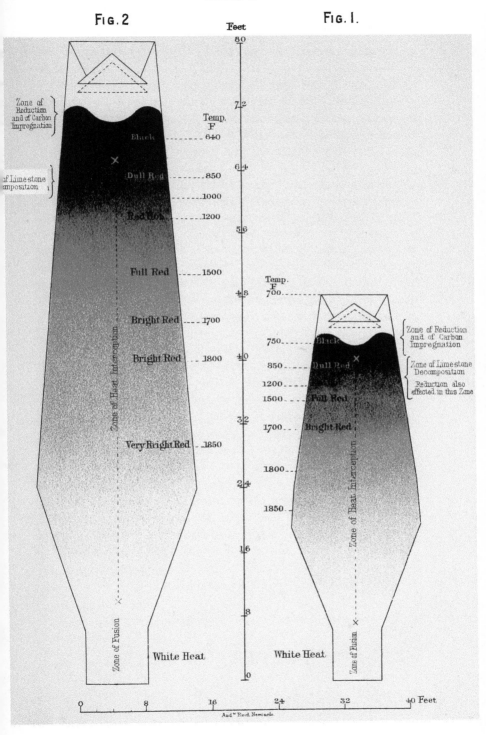

Plate 2

SECTION V.—THE BLAST FURNACE.

and finally where the fourth of the four above-named reactions begin to take effect.

The loss which results, for the reasons given, from the action of the last three of the series, is not all the injury which is sustained by the faulty performance of such a furnace as that in question. The gases finally escape from the throat at a very elevated temperature, forming of itself a very serious waste of heat that might yet do useful work, were opportunity afforded to it.

Now the obvious remedy, for this last mentioned inconvenience at any rate, would be an addition to the height of the furnace; by which means the heated gases, passing through an increased height of column, would have a better opportunity of imparting their heat to the incoming materials. This alteration of capacity, however, requires to be very considerable, owing to the diminished rate at which the gases, as they fall in temperature, impart their sensible heat to bodies cooler than themselves. In practice, when smelting Cleveland ironstone, an increase of about 30 feet to the height, with such other changes in dimensions as raise the capacity from 6,000 to say 12,000 cubic feet, has been found sufficient for practical purposes. Such a furnace is represented Fig. 2, Plate II., drawn also to scale. In it, the upper part, shaded black, will be observed to occupy a much larger area than in the case of Fig. 1. This change of form, as may be easily understood, also permits a very much larger proportion of the ore to be reduced under the first of the four reactions, namely, that due to carbonic oxide.

The results obtained by this mere alteration in the construction of the furnace are such as to produce a saving of nearly 30 per cent. in the fuel employed. This important reduction in the expense of making pig iron is due partly to the more perfect cooling of the escaping gases, and partly to the more complete oxidation of the carbon they contain, as may be ascertained by direct observation.

The following figures indicate the composition of the gases as they left two furnaces, one 48 feet and the other 80 feet in height:—

Composition by weight.	Per Cent.	Furnace of 48 feet.	Per Cent.	Per Cent.	Furnace of 80 feet.	Per Cent.
Nitrogen	58	...	...	57	...	...
Carbonic acid	12	containing carbon	3·27	18	containing carbon	4·91
Carbonic oxide	30	do.	12·85	25	do.	10·71
	100	Total carbon	16·12	100	Total carbon	15·62

The following shews the heat units evolved in the two cases:—

	Furnace of 48 feet. Parts.	Calories.	Furnace of 80 feet. Parts.	Calories.
Carbon to carb. acid	3·27 × 8,000 =	26,160	4·91 × 8,000 =	39,280
Carbon to carb. oxide	12·85 × 2,400 =	30,840	10·71 × 2,400 =	25,704
Total	16·12	57,000	15·62	64,984
Carried off in 100 parts of waste gases		9,650		6,420
Available heat		47,350		58,564

Hence in the smaller furnace we have } $47,350 \div 16\cdot12$ carbon = 2,937 { useful calories per unit of carbon burnt.
In larger furnace we have $58,564 \div 15\cdot62$ do. = 3,749 do. do.

This gives a pretty close approximation to the saving of fuel already mentioned, viz., 30 per cent.[1]

Two questions suggest themselves, on examining the figures as they are set forth above:—1st—is it not possible to reduce still further the loss of the 6,420 units, equal to about 11 per cent. of the whole heat evolved, which are shewn to be escaping in the gases, even from the larger furnace? and 2nd—cannot the proportion of the carbon existing in the gases as carbonic acid be increased, and thus add to the efficiency of the fuel consumed?[2]

With respect to the first of these two suggestions, the subject is one to which much attention has been paid during the course of my enquiries; and in connection with which too little regard has, in some cases, been bestowed on the scientific considerations affecting the process. As soon as it had been demonstrated that increasing the capacity of the furnace from 6,000 to 12,000 cubic feet, and raising the height from 50 to 80 feet, had been attended with so marked a saving in fuel, it was expected, not perhaps that there was no limit to a useful enlargement, but that the dimensions just given still permitted a much further increase, before reaching a point where such increase ceased to be profitable. Although as early as 1869 I had shewn that a furnace 80 feet high, but with a capacity of nearly 25,000

---

[1] At present the heat brought into the account by the hot blast is neglected.

[2] It may also be remarked that the temperature of the escaping gases is affected, more or less, by the warmth of the ironstone as drawn from the calcining kilns. This applies to both furnaces, but this disturbing circumstance has been approximately allowed for.

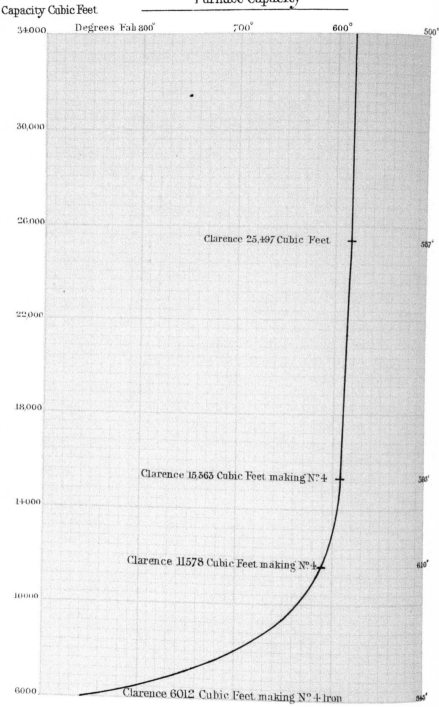

SECTION V.—THE BLAST FURNACE. 75

cubic feet,[1] emitted its gases at as high a temperature as one of half the size, this did not prevent furnaces being built with a height of 103 feet, and a capacity of 41,000 cubic feet, the object stated being a further economy of fuel. A mere glance at a diagram, Plate III., given at the date referred to, is sufficient to shew, from the direction of the curve in its lower, and the want of further change in its upper part, how improbable it was that any benefit in the consumption of fuel could result from the adoption of such extraordinary dimensions—so far at least as a more perfect cooling of the gases was concerned.

It may be asked whether the fact that the cooler materials were thus retained so much longer in contact with the heated gases, without reducing the temperature of the latter, might not have been occasioned by the larger furnaces being driven so much harder than the smaller as to neutralise any advantage they might otherwise have realized. But the very reverse was the fact, as I demonstrated at the time by quoting[2] the then preformance of three furnaces of various dimensions, situate at the Clarence works, as compared with the colossal furnaces just referred to.

The figures (to which I have now added those of a fourth Clarence furnace) are as follows:—

	No. 1.	No. 2.	No. 3.	No. 4.	No. 5.
Furnace—height	48 ft.	80 ft.	80 ft.	80 ft.	103 ft.
Cubic capacity	6,000	12,000	16,000	25,500	41,000
Weekly make, tons	220	260	350	550	550
Weekly make per 1,000 feet of cubic capacity, tons	37	22	22	21·57	13

From this statement it will be perceived that the make of the last named furnace ought to have been close on 900 tons per week, in order to give a result equally favourable, in point of production, to that afforded by the furnace of 16,000 cubic feet; whereas in point of fact it was only 550 tons. It may be convenient here to mention that, whenever it has been attempted to bring up the relative production of the larger furnaces to that of the original furnaces of 6,000 cubic feet, the consumption of fuel, as might be expected, was immediately augmented.

[1] Development of heat and its appropriation in furnaces of different dimensions. Trans. Iron and Steel Institute, 1869.
[2] Limit of economy of fuel in the Iron Blast Furnace. Transactions of Civil Engineers, 1871-72, Part II.

SECTION V.—THE BLAST FURNACE.

The attempts quoted having proved unsuccessful in lessening the the loss of heat, owing to the temperature of the escaping gases failing to fall below that which the enlargement to 12,000 cubic feet had effected, it occurred to me that the chemical action which took place in the reducing zone might be of a heat-producing character; and that to raise the height of the furnace beyond a certain point merely elevated this zone of heat evolution. If so, its distance from the final exit of the gases would remain unaltered, and the power of the incoming minerals to absorb heat from such gases would not be increased.

From the large amount of heat, evolved by the generation of carbonic acid in the reducing zone, has, however, to be deducted that required for dissociating, or tearing away, the oxygen from the oxide of iron. Now the experiments for determining the heat produced by the combinations of different substances are of a very delicate nature, and it was difficult to say, with a narrow margin to work on, how far any calculation respecting the estimated balance between heat generated by burning carbonic oxide to the state of carbonic acid, and heat absorbed by reducing peroxide of iron, might be affected by experimental error. The extent of the margin referred to was known to be small, because the investigations of the best authorities went to show that the heat absorbed, in splitting up, or reducing, peroxide of iron, was a little less, but only a little less, than that generated by the combination of oxygen with the carbonic oxide used to effect the dissociation.

Under such circumstances nothing short of experiment upon the furnace itself could be regarded as a satisfactory solution of the problem; and this I conducted in the following way. It was ascertained that a mixture of flints and blast furnace slag had as nearly as possible the same specific heat[1] as calcined Cleveland ironstone, and hence that there would be little or no disappearance of sensible heat from any difference between the materials in this respect. The rapidity with which the mixture absorbed heat also corresponded pretty nearly with that of the ironstone. The temperature of the upper portion of the furnace was of course far below that required to fuse such materials as

---

[1] That is, to heat up the same weights to the same temperature. the same quantity of heat was absorbed in each case. This consideration is of importance, for the difference in some kinds of material is very great.

those just mentioned; so that there would be an absence, so far as they were concerned, of all chemical action likely to affect the temperature of the gases, and these would therefore be simply cooled by the heat-absorbing properties of the flints and slag.

The ironstone was then withdrawn from the charges of a furnace, and an equal weight of the mixed slag and flints was put on in its room. The mean temperature of the gases gradually fell, as the unreduced oxide remaining in the furnace lost its oxygen; and as soon as the ore was all deoxidized, the mean temperature remained stationary. After continuing it at this point for a short time, the use of the slag and flints was discontinued, and that of ironstone resumed; when the temperature of the gases gradually rose until it reached its original level.

This experiment seems conclusive as to the reducing zone being one of a heat producing character, and that in consequence there was, as I supposed, a point beyond which, for the reasons already given, it was useless to add to the height of the furnaces. This opinion received confirmation from actual experience in the case of the furnaces of 90 and 103 feet in height; for it did not appear that the gases left them at a perceptibly lower mean temperature than those from furnaces of 80 feet in height.

The other alternative alluded to, as a means of obtaining from the fuel consumed in the blast furnace an increase in the quantity of useful heat, consisted in a greater degree of saturation of the gases by oxygen. Any addition from this source, however, must be obtained by the action of the gases on the ore; because it has already been shewn that all which the atmospheric oxygen could effect was the conversion of the carbon at the tuyeres into carbonic oxide.

With a view to supply this increased quantity of oxygen to the burthen of a furnace in good working order, let us suppose an additional quantity of ore to be added. One of two things would happen: either the carbonic oxide, by its combination with the oxygen in this ore to form carbonic acid, would at first raise the temperature of the contents of the furnace in the reducing zone, or else the carbonic oxide, having already taken up as much oxygen as it was capable of holding, would be unable to act further on a fresh supply of ore.

Practically the effect in each case might be the same. In the first the rise in temperature would intensify the action of the carbonic acid in the gases, as explained above in describing the four reactions, and would thus interfere with the process of reduction. The result of this change would be that a portion of the ore would pass down through the reducing zone with a part of its oxygen unexpelled: this part would then have to be driven off either by solid carbon, or if by carbonic oxide, yet under such conditions that the resulting carbonic acid by dissolving carbon, would, immediately revert to the state of carbonic oxide. If, on the other hand, no action took place in the uppermost zone between the carbonic oxide and the unreduced ore, owing to the inability of the former to separate a further quantity of oxygen from the oxide of iron, this separation would have to be performed by solid carbon under precisely the same unfavourable conditions as those just indicated.

Allusion has been made to the power of carbonic oxide to take up oxygen from ore. A few words of additional explanation on this important subject appears necessary for a perfect understanding of the nature of this reaction, for virtually it is the key to the discovery of the extent to which economy in fuel can be carried in smelting iron.

It must be conceded in the first place, that it is quite possible to convert carbonic oxide wholly into carbonic acid by means of iron ore; but this can only be done by using a very large excess of the latter. This is equivalent to our supposing that in reducing peroxide of iron, the oxygen which is separated at first is less firmly held by the iron than those portions which are withdrawn at a later period of the process—a supposition which experiment and observation entirely confirm. Those who may wish to study in greater detail the chemical reactions of the furnace are referred to the separate Section devoted to this particular branch of the subject: here it may be stated that so tenaciously is the last portion of oxygen retained by the iron, that it is believed that the trace of that metal always found in the cinder is due to the impossibility of effecting an absolutely perfect separation of *all* the oxygen from its combination with it.

In the meantime it must be granted that, having regard to the temperature which is set up in the reducing zone of the furnace, and is inseparable from it, a position of equilibrium is necessarily established, when the carbon as carbonic acid bears a certain proportion to the carbon as carbonic oxide. The maintenance of the conditions most favourable for saturating the carbonic oxide with oxygen, as far as the nature of the action permits, in other words for obtaining the highest possible percentage of carbonic acid in the gases, is perhaps a difficult matter. It would indeed be equally difficult to determine by actual observation what is the maximum ratio which the carbon as carbonic acid can bear to the carbon as carbonic oxide; because the determination of the mean amount of each in the gases is not without a certain degree of uncertainty, owing to the circumstance that these amounts are constantly varying, according to the time, after the period of charging fresh ore, at which the samples are taken. Under such difficulties as those referred to, the smelter finds it best to allow himself a little margin in the fuel he supplies to his furnace; or, what amounts to the same thing, if his "burthen" is adjusted with the view of producing No. 3 iron, he must be prepared for having sometimes a higher and sometimes a lower quality than the one he aims at obtaining.

Were it required to name a limit beyond which it would be impracticable, if not impossible, to carry the mean saturation with oxygen of the gases of a blast furnace, the answer probably would be—When the two compounds of carbon and oxygen contain an equal quantity of oxygen. This happens when, out of three equivalents of carbonic oxide, one has been converted into carbonic acid, leaving the remaining two unchanged ($CO_2 + 2CO$). We shall however see later that this is a degree of oxidation of the gases rarely, it may be doubted indeed if ever, attained in actual practice—at all events through any considerable length of time.

SECTION VI.

ON THE USE AND THEORY OF THE HOT BLAST.

In the Section immediately preceding the present, it has been shewn that the heat produced in a blast furnace is that caused, in the first place by burning all the carbon to carbonic oxide, and in the second place by converting something like one-third of such carbonic oxide into carbonic acid. Unavoidable losses of the heat so evolved, are occasioned by some minor sources of escape and by the temperature at which the gases are necessarily permitted to leave the throat. The balance between the heat evolved and the heat so lost is what is actually expended in the operation itself.

If the heat absorbed by the general requirements of a furnace were such as could be supplied by the combustion of a quantity of fuel, which would permit the gases to escape fully oxidized and cooled, so far as is shewn in the last Section to be possible, then all the conditions necessary for a complete utilization of the heating power of the fuel, consistent with the nature of the process, would have been secured. A different state of things however might obtain. A proportion of say 20 units of carbon per 20 units of iron made, might suffice in the furnace, so far as reduction is concerned, supposing a little above 6 units should escape as carbonic acid, and a little under 14 units as carbonic oxide; but the heat evolved by the combustion of the 20 units, might be insufficient to melt the iron and slag, and provide for other minor causes of absorption. If we assume that 24 units of carbon, burnt to carbonic oxide, would suffice for the general requirements of the operation, it is clear we might add the heat represented by the 4 units of carbon by a direct introduction of heat at the hearth, without in any way interfering with the proper relation which the carbonic acid and carbonic oxide must bear to each other in the zone of reduction.

SECTION VI.—ON THE USE AND THEORY OF THE HOT BLAST.

The addition of this heat is performed by means of the hot blast: in other words the chemical composition of the gases is still retained, and they possess a powerfully reducing action on oxide of iron, while their temperature is raised by the amount of heat which is communicated to the air employed in burning the fuel.

This important discovery was originally applied by Neilson under such circumstances, as to make it appear that the heat of a unit of carbon, burnt outside the furnace and conveyed into the hearth by the blast, was many times more efficacious than that of a unit of carbon burnt in the interior. This appeared the more marvellous seeing that, although the coal consumed at the hot blast pipes was fully burnt, *i.e.* to carbonic acid, yet not above 25 per cent. of the heat it produced actually reached the tuyeres ; the remainder being lost at the stove chimneys, by radiation, etc. Many explanations were propounded to account for the seeming anomaly; but with these we need not trouble ourselves at any great length, for the mode of action of the hot blast, is not as it appears to me, in reality difficult to understand.

Before entering upon this explanation, it may be well to dispose of the alleged saving of 3 tons of coal per ton of iron, stated as having been effected at Dunlop and Co.'s works, at the period of the first introduction of the use of heated air some 50 years ago, by merely heating the blast to 615° F. (324° C.) Commercially no doubt this was perfectly correct; scientifically it is simply misleading. As a matter of fact the coke used in smelting the ton of iron at the Clyde works was reduced from 60 to 40 cwts., the excess of gain beyond this being represented by the wasteful mode of coking in the open air, and by a greater consumption at the steam engines due to the smaller make. Still it was not to be denied that a saving of 20 cwts. of coke, by injecting an amount of heat, contained in the blast, represented by the entire quantity capable of being developed out of 2 or 3 cwts. of coal, was sufficiently astounding.

The French, alive then as they are now to the importance of every improvement in iron making, sent over M. Dufrenoy to ascertain the truth of the recommendations urged in favour of heating the blast; and in his work, published in 1834, he mentions the circumstance that, while 20 cwts. of coke per ton of iron were saved in Scotland, only 13 cwts. appeared to be the saving in Wales. Either of these quantities justified its introduction into France, which laboured under the disadvantage of dear fuel; but great must have been the surprise of the owners of the Torteron works in that country, when they found their saving was only 6 cwts., or about half of the lesser quantity just named.

Former attempts to explain the action of the hot blast were chiefly based on some supposed difference in the nature of the combustion, caused by the preliminary heating of the air; but I have been unable to find any analysis of the gases at the hearth, undertaken with a view to determine this question. The actual heat itself, delivered into the furnace with the blast, could not, I apprehend, be expected to perform a higher rate of duty than that obtained from any other source, although this effect has actually been claimed as a consequence of Neilson's discovery.

With respect to the three figures of 20, 13, and 6 cwts., just given as the respective savings of fuel in Scotland, Wales, and France, it ought to be mentioned that in the last two instances the actual consumption of fuel, before the air was heated, was considerably less than what it was in Scotland, where the economy in fuel effected by this means was the greatest. It must further be observed that this difference in the quantity of fuel consumed was independent of any difference in the size of the furnace, or in the richness of the minerals in use, which resembled each other in the three localities.

In the last Section attention was drawn to the importance of having the reduction of the ore effected by carbonic oxide, for reasons which need not be repeated; and it was shown how this end was promoted by the use of enlarged furnaces. Let us now take the case of two furnaces of equal capacity, each fed with a different kind of ironstone, and driven at exactly the same speed. If both varieties of ore lost their oxygen at the same rate, reduction in each case would be performed in the same way, so far as regards its being effected by carbon or carbonic oxide. If on the other hand one of the ores were much less susceptible than the other to the influence of the deoxidizing agency at work in the higher zone of the furnace, then in any given time a larger proportion of such ore would descend unreduced into that region, where reduction would necessarily have to be carried on by solid carbon, and would therefore be accompanied by a waste of fuel, in the manner already explained.

SECTION VI.—ON THE USE AND THEORY OF THE HOT BLAST.

In further illustration of the loss arising from a diminution in the quantity of carbon escaping as carbonic acid, in relation to the iron produced, let us assume the case of an easily reducible ore, the heat requirements of which may be estimated at 96,000 calories per 20 units of pig iron produced.

If in such a case we have 6 units of carbon per 20 units of iron in the gases as carbonic acid, we should have:—

					Calories.
These	6 units of carbon	× 8,000 calories	...	...	= 48,000
Leaving 20	,,	,,	to escape as carbonic oxide, burnt with air at 0° C., × 2,400 ...		= 48,000
Total 26	,,	,,	making together the	... ...	96,000

Included in these 96,000 calories we may assume 12,000 to be carried off in the escaping gases.

Let us now suppose that another variety of ore, resembling the previous one in everything but in readiness with which it parts with its oxygen, to be treated in a furnace of the same capacity as that used in the former case. As a consequence, it is assumed that by the time it is half reduced, a zone of the furnace is reached of such a temperature that carbonic acid is immediately decomposed by heated carbon. Granting that we have now only 3 units of carbon per 20 units of pig iron escaping as carbonic acid, the account will stand thus:—

					Calories.
The	3 units of carbon as carbonic acid × 8,000			...	= 24,000
Leaving 30	,,	,,	,,	oxide, burnt with air at 0° C., to provide the remainder × 2,400 ...	= 72,000
Total 33	,,	,,	making ...	... ...	96,000

The difference, viz. 7 units of carbon, is not however the full measure of the loss; because not only, as will presently appear, is the volume of escaping gas much larger than in the former instance, but it is carried off less perfectly cooled than took place when smelting the more reducible ore. It might therefore easily happen that $1\frac{1}{2}$ to 2 units of fuel might disappear in this way, bringing up the difference to nearly half a ton of coke per ton of iron—and all owing to there being a diminution of 3 units of carbon as carbonic acid in the gases.

The correctness of this surmise respecting differences in reducibility was proved by a series of experiments on artificially prepared oxides, and on natural ores; the full details of which have been explained elsewhere.[1] The former were expressly made with a view to obtain substances likely to be reduced with varying degrees of rapidity; and among other trials they were tested to see if there were any differences in the temperature at which reduction by carbonic oxide commenced on different specimens. The following are the results then obtained:—

	Reduction Perceptible.	Marked Action.
Pure precipitated peroxide of iron merely dried	285° F. (140° C.)	300° F. (149° C.)
Peroxide from calcined nitrate of iron, much more compact than previous specimen	293° F. (145° C.)	310° F. (154° C.)
Peroxide from sulphate of iron, calcined at a higher temperature, still more compact	407° F. (208° C.)	420° F. (215° C.)

Three ores, all with the iron in the state of peroxide, were afterwards enclosed in a vessel containing pure carbonic oxide and maintained for seven hours at a temperature of 770° F. (410° C.) The results were as under:—

	Per Cent. of its Original Oxygen.
Cleveland calcined ironstone lost	20·7
Spathose calcined ironstone lost	28·4
Raw Lancashire hematite lost	57·4

Such variations in susceptibility to reduction by carbonic oxide, as those just enumerated, justify the supposition that a larger quantity of fuel may be expended in smelting Cleveland than is consumed in the treatment of Lancashire ore. In both cases the furnace employed is regarded as having the same capacity, which is moreover considered as being insufficient for economical working. In the furnace working Cleveland stone—the less reducible of the two ores—more oxide of iron arrives at a zone where deoxidation is effected by carbon, thus causing a greater waste of fuel in the manner already described.

In what way, it may be asked, can the heating of the blast produce an effect, resembling in character that consequent upon an enlargement of the furnace? To answer this, let us assume that in a furnace of the olden type, say 50 feet high and containing 6,000 cubic feet, blown

[1] Chemical Phenomena of Iron Smelting.

SECTION VI.—ON THE USE AND THEORY OF THE HOT BLAST.

with air at 32° F. (0° C.), 35 cwts. of carbon per ton of iron was burnt at the tuyeres, while in another, driven with air at such a temperature that 15,000 heat units per ton of iron entered the furnace with the blast, only 25 cwts. of carbon was consumed in the hearth.

The heat development in the cold blast furnace at the tuyeres, when carbonic oxide only is generated, would be 35 × 2,400 = 84,000 units.

	Calories.
The heat development, also at the tuyeres, with hot air is 25 × 2,400 =	60,000
Add heat in the blast	15,000
Total	75,000 units.

But the volume of gases conveying through the materials the 84,000 calories produced by the cold air is 40 per cent. larger than that which is the vehicle of the 75,000 units produced by the hot air.

It is therefore certain that the retarded rate, at which the gases must pass upwards in the latter instance, will enable them to impart a larger quantity of their sensible heat to the cold materials, and will afford the carbonic oxide a correspondingly longer time to act on the ore, than can be the case with the larger volume.

All doubt as to the correctness of this explanation may be regarded as dispelled by the experience gained at a work belonging to Earl Granville, by whose courtesy and that of his then manager Mr. Horton I obtained the fullest information. At the furnaces there, prolonged exposure of the solids to the gases was obtained, not by diminishing the volume of the latter, but by increasing the height of the furnace to 71 feet; when precisely the same economy of fuel was obtained with cold blast in the enlarged furnaces, as if heated air had been used in those of the older type, say of 6,000 cubic feet capacity.

The figures illustrating the results obtained by the use of furnaces of different dimensions, and by the use of hot and cold air, are so pregnant with information bearing upon the present question, that they deserve to be repeated. One ton of metal was made, at the establishment referred to, under the following conditions:—

	Cwts. of Coke.
Cold blast in a small furnace, of say 50 feet, with about ...	40¼
Hot blast in a small furnace } with about	28¼
Cold blast in an enlarged furnace }	

But even 28¼ cwts. of coke represent a quantity of carbon, in which there is so great an excess of carbonic oxide, as compared with the carbonic acid in the gases, that a large portion of carbon may be withdrawn, and heat in the blast substituted in its room, without injuriously impairing their reducing energy. To effect this however, the reducing gas must have an extension of time to perform the work, and this is done by an enlargement of the hot blast furnace, say to 12,000 cubic feet capacity, or 80 feet in height instead of 50 feet.

Taking the coke consumed in the enlarged cold blast furnace, or 28¼ cwts., as being the weight required to furnish the needful heat, when burnt with cold blast, we have the following figures illustrating the use of heated air :—

	Cwts.	Cwts.
Coke required in an enlarged cold blast furnace, per ton of pig		28·250
From which deduct —		
Heat in hot blast, represented in coke burnt to carb. oxide...	5·405[1]	
Saving in heat escaping with gases, due to their volume being lessened, represented by coke burnt to carb. oxide	·165	
		5·570
Leaving		22·680

Now this weight, viz. 22·68 cwts., is a very common weight of coke used for a ton of iron, in a hot blast furnace of the above description.

Let it therefore be remembered that, in two of the cases just described, hot air and increased capacity of furnace mean one and the same thing. This analogy holds good under circumstances of a different order. In the hot blast furnace of 6,000 cubic feet capacity, the air may be considered as having been heated to about 1,000° F. (538° C.); but when the temperature was raised, by means of fire brick stoves, to about 1,400° F. (760° C.), and this was applied to a furnace of 7,500 cubic feet, the consumption of fuel was the same as that in a furnace of about 11,500 cubic feet blown with air at 1,000° F. (538° C.)

In considering the application of the hot blast, the same kind of question suggests itself as that which presented itself upon the occasion of dealing with the capacity of the furnace—viz. to what point will it be profitable to increase the temperature.

[1] This assumes about 12,000 heat units to be communicated by means of the blast.

SECTION VI.—ON THE USE AND THEORY OF THE HOT BLAST.

Much diversity of opinion has prevailed with respect to this very interesting problem. Some formerly held that each successive addition of heat would be attended with the same saving as those preceding it. From this opinion I have always dissented, for reasons which we will now proceed to consider.

The reduction of the ore and carbon deposition arising from dissociation of carbonic oxide limit the carbon escaping as carbonic acid from a Cleveland furnace to 6·58 units per 20 units of pig iron. The reason why any reduction in this quantity is attended with a great loss of heat needs, after what has been said, no further explanation. If however we have to accept as a law, the statement that it is impossible to have more than one-third of the carbon in the escaping gases fully oxidized, it follows that we must have twice as much in the form of the lower oxide, making in all therefore 19·74 units.

The actual weight of coke required to produce the result just referred to is obtained as follows:—

Carbon in gases as carbonic acid 6·58 and as carbonic oxide 13·16 =	19·74
Carbon to combine with the iron	·60
	20·34
Deduct carbon in the limestone say	1·54
	18·80

This weight of carbon (18·80) may be considered as equal to 20·20 of coke. It rarely happens, though, that each unit of carbon as carbonic acid is accompanied with less than 2·2 units, instead of 2·0 units, as carbonic oxide. Hence to preserve the full equivalent of this metalloid (6·58) as carbonic acid, 14·47 units must be present as carbonic oxide. Taking the former figure, however, the actual heat evolved in combustion would be as follows:—

Burnt to carbonic acid		$6·58 \times 8,000 =$	52,640
Do. carbonic oxide	13·16		
Less burnt by carbonic acid in limestone	1·54		
	11·62	$\times 2,400 =$	27,888
Exclusive of ·60 in iron	18·20		80,528

Since the calories required amount to about 86,000, this leaves only 5,000 to 6,000 calories to be supplied by the blast.

As a fact, however, we know that from 10,000 to 12,000 calories are found in the blast of the Cleveland furnaces, working under what may be considered the most favourable average conditions, say with about 20½ cwts. of coke; and this is due to the fact that more or less of the carbonic acid (represented by 6·58 units of carbon) always is wanting. A very common quantity would be 5·58 units; which means a deficiency of 5,600 calories, corresponding therefore exactly with the heat in the blast as commonly used, on the estimate given above. When we come to examine the sources of heat in these two cases, assuming 86,000 calories as having to be provided, we have:—

	Units.		Units.	
Carbon as carbonic acid ...	6·58		5·58	
By oxidation of carbon, calories	80,528	} 86,000 {	74,928	} 86,000
In hot blast do.	5,472		11,072	

This statement is intended to prove that if the carbon fails to perform its full theoretical duty, the deficiency in the instance cited is made up by additional heat of the blast; but it is equally true that the more heat contained in the blast, the larger must be the quantity of carbonic acid which disappears from the gases per 20 units of iron made; because the heat so conveyed into the furnace is intended to displace, and does displace, so much carbon introduced as coke; and as the two oxides of carbon have to be maintained in certain relations to each other, it is obvious that we cannot, when the two are found in these relations, alter the quantity of the one without at the same time producing a rateable effect on the other.

In order to show the manner in which this substitution of coke by hot air acts, I will suppose that the carbon, actually yielding heat, is an integral number of units and that the carbon as carbonic acid to that as carbonic oxide is as 1 to 2, and that there are also required 86,000 calories per 20 units of iron as before. In the last column is placed the temperature the blast must have in order to convey the heat not provided by the combustion of carbon:—

Carbon Burnt by Blast.	Weight of Blast.	Burnt to		Heat Produced by		Total from Carbon.	Blast to Provide.	Total Heat.	Temp. of Blast.	
		CO_2.	CO.	CO_2.	CO.				C. Deg.	F. Deg.
18	104·3	6·00	12·00	48,000	28,800 = 76,800	9,200 = 86,000			372	701
17	98·5	5·67	11·33	45,360	27,192 = 72,552	13,448 = 86,000			576	1,069
16	92·7	5·33	10·67	42,640	25,608 = 68,248	17,752 = 86,000			808	1,486
15	86·9	5·00	10·00	40,000	24,000 = 64,000	22,000 = 86,000			1,068	1,954

SECTION VI.—ON THE USE AND THEORY OF THE HOT BLAST.

In reality, however, these numbers require a little correction; because in practice the proportions of carbon as carbonic acid to carbon as carbonic oxide, instead of being as 1 to 2, more commonly are as 1 to 2·22 or thereabouts. With this correction the figures would stand thus:—

Weight of Carbon furnishing Heat. Units.	Heat produced by Carbon.	Blast to Provide.	Total Heat.	Temp. of Blast.	
				C. Deg.	F. Deg.
18	74,504	11,496	86,000	465	869
17	70,368	15,632	86,000	669	1,236
16	66,232	19,768	86,000	900	1,652
15	62,040	23,960	86,000	1,163	2,152

In the opening remarks of this Section it was mentioned, in the event of 24 cwts. of carbon per 20 units of iron being burnt, that a certain portion of this quantity might be advantageously replaced by heat in the blast. By this was meant that the carbon oxidized being still in sufficient quantity to permit the full amount (6·58 units) to be retained in the form of carbonic acid, no heat was wasted by any action between other portions of carbon and such carbonic acid.

If for simplicity of calculation we assume two units of carbon as carbonic oxide to accompany each unit as carbonic acid, we shall have as our best result 13·16 units of the former against the 6·58 of the latter, making in all 19·74 units. The balance, or 4·26 units of carbon to make up the 24, can be supplied by means of the hot blast without lowering the normal quantity of carbonic acid (6·58 of carbon) in the gases.

The moment, however, that the carbon in the gases is reduced in quantity, so as to prevent the normal amount of carbon as carbonic acid (6·58 units) from holding its proper proportion (one-third of the whole), the carbonic acid partially disappears, so as still to retain its original proportion of the one-third.

The change involved in the alteration of the quantities of carbon, introduced as such into the furnace, will be easily followed by an examination of the circumstances attending the combustion of the 18 and the 15 units of this element referred to in page 88.

There is, in these quantities, a loss of heat on every 20 units of iron represented by the reduction of one unit of carbon in the state of carbonic acid to that of carbonic oxide; but inasmuch as, in each case, the proportions of carbon in the two states of oxidation remain the same, the efficiency of each unit of the fuel is not thereby lowered The alteration of circumstances in question therefore, consists in a mere substitution of heat units in the blast for heat units furnished by the combustion of carbon in the furnace.

In a strictly calorific point of view, the actual gain by the use of the hotter air—1,068° C. (1,954° F.)—were it possible to obtain it at this temperature would be as follows:—

	Carbon.	Units.
a. When using blast containing 9,200 calories:—		
Carbon burnt by the blast		18·00
Blast with 9,200 calories obtained by combustion of fuel in stoves to carbonic acid $\frac{9,200}{8,000}$ equal to	1·15	
Add 15 % the assumed difference between loss of burning carbon in blast furnace and in stoves	·17	
	1·32	
		19·32
b. When using blast containing 22,000 calories:—		
Carbon burnt by the blast		15·00
22,000 calories in blast. Carbon got as before. $\frac{22,000}{8,000}$ equal to	2·75	
Add 15 % loss arising in manner formerly stated	·41	
	3·16	
		18·16

Strictly speaking some small allowance must be made in favour of the superheated air, owing to the diminished volume and somewhat reduced temperature of the escaping gases, consequent on the reduced quantity of carbon burnt by the blast, and the smaller quantity of carbonic acid, per 20 units of iron, which is generated at the top of the furnace. Approximately we may therefore regard the three units of carbon saved in the furnace fuel, to be the result of two units consumed and fully oxidized in the hot air apparatus, instead of being partially oxidized as they would be in the furnace.

When we compare the quantity of fuel burnt either way the difference may appear not very large. This, however, is not the only consideration which determines the conduct of the smelter, who is solely

SECTION VI.—ON THE USE AND THEORY OF THE HOT BLAST. 91

guided by the commercial aspects of the question. The fuel used in heating the blast is the gas which escapes from the furnace, which in many, indeed in most, cases would be wasted; but may at the best be valued as small coal, which in the North of England can be had for threepence per cwt. If then by burning two cwts. of coal worth sixpence, or, still better, by burning furnace gas costing nothing, three cwts. of coke, worth it may be two shillings, could be saved, a great gain would arise from such a change.

The question raised a few pages back was the extent to which the fuel, actually burnt in the furnace, could be lessened by a corresponding amount of heat being communicated to the blast.

It would be difficult perhaps to predict how far this substitution could be carried, in face of the principles which have just been explained; but the real extent is of less consequence, because there is one circumstance which will, in all probability, prove an effectual barrier to much more being done than what has already been achieved.

In the table recently given, page 89, the temperatures of the blast in burning different quantities of carbon were as follows:—

Units of carbon burnt	18	17	16	15
Temperature, degree F.	869	1,236	1,652	2,152
Increase over preceding quantity	—	367°	416°	500°

The nature of the rise of temperature in the air, renders it impossible, as may be easily understood, to continue the increase further; because no apparatus could be made to stand the wear inseparable from a temperature of 2,152° F., this being close to the fusing point of cast iron.

If we adopt 1,600° to 1,700° F. as a practicable point to be attained, the equivalent consumption in coke (using about 12 units of limestone for smelting Cleveland stone), works out to the figures given below:—

Units of carbon per 20 units of pig burnt with blast at 1,652° F.	16·00
Carbon burnt by carbonic acid in limestone	1·44
Carbon in iron	·60
	18·04
Add for impurities to 18·04 carbon to bring it to coke, say 7½ per cent.	1·35
	19·39

It must not be supposed that this weight of coke is unchangeable, as it is susceptible of some trifling modification arising from differences in the richness of the ore or of the quantity of flux required. With this qualification $19\frac{1}{2}$ cwts. of coke may, in my opinion, be accepted as the possible limit with which a ton of Cleveland foundry iron will be produced, using air at a temperature of nearly 1,700° F. If so we have a gain of one cwt. of fuel or thereabouts, as compared with the case when the blast is only heated to 1,000° F.

In connection with the use of superheated air, there are certain considerations, beyond those referred to, which may ultimately lead to its large extension. The difficulty of forcing a large volume of blast through a very lofty column of solid material is well understood by the practical smelter. This difficulty has been notably augmented by the use of the higher furnaces now almost universally adopted in the North of England. The resistance to the passage of the blast did not prove however to be greater than could be dealt with, at the time, by a trifling addition to the pressure applied by the blowing engines.

Of late years the demand for coke for iron works has been so great, in the quarter referred to, that the Durham collieries have been compelled to use inferior seams of coal, to mix with the best in their ovens. This change has not been followed by any material alteration in the purity of the coke, but it has, it is thought, considerably altered its ability to support the heavy load to be carried in a furnace 75 or 80 feet in height.

In some cases, as in the anthracite furnaces of the United States, increased resistance to the entry of the blast is overcome by a mere addition to the pressure; which I found occasionally to be as high as 12 lbs. and even more on the square inch. It may be that similar treatment might prove efficacious in the North of England; although the limited trials which have been made in this direction have not been very encouraging.

The inconvenience, attending such a state of things as that just mentioned, no doubt gives rise to irregularities in the descent of the materials, and in their uniform permeation by and exposure to the reducing gases. This is only met by an increased consumption of coke, which aggravates the evil by calling for a corresponding addition to the quantity of air required for its combustion. These remarks point to

SECTION VI.—ON THE USE AND THEORY OF THE HOT BLAST.

the expediency of diminishing the height of the column, through which the blast has to be forced; and this, we have already seen, can be accomplished by increasing the temperature of the air. It is almost superfluous, however, to repeat that the quantity of carbon present in the furnace, must always be sufficient to maintain a reducing action in the gas produced by the oxidation of the fuel at the tuyeres. It is only therefore when the consumption of coke notably exceeds the limit supposed to be essential for this object, that the use of superheated air can be expected to be beneficial, to any great extent, in saving fuel.

The use of superheated air will be referred to again in a future Section, which will be reserved for the subject of the fuel employed in the blast furnace.

SECTION VII.

ON THE QUANTITY AND QUALITY OF THE FUEL REQUIRED IN THE BLAST FURNACE, USING AIR OF DIFFERENT TEMPERATURES.

IN Sections V. and VI. I have endeavoured to give a concise account of those chemical and physical laws, which are concerned in the reduction of the ore and the fusion of the products in the blast furnace. In the remaining Sections devoted to the smelting process, these laws, at the risk of some repetition, will be examined with greater minuteness, and at the same time some further explanation will be given of the nature of the experimental researches undertaken during the course of my investigations.

Some confusion may arise in considering the subject of this Section, by overlooking the very different conditions which often attend the smelting of iron from different kinds of ore. It is not meant to restrict the observation just made to mere richness of metal contained in the mineral, a circumstance to which undue importance may easily be given, for reasons shortly to be explained.

We have already seen that the economy effected by the hot blast was by no means of an uniform character, nor did the advantages derived from the enlargement of the furnaces, when smelting Cleveland and some other ironstones, always attend a similar alteration, when applied to those engaged on ores of a different kind. In these last mentioned cases, the difference in the results was due to the greater or less facility with which the ore in use parted with its oxygen. But if two minerals are identical in their facility of reduction, but one is poorer in iron than the other, the additional fuel required is that which is needed for melting the additional quantity of slag, and for providing for some minor sources of heat absorption. The reducing power of the gases must always be maintained, and is regulated therefore by the actual quantity of iron to be reduced. Pig metal however, as is well known, is not only impure iron, but the impurity it contains varies in

SECTION VII.—FUEL REQUIRED IN BLAST FURNACE.

amount according to its origin or its mode of manufacture. Such foreign matters as silicon, phosphorus, and some others, are, like the iron itself, reduced from a state of chemical combination, for which reduction, heat and carbon are indispensable: hence it may well happen that similar quantities of pig may require somewhat varying measures of heat, etc., and therefore of fuel, for their treatment in the blast furnace.

The desirability of knowing the actual quantity of heat, necessary for producing a given quantity of pig iron, is obvious; and thanks to scientific research, we have at our command such information, of a sufficiently exact nature for our present purpose.

The following table is an estimate of the number of calories, or heat units, supposed to be required for smelting Cleveland iron of average composition, including certain sources of waste which may be regarded as inseparable from the operation. The quantity of iron taken as a basis of calculation is 20 kilogrammes, which permits of a ready comparison with 20 cwts., the familiar weight to the mind of an English manufacturer :—

	Kilogrammes.		Calories.		Calories.
Evaporation of the water in the coke, estimated at	·58	×	540	=	313
Reduction of 18·60 kilogrammes of iron from state of peroxide	18·60	×	1,780	=	33,108
Carbon impregnation	·60	×	2,400	=	1,440
Expulsion of carbonic acid (CO_2) from limestone	11·00	×	370	=	4,070
Decomposition of carbonic acid from limestone to CO	1·32 carbon	×	3,200	=	4,224
Decomposition of water in blast	·05 hydrogen	×	34,000	=	1,700
Reduction of phosphoric acid, sulphuric acid, and silica, estimated at	...		...		3.500
Fusion of pig iron	20·00	×	330	=	6,600
Fusion of slag	27·92	×	550	=	15,356
Total units usefully applied	...		...		70,311
Transmission through walls of furnace	...	...	...	3,600	
Carried off in tuyere water	...	...	...	1,800	
Expansion of blast, escape into foundations, etc., supposed				3,389	
					8,789
					79,100
Carried off in escaping gases from a modern furnace	...	...	...		7,900
					87.000

SECTION VII.—FUEL REQUIRED IN BLAST FURNACE.

It ought to be observed that this estimate is framed on the supposition that the grade of the metal produced is No. 3 foundry iron. It is necessary to bear this in mind, because, as is well known to iron smelters, the richer, as it is termed, is the quality of the iron, the larger is the consumption of the fuel required in its production; thus, that more coke is needed to run No. 1 than to make other numbers, or white iron, is universally recognised.

Accepting the figures given above as a basis of calculation, and admitting the *possibility* of so oxidizing the carbon of our fuel, that we shall have in the gases one unit of this element escaping as carbonic acid, and two as carbonic oxide—which is the greatest extent to which we may assume it possible to oxidize the carbon as a whole—we can easily compute the coke required for the process.

The heat evolved by one unit of carbon so burnt will be $\frac{1 \times 8{,}000 + 2 \times 2{,}400}{3} = 4{,}266$ calories. Coke of average quality may be regarded as containing, of earthy and volatile matter about 10 per cent. of its weight: hence the actual heat units evolved, per unit of coke, will be something like 3,840. The quantity of heat in the blast raised to a temperature of about 540° C. (1,004° F.), per 20 units of iron, may be calculated at 12,000 units: we have therefore:—87,000 (total absorption) less 12,000 (heat in blast) = 75,000, as the heat units to be provided by the coke; and 70,000 ÷ 3,840 (heat units evolved per unit of coke) = 19·53 as the units of coke per 20 units of metal, or in other words as the number of cwts. of coke per ton of pig iron. Speaking from my own observation, I deem it doubtful whether Cleveland iron has ever been made continuously for any length of time with quite so low a rate of consumption of coke as that just named. At the same time the approximation is sufficiently close to what has been occasionally achieved, to afford strong evidence in favour of the general correctness of the figures quoted for the various items of heat absorption.

Actual experience of the performance of two of the Clarence furnaces, both smelting Cleveland ironstone, but under very different conditions owing to the difference of the capacity of the furnaces, point also to the truth of the statement just given.[1]

[1] Further particulars are given in Sec. XXVIII., "Chemical Phenomena of Iron Smelting."

SECTION VII.—FUEL REQUIRED IN BLAST FURNACE.

SUMMARY OF WORKING OF TWO CLARENCE FURNACES.
COMPARISON OF 80 AND 48 FEET FURNACES.

Height of furnace in feet	80	48
Capacity in cubic feet	11,500	6,000
Materials used—coke per ton of iron	22·32 cwts.	28·92 cwts.
Calcined ironstone do. do.	48·80 „	48·80 „
Limestone do. do.	13·66 „	16·00 „
Weight of blast per ton of iron	103·74 „	128·12 „
Do. escaping gases	138·66 „	170·59 „
Temperature of blast, Centigrade	485°	485°
Do. escaping gases	332°	452°
Heat units evolved by oxidation of fuel, as it was actually effected in the two examples	81,536	89,288
Heat units as contained in the two different quantities of blast used	11,919	14,726
	93,455	104,014
Heat units carried off in the two different weights of escaping gases	8,860[1]	16,409
Heat units absorbed in furnace work	84,595	87,605
Difference between the two ...	3,010 calories.	

The difference between the two sets of figures arises from the additional limestone used in the 48 feet furnace, which required about 1,700 calories for the chemical changes it underwent, and for the fusion of the increase of slag occasioned by its use. This reduces the discrepancy to about 1,300 units, or about 1·5 per cent. on the entire calculation—a difference not in excess of what, from the nature of the case, might have been expected. In the example taken for estimating the theoretical minimum of fuel, there is still less combustible, and therefore less ash to deal with, than in the two examples just given; hence less limestone is required. In like manner the reduced quantity

---

[1] The heat contained in the escaping gases is estimated as follows:—

	Units.
Weight of gases 138·66 × 332° C. × ·24 SH	11,043
Less estimated heat communicated by hot ironstone ...	2,496
	8,547
Latent heat of ·58 units steam × 540	313
	8,860

of coke diminishes the volume of blast required for its combustion, and diminishes at the same time the weight of escaping gases. In this way the amount of heat carried away by the gases is also lessened. The sum of these differences is about 3,800 units in favour of the calculation setting forth the theoretical minimum. Placed in a tabular form, with allowance added or subtracted, as may be required, we have the following figures representing the heat units per 20 kilogrammes of pig iron, smelted from Cleveland stone, in the three given cases :—

	A.	B.	C.
			Supposed minimum consumption.
Character of furnace	80 feet.	48 feet.	
Units of fuel consumed per 20 units of pig ...	22·32	28·92	19·53
Oxidation of fuel—	Heat units.	Heat units.	Heat units.
Carbon to carbonic oxide evolves ...	45,024	58,656	42,000
Carbonic oxide to carbonic acid evolves...	36,512	30,632	33,000
	81,536	89,288	75,000
Blast estimated to contain	11,919	14,726	12,000
	93,455	104,014	87,000
Carried off in escaping gases	8,860	16,409	7,900
Utilized in furnace	84,595	87,605	79,100
Deduct in case of B, to bring down to standard of A for more flux used, etc.	...	1,700	...
Add in case of C, to bring up to standard of A for less flux used at C, etc. ...	...	...	3,800
	84,595	85,905	82,900

In the foregoing estimate of the minimum quantity of heat required for smelting a ton of Cleveland iron, under the circumstances described, it has been assumed that gases of a given composition and temperature, as they leave the furnace, cease to have any useful power in deoxidizing iron. Not only were numerous experiments made to determine what may be considered as the practical point of saturation of pure carbonic oxide by oxygen, when applied as a reducing agent to oxide of iron; but actual trials were conducted, by exposing Cleveland calcined ore to the gases as they left one of the large Clarence furnaces (25,400 cubic feet capacity). Upon one occasion the deoxidizing power of the escaping gas was so very nearly exhausted, that during a period of 24 hours only 3·72 per cent. of the original oxygen, in the oxide of iron exposed, was withdrawn; and on another occasion, during an exposure

SECTION VII.—FUEL REQUIRED IN BLAST FURNACE.

of 48 hours, the weight of oxygen removed was only 5·71 per cent. of its original amount. It will be recollected that the neutral point was supposed to be reached, when in the mixed gases one part by weight of the carbon was converted into carbonic acid, leaving two as carbonic oxide. Speaking volumetrically, this position of saturation is considered as attained when 50 volumes of the former gas (CO_2) are accompanied by 100 volumes of the latter (CO). Now as a fact 45 volumes of carbonic acid per 100 of carbonic oxide was the maximum percentage found in over 14 analyses, made during the trials in question. Further the fact that the weight of oxygen in the two gases, when in the proportion just named, viz. one of CO_2 to two of CO, is the same, renders it yet more probable that it constitutes a neutral position as regards further absorption of this element.

If such a limit as that just described exists, and if, as has been shown, the chemical action in the upper zone of the furnace, from its proximity and position in relation to the exit of the gases, determines the temperature of those gases, it is clear that the full heating power of the fuel suffers a diminution by the existence of natural laws.

The assertion that the theoretical limit previously named, *i.e.* 19·53 cwts. per ton of iron, is only under exceptional circumstances, persistently attained in Cleveland, will be generally accepted by the iron manufacturers of that locality; and it is not difficult to understand why this should be the case.

Let us imagine a blast furnace of the most approved dimensions, in which instead of permitting the entire volume of the intensely heated reducing gas, carbonic oxide, to ascend to the level of the charging plates, one half of this was withdrawn by an aperture made in the lower part of the structure. Dealing only with the carbon gases, with which alone we are concerned in the present argument, such withdrawal would amount to taking 50 volumes out of 150 of the two gases taken together, and would leave, in the remaining 100 volumes, the acid and oxide of carbon (CO_2 and CO) in equal volumes, instead of in the ratio of 1 to 2, the limit to which saturation with oxygen is supposed capable of being carried. To make amends for such disturbance, something like 8 to 10 cwts. of coke, for each ton of iron, would have to be added, in order to restore the 50 volumes of carbonic oxide thus abstracted.

SECTION VII.—FUEL REQUIRED IN BLAST FURNACE.

Such a mode of procedure as the escape of one half of the reducing gas, in exactly the manner supposed, of course never occurs in practice, but disturbances may be occasioned by other causes, such as a faulty method of introducing the materials, etc. The first furnaces under my control were so constructed, that all the charges were thrown in at one opening, over the back tuyere; the result being that all the coarser pieces in point of size rolled to the front. The interior thus became divided vertically into two compartments; through the more open one of these the greater portion of the gases rushed with great rapidity, while they ascended through the other with great difficulty, and of course in smaller quantity, owing to the compactness with which the finer portions of the charge filled it. In the former case they escaped into the air having only imperfectly discharged their proper duty, while in the latter, owing to their deficient quantity, the oxide of iron reached such a level in the furnace that reduction was only imperfectly effected, or was effected exclusively by solid carbon instead of by carbonic oxide, or at all events without the formation of carbonic acid. A mere alteration in the method of charging—using four doors instead of one—by which the inconvenience referred to was in a great measure if not entirely avoided, was followed by a saving of 10 or 12 cwts. of coke on every ton of metal, with an increased daily production of iron amounting to from 30 to 40 per cent. This is quoted as an extreme case; but in ordinary practice derangements similar in principle but smaller in extent may and do take place in a furnace of the most perfect construction, and sometimes in spite of the most active supervision. In such cases, or by the occurrence of scaffoldings as they are termed, a more ready exit is afforded to the gases; and they thus leave the structure before they have time to warm up the materials, and before they have absorbed the full equivalent of oxygen they are capable of holding.

There are many conditions known to furnace managers, which, if present in sufficient force, are invariably followed by derangements in the working of a blast furnace. Irregularities in charging, or the arrival of a large quantity of mineral in a state of fine mechanical division, giving rise to an interruption in the uniform upward flow of the gases, are among the most common of these evils.

Conspicuous among the conditions alluded to, at all events in the Cleveland district, is the quality of the coke. Freedom from ash, of

SECTION VII.—FUEL REQUIRED IN BLAST FURNACE.

course, is at all times a *desideratum*. Not only does the presence of earthy impurity diminish *pro tanto* the amount of carbon in the fuel, but it demands the addition of limestone for its fusion. More heat, and therefore more fuel, has to be provided, on account of the addition of carbonic acid brought by the flux into the furnace, and also on account of the larger weight of slag which has to be fused.

The mechanical texture, however, of the coke is perhaps more important in the eyes of furnace managers than any trifling increase in the percentage of ash. The combustible most in favour with these officials is that variety of coke which is distinguished by hardness, almost metallic lustre, and power of resisting a crushing load: while the lowest in their estimation is the soft, dark-coloured, and easily pulverized variety known as *black ends*.

The prejudicial effect of such coke as this may be twofold. In the first place it may crumble down under the weight of the superincumbent load of a lofty column of material, and thus occasion that diversion of the current of gases already explained. If the belief of competent practical men of large experience is worth anything—and I myself place much value upon it—this constitutes at least one of the objections to friable coke; and they allege in proof that the inconvenience attaching to its use has been intensified since the introduction of lofty furnaces, in which the crushing of the softer material would be increased by the additional weight it has to sustain. The second defect is one of a purely chemical character. The inconvenience of having the coke acted on by the carbonic acid, produced by reduction of the ore, has been already explained, on page 72. It has also been shown in Section V. that this action begins with good coke at about 1,500° F. (815° C.). But experiment in the laboratory, as well as observation on the action of the gases at the furnace itself, led me to the conclusion that the soft black coke, objected to by furnace managers, is affected by carbonic acid at about 800° F. (427° C.) This would mean, at any rate in low furnaces, the existence of a zone, near the top, where the materials reach a region of elevated temperature, and where the carbonic acid, generated by deoxidation of the ore, is being reduced to the state of carbonic oxide by dissolving carbon.

Having regard however to the fact that our furnace managers are so averse to the use of soft coke, I was anxious to make the question one of actual trial in the blast furnace itself; and this was carried out at the Wear Works, on a furnace having a capacity of 17,500 cubic feet. The volumetric percentage of carbonic acid in the gases was as follows:—

Time.	Remarks.	Percentage of CO_2 in the whole.
12 noon	Soon after charging an ordinary round of coke, ironstone, and lime	10·7
12·30	After charging, the carbonic acid gradually rises as the action	11·7
12·50	on the ore progresses. At the same time it is under the	13
1·12	usual average	12·5
1·28	Temperature of gases about 932° F. (500° C.) Charged 52 cwts. of coke	10·6
1·45	Here the effect of the carbon in splitting up CO_2 is evident	4·3
2·10	An increase of carbonic acid is now observable, probably due to those portions of the coke most easily attacked by CO_2 having been previously carried off	13·0
2·30	Volumes of carbonic acid emitted	11·9
2·55	,, ,, (temperature of gases rising) ...	21·0
3·20	Carbonic acid declining, probably owing to reduction of ore	12·0
3·40	having been now largely effected; the first portions of the	15·3
4·5	oxygen being the most readily removed	12·1

In regard to the temperature at which carbonic acid begins to act on coke, the above experiment does not afford any precise information. After introducing the coke, the temperature of the lower part of the charge would rise rapidly. In the laboratory it did not appear that the action was very marked till about 1,200° F. (649° C.) As an evidence, however, of the difference of susceptibility of different kinds of coke to be acted on by carbonic acid, the results are given of several trials, recently made in the Clarence Laboratory on specimens from two different collieries—the one being of known inferiority to the other. The specimens were exposed, side by side, for different periods of time, at a temperature of about 1,500° F. (815° C.)

Exposure.				Percentage of Loss. A, better quality.	B, second quality.	Ratio of A to B.	
2 hours.	Average of 3 samples, A and B	...	1·097	...	2·257	... 100	205
4 ,,	,, 3 ,, ,,	...	1·903	...	3·767	... 100	198
6 ,,	,, 2 ,, ,,	...	2·904	...	5·645	... 100	194
8 ,,	,, 1 ,, ,,	...	3·893	...	7·801	... 100	200
					Average	... 100	199

SECTION VII.—FUEL REQUIRED IN BLAST FURNACE. 103

These trials therefore indicate that, while a ton of coke in the case of A would in 2 hours lose ·778 cwts., the waste on B would be equal to 1·560 cwts.

A second series gave the following results:—

Exposure.					Percentage of Loss. A, better quality.	B, second quality.	Ratio of A to B.	
2 hours.	Average of 2 samples, A and B			...	·758 ...	2·274 ...	100	300
4 ,,	,,	,,	,,	,,	... 1·356 ...	3·152 ...	100	233
6 ,,	,,	,,	,,	,,	... 2·674 ...	5·766 ...	100	215
8 ,,	,,	,,	,,	,,	... 3·892 ...	7·802 ...	100	200
					Average	...	100	237

Both sets of figures indicate that the better quality of coke continued at the end of eight hours less susceptible to the action of hot carbonic acid than the other with which it was compared.

There are other circumstances connected with the quality of coke, besides the presence of ash and proneness to be acted on by carbonic acid, which demand attention. Conspicuous among these are the occurrence of sulphur, usually in the form of sulphide of iron, of volatile matter from imperfect coking, and of oxygen.

It is a fact well known to founders that the use of coke in the cupola, containing an excess of sulphur tends to harden the iron in the process of melting. Speaking in general terms it seems probable, that the presence of this element interferes, in some way, with the separation of carbon in the graphitic state, a condition which appears essential to the formation of soft iron.

In the blast furnace this lowering of the quality of the metal, by the action of sulphur, is sought to be counteracted by the use of an additional quantity of lime beyond that absolutely required for the purpose of a flux. This increased weight of carbonate of lime and the probable necessity of keeping the furnace, when using coke rich in sulphur, at a higher temperature, is the cause of an increased consumption of coke.

To elevation of temperature has to be ascribed those differences in the working of a blast furnace, which constitute the condition required for the production of rich iron; hence anything tending to harden or lower the quality of the metal has to be met by an influence of a contrary character, such as that called into play by a more liberal use of fuel. Besides such effect as that consequent upon a mere increase in the temperature at the tuyeres, I am inclined to think, for reasons to be given in the next Section, that the hotter the iron is, the weaker is the affinity between it and the metalloid in question.

Confining the observation to the produce of the Durham coal-field, the purest samples of coke contain only ·5 per cent. of sulphur while the most impure may have three times this quantity of the objectionable element. A very usual content of sulphur in the fuel consumed at the Middlesbro' furnaces will be about 1 per cent.

It rarely, if indeed it ever, happens in practice that a small portion of those volatile substances existing in raw coal is not left, unexpelled by the process of coking. That variety already spoken of as black ends contains most of the volatile matter referred to, and at the same time from their porous nature they absorb and retain a much greater quantity of water, used in quenching, than coke of a denser character.

The facts just given are set forth in the analyses of specimens from the collieries which supply the Clarence works :—

No.	Carbon.	Hydrogen.	Oxygen and Azote.	Sulphur and Ash.	Moisture.	
1	91·28	·30	·70	7·52	·20	100
2	91·22	·23	1·29	6·86	·40	100
3	76·41	2·41	3·16	7·92	10·10	100
4	92·98	·30	2·11	4·61	...	100

No. 1.—Selected hard pieces from Browney pit.
„ 2.—Average produce from do.
„ 3.—Black ends from do.
„ 4.—Average produce from South Brancepeth pit.

Two specimens brought from Belgium gave the following results in the Clarence laboratory :—

No.	C.	H.	O. and Az.	S. and Ash.		
1	80·85	·51	2·13	16·51	dry	100
2	84·06	·31	·24	15·39	dry	100

In the six samples, of which analyses have been given, the heating power of the coke in the blast furnace may be said to approach very closely to that represented by the first column of figures, viz. the

SECTION VII.—FUEL REQUIRED IN BLAST FURNACE.

carbon they contain. I would however call attention to certain analyses, contained in a recent work by Dr. F. Muck, of Bochum,[1] in which the oxygen and azote in one case reach 7·68 per cent.

	C.	H.	O. and Az.	S. and Ash.	
1. ...	83·487	·737	5·467	10·309 dry	100
2. ...	84·360	·187	6·303	9·150 ,,	100
3. ...	85·060	·860	7·680	6·400 ,,	100

Nos. 1 and 2 are from the Ruhr district. No. 3 is from an English coal.

In these last three examples we have not only a deficient amount of carbon, but we have a portion of it—the heat producing element—already united to a *considerable* quantity of oxygen by which its calorific powers is materially diminished.

Mr. I. Parry's researches on the gases given off by coke heated *in vacuo*, point to the probability that all the oxygen found in coke escapes in combination with carbon and not with hydrogen. In one experiment he obtained a gaseous mixture consisting of

$$2CH_4 - 22H - 23CO - 53CO_2 = 100$$

According to this analysis three-fourths of the carbon present is fully oxidized and therefore incapable of affording any service in the blast furnace.

If we apply this composition of gases to the analysis of No. 3 in the last of the series, it would mean that with the oxygen and hydrogen 3·6 per cent. of carbon would be volatilized bringing the total matter driven off to 12·14 per cent. (·860 + 7·680 + 3·600). This would only leave, even neglecting the probable action of the carbonic acid on carbon, only 81·46 to arrive before the tuyeres. Besides this depreciation in available value some allowance has to be made for the heat absorbed in expelling the volatilized matter, so that about 6 per cent. of the total value of the coke is sacrificed in the case just referred to.

We may assume that 19·53 cwts. represent the theoretical weight of coke, of average good quality, required to make a ton of No. 3 Cleveland iron; and it has often happened that for periods of some weeks an addition of 5 per cent. to this quantity, making 20·50 cwts., has sufficed for keeping a furnace pretty steadily on good grey iron.

[1] Grundzüge und Ziele der Steinkohlen-Chemie.

SECTION VII.—FUEL REQUIRED IN BLAST FURNACE.

This extra allowance of fuel affords a margin to provide against those minor sources of derangement recently spoken of. The result is not a perfectly uniform rate of working; but the furnace so treated makes, when the conditions are all favourable, a somewhat superior quality of iron to No. 3; while on the other hand, if any unfavourable circumstance intervenes to impair the general heat development, and its appropriation, the effect is manifested, not necessarily in an increase in the fuel consumed, but in a lowering of the grade of the metal.

Adopting 87,000 units as the quantity of heat needed for smelting 20 units of iron from Cleveland stone, we have seen that this amount of heat is capable of being obtained by burning 19·53 cwts. of coke with air at 540° C. (1,004° F.) For the performance of this duty, however, a proper saturation of the gases by oxygen is required; while their escape from the furnace must be delayed until they have communicated to the solids all the sensible heat they are capable of imparting. We have seen however that variations in the quality of the coke, and other disturbing causes, have rendered it difficult to produce a ton of Cleveland iron with so low a quantity as 20 cwts., or even more, of good coke. Now that other than the best seam of coal (the Brockwell) is used by coke makers in the County of Durham, 22½ to 23 cwts. of coke per ton of iron is considered as not an uncommon result.

Such a proportion of carbon to the work to be performed leaves of course in the escaping gases a considerable excess of carbonic oxide, as compared with carbonic acid, over and above that shown to be practicable. An examination was made of a furnace using 22·97 cwts. of coke with the following results:—

		Calories.
The requirements for the furnace work ascertained by a similar mode of computation to that already given were found to be for 20 units of iron ...		70,788
Waste by transmission through walls, by heat in tuyere water, etc., etc.	8,957	
Heat in escaping gases, after allowing for that contained in the warm ironstone	8,055	
		17,012
Total calories required		87,800

This heat was derived as follows:—

The actual carbon in the coke was estimated at 20·91 units. Of this

SECTION VII.—FUEL REQUIRED IN BLAST FURNACE. 107

1·30 units was oxidized by the carbonic acid in the limestone, leaving 19·61 to be burnt by the blast; and this was burnt in the following manner:—

	Units.	Calories.	Calories.
To the state of carbonic oxide	14·38 × 2,400 =		34,512
Do. carbonic acid	5·23 × 8,000 =		41,840
	19·61		76,352
Leaving the remainder to be brought in by the blast, viz. 96·99 units heated to about 498 C. (928 F.) ...			11,448
			87,800

It may be mentioned thàt the calculations showing the heat requirements and the heat production corresponded so nearly, as to confirm entirely the general truth of the two sides of the account.

According to the composition of the gases, as given above, it will be perceived that, instead of carbon as carbonic acid to carbon as carbonic oxide bearing anything like the proportion of 1 to 2, the ratio is 1 to 3·24 :[1] in other words, there is a considerable and an unnecessary excess of carbonic oxide. Further, the carbon in the gases as carbonic acid only amounted to 5·23 per 20 units of iron.

The preceding calculations in respect to the consumption of coke are founded on the heat required for smelting Cleveland ironstone, when using the blast at such temperatures as are within the power of stoves in which the air passes through pipes of cast iron. Gradually the form of apparatus known as Cowper's, or the modified form of this principle introduced by my late friend Mr. Thomas Whitwell, is finding increased favour among iron smelters. In these the blast, as is well known, is charged with heat by contact with intensely heated surfaces of fire brick. I propose now to consider to what extent and under what circumstances air heated to 1,400° to 1,600° F. (760° and 871° C.), instead of about 1,000° F. (537° C.), is likely to be really beneficial in the blast furnace.

On referring to an opportunity afforded me of examining the working of two furnaces blown with this superheated air, I find the following results. One of the two, which I will distinguish by the letter A, had a capacity equal to some of the larger Clarence furnaces, and was receiving its blast at a temperature of 854° to 888°, or an average

[1] That is, 5·23 out of 22·21.

SECTION VII.—FUEL REQUIRED IN BLAST FURNACE.

of 871° C. (1,600° F.) The other, which I will call B, was blown with air at 736° to 819°, or an average of 777° C. (1,430° F.); but it had a capacity more than 60 per cent. above that of A. The temperature of the escaping gases from A was 264° C. (507° F.), and of those from B 222° C. (431° F.) Now although A was thus being supplied with blast 94° C. (169° F.) hotter than B, the consumption of fuel was about 20 per cent. more in the case of A (the smaller furnace), than in the case of B. Such a state of things might not unreasonably lead to the inference that the excess of fuel per ton of iron in the case of A was due to its want of capacity.

Before proceeding to consider the justice of such a supposition, let us examine the performance of the furnace B, in order to compare it with that of furnace A. This is desirable; for, as far as consumption of fuel is concerned, the larger furnace B will be found to afford a fair approach to what I have considered the utmost limit of economy. To do this the figures representing the work of the two furnaces are placed side by side:—

	Estimate of Heat requirements.	
	Furnace A. Units.	Furnace B. Units.
Calories calculated as heretofore as usefully employed in the work	74,674	73,987
Waste by transmission, in gases and in tuyere water, loss at foundations, etc., etc.	13,751	11,472
Total calories	88,425	85,459

The heat was produced as under :—

	Furnace A.	Furnace B.
Burning of carbon in coke per 20 units of Iron to carbonic acid:—		
Furnace A 4·76 units × 8,000	38,080	—
Do. B 5·62 units × 8,000	—	44,960
Burning of carbon to carb. oxide including heat brought in by blast	55,481	40,598
	93,561	85,558

Thus while B exhibits a close correspondence between the two sides of the account, A shows the reverse, leading to the belief that in it the coke consumed must be somewhat overstated.

SECTION VII.—FUEL REQUIRED IN BLAST FURNACE.

	Furnace A. Units.	Furnace B. Units.
The weights of the blast upon this occasion were respectively per 20 units of iron ...	93·17	80·48
The escaping gases amounted to respectively ...	130·54	112·61
The average temperature of the blast was...	871° C.	777° C.
The temperature of the escaping gases ...	264° C.	222° C.
Weekly make per 1,000 cubic feet of furnace space ...	24¼ tons.	15 tons.

In addition to the carbon furnished by the coke, there was that brought in by the limestone, so that the total carbon in the gases was:—

	As Carbonic Acid.	As Carbonic Oxide.	Ratio of Carbon as Oxide to one as Carbonic Acid.
Furnace A	4·76	17·64	3·70
Do. B	5·62	12·80	2·27

I am not in a position to name the actual consumption of coke per ton of iron in these cases; but I will adopt the plan which has been recently applied to such calculations as the present, viz., that of estimating the weight of carbon, per 20 units of iron, from the composition of the gases. Although liable to a little fluctuation, when compared with the coke computed to be consumed, according to the charges used, this plan will hereafter be seen to afford a tolerably fair approximation to the truth: and it may be observed that in fact it is a little difficult to say, with any great precision, what weight of carbon is actually delivered to a blast furnace, remembering the somewhat rough practice observed at large iron works, and the varying composition of the coke itself, the impurity of which in South Durham varies from 7½ to 10 per cent.

According to the formula which has been adopted, it would appear that the average consumption of carbon as coke on every 20 units of iron, in the two visits over which my observations extend, was as under:—

	Furnace A.	Furnace B.
Units of carbon ...	20·79	18·61
If to these numbers we add 7½ per cent. for the ash, water, etc., in the coke, we have	1·55	1·28
For the weight of coke used per 20 units of iron	22·34	19·89

Whether these figures represent the actual consumption of coke is unimportant, because the difference between the two sets of numbers will serve quite well to illustrate the argument I wish to found upon the analyses of the gases.

With regard to the performance of A, the lesser furnace, it cannot be maintained, either as a matter of capacity or of heat of blast, that there is here any want of power. This opinion is founded on the fact that there are at the Clarence Works furnaces very little more than half the size of A, doing 20 per cent. more work per 1,000 cubic feet of capacity, and blown with air at a little above 1,000° F. (538° C.); and these furnaces are consuming *very much less* coke than A was doing.

The action of one of the furnaces just referred to proves so unmistakably the correctness of the views just laid down, that I have expressly had a set of careful observations made on its mode of working to serve as a medium of comparison.

It was, according to the Charging Book, consuming, per 20 units of iron, calcined ironstone 46·20 units, coke 20·40 units, limestone 9·38 units. The temperature of the blast was 563° C. (1,045° F.), and that of the escaping gases 262° C. (503° F.)

The heat evolved was estimated as follows:—

	Units.
Coke 20·4, less ash, etc. 1·63 = carbon	18·77
Carbon in limestone, carrying off an equal weight from coke ...	1·12
Leaving to be burnt by blast	17·65

Analyses of the gases proved the carbon to be oxidized in the following way :—

		Calories.
5·48 units as carbonic acid × 8,000 ...	=	43,840
12·17 ,, carbonic oxide × 2,400 ...	=	29,208
17·65		73,048
The blast weighed 87·87 units, heated to 563° C.	=	11,724
Total evolved		84,772

These figures corresponded very closely with the heat requirements, estimated in the way already described in these pages.

The ratio of carbon as carbonic acid to carbon as carbonic oxide is as 1 to 2·22; and the coke, calculated from the quantity of carbon contained in the gases, works out to 20·49, which is an extremely close correspondence with the estimated consumption, viz.: 20·40 cwts.

SECTION VII.—FUEL REQUIRED IN BLAST FURNACE.

The figures showing the results of the two furnaces in question therefore stand thus:—

	Furnace A.	Clarence Furnace— 40 per cent. less capacity than A.
Temperature of blast	853° C. (1,567° F.)	563° C. (1,045° F.)
Temperature of escaping gases	264° C. (507° F.)	262° C. (504° F.)
Make per week per 1,000 cubic ft.	24¼ tons	30 tons
Proportion of carbon as carbonic oxide to one of carbon as carbonic acid	3·70	2·22
Carbon in gases and in 20 units pig iron as furnished by coke, estimated by formula from their composition	20·79 units	18·85 units
Carbon furnished by coke as estimated from charging book	—	18·77 units

The waste of fuel consequent upon irregular passage and distribution of the reducing gases has been already adverted to. If, however, the furnace A were working under such unfavourable conditions as those previously alluded to, there is no reason why carbonic acid once formed should not escape as such. The ratio of carbon as carbonic oxide to carbon as carbonic acid would of course be disturbed, owing to the excessive quantity of coke used in the case of furnace A; because, since the generation of carbonic acid is limited by the oxygen separated from the ore in the process of reduction, all such excess of carbon must find its way out of the furnace as carbonic oxide.

This however constitutes no ground why there should be a greater disappearance of the carbonic acid itself in the one furnace than in the other. The figures representing the difference in this respect may not appear of much importance; but when the results are worked out their significance at once impresses itself upon the mind.

In the two furnaces described as blown with superheated air I ascertained, inclusive of the heat contained in the blast, that each unit of carbon burnt by the air entering the tuyeres, afforded from its superior oxidation nearly 15 per cent. more heat in the case of the furnace B than in that of A.

Not only have we the loss from the difference of combustion above mentioned; but, owing to the larger consumption of coke in the case of the lesser furnace A, the weight of the escaping gases is much larger than in B, and in consequence the loss from this source is correspondingly greater. This last cause of loss was made so conspicuous upon a former occasion as to require no enlarging upon at present. At page 97, in comparing the duty of an 80 feet with that of a 48 feet furnace, in the case of the former 8,860 calories were carried off in the gases, whereas in the lesser furnace no less than 16,409 were so lost, and this in consequence of the greatly increased volume of the gases, escaping also, it is true, at a somewhat higher temperature.

It is perhaps unnecessary to say that it is not only a waste of heat once generated that we have to deal with in furnace A; but with the cause which has interfered with its generation. In other words, the question is in what manner the formation of carbonic acid in the gases has been impeded, or, if not impeded, what has caused its disappearance.

In comparing the duty of the old furnaces of 48 feet with those of 80 feet, the cause of the disappearance of carbonic acid, or the reduction of the ore by solid carbon forming carbonic oxide—it is immaterial which—was considered to arise from reduction being effected in a zone of the furnace where the temperature was too high for carbonic acid to exist in the presence of carbon. The difference between the two furnaces in this respect is shown again in Figs. 1 and 2 Plate IV., in which it will be perceived that the region of lower temperature, coloured dark, is much more extensive in the section of the larger than of the smaller furnace. Let us suppose however that we pour in a vast amount of heat with the blast, a great portion of which is intercepted by the material, and is returned over and over again to the tuyeres in the manner already described; then it is not unreasonable to imagine that the region of less elevated temperature, in an 80 feet furnace, will be so diminished as to approximate in capacity to that of Fig 2, which represents one of 48 feet. In such an event it is clear the same fault which accompanies the reducing action in the lesser furnace, will manifest itself in the larger; viz. that solid carbon will perform this office, or else the carbonic acid generated by the deoxidation of the ore will be resolved into carbonic oxide by dissolving carbon.

There only remains to say a few words on the manner in which the larger furnace B has done its work. The first important point requiring observation is the very small amount of duty performed by

Plate 4

FIG. 1.   FIG. 2   FIG. 3.

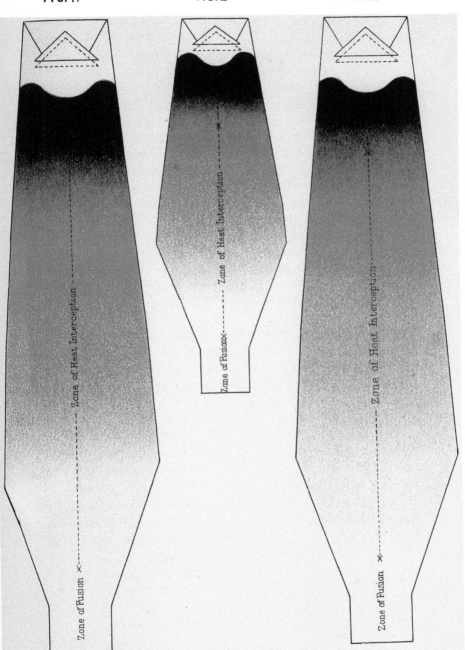

SECTION VII.—FUEL REQUIRED IN BLAST FURNACE.

a furnace of such capacity; in point of fact, for like space, it amounts to only 56 per cent. of the work done by its lesser neighbour A, and only to 50 per cent. of that of some furnaces elsewhere. I admit, in the matter of consumption of coke, that this large furnace is producing a ton of iron for something less coke than has been done for months together by furnaces half its size, making twice as much iron per 1,000 cubic feet of space, and blown with air 300° C. (540° F.) lower in temperature.

An instance has recently been brought under my notice where an 80 feet furnace, probably from defects in the stoves or in the condition of the furnace itself, or for both these reasons, was making 350 tons of forge iron per week, with the high rate of consumption of $25\frac{1}{4}$ cwts. of coke per ton of iron. Fire-brick stoves now deliver the blast at 1,400° to 1,450° F.; and the fuel has been reduced to $21\frac{1}{2}$ cwts., while the make has risen to 500 tons per week. In such a case as this it would of course be unreasonable to give superheated air the credit of saving $3\frac{3}{4}$ cwts. of coke; particularly when the reduced consumption — $21\frac{1}{2}$ cwts. — is in reality by no means an unusually low quantity of fuel for producing forge iron with blast of moderate temperature.

A considerable expenditure of heat, and therefore of fuel, accompanies the use of the unusual quantity of limestone needed in smelting the ironstone of Cleveland; while the slowness with which this mineral is reduced renders the use of large furnaces indispensable for its economical treatment. As a contrast to these results may be adduced the work performed by two furnaces, smelting the easily reduced hematite ore of Bilbao. These examples prove the large amount of work done in a small amount of furnace capacity, and the small amount of heat and therefore of coke required, as compared with Cleveland stone. The figures of the larger furnace also demonstrate the possibility of making iron with almost a theoretical minimum of coke previously insisted on, with blast of moderate temperature. It should be observed that the iron made was for Bessemer steel purposes, and attention is also directed in the case of the loftier furnace to the relation the two oxides of carbon bear to each other.

SECTION VII.—FUEL REQUIRED IN BLAST FURNACE.

	Furnace 57 feet high.	Furnace 70 feet high.
Cubic capacity, feet	7,630	9,550
Temperature of blast	491° C. (916° F.)	522° C. (971 F.)
Do. escaping gases	301° C. (574° F.)	205° C. (401 F.)
Coke used per 20 units of iron	23·21	17·60
Ore do. do.	39·13	38·92
Limestone do. do.	6·90	4·86
Average weekly make	437 tons.	516 tons.
Weekly make of iron per 1,000 ft. of capacity	$57\frac{1}{3}$,,	$54\frac{1}{3}$,,
The heat required for the actual work performed in the furnace was estimated by the same rule as that adopted with Cleveland iron.[1] It amounts to	62,805	60,271
Heat lost by transmission, in gases, etc.	20,361	15,294
Total	83,166	75,565

	Furnace 57 feet.	Furnace 70 feet.
Weight of blast per 20 units of iron	98·67	76·76
The estimated weight and composition of the gases per 20 units of pig were:		
Nitrogen	75·19	58·50
Carbonic acid	15·21 = C 4·15	19·69 = C 5·37
Carbonic oxide	42·23 = C 18·09	26·17 = C 11·21
Water in coke and in ore	7·55	7·32
Hydrogen	·08	·06
	140·26	111·74
Units of carbon as carbonic oxide per unit of carbon as carbonic acid	4·35	2·09
	Calories.	Calories.
Heat evolution by combustion of carbon	73,560	67,944
Heat in blast	11,479	9,495
	85,039	77,439
Appropriation amount was shown to be	83,166	75,565

About 4,000 calories were considered as absorbed in converting the water contained in the ore into steam.

[1] In both cases the temperature of the escaping gases is low, due to the fact that the ore was hematite and contained a certain quantity of water.

SECTION VII.—FUEL REQUIRED IN BLAST FURNACE. 115

I do not possess any better example illustrating the value of sufficiency in the capacity of the furnace than that afforded by the figures just given. The results further confirm much of that which has been advanced in these pages. We have the lesser furnace doing a large amount of work with a comparatively small consumption of coke for its size (23·21 cwts. per ton), owing to the readiness with which the ore is reduced and the small quantity of flux required. In the larger furnace, by running at a somewhat reduced speed ($54\frac{1}{3}$ tons per 1,000 cubic feet) the operation is performed under conditions which, if I am correct in the hypotheses already set forth, form a close approximation to perfect oxidation of the fuel, inasmuch as the carbon as carbonic acid to carbon as carbonic oxide is as 1 to 2·09—1 to 2 being considered as the extreme limit.

The truth of the statement is particularly apparent, when the performance of the furnaces is contrasted with that of a furnace of 6,000 cubic feet capacity, smelting Cleveland stone, given page 97.

With blast at a temperature of 485° C. (905° F.) the gases escaped at 452° C. (846° F.)—a temperature partly due to heat in the ironstone, the consumption per 20 units of pig (No. 3 quality) was as follows:—Coke 28·92, limestone 16·00, ironstone 48·80; weekly make 223 tons, or 37 tons per 1,000 cubic feet.

It may be mentioned that every pains was taken in the hematite furnaces to confirm the correctness of the weighings; but in addition to these precautions, the composition of the gases indicate the general correctness of the weights, and of the analyses; which latter were most carefully performed by Mr. Rocholl at the Clarence laboratory. The samples, like all those in recent experiments, were collected over a period of three hours.

The carbon found by a calculation based on the composition of the gases was:—

	Furnace of 57 feet.	Furnace of 70 feet.
Carbon from coke per 20 units of iron ...	21·14	15·85
Add for impurities in coke 10 per cent.	2·35	1·76
Gives weight of carbon as coke estimated from the analyses of the gases ...	23·49	17·61
The weight of coke estimated by the charge book was	23·21	17·60

Allusion was made at the beginning of the present Section to the undue importance sometimes attached to mere richness of an ore in the metal it contains. This observation had exclusive reference to the quantity of coke required for smelting it; in respect to which a belief sometimes prevails that the amount of fuel consumed bears a direct arithmetical ratio to the richness of the ore. Thus an ironstone yielding 42 per cent. of iron would require, *cæteris paribus*, one sixth less fuel in the furnace than a similar ore yielding 36 per cent. If there be any truth in what has been advanced in these pages, such an opinion is evidently fallacious. The quantity of heat unquestionably will be in excess of that demanded by the richer stone, because the place of metal is taken by earthy matter which must be melted; but not necessarily to the extent suggested.

It is an easy problem to compare the heat requirements of two specimens of ironstone, the one yielding 42 per cent. and the other 36 per cent. of pig iron; and here, in order to avoid complication, we will consider pig iron and pure iron synonymous terms.

	COMPOSITION OF IRONSTONE.	
	42 per Cent. Ore. Per Cent.	36 per Cent. Ore. Per Cent.
Iron	42	36
Oxygen	18	15·43
Earths	40	48·57
	100	100·00
	Cwts.	Cwts.
Hence the earths per ton of iron amounts to exclusive of ash in coke	19·05	26·98
Lime in 10 cwts. of limestone ...	5·60	
Do. in 14·15 cwts., being proportionate increase for additional earths	—	7·92
Cwts. of slag per 20 cwts. of iron	24·65	34·90

By reference to the table of heat requirements given in this Section we can see the additional heat units demanded for the increase in the amount of sterile matter to be dealt with.

Heat Units.

Expulsion of carbonic acid
from additional limestone $(14·15 - 10)$ = $4·15 \times 370$ = 1,535
Decomposition of additional
carbonic acid in limestone = $(1·69 - 1·20)$ carbon, or $·49 \times 3,200$ = 1,568
Fusion of additional slag $(34·90 - 24·65)$ $10·25 \times 550$ = 5,637

8,740

SECTION VII.—FUEL REQUIRED IN BLAST FURNACE.

Now one unit of coke burnt with blast to the state of carbonic oxide at about 1,000° F. (538° C.) may be taken as affording 3,137 calories or units; and $\frac{8,740}{3,107} = 2\cdot 81$ units of coke required in excess by the poorer over the richer ore. But taking 22 units of coke as the requirement of the richer, 3·66 units of fuel would have been the addition to be made, were the ratio of the increase an arithmetical one; making the total 25·66 instead of 24·81 (22·00 + 2·81).

In a following Section, to be devoted to a more minute view of certain chemical reactions set up in the interior of the blast furnace, the grounds will be given for believing that in smelting iron, as a rule, 6·58 units of carbon in the form of carbonic acid, per 20 units of iron, is the extreme limit to which this higher oxide of carbon is generated in the furnace. It is doubtful however whether so large a quantity of carbon as this is very often found in the gases as carbonic acid; there being always a predisposition, when the oxidizing and reducing tendencies of the gases approach the theoretical equilibrium, for carbonic acid to be reduced in its amount. When however there is any considerable weight of slag, as in the case of poorer ironstones, it has to be fused by the combustion of an addition to the coke; and this coke being exclusively converted into the lower oxide of carbon, from the want of oxide of iron to acidify it, the reducing tendency of the gases is *pro tanto* intensified. Thus a nearer approach to the full quantity of carbon as carbonic acid, viz. 6·58 units, is obtained, by complete reduction taking place before the materials reach a zone the temperature of which permits carbon to decompose carbonic acid. In this way more carbon is made to evolve 8,000 instead of 2,400 calories, and thus the general store of heat is thereby increased.

The estimate, it will be observed, has been founded on the assumption that the higher percentage of earthy matter requires for its fusion a corresponding addition of limestone. This by no means follows, for it might easily happen that so great a proportion of the new sterile constituents of the ore consisted of lime, as to render more flux superfluous.

By the kindness of an esteemed neighbour, I possess some interesting particulars of the working of a furnace using ironstone of the Cleveland type, in which the consumption of fuel, looking at the poorness of the mineral, is smaller than I have elsewhere found it in the course of my experience.

The furnaces he uses are 80 feet high, with a capacity averaging 15,500 cubic feet. The calcined ironstone only yields 34·9 per cent. of iron, the additional earthy impurity however being carbonate of lime, the limestone used is only 10·2 cwts. per ton of iron. The average make of the furnaces is 360 tons a week, producing during my observations pig of the average number of 3·5 with 22·7 cwts. of coke per ton, equal to $23\frac{1}{4}$ tons per 1,000 feet of capacity per week.

For all purposes of the furnace, we may assume that the requirements are about the same as those using a similar weight of coke and limestone on the banks of the Tees, and that the only difference in the instance under consideration is in the greater weight of earthy matter associated with the ironstone. Instead of 24·65 cwts. of slag, which accompany the use of a stone of 42· per cent. of iron, we shall have in the present instance about 33·25 cwts., or an increase of 8·6 cwts. Now 8·6 cwts. of slag will require (8·6 × 550) 4,730 heat units for their fusion; and this is really all the additional heat needed for this ironstone over and above that absorbed in smelting ore yielding 42 per cent. of pig metal.

It has, however, been demonstrated that when a furnace is receiving even less than the quantity of coke consumed in smelting this poor ore, the oxygen combined with the iron does not acidify carbonic oxide beyond a point which permits reduction to proceed with suitable rapidity. The generation of carbonic acid being entirely due to the presence of the oxide of iron, no increment of this gas takes place from the fact of a mere increase of earthy substances having to be fused ; and it is therefore a matter of indifference whether the extra heat required for this fusion, proceeds from the combustion of coke or it is brought in with the blast.

This is exactly a case where superheated air is most beneficial. The furnace gases themselves are capable of supplying, free of cost, the necessary heat to the fire brick stoves; and all that is required is to return it in the blast to the materials under treatment.

SECTION VII.—FUEL REQUIRED IN BLAST FURNACE.

It is easy to prove how the additional heat for fusing the additional slag was supplied by referring to the returns sent to me by the friend to whom I am indebted for these particulars.

In one of the Clarence furnaces, of 11,500 cubic feet, using 22·32 cwts. of coke per ton of iron, 103·74 cwts. was the estimated weight of the blast, and it was heated to 905° F. (485° C.) The heat units therefore contained in this volume of air consumed would amount to 11,919. In the case of the poorer stone under consideration, we may take the air entering the furnaces at the tuyeres as weighing 105 cwts., which heated to 662° C. will carry with it 16,473 heat units, or 4,554 more than the less highly heated blast at the Clarence works. This agrees within 176 units with the quantity (4,730) shown to be needed for fusing the extra weight of slag.

The light thrown upon the action of the blast furnace by these figures may be of great value in the future, to the iron smelters carrying on their work on the northern edge of the great Cleveland deposit of ironstone. This seam of mineral, at its outcrop there, is 8 to 10 feet thick, and continues pretty uniform in its content of iron for a distance of from 2 to 3 miles in a southerly direction. Before arriving at the final point a band containing a good deal of shale shows itself, which admits nevertheless of being in a great measure separated. The portion which remains, however, together with other smaller layers of shale, reduces the yield of iron in the furnace, reckoned on the calcined ore, from 42 to about 38 per cent. The larger band gradually thickens and changes into pure shale; it then splits the seam into two distinct beds separated by a distance of several yards, which are worked in Eskdale as the Pecten and Avicula seams, and are of inferior thickness to the main bed further north. The increase of shale bands in both is such, that the raw stone, instead of containing 30 to 32 per cent. of metal, scarcely exceeds 25 per cent.; and the calcined mineral falls in yield from 42 to 35 per cent.

The power of the smelter to deal with an ore, so much poorer than that to which he is at present almost exclusively accustomed, with so small an extra quantity of coke as that indicated, is a circumstance of great importance; and confers a value upon the fire brick stove, which it does not possess to the same extent in the treatment of the richer stone.

All the preceding calculations and statements have been based on the supposition that the fuel employed in the blast furnace was coke. It is now proposed to offer a few remarks on the other combustible substances used for obtaining the necessary temperature in smelting iron.

An amount of fixed carbon, not exceeding that given in the analysis of Durham bituminous coal, does not of itself prevent the produce of certain collieries being advantageously used in the raw state in the blast furnace. The Durham coal owes its power of making excellent coke to its "caking" property; in other words to its becoming sufficiently fused in the oven to run together in one mass. This property, however, is the reverse of being desirable in the blast furnace, because the agglomeration of the coal may bar the access of the reducing gases to the ore; or, by binding together the contents filling the structure, prevents that regular descent of the materials, the importance of which has been dwelt on in these pages.

The following analyses of the furnace splint of Airdrie in Scotland, and of the bituminous coals of Durham as well as of Scotland, will show that something apart from chemical composition renders the first applicable in the raw state as a furnace fuel, while the two latter are regarded as unfit for such work.

	Splint Coal, Scotland.	Bituminous Coal, Scotland.	Bituminous Coal, Durham.
Volatile matter	42·24	38·32	27·78
Sulphur	Not estimated.	Not estimated.	Not estimated.
Ash	3·50	5·78	5·90
Fixed carbon	54·26	55·90	66·32
	100·00	100·00	100·00

Notwithstanding a certain similarity in composition, there is no resemblance whatever between the splint and the bituminous coal as to their conduct in the blast furnace. Fragments of the Scotch splint, instead of showing any inclination to unite when heated, behave almost as charcoal or coke would do under the like circumstances.

The west of Scotland is fortunate in the possession of vast areas of this dry burning coal; hence the iron furnaces there are almost uniformly fed with their fuel in its raw state. The advantages of this

SECTION VII.—FUEL REQUIRED IN BLAST FURNACE.

process are that it avoids the expense of coking, and saves the necessary waste of a portion of the solid carbon, which takes place when the expulsion of the volatile constituents is effected in the presence of atmospheric air. Against these have to be set the providing of the heat required in the blast furnace, for driving off the gaseous portion of the raw fuel. If this be taken at 2,000 calories per unit of gas expelled, and, assuming a Scotch furnace of the old type, say of 45 feet in height, to consume 50 kilogrammes per 20 kilogrammes of iron, there will be 20 kilogrammes of volatile matter to evaporate. It is impossible in the case just cited to calculate, by direct observation, the actual consumption of fuel required to expel the 20 units of volatile matter; because the furnace being open at the top, the flame of the burning gas greatly aids the work of distillation.

The gas emitted by the coal consists chiefly of light carburetted hydrogen; and its quantity and specific gravity are such that, when its volume is added to that of the extra carbon burnt for its expulsion, the volume of the reducing gas is immensely increased, as compared with what it would have been, had coke been used instead of raw coal.

Such a change of course means that over a given period of time, for the same quantity of pig iron, a much larger volume of reducing gas would pass over the ironstone, compared with that given off when coke is the fuel employed. Now I ascertained by direct experiment that oxide of iron lost 2·12 times the quantity of oxygen when 213 litres of the reducing gas were passed over it in 6 hours, as when 65 litres of the same gas were consumed in the same time and at the same temperature: in other words, the addition of 228 per cent. to the volume of gas caused an addition of 212 per cent. to the oxygen extracted.

This great addition to the deoxidizing power of the furnace may thus permit a much more rapid rate of driving, than where coke is employed; and may thus account for the large make in the very small furnaces as used still very largely, in Scotland. In these, with a capacity of only 5,000 cubic feet, (height 42 by $14\frac{1}{2}$ feet) above 200 tons are run, equal to 40 tons per 1,000 cubic feet per week.

Advantageous as is without doubt the increase of reducing energy in these furnaces, the large volume of gaseous substances, rushing upwards with great violence, carries with it a great quantity of sensible heat; but the experience of the Scotch smelters, in adding to the height of their furnaces, had not until recently been such as to induce them largely to follow the example set them in Cleveland.

In the year 1870 however, Mr. William Ferrie of the Monkland works described at a meeting of the Iron and Steel Institute[1] a furnace 83 feet high, in which the coal, as it descended, passed through chambers heated red hot by the combustion in flues, constructed outside of the chambers themselves, of a portion of the furnace gases. By this contrivance the consumption of coal on the ton of iron was reduced from about 51 cwts. to $35\frac{1}{2}$ cwts.; and the weekly make was increased from 185 tons to 224 tons.

It seemed doubtful to me at the time whether the saving was not due entirely, or in great part, to the mere addition to the height of the furnace. Subsequent investigation led me somewhat to modify this opinion, but since then I am informed that by the use of lofty furnaces, without any coking chambers on the top as designed by Mr. Ferrie, the ton of iron is obtained at an expenditure of about 35 to 40 cwts. of coal. Certain it is that a very great reduction in the consumption of fuel, as might have been anticipated, has accompanied this change of capacity; but to Mr. Ferrie is due the credit of having first successfully applied the experience of the tall furnaces of Cleveland to the minerals of Scotland.

Somewhat inconsistent with the disfavour in which the raw coal of the County of Durham is held by furnace managers, is the experience and practice of the North Staffordshire smelters. Their coal is commonly, but not invariably, used with a small portion of coke, say about one-fifth of the latter. The raw coal contains only 56 per cent. of solid carbon, and but 2·7 per cent. of ash. It appears in the laboratory to cake as freely as that of the North of England, and yet the smelters have no difficulty in using it in furnaces having a height of 70 feet. The consumption is about 26 cwts. of raw coal and 6 of coke, equal to 37 cwts. of raw coal, per ton of pig iron. The limestone consumed is about 40 per cent. less than in Cleveland, and the ironstone is richer, so that about 8 cwts. less ore is used, per ton of pig metal, than at Middlesbrough.

Not having sufficient opportunity of examining the facts connected

[1] *Vide* Transactions, 1870.

SECTION VII.—FUEL REQUIRED IN BLAST FURNACE. 123

with the use of raw coal, I can only deal in a somewhat hypothetical manner with the heat required for expelling the gases it contains.

The North Staffordshire example just given was that of a furnace 70 feet high with a closed top, and using super-heated air at a temperature of 785° C. (1,445° F.) In comparing its performance with a Cleveland furnace using 21 cwts. of coke with blast at 1,000° F. I obtain the following figures.

The reduced weight of carbonic acid in the lesser weight of limestone used, and the greater weight of fuel employed in North Staffordshire permit a larger weight of carbon to reach the tuyeres than occurs in the Cleveland case adopted for comparison. This makes a difference in favour of

	Calories.
North Staffordshire in heat evolved at the hearth of ...	10,641
Less CO_2 in limestone to expel and decompose ...	3,632
Lesser weight of slag to fuse 11·76 cwts. × 550 in Staffordshire than in Cleveland	6,468
Leaving for expulsion of gases from raw coal in Staffordshire	20,741

The volatile matter in 26 cwts. of coal at 40 per cent. gives 10·4 cwts. of gas to be expelled, and $\frac{20,741}{10\cdot4}=$ 1,994 calories per unit of gas. When we compare this with the number formerly considered as being absorbed by this function, viz., 2,000 calories per unit of gas, the approximation is a remarkably close one.

The cause of a waste of fuel in furnaces of insufficient capacity has been explained as being partly that the escaping gases, owing to their high temperature, carry away a large amount of heat, and partly that they are charged with less carbonic acid than they are capable of holding. Both these defects, as has been shown, are capable of removal by a suitable increase of capacity, varying in amount with the quality of the ore under treatment.

For the purpose of comparison we will confine ourselves to furnaces using Cleveland stone; and once more I must repeat what has already perhaps been sufficiently insisted on, viz., that as soon as one-third of the carbon in the gases is converted into carbonic acid, and is cooled down to a temperature of about 400° to 500° F. (205° to 260° C.), all further enlargement is unnecessary; because, as I maintain, all further economy of fuel, dependent on these causes, is impossible.

SECTION VII.—FUEL REQUIRED IN BLAST FURNACE.

The drift of these observations and what has preceded, is to prove that something like 12,000 feet of capacity and air at 563° C. (1,045° F.) are very near the useful limits as to size of furnace and temperature of blast; at all events about one cwt. of coke is all the saving by an immensely larger furnace fed with air at 1,600° F.

In the discussions which have taken place at the meetings of the Iron and Steel Institute and elsewhere, the question has often been asked, whether capacity, gained by enlarging the diameter of the furnace instead of adding to its height, would be attended with an equal economy of fuel.

If one may venture to give an opinion on a matter which has not been made the subject of direct experiment, I would say, provided the descending current of materials were so circumstanced that it was exposed to the ascending gases as long and as completely in the furnace where capacity was attained by increased diameter, as in that where the same object was secured by increased height, then a low furnace would do its work as well as a high one of the same cubic capacity.

It is my belief, however, that the conditions referred to cannot be secured in a low furnace of large diameter; and that in consequence no such economy as has attended the loftier furnaces would be obtained. This opinion rests on the following grounds.

We will take the case of a furnace smelting Cleveland stone with a well having a diameter of eight feet. As is well understood, it is necessary to keep the size of a furnace at the tuyeres within moderate dimensions, in order to permit the blast to penetrate to the centre of the materials exposed to its action. Capacity is needed, however, in the upper zones of the structure, in order to permit the solids, while descending to the hearth, to have several hours exposure to the heated gases. To facilitate this, the boshes have an inwards inclination, at a certain angle or slope. I have constructed a diagram, Plate V., in which a furnace of 18,000 cubic feet is comprised in a building of 60 feet from the hearth to the charging plates, the approved angle of the bosh for working Cleveland stone being preserved.

I entertain little doubt, that the practised eye of any furnace manager would condemn, without a trial, such a section as that given in the diagram; and this probably for the same reasons as those which appear to myself to be valid.

Plate 5

Plate to shew imperfect action in a furnace of large capacity but deficient in height.

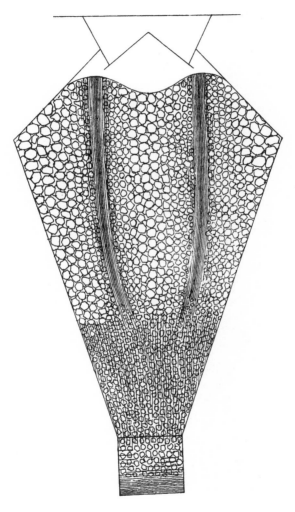

Height . . . 60 feet
Greatest diameter .. .. 32 "
Cubic contents about 18,000 "

And. Reid. Newcastle.

SECTION VII.—FUEL REQUIRED IN BLAST FURNACE.

The inconvenience of having the materials for the smelter in a fine state of division is recognized by all experience, and the obvious cause is their greater impermeability to the heated gases. The more the coarse dust-like matter is divided and distributed by larger blocks the better. Now the quantity of this dust being the same in the low as in the high furnace, but having in the same period of time to pass through 80 feet in the loftier or through 60 feet in the lower furnace, the annulus it forms from the distributing cone will be proportionately thinner in the furnace of 80 feet. Probably however a greater evil would attend the increase of diameter in the furnace itself, from say 20 to 32 feet, in the facility it would afford to the largest pieces of ore, etc., for separating themselves more completely from the smaller fragments. Such separation not only offers very open spaces, through which the gases will rush rapidly upwards from the tuyeres, but the increased horizontal section (two-and-a-half times greater at the top of the bosh in the lower furnace than in the higher) lends itself in an increased degree to the premature escape of the gaseous matters in question. This mode of action possesses the obvious inconvenience of preventing that action between solids and gases in the furnace, which has already received sufficient notice : and therefore no more need be said to prove that a waste of fuel must necessarily accompany the substitution, for increased height, of mere capacity obtained by enlarging the diameter of the furnace.

Allusion was made (Sec. III., p. 41) to the cost at which the ores of iron were freed from the earths in the form of slag, during the process of smelting. To illustrate this a table has been constructed on the following principles.

The heat generated by the formation of carbonic acid has been shown to be but a little in excess of that required for tearing away the oxygen from the iron. We will therefore omit this source of expenditure of heat out of the estimate, and regard all the fuel as burnt to the state of carbonic oxide : the amount we will assume to be 19·42 units per 20 units of iron. In like manner carbon impregnation, being a process of a heat producing character, is also eliminated from the account. We have then :—

19·42 units of carbon burnt to CO × 2,400 = 46,608 calories
Blast containing ... ... 12,000 „
                                          ─────────
                                          58,608

Hence $\dfrac{56,608 \text{ calories}}{19·42 \text{ carbon burnt}}$ = 2,915 calories per unit of carbon,
or say 2,650 calories per unit of coke.

Exclusive of reduction of the oxide of iron, left out for the reason already assigned, we have the following figures, setting forth the coke consumed in each section of the work performed when smelting 20 units of Cleveland iron.

		Calories equal to	Units of Coke.
1.—Evaporation of water in the coke	... ...	313	·118
2.—Expulsion of carbonic acid from limestone	...	4,070	1·536
3.—Decomposition of carbonic acid in limestone		4,224	1·594
4.— Do. water in blast	... ...	1,700	·642
5.—Reduction of phosphoric acid, silicon, etc.	...	3,500	1·321
6.—Fusion of pig iron	... ...	6,600	2·490
7.— Do. slag ...	... ...	15,356	5·794
		35,763	13·495
8.—Waste in gases...	7,900		
9.— Do. other sources ...	8,789		
		16,689	6·294
		52,452	19·789

It will thus be noticed that, in smelting Cleveland stone, fully 5¾ units of coke are expended, for every 20 units of pig produced, in merely melting the slag. Properly speaking, to this must be added the fuel required to effect the change in the limestone, placed under the heads Nos. 2 and 3, and together = 3·130 units: making in all 8·924 units of coke required for offices in connection with the formation and fusing of the slag.

From what has preceded, it follows as a matter of course that a mineral coal, from which there is little or nothing to be expelled, ought, when subjected to a high temperature, to fulfil the conditions required in the blast furnace. That variety of fossil coal known as anthracite is admirably adapted for the purpose in question; for being so free from volatile ingredients, its interior is little changed by the intense heat to which it has been exposed during its passage through the furnace. Owing probably to its tendency to splinter at high temperatures, a very

SECTION VII.—FUEL REQUIRED IN BLAST FURNACE.

strong blast is required for the proper penetration of the materials: I found 9, 10, and even as high as 12 lbs. on the square inch to be in common use in Pennsylvania, and in the States of New York and New Jersey.

The table given below contains particulars of the composition of anthracite, and also of bituminous or gas coal; and from this the superiority of the former as a blast furnace fuel is sufficiently apparent.

Variety.	Anthracite.	Anthracite.	Bituminous.	Bituminous.
Locality.	Pennsylvania.	South Wales.	Co. of Durham.	Pennsylvania.
Volatile matter	6·360	5·86	24·50	37·010
Sulphur	·657	} 1·58	1·22	·761
Ash	5·856		3·84	4·075
Fixed carbon	87·127	92·56	71·66	58·154
	100·00	100·00	101·22	100·00

The best Durham coke may be considered as containing 92 to 95 per cent. of solid carbon; so that the American anthracite referred to above may be regarded as being 6 or 7 per cent. inferior in quality to the finest furnace fuel used in the Middlesbrough district. A kind of anthracite used at the Cedar Point Works, on Lake Champlain, is stated by my friend Mr. Witherbee to contain only 83 per cent. of fixed carbon. Upon the latter basis of computation, $23\frac{1}{2}$ cwts. of anthracite ought approximately to be equivalent to 21 cwts. of Durham coke. In American furnaces of recent construction, from 75 to 80 feet in height, this low rate of consumption, viz. $23\frac{1}{2}$ cwts., has been nearly reached; for 24 to 25 cwts. is not an uncommon figure. The greater number of the furnaces in the United States is however of the old type; and in such, say with a height of 55 feet, 32 to 35 cwts. represent more usually the quantity of anthracite charged into the furnace, as required to produce one ton of foundry iron. Owing to breakage in handling, a portion of small is separated; and this, being used for inferior purposes, entails some addition to the actual cost of fuel at the blast furnace.

In the table given at page 98 it appears that, of the total heat evolved by the oxidation of the fuel, about 45 per cent. (36,512 units out of 81,536) is caused by raising the carbonic oxide, chiefly formed by direct combustion, to the state of carbonic acid. Now seeing that carbonic oxide in composition and in properties is precisely the same, whether it be obtained from burning a diamond, a piece of charcoal, or a morsel of coke, the power of carbon, in the form of carbonic oxide, from whatever source it may be derived, to reduce oxide of iron, must always be the same.

But when the weight of coke actually used in smelting a ton of iron, even from the richest hematite ores and in furnaces of 8,000 to 10,000 cubic feet, was compared with what was done by means of charcoal in the Carinthian furnaces, of not one quarter this capacity, the difference was so marked as either to raise some doubt as to the correctness of the estimates given in these pages, or to lead to the supposition that there was a virtue in charcoal not possessed by coke, or any other variety of mineral fuel. This idea was at one time entertained by my distinguished friend Professor Tunner—himself a resident in Carinthia, and a careful observer of the performance of the furnaces of that province.

In the hope of throwing some light on the apparent anomaly, this able metallurgist kindly consented, at my request, to have the analyses of the gases repeated. The result came out as before, viz. that pig iron was produced at Eisenerz and Lölling with 25 per cent. to 30 per cent. less carbon in the form of charcoal, than was used in this country, in smelting hematite ores, by means of coke.

In estimating the heating power of a given quantity of these two kinds of fuel, the first step is to ascertain the quantity of carbonic acid in the products of combustion. I am bound to say that in the case of charcoal at Eisenerz the proportion of carbon as carbonic acid (CO_2) to that as carbonic oxide (CO), instead of being in the ratio 1 to 2, as with coke furnaces, was stated to be as 1 to 1·72.

In reference to this unusual amount of carbonic acid in the gases, I apprehend that the analyses could scarcely have represented their *average* composition over a long period. It often happens that the proportion of this gas (CO_2) exceeds for a short time the ratio already named (1 CO_2 and $2CO$); but the extra heat so generated is, it is apprehended, lost by being carried off in the escaping gases.

Although I suspected at the time that the secret lay in some difference between the ores of Styria and Carinthia (which were spathose carbonate) and those of Cleveland and Lancashire, or in their respective modes of treatment, I possessed no means of comparing the

SECTION VII.—FUEL REQUIRED IN BLAST FURNACE.

actual facts; which perhaps could be best done by contrasting the results obtained in smelting the same kind of ore, under the same conditions, with charcoal and with coke.

During my visits to the charcoal furnaces on Lake Superior, and in Alabama and Georgia in the United States, I did not meet with any instances in which the consumption of charcoal was as low as that mentioned in connection with the Austrian iron works, 17 cwts. per ton of metal being the lowest of which I have any record. I further learnt that the difference in consumption of carbon between charcoal and coke, when applied to the *same ore*, was imperceptible.

Having obtained from Professor Tunner all the particulars of the materials, etc., used in working the charcoal furnaces which consumed only 14 cwts. of charcoal per ton of iron, I proceeded to estimate the sum of the heat requirements per 20 kilogrammes of pig iron. The figures came out as follows:—

	Calories.
Evaporation of water in the charcoal	300
Reduction of ore and dissociation of carbonic oxide	34,600
Expulsion of carbonic acid from limestone	500
Decomposition of carbonic acid in limestone	500
Do. water in blast	1,500
Do. silica and phosphoric acid	1,000
Fusion of iron	6,600
Do. slag	6,600
Loss by radiation	4,000
Carried off in gases	4,200
	59,800

Now when this amount comes to be compared with 87,000 units, the estimated requirement already given for smelting Cleveland iron, the actual weight of fuel consumed in each case will be found in almost exactly the same proportion as the heat units required in the two cases: for 59,800 : 87,000 :: 14 : 20·37. This statement justifies the idea, that in a suitable furnace, the Austrian iron could be made with the same quantity of coke as of charcoal; for any difference in the figures is easily accounted for by differences in the quality of the product, that of Cleveland being grey, and the other white, iron.

Since the above was written, a very interesting paper has been forwarded to me by the kindness of Mr. John Birkinbine, of the Pine Grove furnace in Pennsylvania, U.S.A.

He has experimented on smelting the same ore with charcoal and with coke, as well as with anthracite. The results given are very instructive, and are as follows; the trials beginning and ending, it will be seen, with charcoal:—

			Charcoal. February, 1879.	Coke. April, 1879.	Anthracite. May, 1879.	Charcoal. August, 1879.
Cwts. of fuel per ton of pig iron			22·59	31·19	34·56	23·66
Do. ironstone do.			52·26	50·00	52·63	50·00
Do. limestone do.			11·50	23·00	24·94	12·20
Quality of iron		No.	2·40	3·00	3·00	2·00
Weekly make		Tons	95	70	58	101

The large additional quantity of limestone consumed, when using mineral fuel, of itself necessitates a higher consumption of fuel, for the reasons already explained. The furnace in which the experiments were conducted was blown with air at only 600° F. (315° C.) It was only 36½ feet high with 9 feet 4 inch boshes—dimensions which will be recognised as being far below those required for successful smelting with mineral fuel. The examples given however do not call for any special explanation in respect to the small weight of charcoal used; which, it will be observed, after making allowance for the yield of the ore, is very near to the weight of coke often used in the Cleveland district. All that can be said is that, for reducing the ore used at the Pine Grove works with coke, the furnace was evidently much too small, and in consequence the consumption of this variety of fuel as well as that of anthracite was wastefully high.

I am indebted to the kindness of my friend Professor Åkerman of Stockholm for some interesting information respecting the performance of 27 charcoal furnaces in Sweden. From these the gas is taken off at two, and occasionally at three, levels; and a mean composition is assumed, based on the three resulting analyses. Unless however the same quantities of gas are given off at each orifice, this mode of computation is liable to error. But the comparison of the heat requirements with the heat evolution agree so closely, that the departure from truth cannot be large.

PARTICULARS OF 27 SWEDISH FURNACES, WORKING WITH CHARCOAL.

	DIMENSIONS IN ENGLISH FEET.		
	Minimum.	Maximum.	Actual Average.
Height	30·3	54·6	45·8
Diameter of bosh	6·8	10·6	8·8
Cubic capacity in feet—supposed	600	2,400	1,400

SECTION VII.—FUEL REQUIRED IN BLAST FURNACE. 131

The heat absorption has been calculated according to the formula applied in the present Section to English furnaces. The average of the 27 furnaces, per 20 units of pig iron, was as follows:—

	Units.		Units.	Heat Units.
Evaporation of water in fuel	3·02	×	540	1,631
Reduction of 19·1 units iron in 20 of pig metal, from peroxide, and partly magnetic	19·10	×	1,663	31,761
Carbon impregnation for carbon in 20 of pig metal	·80	×	2,400	1,920
Expulsion of carb. acid from limestone	3·84	×	370	1,421
Decomposition of do. do.	·46	×	3,200	1,472
Decomposition of moisture in blast ...	·04	×	34,000	1,360
Do. phosphoric and silicic acids	—		—	522
Fusion of pig iron	20·00	×	330	6,600
Do. slag	15·04	×	530	8,270
Transmission of heat through walls, etc., estimated			—	2,543
Carried off in tuyere water do.			—	1,109
Do. do. escaping gases, 93·07 units × 289° C. × ·257 SH,				6,913
Total requirements				65,522

The heat units therefore, necessary for producing iron made at these 27 furnaces, are nearly 25 per cent. less than what was estimated as being absorbed in smelting Cleveland stone.

	Units.
Charcoal used per 20 units of iron was 19·48 units = dry ...	16·46
Limestone do.	3·84
Ore do.	39·56
Temperature of blast was (412° F.)	211° C.
Do. escaping gases (552° F.)	289° C.

To ascertain how the heat is obtained we have—

	Units.
Dry weight of charcoal consumed per 20 units of pig... ...	16·46
Less ash, hydrogen, oxygen, and carbon already combined with oxygen in the charcoal	2·12
Actual available carbon in the charcoal	14·34

The composition of the gases gave 46·51 volumes of carbonic acid for 100 volumes of carbonic oxide; which is equal in weight to 1 unit of carbon as carbonic acid for 2·15 units as carbonic oxide.

The heat evolution is as follows:—

	Units.	Units.	Heat Units.
Carbon in charcoal	14·34		
Less dissolved by carb. acid in limestone	·46		
Leaving to burn to carbonic oxide	13·88 × 2,400	=	33,312
Carbon as CO burnt to CO_2	4·85 × 5,600	=	27,160
Heat in blast, 63·27 units air × 211° C. × ·237 SH,		...	3,163
			63,635
Heat units required as given above		...	65,522
Difference between the two sides of the account		...	1,887

It is worthy of notice, in this estimate, that in connection with the use of limestone and the fusion of the slag, about only 40 per cent. of the heat is required of that needed in the case of Cleveland iron. This of itself is equivalent to a difference of about 3·6 cwts. of fuel upon each ton of iron. It may also be observed that the Swedish iron-master uses blast of a temperature, 412° F., far below that to which it might be easily raised. Experience has probably demonstrated the propriety of this line of conduct. If so, is it possible that a limit to the reduction of the quantity of fuel burnt in the furnace exists, beyond that imposed by a want of ability to heat the air entering at the tuyeres? With a manageable amount of coke or charcoal, the blast easily penetrates the mass of material at or near the hearth, by burning the carbon it meets, but this ready penetration might possibly be interfered with were this carbon, the combustion of which affords access to the air, reduced below a certain point.

One conclusion alone can be drawn from what has preceded as to the use of charcoal, viz. that when it is doing the same kind of work as mineral fuel, the carbon employed appears to be practically the same in quantity. This is no more than might be expected; for the experiments of Andrews, and those of Favre and Silberman, do not indicate any material difference between the quantity of heat evolved by the carbon contained in the two varieties of combustible.

It will be noticed in Mr. Birkinbine's returns, that the weekly make of iron, when using charcoal, is 40 per cent. in excess of that when using coke, and above 70 per cent. over that when using anthracite. In like manner the actual fuel consumed is 35 per cent. more with coke, and 50 per cent. more with anthracite, as compared with charcoal, due most likely, to the smallness of the furnace.

SECTION VII.—FUEL REQUIRED IN BLAST FURNACE.

There is no doubt that, as a rule, furnaces of very moderate dimensions, using charcoal, are able to turn out extraordinary quantities of iron. Thus I visited furnaces in the state of Michigan, 47 feet high and $11\frac{1}{4}$ feet in diameter at the boshes, which had produced upwards of 300 tons per week.

It is difficult, in the absence of the means of continued observation, to give a reason why the two kinds of fuel should vary so much in point of produce, if this really be the fact, seeing that there is by no means a corresponding difference in their heating power.

Of course if we have two furnaces, each burning the same weight of coke and charcoal in the same time, and if there is a real economy of 35 per cent. in weight of combustible, when using charcoal, it is clear that the make will differ in nearly the same ratio. The ore being the same in each case, its reduction, other conditions being equal, ought to proceed in each case at the same rate in point of time. It must be remarked, however, that the other conditions are not equal. We have in the case of charcoal an extremely bulky fuel, and in coke one of a very much denser character—so much so that the same measure weighs fully twice as much when filled with coke as when filled with charcoal. This means that the smaller quantity (22·59 cwts.) of charcoal will occupy about one and a half times the cubic space of that of the larger quantity (31·19 cwts.) of coke, as given in Mr. Birkenbine's paper.

The advantage of a free exposure on all sides of the ore submitted to the reducing gases of a blast furnace may be inferred from what has already been stated in these pages. I have proved that such was the case by direct experiment in the blast furnace itself; and it seems consistent with reason to suppose that a given weight of ore, disseminated in the one instance more or less perfectly through say 30 cubic feet of charcoal, may be better circumstanced, as regards extent of surface exposed to the gases, than the same weight of ore, the various morsels of which were only separated by twenty cubic feet of coke. The mere fact that two surfaces of ore in juxtaposition are both producing carbonic acid, must be less favourable to the progress of deoxidation, than where this action in the one piece is not retarded by the presence of the carbonic acid given off by the other.

This, and other circumstances connected with the relative state of mechanical division in the two cases, render it, perhaps, *possible* that to the mere difference between the spaces occupied by charcoal and coke may be ascribed the larger amount of work done in a given space by the charcoal. In comparing the work really done by each kind of fuel, proper allowance must, of course, be made for the large additional quantity of limestone rendered necessary by the employment of coke or anthracite, which is so much richer in ash than is charcoal. The retardation of the reduction of the ore in the case of coke necessarily involves a larger consumption of fuel, unless a commensurate enlargement of the furnace is provided, in order to compensate for the want of space. Each unit of carbon in the form of carbonic oxide is unable to do its proper amount of duty, either by want of readiness in reaching certain portions of the iron oxide, or from its power to deoxidize being blunted, by having to traverse spaces in which there is an undue proportion of carbonic acid. The consequence of either of these events is the necessity of providing a fresh supply of fuel, to do the duty thus left undone.

This general view of the cause of the higher consumption of mineral, as compared with vegetable fuel, would appear to be confirmed by the figures contained in the paper of Mr. Birkenbine; for it will be seen that when anthracite was substituted for coke the consumption of actual carbon suffered a further increase, due, on the present hypothesis, to the still greater density of the new fuel.

Of course, when these modifications of behaviour are compensated for by suitable enlargements of the furnace itself, each kind of fuel, without reference to its origin, may be assumed to yield a full and therefore equal measure of duty. This however will always be contingent on its suitability in this respect not being impaired by other circumstances—such as friability, in the case of coke, or as an undue tendency to cake in the case of coal, which may render their use impracticable or inconvenient in any form of furnace.

It is scarcely needful to add that, in all comparisons between the different varieties of fuel, regard must be had to the actual fixed carbon in both; and although there is much less ash in charcoal than in coke, yet owing to the facility with which the former absorbs moisture and certain gases, the latter often contains in reality the higher percentage of carbon.

Timber, as it is cut, contains far too much water to be employed as a source of heat in the blast furnace; indeed the cooling effect of the

SECTION VII.—FUEL REQUIRED IN BLAST FURNACE.

mere expulsion of volatile matter, other than water, renders even well-dried wood unfit for the iron smelter. In some charcoal localities, however, a small proportion of the furnace fuel is employed in a partially charred or torrefied state. But with every kind of fuel the heat evolved by the combustion at the tuyeres is that due to the oxidation of the fixed carbon, and to that alone. In other words, any combustible matter of a volatile kind contained in the fuel is expelled long before it reaches the hearth.

The cost of charcoal may be taken to range from 25s. to 40s. delivered at the blast furnace, while that of coke is only about one-half this price. If, as I believe, weight for weight as much work can be done with the one as with the other, it is clear that, as a mere question of economy in the manufacture of pig iron, charcoal has no chance of entering the field against coke. Although these figures dispose of the question, unless in exceptional cases, it is curious to compare the resources of a given area of territory in respect to its power of producing vegetable and mineral fuel.

Dr. Percy[1] gives his authority for stating the annual production of wood from an acre of ground, allowing the ground to continue in the condition of forest. The following figures are calculated from the data contained in the work referred to:—

Locality.	Cwts. of Dry Wood per acre.
Western Slope of the Vosges Mountains	25
Black Forest, Baden, Hornbeam	20
Do. do Silver Fir	31
Lake Superior district	29

The average of these may be taken at 26 cwts. of dry wood per annum. At one place, on Lake Superior, 40 cords or 5,120 cubic feet per acre (equal, according to Dr. Percy, to about 1,160 cwts. of dry wood,) was given me as the produce of a wood after a growth of 30 years. This is equal to 38·6 cwts. of dry wood per acre per annum.

Roughly, the composition of dry wood, exclusive of ash, is as under:—

Carbon 50. Hydrogen 6. Oxygen 43. Nitrogen 1 = 100.
The ash averages 1¾ per cent.

Owing chiefly to the formation of hydro-carbons, but partly to waste in charring, the actual yield of charcoal does not exceed 28 per

---

[1] Metallurgy, volume on Fuel, p. 196.

cent.; and it is sometimes as low even as 15 per cent. Taking it at 25 per cent., we have an acre of land, according to the figures previously quoted, affording from 5 cwts. to 9·65 cwts. per annum of charcoal for each cutting; or if allowed to grow for 35 or 40 years, about $14\frac{1}{2}$ tons per acre.

Now it is a moderate computation which assigns 6 feet thickness of coking coal in any district; and after making allowance for faults, etc., an acre of such a seam ought to yield 8,000 tons of coal. This weight of raw coal is capable of affording 5,040 tons of coke.[1] This being so, it follows that it would require from ten to twenty thousand years for an acre of land to produce as much charcoal as a similar area of ground can give of coke. That the acre of ground can continue to produce timber, while the coal land is exhausted, is too insignificant under the circumstances to be taken into the account—especially as the removal of the minerals does not prevent the surface being used for agricultural or forest purposes.

In the present Section attention has been directed exclusively to those kinds of fuel, which are in common use in the blast furnace. In many localities the variety of coal known as lignite, found in later geological formations than the carboniferous, is somewhat extensively worked. I visited such a mine in Styria, where the mineral was very pure as regards ash, with a fine pitchlike lustre; but it contained an unusually high percentage of volatile matter.

According to Dr. Percy, some of the Austrian lignites only afford about 54 per cent. of coke, and the coal in its raw state contains as much as 12 per cent. of ash. In Styria, where fuel is dear, a small proportion of lignite has been used along with charcoal in the blast furnace.

In 1876 a trial was made with compressed peat at one of the Vordernberg furnaces. Particulars of the experiment are given by Anton Einigl.[2] When about 29 per cent. of the total weight of fuel was peat, the remainder being charcoal, the actual work done was such that one ton of the charcoal was worth nearly three tons of the peat. The low value of peat as a blast furnace fuel is due to the

---

[1] This assumes that all the mechanical work is performed by the waste heat of the coke ovens.

[2] Zeitschrift des berg-und hüttenmännischen Vereins, Feb., 1879.

large quantity of water and volatile gases it contains. Of the really useful constituent, viz. fixed carbon, it only gave in this case 32 per cent.

The statements just made in reference to lignite and peat may be accepted as an indication of the great inferiority of these varieties of fuel, as compared with charcoal or ordinary coke, for the purpose of smelting iron.

It has on certain occasions been proposed, indeed it has been actually attempted, to employ a portion of the fuel required for smelting iron by blowing it in at the tuyeres, and this in the solid as well as in the gaseous and liquid forms. It needs but little consideration to ensure the rejection of all such schemes.

We have already seen that each unit of coke employed in the blast furnace is not burnt until it is charged by the escaping gases with the heat given off by 2·33 units previously consumed. Cold fuel presented to the blast at the tuyeres would tend seriously to refrigerate that region of the furnace where a very intense temperature is indispensable. Raw coal, tried in Belgium, blown in at the tuyeres, was accompanied by an additional inconvenience, viz. the absorption of heat attending the distillation of the gaseous constituents of this description of fuel. After a lengthened trial it was finally abandoned.

In the United States, light carburetted hydrogen occurs in Nature in such quantities and at such pressures, that the opportunity presents itself of an economical application of this gas to the smelting of iron, were such application practicable. However useful the hydrogen, which forms 25 per cent. of the weight of this gas, may be in a reverberatory furnace (in which it is used at Pittsburg), this element is useless or all but useless in the blast furnace—carried off as it is unchanged in the gases. The liquid hydrocarbons no doubt are much better adapted for smelting purposes than those which are gaseous, inasmuch as they are much richer in carbon. Their use however would be attended with the same inconvenience as that which accompanies that of raw coal, viz. a cooling effect upon the zone of fusion. This, and their suitability for purposes of illumination, for which they command higher prices than can be afforded for smelting iron, render it highly improbable that we shall see a repetition of the experiment which was tried in Canada for using petroleum in the blast furnace.

SECTION VII.—FUEL REQUIRED IN BLAST FURNACE.

A claim has been put forward on behalf of the blast furnace, during the course of the present work, for the completeness, as compared with other processes, with which it utilizes the heat which is generated in its interior.

Great loss is almost inseparable from all the various modes in which heat is applied. As might be expected, by far the best results are obtained when a moderate temperature only is required to be communicated to the matter under treatment; such for example as in the evaporation of water. There, a vast volume of flame quickly communicates its heat to the contents of the much colder boiler. As is well known, however, the hot gases, as they cool, impart their heat more and more slowly to the hot water, through the plates of the boilers containing it; and thus even here a considerable amount of the useful effect of the fuel is necessarily lost at the chimney.

It is not an uncommon thing for water to be supplied to steam boilers at nearly its boiling point; but to make allowance for this there is set down in the accompanying table the theoretical quantity of water capable of being evaporated by one lb. of coal, when beginning with water at 0° C. (32° F.) and at 100° C. (212° F.)

Pressure of Steam above Atmosphere.	Lbs. of Water evaporated per Pound of Coal.	
	From 0° C.	From 100° C.
50 lbs.	12·17	14·37
100 ,,	12·04	14·20
150 ,,	11·98	14·11

Now 8½ to 10 lbs. of water evaporated has been returned to me, by one of our best marine engine builders, Mr. G. Y. Blair, as the average duty of 1 lb. of coal in marine boilers; although as much as 11·8 has been obtained in his experience. Adopting 10 as the basis of calculation, and 14 as the possible duty, there is a loss of nearly 30 per cent. The loss is easily accounted for—a certain amount is due to radiation, etc., but the main part is that incurred at the chimney, where the products of combustion are often red hot, or probably at 1,000° to 1,200° F. (538° to 649° C.) The subjoined figures show the loss incurred by the gases leaving at various temperatures, supposing the usual excess of air (100 per cent.), observed as being generally present.

SECTION VII.—FUEL REQUIRED IN BLAST FURNACE. 139

Units of heat evolved by one Kilo. of Coal.	Temperature of Waste Gases.	Units of Heat contained in Waste Gases produced by one Kilo. of Coal.	Loss on Heat evolved per Cent.
7,930 ...	100° C. ...	506 ...	6·4
,, ...	200 ...	1,012 ...	12·8
,, ...	300 ...	1,518 ...	19·1
,, ...	400 ...	2,025 ...	25·5
,, ...	500 ...	2,530 ...	31·9
,, ...	600 ...	3,037 ...	38·3

With locomotive engines matters are even worse than in the marine engine. On the North Eastern Railway, the returns made me by Mr. Edward Fletcher show that a mere fraction above 7 lbs. of water evaporated per pound of coal is the average obtained; which may be taken as indicating a loss of 50 per cent. of the heating power of the fuel used.

Large as the waste of heat is in steam boilers, it is infinitely greater in reverberatory furnaces. In such cases the temperature required is usually very intense, and the structure being small and the walls thin—because otherwise they would melt—the loss from radiation and convection is very considerable. The main loss, however, most commonly arises from the circumstance that, the *whole* of the floor of the furnace requiring to be intensely heated, the gases pass off at the far end almost as hot as they are next the fireplace. Entering the chimney direct from the material under treatment, they of course carry off with them a very large proportion of the heat of the fuel burnt on the grate; and it was estimated by M. Krans of Louvain that, from this and the causes previously mentioned, not above 10 per cent. of the effective power of the coal is rendered available. It is only necessary to look at the chimneys of the ordinary furnaces in our mills and forges, to be satisfied how enormous must be the loss attending their use.

Some improvement has been made in late years in such furnaces, by employing a portion of this waste heat for raising steam; but the actual loss still amounts to a very large percentage of the total power of the coal burnt.

Mention has already been made of the Siemens furnace, in the construction of which the same principle, which secures so large a part of the economy of the blast furnace, has been applied. In the latter we have seen how vast an amount of heat is caught up by the descending solids and brought back to the hearth, instead of being permitted to escape uselessly into the atmosphere. So it is with the Siemens' furnace; the products of combustion leave the working bed at an intensely high temperature, but communicate a considerable portion of their heat to the so-called regenerators; which heat is returned to the furnace, by the gas and air which subsequently enter it. Compared with the blast furnace, however, the apparatus in question is small in its dimensions, and hence the loss from radiation, etc., is much greater. This, added to the circumstance that the coal is all converted into gas, by partial combustion in a separate furnace, or producer, lessens the beneficial effect obtained from the Siemens furnace, which, according to Mr. Krans, does not utilize above 20 to 25 per cent. of the full power of the fuel.

When the conditions under which fuel is burnt in a blast furnace, and those of its appropriation in smelting iron are considered together, it is easy to perceive how immeasurably these are superior to any other form of apparatus, in which heat of great intensity is applied. Radiation and convection depend on time and on surface. The shorter the time, speaking generally, and the smaller the space in which a given quantity of fuel is consumed, the less will be the losses arising from these two causes.

In one of our large furnaces, about a cwt. of coke per minute disappears before the blast at the tuyeres in a space of two or three cubic feet. The walls of the hearth are there of considerable thickness, so that the escape of heat through them, looking at the quantity evolved, is insignificant. In the higher portions of the building, the masonry, although thinner than it is at the well, is much stronger than in any reverberatory furnace. Again, a mass of heated material, filling a cylindrical shaft varying from 12 to 25 feet in diameter, has an infinitely smaller surface exposed to the cooling influence of the atmosphere than happens in the case of buildings of smaller magnitude, such as any reverberatory furnace. The effect is, according to observations I made on such a furnace,[1] that something under 4 per cent. comprises the whole loss from radiation through the walls of the structure. Again, the sensible heat escaping from the blast furnace with the products of combustion is incomparably smaller in proportion

[1] "Development and Appropriation of Heat." Transactions Iron and Steel Institute.

SECTION VII.—FUEL REQUIRED IN BLAST FURNACE.

than that from the general run of furnaces where high temperatures are required. This will be at once admitted on contrasting the waste from this cause—only about 10 per cent. in furnaces of the best construction—with the losses occurring in connection with steam boilers and other forms of furnace.

Against the heat evolved in the blast furnace has to be set the drawback, that, something under one-third of the carbon is capable of being oxidized to the form of carbonic acid; this gives the average heat units evolved per unit of carbon as something under 4,266 instead of 8,000[1], which latter figure obtains with combustion in the ordinary fire-place of a reverberatory furnace.

Nevertheless, the quantity of heat actually appropriated in the blast furnace, per unit of carbon burnt, is of a much higher character than it can be with the more perfect combustion of any ordinary grate; and to this advantage has to be added that of the imperfectly oxidized carbon, leaving as carbonic oxide, being applicable, and being actually applied, to useful purposes, after it has discharged its duty in the smelting of the ore.

Let us follow out the figures illustrating the work performed by a blast furnace, smelting 20 units of iron from Cleveland ironstone, with 11 units of limestone and 21 units of coke, a task which, with fuel of good quality, has often been accomplished.

Assuming the ash and water in coke to amount to $7\frac{1}{2}$ per cent., we have in 21 units of this combustible 19·43 units of actual carbon. From this must be deducted ·60 units of carbon dissolved by the iron itself, leaving 18·83 units which have to serve as the origin of the heat upon which the following calculation is founded.

The total number of heat units capable of being afforded by 18·83 units of carbon, when all is burnt to carbonic acid, is (18·83 × 8,000) 150,640. Taking the amount of carbon which escapes from a coke furnace in Cleveland as carbonic acid, per 20 units of metal, to be 5·85 units instead of 6·58, which latter is the theoretical maximum quantity, we have 5·85 units leaving the furnace fully oxidized, and 12·98 as carbonic oxide.

---

[1] 1 C × 8,000 = 8,000 units evolved when 1 unit of carbon is burnt to CO_2.
$\frac{1 \text{ C} \times 8,000 + 2 \times 2,400}{3}$ = 4,266 do. 1 do. do. $\frac{1}{3}$ to CO_2 and $\frac{2}{3}$ to CO.
Both exclusive of the heat in the blast.

Now 5·85 × 8,000 + 12·98 × 2,400 = 77,952 calories or heat units. This number amounts to 51·75 per cent. of the total heat (150,640 units), capable of being afforded by the fuel were it completely saturated with oxygen, *i.e.*, were it all burnt to the state of carbonic acid. Inasmuch, however, as about 12,000 units, obtained by subsequently burning a portion of the furnace gas to carbonic acid, is returned to the tuyeres in the hot blast, we have (77,952 + 12,000) 89,952 calories as the number really to be placed to the credit of the coke consumed; which brings up the actual development inside the furnace to 59·71 per cent. of its full power.

Of this the only portion which can be regarded as not usefully appropriated in the furnace is that arising from transmission through the walls, and that carried off in the gases. It will suffice for the present calculation if these are assumed as amounting to 12,000 units, or exactly the figure representing the heat supplied in the hot blast. This leaves us therefore with the result that we have applied advantageously 51·75 per cent. of the full heating power of the fuel for furnace work.

We have now to deal with that heat which, having been excited in the furnace, leaves it in a sensible form; to which has to be added that capable of being evolved by the complete combustion of the partially oxidized carbon, escaping from the tunnel head as carbonic oxide. I have caused observations bearing on this question to be made specially for this purpose, on a furnace smelting Cleveland stone, with a consumption of 21·44 units of coke and 10·12 of limestone per 20 of pig.

The gases, in this particular case, were escaping at an average temperature of 630° F. (332° C.)[1] and for 20 units of metal had the following estimated weight, as ascertained by analysis:—

	Units.		Units.	
Nitrogen	73·42	equal to	95·75	of blast
Carbonic acid	22·18	do.	6·04	of carbon
Carbonic oxide	32·31	do.	13·86	do.
Hydrogen ·08 and water ·45	·53			
	128·44			

[1] A small portion of this, it is difficult to say exactly how much, is due to the ironstone being charged warm from the kilns.

SECTION VII.—FUEL REQUIRED IN BLAST FURNACE. 143

The blast (95·75 units weight) was heated to 931° F. (500° C.), and contained therefore 11,345 units; and 128·44 units of escaping gases, at 630° F. (332° C.), would carry away 10,232 units.

The heat evolved by the carbon oxidized is $6·04 \times 8,000 + 13·86 \times 2,400 = 81,584$ calories, or 51·25 per cent. of the full power of the fuel.

The heat placed at our disposal in the escaping gases is as follows:—

	Heat Units.
Sensible heat in the gases, as given above	10,232
Combustion of carbon in carbonic oxide, to state of carbonic acid ... ... $13·86 \times 5,600 =$	77,616
Combustion of hydrogen in gases ... $·08 \times 34,000$	2,720
	90,568

It happens, unfortunately for the calculation we are considering, that there was a considerable excess of gas over and above that required for steam purposes, and for heating the air. This has been estimated in different ways to amount to a proportion varying from 15 to 28 per cent. of the whole. As such estimates can only be regarded as approximate, from the continual changing of one or more of the conditions involved, the escape from this cause may be assumed at an average figure; and then, instead of having to account for 90,568 calories, we may deduct one-fifth, which leaves 72,454 as the number for which appropriation has to be found. This appropriation consists of

		Calories.
A.	Heat in steam supplied to non-condensing engines	28,118
B.	Do. in blast	11,345
C.	Do. in gases entering chimneys	26,779
		66,242
D.	Leaving for radiation, etc.	6,212
		72,454

In respect to A and B, the volume of steam was accurately estimated from the speed of the various engines, and, the weight of blast being known, the quantity of heat for these two sources of consumption is easily ascertained.

The loss from the chimneys (C) is necessarily large (being 37 per cent. of the whole), on account of the very large volume of gaseous matter accompanying the combustion of a fuel such as carbonic oxide, diluted with carbonic acid and nitrogen, and which is capable of affording so small an amount of heating power as compared with solid fuel.

The quantity of free air is considered to be 20 per cent. in excess of that representing complete saturation of the carbonic oxide; and the observed temperatures in the chimneys being 802° F. (428° C.), we have the 26,779 units of loss from this cause.

The heat under D, representing disappearance from radiation, etc., is 8·5 per cent. of the whole, and is simply the difference between the two numbers 72,454 and 66,242.

The various figures given in this last-mentioned example of furnace working present the following result :—

	Units.	Units.
The full equivalent of heat capable of being afforded by 18·83 carbon × 8,000 calories is equal to	—	150,640

The application is as under :—

		Units.	Units.
Useful heat.	(*a*) Furnace work, as per summary already given p. 95	70,311	
	(*b*) Heat in tuyere water	1,800	
	(*c*) Heat in steam	28,118	
	(*d*) Heat in blast, included in *a* and *b*	—	
	(*e*) Available heat in unutilized gas, say	10,837	
			111,066
Waste.	Radiation at furnace, etc., p. 95	6,989	
	Loss at chimneys, at boilers, and hot blast stoves	19,096	
	Do. radiation do. do.	6,212	
	Do. estimated waste in using unutilized gas	7,277	
			39,574
Total heat accounted for			150,640

From this statement we are therefore justified in inferring that about 74 per cent. of the entire heating power of the fuel (111,066 : 150,640) is or may be beneficially employed—an amount of duty not equalled, so far as I know, in any other branch of industry where an elevated temperature is required. In cases where the temperature is very low such as the evaporation of water, an amount of economy

[1] The loss at the chimneys is reduced in this estimate as compared with the amount given in the previous page, otherwise the account would exhibit an excess in the quantity of heat to be dealt with. This alteration, however, does not disturb the percentage of heat usefully employed.

SECTION VII.—FUEL REQUIRED IN BLAST FURNACE.

is practicable which is impossible in the case of any furnace, where the highly heated escaping gases carry off a very large amount of the heat evolved.

Having now dealt at some length with the extent to which fuel has been economised in the past, little remains to be said as to the future possibility of still further reducing this important item in the cost of making pig iron.

So long as the ore and flux remain unchanged in quantity or quality, there are but the two directions, already referred to, in which any relief in this matter need be considered, viz., an enlargement of the furnace, and an increase in the temperature of the blast.

As regards the size of the furnace itself, the fact of there being a zone of heat development close to the upper surface of the materials, producing thus a constancy of temperature in the escaping gases after a certain height is reached, has led me to infer that we cannot hope for any improvement by adding to what has been already done in this respect. It will be apparent from what has preceded that no uniform law can be laid down for determining the height of the blast furnace. This varies greatly according to the different minerals under treatment; because, as has been already explained, there is the utmost diversity in the readiness with which they are reduced. Instances have been quoted in which a particular ore was converted into pig iron in four hours after being charged into the furnace, while the ore of Cleveland, with economical working, requires about seventy-two hours for the same process. Speaking from many observations, I imagine we may assume that for many of the magnetic and peroxide class (or hematites), and for calcined spathose ores a height of 30 to 40 feet would suffice when using charcoal, and perhaps 50 feet were coke used. For the hematites of Cumberland, Lancashire, and Spain 65 feet would probably be ample, while as we have seen, something like 80 feet in height is required for the refractory ironstone of Cleveland.

With regard to the utility of a further enlargement I would merely again refer to the performance of furnaces of 11,500 cubic feet with a height of 80 feet. The coke consumed upon the occasion of a recent examination was 20·40 cwts. and the heat evolved, including 11,724 calories contained in the blast, amounted to 4,155 calories per unit of coke.

146   SECTION VII.—FUEL REQUIRED IN BLAST FURNACE.

After making an allowance for the heat contained in the ironstone, the heat in the escaping gases appeared to be 5,265 calories per 20 units of pig produced. The actual loss in coke represented by this escape of heat was therefore $\frac{5265}{4185} = 1\cdot26$ cwts. of coke per ton of iron, which is the margin of economy any addition to the size of the furnace could effect were it possible in an economical sense.

With regard to the second alternative, viz. the temperature of the blast, I have shown that air used at 1,600° F. to 1,700° F. instead of 1,000° F. was only followed by a saving of about 1 cwt. of coke per ton of iron, and that it would be very difficult, if not impossible, to raise the temperature of the air beyond 1,700° F. in the expectation of effecting further saving of any moment.

Under such circumstances, so far as the consumption of fuel in the furnace itself is concerned, I cannot indulge in the hope, that we shall see any sensible amelioration in the performance of well appointed and well conducted establishments.[1]

In what has been said with regard to using furnaces of unnecessary dimensions it must be borne in mind that the limitation of size is here discussed, simply as a question of economy in fuel. Twelve thousand cubic feet may suffice for a thorough exhaustion of the gases, when smelting Cleveland stone—probably one of the most refractory of the ores of iron; but it does not follow that considerations connected with a more profitable application of labour may not cause a furnace of double this capacity or more to be preferred in point of economy.

[1] Of course I exclude from present consideration any better application which may be made of the waste gases, by the use of different engines or otherwise.

SECTION VIII.

ON THE SOLID PRODUCTS OF THE BLAST FURNACE.

Of the substances, which the title of the present heading is intended to include, the pig iron is of course the most important. The slag, or cinder as it is usually styled, next demands consideration, because it constitutes the principal index of the performance of the furnace itself. A third body which, for our present object, may be classed among the solid products, is the vast volume of white smoke which is seen leaving most furnaces, and which is particularly conspicuous in the treatment of Cleveland ironstone and of some others.

Cast Iron requires the presence of only two elements for its formation, the metal itself and carbon: that is, a compound consisting exclusively of these two substances may be formed, which would be fusible at a moderate temperature. It is strictly true, however, to say that pig iron, as run from the blast furnace, always contains other matter, which may to some extent be regarded as foreign to its essential composition. This impurity, as it may be considered, often includes several different elements, and as Cleveland iron will serve as an illustration of this complex composition, five analyses of this make are subjoined:—

	No. 1.	No. 3.	No. 4.	Mottled	White.
Iron	92·43	93·66	94·64	93·59	93·20
Carbon	3·75	3·41	2·66	3·55	3·20
Silicon	1·70	·88	1·87	·66	·64
Sulphur	·13	·17	Trace	·35	·20
Phosphorus	1·24	1·23	1·00	1·05	1·32
Manganese	·30	·37	·93	·79	·60
Calcium, Magnesium, and Titanium	·62	·46	Trace	·33	1·32
	100·17	100·18	101·10	100·32	100·48

Against iron containing, as in the examples just given, 3 to 4 per cent. of substances which are not essential to its composition, some of which, if not removed, are for certain purposes, absolutely injurious, may be set the analysis of other varieties of pig, in which some of the above-named elements are either wanting or are found in greatly diminished quantities. The following are examples of analyses taken from Dr. Percy's work—a fact which may be accepted as a guarantee for their correctness.

	LOCALITY.		
	Juniata U.S.	Unknown.	Nova Scotia.
Carbon	2·891	4·20	3·27
Silicon	·830	·08	·37
Sulphur	·005	Trace.	·01
Phosphorus	—	·05	·28
Manganese	—	·10	·37
Iron by difference	96·274	95·57	95·70
	100·	100·	100·

It is now above 50 years since Neilson introduced the use of hot air in the blast furnace; and it is a somewhat remarkable fact that even to-day there is an absence of a complete and systematically conducted course of experiments, to prove that cold blast iron is really, as is pretended, superior in point of strength and quality generally to that made with hot blast. On the face of it, there are some obvious reasons why hot blast iron should be purer and therefore stronger, looking at the nature of the impurities, than that made with cold blast. All mineral fuel contains phosphorus and sulphur; in that of one of the most renowned works in Great Britain the former varies from ·025 to ·110 per cent., and the sulphur exists to something like ·5 per cent. If we assume that the consumption of fuel is reduced from 40 cwts. to 30 cwts. per ton of iron, we secure a corresponding reduction in the quantity of phosphorus in the iron; and the sulphur, a portion of which at all events finds its way into the metal, is also reduced in quantity.

There is a kind of vague idea among the advocates of cold blast iron, that the alleged higher temperature of a hot blast furnace affects, in some way they have never explained, the quality of the iron it produces. It is stated by Sir W. Fairbairn and others that the

SECTION VIII.—SOLID PRODUCTS OF THE BLAST FURNACE.

repeated application of heat to solid iron, cast or wrought, does deteriorate its quality; probably owing, I imagine, to a new arrangement of its molecules, to a change in chemical composition, or to an occlusion of oxygen or other gaseous matter. But there does not exist any proof that any such effect takes place, or could take place, in the blast furnace, even were the temperature higher when it is blown with hot air than with cold. But there is really no substantial ground for pretending that the hearth of a cold blown furnace is not quite as hot, taking it as whole, as that of one receiving heated air. The quantity of heat evolved in the two, so far as combustion of fuel is concerned, is actually higher in the cold than in the hot blast furnace. Neglecting the ash in both cases, we have the following figures, representing the carbon used per 20 units of iron:—

Calories.

Cold blast furnace, taking the air even at the freezing point of water } 40 units of carbon × 2,400 = 96,000

Hot blast furnace, receiving its air at about 900° F. (482° C.) ... } 30 units of carbon × 2,400 = 72,000
Calories in blast = 16,500

88,500[1]

Besides this, if we are to believe that the quality (or greyness) of the metal depends on the actual temperature to which it is exposed in the blast furnace, what becomes of the pretended differences of temperature between the use of hot and cold air, both running the same quality of iron?

Again, let us consider the probable effect of an increase of heat on the behaviour of the elements found in combination with iron in its form of pig.

Reasons have already been adduced for believing that an elevation of temperature strengthens the affinity of sulphur for lime or calcium, at all events less is found in iron run from a hot working furnace and I am not aware that there is any information to show that any possible differences of heat can affect the action of phosphorus in its behaviour towards iron during the process of smelting. With regard to calcium

---

[1] These figures are those which set forth the heat evolved in smelting iron in furnaces of small dimensions. The carbon burnt in large furnaces is very much less, and would illustrate the argument in a still more striking way.

and the other metals, it is not improbable that an increase of temperature might be accompanied by a larger amount of these in the pig; but I know of no instances where any ill effects have been connected with their presence.

In the hope of throwing some light on this question, advantage was taken during the blowing out of one of the Clarence furnaces, to reduce the burden and use the blast cold. The iron was white and had the composition given below. Alongside this analysis is placed an example of white iron made with hot air:—

	Cold Blast, White.	Hot Blast, White.
Carbon	3·35	3·20
Silicon	·33	·64
Sulphur	Trace	·20
Phosphorus	1·37	1·32
Calcium, etc.	·30	1·32
Manganese	·28	·60
	5·63	7·28
Iron	95·10	93·20
	100·73	100·48

From these figures it will be seen that the most notable difference is in the silicon, but that most of the other substances are in less quantities in the cold than in the hot blast iron. At the same time the short duration of the experiment does not afford sufficient data to speak with much confidence on the relative quality of the metal. On puddling the pig so made, the malleable iron showed no kind of superiority over that smelted with hot air.

Viewing the question as one of the application of pig iron to castings, the late Sir William Fairbairn, F.R.S., published some experiments undertaken for the purpose of determining the kind of metal to be used in the construction of the High Level Bridge at Newcastle-on-Tyne.

This competent authority gives in a table the following results :—

	Breaking weight applied to bars 1 in. sq., bearings 4½ ft. apart.			Power of the 4½ ft. bars to resist impact.		
	Max.	Min.	Average.	Max.	Min.	Average.
Cold blast—18 Specimens	567	403	457	992	530	746
Hot blast—20 do.	543	353	443	998	532	681

SECTION VIII.—SOLID PRODUCTS OF THE BLAST FURNACE. 151

A second table contains the following :—

	Breaking weight applied as above.			Mean ultimate deflection-inches.		
	Max.	Min.	Average.	Max.	Min.	Average.
Cold blast—5 Specimens ...	598	502	569	·82	·71	·77
Hot blast—9 do. ...	676	472	665	·94	·63	·766
Do. Anthracite—1 Specimen			665			·80

The irons ultimately selected for the Newcastle Bridge were :—

	Parts.	Breaking weight.	Deflection—inches.
Ridsdale No. 3 hot blast ...	40	676	·89
Ystalyfera „ 3 do. ...	40	665	·80
	——80		
Crawshay No. 1 cold blast ...	40	582	·80
Blaenavon „ 1 do. ...	30	502	·82
Coalbrook Dale No. 1 cold blast	30	584	·71
Scrap chiefly do.	30	Not given.	
	——130		

Bars from the above mixture, tested as before, supported 705 lbs. before they broke, and deflected ·89 inches, being superior in strength and power of deflection to Ridsdale, a Northumberland iron, which gave the best results when examined separately.

It would be hazardous for any one to pretend from the above figures, taken as a whole, that cold blast iron is really superior for castings to hot blast; for the information as given above does not settle the question either one way or the other. Ridsdale hot blast may be better, as the trials would indicate it to be, than any of the brands with which it is compared, not because it is hot blast, but on account of the minerals used in its manufacture. Nothing short of the two systems being applied, over a period of time, to precisely the same minerals can satisfactorily settle this question. In the list compiled by Fairbairn there are 3 cases given of the same make of iron smelted with cold and hot air, but it is not positively stated that there were no differences made in the ore, fuel, and limestone, conjointly with the change of blast.

	Cold Blast.		Hot Blast.	
	Breaking weight.	Power to resist impact.	Breaking weight.	Power to resist impact.
Devon No. 3, Scotland ...	448	353	537	589
Carron „ 3, do. ...	444	593	520	710
Coedtalon No. 2, Wales ...	403	600	409	771

As far as these last trials go, they are clearly and unmistakeably in favour of hot blast iron.

At one time it was believed by many that any deterioration in the quality of iron, made with hot air, was due to the use of ores of an inferior description and not to any vital difference between it and that smelted with cold air. I do not believe that there exists any real foundation for this opinion, at the same time it is quite possible, for reasons already given in the Section dealing with the theory of the hot blast, that the semi-vitrified Scotch Black-band, first largely used with hot air, required for its successful treatment a much larger furnace than the more easily permeated clay-ironstones, when lightly calcined, in common use at that time. It might thus have happened that smelting an ore in a furnace too small for its treatment, led to the belief that the mineral itself, and not the furnace, was to blame for any change in the quality of the product.

Of course it is impossible to lay aside the experience of impartial consumers of pig iron, who, paying a high price for an article for special purposes, must have satisfied themselves that it is worth the money. No doubt the iron they prefer is good, and it is cold blast; but is it good because it is cold blast? This is precisely the question which has not perhaps received sufficient attention.

An important element, in considering the strength of a material like cast iron, is the effect produced on it by a continued exposure to the load or strain it has to bear; and also the manner in which it is affected by temperature. Sir Wm. Fairbairn reported to the British Association on both these heads.

It may be urged that the molecular condition of bodies, and their power to withstand a continual series of shocks, are but little understood, particularly when we attempt to connect these attributes with chemical composition. In America, for example, I found that cold blast charcoal iron was infinitely preferred to metal smelted with hot blast, for chilled railway wheels. Wrought iron or steel is rarely used under railway carriages in the United States, cast iron being the substance employed; so that, on nearly one hundred thousand miles of American railroad abundant opportunity has existed over some years for pronouncing an opinion on the relative merits of the two qualities of iron. It was uniformly stated the cold blast iron wheels take a

SECTION VIII.—SOLID PRODUCTS OF THE BLAST FURNACE. 153

deeper "chill," and wear much longer than those made from hot blast; the same charcoal and minerals as I was assured, being used in both cases.

When we direct our attention to pig iron intended for the forge, we are more bewildered than ever in any attempt to connect superiority of quality with cold blast, so far as this superiority is exhibited by any difference of chemical composition. The analyses of the produce of the cold blast furnaces of a well known firm in the West Riding of Yorkshire are as follows:—

Quality of the Pig Iron	No. 1.	No. 2.	No. 3.	No. 4.
Carbon Graphitic	3·421	3·155	3·361	3·308
,, Combined	·583	·581	·393	·319
,, Total carbon	4·004	3·736	3·754	3·627
Silicon	1·708	1·646	1·382	1·381
Sulphur	·073	·070	·063	·081
Phosphorus	·630	·635	·602	·602
Titanium	Trace	Trace	Trace	Trace
Manganese	1·606	1·472	1·475	1·169
	8·021	7·559	7·276	6·860
Iron	92·070	92·644	92·952	93·292
	100·091	100·203	100·228	100·152

The following analyses give the foreign matters associated with other brands, also of cold blast iron:—

	No. 1.	No. 2.	No. 3.	No. 4.
Carbon	3·40	2·88	3·07	3·03
Silicon	1·36	1·09	1·48	·83
Sulphur	·07	·08	·03	·04
Phosphorus	·29	·38	·43	·31
Manganese	·28	·66	·96	·27

The manner in which these five substances affect the quality of pig is now so well understood that the iron is often bought on analysis; and I will venture to say that, for mill purposes, any malleable-iron maker would prefer the last four brands to the four previously given. The only difference which can be alleged in favour of the former is that it contains more manganese, which could be easily and cheaply supplied in the shape of ferro-manganese. Notwithstanding, bars and plates made from the first series sell for at least double the price obtained for those manufactured from the second. No one acquainted with the West Yorkshire irons refuses to admit their great claim in respect to quality; but this in my judgment is more due to the extraordinary care used in the forge and mill, than to any unusual excellence in the pig iron.

The recent improvements in the processes for obtaining steel, either by the pneumatic plan or the open hearth, were made at a time when the exclusive reign of the rule of thumb was beginning to decline. The conditions to secure success were diligently studied, and no ancient doctrine was adopted unless its value was capable of demonstration. So far as I know, neither Bessemer nor Siemens and Martin ever had a word to say against the hot blast. But a more striking instance even than this is that afforded by Sweden. In the manufacture of that quality of Swedish iron which has a world-wide reputation for cutlery steel, no new process is rashly introduced; and the Swedish ironmasters, no doubt for good reasons, adhere to the use of the old Lancashire fire, and, for the highest qualities, have forbidden the introduction of the puddling furnace into their primitive forges.

No iron making community in the world is more dependent for mere existence on a continuance of ancient reputation than the Swedes. They saw nothing in the adoption of the hot blast to impair their traditional renown; and I believe at the present moment the use of hot air is the rule in their country.

In order to ascertain the behaviour of the iron in relation to the four metalloids here considered, viz. carbon, silicon, sulphur, and phosphorus, a portion of slag and iron was withdrawn from one of the Clarence furnaces, just above the tuyeres. Three specimens of the slag, which had a brownish colour, contained respectively ·75 per cent., ·90 per cent., and 1·2 per cent. of iron as oxide. Here evidently there was still unreduced metal in the cinder.

The slag as it ran from the furnace was grey, and only contained ·12 per cent. of iron.

It would appear from the following analyses that the iron, which was partly very hard grey and partly white, had not taken up its full measure of carbon and silicon at the point above the tuyeres. The sulphur, on the other hand, suffered a perceptible diminution in quantity after complete fusion at the tuyeres, probably due to its absorption by the lime in the cinder. The phosphorus also was less

SECTION VIII.—SOLID PRODUCTS OF THE BLAST FURNACE. 155

in the iron run from the furnace, than in the sample taken from above the tuyeres, but the diminution was insignificant.

	Globules from above tuyeres.		Next cast of iron, rich No. 3.
Combined carbon	·763	...	·521
Graphitic ...	2·134	...	3·112
	2·897		3·633
Silicon	·924	...	1·793
Sulphur	·077	...	·050
Phosphorus ...	1·693	...	1·630

We may now proceed to consider some of the circumstances connected with the union of these metalloids with the iron in the blast furnace, and their influence on the quality of the product.

Carbon is well known to possess a certain affinity for iron; but perhaps it would be more correct to designate their union in cast iron as chiefly due to the property on the part of the metal to dissolve the metalloid. The proportion of carbon in this form of solution never approaches the point where the two are present in the ratio of their combining equivalents (28 Fe to 6 C, or 4·66 to 1). The actual relation between the two is nearer 6 equivalents of iron to 1 equivalent of carbon.

When the metal leaves a furnace which is producing rich iron, as it loses heat, it gives off a portion of its carbon, in the form of thin, black, and brilliant flakes, known as "kish." The same thing happens again when the metal solidifies on cooling, the extruded carbon being deposited on the faces of the crystals in the pig iron, in what is known as the graphitic form.

A certain portion of the carbon in pig iron is regarded as existing in combination with the metal; and at one time it was considered that the whole of the carbon in white iron was of this character. It was also held that there was a necessary connection between the "richness" or large sized crystals of the iron and the quantity of carbon it contains in the graphitic form. The extruded kish, is however rarely a pure substance, iron in greater or less quantity being generally present with the carbon. After tapping one of the Clarence furnaces, a considerable quantity of flaky matter was collected from the surface of the pigs. By means of a magnet it was separated into two portions.

SECTION VIII.—SOLID PRODUCTS OF THE BLAST FURNACE.

The part not affected by magnetism gave the following results on analysis:—

	Per Cent.
Carbon	95·00
Iron	1·70
Sand, derived from pig moulds	3·30
	——— 100

The other gave:—

Carbon	54·80
Iron	31·20
Sand, and a portion of finely divided silica ...	14·00
	——— 100

From some pig iron drillings, graphite was separated by levigation in water and all metallic particles were removed by the magnet. This graphite yielded 31·74 per cent. of residue on combustion. It was composed of:—

Peroxide of iron	50·78
Protoxide of iron	28·35
Protoxide of manganese	5·10
Silica	5·08
Phosphoric acid	10·30
	99·61

Assuming that the Mn, Si and P were originally uncombined with O, the following composition is calculated for the unoxidized graphite:—

Carbon	78·32
Iron	18·26
Manganese	1·25
Silica	·75
Phosphorus	1·42
	100·00

At another running, of extremely rich iron, there was collected from the "Sows" a quantity of matter which had the appearance of having been in a state of combustion. It contained as follows:—

	Per Cent.
Carbon	36·00
Silica, probably extruded from the iron as silicon and then oxidized	47·00
Iron	9·10
Manganese	7·21
	——— 99·31

SECTION VIII.—SOLID PRODUCTS OF THE BLAST FURNACE. 157

While speaking of the matters thrown off by iron at the time of tapping, I may mention a circumstance which took place at the Wylam furnace very many years ago. For one or two days in succession the iron was covered on its upper surface with a coating, $\frac{1}{8}$ to $\frac{3}{16}$ of an inch in thickness, of a light grey or brownish matter. It was of a silky fibrous texture, soft to the touch, and yielded easily to pressure. It resembled a sublimate in appearance, was insoluble, and without taste. This happened before the days of constant chemical supervision, and I regret to say that in consequence no further notice was taken of it; but we had no doubt at the time that the substance was silica in a fine state of division, and it was regarded as having been the product of silicon given off by the iron, as it gives off carbon when kish is formed.

In going through the Clarence Laboratory books, the samples containing the highest recorded carbon, had the following composition:—

Clarence Pig Iron	No. 1 Pig.		No. 3 Pig.	
Carbon	4·50		4·82	
Silicon	1·28		1·58	
Sulphur	·08		Slight trace.	
Phosphorus	1·39		1·35	
Calcium	—		·23	
Magnesium	—		·05	
Manganese	·72		·62	
Titanium	·10		·22	
		8·07		8·87
Iron	92·47		92·00	
		100·54		100·87

An exceptionally small quantity of carbon was found to be contained in some No. 1 Clarence iron, which was complained of as turning hard and white on remelting. The analyses gave as under:—

	Original Pig.		Remelted for Castings.	
Carbon	1·61		1·65	
Silicon	·98		·61	
Sulphur	·16		·29	
Phosphorus	1·37		1·25	
Calcium	·40		·81	
Magnesium	·29		·43	
Manganese	·12		·47	
		4·93		5·51
Iron	95·15		94·42	
		100·08		99·93

It is in white iron that carbon, in its combined form, is most generally found in the largest quantity; and although this substance, as graphite, is usually more abundant in very rich iron than in grey iron of a closer grain, yet such is not invariably the case. The following analyses will show that neither the form nor the quantity of the carbon necessarily affects the grade of the iron:—

No. 1 Clarence Pig.			No. 4 Clarence Grey Forge Pig.			White Clarence.		
Graphite.	Combined.	Total.	Graphite.	Combined.	Total.	Graphite.	Combined.	Total.
3·65	·30	3·95	2·45	·26	2·71	1·06	·90	1·96
2·33	·48	2·81	2·72	·22	2·94	·300	3·260	3·560
3·38	·51	3·89	2·491	·606	3·097	·270	3·400	3·670
3·08	·92	4·00	....			·374	2·339	2·713
2·61	1·34	3·95	....			2·209	·993	3·202
2·07	1·76	3·83						
2·536	·668	3·204						
2·156	1·014	3·170						

	Graphitic Carbon.	Combined.	Total.
Specimen of No. 3 Clarence pig	2·256	1·058	3·314
Do. „ White, converted into grey by slow cooling ...	2·296	·644	2·940
Do. „ White	·374	2·339	2·713
Part of the preceding white pig, converted into grey iron by 13 days' exposure to a full red heat ...	1·790	·974	2·764

The condition of the carbon and the quality of the metal appear to be the result of differences between the temperatures at the period when the metal is smelted, or those to which it has been exposed subsequently. Variations in the rate of cooling are also capable of effecting marked changes in the physical character of pig iron: thus, as we have seen, slow cooling alters white iron to grey, and rapid cooling converts grey iron into white.

Specimens of white pig were fused in the slag runners of the Clarence furnaces making No. 3 iron, and the metal so treated also became changed to No. 3. Others were melted in a crucible steel furnace and slowly cooled, and the metal when cooled was No. 4 grey iron.

The following experiments are interesting, as showing the change which lengthened exposure to a high temperature may have on the condition of the carbon.

SECTION VIII.—SOLID PRODUCTS OF THE BLAST FURNACE.

Several bars of grey metal, having a sectional area of 3 × 2 in., were left, during an entire week, in the combustion chamber of the hot air stoves at the Clarence works, the fuel being furnace gas. At the end of that time they were found covered with a coating of oxide of iron, which was removed. The composition of the different parts was:—

	Graphitic Carbon.	Combined Carbon.	Total.
Original pig metal, No. 4 foundry	3·057	·637	3·694
First ⅛ inch of exposed metal underneath crust of oxide after exposure	3·537	Nil.	3·537
Centre of bar after exposure	3·082	·469	3·551

Similar bars of white iron were treated in the same way, but the exposure was continued for 13 days. The analysis gave:—

	Graph.	Combd.	Total. Carbon.
Original metal—foundry No. 4	·374	2·339	2·713
First $\frac{1}{16}$ in. underneath crust of oxide	·031	·129	·160
Centre of bar—after exposure	1·790	·974	2·764

The mere contact of oxide of iron with grey cast iron, both being in the liquid state, changes the latter to white metal. This alteration of quality may be effected, without a diminution beyond 5 or 10 per cent. in the total quantity of carbon in the metal; but the portion which was in the graphitic form is almost entirely converted into the combined variety by the exposure to the fused oxide. This is shown by the following analyses:—

	Before Treatment.		After Treatment.	
	Graphitic Carbon.	Combined Carbon.	Graphitic Carbon.	Combined Carbon.
1.—	3·460	·296	·533	2·636
2.—	3·046	·589	·778	2·645
3.—	Not examined.		·226	3·117
4.—	Do.		·046	3·112

In the last two cases, the graphitic carbon was not black, but consisted of a light brown powder; which however gave carbonic acid, when placed in a combustion tube in a current of oxygen gas.

160   SECTION VIII.—SOLID PRODUCTS OF THE BLAST FURNACE.

In the process of cementation, bar iron is exposed for some days to a high temperature in contact with charcoal. Blister steel is the result, containing various proportions of carbon according to the quality required. As much as 1 per cent. may easily be communicated to the iron in this process.

In the blast furnace, not only have we the iron, in a more or less reduced form, surrounded on all sides by coke, but the metal in its spongy state is permeated in the most complete way imaginable by the carbon already referred to in these pages, as deposited from the carbonic oxide. It has been shown by Dr. Percy and others that iron combines with carbon on being exposed to heat in an atmosphere of carbonic oxide. On repeating this experiment, I found carbonic acid in the gas which had been employed. We may therefore infer that carbon was precipitated by the dissociation of the carbonic oxide ($2CO = C + CO_2$), leaving the precipitated carbon to combine with the metal.

With a view of ascertaining whether it was necessary that the carbon should be in the gaseous form, to enable it to combine with iron, I had a polished disc of wrought iron screwed on to a similar one of cast iron, thus obtaining perfect contact. Around the two plates of metal so secured a casing of cast iron was run, about two inches thick, so as to exclude as far as possible all access of gases. The block of metal with its enclosure was then exposed to a full red heat for four weeks, in one of the hot air stoves at the Clarence works.

The change of composition in the surfaces of the cast and wrought iron discs, which were in contact with each other, is exhibited in the following analyses :—

	Carbon, per Cent.	Silicon, per Cent.	Sulphur, per Cent.	Phosphorus, per Cent.
CAST IRON.				
Before exposure	3·248	1·870	·109	1·542
$\frac{1}{16}$ inch planed off after exposure	2·179	1·890	·104	1·539
WROUGHT IRON.				
Before exposure	·044	·121	·008	·195
$\frac{1}{16}$ inch planed off after exposure	·392	·162	·009	·204
Increase	·348	·041	·001	·009

It may be perhaps open to question whether any of the metalloids, except the carbon, have been affected by the mode of treatment described.

SECTION VIII.—SOLID PRODUCTS OF THE BLAST FURNACE.

The increase in the sulphur and phosphorus is so small as to be almost within the range of experimental error. The addition to the silicon in the wrought iron is more considerable; but there is also a slight increase in the percentage of this substance in the cast iron surface, which renders it somewhat improbable that this substance has moved forwards as it were towards the wrought iron; the excess is probably due to want of homogeneousness in the metal used for the experiment. In this particular experiment, as well as in a previous trial, there seems every reason for concluding that the solid carbon seeks to establish an equilibrium. Hence, as soon as the wrought iron surface has absorbed a portion of this metalloid from the cast iron, the place left vacant in the latter is partially supplied from a stratum of metal immediately behind it.

Notwithstanding this proneness of iron to combine with carbon, I have found the metal in the hearth of a furnace quite free from this element. This was discovered on examining iron infiltrated into pieces of lime, which I had analysed, for the purpose of satisfying myself whether carbonic acid was not brought down to the hearth by means of an imperfect decomposition of the limestone.

The production of white iron is in some districts not the result usually aimed at, but is merely due to imperfect furnace working. This is the case in Cleveland, where grey iron is what is generally demanded. Any circumstance which interferes with the uniform deoxidation of the ore in the upper region of the furnace, or which tends to reduce the temperature of the lower region, causes what is known as a scouring cinder, and with it the production of white iron. Hanging or scaffolding of the materials, or the use of small ore or small fuel, interferes, by an unequal distribution of the gaseous currents, with the process of deoxidation, and may produce a like effect.

Silicon, the base of silicic acid or silica, as it is often designated, is the next constituent in the order in which they stand in the analyses already quoted. Its occurrence in pig iron is easily accounted for, because silica is almost invariably found in all the raw materials delivered to the smelter. Silicon, it is true, has not I believe been hitherto separated from the oxygen, with which it is combined in silicic acid, by the action of carbon and heat alone; but the intense temperature of the blast furnace, aided by the presence of highly-heated iron, appears to suffice for the decomposition of this oxide of silicon. At the same time it must be remembered that in Cleveland iron, which is richer in this metalloid than the general run of pig irons, the total quantity of silicon reduced is not above one-twelfth of that which passes through the furnace. It is however quite possible, by conducting the operation at a very high temperature, to raise the percentage of silicon in the iron from 1¾ or 2 per cent. to 5, 6, or 7 per cent., or even more.

Although, as had been shown, there is less heat evolved at the tuyeres per unit of iron produced, in using hot than in using cold blast, it must be remembered that in a given period of time, owing to the greater make of the hot blast furnace, the rapidity of evolution is about double in this case compared with that of the cold blast furnace. This more rapid generation of the heat of combustion and the influence of the hot air itself, may establish limited *foci* of intense temperatures near the tuyeres which it is impossible to detect by analysis of the products of combustion or by any known means of thermometric measurement. In this way it is possible, in areas of restricted dimensions, silicon may be produced in exceptional quantities.

Iron containing any approach to 6 or 7 per cent. of silicon is absolutely useless in the foundry and forge. It is extremely brittle when in the form of castings, and its richness in silicon renders it perfectly unmanageable, either in the refinery or the puddling furnace, owing to its powerful action on the masonry.

The analysis of two specimens of siliconized or glazed iron, as it is termed, made at the Clarence Works, gave the following results:—

	a.	*b.*
Carbon { (*a* graphitic 2·59, combined ·79) (*b* „ 2·68, „ ·71) }	3·38	3·39
Silicon	5·13	5·13
Sulphur	·17	·23
Phosphorus	1·12	1·12
Manganese	·77	·56
Calcium	·22	·20
Magnesium	·06	·03
Titanium	·26	·18
Iron	88·18	89·70
	99·29	100·54

SECTION VIII.—SOLID PRODUCTS OF THE BLAST FURNACE.

The most expeditious mode of correcting the production of this species of metal is to lower the temperature of the blast, a change which at once makes itself felt in the region where silicon combines with the iron. At the same time the weight of ironstone is increased, and by the time this increased burden comes down to the tuyeres the air is raised again to its normal heat.

Silicon is useful in increasing the fluidity of the metal, and in consequence ought to be sought after, in moderate quantities, for fine castings. Where strength is required it is prejudicial, and in the puddling furnace its presence is the reverse of desirable, for it not only diminishes the yield by its presence, but it increases the waste, on account of the readiness with which the silica it there generates combines with oxide of iron.

It was not until the discovery of the Bessemer process, that any thought was bestowed on aiming at the production of pig iron rich in silicon. In order to maintain the desired temperature in the converter, silicon is considered indispensable; hence a Bessemer steel maker, using the ordinary hematite or analogous pig, requires his metal to contain at least a certain percentage of this element, formerly so justly shunned for most purposes. Indeed cases have arisen in which siliconized iron has been made expressly for the use of steel makers using the open hearth, as well as the Bessemer converter. When however pig iron is of such quality, that it has to be subjected in the converter to the basic treatment recently introduced, the presence of silicon is not desired, and the required temperature is sought to be obtained by the oxidation of the phosphorus.

Sulphur.—In almost every description of clay ironstone sulphur is to be found in greater or less quantity; and the same may be said of almost every variety of pit coal. As a rule, in both minerals the sulphur exists as a bisulphide of iron ($Fe\ S_2$), although in some cases small quantities of sulphuric acid may be detected, generally combined with lime. The calcining of the ore and coking of the coal usually expel half the sulphur from the bisulphide of iron; so that the furnace manager has only to deal with the remaining half of the ingredient— which, so far as my observation goes, is always an unwelcome one in pig iron. For the foundry it hardens the metal; and in the puddling furnace it produces red-shortness, or brittleness when the bar is hot.

164   SECTION VIII.—SOLID PRODUCTS OF THE BLAST FURNACE.

Fortunately lime, and calcium, the metallic basis of lime, share with iron a powerful affinity for sulphur; hence the advantage of using limestone in excess of the quantity absolutely required as a flux. The materials used at an iron work, as well as the products, are so large in volume and so fluctuating in composition, that it is difficult to obtain figures which shall correctly represent the average composition of every thing which passes through the smelter's hands. As nearly however as could be determined, from $2\frac{1}{2}$ to 5 per cent. of the sulphur entering the blast furnaces at the Clarence works appeared in the iron, while the remaining 95 or $97\frac{1}{2}$ per cent. was carried off in the slag; but these figures would require a certain amount of correction, owing to a small quantity of sulphur which finds its way into the fume or smoke, accompanying the gases as they leave the furnace.

A low temperature in the blast furnace seems to favour the absorption of sulphur by the iron; hence it is that white pig iron usually contains more of this metalloid than is found in grey, when the same kind of materials are used in each case. One specimen of white iron, already referred to, was mentioned as containing ·46 per cent. of sulphur; whereas in the grey iron made at the same works from the same materials there is usually only about one-tenth of this quantity.

The fact that sulphur is always found in greater quantities in white than in grey iron, produced from the same materials, entitles us to infer that the ranges of temperature which obtain in the blast furnace suffice to modify the chemical reactions which take place in the hearth. White iron, as is well known is the result of "cold working," and allusion has been already made to the fact that at higher temperatures the affinity of iron for sulphur is weakened—not only is this the case, but it would appear almost as if the affinity of this substance for lime or calcium is strengthened by higher temperatures. I had occasion, some years ago, to try on a large scale the fusion of oxide of iron with soda waste, which, as is well known, is a compound of lime, calcium, and sulphur; the last named substance being present to the extent of about 17 per cent. of the whole. The object was the production of a factitious sulphide of iron. When the temperature of the furnace was kept down, a compound was obtained

SECTION VIII.—SOLID PRODUCTS OF THE BLAST FURNACE. 165

containing sometimes as much as 32 per cent. of sulphur, and about 4 per cent. of oxygen, when, on the other hand, the furnace was worked "hot," nearly all the metal was in the form of cast iron, containing only 1 or 2 per cent. of sulphur.

Phosphorus.—The case is quite exceptional when any material treated in a blast furnace is absolutely free from phosphorus; the fuel, vegetable or mineral, the flux, and the ore, all appear to contain it in greater or less quantities. Beginning with the oxides of iron found in the granite formation, and ending with the lake ore forming at the present day, this element is almost invariably present. It occurs almost always combined with oxygen as phosphoric acid, in the form of phosphates of iron or lime in the ore, or as phosphate of lime in the flux and fuel. When a phosphate of iron or lime is exposed to a high temperature in contact with carbon, the oxygen is separated from these salts, and the product is phosphide of iron or phosphide of calcium. The pig iron then dissolves the phosphide of iron, or decomposes most of the phosphide of calcium.

Thus it unfortunately happens that by far the largest quantity of this substance, as it occurs in the materials, is taken up by the iron. By a similar course of observation to that adopted in the case of sulphur, it was calculated that, of the phosphorus existing in the materials, not less than 90 per cent. found its way into the iron.[1]

Although there is no doubt that the presence of phosphorus in small quantities interferes seriously with the strength of steel, as well as of iron, recent experience rather points to the possibility of very minute proportions not only being admissible, but that it may be advantageous for certain purposes.

Manganese demands an extremely high temperature for its reduction and fusion; but the reducing power of the blast furnace,

---

[1] The following were the quantities of sulphur and phosphorus calculated to be present in the materials, for every 100 parts of pig iron produced from calcined Cleveland ironstone:—

	Sulphur.	Phosphorus.
In calcined ironstone	2·525	1·253
Limestone	·035	·007
Coke	1·896	·318
	4·456	1·578

and the presence of liquid iron, acting as they were described to do in the case of silicon, always effect the reduction of a portion of this metal, which is a very general constituent of iron ores. The exact extent to which this happens depends upon the temperature at which the furnace is being worked. For steel making purposes metallic manganese is now in great demand; and, by the use of a light burthen of ore, ferro-manganese, as it is termed, can be made, containing as much as 70 or 80 per cent. of manganese or even more. The metal however is so difficult of reduction, and the proneness of oxide of manganese to combine with silica is so great, that the cinder produced by the furnaces making ferro-manganese contains a considerable percentage in weight of this metal. A further loss of the manganese arises from the intense temperature employed in the production of ferro-manganese, by which, as I was informed at a Westphalian work, no less than 3 per cent. of this metal, as contained in the ore, is evaporated and carried off in the fume.

Specimens of slag brought from Westphalia, from a furnace making ferro-manganese, contained as follows:—

Silica	29·65
Alumina	9·38
Lime	36·62
Magnesia	2·14
Baryta	1·30
Potash	·99
Protoxide of Iron	·57
Do. Manganese	15·28
Sulphide of Calcium	4·05
Phosphoric Acid	Trace
	99·98

The presence of manganese is sought after in pig intended for malleable iron and for steel purposes. The clay ironstones from which the celebrated makes of Bowling and Low Moor iron are made, contain enough of this metal to give in the cinder from the blast furnaces

SECTION VIII.—SOLID PRODUCTS OF THE BLAST FURNACE. 167

as much as 3·38 per cent. of its protoxide, while the pig itself is often associated with 1½ per cent. of manganese. When the metal used in the Bessemer converters contains a sufficiency of manganese, Spiegel-iron and ferro-manganese may be dispensed with.

Other Metals in Pig Iron.—Little can be said of the influence which any of these, including titanium or its nitride, exercise on the quality of the iron in which they are found. Usually they occur in very minute quantities, and those disappear completely during the process of puddling. Upon one occasion, already mentioned in the present Section, a specimen was found to contain as much as 1·24 per cent. of calcium and magnesium jointly. This, as compared with all other analyses, is an unusually great amount; and had it not been for the peculiarity of the metal, which, with only 1·96 per cent. of carbon, retained great fusibility, I should have set it down to experimental error.

With regard to the last two metals, together with aluminium which also occasionally finds its way into pig iron, no attempts have been successful, so far as I know, in separating them from the earths containing them by the agency of carbon and heat alone. The usual method is to call in the aid of sodium, which at moderate temperatures is able to decompose the chlorides.

Now in the blast furnace chlorine in the form of common salt and compounds of sodium and its kindred metal potassium exist in certain quantities. With this knowledge, and looking to the fact that the temperature and intensely reducing power of the carbon and gases of the lower zone are by no means inconsistent with the production of these two alkali metals, I sought carefully for traces of their presence. The furnace vapours were made to pass through cold mercury, but I entirely failed to obtain a trace of either. The cyanides of potassium and sodium were next suggested as a possible agent, for decomposing the earths containing the calcium, magnesium and aluminium. Trials were made in the laboratory with this object in view; but these hitherto have been entirely unsuccessful.

In the consideration of this subject we must not of course lose sight of the circumstance, that in the blast furnace we have an intense temperature rarely available in the laboratory, that there is an enormous quantity of highly heated iron constantly dropping through the materials, and lastly that the total weight of metallic matter other than iron rarely exceeds one half per cent. of the pig produced.

I incline also to the belief that the presence of silica and other earths in the fume which leaves the furnace show those to be in such a state that it is very possible they may be due to the reoxidation of sodium or potassium. We shall see presently that these earthy sublimates are not found at the tuyeres, and therefore I suspect that at some point above this the alkali metals, or their cyanides, are concerned in the reduction of the silica, lime, etc.

In the event of anything happening to lower the deoxidizing power of the blast furnace, particularly in its lower region, it is only reasonable to expect that substances, which part with their oxygen with so much difficulty, as silicon, calcium, etc. do, or whose oxygen compounds have a great affinity for the unreduced oxide of iron, should be found in diminished quantity in the pig iron. Two cases are appended which seem to point to this conclusion, but the occurrence of these metals in pig iron is not sufficiently uniform to justify its general application.

	Clarence No. 1 Pig.	Clarence Rough White Iron made during a derangement of the Furnace.
Carbon	3·17	1·96
Silicon	2·22	·11
Sulphur	·19	·96
Phosphorus	1·34	·26
Calcium	·24	·15
Magnesium	·06	·06
Manganese	·42	·11
Titanium	·24	—
	7·88	3·61
Iron	91·50	97·30
	99·38	100·91

Cinder.—The quantity of slag or cinder produced per ton of iron varies with the quantity of earthy matter in the ore. At Fullonica I found that in smelting Elba ore, it did not exceed 10 per cent. of the weight of the iron; in other cases it is sometimes nearly 200 per cent. In Cleveland it is about 130 per cent., while in the north west of England, using hematite, it is only about 50 per cent.

SECTION VIII.—SOLID PRODUCTS OF THE BLAST FURNACE. 169

In order to promote the separation of the cinder from the metal, the earths composing the former must constitute a readily fusible substance. A sacrifice however must sometimes be made in respect to the melting point of the cinder in order to free the iron from sulphur. Thus the earths, in the proportions in which they exist in Cleveland stone, melt with facility; but, when an attempt was made at the Clarence works to dispense with the use of limestone, the iron became so highly charged with sulphur as to be useless.

Although in the vast majority of cases lime is a constituent of blast furnace slags, its presence is not indispensable, as may be seen from the following analysis of a specimen obtained in Rhenish Prussia, in which oxide of manganese occupies the place of this earth.

Silica	49·57	48·39
Alumina	9·00	6·66
Protoxide of Manganese	25·84	33·96
Magnesia	15·15	10·22
Sulphur	·08	·08
Protoxide of Iron	·04	·06
	99·68	99·37

Essentially, ordinary blast furnace slags may be regarded as a silicate of lime and alumina, in which more or less magnesia is usually present. M. Berthier pointed out, in his "Essais par la Voie Sêche," that silica, combined with any one of these three earths, melted with great difficulty; and this is the reason why the Cumberland ore, containing only silicious matter, in addition to lime works best with an admixture of aluminous ore, in order to form a double silicate of lime and alumina and thus obtain a fusible slag. Mixtures of different ironstones are frequently recommended on the ground that the iron is better than if they have been smelted singly, the fact probably being that, in the latter case want of fusibility in the cinder interferes with the steady working of the furnace.

Fortunately for the pig iron maker, the composition of slags admits of great latitude, in the relation the earths bear to each other, without interfering seriously with their fusibility. The following numbers show the very different relations, in which the four earths usually present in furnace cinders are met with in different localities; all being sufficiently fusible for easy separation from the metal.

170   SECTION VIII.—SOLID PRODUCTS OF THE BLAST FURNACE.

Composition of Slags.	South Wales. Per Cent.	Staffordshire. Per Cent.	Cleveland Per Cent.
Silica	45	40	29
Alumina	13	15	25
Lime	37	37	40
Magnesia	5	8	6
	100	100	100

In reality however blast furnace cinder is a much more complex body than is exhibited by a mere table of its earthy component parts; as may be seen by the following analyses of specimens from the Clarence works :—

Silica	27·65	30·84
Alumina	24·69	25·71
Lime	40·00	34·03
Magnesia	3·55	6·92
Sulphur	1·95	1·82
Phosphoric Acid	·52	·34
Soda and Potash	1·45	1·30
Protoxide of Manganese	·35	·26
Do. Iron	·72	·23
	100·88[1]	101·45[1]

Notwithstanding the intensely powerful reducing agency at work in the blast furnace, it is extremely doubtful whether the cinder is ever entirely devoid of any traces of iron; and I believe this is due to the fact that there is always retained by iron a last trace of oxygen, so firmly held by the metal that neither carbonic oxide nor even carbon can tear it away.

The quantity of protoxide of iron (Fe O) usually found in the Clarence cinder is ·27 per cent. containing ·21 per cent. of iron; but as small a quantity as ·12 per cent. of iron has been found in these scoriæ.

The increase of iron, already referred to as present in the cinder when making the lower qualities of iron, is apparent in the following analyses of specimens taken from the Clarence laboratory books :—

---

[1] The probable cause of the excess above 100 is that all the calcium is calculated as lime, whereas part of it would be combined with sulphur.

SECTION VIII.—SOLID PRODUCTS OF THE BLAST FURNACE. 171

A specimen of cinder from a furnace making grey forge was
found to contain ... ... 1·03 % of iron.
Ditto,   ditto,   smooth white iron,   ditto 1·40 %   ,,
Ditto,   ditto,   rough white iron,   ditto 2·83 %   ,,

It is therefore from the continuous flow of cinder that the furnace manager is enabled, by its colour, to form an opinion on the condition of his furnace, and on the quality of iron he may expect. By this indication, an experienced eye will predict with great precision the grade of iron being made at the time. Not only so, but a watchful manager learns, by indications of *texture* in the cinder, to foresee impending changes in the furnace; and he is thus enabled to prepare for them accordingly.

In many instances, however, the presence of oxide of iron, as found in the cinder, cannot be the reason of the change in the quality of the metal, owing to the smallness of its amount. In such cases the colour of the slag, due to this presence, must therefore be accepted merely as an indication that the temperature of the furnace is inadequate for the production of grey iron.

When the cinder contains anything like 1 per cent. of protoxide of iron (·77 of Fe), it is black in colour, but this does not indicate a state of things under which a furnace cannot be permanently and conveniently carried on. The extent however to which iron is found in slag, without producing inconvenience, seems to vary somewhat with different ores; this is possibly due to the relation the earths bear to each other, and to the composition of the metal.

In the Duchy of Luxemburg, at Nancy and elsewhere, an ore resembling in geological position that of Cleveland is largely smelted for forge purposes, the product being white metal. This ore is, however, richer in silica than that of North Yorkshire, as may be observed by the relations which this substance bears to the alumina and lime in the furnace cinders of the two districts :—

	Cleveland.	Nancy.
Silica	34	41
Alumina	24	20
Lime	42	39
	100	100

SECTION VIII.—SOLID PRODUCTS OF THE BLAST FURNACE.

Such cinders as run from the furnaces, when producing white pig at Nancy, contain from 2 to 3 per cent. of protoxide of iron. The metal is very suitable for puddling purposes, and has the following composition:—

Carbon	2·339
Silicon	·460
Sulphur	·431
Phosphorus	1·892
Manganese	·092
	5·214
Iron	94·789
	100·003

In instances of extraordinary derangement, the cinder may become so overloaded with iron oxide as to reduce the iron to the condition of a pasty mass, due partly to the reduced temperature of the hearth, and partly to the infusible nature of the metal itself. Under such circumstances it becomes impossible to work a furnace satisfactorily. The slag resembles mill cinder in appearance, and the iron is so thick that it sets on the hearth, giving rise ultimately, if continued, to what is technically known as "gobbing." There is besides so imperfect a separation of the two, that a good deal of the iron is carried away mechanically along the slag runner. In such cases it is more than probable that during the mere passage of the iron through the cinder, it may lose a portion of the carbon with which it had become associated in an upper region of the furnace.

Cooling attends the "overburdening" of the furnace. The carbonic oxide has then imposed upon it more duty than it can perform, unreduced ore descends into the hearth, and the result is a black cinder. The same effect may be brought about by anything which lowers the temperature of the zone of fusion, even when, at the time this occurs, reduction is being suitably carried on in its proper place. A very common cause of this temporary disturbance is the admission of water by a faulty tuyere.

The decomposition of water is attended with a great absorption of heat—no less than 34,000 calories per unit of its combined hydrogen, besides that which is required for its conversion into vapour. If we assume each unit of coke, as it is burnt with hot air, to give 3,300 calories, then, after giving the water credit for the heat it generates by

SECTION VIII.—SOLID PRODUCTS OF THE BLAST FURNACE. 173

oxidation of the fuel, it may be taken that the admission of 15 lbs., *i.e.* 1½ gallons, of water per minute, into a furnace making about 400 tons per week, requires an addition of 3·2 cwts. of coke per ton of iron.

So great indeed is the absorption of heat from the injection of water at the tuyeres, that differences in the extent of moisture in the atmosphere may occasionally be enough to account for variations in the work performed. It has frequently been remarked in winter, at a time of extreme cold, that blast furnaces work more steadily and with less fuel than in summer. Indeed this fact was constantly quoted as a reason against the use of the hot blast in its early days. No doubt the change was due to the comparative absence of moisture in the air; and thus a virtue was attributed to the mere coldness of the air which did not belong to it.

It is I believe capable of demonstration that the reduction of a certain amount of oxide of iron within a very short distance of the tuyeres, and the escape of a trace of the same into the slag, is probably unavoidable; but for many reasons the smaller this amount is the better. If this undecomposed oxide be unduly increased, it is beyond the power of the fuel at the hearth to effect a proper amount of deoxidation; and the unreduced metal is carried off in undue quantity by the slag, giving to it the well known dark colour.

The cinder from blast furnaces, certainly from those in the Cleveland district, possesses a property which, were it possible to reduce all the oxide of iron it contains, ought, one would imagine, to secure that result. This property consists in the occlusion, or power of absorption of a considerable volume of the gases found in the furnace itself, nearly 50 per cent. of these being gas of a highly deoxidizing nature.

For the first time, so far as I know, these have lately been made the subject of analysis at the Clarence works. Two of these were as follows:—

	Furnace 80 feet high making Grey Iron.	
Date of Collection	Oct. 1, 1879.	Oct. 2, 1879.
By measurement—		
Carbonic Acid and Sulphurous Acid	6·28	14·88
Carbonic Oxide	28·80	29·77
Hydrogen	38·98	41·44
Nitrogen	25·94	13·91
	100·00	100·00

174 SECTION VIII.—SOLID PRODUCTS OF THE BLAST FURNACE.

By weight—

Carbonic Acid ...	14·62	32·55
Sulphurous Acid ...	·04	·04
Carbonic Oxide ...	42·72	41·48
Hydrogen ...	4·13	4·13
Nitrogen ...	38·49	21·80
	100·00	100·00

There is no doubt that the volume of gas so absorbed is very considerable, because, being emitted as the cinder cools, it often gives rise to the bursting of the square masses of slag into which it is run. The outer portion of the mass being first solidified, forms a shell which is frequently fractured by the pent up gas. This is often attended with considerable explosive force, not altogether free from danger to those who may be in the immediate neighbourhood. Besides the true gases, which escape from the flowing stream of cinder, a portion of its solid constituents are occasionally emitted in the form of a white smoke; the actual weight of which however is very small. This always happens in smelting iron at a high temperature from Cleveland stone. By condensing and collecting this vapour, probably oxidized after leaving the slag, it was found to consist of :—

Free Sulphuric Acid, SO_3 ...	4·88
Sulphate of Potash ...	85·25
Do. Soda ...	7·46
Do. Magnesia ...	2·41
	100·00

Furnace Fume.—A certain amount of fine dust is propelled mechanically from the blast furnace, by the mere force of the current of gas escaping at the tunnel head. This consists of powder of ironstone, and minute particles of coke, the limestone being rarely in such a condition as to size to permit of its being so ejected. Besides this, there issues from most furnaces a vast volume of a white vapour-like matter or fume, of a varying composition, but differing always widely from any of the materials charged in at the top. A considerable portion of this condenses very speedily; but a part probably from its very minute state of division, passes through the boiler and stove fire places, and floats away for a very great distance from the chimneys of our iron works. Any questions connected with the generation and emission of this sublimed matter will be more con-

SECTION VIII.—SOLID PRODUCTS OF THE BLAST FURNACE.

veniently considered in the following Section, which it is intended to devote more exclusively to the chemical changes, as they occur in the furnace.

As a matter of practical interest this sublimate, if such it be, is of little importance: except that its presence is the source of considerable inconvenience and expense to the furnace manager. It gradually forms a coating on the boilers and hot air apparatus which interferes greatly with the transmission of the heat to their contents. The removal of this, and the emptying of the flues of a large quantity of hot light dust, occupies a considerable time, and is a very disagreeable occupation for the men engaged in the operation.

SECTION IX.

ON THE CHEMICAL CHANGES AS THEY TAKE PLACE IN THE BLAST FURNACE.

In the four preceding Sections the results obtained by the use of the blast furnace have been described. In doing this some general explanation of the chemistry of the process was unavoidable; but in order to render this more complete, I have deemed it desirable to devote a separate Section to a more minute examination of the changes experienced by the materials, during their descent from the top to the lower part of the structure. This mode of dealing with the subject involves some repetition, but, even with this inconvenience, the separation of the chief functions of the furnace from others of a more scientific character, but occupying perhaps a subordinate position as a question of technical interest, has its advantages.

The sequence of chemical actions which takes place in the blast furnace is complicated, and as regards any particular locality is liable to considerable changes in its nature. The immediate action on the materials, viz. ore, coke and limestone, when they enter a close topped furnace, is a mere elevation of temperature. The first to manifest any chemical change is the ore, which quickly commences to lose a portion of its oxygen; then follows the coke, which is attacked to a limited extent by carbonic acid; and lastly the limestone, which does not part with its carbonic acid until it reaches a zone where the temperature is more intense than that required for the changes experienced by the oxide of iron and the coke.

The order in which chemical action is established in the materials, as just set forth, must be understood as relating only to the commencement of the process. A large block of ironstone may pass through the hotter zones, of carbon solution and of limestone decomposition, before

SECTION IX.—CHEMICAL CHANGES IN THE BLAST FURNACE. 177

it is reduced to the full extent to which it is susceptible, owing to the reducing gases not having completely penetrated its mass. Such a state of things as that just mentioned, renders it impossible to judge of the general character of the action by an examination of a particular specimen or specimens of the materials, selected from so large a bulk as that operated on in the furnace. The composition of the gases is therefore a more convenient, if not the only, mode of estimating the progress of the changes, as they occur at different heights of the furnace. Practically we may consider the gases at or a little above the tuyeres to consist of atmospheric nitrogen and oxygen, accompanied, when making grey iron, by as much carbon as converts the latter gas into carbonic oxide. Any addition of oxygen to this normal quantity must be derived from the limestone or ore; and any subsequent increase of carbon in the gases must be contributed by the limestone or coke. This assumption pre-supposes that any carbonic acid in the ore has been expelled by calcination, and neglects taking into account the comparatively small weight of oxygen due to the decomposition of hygrometric moisture.

This pneumatic estimate is subject to many causes of disturbance. The specimen of gas obtained for examination is necessarily drawn from a space of limited dimensions, and may be so affected by local conditions, as not to represent an average of the zone from which it is drawn. From any irregularity produced in the way just described, the gases at the point of final discharge may be considered to be exempt; but these again are liable to very great fluctuations in composition, according to the period after charging at which they are taken. The fillers, as a rule, have their furnace quite full at noon, when they go to dinner; and between this hour and one o'clock no fresh charges are introduced. During this period the carbonic oxide is rapidly acidified by the oxide of iron, and the temperature of the gases rises considerably.

The following table exhibits the volumetric composition of the gases, commencing a little before noon, when the quantity of carbonic acid was much below the usual average—owing probably to its action on the softer portions of coke, in the manner described in the Section on fuel. The furnace supplying the samples had a capacity of 17,500 cubic feet :—

L

SECTION IX.—CHEMICAL CHANGES IN THE BLAST FURNACE.

Hour	11·45		12·20	12·40	1·0		1·20		2·0		2·25
Carb. acid	8·8		10·7	12·3	22·2		10·9		8·3		11·3
Carb. oxide	28·7	Ceased charging.	29·2	25·2	16·9	1st charge after dinner.	28·6	2 rounds charged.	30·4	4th charge.	23·8
Hydrogen	1·1		3·3	3·8	2·1		3·2		2·2		4·0
Nitrogen	61·4		56·8	58·7	58·8		57·3		59·1		60·9
	100·		100·	100·	100·		100·		100·		100·
Vols. CO_2 per 100 vols. of CO	30·66		36·64	48·81	131·36		38·1		27·6		47·4

It may be convenient to observe that giving the volumetric relation of CO and CO_2 to each other, expresses at the same time the ratio of carbon in each by weight.

The large increase of carbonic acid in the gases just referred to, is necessarily accompanied by a considerable evolution of heat; but the heat so generated is practically nearly all wasted, because, owing to the point where the chemical action takes place, viz. near the upper surface of the materials, it is carried off in the escaping gases. In addition to this a considerable quantity of heat rises from the hearth, and being no longer intercepted by the presence of freshly charged materials, finds its way out at the tunnel head. The following observations made in a furnace 80 feet in height with a capacity of 25,500 cubic feet, illustrate the rise in temperature as just stated:—

	Temperature of gases after charging.	Period during which no charge was introduced.		Temperature of gases at end of period.	Rise in temperature.
	° F.	hr.	min.	° F.	° F.
No. 4 furnace	579	1	26	816	237
,,	564	1	20	814	250
,,	593	1	23	797	204
,,	586	1	40	788	202
No. 5 furnace	608	1	30	844	236
,,	644	1	20	834	190
,,	586	2	30	887	301
,,	629	1	42	867	238
		1	36	Average rise	232 = 129° C.

The increase of temperature, while discontinuing the charging, may be thus regarded in this case as amounting to 80° C. (144° F.) per hour.

The increase of carbonic acid in the gases and the rise in their temperature, are chiefly due to the ore parting with its oxygen and thus acidifying the carbonic oxide and partly to a dissociation of carbonic

SECTION IX.—CHEMICAL CHANGES IN THE BLAST FURNACE.

oxide. A precisely opposite effect is produced when coke alone is introduced—the carbon, particularly the softer portions, as described in Section VII. are attacked by the hot carbonic acid and the relations in the volume of the two carbons undergo an immediate change.

During 50 minutes no materials had been charged into a furnace of 80 feet having a capacity of 17,500 cubic feet.

		Carb. Oxide. Vols.		Carb. Acid. Vols.
At end of this time the carbonic oxide and carbonic acid gases were found to exist in the ratio of	...	100	to	105
52 cwts. of coke were then introduced, immediately after which the escaping gases contained for	...	100	...	58
In 16 minutes after this the proportions were	...	100	...	44
42 ,, after last trial		100	...	54
45 ,, ,, ...		100	...	128
55 ,, ,, ...		100	...	96
25 ,, ,, ...		100	...	44

The lesson taught by the figures just given is, that those portions of the coke more easily attacked by carbonic acid give rise to a large decrease of this gas by its conversion into carbonic oxide. This goes on until these are gasified, when the carbonic acid commences to increase in quantity by not finding any more carbon which it is capable of dissolving. Ultimately the proportion of carbonic acid again falls, owing to the fact that the greater portion of the ore is reduced and is incapable of furnishing the right equivalent of oxygen, for a proper saturation of the gases with this element. As might be expected the immediate effect upon the escaping gases by the introduction of coke, limestone, and ore is one of a cooling character, arising from the mere fact of their passage through substances cooler than themselves. I determined, in the laboratory, that weight for weight, the power of intercepting heat of the three materials used in the Cleveland district was about the same. Bulk for bulk however, the ironstone possesses double the cooling capacity of coke and in reality it was found, in the blast furnace that 24 cwts. of coke reduced the temperature of the gases 21° C. (38° F.) while 50 cwts. of ore and limestone cooled them 13° C. (23° F.).[1]

---

[1] The difference in the laboratory results and those afforded by the furnace is due to the warmth contained in the ironstone as it is brought from the kilns. the

SECTION IX.—CHEMICAL CHANGES IN THE BLAST FURNACE.

In each case the observation was made five minutes after charging and before the materials were considered to have become sufficiently heated to start chemical action. Of course on introducing a complement of the materials, or a *round* as it is termed, the cooling action becomes complicated—the first contact of the gases with the solids is accompanied by a fall of temperature of the former, which is then over powered by the formation of carbonic acid, consequent upon the reduction of the ore—further cooling takes place by the action of the newly formed carbonic acid on the coke and next by the dissociation of the limestone.

When the materials are rapidly charged, as happens immediately after the dinner hour, the net result of the two antagonistic tendencies just spoken of, was ascertained at the Clarence Works to be one of a cooling nature. On resuming the duty of charging, it requires something like one hour for the men to fill the furnace to its proper level, during which time, about 20 tons of materials are introduced. The average of eight observations indicated a fall in the temperature of the gases of 233° F. (129° C.) *i.e.* they fell from 830° to 597° F.

In order to complete the observations on the temperature of the escaping gases from an 80 feet furnace, as it is affected by charging, the table is inserted, containing the changes in temperature five minutes before and five minutes after charging.

In conducting any experiments to determine the action of the furnace gases on oxide of iron, the constant changes which occur in the temperature and composition of the former must be borne in mind.

Temperature before charging.		Nature of the charge introduced.	Temperature after charging.		Maximum difference between each charge.	
° F.	° C.		° F.	° C.	° F.	° C.
655	346	24 cwts. coke.	640	338	—	—
525	274	50 ,, calcined ironstone and limestone.	495	256	− 130	72
495	256	24 ,, coke.	450	232	− 30	17
435	224	50 ,, ironstone and limestone.	415	213	− 60	34
420	215	24 ,, coke.	400	204	− 15	8
380	193	50 ,, ironstone and limestone.	365	185	− 40	22
475	246	24 ,, coke.	430	221	+ 95	53
405	207	50 ,, ironstone and limestone.	405	207	− 70	39
395	202	24 ,, coke.	375	191	− 10	5
370	188	50 ,, ironstone and limestone.	365	185	− 25	14

SECTION IX.—CHEMICAL CHANGES IN THE BLAST FURNACE. 181

Furnace was now full—ceased charging for one hour. Charges sank 4 feet below the charging bell after which the temperature rose to 700° F. (371° C.) when charging was resumed.

Temperature before charging.		Nature of the charge introduced.	Temperature after charging.		Maximum difference between each charge.	
°F.	°C.		°F.	°C.	°F.	°C.
700	371	24 cwts. coke.	625	329	+ 330	183
		Furnace charges, 4 feet down.				
565	296	50 ,, ironstone and limestone.	535	268	− 135	75
530	277	24 ,, coke.	510	264	− 35	19
460	238	50 ,, ironstone and limestone.	445	229	− 70	39

It is almost needless to say that the temperatures, given as those taken 5 minutes after charging, do not indicate the entire cooling influence of the charge; for it will be seen that the heat of the gases continues to fall up to the period at which the succeeding charge is introduced.

Upon another occasion at a furnace 80 feet high with a capacity of 11,500 cubic feet the following readings were obtained:—

Hour. A.M.	Temperature.	
	°F.	°C.
10· 5	575	302
10·10	605	318
10·15	670	354
10·20	700	371
10·25	750	399
10·30	800	427
10·35	835	446
10·40	875	468
10·45	935	502
10·50	975	524
10·55	1,015	546
11· 0	1,025	552
11· 5	1,075	579
11·10	1,140	615

No materials were introduced during these 65 minutes. Furnace by this time was three complete rounds down, equal to 271½ cwts. Charged 25 cwts. of coke.

182   SECTION IX.—CHEMICAL CHANGES IN THE BLAST FURNACE.

Hour. A.M.		Temperature. °F.	°C.
11·15	... ...	1,140	615
	Charged 65½ cwts. of ore and flux.		
11·20	... ...	1,025	552
11·25	... ...	930	499
11·30	... ...	930	499
	Charged 25 cwts of coke.		
11·35	... ...	930	499
	Charged 65½ cwts of ore and flux.		
11·40	... ...	930	499
	Charged 25 cwts. of coke.		
11·45	... ...	800	427
11·50	Charged 65½ cwts. of ore and flux.		
11·55	... ...	750	399
	Charged 25 cwts. of coke.		
12· 0	... ...	725	385
12· 5	Charged 65½ cwts. of ore and flux.		
12·10	... ...	680	360
1·10	... ...	1,070	577

In this instance the temperature of the gases has, from the mode of charging, risen considerably above the average of that at which they leave this particular furnace, which does not usually exceed 630° F. (332° C.)

In making any estimate of the quantity of heat thus carried away in the gases, an allowance must be made for that contained in the iron-stone which is used warm from the kilns. It was found that the temperature of the ore was often considerably above 212° F. (100° C.) so that when cold ironstone was being used the gases escape 220° F. (122° C.) cooler than when hot ore was being charged.

Previous to commencing an examination of the action going on in the furnace itself, much care was bestowed on observing, in the laboratory, the behaviour of the materials when exposed to the influence of the oxides of carbon alone; those gases being employed in different proportions, and at different temperatures. In one or two cases these results have been anticipated in page 65 and following pages; but for the sake of rendering this Section more complete in itself, some of them will be repeated upon the present occasion, accompanied by certain details as to the experiments themselves.

SECTION IX.—CHEMICAL CHANGES IN THE BLAST FURNACE. 183

The first trials were made to ascertain the temperature at which an ore of iron commences to lose its oxygen. In the case of Cleveland ironstone it was found that:—

Faint action commenced at	... ...	392° F. (200° C.)
Marked action at ...	... ...	410° F. (210° C.)

With pure precipitated oxide of iron, faint action manifested itself at 285° F. (141° C.) and marked action at 300° F. (149° C.)

The next observations were directed to learn the rate at which ores of various kinds parted with their oxygen, at different temperatures, when subjected to the action of carbonic oxide. The following numbers give the percentage of the oxygen originally existing in the oxide of iron, which was removed during seven hours exposure:—

	Temperature.	Duration of Exposure. Hours.	Cleveland Calcined Stone. Per Cent.	Lancashire Red Ore. Per Cent.
1st trial	782° F. (417° C.)	7	9·4	—
	770° F. (410° C.)	7½	—	37·1
2nd ,, ...	770° F. (410° C.)	7½	20·7	57·4

The differences perceived in these two experiments were suspected to depend on the rapidity with which the current of reducing gas was made to pass over the oxide of iron. Two other trials were made, preserving the temperature at the same point in each, viz. 410° C. (770° F.), and both were continued during 6 hours. In the first experiment 10·5 litres per hour and in the second 40·8 litres were employed. With the smaller quantity it will be found that, for every unit of oxygen removed from the ore, the proportion of carbonic acid in the gas, after the duty is performed, must be more than six times that which exists when the larger volume is employed. In other words, the reducing gas being less saturated with oxygen in the latter than in the former case, the rapidity with which it withdraws this gas from the ore is proportionately greater. The following figures give the rates at which reduction proceeded, during an exposure of six hours, in the two experiments in question:—

	Temperature.	Calcined Cleveland Ironstone. Oxygen removed. Per Cent.	Lancashire Ore. Oxygen removed. Per Cent.
Slow current of carbonic oxide	770° F. (410° C.)	37·3	35·9 [1]
Rapid ,, ,,	770° F. (410° C.)	50·6	70·9

[1] In this instance the specimen of Lancashire ore from some cause lost less oxygen than the Cleveland ironstone, but as a rule the reverse was the case.

When the temperature at which the ore is subjected to the action of carbonic oxide is low, the action is very languid; and it may be doubted whether any approach to complete reduction under such circumstances would ever ensue. Thus an exposure of calcined Cleveland stone to a current of this gas for $20\frac{1}{2}$ hours, at a temperature of 420° F. (216° C.), failed to withdraw more than 13·65 per cent. of the original oxygen which the mineral contained in combination with iron. On the other hand, as the temperature is raised above that which obtained in the first experiments (770° F. or 410° C.), reduction proceeds at a vastly accelerated speed. Oxide of iron in Cleveland calcined stone lost oxygen as under:—

At a heat visibly red in daylight ... 63 per cent. in 8 hours.
At a bright red heat ... ... 90 „ $3\frac{3}{4}$ „

The change experienced by carbonic oxide (CO) when passed over oxide of iron, is its conversion into carbonic acid (CO_2). When this latter however, at certain temperatures, is passed over metallic iron—particularly iron in a fine state of division, such as what is known as iron sponge—it rapidly loses the equivalent of oxygen it has previously absorbed from the oxide of iron, and is again reduced to the condition of carbonic oxide.

The temperature required for this reoxidation of metallic iron by carbonic acid is notably higher than that necessary for the reduction of the original oxide. At 572° F. (300° C.) there was no action whatever; but at 782° F. (417° C.) a slow current of this gas (CO_2) was almost entirely resolved into carbonic oxide.

We have thus to consider two antagonistic forces, due to the behaviour of the two oxides of carbon (CO and CO_2)[1] in the presence of oxide of iron and of metallic iron: for when oxide of iron is being reduced by carbonic oxide, the presence of carbonic acid retards reduction, and when it is attempted to oxidize metallic iron by carbonic acid, the presence of carbonic oxide in like manner impedes the action.

Under those circumstances it is easy to comprehend that there must be a position of equilibrium, in which the two gases acting in reverse directions, one will neutralize the other. Hence there are certain

[1] To save space the chemical symbols of the substances under consideration will frequently be made use of. Even to those who are not familiar with this mode of expression, its use, from the context, will present no difficulty.

SECTION IX.—CHEMICAL CHANGES IN THE BLAST FURNACE. 185

mixtures of CO and CO_2 which at certain temperatures are without any reducing action on oxide of iron; and others where oxidizing action on metallic iron is suspended.

As the two forces in question are not called into existence by temperatures of the same intensity, nor are influenced in the same degree by each increment of temperature, it follows as a matter of course that the position of equilibrium referred to is altered, as the heat is increased or diminished. The position of equilibrium is further materially affected by the relative quantity of oxide which is exposed to the influence of the mixed gases.

As an illustration of what has just been said, carbonic oxide may be passed over an ore, and be almost, if not entirely, converted into carbonic acid; but then it is only the first and apparently the most loosely held portions of the oxygen which are torn away from the iron. This happened when Cleveland calcined stone in large excess was submitted to a current of carbonic oxide at a temperature of about 800° F. (427° C.); but in $1\frac{3}{4}$ hours only 5·25 per cent. of the oxygen combined with the iron was removed, although the gas on leaving the apparatus was almost wholly composed of carbonic acid.

When the two gases, in the proportion of 2 volumes of carbonic oxide to 1 of carbonic acid, were passed over calcined Cleveland ore, at about the same temperature as in the preceding case, in one experiment only ·9 per cent. of oxygen was lost by the iron oxide in $5\frac{1}{2}$ hours, and in another trial extending over $11\frac{1}{2}$ hours the loss of original oxygen was 4·3 per cent. Here again it was only the first portions of oxygen which were withdrawn. No trace of metallic iron appeared in the ore under treatment. These facts have an important bearing on the actual work of the blast furnace; because the ratio given contains above the maximum average proportion of carbonic acid, as found in the gases escaping from a modern Cleveland furnace, however large its dimensions may be.

We have already seen that at 782° F. (417° C.) spongy iron was rapidly oxidized by carbonic acid. The presence of an equal volume of carbonic oxide however completely stops this action.

If pure carbonic acid (CO_2) be passed over the sponge at a low red heat, it is reduced to the condition of carbonic oxide; but as soon as the two gases arrive at the proportion of 100 CO to 150 CO_2, further change of composition is suspended.

Perfectly reduced ore was exposed to a mixture consisting of about 100 volumes of carbonic oxide to 60 of carbonic acid, at a bright red heat. As soon as the metallic iron had absorbed 8·2 per cent. of the oxygen required to convert it into peroxide of iron, and when the carbonic oxide and carbonic acid were as 100 to 47, all further action ceased; although the experiment was continued for $9\frac{1}{2}$ hours after this position of equilibrium had been reached.

The same experiment was repeated, but with the reduced iron raised to a temperature approaching whiteness. A point of equilibrium was attained, when 100 volumes of carbonic oxide were accompanied by only 11 of carbonic acid, and when the iron had absorbed 11·9 per cent. of the oxygen required to constitute peroxide of iron.

A mixture of equal volumes of the two gases was passed over different specimens of ores, and over artificially prepared peroxide of iron, at a bright red heat, until they ceased respectively to be further affected. This happened when about one-third of the oxygen of the latter was removed, leaving two-thirds still combined with the iron.

A precisely similar trial was made with reduced iron; in this instance oxidation began at once and ceased as soon as the metal had absorbed two-thirds of the oxygen needful for constituting peroxide of iron (Fe_2O_3). In both cases the product was protoxide of iron (FeO).

It would appear then that a considerable excess of carbonic oxide is indispensable for the reduction of the oxides of iron, so that the formula representing the change on the peroxide may be thus expressed :—

$$Fe_2O_3 + 9CO = 2Fe + 3CO_2 + 6CO.$$

in which it will be seen, that there is exactly as much oxygen in the resulting carbonic acid, as there is in the carbonic oxide which escapes unchanged.

The reduction of oxide of iron by carbonic oxide is, however, not quite so simple an action as might be inferred from what has preceded: *i.e.* something more takes place than mere loss of oxygen by the ore, and conversion of a portion of the gas effecting this change into carbonic acid.

SECTION IX.—CHEMICAL CHANGES IN THE BLAST FURNACE. 187

While one portion of the carbonic oxide undoubtedly performs its office as described, another acts in a totally different manner; for it would appear that a portion of this gas is always split up into carbonic acid and carbon. We must suppose that two equivalents of the oxide (2CO) containing 2C and 2O are required for the change, which is represented by the equation

$$2CO = CO_2 + C$$

In the description given of this curious rearrangement of the elements[1] I suppose that the reducing agent must be a lower oxide of iron and that the action was probably of the nature expressed in the formula,

$$Fe_xO_y + zCO = zC + Fe_xO_{y+z}$$

in which it is considered that x is greater than y, in other words that Fe_xO_y is a suboxide of iron, containing therefore less oxygen than the protoxide FeO.

It has been since urged by other writers that the presence of iron in the metallic state is essential for the reaction in question; but my own experiments did not lead me to this conclusion, at the same time it is clear that when the action of carbonic oxide on the oxide of iron has gone on for some time, metallic iron will make its appearance. We shall then have a mixture of the three substances, viz:—

Metallic iron.   Unknown oxide of iron Fe_xO_y.   Carbon.

There will thus be set up among these, together with the carbonic oxide in the furnace, a tendency towards each of the five following reactions:—

A.—Action of CO on Fe_xO_y, producing CO_2 and some lower oxide.

B.—Reaction of this lower oxide on CO, reproducing Fe_xO_y and setting free carbon.

C.—Action of carbon on Fe_xO_y, forming metal and lower oxide, together with CO and CO_2.

D.—Action of metallic iron on CO, reproducing Fe_xO_y or lower oxide, and setting free carbon.

E.—Action of CO_2, set free in above reactions, on carbon, metallic iron, and lower oxide of iron.

It has to be observed that in none of the numerous experiments performed with great care ten years ago, by my then assistant Dr. C. A. Wright, was a product obtained in which oxygen was

[1] *Vide* Transactions Iron and Steel Institute.

entirely absent. At a temperature of about 800° F. (399° C.) carbon is largely deposited, indicating that under these circumstances the tendency towards carbon deposition is much in excess of that towards gasification.

At a bright red heat a different state of things prevails. The known tendency of carbonic acid to reäct on carbon and produce carbonic oxide prevents any large deposit of carbon; but even here a position of equilibrium among the five tendencies referred to is attained, and both carbon and oxygen are present in the compound, which is of such a nature that carbonic oxide has no action on it.

Peroxide of iron (Fe_2O_3), exposed to carbonic oxide at the temperature just mentioned, gave a substance which contained:—

$$Fe\ 99\text{·}33 + O\ 0\text{·}43 + C\ 0\text{·}24 = 100$$

Similar treatment applied to iron *perfectly* reduced by hydrogen from peroxide of iron gave:—

After one hour's exposure ... $Fe\ 99\text{·}20 + O\ 0\text{·}48 + C\ 0\text{·}32 = 100$
After four ,, ... $Fe\ 99\text{·}34 + O\ 0\text{·}30 + C\ 0\text{·}36 = 100$

It appears then, that by the action of pure CO on metallic iron some oxide of iron is formed and free carbon is set free, thus:—

(A.) $\quad a\,Fe + b\,CO = Fe_a\,O_b + b\,C$

On the other hand, a mixture of oxide of iron lower than Fe_2O_3 and carbon furnishes by heat alone metallic iron, CO and CO_2.

(B.) $\quad Fe_m\,O_{2n} + 2n\,C = m\,Fe + 2n\,CO$
(C.) $\quad Fe_m\,O_{2n} + n\,C = m\,Fe + n\,CO_2$

Again it was ascertained that Fe_2O_3 exposed to the action of CO gave CO_2, and a lower oxide, thus:—

(D.) $\quad p\,Fe_2\,O_3 + q\,CO = Fe_{2p}\,O_{3p-q} + q\,CO_2$

in which it was imagined that $3p-q$ was always a positive quantity (*i.e.* is always less than 3 p); or, in other words, that carbonic oxide *alone* is not able to produce metallic iron, that office, if completely performed, is accomplished by deposited carbon in accordance with equations B and C, probably the lowest point to which CO *alone* can reduce Fe_2O_3 is expressed by the composition Fe_2O.

Experiments showed that, up to a bright red heat, the tendencies expressed by equation A predominate over the others, metallic Fe

SECTION IX.—CHEMICAL CHANGES IN THE BLAST FURNACE.

under the influence of CO, at all temperatures up to, and including a bright red heat, becoming more or less oxidized and impregnated with carbon.

On the other hand it seems pretty sure that considerable quantities of carbon may be deposited without the formation of any metallic iron whatever.

Whatever uncertainty may attach to the precise mode of action by which carbon is thus recovered from carbonic oxide, there is no difficulty in ascertaining the conditions favourable for its deposition.

M. Caron apparently imagined that something approaching a red heat was most favourable for the reaction, and M. Schinz sets it down as requiring 800° to 900° C. (1,472° to 1,652° F.) My own observations prove beyond any doubt that it commences with the commencement of reduction, *i.e.* at or a little above 400° F. (204° C.) At this temperature however the deposition of carbon is very slow; but at 469° F. (243° C.) it is very marked, pure peroxide of iron throwing down nearly 1 per cent. per hour, of the weight of iron present in the oxide. And at about 788° F. (420° C.) each 100 parts of iron present were found to be associated on one occasion with 144 of carbon, after an exposure of 7 hours; being a deposition equal to 20 parts per hour.

Different oxides however vary considerably in their power to dissociate carbonic oxide; thus after 6 hours exposure at a temperature of 770° F. (410° C.):—

	Slow Current. 63 Litres.	Rapid Current. 245 Litres.
100 units of iron present in Cleveland calcined stone precipitated of carbon	12·6	22·3
100 units of iron present in Lancashire ore precipitated of carbon	60·1	270·8
100 units of iron present in pure precipitated peroxide of iron	78·7	335·4

As the temperature is raised, the extent to which carbon deposition takes place rapidly diminishes. After 4½ hours exposure of Cleveland calcined stone, at a temperature below redness, to a current of this gas (CO), the carbon thrown down amounted to 64 per cent. of the iron present. After 21 hours exposure, at a heat beginning at red and ending at bright red, the carbon found in a specimen of the same ore only amounted to 2·3 of the metal present in the oxide of iron. And in another trial, when the temperature was increased to very bright redness, the carbon only amounted to ·3 per cent. of the iron present after $4\frac{1}{2}$ hours exposure.

The general result of many experiments pointed to 752° to 842° F. (400° to 450° C.) as being the most favourable range of temperature for this phenomenon of carbon deposition.

The presence of carbonic acid in the gas passed over the specimens, as might be supposed, interferes greatly with the dissociation of carbonic oxide, although reduction may proceed pretty rapidly. The following experiments were performed on calcined Cleveland stone:—

Temperature.	CO Vols.	CO_2 Vols.	Hours of Exposure.	Percentage of O removed.	Carbon per 100 Fe.
525° F. (274° C.)	100	100	6	8·8	Nil.
832° F. (444° C.)	100	100	3	28·7	Nil.
Bright red	100	100	$5\frac{1}{2}$	33·3	Nil.
770° F. (410° C.)	100	50	$10\frac{1}{4}$	10·0	Nil.
Dull red	100	50	6	13·5	1·5
770° F. (410° C.)	100	33	5	4·4	Nil.
Dull red	100	33	5	70·0	1·5

It may be well to state that the trials about to be referred to were undertaken in consequence of the discovery, that not only did different samples of peroxide of iron, according to origin, vary greatly in their power to dissociate carbonic oxide, but that the same variety of ore possessed this property in very different degrees. This, it was conceived, was due to some physical differences among the specimens; because in each the metal was in the state of peroxide. The samples (Cleveland ironstone), submitted to the action of pure carbonic oxide, had been calcined to different degrees of hardness, as a ready means of obtaining variations in their physical structure. The same quantities (2 grammes of each specimen) were placed simultaneously in an iron vessel immersed in melted lead, and a pretty uniform temperature, viz. 770° F. (410° C.), was maintained for 6 hours. During this time 65 litres of the gas were passed through the apparatus in the first experiment, and 213 litres in the second.

The specimens are distinguished by the letters of the alphabet, *a* being the least calcined and *f* the most calcined.

SECTION IX.—CHEMICAL CHANGES IN THE BLAST FURNACE.

	Slow Current.			Quick Current.		
	Oxygen removed. Grammes.	Carbon deposited. Grammes.	Ratio of Carbon deposited to Oxygen removed.	Oxygen removed. Grammes.	Carbon deposited. Grammes.	Ratio of Carbon deposited to Oxygen removed.
a	·047	·160	3·40	·091	·572	6·28
b	·018	·354	3·28	·148	1·005	6·79
c	·104	·326	3·13	·121	1·562	12·89
d	·061	·211	3·46	·201	2·868	14·24
e	·063	·131	2·09	·200	1·125	5·62
f	·025	·006	·13	·104	·659	6·33
Total	·408	1·118	2·58	·865	7·791	8·69

It is not of course to be wondered at, that the amount of reduction and weight of carbon deposited should be more marked in the rapid than in the slower current of the carbonic oxide; but it happens that not only do the 213 litres of resulting gas contain more CO_2 than the 65 litres, but the former was found to contain more CO_2 than the latter per unit of volume. The composition of resulting gases by volume was as follows:—

When 65 litres were passed over samples in 6 hours $4·29 CO_2 + 95·71 CO = 100$ vols.
„ 213 „ „ „ $8·22 CO_2 + 91·78 CO = 100$ „

Other trials afforded similar results as regards the much larger amount of oxygen removed and carbon deposited in the way described; although it did not always happen that the carbonic acid (CO_2) in the rapid current was so much in excess of that in the slow current.

There is no doubt that a very moderate elevation of temperature suffices to expedite the action of carbonic oxide on oxide of iron: thus we have the Cleveland calcined stone losing its oxygen at the following rates:—

Temperature.		Loss of Oxygen.
° F.	° C.	
392	(200)	... First indication.
410	(210)	... Marked action.
440	(227)	... Loss ·28 per cent. per hour of original O combined with Fe.
515	(268)	... „ 1·31 „ „ „
779	(415)	... „ 5·80 „ „ „

On the whole I am inclined to the belief that some slight elevation of temperature accompanies the rapid passage of the gas, owing to a more energetic action than takes place with the slower current, and that this rise of temperature in its turn has promoted chemical action. In the above figures it will be noticed, in the third and fourth examples, that a rise of 75° F. (42° C.) or from 440° F. to 515° F., has increased the reduction above four-fold.

There is no regular correspondence of the amount of carbon deposited with the quantity of oxygen removed, as will be perceived on reference to the table already given. In it the carbon deposited by the slow current varies from ·13 (f) to 3·46 (d) per unit of oxygen; in the rapid current the fluctuation is from 5·62 (e) to 14·24 (d). This is probably determined by some minute differences in the physical condition of the oxide under treatment; for it was distinctly ascertained that such differences greatly affected the results manifested by the simultaneous exposure of different specimens to a current of carbonic oxide, in the same vessel, and all heated alike. The results were :—

	Carbon per 100 of Iron present.	Original Oxygen removed. Per Cent.
Peroxide of iron obtained from calcining sulphate of iron	476·5	72·6
Precipitated peroxide of iron	335·4	80·6
Lancashire hematite	270·8	71·0
Calcined Cleveland stone	22·3	50·7

So far then as my own examination goes, I see no reason to infer that there is necessarily any connection between the rate of reduction and that of carbon deposition. The two actions are probably quite distinct in their operation; and reduction, at least when the mixture of the two gases contains one of carbonic acid to two parts in volume of carbonic oxide, can be effected without any carbon deposition being manifested.

In the experiment performed to prove this, the temperature was 410° C. (770° F.) and the exposure lasted for 5½ hours. A specimen of Cleveland stone lost 2·2 per cent. and one of artificially prepared peroxide of iron lost 7·9 per cent. of their original oxygen, without there being any appearance of deposited carbon.

As already mentioned a question has been raised respecting the conditions needed for this dissociation of carbonic oxide, some maintaining that the presence of metallic iron is indispensable. There is no doubt that reduced iron, when exposed to a current of carbonic oxide, rapidly decomposes the gas in the manner described.

SECTION IX.—CHEMICAL CHANGES IN THE BLAST FURNACE.

At a temperature of 770° F. (410° C.) perfectly reduced iron threw down in 4½ hours, 20·3 per cent. of its weight of carbon.

At a dull red heat, 23·8 per cent. of carbon was found in iron sponge, after being exposed for one hour to carbonic oxide.

It has been demonstrated that, whenever this gas (CO) is passed over metallic iron, a certain amount of the metal is reoxidized. On the other hand I am not disposed to admit that metallic iron is indispensable for setting up the action under consideration, viz. the deposition of carbon. Of course, when a given quantity of oxide of iron has lost, even the first portions of its oxygen, one of two things may have taken place—either there may be some metallic iron present, which has lost the whole of its oxygen or, as is more likely, there has been a partial reduction of a large quantity of oxide of iron, without any being reduced to the state of metal.

I am led to the belief that metallic iron is not indispensable for determining the dissociation of carbonic oxide, from having observed free carbon in oxide when only 7·4 per cent. of the oxygen originally present was removed. It had further been ascertained by Dr. Wright that an aqueous solution of iodine was capable of dissolving metallic iron, leaving the oxides untouched. This treatment enabled us to ascertain, that in many cases of carbon deposition no iron in the metallic form was present. Again, if iron in its perfectly reduced form were needed, it might be expected that spongy iron would split up carbonic oxide more readily than the oxide. The following experiment proves the contrary, although it was performed simultaneously on specimens in the same vessel, which were exposed during a period of nine hours to a temperature of 800° F. (427° C.):—

	Carbon deposited per 100 of Iron present.
Pumice stone saturated with peroxide of iron	808
Anhydrous precipitated peroxide of iron	207
Peroxide of iron from calcined sulphate	206
Spongy iron from peroxide reduced by hydrogen	158

Before leaving the subject of dissociation of carbonic oxide, it may be remarked that the power of certain other substances to split up the gas was also made the subject of experiment. The results are classified as under:—

SECTION IX.—CHEMICAL CHANGES IN THE BLAST FURNACE.

Substances capable of precipitating Carbon from Carbonic Oxide.
Oxide of nickel (NiO).
  ,,   ,, (Ni_2O_3).
Metallic nickel.
Oxide of cobalt (Co_3O_4).
Metallic cobalt.
Titanic acid.

Substances devoid of this power.
Dioxide of manganese (MnO_2).
Manganoso-manganic oxide (Mn_3O_4).
Oxide of copper (CuO).
Metallic copper.
Oxide of zinc (ZnO).
Oxide of tin (SnO).
Oxide of chromium (CrO).
Oxide of lead (PbO).
Platinum.
Silicon doubtful.

The experiments setting forth the results just given were carried on at various temperatures, from that of melting zinc to a red heat, and during different periods of time, often for six hours and more. It may be remarked that nickel and cobalt possess this power of splitting up carbonic oxide in a remarkable degree; and it was proved that a mixture, containing carbonic oxide and carbonic acid in equal volumes, possessed the property of completely reducing both these metals from their oxides at a low red heat.

The next item which demands attention is the power which carbon possesses at elevated temperatures of splitting up carbonic acid: and it is by no means an unimportant one in the economy of the blast furnace. As already stated, one unit of carbon burnt to carbonic acid affords 8,000 calories; instead of which in the action in question (viz. $CO_2 + C = 2CO$) the two units of carbon involved in the decomposition give only $2 \times 2,400$ or 4,800 units, being a direct loss of $\left(\dfrac{16,000 - 4,800}{2} \right)$ 5,600 heat units.

The extent to which so great a loss as that just indicated can take effect is limited of course by the quantity of carbonic acid introduced as such or which can be generated in the furnace. This on 20 units of Cleveland pig iron may be taken to be as follows:—

	Carbon Units.
Carbon raised to condition of carbonic acid by deoxidation of peroxide of iron, regarding 20 units of pig as containing 18·6 units of actual iron	5·97
Carbon raised to condition of carbonic acid by dissociation of carbonic oxide, a similar weight being dissolved in pig iron ...	0·60
Carbon as carbonic acid in the usual quantity of limestone, say 10½ units per 20 units of Cleveland pig iron	1·26
	7·83

SECTION IX.—CHEMICAL CHANGES IN THE BLAST FURNACE. 195

Now the largest average quantity of carbon as carbonic acid I can recollect meeting with in the gases from the Clarence furnaces was 6·52 units per 20 units of pig iron.[1] There was thus a diminution in the carbonic acid of 1·31 carbon units, caused no doubt by the solvent power of carbonic acid on carbon in the fuel.

It is very easy from what has preceded, to estimate the loss, in a calorific point of view, which is entailed by the disappearance of carbonic acid as represented by the 1·31 units of carbon.

Let us assume that the coke consumed per 20 units of iron in the case referred to, was $21\frac{1}{2}$, which may be regarded as equivalent to 20 units of pure carbon. It will hereafter be shown that the carbonic acid contained in the limestone is given off under conditions which render it impossible that it can escape decomposition. The theoretical evolution of heat by the 20 units of carbon is as follows:—

20·00 units of carbon contained in the coke.
1·26 „ deducted as dissolved by carbonic acid of limestone.

		Calories.
18·74 „ burnt in hearth to state of carbonic oxide × 2,400 =		44,976
6·57 „ of the 18·74, the limit to which it is considered as possible for carbon to escape as carbonic acid × 5,600 =		36,792
Total evolution by the extent to which oxidation of carbon can be carried		81,768

For the oxidation as it was in reality effected we have the following figures:—

20·00 units of carbon.
1·31 „ as carbonic acid which have disappeared from the gases carrying with it the same weight of carbon.

		Calories.
18·69 „ burnt to state of carbonic oxide in hearth × 2,400 =		44,856
6·52 „ of the 18·69 burnt to carbonic acid × 5,600 =	...	36,512
0·05 „ burnt in reducing zone to state of carbonic oxide by action of carbonic acid × 2,400 = ...		120
		81,488

---

[1] Care must be taken to discriminate between the *occasional* amount of CO_2 in the gases and the *average* quantity. This subject has been already referred to at the beginning of the present Section.

Showing a loss of under one-half per cent. as compared with the full theoretical quantity of heat considered capable of being evolved.

Soft coke has been already mentioned in Section VII. as an unpopular form of fuel with furnace managers. Certain experiments proving this were set out in detail and need not be repeated in the present place.

This is probably to some considerable extent due to its tendency to be crushed, which by directing the reducing gases into channels prevents their becoming as completely saturated with oxygen as otherwise would happen. There is at the same time no doubt that the power of coke of this description to split up carbonic acid into carbonic oxide exceeds that of the hard silvery description, which is in such favour at our iron works. In addition to the experiments already given specimens of each of the above-named varieties of coke were exposed at a bright red heat (1,560° to 1,580° F.), for the same time, to the action of a current of carbonic acid. In the case of the soft coke 18·7 per cent. of the carbon in the carbonic acid was reduced to the state of carbonic oxide, while in that of the hard coke the proportion of carbon so altered was only 7·3 per cent. of the total quantity.

In addition to the example given in Section VII. upon another occasion, after lighting a large number of new ovens, a considerable quantity of soft coke was delivered to the Clarence works. The quantity of carbonic acid found in the gases, per ton of iron made, fell considerably, and the consumption of coke rose in proportion. Ten analyses of the escaping gases gave the following results by volume:—

	Carbonic Acid.	Carbonic Oxide.	Hydrogen.	Nitrogen.
	9·6	30·6	1·5	58·3
	9·5	30·5	1·5	58·5
	10·5	27·6	1·5	60·4
	10·6	28·4	·9	60·1
	11·0	30·8	·8	57·4
	11·2	30·6	1·0	57·2
	10·0	29·3	1·5	59·2
	10·0	31·3	1·4	57·3
	10·8	27·0	1·4	60·8
	10·8	27·4	1·6	60·2
Average...	10·4	29·35	1·31	58·94 = 100 vols.

SECTION IX.—CHEMICAL CHANGES IN THE BLAST FURNACE.

These figures indicate that the volume of carbonic acid to that of carbonic oxide, instead of being 40 or 45 to 100, as it might easily be, exists only in the proportion of 35·4 to 100, and that the weight of carbon fully oxidized bears a ratio to the lower oxide of 1 to 2·81, instead of a little below 1 to 2 or 1 to 2·2. This difference in the proportions is of course not wholly due to the reduction of carbonic acid to the state of carbonic oxide, but in considerable part to the mere addition of carbon in the extra quantity of coke consumed.

This same loss of heat, consequent upon carbon splitting up carbonic acid into carbonic oxide, also ensues whenever carbon is the reducing agent; because, as in the former case, the heat units evolved are those resulting from the burning of carbon to carbonic oxide. Fortunately the temperature at which this reaction (FeO + C = Fe + CO) takes place, as we have already seen in a previous Section, is considerably higher than that at which reduction of oxide of iron is effected by carbonic oxide, with generation of carbonic acid. In the case therefore of a furnace of sufficient dimensions, working properly with suitable materials, deoxidation of the ore by this gas is almost entirely completed, before its arrival at that region where the temperature is sufficiently intense to enable carbon to reduce oxide of iron. The power of carbon to act on oxide of iron appears, like that of carbon on carbonic acid, to be greatly modified by the physical condition of the coke containing it.

Specimens of calcined Cleveland ironstone were mixed with carbon in the form of coke, graphite, deposited carbon and charcoal; and placed, enclosed in tubes with a capillary opening, in the hot blast pipe of one of the Clarence furnaces, the temperature of which varied from 1,000° to 1,080° F. (538° to 582° C.) After 48 hours of such exposure, the following were the results:—

	Metallic Iron.	Peroxide reduced to Protoxide.	Per Cent. of original Oxygen removed.
Coke including hard and soft, six specimens gave	Nil.	Nil.	Nil.
Graphite	Nil.	Nil.	Nil.
Carbon deposited from carbonic oxide dried	Nil.	2·3	2·49
Deposited carbon, previously ignited	Nil.	·3	·32
Charcoal dried	Nil.	6·00	6·50
,, previously ignited strongly	Nil.	1·60	1·72

In an earlier page of the present Section it was explained that a current of carbonic oxide at a temperature of 770° F. (410° C.) removed 37·3 per cent. of the oxygen contained in Cleveland calcined stone in 6 hours; and that a more rapid current of the gas separated as much as 50·6 per cent. in the same period of time. On the other hand we have just seen that coke, when exposed in contact with the same ore for 48 hours at a temperature of above 1,000° F. (538° C.), is not affected. This difference of results is probably due to the easy penetration by the gas into the interstitial spaces of the ore, which does not happen when solid carbon is the reducing agent. Otherwise it is only reasonable to suppose that carbon, having its power to combine with oxygen in no way blunted by any previous saturation with this element, as is the case with carbonic oxide, must be the more ready of the two to tear it from the oxide of iron. These differences of temperature, at which the two series of actions are set up, explain how a closer approximation to complete reduction may be obtained by carbonic oxide, in a properly constructed furnace, before the materials reach that zone where the heat is intense enough to enable carbon to act on oxide of iron or on carbonic acid. Why one or both the last-mentioned changes may take place in a less perfect form of furnace will be presently explained.

It is however a much more difficult matter to avoid the loss which ensues from the presence of the carbonic acid set free from the flux employed; because, although the generation of the carbonic acid by reduction of the ore is effected at a comparatively low temperature, it is otherwise with the expulsion of the carbonic acid from the limestone. It had been ascertained that caustic lime, when exposed to a red heat in an atmosphere of carbonic acid, became speedily converted into carbonate; it is highly improbable therefore that the limestone, until it reaches a depth in the blast furnace where there is little or no carbonic acid generated, or where the temperature is considerably above a red heat, will part with any notable quantity of the carbonic acid it contains.

It was proved experimentally that, when this acid (CO_2) was passed over a mixture of hard coke, from which all volatile matter had been expelled, with limestone, then the carbonic acid employed suffered partial decomposition before the lime parted with any of its carbonic acid.

SECTION IX.—CHEMICAL CHANGES IN THE BLAST FURNACE.

The facts, as thus stated, point to the possibility of the carbonic acid of the flux being retained by the lime, until it reaches a point where temperature and the presence of carbon render it impossible for this higher oxide of carbon to exist. If this be so, we must regard the loss of heat accompanying the reduction of the carbonic acid of the limestone to carbonic oxide, to be in a great measure, if not entirely, unavoidable. Experience indeed has not demonstrated that any perceptible economy results from the use of quick lime in a Cleveland furnace of modern construction. This is probably due to the readiness with which quick lime absorbs carbonic acid, of itself no doubt a source of heat evolution. The resulting elevation of temperature however will take place so near the top of the furnace that its beneficial effect in heating the materials is lost, thus leaving the absorption of heat, due to the decomposition of the carbonic acid, absorbed by the caustic lime, to cool the contents of the furnace at a point where such cooling is prejudicial.

Having thus considered the circumstances which attend the action of carbon and that of the two carbon gases on the ore and on the reduced iron, as well as on the fuel and flux, we may now proceed to apply the information thus acquired to the blast furnace itself; and to compare the experimental results with those furnished by the actual smelting process.

There does not appear any reason for thinking that, if highly heated carbon meets its equivalent of oxygen, in the proportion to form carbonic oxide, this gas may not be the sole product of combustion. In the hearth of a blast furnace, receiving something like 100 cubic feet of air per second, an instantaneously complete admixture of fuel and oxygen is of course impossible; and accordingly a certain quantity of carbonic acid may be, and is generally, found at or about the spot where the blast enters. Indeed at certain points free oxygen may be occasionally detected.

Samples of the gas were taken on five occasions, at a tuyere from which the blast was shut off, so that roughly the specimens were collected at a distance of $4\frac{1}{2}$ feet in a horizontal line from the tuyeres at which the air continued to be admitted.

SECTION IX.—CHEMICAL CHANGES IN THE BLAST FURNACE.

These samples exhibited the following composition by volume:—

	No. 1.	No. 2.	No. 3.	No. 4.	No. 5.
Carbonic acid ...	·76	1·1	0·8	1·8	1·7
Carbonic oxide ...	37·60	31·7	33·3	35·2	35·0
Hydrogen ...	—	·4	1·0	—	·3
Nitrogen	61·64	66·8	64·9	63·0	63·0
Volumes ...	100	100	100	100	100

On recently repeating the analyses of gases collected near the entry of the blast, a higher percentage of carbonic acid than the above was occasionally obtained.

The following statement contains the results from a furnace of 25,500 cubic feet, producing grey iron:—

	A. At one Tuyere, at which the Blast was shut off.	B. 2 feet higher up.	C. Same as B.
Carbonic acid ...	5·98	5·30	—
Oxygen	·85	—	—
Carbonic oxide ...	33·33	35·60	42·33
Hydrogen	·85	·76	·20
Nitrogen	58·99	58·34	57·47
Volumes ...	100	100	100

Six feet above the tuyeres all trace of carbonic acid, generally speaking, disappears; as may be seen from analyses made at the same time as the above:—

	No. 1.	No. 2.	No. 3.	No. 4.
Carbonic acid ...	0·0	0·0	1·5	0·0
Carbonic oxide ...	35·4	34·4	36·5	34·8
Hydrogen ...	·2	1·3	1·3	·7
Nitrogen	64·4	64·3	60·7	64·5
Volumes ...	100	100	100	100

These observations, respecting the speedy disappearance of carbonic acid from the gases in the vicinity of the tuyeres, have especial reference to a furnace making grey iron. In such cases it seems essential that the reducing energy, as the hearth is approached, should be impaired as little as possible by the presence of an oxidizing substance like carbonic acid. Imperfect reduction, as is well known, gives a black slag which is incompatible with the production of the above-named quality of metal.

SECTION IX.—CHEMICAL CHANGES IN THE BLAST FURNACE. 201

In the charcoal-smelting works of Styria and Carinthia, engaged in the manufacture of *white iron,* carbonic acid appears to occur, in greater or less quantity, in every zone of the furnace. In these, however, the slag is always very much richer in unreduced metal than that accompanying the smelting of grey iron. From this it may be inferred, that the deoxidizing action is less powerful in the hearth of these charcoal furnaces, when making white iron, than it is in a Cleveland furnace, producing grey metal.

In all cases where hydrogen is detected at the depths just given, it may be assumed that it is the result of a decomposition of hygrometric moisture in the atmospheric air entering the furnace or of water accidentally admitted by leakage from damaged tuyeres. For our present purpose however we will consider the intensely heated gas ascending through the furnace, to consist of carbonic oxide mixed with nitrogen, and to communicate, during its passage upwards, by far the greater quantity of its heat to the descending train of solids. Soon of course, the ascending gas has its composition materially altered by its chemical action on the materials it encounters on its way to the point of exit. Although there are considerable fluctuations both in composition and in temperature, we will assume that from a furnace of 80 feet in height the gases leave at an average mean temperature not exceeding 326° C. (620° F.), with carbonic acid and carbonic oxide in the ratio of 42 volumes to 100. The experiments already quoted would prepare us for expecting that the reduction of the ore under such circumstances would, at the point of exit, be very slow and very imperfect, and that carbon deposition would barely manifest itself. The following observations to determine this were made on a furnace having a capacity of 25,500 cubic feet, by exposing calcined Cleveland stone to the action of the escaping gases, under the above-named conditions as to temperature and composition:—

Hours of exposure.	Loss of original Oxygen. Per Cent.	Deposited Carbon.
24	2·0	Trace.
48	2·9	,,
72	13·0	,,
96	13·0	,,

As might be looked for, reduction is accelerated as the temperature increases and the volume of carbonic acid diminishes—two conditions which are secured by exposing the ore to the action of the gases lower down in the furnace, which in this case was one of 48 feet in height:—

Hours of exposure.	Distance from top.	Approximate Temperature. ° F. ° C.	Vols. CO_2 per 100 Vols. CO.	Loss of original Oxygen. Per Cent.
2	4·3 feet	800 ... 426	23	11·8
2	9·9 „	1,250 ... 676	7	76·2
2	15·7 „	1,600 ... 865	2	76·3
2	21·5 „	1,800 ... 982	3	81·7

Carbon deposition took place upon two different occasions, under the circumstances given below, on exposing calcined Cleveland stone to the action of the escaping gases:—

Hours of exposure.	Approximate Temperature. ° F. ° C.	Vols. CO_2 per 100 Vols. CO.	Carbon per 100 of Fe present.
24	700 ... 371	19	1·96
24	800 ... 426	19	2·40

The intensely heated gases, which ascend from the hearth, having to perform the duty of reducing the ore as well as to heat the materials, the point which has to be kept in view in the construction of the blast furnace is, that its capacity should suffice for the heated gases to be retained among the solids long enough to communicate to the latter as much of their sensible heat as is possible, and to become as completely saturated with oxygen, as the nature of the chemical action will permit. These two objects seem to be attained in the Cleveland district, when the furnace has a height of 80 feet, with an interior capacity of about 12,000 cubic feet. These dimensions enable the gases to be cooled down to such a point, and to be so saturated with carbonic acid, as to retain little or no further power of reducing the ironstone of the country. The correctness of these views has been made apparent by the fact that furnaces, mentioned as having been constructed upwards of 103 feet high, and others nearly as lofty, with a capacity of 40,000 cubic feet, have failed to show any marked advantage, so far as economy of fuel is concerned, over that which has been frequently obtained in those of more moderate dimensions.

Let us now consider the structural error in a furnace to be in the opposite direction; *i.e.*, that instead of being unnecessarily high and capacious, it is insufficient in both respects.

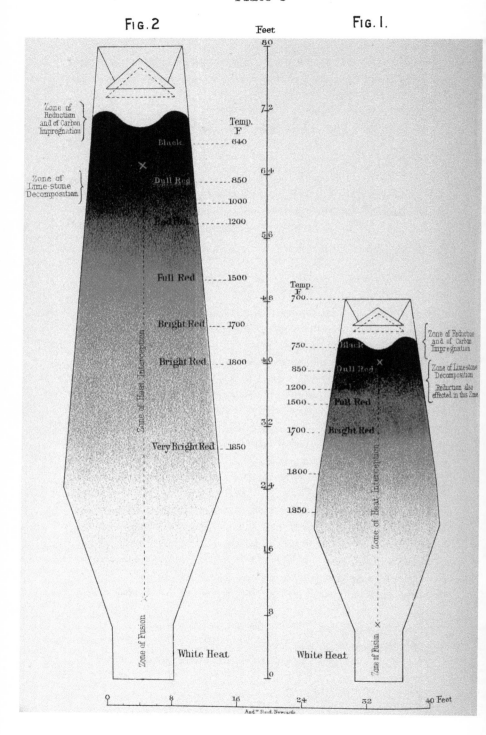

SECTION IX.—CHEMICAL CHANGES IN THE BLAST FURNACE. 203

The accompanying sketches are intended to afford a general idea of the temperatures of the materials filling two furnaces—one 80 feet high with boshes of 20 feet, and the other 47 feet in height with boshes of 16 feet. The capacities are 15,400 and 6,000 cubic feet respectively.

By drilling holes in the sides of the two structures the temperature of each was roughly ascertained, and the figures denoting these are inscribed along-side of each of the drawings. The indications of temperature given in the sketches must only be considered as comparative; for it is very possible that the contents of the furnaces at a distance from the walls may be hotter than is indicated by the figures. Indeed, looking at the rate at which reduction is carried on in the first $16\frac{1}{2}$ feet of the height of the larger furnace, of which space not above one-half on an average is actually occupied by the materials, it seems very probable that the interior of the mass is somewhat more highly heated than I have supposed.

The advantages possessed by the larger furnace are firstly, that the gases pass away cooled, as far as it is practicable to effect this; and secondly that the deoxidation of the oxide of iron is performed in a portion of the furnace where the temperature is so low, as to avoid as much as possible the carbon acting on the carbonic acid generated by the act of reduction. The comparative magnitude of this zone of moderate temperature in each furnace can easily be appreciated by an inspection of the two sketches. In the larger one the contents do not exhibit a dull red heat until a depth of 16 or 17 feet is reached, whereas in the other this temperature manifests itself at a depth of about 9 or 10 feet. In each case the distance is reckoned from the charging plates.

It might be supposed at first sight that the conditions of the two furnaces could be brought into harmony by diminishing the rate of driving in the lesser. The increased period of time however, during which the ore would then be exposed to the action of the heated gases, only brings the hotter zone nearer the top of the furnace, and an actual trial of slower driving, extended over some time, induced me to think that there was no gain to be expected from a change of that kind.

A comparison between two furnaces, of 11,500 and 6,000 feet respectively, taken from actual experience, is perhaps the simplest mode of pointing out the points of dissimilarity.

Height of furnace, feet ...	80 ...	48
Diameter of boshes, do. ...	17½ ...	16
Capacity, cubic feet ...	11,500 ...	6,000
Coke per 20 units of iron ...	units 22·32 ...	28·92
Ironstone do. ...	„ 48·80 ...	48·80
Limestone do. ...	„ 13·66 ...	16·00
Weight of blast	„ 103·74 ...	128·12
Do. escaping gases ...	„ 138·66 ...	170·59
Temperature of blast... ...	485° C. ...	485° C.
Do. escaping gases	332° C. ...	452° C.

From this statement it will be perceived that in the smaller furnace there is a greater weight of blast consumed for a given quantity of metal, and a larger volume of gases emitted, owing to the larger weight of coke burnt as compared with the larger furnace. The use of more coke demands more limestone, to flux the ash it contains; which necessitates further increase of coke, from the heat required to decompose the carbonate of lime, as well as to provide carbon to split up the further quantity of carbonic acid liberated. Besides this, additional fuel is rendered necessary, in order to melt the larger quantity of slag thus formed.

The following statement exhibits the mode by which the performance of the larger blast furnace was ascertained:—

Seven specimens of gases were collected during 3 hours and 20 minutes, so as to avoid the fluctuations in composition already spoken of; and the analyses, to ensure accuracy, were all made in duplicate.

By Volume	CO_2.	CO.	H.	N.
Minimum proportion of CO_2	10·	28·1	·9	61·
Maximum do.	12·9	28·8	·7	57·6
Actual average of the 7 trials	11·7	26·7	·7	60·9

From these figures the average composition by weight was calculated, as under:—

Composition of Gases by Weight.			Carbon.	Oxygen.	Nitrogen.	Hydrogen.
CO_2	17·30	containing	4·72	12·58	—	—
CO	25·20	do.	10·80	14·40	—	—
H	·10		—	—	—	·10
N	57·40		—	—	57·40	—
	100	containing	15·52	26·98	57·40	·10

SECTION IX.—CHEMICAL CHANGES IN THE BLAST FURNACE.

The hydrogen, being insignificant in weight, is neglected in the subsequent calculations bearing on the subject. The heat developed by the oxidation of the carbon is found as follows:—

	Units.		
Coke used per 20 units of iron	22·32		
Less ash, etc., 1·92, gives actual carbon	20·40	Carbon in coke and limestone	22·04
Carbon in limestone, carrying off an equal weight of C from coke[1]	1·64		
	18·76		

Heat Units.
Giving from carbon burnt to CO ... ... 18·76 × 2,400 = 45,024
Carbon of this CO burnt to CO_2 ... ... 6·52 × 5,600 = 36,512

Total development by combustion of carbon 81,536

The weight of carbon contained in the gases *per* 20 units of iron is equal to the total carbon made use of 22·04, less ·60 dissolved in iron, = 21·44 units in gases.

Referring this weight (21·44) to the above analysis, we obtain:—

		Carbon.		Oxygen.
Total nitrogen per 20 units of iron (15·52 C : 57·40 N :: 21·44 C)	79·30			
CO_2 (57·40 N : 17·30 CO_2 :: 79·30 N)	23·89	= 6·52		17·37
CO (57·40 N : 25·20 CO :: 79·30 N)	34·81	= 14·92		19·89
Water in coke ...	·58			
H in moisture of blast ...	·08			
Total weight of gases ...	138·66	containing 21·44		37·26

The weight of blast is : 79·30 N + associated atmospheric oxygen 23·70 + moisture ·74 = 103·74.

Calories or Heat Units.
The heat units in the heated blast are:—
103·74—weight of air × temperature 485° C. × specific heat of air ·237 = 11,919

The heat in the escaping gases is:—
138·66—weight of gases × temperature 332° C. × average specific heat ·24 = 11,043

From this an allowance was made for heat of ironstone fresh from the kilns, and latent heat of steam, reducing the figure to 8,860 calories in the escaping gases.

---

[1] This arises, as formerly explained, by the carbonic acid of the limestone, not being liberated until it reaches a zone of the furnace, where it dissolves and carries off carbon as carbonic oxide, $CO_2 + C = 2CO$.

SECTION IX.—CHEMICAL CHANGES IN THE BLAST FURNACE.

The factors of the work obtained at the lesser furnace were calculated in the same way; it will therefore be sufficient to call attention to the higher temperature of the escaping gases—452° C. against 332° C.—and to the fact that they contained much less carbonic acid than those of the larger furnace. Their composition was, by weight,

CO_2.	CO.	H.	N.
11·8	30·5	·1	57·6 = 100

The heat developed by the combustion of carbon was calculated at 89,288 calories on 20 units of iron. The heat contained in the gases was 16,409 calories. From these figures it will be at once apparent how much less perfectly the heating power of the fuel is utilised in the smaller furnace.

	80 Ft. Furnace.	Calories.	48 Ft. Furnace.	Calories.
Coke used per 20 units of iron	22·32		28·92	
One unit of coke develops	$\dfrac{81{,}536}{22{\cdot}32}$	= 3,653	$\dfrac{89{,}288}{28{\cdot}92}$	= 3,087
Sensible heat of escaping gases per unit of coke	$\dfrac{8{,}860}{22{\cdot}32}$	= 397	$\dfrac{16{,}409}{28{\cdot}92}$	= 567
Leaving calories utilized in the furnace per unit of coke consumed		3,256		2,520

When proper allowance is made for the different conditions alluded to, as obtaining in the two furnaces, the ultimate figures will be found to correspond very fairly. Per ton of pig iron we have :—

	Furnace of 80 feet. Calories.	Furnace of 48 feet. Calories.
Heat evolved by oxidation of carbon...	81,536	89,288
Do. contributed by heat in blast	11,919	14,724
	93,455	104,012
Less carried off in the gases ...	8,860	16,409
Net heat units left for furnace requirements	84,595	87,603
Extra heat required in lesser furnace for fusion of slag, decomposition of limestone, etc, the subtraction of which brings both sets of figures as nearly to the same value as can be expected in such a calculation		3,858
		83,745

SECTION IX.—CHEMICAL CHANGES IN THE BLAST FURNACE. 207

Connected with all the changes which take place in the composition of those constituents of the ore which are found in the pig iron, deoxidation may be regarded in the light of a preliminary process; and by estimating the quantity of oxygen beyond that brought in by the atmospheric air, an opinion can be formed of the extent of the change effected at any particular point of the furnace from which the specimen of gas may be taken.

Such calculations as those now under consideration can at best only be approximate, from the unavoidable differences arising from time to time in the composition of the materials employed; as well as from trifling variations in the constitution of the products, and in the character of the chemical action in operation, at any particular period as compared with another.

The following calculations apply to a furnace 80 feet high containing 17,500 cubic feet, and consuming, per 20 units of pig iron, 12·8 units of limestone and 23·5 units of coke containing 90 per cent. of carbon. The full quantity of carbon delivered to the furnace may be assumed to have been:—

	Units per 20 Units of Iron.
That in 23·5 of coke less ash, etc., 2·35 ... ... =	21·15
Do. 12·80 of limestone	1·53
	22·68
Less dissolved in the iron	·60
Leaving in the gases	22·08

The total oxygen leaving the throat of the furnace in the gaseous form per 20 units of Cleveland iron was as follows:—

		Units.
Oxygen brought in by the blast, including ·66 in the atmospheric moisture		23·91
Separated from oxide of iron, chiefly at moderate temperatures, 7·97, less ·07 considered as inseparable and passing into the slag		7·90
Separated from the limestone in combination with carbon...		4·09
Separated from phosphoric, sulphuric, and silicic acids, the sources of the P, S, and Si found in the pig	·75	—
Oxygen separated from lime to form sulphide of calcium found in the slag	·28	
		1·03
Total oxygen in escaping gases by calculation ...		36·93

208   SECTION IX.—CHEMICAL CHANGES IN THE BLAST FURNACE.

The figures which follow, exhibit the quantities of oxygen per 20 units of iron, found in the gases at different levels of the furnace. They were obtained by ascertaining, from analysis of gas specimens, with how much oxygen the same quantity of nitrogen was associated, which latter was estimated as being present in the escaping gases per 20 units of iron, and may be considered, for purposes of calculation, as being a constant quantity throughout.

EXAMPLE No. 1.

Per 20 Units of Iron. Units.

A.—The actual quantity of oxygen estimated in the escaping gases by analysis was found ... ...   36·87
Thus showing a very trifling difference between it and the previous number, viz. 36·93.
B.—At a depth of 16½ feet from top, the quantity of oxygen was...   25·09
C.—   Do.   26   do.   do.   ...   24·71
D.—   Do.   39   do.   do.   ...   24·69
E.—   Do.   52¼   do.   do.   ...   24·72
F.—   Do.   65   do.   do.   ...   24·14
G.—   Do.   70½   do.   do.   ...   23·74
H.—   Do.   76½   do.   do.   ...   26·97

The ascertainment of carbon at the various levels in the furnace, was also calculated by a reference to the nitrogen present in the gases.

Units.

A.—The quantity of carbon in escaping gases per 20 units of iron was   22·08
B.—   Do.   16½ feet from top   do.   17·29
C.—   Do.   26   do.   do.   16·33
D.—   Do.   39   do.   do.   17·42
E.—   Do.   52¼   do.   do.   18·09
F.—   Do.   65   do.   do.   17·98
G.—   Do.   70½   do.   do.   17·80
H.—   Do.   76½   do.   do.   19·35

It will be observed that in both sets of figures just given there is a decrease of oxygen and carbon, below the original quantities estimated as present in the escaping gases, down to a certain point, somewhere between F and G. After this point both oxygen and carbon commence to increase, and they rise to a greater quantity than at any of the

SECTION IX.—CHEMICAL CHANGES IN THE BLAST FURNACE.

higher levels of the furnace except the point of final escape. This was found invariably to be the case, as may be seen from the following additional examples from an 80 feet furnace :—

Depth from Top.	Oxygen Present per 20 Units of Iron.			Carbon per 20 Units of Iron.		
Example	No. 2.	No. 3.	No. 4.	No. 2.	No. 3.	No. 4.
Flux used..	(Raw Limestone.)	(Calcined Limestone.)	(As No. 3.)	(Raw Limestone.)	(Calcined Limestone.)	(As No. 3.)
A { Escaping gases. }	36·87	33·40	33·40	22·08	20·96	20·96
	Feet.					
B	16½	Not given.	26·19	25·43	Not given.	18·53	17·94
C	26	28·21	24·55	24·43	19·28	18·15	17·96
D	39	29·65	25·21	26·01	20·92	18·36	18·35
E	52¼	23·81	26·24	Not given.	17·88	18·92	Not given.
F	65	24·14	25·13	25·23	17·75	18·62	18·52
G	70½	28·26	25·07	25·70	20·43	18·79	19·55
H	76½	27·33	27·56	27·74	19·55	20·27	20·43

The data upon which the preceding calculations were based were obtained about ten years ago, the analyses being made by Dr. C. A. Wright on the gases in a furnace containing 17,500 cubic feet.

Further experiments were undertaken by Mr. Rocholl, the present chemist at the Clarence works, in January, 1880. The gases were drawn from nine levels at four different periods during one and the same day, and the averages of the four different specimens taken at each of the nine points were carefully analysed by that gentleman.

The furnace from which these last named samples were taken is 80 feet high, with a content of 25,500 cubic feet. A space at the top of not less than 9 feet 2 inches is occupied by the cup and cone, and the room required for lowering the cone. The furnace therefore, when "full," contains a column of material, including the pig iron and slag, 70 feet 10 inches in height.

The furnace was making No. 3 iron. The consumption per 20 units of iron was as follows :—

	Coke.	Calcined Ironstone.	Limestone.
Units	22·44	47·45	10·91

SECTION IX.—CHEMICAL CHANGES IN THE BLAST FURNACE.

EXAMPLE 5.

Composition by weight of gases:—

Distance from Top	At Escape Pipe.	11 Ft. 6 Ins.	17 Ft. 9 Ins.	24 Ft. 1 In.	30 Ft. 5 Ins.	36 Ft. 8 Ins.	42 Ft. 11 Ins.	49 Ft. 9 Ins.	74 Ft.
Carbonic acid	16·07	11·71	10·03	8·17	6·12	—	—	·72	3·01
Carbonic oxide	27·34	29·71	31·39	31·40	32·79	35·27	36·00	36·02	39·47
Hydrogen	·11	·10	·07	·14	·28	·10	·11	·08	·14
Nitrogen	56·48	58·48	58·51	60·29	60·81	64·63	63·89	63·18	57·38
	100·	100·	100·	100·	100·	100·	100·	100·	100·
Carbon	16·09	15·94	16·18	15·69	15·72	15·12	15·43	15·44	17·72

Units per 20 of iron:—

Oxygen	36·48	32·86	32·53	29·88	28·76	23·52	24·29	25·20	32·55
Carbon	21·49	20·56	20·86	19·62	19·50	17·64	18·22	18·43	23·29
Nitrogen	75·43	75·43	75·43	75·43	75·43	75·43	75·43	75·43	75·43

The factors for this series were obtained as follows:—

Carbon per 20 units of iron, coke 22·44 less 7 per cent.
 ash, etc. ... ... 20·87
 Do.   do.   in limestone ... ... 1·30
                                                          ——— 22·17
Less dissolved in iron, containing 3·4 per cent. ... ...   ·68
      Weight of carbon in gases ... ... 21·49

Weight of carbonic acid, carbonic oxide and nitrogen per 20 units of iron:—

Nitrogen   (16·09 : 56·48 :: 21·49 :)   ...   ...   75·43
Carbonic acid (56·48 : 16·07 :: 75·43 :)   ...   ...   21·46
Carbonic oxide (56·48 : 27·34 :: 75·43 :) ...   ...   36·51

	Carbon.	Oxygen.
Units in escaping gases per 20 units of iron, in CO_2 ...	5·85	15·61
Do. do. do. in CO ...	15·64	20·87
	21·49	36·48

Oxygen due from various sources as follows:—

Brought in with 75·43 of atmospheric nitrogen   ...   22·82
From moisture in blast ... ...   ·65
From oxide of iron, phosphoric acid, silica, etc., in ore...   9·07
In carbonic acid of limestone ... ...   3·49
                                                          ——— 36·03
      Difference for experimental error ... ...   ·45

SECTION IX.—CHEMICAL CHANGES IN THE BLAST FURNACE.

The oxygen contributed by the calcined ironstone, 9·07, is computed from an average of several recent analyses, as follows:—

Combined with 18·64 units of iron in 20 units of pig	… …	7·98
Do.	·33 of phosphorus as phosphoric acid … …	·43
Do.	·34 of silicon as silica … … … …	·39
Do.	·008 of sulphur as sulphuric acid … …	·01
Do.	calcium which forms sulphide of calcium in slag..	·26
		9·07

It will be observed that in this instance (*Example No.* 5) the increase of oxygen and carbon at the tuyeres exceeds that in any of the previous experiments. The four specimens from the lowest point which served for the analysis were taken from a perforation made through the *tymp*, and represented therefore the composition of the gases next to the wall of the furnace. In order to ascertain whether the blast, rushing as it does towards the centre of the hearth, generated more carbonic acid there than was found nearer the exterior of the minerals, a tube was thrust into the mass, and samples rapidly collected before the tube was melted by the extreme heat.

The following was the information obtained in this way, in five separate trials a, b, c, d, and e:—

EXAMPLE 6.—IN HEARTH.

	Composition of Gas by Weight.					Units of Oxygen and Carbon per 20 Units of Iron.		
	O.	CO_2	CO.	H.	N.	O.	C.	N.
a …	·95	9·14	32·44	·06	57·41 = 100 …	33·09	21·53	75·43
b …	—	16·68	30·61	1·25	51·46 = 100 …	43·41	25·88	75·43
c …	—	8·14	34·79	·05	57·02 = 100 …	34·13	22·66	75·43
d …	—	—	42·41	·01	57·58 = 100 …	31·75	23·80	75·43
e …	—	—	36·71	·14	63·15 = 100 …	25·05	18·79	75·43
				Average … …		33·48	22·53	

	Oxygen.	Carbon.	Nitrogen.
The computed weights of oxygen and carbon in the escaping gases from the furnace being as formerly stated … … … … … …	36·48	21·49	75·43

The construction of the furnace whose gases were examined in *Examples* 5 and 6 did not permit the samples being taken off precisely at the same points as those formerly supplying the specimens. Both furnaces were 80 feet in height, but of different diameters. The specimens E (*v.* p. 209) taken at a place about 52 feet from the top in the first four examples, Nos. 1 to 4, and in the fifth (*v.* p. 210) the sample obtained from an aperture 49¾ feet below the same point may fairly be considered as representing corresponding zones of the two furnaces. The lower one, from which the gas was collected, was 22 to 24 feet below the higher one.

The comparison of the quantities of oxygen and carbon per 20 units of metal at this point, with those present 22 to 24 feet lower down, in the neighbourhood of the tuyeres, is as under:—

	OXYGEN.			CARBON.		
	Higher Point.	Lower Point.	Increase at Lower.	Higher Point.	Lower Point.	Increase at Lower.
Examples 1 to 4 average	24·92	27·40	2·48	18·29	19·95	1·66
Example No. 5 ...	25·20	32·55	7·35	18·43	23·29	4·86
Example No. 6 average	—	33·48	—	—	22·53	—

Any one who gives himself the trouble to examine carefully the analyses of the gases in the six examples selected for illustration, will perceive considerable irregularities in the results they disclose. This is no more than might be expected, looking at the everchanging action going on in particular localities of the interior of a blast furnace. It is however perfectly clear, looking at the whole of these experiments as well as those of former authorities, that oxygen does appear at the hearth in excess of that computed to be due by the atmospheric air; and that both this gas and carbon are at a minimum at a point about 40 feet below the top of the furnace experimented on, above which they gradually increase.

It is worthy of remark that Ebelmen gives three analyses, and Tunner and Richter one, which confirm the observations just given. The first named authority, and after him my friend Dr. Percy, made this increase in the oxygen at the tuyeres above that corresponding with the nitrogen of the blast the object of close investigation, without however arriving at any definite conclusion in reference to the cause.

Ebelmen had noticed that the oxygen at the tuyeres was in excess of that corresponding with the nitrogen of the blast. To some extent this is accounted for by:—

1st.—The moisture in the atmospheric air.

2nd.—The decomposition of silica, phosphate of lime, etc., which probably takes place in the hottest part of the furnace.

SECTION IX.—CHEMICAL CHANGES IN THE BLAST FURNACE. 213

After allowing however for both these sources of oxygen, I have found a further excess, which cannot be accounted for by any of the above mentioned causes.

It had also been observed by Ebelmen that this excessive amount of oxygen was not to be found in the gases taken at a short distance above the tuyeres; but when I came to examine the general composition of the gases taken from an 80 feet furnace I noticed that there was, at certain points, also a very material difference in the quantity of carbon, in relation to the nitrogen, from that which calculation would lead us to expect should be present. Of this circumstance I have not found any mention in the writings of previous observers.

The total quantity of oxygen which can be accounted for as an increase at the hottest part of the furnace, is that separated from the sulphur, phosphorus, silicon and calcium, together with that in the minute quantity of oxide or iron in the slag, ascertained to arrive very near the tuyeres in an unreduced form.

In one instance the oxygen from the sources just quoted was estimated as follows, per 20 units of iron:—

From P_2O_5 ·43, SiO_2 ·39, SO_3 ·01, CaO ·26	= 1·09
If there is 1·20 per cent. of iron in the slag as it arrives in the hearth, this has to be reduced, and will supply ·34 of oxygen, giving per 20 of pig say	·51
Total oxygen separated at or near the hearth ...	1·60

This calculation of course only assigns a reason why the oxygen in the gases should show an increase *below* the point where none of the causes just enumerated are supposed to have come into active operation; and does not in any way account for any apparent excess, which uniformly seems to be present at this point of the furnace, and which can scarcely be accounted for by accidental admissions of water. Neither does it in any way explain the excess in reference to the carbon which invariably manifests itself in the gases below a certain level.

A reference to one of the examples formerly quoted will show the amount of carbon and oxygen in the gases taken at the tuyeres, for every 20 units of iron produced. This, it will be observed, is considerably larger than that obtained from the gases, produced for a similar quantity of metal, taken at a point 6 feet higher in the furnace.

Example No. 1 (formerly given) contained of carbon and oxygen at different levels of an 80 feet furnace per 20 units of iron:—

Depth—Feet	A, at Exit. 16½	B. 26	C. 39	D. 52¼	E. 65	F. 70½	G.	H, at Tuyeres. 76¼
Carbon	22·08	17·29	16·33	17·42	18·09	17·98	17·80	19·35
Oxygen	36·87	25·09	24·71	24·69	24·72	24·14	23·74	26·97

Now in this particular case the carbonic acid which ought to be found in the gases would be (exclusive of any due to carbon deposition) as follows:—

	Units.	Units.
Formed by deoxidation of the peroxide of iron ...	21·92	
Separated from the limestone	5·62	
		27·54
Whereas there was only		20·38
Showing that there was of carbonic acid decomposed		7·16

This quantity (7·16) of carbonic acid contains 1·95 of carbon, which would, by its reduction to carbonic oxide, carry off a similar weight of carbon. The weight, per 20 units of iron, which we should expect to find at the tuyeres, would be therefore:—

$22·08 - (2 \times 1·95) \; 3·90 \quad ... \quad = 18·18$
Whereas there was in reality ... $\quad 19·35$—Excess 1·17 units.

The oxygen which ought to be at the tuyeres, in the absence of any disturbing cause, is that supplied by the blast and reduction of silica, etc., together 25·51 units; but there were really 26·97 or an excess of 1·46 units.

Now, without pretending to describe the exact manner in which the subsequent absorption of this 1·17 of carbon and 1·46 of oxygen takes place, we may suppose that a portion of the former is due to a deoxidation of a small quantity of alkali; the following equation would then account for the disappearance of the two elements:—

$1·17C + 1·46O + xNaOKO + yNaK[1] =$
$·55C + 1·46O + xNaKO$ (forming carbonates of potash and soda).
$·62C + ·72N + yNaK$ (forming cyanides of potassium and sodium).

In the absence of anything interfering with the oxygen, the moment this element falls in quantity to that represented by the blast and metalloids, etc., viz. 24·81 units, we may assume that the oxygen

---

[1] The weights of NaOKO and NaK, are not given because they vary with the relative quantities of potassium and sodium and their oxides.

SECTION IX.—CHEMICAL CHANGES IN THE BLAST FURNACE.

supplied by the ironstone and the carbonic acid of the limestone has been all expelled. At C, 26 feet from the top, the oxygen is 24·71; showing that a trifle less of oxygen (·10) is in the gases here than the equivalent just named, *i.e.* 24·81.

At B however it will be remarked that, while the oxygen is ·28 in excess (25·09 — 24·81), the carbon is actually deficient by ·89 units (18·18 — 17·29). Taking 18·18 and 24·81 as the standard quantities of carbon and oxygen respectively, the following shows the position of both at the various points per 20 units of iron :—

EXAMPLE 1.

Depth from Top, feet	B. 16¼	C. 26	D. 39	E. 52¼	F. 65	G. 70½	H. 76¼
Carbon deficient	·89	1·85	·76	·09	·20	·38	—
Do. excess	—	—	—	—	—	—	1·17
Oxygen deficient	—	·10	·12	·09	·67	1·07	—
Do. excess	·28	—	—	—	—	—	2·16
Units of carbonic acid per 20 units of pig iron	5·13	7·97	4·23	1·67	·59	·00	3·22

In so large a mass of material as that which fills a blast furnace, with ever varying currents of the reducing gases passing through its various portions, it is impossible to do more than estimate generally, from a great number of analyses, the character of the action which is going on at any moment of time. As an illustration of the changes of composition in the gases, a similar Table to that last given has been constructed from the figures of Example 2 already quoted and also taken on 20 units of pig iron.

EXAMPLE 2.

Depth from Top, feet	B. 16½	C. 26	D. 39	E. 52¼	F. 65	G. 70½	H 76½
Carbon deficient	—	—	—	·30	·43	— —	—
Do. excess	—	1·10	2·74	—	—	2·25	1·37
Oxygen deficient	—	—	—	1·00	·67	—	—
Do. excess	—	3·40	4·84	—	—	3·45	2·52
Units of carbonic acid per 20 units of pig iron	—	6·89	4·88	·00	1·31	2·91	3·46

Thus it will be seen that, while the whole of the carbon brought in by the limestone, together with the oxygen combined with it and that combined with the iron in the iron ore, has disappeared soon after passing B in Example I., both are present in considerable quantity in

Example II., until after we pass D. The differences in detail do not however affect the principle viz., that after the oxygen combined with the iron, and the carbon and oxygen forming the carbonic acid in the limestone, or their equivalents in point of quantity, disappear from the gases, the amount of these two elements falls below that which ought to be found there, even after making allowance for the carbonic acid of the limestone carrying off a weight of carbon equal to its own. After this diminution in point of amount, we have the excess of carbon and oxygen reappearing at the tuyeres, to the extent given below in the two cases under consideration, per 20 units of iron as before:—

	Carbon.	Oxygen.
Example 1	1·17	2·16
„ 2	1·37	2·52

In these numbers it will be seen that the proportion of carbon to oxygen is almost precisely the same; for

$$1{\cdot}17 \ : \ 2{\cdot}16 \ :: \ 1{\cdot}37 \ : \ 2{\cdot}529$$

These figures indicate a ratio of carbon to oxygen of 1 to 1,846 and 1,839 respectfully, which is something under 1½ equivalents of oxygen for each equivalent of carbon.

Many years ago Bunsen and Playfair mentioned the fact that they had detected 2·6 grammes of cyanide of potassium per cubic metre of the gases of a blast furnace; and at the same time stated their belief that four times as much as this quantity was condensed in the pipes of their apparatus.

The presence of this compound of carbon, nitrogen and potassium which subsequently has been proved to be a never failing accompaniment—in the form of vapour—of the gaseous current up to, or within a short distance of the charging plates, is not indicated in the customary analyses of blast furnace gases, which comprise only the constituents of a true gaseous nature. We might therefore infer that a quantity of nitrogen, equivalent to the oxygen which was in excess, had disappeared from among the gases, carrying with it its equivalent of carbon (14 N. to 12 C.) to form the cyanogen compounds. Again, this very action would of itself account for a portion of the excess of oxygen, because it is not free cyanogen that we have to deal with, but this substance combined with potassium and sodium. Every 14 parts

SECTION IX.—CHEMICAL CHANGES IN THE BLAST FURNACE.

therefore of nitrogen, so disappearing, would liberate 8 of oxygen from potash or soda, these alkalies, as is well known, being present in the minerals used in the furnace. The oxygen, so set free, also accounted, in its turn, for a part of the excess of this element; thus bringing about the disturbances in the relative quantities of carbon and nitrogen.

The quantity of cyanides however, mentioned by Bunsen and Playfair, seemed to me insufficient to account for the discrepancies just referred to; and I had numerous experiments made to ascertain the amount of these alkaline salts in the very furnaces upon the action of which my calculations are founded. No doubt there was some irregularity in the quantities so determined on different occasions, but I consider that the results of the investigations justify the conclusion that the cyanogen compounds, at all events in the furnaces which were examined, greatly exceeded in amount that mentioned by those distinguished chemists.

The following Table gives the weight of these salts in grammes per cubic metre of gas, taken at a point about 8 feet above the tuyeres, and $68\frac{1}{2}$ feet from the top of the Wear furnace, of 17,500 cubic feet:—

1870, July	14th.	15th.	19th.	22nd.	26th.	28th.	Average.
Potassium and sodium in combination with carb. acid, oxygen, or cyanogen	46·69	30·17	33·15	21·09	31·65	11·88	29·11
Cyanogen	19·00	12·93	17·32	11·34	20·61	9·16	15·06

In the escaping gases the quantity of these substances on some of the days was as follows per cubic metre of gas:—

July, 1870.	15th.	19th.	22nd.	26th.	28th.	Average.
Potassium and sodium combined as before (grammes)	11·20	15·30	8·68	5·89	4·29	9·07
Cyanogen	4·00	6·60	3·57	2·91	1·79	3·77

The data upon which the quantities just enumerated were calculated were obtained from two hundred litres of gas withdrawn in 45 minutes, and from this the whole of the solids was condensed.

Subsequent experiments on one of the Clarence furnaces of 25,500 cubic feet gave a considerably less quantity of alkaline salts, the weight per cubic metre not having exceeded 13·61 grammes. Potash and soda were detected in the escaping gases, but the cyanogen appeared to have been decomposed before reaching this point. In the later trials the period of collection was extended over 5 hours, but the quantity of gas collected was only 60 litres.

The gradual decrease in the average quantity of cyanides in the gases, during their ascent in the furnace, is apparent from the following series of observations:—

Alkaline Salts collected from the gases (60 litres) of one of the Clarence furnaces:—

Height above Tuyeres	8 Feet.	21¼ Feet.	50¼ Feet.	60 Feet.	In escaping Gases.
Cyanogen (grammes)	15·06	15·76	7·67	5·94	3·77
Potassium and sodium (grammes)	29·11	31·99	25·87	24·14	9·07
Number of Experiments	Six.	One.	One.	One.	Five.

The alkaline metals are in quantity considerably above that required to combine with the cyanogen, the excess being combined with carbonic acid.

It has to be remarked however that there are many practical difficulties which interfere with accuracy in determining the true quantity of alkaline matter contained in the furnace gases, particularly in the hotter regions, where it is trickling down in a fused or semi-fused condition. The choking of tubes employed for conveying the gas, and the occasional projecting of a fragment of unusual dimensions, composed of matter, largely impregnated with alkalies, are cases of such difficulties. Having regard however to the number of experiments given above, fourteen in number, it is obvious that there must have been present a considerable quantity of cyanides and carbonates of potassium and sodium.

A constant return, from the upper to the lower zone of the furnace, of the alkaline carbonates and cyanides present might no doubt account for such discrepancies in the quantities of carbon and oxygen as have been already described, but the same results might be due to carbonic acid brought down to the tuyeres in combination with the lime in the flux. Carbonate of lime is well known to retain its carbonic acid even when exposed to very high temperatures, if free access of steam or other gases is excluded. The presence of alkaline and other fusible matter might hermetically seal fragments of limestone, and prevent the displacement of their carbonic acid until they reached the tuyere partly or wholly undecomposed. Pieces of this mineral were drawn out,

SECTION IX.—CHEMICAL CHANGES IN THE BLAST FURNACE.

and were found, as was anticipated, to be coated with a fused crust. Generally speaking, the fragments then examined were entirely free from a trace of carbonic acid. Upon rare occasions however, when a large piece came down to the hearth, the centre seemed to have been but slightly affected by the heat during its passage through a furnace having a capacity of 25,500 cubic feet. Recently such a piece of limestone contained a kernel 2 in. diameter, which had only lost about 10 per cent. of its carbonic acid, the remainder of the block containing only a trace. Undecomposed limestone however does not appear to afford any sufficient explanation for any irregularity in the ratio of carbon and oxygen to the nitrogen.

It is quite clear from what has preceded that it is impossible to separate the respective causes of the two changes in the gaseous contents of the blast furnace.

To avoid as far as is possible all needless complication, the results obtained by the group of analyses designated as Example No. 1 will be taken as a means of illustrating the present subject, which is to attempt to show that in all probability the dissociation of carbonic oxide also contributes to the irregularities in the composition of the gases, from the top downwards.

Commencing at the top, we know that as soon as the oxide of iron in Cleveland calcined stone is warmed up to about 400° F. (204° C.) carbonic acid is generated; and that when the ore passes the point where, having regard to temperature, the quantity of the gas permits the deposition of carbon, we shall have carbonic acid proceeding from two sources, viz., largely from reduction of the peroxide of iron, and, to a certain extent, from dissociation of carbonic oxide. The actual quantity due to the latter cause will at first no doubt be very small, for at a temperature of 770° F. (410° C.) it was mentioned that when the gases contained 33 volumes of carbonic acid to 100 of carbonic oxide no carbon was deposited. As the heat approaches a dull red, which it does at 12 to 15 feet from the top, this action ought to become marked, because at that point 100 volumes of CO are only accompanied by 13 of CO_2.

At some point not far below B the temperature has become sufficient to expel the chief part of the oxygen from the ore, as well as the carbonic acid from the limestone; because at this level the oxygen in the gases is only equal to that injected or separated at the tuyeres (24·81 units) as already described. At the same time, while this is going on, no doubt carbon deposition must be in active operation, and by it carbonic acid is formed. If it happens that this carbonic acid, along with some of that furnished by the limestone, escapes decomposition, such a position of equilibrium will be established at that level between CO and CO_2, that reduction of the ore by carbon direct will be set up.

Leaving the carbon and oxygen, which are considered to have been withdrawn from the gases by the formation of alkaline carbonates and cyanides, to pursue their way upwards, we will proceed to consider the state of affairs at 6 feet above the tuyeres or at the point G. Here that which was regarded as the normal composition of the gases, viz., 18·18 carbon and 24·81 of oxygen, has been interfered with to the extent of ·38 of the former and 1·07 of the latter.

The only way I can account for this is by the withdrawal of carbon owing to the dissociation of carbonic oxide, while at the same time the carbonic acid formed ($2\ CO = C + CO_2$) has been decomposed by the iron having seized one of its equivalents of oxygen ($xFe + CO_2 = Fe_xO + CO$). The quantity of oxygen, 1·07, may appear large compared with the carbon, ·38; but at this very point reduced iron was, on one occasion, exposed to the gases, and for 1·50 of carbon deposited, as much as 9·53 of oxygen was absorbed.

Of course such a fixation of oxygen as that suggested means that in a zone essentially reducing in its character, oxidation can be effected. Dr. Percy points out this anomaly, and does not understand how oxidation and reduction can occur alternately in so limited a space. My own experiments however, many times repeated, prove that oxidation really does go on when the pure reducing agent alone is employed. Thus, for example, when spongy iron was subjected to a current of carbonic oxide at a low red heat, for each 100 of carbon deposited 6·6 of oxygen had united with the iron; while at a temperature of melting zinc the oxygen was only 10 per cent. of the carbon deposited.

At a bright red heat the quantity of carbon separated by metallic iron is not large, for then, as it would appear, the oxygen absorbed

SECTION IX.—CHEMICAL CHANGES IN THE BLAST FURNACE. 221

by the iron reacts on the carbon. Thus perfectly reduced spongy iron was exposed at the above-mentioned temperature to a current of pure carbonic oxide for one hour, and afterwards for three hours more without any further addition to the carbon or oxygen taking place. The results were as follows:—

	Iron.	Carbon.	Oxygen.	
For one hour	99·20	·32	·48	= 100
Continued for three hours more ...	99·34	·30	·36	= 100

In these instances for 100 of carbon received we have 120 and 150 of oxygen respectively. To secure more favourable conditions for carbon deposition, calcined ironstone was placed in the tubes and exposed during two hours to the gases of a furnace 48 feet high.

Distance from Top. Ft. In.	Vols. CO_2 per 100 CO.	Temperature.	C deposited per 100 Fe present.	Per Cent. Original O removed.
4 3	21	No signs of redness.	3·07	13·7
9 9	5	Cherry red.	·56	76·2
15 9	0	Bright red.	·32	76·3
21 5	6	Very bright red.	·55	68·2
27 3	9	Do.	·32	68·1

Upon another occasion a specimen of spongy iron was immersed in the gases of an 80 feet furnace, taken at a level 53 feet from the top. The temperature was bright red, and the composition of the gases as observed before and after the experiment, which lasted for $4\frac{1}{2}$ hours, was by weight as follows:—

	Before.	After.
Carbonic acid	2·7	4·9
Carbonic oxide	31·5	28·5
Hydrogen	Nil.	1·4
Nitrogen	65·8	65·2
	100	100

The iron present had acquired ·99 per cent. of carbon and ·9 of oxygen.

The largest amount of carbon thus deposited in an ore of iron took place when the latter was placed during 24 hours in the gases as they left a furnace 48 feet in height, the temperature being between 800° and 900° F. (427° and 482° C.) The quantity of carbon which had been precipitated from its oxide, amounted to 33·04 per cent., reckoned on the amount of iron present.

It is obvious from the facts just given that there is abundant opportunity for a very large quantity of finely divided carbon being thus infiltrated, as it were, throughout the ironstone; by means of which deoxidation can be most readily effected. In confirmation of the idea that the quantity is a large one, may be adduced the fact that, while the coke and lime do appear at the tuyeres in good sized pieces, the ironstone being torn asunder by the act of carbon deposition, is rarely so seen in pieces of any size. Moreover, when a furnace is blown out, large accumulations of the precipitated carbon are found in the masonry and among the materials.

Nor does the value of this substance appear to be confined to the reduction of the ore. It would appear to form with the slag a species of conglomerate, which probably acts as a protection to the masonry in the region where the temperature is highest. It would be difficult indeed to conceive how the fire brick of the walls surrounding the hearth could resist the corroding action of the slag without this or some analagous mode of defence; because a plain fire brick, when placed in the current of slag as it runs from the furnace, melts away almost like sugar in water. Experience has already proved that the 80 feet furnaces are resisting the extraordinary demand constantly exercised on their powers of endurance, with greater effect than did the smaller furnaces. Now in the former, as is clear from what has preceded, the opportunities for carbon deposition are greater than in the latter; which may account for the increased durability which has accompanied the use of the larger dimensions.

The simultaneous deposition of carbon and absorption of oxygen appears a reasonable mode of accounting for that excess of both, which manifests itself when the iron falls as fluid metal into the hearth, and which disappears, at all events to some extent, in the manner supposed, by the formation of alkaline carbonates and cyanides.

We may now return to our two examples (pp. 214, 215) and consider what goes on at the other levels. An action similar in principle to that described occurring at G goes on as we pass the levels designated as F, E, D, and C; but here probably owing to the heat being less intense the carbonic acid or a part of it is not decomposed by the iron, and hence the absorption of oxygen by the metal is much less. To facilitate

SECTION IX.—CHEMICAL CHANGES IN THE BLAST FURNACE. 223

comparison, the units of carbonic acid per 20 of pig are given in *Example* 5, and it will be observed that as the cooler portions of the furnace are approached the quantity of carbonic acid increases.

We have now to imagine the condition of things in what is generally regarded as the true zone of reduction at and above B.

The cyanides will melt when in contact with partially deoxidized ore—let us say with the protoxide, FeO. The action in that case will probably be $2(KCN) + 3 FeO = K_2O + 2 CO + 3 Fe + 2 N$; where it will be observed that all the carbon and nitrogen are restored to the gases. The alkaline metal will immediately be oxidized: part will escape, and part, with a portion of the alkaline carbonates, will condense on the cooler materials, and be carried back to form a fresh supply of cyanide. The carbonates of potash and soda, during their passage through incandescent iron, will probably be entirely decomposed ($Na_2CO_3 + Fe = Na_2O + FeO + CO$)—a form of action which also ultimately restores the oxygen and carbon absorbed by the soda to the gases.

It would, of course, be difficult to say what is really the precise nature of the action, which leads to the disappearance of the cyanides as they ascend through the materials.

It would also be difficult in a mere laboratory experiment to imitate the conditions as they exist in the blast furnace. Failing this, some experiments were undertaken, to ascertain the reducing power of cyanogen when in the presence of carbonic acid. The absence of all oxygen in the former gas rendered it pretty certain that its deoxidizing power would be superior to that of carbonic oxide, with which it was the object of the trial to compare it. The quantity of carbon in a given volume of cyanogen and carbonic oxide, it may be observed, is the same.

Vols. Cy.	Vols. CO$_2$.	Duration of Expt. Hours.	Temp. C.	Exposed Oxide contained per 100 of Total Fe.			
1	6	2.75	698°	Met. Fe 56.3.	Oxidized Fe 43.7.	Comb. with O 9.1.	Carbon 28.5
1	6	3.1	Bright red.	,, 22.7	,, 77.3	,, 17.4	,, 13.8
1	15	2.5	806°	,, 6.5	,, 93.4	,, 32.2	,, 1.3
1	30	3.	775°	,, .9	,, 99.0	,, 33.82	,, 2.52
1	15	2.8	Bright red.	,, none.	,, 100.	,, 28.90	,, .5

All these experiments tend to show that, both in reducing power and in the facility of depositing carbon, cyanogen is greatly superior to carbonic oxide; for it will be remembered that equal volumes of the latter gas and carbonic acid, at a bright heat, far from affording any metallic iron, oxidized the metal, and at no temperature was any carbon thrown down from this mixture of the two gases.

Notwithstanding the difficulty referred to, in explaining the exact nature of the reäctions which take place between the alkaline cyanides and the unreduced ore, so much was ascertained as to make it appear that, as they passed through the contents of the furnace, they were gradually converted into carbonates.

Feet above Tuyeres.	Cubic Metre of Gas Contained Alkaline Bases.	
	As Cyanides. Grammes.	As Carbonates. Grammes.
24	19·75	12·24
50½	10·68	15·19
60	9·14	14·90

Finally, as the gases pass out of the furnace, cyanogen as has been already observed, is often entirely absent, the alkalies being found entirely in the form of carbonates.

With regard to the gradual increase of alkaline matter, I had the gases of a furnace examined soon after it was "blown in." As might be expected, the quantity of potassium and sodium was very small, no opportunity having been afforded for its accumulation.

	Gases 8 Feet above Tuyeres.		Escaping Gases.
	1st Expt. Grammes per Cubic Metre.	2nd Expt. Grammes per Cubic Metre.	Grammes per Cubic Metre.
Potassium	6·04	Not estimated.	·99
Sodium	·32	,,	·06
Cyanogen	4·25	4·34	Nil.

Ultimately it may be that this cumulative action is brought to an end and the excess finds its way partly into the escaping gases and partly into the slag.

In the expectation that an examination of the fume at different heights of the furnace might throw some light on the nature of the reactions going on at different levels, a careful analysis was made of this substance, beginning with the small quantity of vaporous matter which is given off by the slag as it flows from a furnace producing grey iron, and terminating with that contained in the escaping gases.

SECTION IX.—CHEMICAL CHANGES IN THE BLAST FURNACE.

COMPOSITION OF FUME AT CLARENCE FURNACES, 80 FEET HIGH 25,500 CUBIC FEET, SMELTING CLEVELAND IRONSTONE.

Height above tuyeres	A. Emitted from Slag.	B. 2 Feet.	C. 26¼ Feet.	D. 39¾ Feet.	E. 45¼ Feet.	F. 58¼ Feet.	G. Escaping Gas. 76 Feet.
	Per Cent.	Per Cent.	Per Cent.	Per Cent.	Per Cent.	Per Cent.	Per Cent.
Sulphuric acid (free)	5·00	...	...	...	...	...	...
Phosphoric acid	...	...	...	...	...	...	1·57
Sulphate potash	85·00	...	...	...	...	...	1·00
,, soda	7·50	...	...	...	...	...	·23
,, magnesia	2·50	...	...	...	...	...	...
,, lead	...	...	...	...	...	...	1·87
Cyanide potassium	...	75·98	89·20	71·21	...	...	...
,, sodium	...	...	3·51	·07	...	...	...
Carbonate potash	...	3·95	...	...	...	...	} 2·61
,, soda	...	20·70	3·52	3·91	...	...	
,, calcium	...	...	...	...	...	...	? 2·39
Chloride ammonium	...	...	...	...	...	...	...
,, potassium	...	·55	...	...	5·42	1·80	5·92
,, sodium	...	...	2·48	6·20	2·93	1·00	1·48
,, magnesium	...	...	...	...	...	...	1·56
Oxide of zinc	...	tr.	tr.	6·71	15·27	17·17	13·39
Carbonate ,,	...	...	...	11·88	...	...	...
Sulphide ,,	...	...	...	...	...	...	7·70
Metallic ,,	...	...	...	...	50·75	13·66	...
Chloride of zinc	...	...	...	...	7·65	13·16	4·95
,, calcium	...	...	...	...	...	...	2·06
,, lead	...	...	...	...	1·47	8·77	...
Sulphate calcium	...	...	...	...	tr.	·57	3·60
Silica	...	...	...	...	6·38	15·00	10·25
Lime	...	...	...	...	6·70	14·37	·56
Alumina	...	...	...	...	1·84	6·11	3·66
Magnesia	...	...	...	...	·31	4·33	tr.
Protoxide iron	...	...	...	...	1·03	2·63	...
Peroxide ,,	...	...	...	...	...	...	15·00
Iodide of potassium	...	...	...	...	...	...	·12
Carbon	...	...	...	...	·27	·41	13·66
Water	...	...	...	...	...	...	6·42
	100·00	101·18	98·71	99·98	100·02	98·98	100·00
Approximate temperature F.	—	2,000°	1,800°	1,600°	900°	600°	
Grammes of fume per cubic metre	13·61	7·74	2·27	1·39	1·65	1·903[1]	

Composition, by vol., of the gases at :—

	B.	C.	D.	E.	F.	G
Carbonic acid	1·90	·46	·00	3·84	6·56	10·69
Carbonic oxide	39·18	35·80	34·82	32·33	32·25	28·58
Hydrogen	1·97	1·14	1·41	3·88	1·08	1·70
Nitrogen	56·95	62·60	63·77	59·95	60·11	59·03
	100·	100·	100·	100·	100·	100·

[1] Three trials were made of this—they gave respectively 2·610, 1·435, and 1·666, average 1·903 grammes per cubic metre.

Commencing with the vaporous matter emitted by the slag (A), it has been shown that there was always a certain quantity of sulphur, together with some soda and potash, in the cinder produced in smelting Cleveland stone. It looks as if a portion of the sulphur were converted, when it comes in contact with the air, into sulphuric acid; after which it combines with potash, soda, and magnesia, a small quantity escaping as free acid.

At B, 2 feet, and at C, $26\frac{1}{2}$ feet above the tuyeres, the whole of the sublimate consists of alkaline salts. At C almost the whole of these consist of cyanides, which $24\frac{1}{2}$ feet higher up are partially converted into carbonates.

The carbon and oxygen, which go to form the cyanides and carbonates just mentioned, appear to be withdrawn from the gases by condensation between B and C, to be returned to the gases either at the tuyeres or where the sublimate is melted and carried off in the cinder. It is the portion returned to the gases which is supposed to account in part, for the increase of carbon and oxygen in the gases at a point below B.

At D, $13\frac{1}{4}$ feet above C, a large percentage of the fume (fully 75 per cent.) consists of alkaline salts, chiefly cyanides. These seem to have accumulated after the materials passed E, situate $5\frac{3}{4}$ feet higher up; the heat at the latter point apparently not having sufficed for their volatilization. At D vaporization must have fairly started and have carried upwards the salts in question, which must have been condensed and brought back to the lower levels, where the fume is exclusively composed of alkaline cyanides and carbonates.

At E it will be perceived that one half of the sublimate consisted of metallic zinc, which sufficed to cover the inside of the tubes employed with a lustrous lining. At the same time that there was a large quantity of zinc in the metallic form, 15·29 of the fume consisted of oxide of zinc. A reference to D, $5\frac{3}{4}$ feet further down in the furnace will show that all this metal existed there as carbonate or oxide, the two amounting to 18·59 per cent. of the whole.

To form an idea of the reducing power of these two regions of the furnace on oxide of zinc, we must consider their temperature and the nature of the atmosphere which pervades them. These, obtained from a furnace of 80 feet in height, are given below:—

SECTION IX.—CHEMICAL CHANGES IN THE BLAST FURNACE.

	D (39¾ feet above Tuyeres).	E (45½ feet above Tuyeres).
Temperature about	980° C. (1,792° F.)	870° C. (1,598° F.)
Composition of gases by volume:—		
Carbonic acid	·00	3·84
Carbonic oxide	34·82	32·33
Hydrogen	1·41	3·88
Nitrogen	63·77	59·95
	100·	100·

The first step was to ascertain the nature of the action of carbonic oxide on oxide of zinc; which was done by passing the perfectly pure gas over the oxide in an apparatus from which all air had been expelled. The first indication of carbonic acid in the current of gas was when the temperature reached 800° F. (427° C.)—a much higher temperature therefore than that at which peroxide of iron first shews marked action, which, it will be remembered, is 400° F.

On increasing the heat to the softening point of the glass tube employed (about 815° C. or 1,500° F.), metallic zinc began to sublime, mixed with oxide. Inasmuch, however, as oxide of zinc is not volatile, its presence among the sublimate is probably due to a reoxidation of the newly formed metal, through contact with the highly heated carbonic acid formed by the oxidation of the carbonic oxide.

An experiment was made to ascertain the behaviour of metallic zinc in contact with carbonic acid at certain temperatures. This gas readily converted the zinc into oxide before the metal was melted. At a higher temperature, 800° to 850° C. (1,472° to 1,562° F.), the carbonic acid was entirely converted into carbonic oxide, and the zinc was oxidized in proportion.

The composition of the gases may vary so much during the time of collecting the sublimed matter they contain that it is useless to speculate on the reactions which are taking place in the interior of the large mass of heated substances which fill the furnace. No doubt much of the metallic zinc reduced in the strongly reducing zone is reoxidized; for it was ascertained that on exposing peroxide of iron to a current of zinc vapour, in an atmosphere of nitrogen and at a full red heat, for 2½ hours, one-half the iron was found completely reduced. Any unreduced oxide of iron therefore at D would also be able to reoxidize the metallic zinc.

228   SECTION IX.—CHEMICAL CHANGES IN THE BLAST FURNACE.

Although in the table of analyses of the fume the acids and bases are mentioned as existing in a certain form of combination, it is impossible to speak with any degree of confidence as to their precise condition in this respect.

Before leaving the phenomena connected with the presence of zinc, it should be stated that this metal is frequently collected in considerable quantities, after running out in its metallic state through the brickwork near the tuyeres. As the temperature in that region is inconsistent with the presence of the metal in its liquid form, this efflux is probably due to zinc being distilled higher up the furnace and accumulating behind the masonry, where from a subsequent increase of temperature it melts, and then escapes in the manner described.

The presence of lead in the vapourized substances which are found in the blast furnace, indicate the frequent existence of this metal in the material employed by the iron smelter. This is particularly the case when the ore is obtained from veins, more or less productive of galena, in which event metallic lead sometimes runs from the tappinghole into the pig moulds. I have not met with this during my own experience; but the circumstances attending its occurrence, no doubt, are analogous to those which accompany the reduction of zinc.

Whether ammonia is brought in by traces adhering to the coke, or is the result of a direct combination of nitrogen with hydrogen in the furnace, or occurs through the instrumentality of the cyanogen compounds, I have no means of determining. About 10 years ago, when using raw Cleveland ironstone in the furnaces, a large quantity of water was given off; and this dissolving the deliquescent chlorides formed stalactites outside the gas tubes. These consisted of:—

		Per Cent.
Insoluble matter, probably peroxide of iron	... ...	1·0
Chloride of ammonium	... ...	52·9
Do. iron	... ...	14·7
Do. zinc	... ...	6·6
Do. sodium and calcium	... ...	trace.
Water		24·8
		100·

SECTION IX.—CHEMICAL CHANGES IN THE BLAST FURNACE.

In connection with the presence of the alkalies in the vapours emitted during the smelting of iron, it should be stated that the comparatively rare alkali lithia, was observed, by means of the spectroscope, in the flame burning at the tymp of one of the Clarence furnaces, by Mr. Norman Lockyer and myself; and subsequently this substance was detected as a trace in an analysis of the fume.

The substances contained in the fume which now claim attention are the silica, lime, alumina, and magnesia. These four earths occur in large quantity in the ironstone; they constitute the ash of the coke, and the limestone in use, the last mentioned although essentially a carbonate of lime, contains also a small amount of silica, alumina, and magnesia.

No doubt the mere current of gas passing through the materials when charged, will mechanically propel a certain quantity of the fine dust, which occurs largely in the calcined Cleveland stone and to a less extent in the coke. The presence of dust of peroxide of iron and of carbon in the fume deposited by the gases as they escape, is a proof of a certain portion of matter being carried over mechanically in the way suggested. Such portions are however coarse in their texture, and form no part of the chief mass which has been referred to as a sublimate.

This appellation has been bestowed on the true fume, because it bears all the physical attributes of matter which has been in vapour, and, condensing from that state, has adhered to any surface exposed to it.

The circumstance that four substances, of such a fixed nature as silica, lime, alumina, and magnesia, are evaporated by heat, and by a comparatively moderate heat, seems the reverse of what might be expected; especially as some of these substances are employed in the construction of the hottest furnaces.

In searching for the probable origin of this sublimed matter, I was formerly led to look for metallic potassium and sodium in the furnace, believing that they might reduce the bases of the earths, and that these, being carried mechanically upwards in a fine state of division, might be reoxidized in the upper region; but I failed to discover a trace of these metals.

Attempts were then made to obtain metallic aluminium in the laboratory, by exposing to a very high temperature, in a graphite crucible, mixtures of alumina and cyanide of potassium, with and without carbon. In no case was there a trace of the metal discovered.

There is one fact in connection with the properties of this fume which deserves attention. The silica in all the raw materials of the furnace is almost if not entirely insoluble in water, after treatment by hydrochloric acid; whereas in the fume or sublimate this substance is all soluble in that acid. This proves the fact that chemical action has been at work, and the most probable cause of such action lies in the alkaline constituents of the fume.

It is to be observed that the gases contain no trace of any of these four earths, until we have reached a point about 45 feet above the tuyeres, in a furnace having a height of 80 feet. I apprehend that this is due to the condensation of the alkalies, and their existence in a fused state among the earths. Under such circumstances nothing more likely than that we should have a fused mass of alkalies, with silica, lime, alumina, and magnesia. This mode of action does not necessarily account for the formation and possible evaporation of the metallic bases of earthy substances by the intervention of potassium or sodium, for these metals appear to exist in the fume obtained from the upper parts of the furnace chiefly as potassium and sodium chlorides. The earths at E ($45\frac{1}{2}$ feet above the tuyeres) bear the following relation to each other:—

Silica.	Lime.	Alumina.	Magnesia.	
42 ...	44 ...	12 ...	2 =	100

Looking at the fixed nature of these bodies under the most intense heats we can command it would seem as if the only means of accounting for the sublimation, is by the power which currents of highly heated gases are known to possess of carrying away substances otherwise fixed in their nature.

During the passage of this earthy sublimate, if such it really be, through the colder portions of the materials, a good deal will no doubt be condensed, and carried back in the solid form to the lower regions of the furnace; and it would appear that, while the proportion

SECTION IX.—CHEMICAL CHANGES IN THE BLAST FURNACE. 231

of silica is not very sensibly affected in quantity by this cumulative process, the lime certainly, as shown in the table, is considerably diminished in amount; its place being taken by alumina and magnesia.

The following figures exhibit the changes just referred to; and to them is added an example, showing the composition of the slag from a furnace smelting Cleveland stone; by which it will be seen that there is little analogy between the fume and the slag, as to the relations which the earths bear quantitively to each other. Where sulphates and chlorides occur, their equivalent of oxide is taken into account:—

	Silica.	Lime.	Alumina.	Magnesia.		
E $45\frac{1}{2}$ feet above tuyeres ...	42	44	12	2	=	100
F $58\frac{1}{4}$ do. do. ...	37	37	15	11	=	100
G 76 do. escaping gases...	28	17	21	4	=	100
Slag	29	38	24	9	=	100

The composition of the fume deposited from the gases after they leave the furnace, cannot be relied on as correct; for it is contaminated with mechanically projected dust, as evidenced by the large quantity of peroxide of iron and carbon it contains.

Looking at the small proportion of the earths which leave the furnace in the form of this fume, the composition of the slag necessarily represents the relations they bear to each other in the materials charged. Assuming this, and comparing the composition of the slag with that of the fume, before it is contaminated with the dust mechanically carried over from the ironstone and coke, we have the following numbers:—

	Silica.	Lime.	Alumina.	Magnesia.		
Fume $58\frac{1}{4}$ feet above tuyeres	37	37	15	11	=	100
Normal relation as in slag ...	29	38	24	9	=	100

From these we may infer that the silica is most affected by the causes in operation in the furnace, which conduce to sublimation.

Compared with the slag, 37 parts of silica ought to have been accompanied by:—

	Lime.	Alumina.	Magnesia.
	48 ...	31 ...	11
Whereas there was only ...	37 ...	15 ...	11
Deficiency ...	11 ...	16 ...	0

Hence it appears that the lime and alumina are the least affected by the sublimating influences.

The fume, during its passage from the furnaces to the fireplaces, where the gas containing it is burnt, deposits its grosser particles. After combustion however there accumulates in the boiler and stove fireplaces a large quantity of a fine powdery matter, having a dirty yellow colour.

From hence the burnt gas, diluted with the air required for its combustion, and bearing with it a large quantity of the finest portions of the sublimate, is carried forward towards the chimney. At the base of one of the boiler chimneys at the Clarence works, the amount of the fume contained in each cubic metre of the diluted gas was ascertained to amount to 1·07 grammes.

I will add to what has been said on this subject the composition of the deposit gathered in one of the boiler fireplaces, and that of the fume as it enters the chimney.

	Dust from Fireplace. Per Cent.	Fume at Chimney. Per Cent.
Silica	23·09	32·85
Lime	11·87	21·23
Alumina	18·48	16·66
Magnesia	4·45	13·26
Potash and soda	2·66	2·54
Oxide of zinc	11·12	3·34
Do. lead	3·08	2·55
Peroxide of iron	8·29	6·57
Sulphuric acid combined with lime, etc....	16·68	·55
Phosphoric acid	—	·69
	99·72	100·24

Although the earthy constituents which form the bulk of this vaporous or finely divided matter appear to take their origin in a part of the furnace where the temperature is by no means very intense—probably not above 1,093° C. (2,000° F.)—I have remarked that when the furnaces in the Middlesbrough district are "working cold," which happens when they are making white iron, little or no fume is given off. On the other hand its comparative absence is to be observed in the Scotch furnaces smelting rich foundry iron, where it might well be

SECTION IX.—CHEMICAL CHANGES IN THE BLAST FURNACE. 233

expected to occur, seeing that the slag much resembles in composition that of the works using Cleveland stone. The slag from one of the Coltness furnaces contains the four earths in the proportions given below, which, it will be seen, closely resemble those at the Clarence works:—

	Silica.	Lime.	Alumina.	Magnesia.
Coltness iron works ...	30	38	20	12
Clarence do. ...	29	38	24	9

At Coltness however the coal is used raw, which gives rise to the distillation of a large quantity of tar in the upper region; and it may be that the presence of this substance may serve to intercept the escape of the sublimate in question.

There are no doubt many circumstances which may conduce to changes in the composition of the fume from blast furnaces. The more prominent of those are alterations in the contents of the materials; this observation is chiefly applicable to the ironstone, which is much more subject to changes in composition than the coke or limestone.

If this remark be appropriate in the case of ironstone, taken from one district, which is nevertheless variable in its composition, it is still more so in the case of furnaces employing ironstone from other districts and of a totally different character.

There is to be found in the present Section a good deal of matter of a more or less speculative character. This observation applies particularly to the alterations in the relative proportions which the carbon, oxygen and nitrogen bear to each other in different levels of the blast furnace. It is also applicable to the formation and behaviour of the cyanogen compounds, and other substances found in the fume given off during the process of smelting. Whatever value the speculations themselves may have, it is hoped that the facts, mentioned in connection with the attempted explanations, may justify the space devoted to their description.

SECTION X.

ON THE EQUIVALENTS OF HEAT EVOLVED BY THE FUEL IN BLAST FURNACES.

SINCE the matter relating to the subject of the present Section was printed in this work, further consideration has induced me to add a few more pages on a question of so much practical importance to the iron smelter. This appeared to me especially desirable because at the present time the use of fire brick stoves is engaging increased attention. I have also been led to give the question further notice by having had a recent opportunity of again examining the performance of furnaces which are exclusively engaged in the manufacture of charcoal iron. Besides this, Mr. Charles Cochrane has published his experience with superheated air in his very capacious furnaces at Ormesby;[1] so that I am now able to supplement the data given at p. 107, *et seq.*, in reference to the interesting problems connected with this branch of my work.

The value of every description of fuel, so far as its use *in* the blast furnace is concerned, I repeat may be primarily regarded as dependent on the fixed carbon it contains.

If the fuel used in smelting iron is uncharred, the gaseous matter, instead of being previously burnt in the coke ovens or charcoal pits, is rendered available for purposes outside the furnace; but it is next to useless inside it. Of course some sacrifice of solid carbon must be made in volatilizing the gases, when using raw fuel; but this does not amount to the waste of the fixed portion of the coal generally incurred during the process of charring.

In practice a coal containing 72 per cent. of coke only yields about

[1] Proceedings of Institution of Mechanical Engineers, August, 1882.

62 per cent. in the ordinary so-called bee hive oven, which means a loss of 10 units. Allowing for ash, these 10 units of coke, if burnt to carbonic acid, will afford 72,000 Centigrade calories. Supposing on the other hand, that 18 units of volatile matter have to be expelled, 36,000 calories will be absorbed; or about one half the quantity wasted in the oven, by the needless combustion there of solid carbon.

The only value of the hydrocarbons which are given off by raw fuel in the blast furnace must, I imagine, be in the power they possess of maintaining a deoxidizing agency in the reducing zone. If this be true, then they may save so much carbonic oxide, which otherwise would be required to be furnished by the combustion of solid carbon.

The same observations as those made in reference to coke apply, in point of principle, to the manufacture of charcoal; but the large quantity of water present in wood, renders its use in the raw state, except in very limited proportions, impracticable.

The heating power of the fixed carbon, in fuel used in blast furnaces, is affected :—

(*a*) By the quantity and nature of the foreign matter with which it is associated :

(*b*) By the temperature and condition of the air with which it is burnt:

(*c*) By the state of oxidation in which it leaves the furnace:

(*d*) By the quantity of heat which escapes unutilized in the blast furnace gases.

I propose to examine these items in the order in which they stand.

(*a*) *Effect of quantity and nature of foreign matter.*—An obvious inconvenience of foreign matter, in coke or charcoal, is that it affords no heat in the furnace; for even if a portion is in the form of combustible gas, it is evaporated long before it reaches the tuyeres, where the solid carbon is burnt. The extent to which fuel is injured, under this aspect of the case, depends on the quantity of foreign matter present. Supposing it to be 10 per cent., the co-efficient of heat, for one unit of coke burnt, is as under :—

To carbonic oxide $1 \times 2,400 - 10$ per cent. impurity $= 2,160$.
To carbonic acid $1 \times 8,000 - 10$ per cent. impurity $= 7,200$.

When these two gases are found in the proportion of 1 vol. CO_2 to 2 vol. CO, the heat equivalent of coke containing 10 per cent. of foreign matter will be :—

$$\frac{1 \times 7{,}200 + 2 \times 2{,}160}{3} = 3{,}840 \text{ per unit.}$$

The heating value of the combustible is further materially affected by the nature of the foreign matter.

Uncombined water absorbs as much heat as is required to evaporate it, *plus* that required to raise it to the temperature of the escaping gases. The last-mentioned portion of the loss is however included under heading (d).

Coke is frequently quite dry, at other times it may contain 2 per cent. of water: whereas charcoal sometimes contains as much as 15 per cent. to 20 per cent. of moisture. At 15 per cent., each unit would contain ·15 unit, and this multiplied by 540 calories (required to convert a unit of water into vapour) gives

$$·15 \times 540 = 81 \text{ calories.}$$

This is equivalent to a loss of something above $3\frac{1}{2}$ per cent. when the charcoal is burnt to the state of CO, and of a little more than 1 per cent. when CO_2 is the product of combustion. Oxidized as carbon usually is in blast furnaces, say at something like 2 volumes of CO to 1 volume CO_2, the loss represents about $2\frac{1}{4}$ per cent. of the heating power evolved by charcoal. A common ingredient in coke and charcoal is unexpelled volatile matter. This, consisting of CO_2, CO, CH_4, H and N, exceeds sometimes, in the case of charcoal, 6 per cent. of the total weight. Adopting this number, each unit contains ·06 of such matter, which, multiplied by 2,000 calories, gives an absorption of 120 calories, or a loss about one half greater than when the charcoal was associated with 15 per cent. of water; say therefore a loss of about 3 per cent. of the heat evolved by the carbon as it is usually burnt in the furnace. In well made coke the volatile matter besides water rarely exceeds $1\frac{1}{4}$ per cent., and occasions therefore a loss of only 25 calories per unit.

When the volatile constituent is carbonic acid, a further loss attends its presence beyond that represented by its own weight; because, not being gasified until it reaches a position in the furnace, where the temperature enables it to carry off a quantity of solid carbon equal to that it contains; this quantity is thus oxidized without affording any useful heat in the furnace.

Recently Dr. F. Muck has given, in a very useful publication,[1] the particulars of certain analyses, in which as much as 7·68 per cent. of oxygen appears as one of the constituents of coke. According to the researches of Parry, when coke was heated *in vacuo*, the oxygen was gasified, producing CO and CO_2 in the ratio of one-third in the former and two-thirds in the latter state of combination. Applying these figures to a coke containing 7·68 per cent. of oxygen, this means that 3·84 per cent. of its carbon would disappear as CO_2, and be wasted before it reached the tuyeres.

The last kind of the foreign matter requiring notice is the solid ash; which acts hurtfully, not only because it must itself be fused, but because additional limestone has to be used, partly as a flux and partly, I imagine, to free the iron from the additional sulphur often brought in with the ash of the fuel.

It may be considered roughly, that when 30 units of coke are used to produce 20 units of Cleveland pig iron, 16 units of limestone are required; and this, when the consumption of coke is reduced to 22 units by the use of large furnaces 80 feet in height, is brought down to 11 units.

If we take the ash and sulphur at 7 per cent., we should have to deal with 2·10 units for the 30 units of coke, and 1·54 for the 22 units. This is equivalent to an increase of (2·10 − 1·54), or ·56 of ash; to which has to be added the lime in 5 units of limestone, say 2·8 units united with 2·2 of carbonic acid. We have therefore 3·36 units more of slag to fuse in the one case than in the other.

The total loss upon the 20 units of iron will stand thus:—

		Calories.
Expulsion of CO_2 from 5 cwts. of limestone	5 × 370 =	1,850
Decomposition of the 2·2 CO_2	·6 carbon × 3,200 =	1,920
Fusion of ash and lime	3·36 cinder × 550 =	1,848
Total loss		5,618

[1] Steinkohlen Chemie. Translated notice is given in Transactions Iron and Steel Institute, 1882.

From the facts just enumerated, it may be estimated that from causes connected with the composition of the coke, its value as a source of heat may be affected to the extent of from about 8 per cent. to about double this amount. This diminution of power is frequently aggravated by a still larger quantity of ash than that considered as present in the above calculations. On the Continent I have met with instances in which coke used for smelting iron contained as much as 12 to 15 per cent. of earthy matter.

An excessive use of fuel in the blast furnace is thus accompanied by a notable waste; for it has just been shown that 5,618 heat units may be devoted to the formation of the additional slag incidental to an increase of 8 cwts. of coke consumed per ton of iron.

(*b*) *The temperature and condition of the air with which the coke is burnt.*—Each addition to the temperature of the blast increases by so much the number of calories afforded by the coke. If the blast contained 500 calories for each unit of the fuel, and if the heat afforded by the combustion of the fuel were 3,840 calories, the equivalent of fuel so burnt would be $3,840 + 500 = 4,340$ calories. On the other hand, if the air contains 1 per cent. of moisture, which it occasionally does, 160 calories per unit of coke burnt might be absorbed from this cause. This however is an extraordinary amount of saturation: 100 to 120 calories is therefore a more common amount of heat absorption from moisture equal to about $2\frac{1}{4}$ or $2\frac{3}{4}$ per cent. of the total quantity of heat evolved under the conditions just mentioned.

(*c*) *The state of oxidation in which the carbon leaves the furnace.*— Reference has frequently been made in the present work, to an apparent limit beyond which the oxidation of the carbon in the gases could not be carried. The observations upon which this point was approximately determined were carried on by means of furnaces smelting Cleveland ore with Durham coke. At the same time I have no reason for thinking, that the conclusions I arrived at, under the conditions just named, are not equally applicable to all furnaces using mineral fuel.

It is needless to refer again at any length to the methods employed, not only at the furnaces themselves but in the laboratory, to demonstrate what seems to be a chemical law, viz., that when the gases have about one-third of their carbon in the form of carbonic acid, they are unable to reduce iron. I would explain this law in the following way:—

The combining equivalent of peroxide of iron Fe_2O_3 is 80, composed of 56 of iron and 24 of oxygen. Of these 24 parts of oxygen, 12 are so feebly held by the iron that, $4\frac{1}{2}$ parts of carbon suffice for separating this half of the combined oxygen, giving $16\frac{1}{2}$ parts of carbonic acid. The remaining half of the oxygen is however, so firmly retained, that it requires twice as much carbon, or 9 parts, to tear it away from the iron, producing by this union 21 parts of carbonic oxide. These figures it will be perceived give us, one-third of the carbon as carbonic acid, and two-thirds as carbonic oxide.[1]

Saturation of the carbon with oxygen, in furnaces using coke, never reaches the limit just named, *i.e.* one-third as carbonic acid and two-thirds as carbonic oxide, at least such is the result of my own experience and observation. The simplest way to show this will be by quoting the results actually obtained in different cases which have been examined, assuming the air to have been at 32° F. There it will be seen that a unit of coke gives, by its combustion with air at 0° C. (32° F.), in one case 3,087 calories, and in another 3,860, showing a difference of fully 25 per cent.[2] due entirely to differences in the extent of its oxidation.

For this purpose Tables have been constructed, containing examples of a varied character both as regards the development of heat and its appropriation. In the comparison of the two sides of the account is given, the number of calories produced by the coke, modified by the extent to which the carbon is oxidized, and by the temperature of the blast. On the other side is inserted the requirements of the furnace as affected by differences in the ores and flux under treatment, as will be explained in its proper place.

---

[1] As explained in its proper place. Sec. V., this is not the way in which reduction is actually performed in the blast furnace. The carbon is considered as first entirely converted into carbonic oxide, and when one-third of it, in this new form, is converted into carbonic acid further action ceases, leaving two-thirds unchanged. The final result is, of course, the same in each case, viz. $CO_2 + 2CO$ in which the oxygen in the two forms of oxidation is the same, but the C in the CO_2 is only half that in the 2 CO.

[2] *Vide* Examples B and I. pages 244 and 246.

The heat produced by the actual combustion of the coke is necessarily affected by the amount of impurity it contains. This is very variable and uncertain; but as the difference will not exceed $2\frac{1}{2}$ per cent. in any case, this cause of disturbance is not, generally speaking, taken into the account.

(*d*) *The quantity of heat which escapes unutilized in the gases.*— This is estimated by observations on their temperatures; and the composition and weight for a given quantity of iron made being known, it is easy to calculate the amount of heat thus carried off. Care must of course be taken to extend the observations over a sufficient period of time to secure an average representation of the whole; because, as has been explained in Section IX., page 180 *et seq.*, the temperature varies very considerably according to circumstances.

Beyond any irregularity in the composition of the coke, there are many other circumstances which render it impossible to claim perfect correctness in the estimates of the performance of a blast furnace.

The ironstone may be, and occasionally is, imperfectly calcined. This mineral, as it occurs in the Cleveland hills contains about 22 per cent. of carbonic acid or 6 per cent. of carbon. If we assume 65 units of raw stone to be consumed for each 20 units of iron we have 3·9 units of carbon in this weight of ore. Supposing 5 per cent. of this carbon to remain unexpelled we may have ·19 of a unit of this element derived from this source instead of being contributed by the coke. The retention of this carbon as carbonic acid would, were it combined with the iron, mean that a corresponding quantity of the metal remains in the form of protoxide, instead of being peroxidized as happens in roasting the ore. The ore for each 20 units of pig iron might therefore afford ·26 units less oxygen than is estimated for in my calculations.

The unexpelled carbonic acid however may be partly or possibly wholly retained by lime and magnesia which earths constitute about 12 per cent. of Cleveland raw ironstone. We should then have the relation between carbon and oxygen as they exist in the imperfectly calcined ore disturbed from this cause.

The coke itself may contain oxygen, generally in very minute

SECTION X.—BY FUEL USED IN BLAST FURNACES.

quantities; but I have recently met with a case in which $2\frac{1}{2}$ per cent. of this substance was found in a quantity furnished by a mine in the county of Durham.

Again the air blown into the furnace is constantly altering in the quantity of moisture it contains. Kämtz of Halle found over a series of years that the mean quantity of water in the atmosphere at that place varied from 61 to 86·2 per cent. of the quantity required for complete saturation. Taking 100 units of air as the consumption for 20 units of iron we should have ·34 units of water entering the furnace were 61 per cent. of what is required for complete saturation and ·48 units of water were it 86·2 per cent., or nearly one-half more of the first-named quantity.

Now when it is remembered that a large Cleveland furnace may contain at one time 1,000 tons of material, to smelt which about 1,000 tons of air is required it is easy to conceive the nature of the obstacles which may prevent perfect accuracy in any calculation founded on the process itself.

This difficulty of a correct appreciation of the various factors which have to serve as a basis of calculation makes us rarely able to produce an exact correspondence between the quantity of heat estimated to have been produced and that really required. In all cases, the coke consumed appears to give a little more heat than is actually needed, according to the figures taken for the rate of absorption. In the nine examples, contained in the two tables, the average discordance between the two sides of the account is a trifle under 5 per cent.; but in each case the weight of coke itself has been taken as the basis of computing the consumption of fuel for the process.

The calculations which follow apportion to each function of the furnace its own equivalent of coke, together with the heat in the blast—the latter reduced to an equivalent of coke, so as to give at a glance the actual value of the hot blast in each case. These functions are divided into three sections, as follows:—

Class I.—The elements requiring heat which depend upon the quantity of slag to be melted, the weight of limestone requiring dissociation, the quantity of water to be evaporated from the materials, and the dissociation of the moisture in the blast.

Class II.—Those requirements of a furnace which comprise reduction and fusion of iron, dissociation of carbonic oxide, reduction of metalloids absorbed by the pig, loss of heat through walls of furnace, and heat carried off by tuyere water in the case of hot blast furnaces.

Class III.—Heat leaving the furnace in the escaping gases.

The first example, A, Table I., represents an ideal case of smelting Cleveland stone with cold blast, the particulars of which are sufficiently close to the truth to enable any one to compare the sources which make up the saving effected by the use of hot blast.

In this case, the mode of dividing the items of heat appropriation is set forth at length; but to save space the details are omitted except in two other examples, the elements being estimated according to the yield of the minerals and the quantity of flux required.

The work of the cold blast furnace—one of old construction, say of 6,000 cubic feet capacity—is estimated on a consumption of 45 cwts. of coke, requiring as much as 25 cwts. of limestone, with ironstone of the same richness as the other examples of Cleveland iron mentioned in the table—say 42 per cent. or 47·6 cwts. of calcined stone per ton of metal.

The three items for this furnace are thus calculated.[1]

Class I.—Elements of heat absorption:—

			Calories.
Fusion of slag	36 cwt. ×	550	19,800
Decomposition of water in blast	·16 (H) × 34,000		5,440[2]
Expulsion of carbonic acid from minerals 25 $(CaCO_3)$ ×		370	9,250
Decomposition of carbonic acid in do.	3 (C) ×	3,200	9,600
Evaporation of water in coke	1·16 ×	540	626
Forward			44,716

[1] The elements of absorption have been formerly (Chem. Phen., p. 163) divided into variables, constants and absorption by gases. The variables were subject to great fluctuations and are given under Class I. The constants were less liable to change but not being absolutely constant are viewed under Class II.

[2] The large consumption of coke necessitates a great increase in the weight of air required for its combustion as compared with a furnace using hot blast: hence the excessive absorption of heat under this item.

SECTION X.—BY FUEL USED IN BLAST FURNACES.

Class II.—Elements of heat absorption :—

Brought forward		44,716
Reduction of iron	18·6 × 1,780 =	33,108
Carbon impregnation	·60 × 2,400 =	1,440
Reduction of silicon, sulphur, phosphorus from acids...	...	4,174[1]
Transmission through walls, etc.	...	3,658
Fusion of pig iron...	20 × 330 =	6,600
Carried off in tuyere water	...	Nil.
		48,980

Class III.—Balance, regarded as carried off in escaping gases 28,304

122,000

The evolution of heat is estimated as follows:—

Coke consumed—45 units per 20 of pig less 4 of ash, &c. ...	...	41
Less carbon in limestone, carrying off an equal weight from coke	...	3
Leaving carbon as source of heat	...	38

Assume combustion ... 5·50 as CO_2 × 8,000 = 44,000
32·50 as CO × 2,400 = 78,000

38·00   122,000 = 2,711 calories per unit of coke burnt.

Particulars of heat absorbed in a large Cleveland furnace blown with hot blast D, Table I. :—

Class I.	Calories.	Class II.	Calories.	Class III.	Calories.
Fusion of slag	13,475	Reduction iron	33,108	Escaping gases	7,537
Decomp. H_2O in blast	2,380	Carb. impreg.	1,440		—
Expulsion of CO_2	3,471	Reduction Si., S., P.	4,174		—
Decomp. of do.	3,584	Transmission	3,658		—
Evap. H_2O coke	286	Fusion pig	6,600		—
		Tuyere water	1,818		
	23,196		50,798		7,537

[1] These items vary considerably, hence the figure inserted here can only be regarded as a rough approximation.

TABLE I.—SMELTING CLEVELAND STONE.

	A.	B.	C.	D.	E.	F.
Height of furnace	48	48	80	80	76	90
Capacity, cubic feet	6,000	6,000	11,500	11,500	20,642	35,016
Weekly make per 1,000 c. ft.	20	36¾	30	30	24¼	14$\frac{1}{10}$
Coke per ton of iron, cwts.	45	28·92	22·32	20·40	24·14	19·69
Limestone do. do.	25	16·00	13·66	9·38	12·61	12·62
Quality of iron, No.	3·0	3·0	3·0	3·5	3·0	3·0
Temp. of blast, degrees C...	0	485	485	563	888	819
Do. escaping gases do. C....	496	452	332	262	329	222
Units of C as CO to 1 of C as CO_2 in escaping gases	5·91	4·06	2·28	2·52	3·07	2·28
Sources of heat, per unit of coke, in centigrade calories—						
Combustion of coke, calories	2,711	3,087	3,653	3,580	3,258	3,551
Contained in blast do.	...	509	534	574	906	793
Total calories	2,711	3,596	4,187	4,154	4,164	4,344
Less carried off in gases	629	567	397	369	365	189
Useful equivalent of heat	2,082	3,029	3,790	3,785	3,799	4,155
Comparison of heat developed and heat absorbed [1]—						
Heat evolved by fuel, including heat in blast, calories	122,000	104,012	93,455	84,772	100,504	85,538
Do., absorbed	122,000	101,120	89,712	81,531	89,165	83,162
Difference	nil.	2,892	3,743	3,241	11,339	2,376
	122,000	104,012	93,455	84,772	100,504	85,538
Percentage of difference calculated on absorption	...	2·86	4·17	3·97	12·72	2·85
Calories in blast	nil.	14,724	11,919	11,724	21,864	15,618
Elements of absorption corrected to cover differences between two sides of the account—						
Class I. [1]	44,716	34,884	31,308	24,118	33,324	29,477
Class II.	48,980	52,249	52,916	52,818	57,258	52,236
Class III., gases	28,304	16,879	9,231	7,836	9,922	3,825
Calories	122,000	104,012	93,455	84,772	100,504	85,538
Elements of absorption and heat in blast converted into coke[2] per 20 units of iron—						
Class I., units	16·49	9·70	7·47	5·80	8·00	6·78
Class II. ,,	18·07	14·53	12·65	12·72	13·76	12·03
Class III., gases, units	10·44	4·69	2·20	1·88	2·38	0·88
Total units of coke	45·00	28·92	22·32	20·40	24·14	19·69
Heat in blast, equivalent value in coke as burnt with hot air, units	nil.	4·09	2·84	2·82	5·25	3·59
Sum of coke and equivalent of blast reckoned as coke	45·00	33·01	25·16	23·22	29·39	23·38

Foot notes *vide* p. 245.

SECTION X.—BY FUEL USED IN BLAST FURNACES.

It may be here remarked, in reference to the example E, that the fuel stated as having been consumed (24·14 cwts. per ton of iron) seems to be overstated; for it will be perceived that there is an unappropriated excess of 11,339 calories. This is 7,000 to 8,000 calories above the difference exhibited in the other examples. This fact, together with a computation based upon the composition of the gases, induces me to think that the consumption of coke is exaggerated by about $1\tfrac{3}{4}$ cwts.

Table II—I. Elements of absorption are thus classified :—

Class I.	Calories.	Class II.	Calories.	Class III.	Calories.
Fusion of slag	4,240	Reduction iron	33,464	Gases	5,914
Decomp. H_2O blast	2,040	Carb. impreg.	1,775		—
Expulsion of CO_2	2,334	Reduction Si., S., P.	3,200		—
Decomp. of CO_2	2,432	Transmission	3,658		—
Evap. H_2O coke	3,954	Fusion pig	6,600		—
Do. H_2O ore	232	Tuyere water	1,818		—
	15.232		50,515		5,914

[1] The divisions of the calories absorbed are corrected by adding to each head, the percentage of difference between the heat calculated to be produced and that absorbed.

[2] The estimates are obtained by adopting the calories evolved by combustion of the coke as a basis of computation. Thus in the case of D 4,154 calories represent the value of a unit of coke.

Then for Class I. we have ... $24,118 \div 4,154 = 5·80$ coke
„ Class II. „ ... $52,818 \div 4,154 = 12·72$
„ Class III., gases, „ ... $7,836 \div 4,154 = 1·88$

                                                                     20·40

Calories in blast ... ... $11,724 \div 4,154 = 2·82$

                                                                     23·22

246   SECTION X.—EQUIVALENTS OF HEAT EVOLVED

TABLE II.—SMELTING HEMATITE.

	G.	H.	I.	K.
Height of furnace	57	55	70	55
Capacity, cubic feet	7,630	9,400	9,550	10,300
Weekly make per 1,000 cubic feet	57½	31	54½	38
Coke per ton of iron, cwts.	24·43	22·75	18·38	18·00
Limestone do. do.	6·90	8·25	4·86	8·12
Quality of iron, No.	Bessemer	Forge 4·9	Bessemer	Forge 4.
Temperature of blast, degrees C.	491	454	522	718
Do., escaping gas. ,, C.	301	477	205	248
Units of C as CO to 1 of C as CO_2	4·37	3·13	2·04	2·52
Sources of heat per unit of coke—				
Combustion of coke	3,011	3,337	3,696	3,532
Contained in blast	469	463	517	671
Total calories	3,480	3,800	4,213	4,203
Less carried off in gases	442	608	319	283
Useful equivalent of heat, calories	3,038	3,192	3,894	3,920
Heat developed and heat absorbed, compared—				
Heat developed by fuel, including heat in blast, calories	85,039	86,472	77,439	75,665
Do., absorbed, calories[1]	79,089	84,446	71,661	75,074
Difference, do.	5,950	2,026	5,778	591
Calories	85,039	86,472	77,439	75,665
Percentage of difference reckoned on absorption	7·53	2·40	8·06	·79
Calories in blast	11,479	10,528	9,495	12,081
Elements of absorption corrected to cover differences between two sides of the account—				
Class I.) Corrected to bring up	19,860	20,284	16,401	19,332
Class II. (absorption to agree	53,576	52,017	54,682	51,198
Class III. (with coke alleged to be				
in gases.) burnt.	11,603	14,171	6,356	5,135
Calories	85,039	86,472	77,439	75,665
Units of coke consumed per 20 units of iron—				
Class I.	5·70	5·34	3·92	4·60
Class II.	15·40	13·68	12·96	12·18
Class III., in gases	3·33	3·73	1·50	1·22
Total coke	24·43	22·75	18·38	18·00
Heat in blast, equivalent value in coke as burnt with hot air	3·30	2·77	2·44	2·88
Sum of coke and equivalent of blast reckoned as coke	27·73	25·52	20·82	20·88

[1] A new element of absorption appears in G and I, that due to liquefaction of combined water in the ore which is brown hematite. The heat however so applied only amounts to 239 and 232 calories respectively.

SECTION X.—BY FUEL USED IN BLAST FURNACES.

The first point to which I would direct attention, is the causes which appear to affect the oxidation of the carbon of the coke. As already explained in a previous section, time is required for this function, and where the furnace is of insufficient capacity and height it is more or less imperfectly performed. Thus in the case of B, a furnace of 6,000 cubic feet, out of 5·06 units of carbon only 1 (=19·74 per cent.) is converted into carbonic acid: whereas with C of a furnace of 11,500 cubic feet, the carbon as carbonic acid amounts to 30·48 per cent. of the whole, *i.e.* 1 unit out of 3·28.

Although I have occasionally met with a little higher oxidation than that set forth in the last named instance, when smelting Cleveland stone, yet the Tables themselves—as well as many other analyses, besides those upon which the Tables have been constructed—justify the belief that in a furnace of 11,500 cubic feet this operation is carried to nearly its utmost practical limit, at least when grey iron is the object of the smelter. In no case have I found 33 per cent. of carbon in the highest state of oxidation. In one of the Ferryhill furnaces, 103 feet high, and containing 33,300 cubic feet, it only amounted to 25·13 per cent.; and in the example F in Table I., which is a record of a 90 feet furnace at Ormesby, containing 35,016 cubic feet, the proportion is 30·48 per cent.

The same Ormesby furnace however afforded at another time one of the best examples I have met with, viz. 31·94 per cent., but this was when using blast of 83° C. (150° F.) lower temperature than that mentioned in example F.

I have maintained, on many occasions, that within certain limits, capacity of furnace and temperature of the blast are convertible terms, the same advantages being derived from an increase in either.

The advantage of space is exemplified in G and I, both smelting the same hæmatite ore in South Wales. Here G is only 20 per cent. smaller than I yet 27 per cent. less coke is required for the same quantity of pig iron in the larger furnace than in the smaller. The result is that we have a larger percentage of the carbon as carbonic acid in I than in G, and the gases are much better cooled in the former.

Next in importance to a high state of oxidation of the carbon in the gases, comes the avoiding of waste of the heat so obtained. Practically, supposing the furnace to be in good working order, the escaping gases are the only source of loss which need occupy our attention in this respect. The same observations, made with regard to oxidation of carbon in small and large furnaces, hold good in respect to the item just mentioned. In the former, the larger volume of gas, consequent upon the greater consumption of coke, has not the time afforded it for parting with its heat to the descending train of cold material introduced at the top. In consequence the gases escape at a higher temperature: besides which they are much greater in point of quantity. These two circumstances combined give rise to a loss of about 16 per cent. of all the heat evolved in a furnace of 6,000 cubic feet (B Table I.), as against 9·9 and 9·2 per cent. respectively, from those of 11,500 cubic feet (C and D Table I.)

An example of high temperature of blast producing the same effect as increased capacity is afforded in two examples in Table II. Furnace K, only 10 per cent. larger than H, (both being of moderate size), but blown with air at 718° C. against 454° C. in H, loses only 6·78 per cent. of its heat in the gases, against 16·75 per cent. in the furnace H (Table II.)

I have to remark, in reference to the quantity of heat carried away in the escaping gases, that my own observations on furnaces at the Clarence works, ranging from 11,500 to 25,500 cubic feet, have induced me to consider that no material diminution of the loss, arising from this cause, accompanied any increase of capacity after that of the 11,500 cubic feet; the height in each case being 80 feet. This opinion was confirmed by an examination of a furnace of a much larger capacity, having a height of 103 feet.

In example D, Table I., out of 4,154 calories, the loss by the escaping gases amounted to 369 calories, equal to 8·88 per cent. The more recent performance of an Ormesby furnace, 90 feet high with a capacity of about 35,000 cubic feet, as given under F in Table I., shows the heat in the escaping gases to be only 189 calories out of 4,344, equal therefore to 4·35 per cent.[1]

---

[1] The temperatures of the escaping gases are given as observed. These however include the heat brought in by hot ironstone, which, from a series of observations was equal to about 2,000 to 2,400 calories per 20 units of iron. This in a furnace producing 130 to 140 cwts. of gas for this quantity metal would be equivalent to 64 to 66° C. which ought therefore to be deducted from the temperatures given to obtain. In the computation of the heat carried away in the gases an allowance is made for that considered as brought in by the hot ironstone.

SECTION X.—BY FUEL USED IN BLAST FURNACES.

I am not quite sure that a part of this last mentioned difference may not be due to differences in the temperature of the ironstone. Whether this be so or not, it must be remembered that the Ormesby furnace was driving so slowly, that its rateable produce was only half that of the furnaces to which my observations applied.

The general tenor of these remarks on this large furnace at Ormesby is, I admit, inconsistent with my belief, that raising the height or increasing the capacity of a furnace, only raised the zone where heat, generated by reduction, was produced. The difference however in the results, between one furnace and another of three times the size, is not a large one—only 4·7 per cent. reckoned on the total heat—so that the causes just alluded to may partly account for its occurrence.

I now come to the much debated question of the value of successive additions to the temperature of the blast; and for greater clearness it will be assumed, in each case, that we are dealing with furnaces of such dimensions that the carbon does not exhibit any striking difference, in point of oxidation.

Unfortunately, I do not possess any data from which I can compile the performance of a cold blast furnace having a capacity of 11,500 cubic feet or more. I would however recall the fact, mentioned page 85, of an enlarged cold blast furnace producing iron with the same weight of fuel per ton as a smaller furnace of the old type, blown with hot air. We will therefore assume the performance in respect to coke, of a furnace such as C in Table I. (11,500 cubic feet), if blown with air at 0° C., to resemble that of the small furnace B (6,000 cubic feet), Table I., when blown with air at 485° C.—say 28·92 cwts. of coke per ton of iron.

Examples of the effect on consumption of coke by raising the temperature of the blast from 0° to 819° C.

	Temperature. °C.	Class I.	Class II.	Class III. Gases.	Total Units Coke.
Small furnace. coke for 20 units of iron, 48 feet high, 6,000 cubic feet	0	16·49	18·07	10·44	45·00
	485	9·70	14·53	4·69	28·92
Large furnace, coke for 20 units of iron, 80 feet high, 11,500 cubic feet	0	9·70	14·53	4·69	28·92
	485	7·47	12·65	2·20	22·32
	563	5·80	12·72	1·88	20·40
Large furnace, coke for 20 units of iron, 80 feet high, 35,016 cubic feet	819	6·78	12 03	·88	19·69

In all these cases, where heated air is used, the data are taken from actual experience in smelting Cleveland stone. In those of cold blast the estimate is also founded on using the same quality of ore.

From these numbers the saving of coke, at each increment of temperature in the blast, appears to be as follows:—[1]

	Class I. Cwts.	Class II. Cwts.	Class III. Gases. Cwts.	Total. Cwts.
Small furnaces, blast raised from 0° to 485° C. (32° to 905° F.) ...	6·79	3·54	5·75	16·08
Large furnaces, raising blast from 0° to 485° (32° to 905° F.) ...	2·23	1·88	2·49	6·60
Large furnaces, raising blast from 485° to 563° (905° to 1,045° F.)...	1·67	—·07	·32	1·92
Large furnaces, raising blast from 563° to 819° (1,045° to 1,506° F.)...	—·98	·69	1·00	·71

These figures are not given under the idea that they accurately set forth the saving of coke due to the different additions to the temperature of the blast. To do this it would be necessary that the furnaces in all respects were working under precisely the same conditions. All the figures are meant to convey is that the saving, large as it is with the first additions of heat in the blast, falls off afterwards very rapidly in its amount.

It will be noted that the saving in coke is by far the largest at the first raising of the temperature of the blast. This, it is almost needless to remark, does not arise from any difference in the value of a given amount of heat contained in air between 0° C. and 400° C., and the same amount contained in air between 400° C. and 800° C., or any other numbers. The real cause is due, partly to differences in the quantity of work to be performed, and partly to a difference in the nature of the combustion of the fuel, the heat from which has to be replaced by heat contained in the blast.

As regards the relative quantity of work to be performed, we have only to inspect the Tables already given, to perceive, that out of 122,000 calories, in the case of A, 73,020 are accounted for under the heads of Class I., and Class III. or in escaping gases. This large expenditure of heat, as has been already mentioned, is due to the excessive quantity of flux required, and to the large volume of highly heated gases which

---

[1] *Minus* sign means increase.

escape. With hot blast having a temperature of 485° C., in a furnace of the same size (B, Table I.—6,000 cubic feet), the absorption from these two causes has fallen to 51,763 calories. After this, absorption under Class I. and escaping gases instead of falling proportionately with further increments in the temperature of the blast, only shows a little above half the saving affected by the first additional heat communicated to the air. Thus between 0° and 485°, the absorption under these two heads was reduced, in small furnaces (A and B 6,000 cubic feet) by 21,357 calories, whereas between 485° C. and 819° C., in enlarged furnaces (C and F), the difference was ascertained to be under 7,300 calories. Thus while each degree (Centigrade) in the blast has saved 43 calories in the first case the saving in the second only amounts to 22 calories per degree of temperature.

The economy observed in the escaping gases in C, D, E and F is, to some extent, due to the reduced quantity of coke burnt, giving a diminished volume of gases leaving the furnace; but it is also due, in some degree, to their reduced temperature, arising from enlarged capacity of the structure.

The manner in which the nature of the combustion, *i.e.* the degree of oxidation of the carbon, affects the amount of economy of fuel has been already sufficiently explained to need much further notice.

Practically the furnace gases may by their combustion be regarded as the ordinary means of heating the blast. At page 143 it was estimated that out of 72,454 calories obtained in this way there was a waste of 45 per cent.: hence the last 1,000 calories in the air required the expenditure of 1,818 calories in the hot air stoves containing metal pipes. The loss in fire brick stoves is stated to be only about half this amount, in which case an expenditure of something like 1,350 calories will afford 1,000 calories in the air.

These figures, obviously have no reference to the money economy attending the use of hot air; because the coal, and latterly the escaping gases, required to heat the blast, were of little, indeed in some cases of no, value.

It is, in our present state of knowledge, a very easy matter to calculate, with a fair approach to correctness, the quantity of coke required to produce a given number of calories, when burnt with air at a given temperature. Let us select the most favourable case in the Table for the purpose of illustration: viz. D, where 84,772 calories sufficed to produce 20 units of pig iron.

For the purpose of calculation, it is indispensable that we should know before-hand, what proportion of the carbon in the gases is found in the form of carbonic acid. If we take it at 2·10 as CO to 1 as CO_2, the heat evolved will be:—

$$\frac{2 \cdot 10 \times 2{,}400 + 1 \times 8{,}000}{3 \cdot 10} = 4{,}206, \text{ less } 7\tfrac{1}{2} \text{ per cent. for impurity} = 3{,}891$$

calories per unit of coke.

To obtain the heat required in the blast, when using different weights of coke, the following figures are given:—

	Calories.		Calories.	Calories.
18 units of coke	× 3,891 = 70,038,	leaving for blast	14,734	= 84,772
17 ,,	× 3,891 = 66,147,	,, ,,	18,625	= 84,772
16 ,,	× 3,891 = 62,256,	,, ,,	22,516	= 84,772

Now to burn these respective quantities of coke, supposing 9·38 units of limestone to be used, as was done in the case of D, we should require as under:—

						°C.	°F.
For 18 units of coke,	78,91 units of air having a temperature of	788	= 1,450
,, 17 ,, ,,	71,87	,,	,,	1,095	= 2,003
,, 16 ,, ,,	65,30	,,	,,	1,455	= 2,651

Judging by past experience, with brick stoves provided for superheating the air, the first of the three examples alone can enter into our calculation; for it would, I apprehend, be practically impossible to heat the blast continuously to 1,095° C.

In such a calculation, all depends on the possibility of maintaining the carbon at the degree of oxidation supposed, *i.e.* 1 out of every 3·1 units, equal to 32·26 per cent. as carbonic acid; which in my experience, is quite exceptional at all events in furnaces using coke.

We are thus, after all, constrained to consider what has been effected in practice. Now at the Ormesby furnaces, certain details of which are given at page 108 of this work, 19·69 cwts. of coke was named in 1881, as the then rate of consumption in a furnace of 35,016 cubic feet, driven with air ranging from 736° to 819° C. In 1882 this had not only been maintained, but Mr. Cochrane, the senior partner in the concern, mentioned at the meeting of the Institution of Mechanical Engineers in Leeds that the coke had been reduced to 18·70, which the above calculation shows to be *possible*.[1]

No one can dispute the proposition, that each addition to the calories contained in the blast, must place at the disposal of the smelter a corresponding increase of heat. Failure to render this heat available may however arise from an undue quantity being carried away out of the throat of the furnace in the gases or from the carbon, by its comparative imperfect oxidation, not affording the heat it is capable of developing.

At page 112 of this work, in the absence of any information as to the interior condition of one of the Ormesby furnaces (E, Table I.), I hazarded the opinion that possibly the very high temperature of the blast, by elevating the temperature of the reducing zone of the furnace, caused carbonic acid to act on carbon, and thus to be lowered to the condition of carbonic oxide. Mr. Cochrane, in his paper, mentions the discovery of scaffoldings in the furnace in question; which no doubt would produce the same effect as that just named, although in a somewhat different way, viz., by allowing part of the reducing gas to escape before it had taken up the full equivalent of oxygen it is capable of separating from the ore.

The figures already given indicate quite clearly the deficiency of carbonic acid; and whatever may be the immediate cause of this, the practical result is the same, viz., a waste of fuel, due to its imperfect oxidation.

If Mr. Cochrane's hypothesis be the correct one, it follows, either that the particular shape of these furnaces, or the fact of their being blown with superheated air, has rendered the formation of scaffolds an evil of a very persistent kind. This is inferred from the fact that in Mr. Cochrane's communications respecting their performance, extending over 10 or 12 years, furnaces of 20,642 cubic feet never appear to have done as well as many other furnaces, even of much smaller dimensions and blown with air at only 500° C. (932° F.)

---

[1] It must be remembered however that the produce, per 1,000 cubic feet of furnace space, was only half that of the general run of furnaces making Cleveland iron.

One example of the co-efficient of useful heat at a furnace of 90 feet has already been given at F in Table I. Observations of the same furnace on another occasion gave a somewhat superior figure, although the temperature of the blast was somewhat lower—736° C. instead of 819° C. The details founded on an analysis made at the Clarence Works were as follows, per unit of fuel consumed:—

	Calories.
Heat due to combustion of coke	3,637
Do. contained in blast	755
	4,392
Less in escaping gases	111
	4,281 calories per unit of coke.

In this case, however, I suspect that the temperature of the escaping gases was not a correct average; for notwithstanding that the total heat produced was superior to the example given in the table—4,392 instead of 4,344—the consumption of coke per ton of iron was more, viz., 20·08 cwts. instead of 19·69, as shown by the charging-book account.

In both cases the results in point of consumption of fuel may be regarded as satisfactory; but there is the inconvenience, already alluded to, *quantum valeat,* of the make being only half that of the furnaces with which it has been compared. This, however, is perhaps not a serious matter, as Mr. Cochrane justly observed in the discussion on his paper at Leeds; for any such loss amounts to little more than the interest on the extra cost of the furnace itself. Supposing the additional outlay to involve £300 a year for this item, the saving of 1 cwt. of coke per ton of iron on 30,000 tons per annum would, if it can be maintained, far more than cover it.

We must not however lose sight of the danger of attaching too much importance to the performance of a few individual cases. I have given 20·4 cwts. as the consumption of coke in a furnace of 11,500 cubic feet, blown with air at 573° C. This is one of those at Clarence (D, Table I.); but 22 cwts. or even more is nearer the average consumption of the entire establishment. Again, I have quoted the case of a hematite furnace of 9,950 cubic feet, blown with air at 522° C.

(I, Table II.) and making Bessemer pig with as little as 18·38 cwts. of coke; but every one knows perfectly well that 20 to 21 cwts. is nearer the average consumption of furnaces producing this quality of metal from hematite ores, even in furnaces of like dimensions to that in question.

What after all we have to deal with is actual experience, acquired at furnaces generally. Guided by this, my belief is that up to this time, the application of superheated air to furnaces under 25,000 cubic feet at all events, has not been attended with any very marked economy in coke, when smelting the usual run of clay ironstones.

As has been already remarked, when viewed as a mere question of heat development, each calorie delivered along with the blast, into a furnace, *in which the limestone, and the loss in the escaping gases, have been reduced to their lowest quantities,* ought unquestionably, other circumstances being the same, to save one calorie previously supplied by the fuel. Parenthetically it may be observed that stress is here laid on these two items; because, as will be seen on reference to the tables contained in the present section, the use of hot blast in small furnaces led to a considerable diminution in both. In such a case these changes were regarded as having resulted in a better oxidation of the gases, and having consequently raised the heating power derived from the combustion of the carbon from 2,711 to 3,087 calories. At the same time 104,012 calories in the furnace were enabled to do the work of 122,000.[1] These 104,012 calories consisted of 89,288 furnished by the fuel, and 14,724 contributed by the blast. We thus have, in the case of cold blast, a requirement of 122,000 calories, to the production of which the more expensive fuel consumed in the furnace itself was devoted. On the other hand, out of the 104,012 calories developed in the hot blast furnace, 89,288 only were due to the coke and 14,724 to the blast. By this account it appears that 32,712 calories obtained by burning coke—the difference between 122,000 and 89,288—have been replaced by the application of 14,724 calories contained in the blast; so that by means of an alteration in the heat required for the elements of absorption under Class I., by the better oxidation of the carbon, and by the reduced loss attending

[1] *Vide* Table I., A and B.

the diminished volume and temperature of the escaping gases, a saving of (32,732 − 14,724) = 18,008 calories has been effected. In this way each calorie in the blast has been enabled to render the same service as 2·22 calories contributed by the coke which it has displaced.

Returning to the question before us, anterior to this short diversion, it would appear, so far as I have been able to learn, that the experience of Bessemer-iron makers corresponds with what has been laid down, as being that of smelters of Cleveland and other clay ironstones.

Soon after the meeting at Leeds above referred to, I visited some of the principal works engaged in smelting hematite ores for steel making, on the West Coast of England. At the furnaces there they were all using pretty much the same quality of ore, and the same quantity of limestone, and all were producing nearly the same description of iron for Bessemer steel. From the experience supplied by the performance of furnaces having a height of 70 feet, it appears that the results as to consumption were practically the same with the two kinds of hot air apparatus, viz., pipe stoves and fire brick stoves. With the blast heated in metal pipes, the coke used was 20·43 cwts., against 20·36 cwts. for fire-brick stoves. It has to be observed, however, that the difference between the temperature of the blast was only given at about 130° F. in favour of the brick stoves; and also that the make was higher—in one case by 16 per cent.—from furnaces using fire-brick stoves and blast at the higher temperature than from those using pipe stoves, and air at about 1,000° F. This, it need hardly be said, is of itself an important advantage on the side of the fire-brick stoves.

With regard to the small difference between the temperature of air received from the two kinds of stoves, as observed on these particular furnaces, I have been informed that fire-brick stoves are apt to fall back after some years use, in the temperature at which they deliver the air. Whether this arises from the difficulty of removing the coating deposited by the furnace gases used as fuel, or from a want of care in working I am unable to say. The metal pipes in hot air stoves, of the old kind, are liable to the same inconvenience; but, as they are more accessible, and do not present above one-fourth the surface to be cleaned, it is one of much less magnitude. I would however point out, in reference to the performance of the furnaces in question, that the saving in coke does

SECTION X.—BY FUEL USED IN BLAST FURNACES.

not quite correspond with even the small difference in the temperatures of the blast, viz., 130° F.; the saving being still something below what it ought to be.

The weight of blast per 20 units of pig would be about 88 units which heated to 130° F. (72° C.) ought to give 88 units × ·237 (sp. heat) × 72° C. = 1,498 calories. This number, taking 4,000 calories as the heat evolved per unit of coke, will represent $\frac{1498}{4000}$ = ·375 cwts.; whereas the saving only appeared to be ·07 cwts.

At a pair of furnaces on the East Coast of England, also making Bessemer iron, from Spanish ore, the average consumption of coke is about 20·5 cwts. Their dimensions are 60 feet high, with a capacity of 11,650 cubic feet. They are supplied with blast heated in firebrick stoves to 1,150° F. (621° C.)

The necessity of caution in forming any opinion on the matters under consideration, save that acquired by actual experience, is confirmed by the information I have received from friends in South Wales.

The duty of Example I., Table II., is that of a furnace making a fair proportion of Bessemer iron in the Principality. It is 70 feet high, with a capacity of 9,550 cubic feet, and its consumption is given at 18·38 cwts. of coke per ton of iron, blown with air heated in metal pipes, to 972° F. (522° C.)

At a recent meeting of the South Wales Engineers, Mr. E. P. Martin, now of Dowlais and until recently of Blænavon, gave the consumption of coke at No. 9 furnace, using fire-brick stoves, at the latter place at 18·59 cwts. I learn from another source that the temperature of the blast is 1,600° F. (871° C.), and that the consumption of coke is now less than the figure named by Mr. Martin. The ores in use are pretty nearly the same in each case at the two works (I, Table II. and Blænavon); and assuming the quantity of coke consumed also to be the same, we have merely to consider the difference in the quantity of heat in the blast at the temperatures of each.

The difference amounts to 349° C., and taking the weight of air used at 77 units per 20 of iron we have:—

77 blast × ·237 (sp. heat) × 349° C. = 6,369 calories.

Now the net equivalent of the fuel in the case of I, Table II., was 3,894 calories per unit of coke; so that according to this calculation, the work performed by the furnace using superheated air represents a loss in heat, as compared with example I with its metal pipe stoves, equal to 6,369 ÷ 3,894 = 1·63 cwts. of coke per ton of iron.

This must be due to the less perfect oxidation of fuel; but whether this is owing to the difference in height—the furnace I being 70 feet and that at Blænavon 60 feet—or some other cause, I am unable to say. In the matter of capacity the Blænavon furnace, however, is the larger of the two.

From another work in South Wales I am informed that air at 1,500° does its work with 10 per cent. less fuel than air at 1,000° in furnaces of the same size. The difference here is 500° (278° C.) and the weight of blast was about 100 units per 20 of iron. We have therefore :—

100 units × ·237 (sp. heat) × 278° C. = 6,588 calories.

The net useful effect of the coke used, which is of second-rate quality, may be taken at 3,400 calories; and $\frac{6588}{3400}$ = 1·94 units of coke, which closely corresponds with the saving stated to result from the use of the more highly heated air.

A second furnace at the same establishment is making No. 4 iron from hematite, chiefly Spanish, yielding 56·4 per cent. metallic iron with a consumption of 4·64 cwts. of limestone. The coke required is 21·19 cwts., burnt with air at 1,500° (815° C.), for "tinplate" pig, average No. 4. Having regard to the richness of the ore and the small quantity of limestone, this is by no means an unusually low rate of consumption for a furnace receiving blast even at only 500° C. I am however, not in a position to form an opinion as to the cause. The furnaces at the work in question have a capacity of 9,722 cubic feet, with a height of 60 feet. Whether the height or the difference in the quality of the fuel has produced so much less favourable results than those given in I, Table II., I am unable to say.

SECTION X.—BY FUEL USED IN BLAST FURNACES.

Returns from another work in Wales exhibit the following results in making Bessemer iron:—

	Capacity, Cubic Feet.	Make per Week per 1,000 Cubic Feet. Tons.	Stoves.	Temperature Blast, ° F.	Coke per Ton.	Cwts. Ore and Limestone per Ton Iron.
a	8,593	64½	Iron.	900	21·67	44·55
b	9,401	72¼	Fire-brick.	1,300	20·59	42·65
c	10,543	54¾	,,	1,450	20·11	46·87
Average of b and c	9,972	63½	,,	1,375	20·35	44·76

It will be observed that the two furnaces using superheated air, supplied by fire-brick stoves, are considerably larger than the furnace where the blast is heated in pipes. In one case also the ore smelted in the former was richer. Taking the average of the two furnaces, b and c, fed with superheated air, the saving of coke as compared with a is 1·32 cwts.: so that making allowance for the extra capacity of b and c, one cwt. of coke is probably something like the saving obtained by raising the temperature of the air 475° F. above that supplied to a.

It is scarcely necessary to pursue the subject at any greater length. The facts and figures clearly prove that there are many disturbing causes which interfere greatly with the action of blast furnace work. These render it impossible for any one to speak unhesitatingly on the questions involved in such comparisons as are contained in the present section, unless he has the means, under his own control, of *carefully and continuously* studying the action of the furnaces themselves.

Of this I think I may speak with some confidence, viz., that the question is capable of being considered on the plain and simple lines I have laid down, by any one who will give himself the trouble of pursuing it. Some misconception, I take it, has arisen from neglecting a proper appreciation of the circumstances of particular cases. Superheated air is often applied to furnaces worn out themselves, and blown with air insufficiently heated by dilapidated metal pipe stoves. Of course the change is accompanied by a great economy of fuel, sometimes stated to be five cwts. on the ton of iron.

The fact may be accepted as perfectly true; but if it is pretended that this weight of coke represents the difference between the performance of a furnace, in good condition and carefully attended to, blown with air at 1,000°; and that of a similar furnace blown with air at 1,500°, such an interpretation is a complete and obvious fallacy.

I would repeat that, in furnaces of *sufficient capacity*, the oxidation of the fuel and loss of heat in the escaping gases are mainly if not entirely independent of the temperature of the air. The heat necessary for the process is sufficiently understood, so that no difficulty need arise in connection with its correct estimation. Admitting these two premises, it is, I would submit, quite idle to imagine that the economy resulting from the use of superheated air can exceed in value the co-efficient of coke when properly burnt.

If the combustion of the fuel is such that the volume of carbonic oxide is as 100 to 50 of carbonic acid, the values of two units of coke containing 10 per cent. of impurity, when using air at such temperatures, that in the one case we have 560 calories, and in the other 860 calories, in the blast per unit of coke burnt, is as follows:—

	Calories evolved per Unit of Coke burnt.	
	A.	C.
Combustion of coke	3,840	3,840
Contributed by the blast	560	860
	4,400	4,700

showing an increase of heating power of 6·82 per cent. in C.

If on the other hand there are 45 volumes of carbonic acid per 100 of carbonic oxide in the gases, we have:—

	B.	D.
Combustion of coke	3,722	3,722
Contributed by the blast	560	860
	4,282	4,582

or an increase of heating power equal to 7 per cent. in D.

The blast in these two sets examples A B and C D, may be taken at 540° C. (1,004° F.) and 870° C. (1,598° F.), respectively.

Beginning with a consumption of 21 cwts., as a starting point in the case of A, we have the following results per ton of iron:—

	Air at 540° C. Cwts.	Air at 870° C. Cwts.	Cwts.
Carbon oxidised as in A	21·00	as in C ... 19·66—saving	1·34
Do. B	21·58	„ D ... 20·17— „	1·41

SECTION X.—BY FUEL USED IN BLAST FURNACES.

These figures contain an instance of the fuel in an inferior condition of oxidation, having the defect more than compensated for by the superior amount of heat in the air—see A and D.

An obvious question will suggest itself to the furnace manager, in connection with the figures made use of in the present section, viz., their practical bearing on the quantity of coke required to make a ton of pig iron.

It is clear that the answer is dependent upon a great variety of conditions, the whole of which must be clearly understood and accurately stated. These conditions are comprised under two heads :—first, the quality of the fuel and the manner in which it is burnt; and second, the amount of heat required to produce the iron manufactured at any particular furnace.

For our present purpose it will be assumed that the plant is one of the modern type now commonly used in the Cleveland district: the fuel employed is good Durham coke containing 6 per cent. of ash and 1½ per cent. of other foreign matter, free from any combined oxygen, and burnt to two different conditions of oxidation, by air at two different temperatures, viz., 485° C. and 800° C. (905° and 1,472° F.)

Blast having a temperature of 485° C.—
a Gases containing 45·45 vols. CO_2 per 100 vols. CO.[1] Calories from coke 3,839, calories in blast 529. Heat evolved per cwt. of coke   **4,368**
b Gases containing 47·62 vols. CO_2 per 100 vols. CO.[2] Calories from coke 3,890, calories in blast 529. Heat evolved per cwt. of coke   **4,419**

Blast having a temperature of 800° C.—
c Gases containing 45·45 vols. CO_2 per 100 vols. CO.[1] Calories from coke 3,839, calories in blast 872. Heat evolved per cwt. of coke   **4,711**
d Gases containing 47·62 vols. CO_2 per 100 vols. CO.[2] Calories from coke 3,890, calories in blast 872. Heat evolved per cwt. of coke   **4,762**

In estimating the heat required, the factors given at page 95, for smelting Cleveland iron, will be adhered to. They are based on a

---

[1] By weight this is equivalent to 1 of carbon as carbonic acid to 2·2 as carbonic oxide.

[2] By weight this is equivalent to 1 of carbon as carbonic acid to 2·1 as carbonic oxide.

consumption of 11 cwts. of limestone per ton of pig iron, accompanied by a production of 27·92 cwts. of slag. These conditions brought out the various elements of absorption to 87,000 calories. On referring however to Table I., p. 244, it will be seen, on comparing the heat required with the heat evolved, that the former is deficient in quantity by about 3 per cent. This may be due to trifling errors in the factors made use of, or there may be an absorption of heat in vaporizing the earthy matters carried off in the escaping gases. To cover this difference 2,600 calories are added, making the total requirement 89,000 calories.

With the elements for calculation just stated, we have the following weights of coke needed for each ton of pig iron:—

					Cwts. Coke.
a,	45·45 Vols. CO_2,	100 Vols. CO.	Blast 485° C, ...	$\dfrac{89,600}{4,368}$	$= 20\cdot51$
b,	do.	do.	do. 800° C, ...	$\dfrac{89,600}{4,419}$	$= 20\cdot28$
c,	47·62 Vols. CO_2,	100 Vols. CO.	Blast 485° C, ...	$\dfrac{89,600}{4,711}$	$= 19\cdot02$
d,	do.	do.	do. 800° C, ...	$\dfrac{89,600}{4,762}$	$= 18\cdot82$

In all these cases the quality of the iron is considered to be between No. 3 and No. 4, say an average of No. 3·25. In each instance it is, of course, assumed that the supposed conditions are maintained constant; but every furnace manager knows that this is impossible. The air varies in the amount of moisture it contains, the minerals vary in composition as well as condition, and the workmen in the attention they give. If then it were essential to produce a large quantity of iron not falling below the quality stated, some trifling addition would, no doubt, be required to the quantities given in these estimates, and I propose to add a few lines on this head in connection with the figures contained in the tables given in the present section.

Let us take the example of a furnace of dimensions insufficient to permit that prolonged contact between the gases and solids, which is needed for securing the saturation of the former with oxygen, and the proper heating of the latter. A waste of fuel is the consequence;

because each unit affords less heat from its imperfect oxidation, and more heat is also carried off in the escaping gases, partly from their high temperature, and partly from their larger volume caused by the combustion of an increased quantity of coke for a given quantity of iron.

A specimen of this defective working is seen on comparing B and C in Table I., which contains the following figures:—

	B.	C.
Fuel consumed per 20 units of iron	28·92 units	22·32 units.
Carbon as carbonic oxide to one of carbon as carbonic acid	4·06 ,,	2·28 ,,
Cubic feet in furnace	6,000	11,500
Temperature of escaping gases	452° C.	332° C.
Do. blast	485° C.	485° C.
	Calories.	Calories.
Heat from combustion per unit of coke	3,087	3,653
Heat in blast per unit of coke	509	534
	3,596	4,187
Carried off in escaping gases per unit of coke	567	397
Net heat equivalent	3,029	3,790

The result of the differences given above is such that, with the blast at the same temperature in each case, there is a higher useful effect, by about 25 per cent. in the heat evolved from C than from B, accompanied by a corresponding economy in the fuel consumed.

It would be useless to attempt to obtain the net result given in C, viz. 3,790 calories, by a mere increase in the temperature of the blast; because it will be seen that the air employed for burning the coke is too small in weight to contain the necessary quantity of heat. Thus:—

	Calories.	
Heat from combustion of coke as in B	3,087	
Heat in blast required to secure the net result of C	1,100	(Temp. 1,009° C.)
Total heat	4,187	
Heat carried off in escaping gases, taken as in C	397	
Net heat	3,790	

After making the necessary allowance for oxygen supplied by the ore and flux and for the carbon which combines with the iron, 4·60 units of air are required for the combustion of the coke. We have therefore 1,100 calories to divide by 4·60 (units of blast) × ·237 (sp. heat of atmospheric air): giving 1,009° C. or 1,848° F., which represents the temperature of the blast.

Experience with fire-brick stoves does not hold out any expectation at present of our being able to heat air to the point just named. A compromise therefore will require to be made, *i.e.* instead of seeking to raise the power of the unit of coke to 3,790 from 3,029 calories simply by raising the temperature of the blast, an enlargement of the furnace must be provided, in order to obtain a higher degree of saturation of the gases with oxygen. In this calculation we will assume that the ratio of carbon as carbonic oxide to carbon as carbonic acid is 3·5 to 1, and that the loss in the escaping gases is midway between B and C, viz., 482 calories per unit of fuel instead, of 567 or 397 calories.

	Calories.
Heat by combustion of coke—3·5 of C as carbonic oxide to 1 of C as carbonic acid	3,465
Heat in blast required to secure the net result of example C	807 (Temp. 740° C.)
Total heat	4,272
Carried off in escaping gases—mean between B and C	482
Same net result as C	3,790

Probably the results obtained in this case would be secured in a furnace containing 8,000 or 8,500 cubic feet, and receiving its blast at 740° C., or 1,364° F.

It is obvious however that it is much simpler to obtain as much as possible of the heat required by a high state of oxidation of the fuel, rather than by having to expend fuel and provide apparatus for heating the blast. I would quote as an instance of this the example C in Table I. Thus:—

SECTION X.—BY FUEL USED IN BLAST FURNACES.

	C.
Carbon as carbonic oxide to 1 of carbonic acid	2·28
Temperature of blast	485° C.

	Calories.
Heat from combustion of coke	3,653
Heat contained in blast	534
	4,187
Carried off in gases	397
Net result	3,790

As has been already stated in Section VI. p. 84, the value of the hot blast depends, up to a certain point, on effecting what is equivalent to a practical enlargement of a furnace, by diminishing the volume of gases produced in the manufacture of a given weight of iron. This it does by making the air the vehicle of heat, instead of having recourse to the combustion of fuel.

In addition to what may have been already said on the subject I would submit a few observations on the nature of the saving effected by successive additions to the temperature of the air blown into the furnace.

Let us admit the possibility of so oxidizing the carbon that the gases as they escape contain 1 of carbon as carbonic acid to 2·1 as carbonic oxide. So long as the carbon in the latter form is in excess of the ratio just named, viz. 2·1 to 1 as carbonic acid, the introduction of heat in the blast means the saving of heat afforded by burning fuel to the state of carbonic oxide only. Now to cover loss from impurities in the coke, say $7\frac{1}{2}$ per cent., it has been assumed that each unit of this description of fuel gives 2,220 calories when burnt to the form of carbonic oxide, and 7,400 calories when burnt to carbonic acid. If thus 1,000 calories are contributed by the blast, and we are saving fuel which only goes to form carbonic oxide, we have $\frac{1000}{2220} = \cdot 450$ units of coke.

If on the other hand the saving of coke which only serves to generate carbonic oxide has been carried to a point where the gases contain only 2·1 of carbon in this form to 1 in that of carbonic acid, or 100 volumes of CO to 47·6 of CO^2, then, the ratio of the two carbon oxides remaining unaltered, the saving is found as follows:—

```
1   of carbon as carbonic acid  × 7,400 = 7,400 calories.
2·1   do.    carbonic oxide    × 2,220 = 4,662     ,,
___
3·1 units of carbon give ...    ...     ... 12,062
                                            _____ = 3,890 cals. per unit of coke.
                                              3·1
```

Then we have $\frac{1000}{3890}$ = ·257 units of coke per 1,000 calories.

Thus, so long as there is an excess of carbonic oxide in the gases, the saving for each 1,000 calories in the blast amounts to nearly half a unit of coke (·450); whereas, when the gases arrive at the supposed point of static equilibrium, the saving is only one-quarter of a unit (·257). At this rate the saving will remain constant *so long* as the ratio of the two oxides of carbon remains constant.

The limit to further economy under these conditions will be the temperature to which it will be practicable to raise the blast. In the case of E and F, Table I., the temperatures were 888° C. and 819° C. respectively. If we assume 850° C. (1,562° F.) as an average practical temperature, then 4·60 units of atmospheric air would contain 981 calories, and taking the loss on the escaping gases at 258 calories, we have:—

	Calories.
Heat from combustion of one unit of coke—2·1 as carb. oxide to 1 as carb. acid, ...	3,890
Heat contained in the blast (4·6 × ·237 sp. heat × 850°)	925
	4,815
Less carried off in the gases	258
Net result	4,557

In the estimates just given the proportion of carbon as carbonic acid is in reality in excess of what I have usually found in the gases; 1 to 2·20 or 2·25 as carbonic oxide being a more common ratio. In the cases of C and D, already particularly referred to, the figures were 2·28 and 2·52 respectively, and then the net results were 3,790 and 3,785 calories.

SECTION X.—BY FUEL USED IN BLAST FURNACES.

Applying the above ratio, say 2·20 to 1, to a furnace blown with air at 850° C., the effect would be as follows:—

	Calories.
Heat from combustion of coke (2·2 as carb. oxide to 1 as carb. acid)...	3,839
Heat contained in blast (4·60 air × ·237 sp. heat × 850° C.) ...	925
	4,764
Less carried off in gases	258
	4,506

It is a remarkable fact however that, with one exception, I have invariably found the proportion of carbon as carbonic acid, in the gases from furnaces blown with superheated air, below that in furnaces using air of more moderate temperatures. The exception referred to is the fourth in the subjoined list (F in Table I.). It contains the particulars of the large Ormesby furnace, which it will be remembered was running only half the quantity of metal usually produced for a given capacity from a Cleveland furnace. In the following list the performance of three other furnaces is added in order to obtain an average result.

Equivalents of Heat per Cwt. of Coke in furnaces using air superheated in Firebrick Stoves.

Work.	Capacity.	Temp. of Blast °C.	C as Co to 1 C as CO_2.	Calories from combus. of Coke.	Calories in blast.	Total.	Less in Gases.	Net result.
Ormesby	20,642	780	2·89	3,442	747	4,189	464	3,725
Consett, K, Table II. ...	10,300	718	2·52	3,532	671	4,203	283	3,920
Ormesby, E, Table I. ...	20,642	888	3·07	3,258	906	4,164	365	3,799
Do. F, Table I. ...	35,016	819	2·28	3,551	793	4,344	189	4,155
Average	21,650	801°	2·69	3,445	779	4,224	325	3,899

Against these I would place the figures appertaining to furnaces using air heated by means of iron pipes :—

Work.	Capacity. c. feet.	Temp. of Blast. C.	C as CO. to 1 C as CO_2	Calories from Combst. of Coke.	Calories in Blast.	Total.	Less in Gases.	Net result.
Clarence, C. Table I....	11,500	485°	2·28	3,653	534	4,187	397	3,790
Do. D., Do. ...	11,500	563°	2·52	3,580	574	4,154	369	3,785
Forest, I., do. II....	9,550	522°	2·04	3,696	517	4,213	319	3,916
Average ...	10,850	523°	2·28	3,643	542	4,183	362	3,830

Without implying any distrust whatever in the value of fire-brick stoves, I would point out that with air of an average temperature of 523° C., used in furnaces of dimensions only half those of other furnaces using blast heated to an average of 801°, results if anything rather better than in the larger furnaces have been obtained. It must not be overlooked, however, that the question of the kind of stove to which the preference has to be accorded can only be settled by the average performance over long periods of a great number of furnaces. All I intend at present is to shew by the figures I have made use of, that as great an average heating power has been obtained from coke, in furnaces of moderate size driven with air at comparatively moderate temperatures, as in those of much larger dimensions driven with superheated air.

There is one other matter in connection with the economy of fuel raised by a most competent authority in the manufacture of iron, I mean Mr. Edward Williams. The importance of attending to the *shape* of a furnace as well as to its capacity, was insisted on, in July, 1882, by this gentleman, at a meeting of the South Wales Institute of Engineers, of which he is president.

The question has already been alluded to in these pages (page 124), but without insisting, to the extent Mr. Williams does, on the superior advantages gained by additions of height in preference to enlarging the capacity of a furnace by increasing its width. This, like most other branches of the enquiry, can be determined by experience alone; for we not only have to design a form, with which gases, in various degrees of dilatation, may find their way most equably through the solids, but we have also to study the conditions least favourable to the "scaffolding" of the materials.

This is a subject to which practical smelters would do well to turn their attention, recording for mutual instruction the results of their experience. My own observations on the subject are too limited to speak with a sufficient degree of confidence. This much I may be permitted to state, viz., that the furnaces which seem to have done the best work have boshes sloping at an angle of 73 or 75 degrees, instead of about 60 degrees, as was formerly the common practice in Cleveland.

SECTION X.—BY FUEL USED IN BLAST FURNACES.

Allusion has been made to the possibility of error in stating the quantity of carbon and therefore of coke required in producing one ton of iron. As an example may be quoted that of one of the Ormesby furnaces (v., page 245) in which 24·14 cwts. of coke, containing 90 per cent. of carbon, was given as the consumption per ton of pig. It was pointed out at the proper place that this quantity seemed excessive, having regard to the quantity of heat needed in this particular case. The excess was shown to be 7,000 to 8,000 calories. Calling it 7,500 calories this number divided by 4,164 (the calories developed per unit of fuel) represents an overstatement of 1·79 cwts. of coke per ton of metal.

Now there is really no difficulty (supposing all the ore to be so calcined as to have all its carbonic acid driven off, as is often the case) in calculating from the average composition of the gases, how much carbon has been derived from the fuel.

To prove this I will assume a case in which the quantity of carbon burnt in the furnace is known, and thereby adopting a composition of the gases based on this knowledge, we shall find that the coke used corresponds in the calculation with the assumed quantity. For my present purpose I will take an example where each ton of iron is supposed to have required as follows:—

22 cwts. of coke containing 92·5 per cent. of carbon.
10 " limestone " 12·0 " " united to 32·0 per cent. of oxygen to form CO_2.

For simplicity's sake I will assume that the blast was absolutely free from moisture, and that the gases contained 6·58 units of carbon as CO_2, and the remainder as CO, per 20 units of iron.

By calculation, the carbon in the escaping gases comes out to be as follows:—

	Units.
Coke, 22 units containing 92·5 per cent. C. = ...	20·35
Limestone contains	1·20
	21·55
Less absorbed by iron	·60
Total carbon contained in gases	20·95[1]

[1] The possible extent to which carbon may be derived from the ironstone has been referred to, page 240.

CONDITION OF THE CARBON IN THE GASES.

From our assumption that 6·58 carbon existed in the form of CO_2 it follows that 14·37 are present as CO, making together the 20·95 units of carbon given above. Although the condition of the carbon in the gases as regards its state of oxidation is given, it must not be understood that this condition affects the calculation we are considering.

OXYGEN IN THE GASES.

Analysis of the iron indicated that (including the oxygen originally combined with the silicon, sulphur, and phosphorus therein contained) 9·00 units of oxygen were derived from the ore. This together with 3·20 units derived from the limestone as already mentioned, makes 12·20 units of oxygen, contributed by the solids entering the furnace for each 20 units of iron produced. The oxygen due to the blast, for this weight of metal, is thus estimated:—

	Units.
Carbon as carbonic acid = 6·58, united with oxygen ...	17·55
„ carbonic oxide = 14·37, „ „ ...	19·16
Total oxygen in the gases	36·71
Deduct oxygen in ore and flux, as above mentioned ...	12·20
Oxygen due to the blast, considered as dry	24·51

NITROGEN IN THE GASES.

The analysis of the gases gives the exact relation which the nitrogen they contained[1] bears to the carbon and oxygen, consequently the exact quantity of atmospheric air employed (taken as dry in this particular case) *per* 20 units of iron is then easily calculated. The correctness of the analysis can be judged by this nitrogen corresponding with the quantity estimated to be brought in by the oxygen just mentioned, viz. 24·51 units of oxygen for 20 units of iron. This of course is not attempted in the present case, because the assumed composition of the gases and solids ensures an exact correspondence between the two sets of numbers. Following the figures adopted we have :—

---

[1] All reference to the minute quantities of N which go to form NH_3 and Cy is neglected in this calculation.

SECTION X.—BY FUEL USED IN BLAST FURNACES.

23·25 (O. per 100 air) : 76·75 (N. per 100 air) :: 24·51 (O. in gases per 20 units of iron derived from blast) : 80·91 (the N. per 20 units of iron).

The escaping gases, adopting the figures assumed in the present case, will have the following composition:—[1]

CO_2	...	17·41 =	C. 4·748	...	O. 12·662
CO	...	24·19 =	10·367	...	13·823
N	...	58·40	—	...	—
		100·	15·115		26·485

Having determined in the manner described above, the actual weight of nitrogen (80·91 units) accompanying the production of 20 units of metal, we have simply to estimate the carbon and oxygen which are found associated with this quantity of nitrogen according to the proportions in which they are found by analysis. According to such computation the figures will stand thus:—

Carbonic acid...	...	24·13 =	Carbon 6·58	and oxygen	17·55
Carbonic oxide	...	33·53 =	„ 14·37	„	19·16
Nitrogen	...	80·91	—		—
Total weight of gases per ton of iron	...	138·57	Total carbon 20·95	Total oxygen	36·71

From these figures the oxygen in minerals and in air is thus calculated :—

76·75 (N. in air) : 23·25 (O. in air) :: 58·40 (per cent. N. in gases) : 19·69 (O in per cent. O in gases derived from blast.

But the oxygen in the gases associated with the 58·40 of nitrogen weighed 26·485; and 26·485 less 19·69 (oxygen in air) gives 8·795 as due to the minerals.

Now the total weight of oxygen derived from the minerals per 20 units of iron has been shown to be 12·20 units, and the weight of carbon in 100 units of the gases was ascertained to be 15·115 units: hence we have:—

8·795 (O. from minerals per 100 of gases) : 15·115 (C. in 100 of gases) :: 12·20 (oxygen in minerals per 20 of iron) : 20·96 (total C. in gases per 20 of iron).

---

[1] The composition of the gases in the analyses is an estimated one and corresponds exactly with the carbon, &c., delivered to the furnace. Of course in actual practice the analysis itself constitutes the basis of computation.

The total carbon in the gases per 20 of iron is thus shown to be	20·96
Of this there was contributed by the limestone	1·20
Difference	19·76
To which add carbon absorbed by pig iron	·60
Actual carbon furnished by coke	20·36
To convert 20·36 into coke add for 7½ per cent. impurity ...	1·64
Making together and corresponding with the assumed quantity of coke consumed	22·00

The calculation just given is of course liable to be constantly affected by the ever changing amount of moisture in the air.

To avoid complicating my calculations unnecessarily, I have adopted in all the estimates given in these pages what I consider to be a mean, viz., ·72 per cent. of water in the blast. Every 100 parts of such blast is therefore composed of:—

Nitrogen.		Oxygen.		Hydrogen.	
76·20	...	23·72	...	·08 =	100
Instead of 76·75	...	23·25	...	— =	100

as assumed in the previous calculation, where the blast was regarded as absolutely dry.

Suppose on analysis 100 parts of the escaping gases are ascertained to consist of:—

Carbonic acid	...	...	17·68	=	C. 4·82	+	O. 12·86
Carbonic oxide	...	...	24·57	=	C. 10·53	+	O. 14·04
Hydrogen	...	...	·05		—		—
Nitrogen	...	...	57·70		—		—
			100·00		15·35		26·90

We have then as before :—

76·20 N. : 23·72 O. :: 57·70 : 17·96 = O in blast per 100 of gas.

But O. in 100 of gases is 26·90; and deducting 17·96 we have per 100 gas, 8·94 O. derived from the ore and limestone instead of 8·795 as formerly.

This modifies the figures to be made use of as follows:—

8·94 (per cent. O. in gas from minerals) : 15·35 (per cent. C. in gases) :: 12·20 (weight of O. from minerals per 20 units of iron) : 20·95 (the total weight of carbon in the gases per 20 units of iron).

SECTION X.—BY FUEL USED IN BLAST FURNACES.

We have then as before, on deducting the carbon supplied by the limestone and adding the carbon absorbed by the pig iron together with the ash to form coke, a net result of 22 units, which was the assumed weight of coke employed per 20 units of metal produced.

Abbreviated, the following formula will serve to estimate the quantity of carbon derived from the coke used for 20 units of pig iron:—

Let x be the total weight of carbon in the gases per ton of iron.
  O the total percentage of oxygen in the gases.
  N   do.   nitrogen   do.
  O − ·3112 N. the percentage of oxygen in the gases as far as it originates from ore and limestone, *i.e.* allowing for that derived from ·72 per cent. of moisture—the assumed quantity in the blast.
  C the percentage of carbon in the gases.
  O_M the total oxygen brought by the minerals per 20 units of pig.

Then $x = \dfrac{C \times O_M}{(O - ·3112\,N.)}$

Allowing then, as shown above, for carbon in limestone and pig iron, and ash in coke, the value obtained for x readily gives the weight of coke per 20 units of iron. On applying this mode of computation to the case of the Ormesby furnace, given in page 244, the coke came out 22·35 instead of 24·14 cwts.—showing thus an excess of 1·79 cwts. and corresponding exactly with the amount computed from the quantity of heat required.

I subjoin five other examples as a means of comparing the quantity of fuel said to be consumed with that appearing as carbon in the escaping gases:—

	Ormesby Furnace of 35,000 cubic feet.	Ormesby Furnace of 35,000 cubic feet.	Clarence Furnace of 24,000 cubic feet.	Clarence.	Clarence Average 4 Furnaces 1 Week.
Coke per charging book ...	20·03	19·09	22·97	22·44	21·22
Coke estimated from C in gases by the formula ...	21·19	19·89	22·36	21·85	21·06
Coke as consumed understated	1·16	·20	—	—	—
Coke as consumed overstated	—	—	·61	·59	·16

R.

ON THE HEAT EQUIVALENT OF CHARCOAL IN THE BLAST FURNACE.

Already in Section VII., on fuel, a general comparison was attempted to be drawn between the amount of duty obtained from charcoal and from coke. The application of vegetable fuel to the smelting of iron derives much interest from the small quantities occasionally consumed in the blast furnace to produce a ton of iron. A more minute enquiry into the causes which lead to this result induces me to supplement what has been already said on the subject. My present opinions are mainly founded on information obligingly supplied to me by the owners and managers of different works, during a recent visit to the Vordernberg Valley and elsewhere.

As examples of a remarkably low consumption of charcoal, I append the following particulars, obtained during the late meeting (1882) of the Iron and Steel Institute in Austria:—

Work.	Quality of Iron.	Height and Capacity of Furnace.		Temperature of Blast.		Weekly Make. Tons.	Charcoal used per 20 Units of Pig. Units.
		Height.	Capacity.	° C.	° F.		
Trofaiach ...	White	52	2,030	340	644	140	13·50
Vordernberg No. 14	,,	43½	1,190	313	595	154	14·20
,, ,, 13	,,	29	430	cold blast.		50	16·70
,,[1] ,, 2	,,	35½	1,120	300	572	105	14·80
,,[1]· ,, 3	,,	53	3,690	450	842	269	12·60
Neuberg	Bessemer	44	—	300	572	115	18·00
Heft.[2] Corinthia...	,,	43	—	375	707	—	16·80

From a list of furnaces given further on (pp. 276, 277) it will be seen that a ton of grey forge iron is made in Sweden with quantities of charcoal varying from 15·44 to 23·94 cwts.; and Bessemer iron, also with charcoal, from 17·70 to 25·56 cwts.

There is, as might be expected, a certain, indeed in some cases, a considerable amount of variation perceptible in the figures just given; but on the whole the quantity of charcoal required to produce a ton of metal is considerably below that usually consumed by smelters using coke as their fuel.

The most rational mode of investigation into the cause of this difference is to consider, in the first place, to what extent any variations

---

[1] Paper by M. Friderici.   [2] Letter from Ritter v. Tunner.

SECTION X.—BY FUEL USED IN BLAST FURNACES.

in the work to be performed may account for this apparent superiority of charcoal over coke. One example was given at page 129, by which 59,800 calories appeared to be absorbed per 20 units of *white* iron. Underneath is inserted a comparative statement of the requirements for smelting *grey* iron in two furnaces, the one using coke and Cleveland ore, and the other using charcoal and spathose ore :—

Heat Requirements:[1]—

	Charcoal. Calories.		Coke. Calories.	
CLASS I.—				
Fusion of slag	8,811		14,520	
Decomposition of water in blast	1,700		2,444	
Expulsion of carbonic acid in minerals	962		4,013	
Decomposition of ,, ,,	992		4,160	
Evaporation of water in fuel	1,674	14,139	324	25,461
CLASS II.—				
Reduction of iron in oxide of iron	32,100		33,108	
Carbon impregnation	1,920		1,440	
Decomposition of silicic, phosphoric and sulphuric acids	522		4,174	
Transmission through walls	2,615		3,658	
Fusion of pig iron	6,600		6,600	
Carried off in tuyere water	1,176	44,933	1,818	50,798
CLASS III.—				
Carried off in gases		3,480		7,542
		62,552		83,801
Estimated heat produced in the furnace		63,896		84,772

The close correspondence between the two sides of the account, in both these examples, affords a reasonable ground for a belief in their general correctness. Admitting this, it would appear that iron made under the assumed conditions requires 33 per cent. more heat in the case of the coke than in that of the charcoal furnace. There are, however, differences in the composition of the two kinds of fuel, as well as differences in the character of their combustion, which prevent any direct comparison being made by having regard only to the

---

[1] The same classification of the heat requirements has been retained as that given at page 242. In comparing different qualities of charcoal and coke, as well as charcoal iron and coke iron, care must be taken to make proper allowance for differences of composition.

quantity of heat required. This will be best understood when the heat equivalents of each kind of fuel come to be considered.

At the risk of appearing to overload these pages with examples of the performance of blast furnaces using charcoal, I annex extracts from a list compiled by Professor Åkerman. This is done because, as is well known by furnace managers, it would be easy to select a few cases which might serve to illustrate conditions of a very opposite nature. The iron produced in all the examples given may be considered as grey in quality; but that used in the Lancashire hearths for making malleable iron, being run in iron moulds, is white at the outside, where it comes in contact with the cold iron. The metal intended for Bessemer steel is of course "richer" than that used for forge purposes. The consumption of fuel being affected by the richness of the ore, and the quantity of limestone charged, I have added the weight of ore and limestone per 100 units of fuel, together with the temperature of the blast, and the percentage of carbonic acid to carbonic oxide.

I.—*Swedish works making grey forge iron:—*

Work.	Height. English Feet.	Units of Charcoal per 20 units of Pig.	Units of Ore and Flux per 100 units Fuel.	Temperature of Blast. °C.	°F.	Vols. of CO_2 per 100 Vols. CO.
Hagforsen	54½	19·56	198	200 =	392	38·4
Gustavfors	46¾	23·94	199	140	284	28·8
Finshyttan	43¾	23·80	224	100	212	31·1
Söderfors	50½	18·86	199	230	446	34·1
Gammelkroppa	42¾	20·96	204	180	356	30·6
Hofors, No. 1	51	22·50	247	150	302	41·0
Do. No. 3	52½	19·56	220	250	482	35·9
Harnäs	41	19·30	195	80	176	32·1
Vikmanshyttan	30¾	20·18	195	170	338	35·6
Starbo	39	17·01	219	300	572	60·9
Degerfors	50¾	18·52	223	200	392	56·3
Högfors	37	19·36	235	230	446	50·4
Björnhyttan	42	15·44	204	200	392	85·3
Norn	43	19·60	236	200	392	58·4
Bofors	52	20·04	236	170	338	60·4
Seglingsberg	48¾	17·26	224	200	392	61·2
Dalkarlshyttan	43¼	15·70	237	100	212	68·3
Do.	43¼	15·66	253	100	212	69·1
Klenshyttan	31½	18·28	188	250	482	44·4
Average for grey forge	44½	19·23	217·68	182	359	48·5

SECTION X.—BY FUEL USED IN BLAST FURNACES.

II.—*Swedish works making Bessemer iron:*—

Work.	Height. English Feet.	Units of Charcoal per 20 units of Pig.	Units of Ore and Limestone per 100 units of Fuel.	Temperature of Blast ° C.	° F.	Vols. of CO_2 per 100 Vols. of CO.
Nykroppa	50½	23·42	208	250	482	27·3
Långshyttan	51¾	17·70	212	300	572	41·5
Domnarfvet	53½	21·00	221	280	536	40·8
Ulfshyttan	40	17·74	228	300	572	48·0
Sandviken	50½	25·56	229	280	536	27·0
Bångbro	53½	19·02	220	300	572	56·2
Guldsmetshyttan	50½	17·50	223	230	446	75·1
Westanfors	41	22·12	183	250	482	36·7
Average for Bessemer	49	20·53	215·5	273·7	525	44·1

It may be observed that the average weight of ore and limestone carried on 100 lbs. of charcoal is about 218 lbs. for Lancashire hearth iron, and about 215 lbs. for Bessemer. Reckoned on dry charcoal, these numbers will be 257 lbs. and 254 lbs. respectively. In British furnaces 100 lbs. of good Durham coke carry about 252 to 272 lbs. of ore and limestone for grey forge, and about 222 lbs. for Bessemer pig.

Dealing with the furnaces just enumerated as a whole, there will be found nothing to excite any surprise in reference to the consumption of fuel. It is true that the charcoal usually contains about 15 per cent. of water, so that in 19·23 units and 20·53 units we have in reality only 16·35 and 17·45 units respectively of dry fuel to account for. Now if Cleveland coke iron of forge quality requires, as appears in the statement given at page 275, 33 per cent. more heat than the charcoal pig made in these furnaces, we have (16·35 + 33 per cent.) 21·80 units as the proper quantity of coke required for grey forge iron. Bessemer iron, as made in England, usually absorbs about 75,500 calories per 20 units, equal therefore to 21 per cent. in excess of that produced in the charcoal furnaces of Sweden. If then to 17·3 cwts. the weight of dry charcoal, we add 21 per cent., we have 20·93 cwts. as its equivalent in coke at an English Bessemer furnace. Now when we consider that Durham coke contains 6 per cent. of ash, or three times the quantity of that present in charcoal—also that 21½ cwts. of the former suffice to make a ton of grey forge iron and under 21 cwts. a ton of Bessemer iron—the figures just given prove that so far as grey iron is concerned the average results in these Swedish furnaces, in point of the consumption of fuel, are in no way superior to those commonly obtained in modern furnaces using coke.

On contrasting the weight of charcoal, given in page 274, as employed in the manufacture of white iron in Styria, with that consumed in the Swedish furnaces for producing grey iron, there is, it will be seen, a difference of above 6 cwts. to the ton; and even in making Bessemer pig, the smelters of Austria appear to be able to do the work with less fuel than those of Sweden. I propose now to examine the work of the furnaces in the two countries, by means of the data in my possession, in order that a comparison may be drawn as to the quantity of heat required in each case and the means employed for obtaining it. With this view I have prepared Tables containing the needful factors, together with the estimates based thereon; selecting cases where a maximum, a minimum, and something like a mean quantity of fuel has been consumed.

An examination of the Tables III. and IV. points to the supposition that the weight of charcoal used may be somewhat understated in some of the instances under examination. This observation is particularly applicable to the furnace of Seglingsberg and Guldsmetshyttan, the former showing a deficiency of about 2 cwts. and the other of 1 cwt. to the ton of iron. This is not surprising; because, owing to the readiness with which charcoal absorbs moisture, it is the practice of those using this description of fuel, to deliver it to the furnaces by measure, and it may easily happen that a given bulk may contain a greater weight of dry charcoal than the figure assumed for the estimates given in the Tables. No allowance has been made for the heat contained in the ore, usually freshly drawn from the kilns; but this is considered as varying from 1,600 to 1,800 calories per 20 units of iron.

My want of sufficient experience with charcoal furnaces renders it however inexpedient that I should do more than point out the discrepancies alluded to ; and the numbers given by the authorities, quoted as the source of my information, will be adhered to in the calculations which follow.

Although it has been demonstrated (page 277) that the *average* consumption of charcoal in making grey iron in Sweden shows, on proper

SECTION X.—BY FUEL USED IN BLAST FURNACES.

TABLE III.—CHARCOAL FURNACES.

Name of work	Duty Performed in Swedish Furnaces using Charcoal.							Coke.	
	Hagforsen. Grey Forge. 54¼	Björnhyttan. Grey Forge. 42	Seglingsberg. Grey Forge. 48¾	Sandviken. Bessemer. 50¼	Bångbro. Bessemer. 53¼	Guldsmets-hyttan. Bessemer. 50¾		Cleveland Forge. Clarence. 80	South Wales. Bessemer. 70
Quality of Iron									
Height of furnace, English feet									
	Units.	Units.	Units.	Units.	Units.	Units.		Units.	Units.
Materials used per 20 units of pig:—									
Fuel	19·56	15·44	17·26	25·56	19·02	17·50		20·40	18·38
Ore	36·36	36·44	39·71	42·68	41·40	39·13		47·62	
Carbonate of lime in ore and flux	3·24	4·36	5·09	3·12	2·60	5·47		9·38	4·86
Estimated weight of slag per 20 units of pig	12·26	12·01	15·71	17·86	16·02	15·15		24·50	7·71
Temperature of blast	200° C. = 392° F.	200° C. = 392° F.	200° C. = 392° F.	280° C. = 536° F.	300° C. = 572° F.	230° C. = 446° F.		563° C.	529° C.
Do. escaping gases	400	222	182	270	145	225 437		262	205
Heat produced, after allowing for H₂O and ash in charcoal together with C removed by CO₂ in flux:—	Units. Cal.	Units. Cal.	Units. Cal.	Units. Cal.	Units. Cal.	Units. Cal.			
From C burnt to CO (2,400 cal. per unit)	14·59 = 35,016	11·56 = 27,740	11·63 = 27,912	17·68 = 42,432	14·03 = 33,672	12·26 = 29,424			
C of this CO burnt to CO₂ (5,600 cal. do.)	4·24 = 23,744	5·83 = 32,648	4·89 = 27,384	3·91 = 21,896	4·99 = 27,944	5·47 = 30,634			
Less heat of combination for O in charcoal	58,760 2,600	60,388 2,080	55,296 2,170	64,328 3,120	61,616 2,490	60,058 2,240			
Net heat from combustion of charcoal	56,160	58,308	53,126	61,208	59,126	57,818			
Heat in blast	3,128	2,607	2,337	5,119	4,770	3,176			
Grand total of heat evolved	59,288	60,915	55,463	66,327	63,896	60,994			
CO₂ per 100 vols. CO	38·4	85·3	61·2	27·0	56·4	75·1			

To serve as a means of comparing the work of coke-furnaces with these of charcoal I have added some of the figures already given in this work, pp. 244 and 246.

TABLE IV.—CHARCOAL FURNACES.

DUTY PERFORMED IN SWEDISH FURNACES USING CHARCOAL.

Name of work	Hagforsen. Grey Forge.	Björnhyttan. Grey Forge.	Seglingsberg. Grey Forge.	Sandviken. Bessemer.	Bangbro. Bessemer.	Guldsmetshyttan. Bessemer.
Elements of heat absorption	Calories.	Calories.	Calories.	Calories.	Calories.	Calories.
CLASS I.—						
Evaporation of H_2O in fuel	1,577	1,090	1,966	2,059	1,674	1,696
Reduction of Fe from oxide	30,098	32,565	32,726	31,958	32,100	31,960
Carbon impregnation	1,920	1,920	1,920	1,920	1,920	1,920
Expulsion of CO from carbonate of lime	1,199	1,636	1,883	1,154	962	2,024
Decomposition of CO_2 of do.	1,241	1,664	1,955	1,216	992	2,112
Do. of H_2O in blast	1,720	1,360	1,360	1,720	1,700	1,360
Reduction of P, Si, and S	522	521	522	522	522	522
Fusion of iron	6,600	6,600	6,600	6,600	6,600	6,600
Do. slag	6,743	6,605	8,643	9,823	8,811	8,332
Total requirements, Class I.	51,620	53,961	57,575	57,872	55,281	56,526
CLASS II.—						
Transmission through walls	2,337	2,341	2,592	2,747	2,615	2,550
Carried off in tuyere water	1,083	936	839	1,309	1,176	1,024
Total, Class II.	3,420	3,277	3,431	4,056	3,791	3,574
CLASS III.—						
Carried off in escaping gases[1]	9,560	4,121	3,677	8,099	3,480	4,920
Total requirements, Class II. and III.	12,908	7,398	7,108	12,155	7,271	8,494
Grand total of heat required	64,600	61,359	64,683	70,027	62,552	65,020
Heat evolved as given in page —	59,288	60,915	55,463	66,327	63,896	60,994
Heat deficient for requirements	5,412	444	9,220	3,700	1,344	4,026
Do. in excess of do.	..	..	..	..	..	..

[1] The heat conveyed away in the escaping gases must only be considered as an approximate estimate, the mode of taking their temperatures not admitting of a very accurate calculation.

SECTION X.—BY FUEL USED IN BLAST FURNACES.

allowances being made, scarcely as good results as those obtained in the coke furnaces of Great Britain, yet there are in the list some remarkable examples which lead to a contrary conclusion. Some of these have been given at 274 and following pages, from which it will be perceived that in certain cases there is an extraordinary proportion of carbonic acid[1] as compared with carbonic oxide.

In order to compare the available heat developed per unit of charcoal with that produced from each unit of coke, the following estimates have been compiled:—

	Björnhyttan. *a.*	Seglingsberg. *b.*	Sandviken. *c.*	Guldsmetshyttan. *d.*	Vordernberg No. 2. *e.*	Vordernberg No. 3. *f.*	Average.
Combustion of charcoal ...	3,776	3,077	2,394	3,303	3,626	3,748	3,321
Contained in blast	169	135	200	181	268	363	219
	3,945	3,212	2,594	3,484	3,894	4,111	3,540
Carried off in gases... ...	266	213	317	281	242	185	251
Useful equivalent of heat per unit of charcoal	3,679	2,999	2,277	3,203	3,652	3,926	3,289

On referring to the estimates of the quantities of heat available, as produced in coke and charcoal furnaces respectively it will be seen that, at first sight, there is a very close correspondence, in this respect, between the two. Taking C, D and E, given page 244, as fair examples of the former, we have :—

Per unit of Coke.	C.	D.	E.	Average.
Combustion of coke ...	3,653	3,580	3,258	3,497
Contained in blast ...	534	574	906	671
	4,187	4,154	4,164	4,168
Carried off in gases ...	397	369	365	377
Useful equivalent ...	3,790	3,785	3,799	3,791

[1] The samples of gas were not taken in the manner now practised in the Clarence laboratory, viz. over a period of a couple of hours, but over very short periods only. Professor Åkerman, in a private letter, expresses his belief in their general correctness, and points out that the quantity of fuel consumed corroborates this view. In this I entirely agree; because, if the weight of charcoal for a given quantity of pig is truly stated, the quantity burnt to the state of carbonic acid is necessary, in order to provide the quantity of heat required for the operation.

In like manner we may take a, e, f as illustrating the best examples of the charcoal process:—

Per unit of Charcoal.	a.	e.	f.	Average.
Combustion of charcoal...	3,776	3,626	3,748	3,717
Contained in blast ...	169	268	363	266
	3,945	3,894	4,111	3,983
Carried off in gases ...	266	242	185	231
Useful equivalent ...	3,679	3,652	3,926	3,752

Practically coke, as supplied by the Durham collieries, may be considered dry, while the samples of charcoal chosen for illustration contained 9 per cent. of moisture. With this allowance the two sets of figures stand thus:—

	Dry Coke as above.	Charcoal Dry weight.
Combustion of the dry fuel, average ...	3,497 ...	4,084
Contained in the blast ,, ...	671 ...	292
	4,168 ...	4,376
Carried off in the gases	377 ...	253
Useful equivalent...	3,791 ...	4,123

Viewed in this way, charcoal as burnt affords about 8·5 per cent. more available heat than is obtained from Durham coke of fair quality. There is however no proof whatever that the heating power of vegetable exceeds that of mineral carbon; but there does appear ample reason, accepting the correctness of the quantities of charcoal stated to be consumed in these examples, and of the analysis of the gases as representing an average composition, for believing that the circumstances attending the combustion of charcoal differ from those of coke.

There is not any material discrepancy in the loss by the escaping gases between the two kinds of fuel—nothing beyond what might be expected from the fact that charcoal contains so much water, the presence of which must necessarily cause considerable cooling of the volatile substances as they leave the furnace. The great difference in the calorific power, as between coke and charcoal, arises, as has been already intimated, from the high ratio of carbonic acid in relation to

SECTION X.—BY FUEL USED IN BLAST FURNACES. 283

carbonic oxide, which is generally found to prevail, when using charcoal. In consequence of this it appears, although the same numbers have been applied in estimating the calories developed by both kinds of fuel, that one unit of charcoal taken as dry affords by its combustion with air at 0° C. (32° F.), fully 9 per cent. more heat than coke, viz. 4,373 calories instead of 3,497.

Having regard to the high price of charcoal, it is somewhat remarkable that more has not been done in raising the temperature of the blast. On referring to the list of the Swedish furnaces, pp. 276 and 277, it will be noticed that for grey forge iron, in one instance, it is as low as 80° C. (176° F.) and that the average is only 184° C. (363° F.) In making Bessemer iron the highest blast heat is 300° C. (572° F.) and the average 273·7° C. (524° F.) It is true that in most cases all the escaping gases are utilized either in heating the air or in calcining the ore—the blowing power being usually water—so that there is usually no gas to spare. On the other hand, notwithstanding the loss of heat in all special air heating arrangements, yet a much less expensive fuel than charcoal can be used for the purpose; and therefore it seems strange that more attention is not paid to so important a question as that referred to. These observations have particular application to the fact that in the coke furnaces (in the examples given at page 281) for each unit of fuel burnt there are 671 calories due to the blast against 292 per unit of dry charcoal.

In seeking to secure any great economy of charcoal by increasing the temperature of the blast, it is necessary to consider the difficulties which have to be encountered. In the instance given of the Vordernberg furnace, the charcoal consumed is 12·60 units per 20 of iron. The weight of the blast is about 46 units, having a temperature of 450° C., and containing therefore 4,770 calories. Now this exceeds considerably the value of one unit of the fuel, in the state in which it is delivered to this Vordernberg furnace when it only gave 4,111 calories. If we so reduced the consumption of charcoal as only to need 40 units of blast, and heated that blast to 1,000° C. (1,832° F.), the heat it would contain would be represented by 9,480 calories or say 4,710 more than when the air had a temperature of 450° C. Now 4,710 calories is equal to about ($\frac{4710}{4111}$) 1·14 units of charcoal; which represents the utmost economy to be obtained by raising the temperature of the blast to the probably unattainable temperature of 1,000° C.

It may be remarked that any decrease in the quantity of vegetable fuel is not, as in the case of coke, attended with any material decrease in the weight of limestone, the earthy impurity in charcoal being basic in its nature and insignificant in point of amount.

The arguments employed up to this time, to prove that charcoal as used in timber-growing countries does not differ materially, in point of quantity required for a given amount of work, from coke as used in England, have been chiefly founded on the experience of furnaces making rich pig iron for the converter. Before leaving the subject, it may be expedient to say a few words on the practice of charcoal furnaces when making white iron. For this purpose, the two Vordernberg furnaces Nos. 2 and 3, described in great detail by M. Friderici, are selected; the former of these is stated, page 274, to be producing a ton of pig with 14·80 and the latter with 12·60 cwts. of charcoal.[1]

The heat equivalent of one unit of dry charcoal compared with coke and calculated in the manner adopted, page 279 (*i.e.* allowing it as stated by Mr. Friderici to contain 7 per cent. of water), stands thus:—

	Vordernberg White Iron.			Coke of Durham.
	No. 2.	No. 3.	Average.	
Combustion of dry fuel ...	3,900	4,030	3,965	3,497
Contained in the blast ...	288	390	339	671
	4,188	4,420	4,304	4,168
Carried off in the gases ...	260	199	230	340
Useful coefficient	3,928	4,221	4,074	3,828

All perhaps that need be said as to these two instances of charcoal working, is that the high state of oxidation of the carbon is still more conspicuous than in the production of Bessemer iron; while the extent to which the general fund of heat has been aided by the heat in the blast is also somewhat higher—viz. 339 calories instead of 211, which latter is the average of the Sandviken, Bångbro, and Guldsmetshyttan furnaces, reckoned upon charcoal in its dry state.

Experience with modern Cleveland furnaces in the manufacture

[1] There are some grounds, for supposing that this (12·60) may be understated.

of white iron is not sufficient to enable me to speak with much confidence on the heat required for its production. Taking it at 84,000 calories, as against 58,000 for that made in the Styrian furnaces, and taking 14 units of charcoal as sufficing to provide this latter quantity of heat, we should require 20·27 units of charcoal to give the 84,000 calories supposed to be required for producing white pig from Cleveland ore. This is probably more than the outside quantity of coke which would be needed to produce this quality of metal in a furnace of 15,000 cubic feet.

Apart however from the mere quantity of fuel required of the two kinds, there are differences in the mode of working furnaces using charcoal and coke, and differences in the way in which the heat is produced from each, which deserve especial attention.

We have to begin with the fact that usually the make in a charcoal furnace is much larger, in relation to the capacity, than it is in a furnace using coke as the fuel. In most cases—certainly it is so in the Styrian furnaces—the ore to be treated is much more rapidly reduced than many with which it may be compared. Professor v. Tunner some years ago kindly selected for me a group of specimens from the great mine at Eisenerz, which were subjected to examination. All were carefully calcined to the same degree, so as to have the iron all peroxidised. The whole were then placed in a closed vessel of iron, immersed in melted lead so as to ensure, as far as possible, the exposure of all to the same temperature. This temperature was about 400° C. (752° F.); for a morsel of pure zinc, placed alongside them, softened but did not melt.

No. 1 was calcined unaltered spathose ore ($Fe\ CO_3$).
,, 2 calcined spathose ore slightly altered by atmospheric influence.
,, 3   ,,   ,,   changed to brown hæmatite.
,, 4   ,,   ,,   ,,   brauner glaskopf.
,, 5   ,,   ,,   ,,   plauerz.
,, 6   ,,   Cleveland ore.

The last was inserted to serve as a standard of comparison, being the mineral with whose properties I was most familiar.

Two experiments were made, in which pure carbonic oxide was passed through the vessel, each lasting eight hours, with the results given below:—

Number of Specimens			1.	2.	3.	4.	5.	6.
O removed per cent. of original oxygen.	Exprt.	1......	77·73	91·31	29·50	42·92	17·18	41·91
	„	2......	65·30	74·10	30·30	40·80	17·00	39·30
Carbon deposited per 100 of Fe.	„	1......	1·00	2·10	7·41	18·01	4·71	2·32
	„	2......	1·55	2·00	7·20	18·50	4·70	2·80

A fresh lot of the above specimens was then exposed for 8 hours, under similar conditions, to a mixture of equal parts by volume of CO and CO_2. The results were as under:—

Number of Specimens	1.	2.	3.	4.	5.	6.
Original O removed, per cent.	29·20	31·20	12·89	9·71	6·48	9·35
C deposited per 100 of Fe	Nil.	Nil.	Nil.	Nil.	Nil.	Nil.

In order to imitate more nearly the conditions of the blast furnace, 20 parts of calcined spathose ore and 7 parts of bruised charcoal were placed in one hard glass tube; and the same quantity of calcined Cleveland ironstone, with 8 parts of pounded coke, were introduced into another. Both were exposed to a good red heat in a Hofman's double gas-furnace for 45 minutes, while a current of oxalic gas (equal volumes of CO and CO_2) was conducted over the mixture. The following changes were noted:—

	Percentage of Carbon removed.	Percentage of Oxygen removed.
In the case of Eisenerz ore and charcoal ...	23·90 ...	31·00
„ „ Cleveland ore and coke ...	4·90 ...	17·40

These results, together with the fact that a much smaller weight of carbon has to be burnt to supply the lesser quantity of heat needed for smelting Nos. 1 and 2 (the usual description of Eisenerz ore treated in the Vordernberg blast furnace), render it quite intelligible that the reduction of such a mineral as this spathic ore should proceed much more rapidly than in the case of Cleveland ore, and that a given capacity of furnace should be able to do much more work.

The trials just described indicate that the softer charcoal acts more readily both on the oxide of iron of Eisenerz and on carbonic acid, than the harder coke can act either on carbonic acid or on the more stubborn ore of Cleveland.

SECTION X.—BY FUEL USED IN BLAST FURNACES.

A series of experiments was then undertaken, with a view to ascertain the behaviour of charcoal, as compared with coke, when exposed to heated carbonic acid.

Hard coke, soft coke and charcoal, pounded as nearly as possible to the same size, were placed in hard glass tubes which they filled, and were then raised to a good red heat in a Hofman's double gas-furnace. During the space of 30 minutes 800 c.c. of carefully dried carbonic acid was passed over each specimen. The issuing gases had the following volumetric composition:—

	Hard Coke.	Soft Coke.	Charcoal.
Carbonic acid	94·56	69·81	35·2
Carbonic oxide	5·44	30·19	64·8
	100·00	100·00	100·00

As in former experiments, the soft coke, it will be seen, reduced a much larger quantity of CO_2 to the state of CO. In the present trials the soft coke has produced about 6 times and the charcoal about 12 times as much carbonic oxide as the hard coke; and it has to be noted that, owing to its lightness, the weight of charcoal filling the tube was only half that of the coke.

A mixture of 20 grains of calcined Austrian spathose ore with 7 grains of charcoal, and another of 20 grains of calcined Cleveland ore with 8 grains of hard coke, all pounded in a similar manner, were placed in two hard tubes and heated to a bright red heat. Three litres of a mixture of 100 volumes of CO and 56 volumes of CO_2 were passed over each in 45 minutes.

The loss of carbon and oxygen was as follows:—

	Loss of C. Per Cent.	Loss of O. Per Cent.
Experiment 1.—Calcined Austrian ore and charcoal	22	46·2
,, 2.— ,, Cleveland ore and coke	5	22·3

The issuing gases had the following volumetric composition:—

	Austrian Ore and Charcoal.		Cleveland Ore and Coke.	
	CO.	CO_2.	CO.	CO_2.
10 minutes after commencing	100	180	100	177
30 ,, ,,	100	41	100	76
45 ,, ,,	100	38	100	82

The effect of a previous strong heat, in rendering carbon less easily acted on by heated iron ore, has been demonstrated at page 197 of the present work. The specimens submitted to trial were exposed for 48 hours, together with Cleveland calcined ore, to a temperature of from 538° to 582° C. (1,000° to 1,080° F.), and the percentage of original oxygen removed was noted:—

	Carbon from dissociated CO. O removed from Ore. Per Cent.	Charcoal dried. O removed from Ore. Per Cent.
Before ignition...	2·49	6·50
After previous ignition of carbon for 48 hours	·32	1·72

Results similar in character are described by my friend Professor Åkerman as having been obtained in the laboratory of the Stockholm[1] School of Mines. Specimens of charcoal and Durham coke were then submitted to a current of carbonic acid at different temperatures with the following results:—

	Length of Stratum. Lines.	Temperature during Experiments. ° C.	° F.	Percentage of CO after exposure to Carbon.
Charcoal	100	319	606	0·0
,,	100	393	740	0·4
,,	100	918	1,684	13·0
Durham coke	100	332	630	0·0
,,	100	485	905	0·3
,,	100	906	1,663	2·5

Previous ignition of the specimens greatly lessened their action on the CO_2, as will be perceived in the following Table:—

	Length of Stratum. Lines.	Temperature during Experiments. ° C.	° F.	Percentage of CO in Gas after exposure to Carbon.
Charcoal	80	515	959	0·5
,,	80	741	1,366	·8
,,	80	909	1,668	1·6
Coke	100	814	1,497	0·1
,,	100	906	1,663	1·0

Professor Åkerman proceeds to observe, very truly, that the action of the blast furnace is such, that the fuel has to reduce the ore as well

---

[1] Translation of Report on Manufactures of the United States in "Iron." 22nd September, 1877.

as to provide the needful heat for fusion; and he also remarks that the more speedily any CO_2 formed at the tuyeres is converted into CO the better.

I would point out that although the power of carbon in charcoal and coke, as both are delivered to the blast furnaces, to reduce carbonic acid to the state of carbonic oxide, is expressed by the ratio 13 to 2·5 (according to the Stockholm experiments), yet after previous ignition it is altered (according to the same experiments) to the ratio of 1·6 to 1·0. Under these circumstances it may fairly be considered whether, after charcoal has been exposed to previous ignition in the blast furnace, the temperature being far above those mentioned in Professor Åkerman's experiments, there is any difference between coke and charcoal in their power to resist the action of carbonic acid. Indeed it seems possible that, at the higher temperatures which obtain in smelting iron, charcoal may not be rendered less liable to attack by carbonic acid than coke is. It may be also that the easily reducible ores of Austria part with their oxygen in an atmosphere possessing a less energetic deoxidizing power than that required for such ores as are commonly treated in the furnaces of the United Kingdom.

Whatever inference the experiments just described may point to as regards the action of carbonic acid, generated by the reduction of the ore, on charcoal and on coke, there is no question that in furnaces using charcoal the ratio of carbonic acid to carbonic oxide is generally much larger than in furnaces using coke. It is indeed this very excess of carbonic acid in the gases which enables charcoal, with less heat contributed by the blast, to give better results, in a heat producing point of view than coke.

At page 282 figures were given to illustrate the nature of the difference in question. These were as follows:—

	Coke.	Dried Charcoal.
Combustion of fuel in dry state...	3,497	4,084
Contained in blast	671	292
	4,168	4,376
Carried off in gases	340	253
Useful equivalent in furnace	3,828	4,123

Now it will be perceived that so far as the analyses enable us to judge there appears to be a larger proportionate quantity of carbonic acid in the gases of these charcoal furnaces making white iron, as well as in some of those producing Bessemer iron, than in any of the analyses of the gases of the coke furnaces made at the Clarence works.

It is of course this larger ratio of carbonic acid as compared with the carbonic oxide which raises the co-efficient of heat in the case of charcoal—4,123—so much beyond that of coke—3,791.

It has been shown at page 287 that soft coke is more readily affected by CO_2 than hard coke; and the experience of the laboratory is confirmed by that of the furnace, as has already been demonstrated, pages 101 and 196, by the disappearance of carbonic acid from the gases, when using soft coke, and by the increased weight in the fuel consumed for a given weight of iron.

Professor Åkerman observes that, anthracite being denser than coke, more of the former is required than of the latter to smelt a ton of iron. By a parity of reasoning he supposes that the superior density of coke is the cause of its inferiority to charcoal as a fuel in the blast furnace. I attach much more importance however to the other ground, mentioned by this very high authority, for any difference there may be in the value of anthracite as compared with coke, viz., the tendency it has to splinter, thereby blocking up the air-passages, and interfering with the proper access of the reducing gas to the ore. In like manner the more open character of large pieces of coke affords a more ready exit for the ascending gases—a defect which does not obtain in the more closely packed contents of a charcoal furnace. This inconvenience is remedied, in the case of anthracite and coke, by a sufficient addition to the height and cubical capacity of the furnace. Mr. Birkinbine, in his paper[1] referred to page 129, describes the use of all three varieties of fuel in a charcoal furnace $36\frac{1}{2}$ feet high, with boshes of $9\frac{1}{3}$ feet. The following are his figures per ton of iron, using an ore giving about 40 per cent. of pig iron fed with air, at 315° C. (600° F.):—

Charcoal 22·50 cwts. Coke 31·19 cwts. Anthracite 34·56 cwts.

Now it is well known that in a suitably constructed furnace, instead of these excessive quantities, 22·50 cwts. of coke and about 24 cwts. of

[1] Transactions of American Institute of Mining Engineers, September, 1879.

anthracite are able to produce a ton of pig iron from such an ore as that referred to.

My opinion, already expressed in these pages, is that carbonic oxide is incapable of completely reducing oxide of iron. To ascertain how far conditions which obtain in the laboratory are reproduced in the blast furnace, a portion of slag was withdrawn from a few feet above the tuyeres at a furnace running a grey cinder and making foundry grey iron. The specimen contained 1·22 per cent. of protoxide, while that running away at the level of the dam had only ·15 per cent.

It is worthy of observation that even when a furnace is producing rich Bessemer iron, the cinder is not absolutely free from unreduced oxide of iron. During my recent visit (September, 1882) to Neuberg in Styria, I brought a snow-white piece of slag from a furnace making charcoal Bessemer pig, and found that it contained ·19 per cent. of FeO.

Our present enquiry however is not directed to mere traces of iron oxide in the slag but to a quantity of from 3 to 5 per cent., which renders the latter dark in colour, and is the cause of the iron being white instead of grey. All my observations go to prove that the last portions of oxygen, short of the trace which is irremovable by carbonic oxide, hold to the iron with great tenacity. It is therefore easy to comprehend that the hearth of a furnace, from insufficiency of heat or perhaps from an excess of carbonic acid, may arrive at a position of equilibrium before all the oxide can possibly be reduced to the metallic state. This insufficiency of heat proceeds no doubt from the charcoal having to carry a load of ore, when making white metal, beyond that which it is capable of raising to the temperature required for complete reduction.

Notwithstanding the very high percentage of carbonic acid given as the occasional product of some of the charcoal furnaces we are examining, the quantity of this gas is much below that actually produced by the action of carbonic oxide on the ore. This quantity was mentioned, page 87, to be equivalent to 6·58 units of carbon as carbonic acid for every 20 units of pig iron produced. In the six Swedish furnaces examined in detail, the maximum weight of carbon for the same quantity of iron was 5·83 and the minimum 3·91, the average of the whole being 4·88. At the two furnaces of Vordernberg the average was 4·908. In the gases from modern furnaces in Cleveland, using coke, I have met with as much as 6·52 units of carbon per 20 of iron in the form of carbonic acid; and as a rule it is not much under 6 units, in furnaces of modern construction in that district. It would thus appear that we have to account for an excessive ratio of carbonic acid in the escaping gases, at the same time that 25 per cent. of the carbon, which might have been present in this form, has disappeared.

In the case of a Cleveland furnace I have already pointed out that, having regard to the temperature of the escaping gases, an average position of equilibrium is arrived at when the two oxides of carbon are found in the ratio of a little under one volume of the higher to two of the lower. I have also shown experimentally elsewhere that, as the temperature was raised, the reducing energy of a mixture containing carbonic acid was lowered.[1]

It would appear however that in all the charcoal furnaces whose gases have been examined, carbonic acid exists in considerable quantity at depths below the throat which I have never met with in any furnaces in Cleveland, either of the older or of the more modern type. In illustration of this I annex the following examples, giving the volumes of carbonic acid accompanying 100 volumes of carbonic oxide:—

Furnaces.	Height. Feet.	Examined.	Blast.
A.—Veckerhagen	21	Bunsen	—
B.—Clerval	33	Ebelmen	Cold blast.
C.—Bärum	28	Sheerer and Langberg	Hot blast.
D.—Wrbna	36	Tunner and Richter	—
E.— ,,	36	,, ,,	Hot blast 400° C. (752° F.)
F.—Cleveland	48	Clarence laboratory	Hot blast.
G.— ,,	80	,, ,,	,,

The following Table gives the distances reckoned below the throat, and the volumetric quantities of CO_2 per 100 volumes of CO:—

A.—Veckerhagen	3′	$4\frac{1}{2}$′	6′	$7\frac{1}{2}$′	9′	12′	15′	—
Vols. CO_2	36	50	13	11	12	28	23	—
B.—Clerval	Throat.	$4\frac{1}{2}$′	$6\frac{3}{4}$′	13′	$17\frac{1}{2}$′	$18\frac{1}{4}$′	$25\frac{1}{4}$′	$27\frac{1}{2}$′ tymp.
Vols. CO_2	55	62	61	31	6	0	$\frac{3}{4}$	0
C.—Bärum	2′	$4\frac{1}{2}$′	7′	$9\frac{1}{4}$′	12′	15′	—	—
	277	118	67	16	42	29	—	—

[1] Chemical Phenomena of Iron Smelting, page 29.

SECTION X.—BY FUEL USED IN BLAST FURNACES.

D.—Wrbna	...	...	11′	17′	23′	27′	34′	—	—	—
	Vols. CO_2		125	163	44	7	50	—	—	—
E.—	,,	...	Throat. 18′	$25\frac{1}{2}$′	28′	32′	$34\frac{1}{2}$′	—	—
	Vols. CO_2		57	$48\frac{3}{4}$	$51\frac{3}{4}$	$44\frac{3}{4}$	$42\frac{1}{4}$	$7\frac{3}{4}$	—	—
F.—Cleveland		...	Throat. $9\frac{3}{4}$′	$15\frac{1}{2}$′	$21\frac{1}{4}$′	$27\frac{1}{4}$′	—	—	—
	Vols. CO_2		$22\frac{3}{4}$	$7\frac{1}{4}$	2	$3\frac{1}{2}$	$5\frac{1}{2}$	—	—	—
G.—	,,	...	Throat. $10\frac{1}{2}$′	26′	42′	$52\frac{1}{4}$′	65′	$70\frac{1}{2}$′	Tuyeres.
	Vols. CO_2		34	$6\frac{1}{4}$	$1\frac{3}{4}$	3	$4\frac{1}{4}$	$1\frac{1}{4}$	0	2

On referring to the examples F and G, which contain the analyses of the gases taken at different depths of furnaces smelting with coke, it may be said that although certain quantities of carbonic acid are found at a considerable depth in the furnace, reduction of the ore is practically completed within a distance of 25 feet from the throat. Thus, on referring to the quantity of oxygen in the gases of an 80 feet furnace, page 208, it will be seen that there is no increase of oxygen in the gases after passing a depth of 26 feet until the depth of $70\frac{1}{2}$ feet is reached, and there the small increase is considered as being due to a deoxidation of silica, etc., together with other minor causes as described, in Section IX.

A striking contrast to the state of things just described is that afforded by the list of charcoal works also given above. In these not only are large quantities of carbonic acid, as compared with carbonic oxide, found in the gases as we descend, until the region near the tuyeres is reached, but the reduction itself seems not to be effected until the materials reach this the hottest zone of the furnace. This is proved as follows:—

Ritter v. Tunner in 1871 made at my request a careful examination of the gases of the Wrbna furnace (E in the list), $37\frac{1}{2}$ feet high, when smelting the spathose ore of Eisenerz, with the following results by weight:—

	N.	H, etc.	CO_2.	CO.			Total per 100 Vols. CO Vols. CO_2.
Escaping gases	54·46	·67	21·47	23·40	=	100	57
,, ,, second trial	54·90	·39	20·90	23·81	=	100	
18 feet from throat ...	53·58	1·01	19·79	25·82	=	100	$48\frac{3}{4}$
$25\frac{1}{2}$,, ,, ...	54·57	·27	20·29	24·87	=	100	$51\frac{3}{4}$
28 ,, ,, ...	55	·24	18·50	26·26	=	100	$44\frac{1}{2}$
32 ,, ,, ...	54·34	·20	18·14	27·32	=	100	$42\frac{1}{4}$
$34\frac{1}{2}$,, ,, ...	57·35	·11	4·61	37·93	=	100	$7\frac{3}{4}$

From these figures has been estimated, as nearly as my information permits, the weight of the oxygen, separated from the minerals and supplied by the blast, per 20 units of iron, at the various depths just given. In the Table inserted below the weight of this oxygen is stated, together with the weight of carbon in the two forms of carbonic oxide and carbonic acid, also per 20 units of iron.

	A. Average Escaping Gases.	B. 18 ft.	C. 25¼ ft.	D. 28 ft.	E. 32 ft.	F. 34½ ft. (Tuyeres.)
Carbon as CO	10·11	11·06	10·66	11·25	11·71	16·25
,, CO_2	5·78	5·40	5·53	5·05	4·95	1·26
Total carbon	15·89	16·46	16·19	16·30	16·66	17·51
Oxygen as CO	13·48	14·74	14·21	15·00	15·61	21·66
,, CO_2	15·41	14·40	14·74	13·46	13·20	3·36
Total oxygen	28·89	29·14	28·95	28·46	28·81	25·02
Ratio of C as CO to 1 as CO_2 by weight	1·76	2·05	1·93	2·23	2·37	12·89
Per 100 vols. CO = vols. CO_2	57	48¾	51¾	44½	42¼	7⅝

It may be well to state, in reference to the ratio which the reducing gas (CO) bears to the oxidizing gas CO_2, that in each case the presence of other reducing gases, such as hydrogen and marsh gas, has been neglected. This has been done purposely, to avoid complication; but the subject will receive attention in the next section, which will be devoted to a consideration of hydrogen and certain compounds of hydrogen in the blast furnace. In the meantime it may be mentioned that these two gases often constitute about one-tenth of the bulk of the carbonic oxide; so that, in taking a case where the carbonic acid was 57 volumes per 100 volumes of carbonic oxide, if the hydrogen and marsh gas are included, the ratio is reduced to 52 volumes of CO_2 per 100 of CO.

These figures indicate a very moderate increase of carbonic acid between the points C and A and practically what may be considered as no increase of oxygen between a point 2½ feet above the tuyeres and the escaping gases; in other words that the reduction of the ore has been delayed until it reached the zone of fusion.

In a furnace using coke, the upper portion of the furnace has such a temperature that reduction is able to be practically completed at no

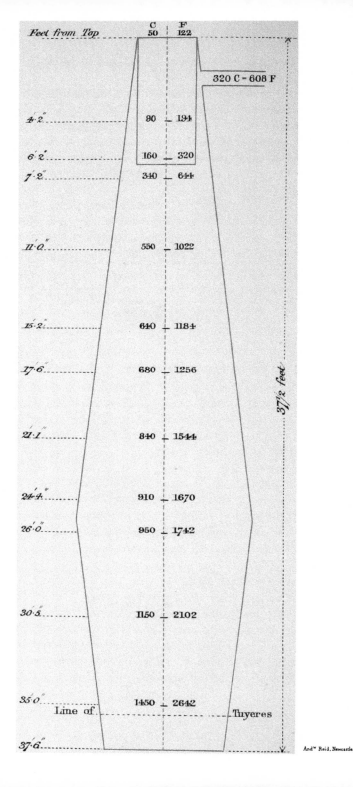

Diagram shewing temperatures of a Styrian Charcoal Blast Furnace at different levels as ascertained by Ritter V. Turner.

great distance from the throat, say within one fifth of the entire height in a furnace of 80 feet. A position of neutrality, as regards deoxidation by the gases is here regarded as being attained when one-third of the volume of carbon gases is fully saturated with oxygen.

It becomes interesting and important to know the rate at which the temperature increases, as the lower portion of a charcoal furnace is reached. Ritter v. Tunner in his paper, already referred to,[1] describes a mode, in which he expresses great confidence, of obtaining this information, in the case of a furnace having a height of about $37\frac{1}{2}$ English feet. His plan was to place in a properly constructed vessel alloys of lead with silver, with gold, and lastly with platinum. This vessel was attached to a chain of iron, and permitted to descend with the materials to a given level, when it was withdrawn by being drawn up through the charges. The temperature was then ascertained by finding that some alloy, the melting point of which was known, had been fused in the operation.

The gases are withdrawn from this furnace by a lateral tube inserted in the side of the furnace near the top; escape upwards being prevented in a great measure by a cylinder of sheet iron, which is kept filled with ore and fuel. The chief part of the gas therefore leaves the furnace at a temperature of 320° C. (608° F.) and only a small part at 50° C. (122° F.)

If we compare this furnace with that shown Fig. 1, Plate II, page 72, a remarkable difference is perceptible. In the latter, which is a furnace having a height of 48 feet, the temperature at a height of 20 feet from the hearth is given at 1,850° F.; whereas in the charcoal furnace shown in the diagram, a zone of this temperature is not found until a height of only 10 to 12 feet from the hearth is reached.

Experiments of my own, and others by Professor Åkerman, referred to page 288 clearly show that both coke and charcoal are less liable to be acted on by carbonic acid after previous exposure to even moderate temperatures. Both sets of experiments however clearly demonstrate that the resisting power of charcoal was increased to a much greater extent than that of coke. It is very probable that this difference, in point of susceptibility to dissolution by carbonic acid, goes on increasing

---

[1] Theorie des Hauts fourneaux. Revue Universelle, 1860. 1re Semestre, p. 132.

with increase of temperature; and, if so, it is intelligible enough that a position of equilibrium between heated charcoal and an atmosphere containing carbonic oxide and carbonic acid may be established, when the latter gas exists in a much higher ratio than obtains where heated coke is brought in contact with an atmosphere containing the same two gases.

There appears no doubt whatever that the analyses of the gases, quoted at page 294, indicate unmistakeably that they were then charged with oxygen in the form of CO_2 so low down as the point E, 32 feet from the throat, and that between this point and the exit they have been unable to withdraw any further quantity of this element from the ore under treatment.

This view, as to the delayed action of the gases on the ore (spathose), which thus is not reduced until it is close to the tuyeres, is amply confirmed by the experiments of Professor v. Tunner.[1] In the same piece of apparatus, by means of which the temperatures at different levels were determined, samples of ore were placed, and the rate of deoxidation was noted. The results of his experiments were as follows:—

Depth from Top. Feet.	Temperature. ° C.	° F.	Time of Exposure. Hours.	Progress of Reduction.
$15\frac{1}{2}$ to $17\frac{1}{2}$	650	(1,202)	1	First signs.
$25\frac{1}{4}$ to 26	850 to 900	(1,562 to 1,652)	2	First signs of metallic iron.

These figures do not correspond in any way with the temperatures at which reduction first manifests itself in the various samples of ore, etc., given in Section V. of the present work. Upon this latter occasion we had to deal with pure carbonic oxide, whereas in Professor v. Tunner's trials his specimens were being exposed to a mixed atmosphere, containing for every 100 volumes of the reducing gas (CO) about 50 of one having an antagonistic tendency (CO_2).

The delay, referred to above, in the reduction of the ore is however one of position and not one of time. It is true that when the ore in a coke furnace, smelting Cleveland stone, has reached a level of from one-fifth to one-third of its height, deoxidation may be regarded as complete; whereas, according to v. Tunner's experiments, at a distance of one-half the height of the Wrbna furnace, it has barely begun. But it will be seen that while in the Cleveland furnace, the ore has

---

[1] Revue Universelle, 1860. 1re Semestre, page 460.

been exposed for at least 12 hours to the gases, in that of Wrbna it travels half the height of the furnace in one single hour.

This lowering in level of the zone of reduction gives rise to the generation of carbonic acid, at depths when little or none is formed in the coke furnaces with which we are comparing that of Wrbna. Further, it may be observed that, were the process of reduction effected, in a furnace using coke, at the temperature at which it is carried on in charcoal furnaces, it is most probable that the carbonic acid produced by the act would be resolved into carbonic oxide by the highly heated carbon of the coke. Under these circumstances I am led to the conclusion, that charcoal has its power of resisting the solvent action of carbonic acid increased, by previous ignition at elevated temperatures, *in a higher degree than coke.*

In seeking to assign a cause for the different modes of action exhibited by charcoal and coke, I labour under the disadvantage of not being able to test the accuracy of my conclusions (as I have so often done from time to time, in smelting with coke), by an appeal to the furnace itself. All I can do is to consider each point as it occurs, and examine it by the information in my possession. As an example, one may conceive that as time is an element in the deoxidation of the ore, time also may be an element in the deoxidation of carbonic acid by carbon. Having regard to the make of a Cleveland furnace, with a capacity of 25,000 cubic feet, the weight of gases escaping at the throat may be taken at 410 cwts. per hour; whereas from the Vordernberg furnace, No. 2, already referred to, with a capacity of only 1,120 cubic feet, 51 cwts. of gases are passing away per hour. This means that the current of gas through the smaller furnace has about three times the velocity of that passing through the greater.

When however we come to compare a coke furnace, of a capacity of 6,000 cubic feet, with the Vordernberg No. 3, of 3,664 cubic feet, the case assumes a different aspect. The coke furnace of 6,000 cubic feet emits about 222 cwts. of gases per hour against 101 cwts. given off by the Vordernberg No. 3, of 3,664 cubic feet. Here we have the current of gases passing through the coke furnace at a speed about one and one-third times greater than those of the charcoal furnace.

With the facts as just stated, it may be considered as proved that the escape of so much carbonic acid from a charcoal furnace cannot be set down in any way to its want of opportunity of surrendering one equivalent of its oxygen to the heated carbon, with which it is brought in contact during the process of smelting.

Whatever the cause may be, the generation and retention of a large quantity of carbonic acid near the tuyeres renders the gases containing it incapable of acting on the freshly charged ore, exposed to their influence on their way to the point of exit. This altered condition of things confines more than four-fifths of the entire height of the furnace to the simple office of intercepting the heat which remains unutilized, after reduction is completed.

There is at Schwechat near Vienna a pair of furnaces using mineral fuel for smelting the ore of Eisenerz which might have afforded an opportunity of testing the soundness of the conclusions I have been led to form on the subject of charcoal smelting. Unfortunately I am not in possession of all the data required for undertaking such an examination. At these works a mixture is employed of coke, of a somewhat inferior character, with raw coal. The consumption per 20 units of pig iron was given at 22 units, say 16 of coke and 6 of raw coal. The impurity of the coke necessitates the use of limestone, which, together with an increased weight of slag and other sources of heat absorption, etc., may easily account for the increase in the consumption of fuel.

Having regard to the low temperature to which the blast is heated for charcoal furnaces, we are naturally led to ask what the effect would be, as regards the relation which the two oxides of carbon bear to each other, of the application of superheated air; for this is the question whose answer must necessarily determine the possibility of any economy in the charcoal used. My enquiries at Vordernberg rather tended to the conclusion that, when 13 or 14 units of charcoal were being used per 20 units of iron, there had not been any notable saving effected by raising the temperature of the blast. Ritter v. Tunner on the other hand expressed to me his belief that, by suitably increasing the heat of the air, Bessemer iron might be produced with as small a quantity of fuel as is now required in making white iron. He does not however state that, by the same means, white iron should be made with something less of fuel, although this inference may fairly be drawn.

I apprehend nothing short of actual experience can settle this point

SECTION X.—BY FUEL USED IN BLAST FURNACES. 299

—all I would do at present is to express an opinion of what might happen.

Referring to what has preceded on this subject, we may assume that charcoal, previously ignited in the furnace at a temperature of 1,850° F., arrives at a position of neutrality, as regards carbonic acid, when this gas exists in the ratio of 50 volumes to 100 volumes of carbonic oxide. The effect of introducing a further quantity of heat in the blast would be to elevate the temperature of the lower zone, say that between E and F, page 294. Such a change might at once alter the power of the gases to retain so large a ratio of carbonic acid as they had hitherto held. Let us assume that 5,600 calories per 20 units of iron had been so added, and that the addition had displaced carbon existing as carbonic acid, to the extent of one unit. By this transfer 5,600 calories would be absorbed; and the sum of the heat would thus remain unchanged not only at or near the point e but presumably upwards to the point of exit.

One of two things would happen; either the ascending gases, enriched by the addition of the unit of carbon as carbonic oxide referred to, would be enabled to separate a corresponding equivalent of oxygen from the ore; or, their solvent power over carbon being promoted by the increased temperature derived from the superheated air, an absorption of carbon would take place, which would reconvert the carbonic acid generated by the act of reduction into carbonic oxide.

In the first case there would be an economy of fuel represented by 5,600 calories. In the second, the heat contributed by superheating the blast would be reabsorbed by the diminution in the quantity of carbonic acid contained in the gases: in other words, there would be no economy in the charcoal, and the heat generated to superheat the air would be so much loss.

A third contingency might arise, viz. the quantity of carbonic acid in the gases might suffer a diminution short of that required to absorb the 5,600 calories: in that case there would be a corresponding economy of fuel; but to what extent this would happen, my information on the subject does not enable me to conjecture.

As bearing upon this question, it may be well to mention that Ritter v. Tunner, in his paper on the theory of the blast furnace written about 1860, gives the consumption of charcoal in the Wrbna furnace at about 14 units per 20 of pig, the blast being heated to

200° C. (392° F). Subsequently the temperature of the blast was raised to 400° C. (752° F.), when the consumption of charcoal was given as being 13·20. Whether the saving was solely due to the change in the temperature of the air I am unable to say.

It may be well to work out approximately the temperature of the blast required to effect the whole of the saving just referred to, supposing the same to be possible. This I propose doing upon the basis of the estimate applied to the Vordernberg furnace No. 3, page 284; in which however it appears possible that the consumption of charcoal was somewhat understated.

Here the weight of charcoal assumed in the figures of M. Friderici was 12·6 units per 20 of pig iron, requiring about 46 units of air. If this quantity of fuel is reduced by 1 unit of carbon, or say 1·2 units of charcoal, the air required for its combustion, oxidised as it was in this case would be about $41\frac{1}{2}$ units.

The blast in the estimate (weight 46 units) was heated to 450° C. (842° F.) and contained 4,950 calories. Under the altered conditions, the weight of the air will be reduced to $41\frac{1}{2}$ units: and to load this quantity with 10,550 calories (4,950 + 5,600), a temperature of 1,063° C. (1,945° F.) would be necessary—a temperature, it is conceived, which would be found impracticable to generate or maintain.

POSTSCRIPT.

Since writing the preceding pages of the present Section, No. 6, Vol. III. of the Journal of the "United States Association of Charcoal Iron Workers" has come to hand. It contains the first account I have met with of the application of superheated air to charcoal furnaces, as well as some particulars of similar furnaces blown with air of more moderate temperatures. The descriptions unfortunately do not embrace all the data required for a proper estimate of the nature of the work performed. So far however as my information permits I purpose instituting a comparison among the furnaces in question, and between them and some of those already referred to in these pages.

The ore smelted in these American furnaces is very rich, and contains so little impurity as to require a very small addition of limestone.

The following particulars have been abstracted from the journal referred to. The whole produce was grey iron, equal to an average of something above No. 3 :—

SECTION X.—BY FUEL USED IN BLAST FURNACES.

Name of Furnace	Height of Furnace. Feet.	Temperature of Blast. °F.	°C.	Make per Day. Tons.	Used per Ton of Pig.		
					Ore. Cwts.	Limestone. Cwts.	Charcoal. Cwts.
Detroit	53	Not given.		41¾	33·50	1·28	17·05
„ One month's make ...	53	Not given.		47	34·18	1·28	15·39
Bangor	43	825	440½	40⅓	32·80	1·35	17·50
Midland	53	900	482	42	34·83	3·2	15·54
Martel	53	1,454	790	38¼	33·55	·80	13·22

From the yield and quantity of limestone, and assuming 2 per cent. of ash in the charcoal, I have calculated the weight of slag produced, and the heat required for its fusion for the three furnaces of which the required data were given:—

Bangor, 5·51 cwts. of slag per ton of pig needs 3,030 calories for fusion of the slag.
Midland, 9·20 „   „   „   „   5,060   „   „
Martel, 6·34   „   „   „   „   3,490   „   „

I have also estimated approximately the following factors:—

Furnace	Bangor.	Midland.	Martel.
Weight of blast per 20 units iron, cwt. ...	51	48	42
„ Escaping gases „ „ ...	87	77	66
Calories in blast „	5,318	5,492	7,864
„ escaping gases „ (taken at average temperature for charcoal furnaces)	4,176	3,696	3,168

In like manner the approximate quantity of heat absorbed is as follows:—

Class I.—

	Bangor.	Midland.	Martel.
Fusion of slag	3,030 ...	5,060 ...	3,490
Decomposition of H_2O in blast	1,360 ...	1,360 ...	1,020
Expulsion of CO_2 in flux ...	500 ...	1,184 ...	296
Decomposition of CO_2 in flux ...	512 ...	1,216 ...	307
Evaporation of H_2O in fuel, assuming 15 per cent. of moisture in all	1,398 ...	1,242 ...	1,022

Class II.—

Reduction of iron	33,000 ...	33,000 ...	33,000
Carbon impregnation	1,440 ...	1,440 ...	1,440
Reduction of metalloids	2,000 ...	2,000 ...	2,000
Transmission through walls ...	1,800 ...	1,800 ...	1,800
Lost in tuyere water	900 ...	900 ...	900

Class III.

Lost in gases	4,176 ...	3,696 ...	3,168
	50,116	52.898	48,443

The heat afforded by the fuel is thus estimated:—

	Bangor.	Midland.	Martel.
Heat required as per estimate of requirements	50,116	52,898	48,443
Heat estimated in blast	5,318	5,492	7,864
Derived from charcoal	44,798	47,406	40,579

From the figures contained in the above statements, the heat afforded by the fuel consumed per 20 units of pig (burnt with air at 0° C. (32° F.) is arrived at as follows:—

	Units Fuel.		Calories per Unit of Charcoal.
Bangor	44,798 ÷ 17·50 =		2,519
Midland	47,406 ÷ 15·54 =		3,088
Martel	40,579 ÷ 13·22 =		3,069

If we refer to the heat evolved in the case of the Swedish and Styrian furnaces given at pages 274, 276, and 277, it will be found the heat equivalent is higher when the blast is heated to more moderate temperatures than that indicated by the figures just given:—

	Björn-hyttan.	Seglings-berg.	Sand-viken.	Gulds-meds-hyttan.	Vordernberg. No. 2.	Vordernberg. No. 3.	Average.
Temperature of blast, in degrees C.	200	200	280	230	300	450	332 = 629° F.
Percentage water in charcoal, average Swedish 15¾ %	13	18	18	16	7	7	11
Heat from combustion of charcoal, calories	3,776	3,077	2,394	3,303	3,626	3,748	3,321
Heat contained in blast, calories	169	135	200	181	268	363	219
	3,945	3,212	2,594	3,484	3,894	4,111	3,540
Less carried off in gases, calories	266	213	317	281	242	185	251
Net heat equivalent of fuel	3,679	2,999	2,277	3,203	3,652	3,926	3,289

Estimated in this way the three American furnaces give the following figures per unit of charcoal, as it is received (moist weight):—

SECTION X.—BY FUEL USED IN BLAST FURNACES.

	Bangor.	Midland.	Martel.	Average.
Temperature of blast ° C.	440½	482	790	571 = 1,060° F.
Combustion of charcoal...	2,519	3,088	3,069	2,892
Contained in blast ...	304	353	597	418
	2,823	3,441	3,666	3,310
Carried off in gases ...	238	238	239	238
Net equivalent ...	2,585	3,203	3,427	3,072

On comparing the actual quantity of heat evolved by combustion with air at 0° C. (32° F.) it would appear that the equivalent of heat actually obtained from one unit of charcoal has in these cases declined when the blast is increased in temperature; in other words that the proportion of carbonic acid in the escaping gases seems to have fallen off on the application of superheated air. Of course, in dealing with the figures before us we are labouring under the disadvantage of not being informed as to the percentage of water in the charcoal used in the furnaces of the United States. It is not likely however that it would contain more moisture than that given for the first four of the Swedish furnaces, with which we are comparing them. As stated above, the average for these four amounted to 15¾ per cent.

The average heat equivalent per unit of charcoal of these four Swedish furnaces works out as follows:—

Average temperature of blast	227½° C = 441° F.
Combustion of charcoal	3,137 Calories.
Contained in blast	171
	3,308
Loss in escaping gases	269
Net equivalent	3,039

So far as these examples, founded on the information contained in the "American Journal," enable us to form an opinion, the introduction of "superheated"[1] air, as at Martel, or even of air less intensely heated, as at the Bangor and Midland furnaces, has been attended with a diminution of the heat evolved by the mere oxidation of the charcoal.

[1] By the word superheated I mean air that is heated above the temperature at which it can safely be done in metal pipes.

This is plainly perceptible on comparing the figures of the Swedish works, where 3,137 calories are afforded for each unit of charcoal, with 3,069 calories at Martel and with 2,519 and 3,088 at Bangor and Midland respectively. The deficiency of heat afforded by the combustion of the charcoal is, on the whole, nearly compensated for by the extra heat contained in the blast; and that which is wanting in this respect is made up by the smaller quantity carried off in the escaping gases. The extent of this last-named item is, in the absence of any data, an assumed one, the temperature being taken as the same as that of the Swedish furnaces and the above saving is merely the result of the calculated diminished weight of the gases themselves.

On comparing the Midland and Martel furnaces alone with the four in Sweden, the use of air heated to 636° C. (1,177° F.) in the former instead of $227\frac{1}{2}$° C. (441 F.) in the latter has been attended with an improvement of nearly 9 per cent. in the net useful effect of the fuel.

The pig made in the Martel furnace was good grey iron, averaging about 2·50; and at first sight the production of 20 units of such quality with 13·22 units of fuel seems extraordinary work. The value of the carbon in the charcoal used is further enhanced, after deducting 15 per cent. of water and 2 per cent. of ash, leaving only 10·98 units of actual carbon to perform the work of the furnaces.

To compare this with a coke furnace using 22 cwts. of coke per ton of iron, we will assume 87,000 calories to represent the heat required in a furnace producing Cleveland iron.[1] Now 22 of coke containing 91 per cent. of carbon represents 20·02 of the latter element. Adopting these factors, to ascertain the heat equivalent of the 10·98 units of carbon in the charcoal as compared with that in coke, we have 20·02 : 87,000 :: 10·98 : 47,212. On referring to the heat considered approximately (page 301) as being required for the iron produced in the charcoal furnace it will be seen that it amounted to 48,443 calories, so that according to these calculations one unit of the vegetable carbon, burnt with superheated air, has afforded a trifle less heat than the mineral carbon burnt with blast at about 1,000° F.

[1] For the causes which render so much more heat being necessary to produce a ton of pig from Cleveland ore than from the richer ores treated in these charcoal furnaces the reader is referred to the elements of absorption given page 275 and elsewhere in this work.

SECTION XI.

ON HYDROGEN AND CERTAIN HYDROGEN COMPOUNDS IN THE BLAST FURNACE.

In those divisions of this work which have dealt with the reduction of iron ore in the blast furnace, carbonic oxide has been alone considered as the reducing agent. This gas, indeed, has been regarded as sufficing to account for all the phenomena which take place in the reducing zone. At the same time, on referring to the various analyses given in previous sections, it will be found that hydrogen appears as a constituent in most cases, and when it does not so appear, its omission from the list must not be construed as an indication of its entire absence; for in point of fact, this element, in greater or less quantities, is always to be found in the gases which are generated in smelting iron.

In addition to the reason given above, for not complicating the description of the process by taking any account of the small weights of hydrogen which are present, there is the further possibility, that this gas may be nearly, if not entirely, inert so far as the direct office of reduction is concerned. This statement however must be given with some little reserve, owing to the difficulty of determining the behaviour of such small quantities of this gas as are usually found in furnaces where the fuel is coke or charcoal.

It is pretty clear that the moisture in the air or the introduction of water in some way or another, as by accidental leakings from the tuyeres, will account for the presence of hydrogen in the hearth. Nearer the throat imperfectly charred wood or coal will be a frequent source of hydrogen, as well as of its carbon compounds. It therefore becomes impossible to say whether those variations which occur in the

SECTION XI.—HYDROGEN IN THE BLAST FURNACE.

quantities of this element found in the escaping gases, are due to changes in the quantity of water decomposed in the zone of fusion; or to alterations in the amount of hydro-carbons liberated in the upper zone of the furnace; or to a certain portion of these hydro-carbons having disappeared, by conversion into water either in the deoxidizing of the ore or through other reactions to be presently mentioned.

As an example of the varying quantities of hydrogen passing off from a coke furnace, I would refer to the volumetric analyses, at page 178, of gases taken at different times of the same day from the Wear furnace having a capacity of 17,500 feet:—

Hour					11·45	12·20	12·40	1·0	1·20	2·0	2·25
Vols. of H per 100 of escaping gases...					1·1	3·3	3·8	2·1	3·2	2·2	4·0

Ten trials at one of the Clarence furnaces (page 196) gave an average of 1·31 vols. of hydrogen per 100 of gases. The different figures were:—

											Average.
Number of trial	...	1.	2.	3.	4.	5.	6.	7.	8.	9.	10.	
Per cent. of hydrogen ...		1·5	1·5	1·5	·9	·8	1·0	1·5	1·4	1·4	1·6	1·31

For one of the Clarence furnaces, of 25,500 cubic feet, the following analyses are given at page 200:—

Position in Furnace.	At Tuyeres.	Two Trials 2 feet higher up.	
Vols. of H per 100 vols. of gases ...	·85	·76	·20

Four experiments on samples of gases, taken at different times six feet above the tuyeres at the same furnace, gave the following figures:—

	No. 1.	No. 2.	No. 3.	No. 4.
Vols. H per 100 vols. of gases ...	·2	1·3	1·3	·7

In all the instances just cited, the analyses were made in the ordinary way, by explosion with oxygen in eudiometers. Want of accuracy in the tubes themselves, and the occasional smallness of the quantities to be dealt with, induced me to try a different plan; for I was anxious to ascertain, as correctly as possible, the proportion borne by hydrogen to carbonic oxide in the ordinary reducing gas of the furnace.

Three litres of the escaping gases were drawn from the Wear furnace, and passed through tubes, containing respectively potash,

SECTION XI.—HYDROGEN IN THE BLAST FURNACE.

sulphuric acid and chloride of calcium, whereby all carbonic acid and moisture were removed. The gas, consisting then of nitrogen, carbonic oxide and hydrogen, was conducted over red-hot oxide of copper, which converted the carbonic oxide into carbonic acid and the hydrogen into water. By this treatment the following quantities of hydrogen were obtained per 100 of carbonic oxide present:—

Locality from which gases were withdrawn.	By Weight.	Average.	By Volume.	Average.
A.—Escaping gases mean of two trials	1·01		14·10	
B.— Do. do.	·86	·903	12·04	12·633
C.— Do. do.	·84		11·76	
D.—6½ feet above tuyeres	·63	·585	8·82	8·190
E.— Do.	·54		7·56	

Let us suppose the case of a furnace emitting, per 20 units of iron, gases having the following composition by weight:—

Carb. Acid.	Carb. Oxide.	Hydrogen.	Nitrogen.		Total Units.
23·	35·	·316	80·	=	138·316

At the tuyeres there would be a mere trace of oxygen derived from the ore, nor would there be any carbonic acid separated there from the limestone. Approximately therefore each 20 units of iron would be accompanied by the following weight of gases at the tuyeres:—

Carb. Acid.	Carb. Oxide.	Hydrogen.	Nitrogen.		Total Units.
Nil.	47·	·275	80·	=	127·275

In both cases the weight of the hydrogen is computed from the weight of water given as having been collected after passing the gases over the oxide of copper.

We then have:—

	Units H.
Hydrogen for 20 units of iron in the escaping gases	= ·316
,, ,, ,, at the tuyeres as above	·275
Increase in the escaping gases	·041

It is very probable that part or even all this apparent increase may be due to hydrogen given off by the coke; but what seems very evident is that any amount of reduction effected by this gas is of a very insignificant character, so far at least as the figures just given are capable of affording the necessary information.

I have endeavoured to ascertain from the labours of other enquirers the changes which hydrogen and its carbon compounds experience during their passage through the column of materials which fills a blast furnace. These gases appear in the following quantities, per 100 volumes, in the different furnaces below, as quoted in a paper by M. Ebelmen.[1]

VECKERHAGEN (CHARCOAL).		CLERVAL (CHARCOAL).		BARUM (CHARCOAL).		SERAING (COKE).	
Depth from Throat. Feet.	Hydrogen and Hydro-carbons.	Depth from Throat. Feet.	Hydrogen and Hydro-carbons.	Depth from Throat. Feet.	Hydrogen and Hydro-carbons.	Depth from Throat. Feet.	Hydrogen and Hydro-carbons.
2	4·69	At throat.	5·82	3	4·33	1	2·94
4·5	4·37	2¼	6·90	5½	4·81	4	2·45
6	4·62	6¾	5·44	8	5·78	9	2·12
7·6	4·01	13	3·82	9½	2·28	10	2·15
9·0	4·23	17½	3·50	13	4·10	12	2·26
12·0	3·99	18¼	1·92	—	—	45	·32
15·0	3·94	25½	1·42	—	—	—	—
—	—	27	1·25	—	—	—	—

Speaking generally, it cannot be said that any of the examples given by this early enquirer into the action of the blast furnace indicate any notable deoxidation having been effected by means of hydrogen.

I have not been able, in dealing with the well-known paper of Messrs. Bunsen and Playfair, to make out a satisfactory statement from which I can calculate the weight of hydrogen which enters the furnace, with a view to compare it with the various analyses given by these chemists. According however to the tabulated results given below,[2] there would appear to be a notable diminution of hydrogen and hydro-carbons between the point d and the point a, the latter being close to the point of final escape.

---

[1] Annales des Mines, Tome XIX., 1881, p. 89.
[2] Transactions of British Association for Advancement of Science, 1845, p. 142.

SECTION XI.—HYDROGEN IN THE BLAST FURNACE.

	From Throat. Feet.	Nitrogen.	Cyanogen.	Carb. Acid.	Carb. Oxide.	Hydro-carbons and Hydrogen.		Vols.
a	5	55·35	—	7·77	25·97	10·91	=	100
b	8	54·75	—	9·42	20·24	15·59	=	100
c	11	52·57	—	9·41	23·16	14·86	=	100
d	14	50·95	—	9·10	19·32	20·63	=	100
e	17	55·49	—	12·43	18·77	13·31	=	100
f	20	60·46	—	10·83	19·48	9·23	=	100
g	23	58·28	trace	8·19	26·97	6·56	=	100
h	24	56·75	trace	10·08	25·19	7·98	=	100
i	34	58·05	1·34	0·00	37·43	3·18	=	100

It would not however be safe to connect these last analyses with the idea that hydrogen has actually served the purpose of reducing any oxide of iron in the furnace. Among a vast amount of valuable information, contained in the paper, there is an absence of certain factors required for making a close comparison of the behaviour of the various gases during their progress upwards among the materials. Here are indeed certain irregularities in the composition of the gases, taken at different levels of the furnace, which far exceed anything I have met with in my own researches. Thus at a depth of 24 feet from the throat there appear to be, taking 79·2 units of nitrogen as a standard of comparison, 31·6 units of oxygen; as against only 29·7 units of oxygen at a point 5 feet below the top of the furnace. The oxygen therefore instead of increasing as we approach the outlet is less than it was 24 feet from the throat. The authors of the paper admit the anomaly; alleging that the mean composition of the gases *cannot* (*sic*) be determined where the evolution of gas by distillation is at its maximum. This may be true as regards the proportion which hydrogen and its carbon compounds bear to the joint quantity of nitrogen and oxygen; but it is not very clear why the relation which these last mentioned elements bear to each other should be so much disturbed.

Viewing the instances already described as a whole it cannot be said that there is any good ground for believing that the hydrogen of the gases actually assists directly in reducing the ore. At the same time it can, one would suppose, hardly fail that a powerfully reducing gas like hydrogen or its carbon compounds, must assist as a restraint on the oxidizing tendency of carbonic acid. By this I mean that if 50 volumes of carbonic acid were found in the gases, accompanied by 100 volumes of carbonic oxide and 10 of hydrogen, we should in point of fact have 110 volumes of the reducing gases present with 50 volumes of the oxidizing gas, bringing the ratio of the former to the latter up to 100 : 45·4. In such a case as that in question, I am supposing it possible that the process of reduction may be entirely effected by means of carbonic oxide; but that at the same time the 10 volumes of hydrogen are acting as a restraining influence to the carbonic acid, in the same way as if they were so much carbonic oxide.

Some twelve years[1] ago I undertook a series of experiments in order to compare the action of hydrogen with that of carbonic oxide on oxide of iron as it exists in Cleveland stone. The ore in a divided state was heated in a glass tube and exposed to a current of the gas; and the amount of reduction was ascertained by determining the weight of the water formed.

Sample.	Temperature during the Experiments.	Time of Exposure.	Effect.
A.	104° to 127° C. (220° to 260° F.)	Half an hour.	No sign of action.
B.	199° to 227° C. (390° to 440° F.)	Do.	Small signs of reduction.
C.	Good red heat, maximum of single Hofman's gas furnace	4 hours.	Not completely reduced.
D.	Sample C, after exposure for four hours in the Hofman's gas furnace, was heated to a white heat in a coke fire ...	1 hour.	Perfectly reduced.

These trials indicate that reduction by hydrogen commences at about the same temperature as that effected by carbonic oxide (*vide* p. 71).

A mixture of 100 volumes of carbonic oxide and 12 volumes of hydrogen was then passed over 63·65 grains of Cleveland stone (calcined) in a tube from which all air had been carefully expelled. The tube was maintained for *one-and-a-half hours* at a temperature short of a red heat, zinc remaining melted all the time. The current of mixed gases was continued until the tube and its contents were cold.

[1] Vide Chemical Phenomena of Iron Smelting, p. 118.

SECTION XI.—HYDROGEN IN THE BLAST FURNACE.

The ore was found to contain 1·20 grains of deposited carbon, a weight equal to 4·61 per cent. of that of the iron present in the ore. The quantity of oxygen removed was about 68 per cent. of that originally combined with the iron in the mineral, being in all 7·52 grains.

	Grains.		Grains of Oxygen.
The total water collected was	0·95	=	·85
The total carbonic acid collected was ...	22·75		
From which deduct that due to the dissociation of carbonic oxide (2 CO = CO$_2$ + C) ...	4·40		
	18·35	=	6·67
Total oxygen separated from the ore equal to 45·2 per cent. of that required to form Fe$_2$O$_3$			7·52

Experimentally it had been ascertained that pure carbonic oxide, at a temperature probably about 800° F. (427° C.), only removed 9·4 per cent. of the total oxygen in the Fe$_2$O$_3$, in *seven hours* from calcined Cleveland ore, equal therefore to 1·34 per cent. per hour.

The same mixture (of 100 volumes CO and 12 volumes H) was then passed over 68·85 grains of calcined Cleveland ore at a very bright red heat, during *one hour*, with the same precautions observed as in the previous experiment.

About 70 per cent. of the original oxygen, viz. 8·42 grains, was expelled, and ·38 grain of carbon was precipitated. Calculated in the same way ·58 grain of oxygen had been removed by hydrogen, and 7·84 grains by carbonic oxide.

A third trial was conducted at the white heat of a coke furnace for 40 *minutes*, the ore weighing 61·8 grains. The oxygen removed was 6·46 grains, or about 60 per cent. of the whole, and carbon deposited only ·08 grains.

				Grains.
The hydrogen had removed of the oxygen ...	...	...	1·42
The carbonic oxide do. do. ...	...	...	5·04
Total	...	...	...	6·46

The amount of reduction in this trial would no doubt have been greater, had the ore not fused under the intense temperature to which it had been exposed.

Taking the results per 100 parts of ore used, and over similar periods of time, we get the following figures:—

Temperature	Experiment 1. Melting Zinc. Grains.	Experiment 2. Bright Red. Grains.	Experiment 3. White Heat. Grains.
Loss of oxygen per hour	7·86	12·23	15·68
Carbon deposited do.	1·26	·55	·19
Loss of oxygen do. due to H.	0·89	·84	3·44
Do. do. due to CO.	6·97	11·39	12·24
Total loss of oxygen	7·86	12·23	15·68

The whole of these experiments seem to indicate that hydrogen, having regard to its volume, is the more powerful reducing agent of the two gases.

Notwithstanding the apparent superior affinity of hydrogen for oxygen, as compared with that of carbonic oxide, it cannot be considered as satisfactorily proved that the water formed by the hydrogen is the consequence of its union with the oxygen of the ore. I have arrived at this conclusion, because I ascertained that when hydrogen was passed over carbonate of lime the carbonic acid was decomposed—the result being carbonic oxide and water. The carbonate of lime employed was Carrara marble containing 44 per cent. of carbonic acid, and mountain limestone containing 43·61 per cent. of the same. The results per 100 parts of carbonate of lime were as follows:—

Substance used.	Temperature.	Time of exposure. Minutes.	CO_2 lost.	CO_2 not reduced.	CO_2 reduced.	Water found.
Marble	melting zinc	40	1·03	0·19	0·84	0·55
Mountain limestone	dull red heat	30	3·21	0·38	2·83	1·07
Do.	bright red	30	40·56	21·42	19·14	7·44
Marble	do.	20	43·47	23·47	20·00	8·67

Not only, as has just been shown, is hydrogen capable of decomposing carbonic acid, but water can be formed at the expense of

SECTION XI.—HYDROGEN IN THE BLAST FURNACE.

oxygen in carbonic oxide. Equal volumes of the two gases were passed for two hours through pumice-stone heated to a dull red. The water collected weighed 1·2 grains and the carbonic acid 0·1 grain.

A similar experiment was performed with the pumice-stone heated to full red. The water collected weighed 4·6 grains and the carbonic acid 1·2 grains. In this case the pumice-stone was blackened; and when it was heated in oxygen 1·59 grains of carbonic acid were obtained, equal to ·43 grains of carbon; which must have been deposited by the dissociation of the carbonic oxide.

Enough has been advanced in the present section to prove the difficulty, perhaps the impossibility, of demonstrating the precise nature of the part played by any hydrogen which may be present in the blast furnace. So far, however, as the promotion of reduction is concerned, it is probably immaterial whether this gas deoxidizes the ore direct, or whether it decomposes carbonic acid with the liberation of carbonic oxide. In either case the presence of hydrogen must, it may be supposed, help to maintain the reducing energy of the gases.

All my experiments tend to prove not only that this energy is intensified by the co-operation of hydrogen with carbonic oxide; but that a mixture of the two oxides of carbon (CO and CO_2) had its power of deoxidizing iron ore increased by the presence of this third gas, perhaps beyond that point which might be expected from its relative volume.

Experiments were first made to ascertain the mode of action of mixtures of the two oxides of carbon on the ore, with the following results:—

Volumes.		Temperature.	Ore.	Hours of Exposure.	Action.	Percentage of O removed per hour.
CO	CO_2					
100	46	melting Zn., say 800° F. (427° C.)	Calcined Cleveland	15	none	nil
100	31	do.	do.	10½	6·8 per cent. of original oxygen lost	·65 per cent.
100	50	low red heat	do.	6	13·55 do.	2·26 ,,

A mixture was then made resembling somewhat in composition blast-furnace gas; and into this some hydrogen was introduced. It then contained CO_2 15·9, CO 33·5, H 4·2, N 46·4 = 100 volumes.

In this, 100 volumes of reducing gases consist of CO 88·86, H 11·14; and the proportion of CO_2 to CO is 47 volumes of former to 100 of the latter, or taking the CO and H conjointly (57·7 volumes) we have 42 volumes of CO_2 to 100 of reducing gases. After an exposure of $10\frac{1}{2}$ hours to a temperature of say 842° F. (450° C.), antimony melting, the loss of oxygen amounted to 11·7 per cent. of that originally combined with the iron in the ore.

At a later period in the present section when the use of raw coal in the blast furnace is being considered we shall be better able to discuss the practical value of hydrogen as a reducing agent.

Messrs. Baird & Co. of the Gartsherrie works in Lanarkshire have recently introduced a new industry at this well known establishment, which promises to become one of immense importance even in a national point of view. The gases from two of their blast furnaces fed with raw coal are aspirated by means of a large ventilating fan through huge chambers of sheet iron, in which they are cooled by suitable means, and the tar and ammonical water are thereby condensed. From the former the usual oils and pitch are obtained, and the latter, being used for the manufacture of sulphate of ammonia, will be able to render valuable service to agriculture.

The average quantity of oxygen in the coals used at Gartsherrie is stated to be 11·06 per cent., and of hygrometric moisture 7·77 per cent.—or per ton of coal 2·212 cwts. of oxygen and 1·554 cwts. of moisture.

Now this quantity of oxygen if united with hydrogen and added to the moisture would be equal to 451 lbs. or 45·1 gallons of water.

I am informed by my friend Alderman Thomas Hedley, chairman of the gas works at Newcastle-on-Tyne, that 30 gallons of water are condensed per ton of coal distilled. A similar result is obtained at the coke works of Messrs. Pease & Partners, in which the coal is coked in closed ovens. The Durham coal contains on an average under 1 per cent. of water with from 10 to 11 per cent. of oxygen, which latter, by combining with hydrogen in the coal, will afford just about the 30 gallons of water referred to. According to Mr. Cosh, of the Gartsherrie works, 30 gallons also is the amount of water condensed in the apparatus referred to above; but the Gartsherrie coal, contain-

SECTION XI.—HYDROGEN IN THE BLAST FURNACE.

ing 7·77 per cent. of moisture and fully as much oxygen as the Durham coal, it follows, irrespective of any water formed by the action of hydrogen as a reducing agent, that a good deal must escape condensation in the apparatus used for collecting the tar and ammonia.

In the Gartsherrie coal hydrogen exists to the extent of 5 per cent. of the weight of the coal; equal therefore to 2 cwts. for every ton of pig iron made. Assuming each ton of metal to contain 18·6 cwts. of iron (Fe), there will be 7·97, say 8 cwts. of oxygen, to be removed in the process of its reduction. For this 1 cwt. of hydrogen would suffice; so that there is twice as much of this element present as is required to withdraw all the oxygen from the oxide of iron.

The want of any authentic information on the office performed by hydrogen in the blast furnace induced me to request permission, from a friend who is a furnace owner in Scotland, to extend my observations on smelting iron to cases when, instead of coke, raw coal is the fuel employed. Every possible facility was extended to my assistant, Mr. Rocholl, and I now propose to examine the results of his analyses and observations.

The fuel employed at the furnaces in question is the well known splint of the Lanarkshire coal-field. In lumps it burns without any approach to fusion, or one piece adhering to another, but when powdered and heated in a crucible it coheres and forms a dense coke without any apparent change of volume. The coal appears perfectly dry to the eye and touch, nevertheless when dried at 100° C. (212° F.) it parts with more than 11 per cent. of water. Its constituents were found to be:—

	Per Cent.		Volatile.
Water, given off at 212° F.,	11·62		
Carbon	66·00,	viz., 53·41 fixed and	12·59
Hydrogen	4·34		
Oxygen	11·09	} Total	16·37
Nitrogen	·94		
Sulphur	·59		
Ash	5·42		
	100·00	Total gaseous constituents	28·96

SECTION XI.—HYDROGEN IN THE BLAST FURNACE.

The ironstone employed was partly calcined black-band, containing 1·34 per cent. of iron as protoxide and 48·63 per cent. as peroxide, together 49·97 per cent.; partly clay-band with ·48 per cent. of iron as protoxide and 40·89 as peroxide, together 41·37 per cent.; and partly Spanish hematite ore containing 14·6 per cent. of water.

The pig iron on analysis gave as follows:—

		Oxygen in Ore per 20 units of Pig equivalent to
Carbon	3·51	—
Silicon	2·16	·49
Sulphur	·004	·00
Phosphorus	·85	·22
Titanium	·10	·01
Manganese	1·46	·11
Iron	91·00	7·66
	99·084	Total oxygen per 20 of metal...8·49

The particulars of the two furnaces which were examined are contained in the following Table:—

	Furnace A.	Furnace B.
Height	51 feet.	74 feet.
Cubic capacity	6,500 c. feet.	10,000 c. feet, of which only about 8,500 were filled.
Weekly make, chiefly No. 1 iron	220 tons.	220 tons.
Temperature of blast—	°C. °F.	°C. °F.
Minimum	405 762	412 773
Maximum	568 1,073	472 881
Average of 7 observations (A) / Average of 8 observations (B)	475 887	427 800
Temperature of escaping gases—		
Minimum	155 310	138 280
Maximum	298 568	254 490
Average (12 observations over 4 hours) (A) / Average (14 observations over 3 hours) (B)	200 392	190 374
Raw coal used, per ton of iron... cwts.	42·39	42·39
Ironstone, do. do.	37·46	37·46
Limestone do. do.	10·93	10·93

SECTION XI.—HYDROGEN IN THE BLAST FURNACE. 317

Volumetric analyses of gases—specimens collected over three hours:—

	Furnace A.		Furnace B.	Furnace A, Gas from above tymp.
	a.	b.	c.	d.
CO_2	5·84	5·70	6·27	1·40
CO	30·77	30·31	29·11	32·96
CH_4	3·40	3·45	2·84	—
C_2H_4	·38	·14	·24	—
H	6·65	5·98	6·84	2·60
N	52·96	54·42	54·70	63·04
	100·00	100·00	100·00	100·00
100 vols. CO accompanied by vols. CO_2	18·80	18·98	21·54	4·25
100 vols. CO, H and hydrocarbons accompanied by vols. CO_2	14·29	14·15	16·06	3·93

Analyses of escaping gases by weight:—

	Furnace A.		Furnace B.
	a.	b kept full.	c.
Carbonic acid	8·95	8·52	9·66
Carbonic oxide	30·02	28·82	28·36
Marsh gas	1·89	1·87	1·59
Olefiant gas	·37	·13	·23
Hydrogen	·46	·41	·48
Nitrogen	51·67	51·75	53·34
Ammonia	·07	Not estimated.	·07
Water	6·57	8·50	6·27
	100·00	100·00	100·00
Carbon in carbonic acid	2·44	2·32	2·63
Do. do. oxide	12·86	12·35	12·15
Do. do. acid taken as unity	1	1	1
Do. do. oxide	5·28	5·32	4·62

The analyses marked *a* and *c* will be employed in the estimates which follow. The furnaces were charged, when *a* and *c* were taken, at short intervals, according to the usual plan followed at most ironworks; whereas when the sample of gas *b* was collected the furnace

318   SECTION XI.—HYDROGEN IN THE BLAST FURNACE.

was constantly kept full to the charging plates. It has also to be observed that furnace B, being found too high to work satisfactorily with raw coal, was only filled to within 12 feet of the top.

The total weight of carbon delivered to the furnace per 20 units of iron is as follows:—

	Furnace A.	Furnace B.
Fixed carbon in raw coal at 53·41 per cent. ...	22·65 ...	22·65
Carbon of vol. matter in do. 12·59 „ ...	5·34 ...	5·34
Fixed carbon in limestone	1·31 ...	1·31
	29·30	29·30
To obtain the carbon in the permanent escaping gases, deduct:—		
Carbon absorbed by iron	·70	·70
Do. in tarry matter	1·33	1·38
	2·03	2·08
Carbon in escaping gases	27·27	27·22

The weights of the escaping gases per 20 units of iron, computed in the manner already described in this volume, are as follows:—

FURNACE A.—SAMPLE a.

		C.		O.		H.		N.
CO_2 ...	... 14·32 =	3·91	+	10·41	...	—	...	—
CO ...	... 48·03 =	20·58	+	27·45	...	—	...	—
CH_4 ...	... 3·02 =	2·27	...	—	+	·75	...	—
C_2H_4 ...	... ·59 =	·51	...	—	+	·08	...	—
H ...	... ·74 =	—	...	—	...	·74	...	—
N ...	... 82·68 =	—	...	—	...	—	...	82·68
NH_3 ...	... ·11 =	—	...	—	...	·02	+	·09
H_2O ...	... 10·51 =	—	+	9·34	+	1·17	...	—
Tar ...	... 1·55 =	1·37	+	·03	+	·15	...	—
Totals	... 161·55	28·64		47·23		2·91		82·77

FURNACE B.—SAMPLE c.

		C.		O.		H.		N.
CO_2 ...	... 16·26 =	4·43	+	11·83	...	—	...	—
CO ...	... 47·73 =	20·45	+	27·28	...	—	...	—
CH_4 ...	... 2·67 =	2·01	...	—	+	·66	...	—
C_2H_4 ...	... ·38 =	·33	...	—	+	·05	...	—
H ...	... ·81 =	—	...	—	...	·81	...	—
N ...	... 89·77 =	—	...	—	...	—	...	89·77
NH_3 ...	... ·11 =	—	...	—	...	·02	+	·09
H_2O ...	... 10·55 =	—	...	9·38	+	1·17	...	—
Tar ...	... 1·56 =	1·38	+	·03	+	·15	...	—
Totals	... 169·84	28·60		48·52		2·86		89·86

SECTION XI.—HYDROGEN IN THE BLAST FURNACE.

The items of absorption are thus computed:—

Class I.

	Calories.	Totals.	Calories.	Totals.
Fusion of slag	10,054		10,054	
Decomposition of H_2O in blast	2,890		3,162	
Expulsion of CO_2 in minerals	4,044		4,044	
Decomposition do.	4,192		4,192	
Evaporation H_2O in coal Do. in hematite	3,051		3,051	
Expulsion of volatile carbon gases (H, O, and N) = for 12·28 units	24,560		24,560	
		48,791		49,063

Class II.

Reduction of Fe_2O_3	32,710		32,710	
Carbon impregnation	1,680		1,680	
Reducing SiO_2—P_2O_5 and SO_3	4,266		4,266	
Transmission through walls	3,658		5,487*	
Fusion of pig iron	6,600		6,600	
Carried off in tuyere water	1,818		1,818	
		50,732		52,561

Class III.

Carried off in escaping gases		9,856		8,953
Grand total		109,379		110,577

In the elements of absorption we have a very large amount of heat devoted to the expulsion of the volatile constituents of the coal; an absorption with which the low temperature of the escaping gases, under 400° F., is connected. On the other hand there is a not inconsiderable contribution to the supply of heat through a certain amount of hydrogen which appears to have been oxidized. This is inferred from the fact that there appears more water in the gases than was introduced as such in the materials. In the coal and ore used per 20 units of iron the water amounted to 5·65 units, whereas there were collected for this weight of metal 10·51 units of water at furnace A and 10·55 units at furnace B. This exhibits an increase of 4·86 and 4·90 units of water, equal to ·54 and ·55 unit of oxidized hydrogen respectively.

* This furnace is much larger and yet makes no more iron than No. 1, hence the extra allowance for loss by transmission through the sides.

SECTION XI.—HYDROGEN IN THE BLAST FURNACE.

The heat evolved is therefore thus estimated:—

Carbon delivered to both furnaces per 20 units of iron, viz. 42·39 of coal containing 53·41 per cent. of fixed carbon...		22·65
Less carbon carried off with an equal weight of carbon in limestone		1·31
		21·34

Heat Evolved.	Furnace A. Sample a. Calories.		Furnace B. Sample c. Calories.
Burning fixed carbon 21·34 to CO ...	51,204	21·34 to CO ...	51,204
Of this C as CO oxidized 3·93 to CO_2	21,896	4·43 to CO_2 ...	24,808
	73,100		76,012
Combustion of hydrogen ·54 ...	18,360	·55	18,700
Contained in blast	12,160		11,920
	103,620		106,632

Having regard to the complicated nature of the analysis, etc., the figures arrived at in this computation of heat, as referring to the use of raw coal, seem fairly satisfactory on both sides, the difference being under 5 per cent.

To test the general accuracy of the analyses, the sources of the oxygen and hydrogen are compared below with their respective amounts in the escaping gases, both taken from furnace A, sample a:—

	Units.
Sources of oxygen: In the blast including its moisture...	25·73
,, ore ,, ,, ...	9·24
,, carb. acid of the limestone...	3·50
,, coal including its moisture...	9·07
	47·54
Weight of oxygen as per analysis given at p. 318 ...	47·23
Difference ...	·31
Sources of hydrogen: In moisture of blast ...	·08
,, ore ...	·08
coal including its moisture ...	2·39
	2·55
Weight of hydrogen in various forms given at p. 318 ...	2·91
Difference ...	·36

SECTION XI.—HYDROGEN IN THE BLAST FURNACE.

The correctness of the hydrogen estimate may be easily affected by differences in the composition of the coal, changes in the humidity of the air, etc., or by leakages at the tuyeres.

Although, as far as the present estimates show, there is a distinct oxidation of hydrogen, it is by no means proved, nor is it probable, that this oxidation is due to the direct action of this element on the ore. On referring to the analysis given in p. 315 it will be found that the coal contained 11·09 per cent. of oxygen, equal therefore to 4·70 units in the coal required to smelt 20 units of iron. Now this quantity of oxygen would combine with ·58 units of hydrogen, which almost exactly corresponds with the excess supposed to have been converted into water, viz.—·54 and ·55 (v. p. 319).

There is of course no doubt that the moment water, in the form of steam, meets with highly heated iron or carbon, the latter substances are instantly oxidized, hydrogen being liberated.

Being wishful to compare the rapidity of this action with that of reduction I had placed in a porcelain tube 10 grammes of very pure hematite ore which filled a certain length of the tube; and immediately beyond it 7·061 grammes of spongy metallic iron, reduced from other 10 grammes of the same ore by means of hydrogen gas; this iron also filled the diameter of the tube.

This apparatus was filled with hydrogen and maintained at a temperature approaching that of melting silver, therefore probably a little below 1,000° C. (1,800° F.) Two litres of hydrogen were passed over its contents during a period of one hour. The gas coming in contact with the ore effected a certain amount of reduction; and then, the resulting vapour of water, together with any unchanged hydrogen, meeting with the heated iron sponge, the latter was partially oxidized.

We had thus two equal quantities of iron (7·061 grammes) acting in presence, the one of a reducing and the other of an oxidizing atmosphere at the same temperature. The results were as follows:—

Experiment.	10 Grammes of Hematite Ore. Lost Oxygen. Grammes.		7·061 Grammes of Spongy Iron, obtained from 10 grs. of the same Hematite. Absorbed Oxygen. Grammes.
1	1·2799	...	·5790
2	1·6062	...	·6475
3	1·0964	...	·6614
Average	1·3275	...	·6293

U

SECTION XI.—HYDROGEN IN THE BLAST FURNACE.

Although there is not entire uniformity in the results, we are justified in inferring that reduction proceeds much more rapidly than oxidation; the average relation being nearly as 2 to 1, $\frac{1\cdot 3,275}{\cdot 6293}=2\cdot 11$.
This perhaps is no more than might be expected; because we have the pure hydrogen, as it meets oxide of iron, sweeping out the watery vapour before it, whereas the metallic iron has its tendency to be oxidized always restrained, so long as both actions are in operation, by the presence of free hydrogen.

In the blast furnace itself, the small quantity of water to be decomposed in the hearth and the immense surface of iron it meets with render decomposition the work of an instant, leaving the oxidized metal to be reduced by the highly heated carbon.

Whether hydrogen promotes reduction in the upper zone of a blast furnace or not, there can, from what has preceded, be no question of its energy as a deoxidizing agent. Under these circumstances it becomes necessary to consider the cause of there being, in the case of these furnaces burning raw coal, so much less carbonic acid for a given weight of iron than is found in the gases of a furnace using coke.

If we refer to the performance of a furnace having a capacity of 6,000 cubic feet, $v.$ page 244, it will be found that $28\cdot92$ cwts. of coke sufficed to produce a ton of iron. For this quantity of iron there was $5\cdot47$ cwts. of carbon in the form of carbonic acid. On the other hand in one of the Scotch furnaces, using raw coal, there was only $4\cdot43$ cwts. of carbon in this condition.

Let us first contrast the circumstances of the two furnaces:—

	Coke Furnace.	Raw Coal Furnace.
Capacity ... ... cubic feet	6,000	10,000[1]
Fuel used per ton of metal cwts.	28·92	42·39
Volumes of carbonic acid per 100 volumes carbonic oxide ...	31·31	21·54
Volumes of carbonic acid per 100 volumes CO, H, &c.	31·18	15·81
Carbon as carbonic oxide per one C as CO_2	4·06	4·62
Cubic feet of gases per ton of iron[2]	237,146 (170·59 cwts.)	247,837 (160 cwts.)

[1] Only about 8,500 cubic feet filled.
[2] Barometer at 30 inches and thermometer 60° F.

SECTION XI.—HYDROGEN IN THE BLAST FURNACE.

These figures indicate that in the furnace using raw coal there passed over the ore about $4\frac{1}{2}$ per cent. greater bulk of gases possessing for an equal volume a higher reducing power by about 34 per cent. (21·54 volumes CO_2 instead of 31·31) than in the coke furnace. Nevertheless there is per ton of iron 23 per cent. less carbon as carbonic acid in the raw coal furnace than in the other (4·43 instead of 5·47).

It has been shown (page 190) that strongly roasted ore, like black band, is less susceptible to the reducing action of carbonic oxide than ironstone in a more moderately calcined condition, like the ironstone of Cleveland as it commonly comes from the kilns. If this were so, it is possible, notwithstanding the more energetic nature of the deoxidizing zone in the furnace using raw coal, that a quantity of semi-fused black band might pass down unreduced, until it reached a position where the temperature determines the decomposition of carbonic acid by carbon.

To remove this uncertainty, the same friend supplied me with the particulars of a furnace smelting Spanish (Bilbao) ore with raw coal, in order that I might compare it with the same furnace using the Scotch ironstone.

In the former estimate, the heat absorbed per 20 units of iron to smelt a mixture consisting chiefly of argillaceous ores was 110,577 calories; and the coal consumed being 42·39 units, we have $\frac{110,577}{42·39}$ or 2,609 calories per unit of raw coal.

Employing the same data as before for the means of computation, I find that in smelting hematite the absorption to be 94,100 calories; and the raw coal used being 36·62 units per 20 of iron, we have $\frac{94,100}{36·62}$ or 2,569 calories per unit of fuel.

I do not possess any analysis of the gases from this last named furnace; but, the heat-equivalent of the fuel agreeing so closely with the previous one, it is only reasonable to infer that the quantity of carbon as carbonic acid must have been the same in each case. It is obvious therefore that any peculiarity in the ironstone is not the cause of the disappearance of carbonic acid; for an example was given, p. 246, Table II. I., where the carbon in this form, with similar ore (hematite), amounted to 1 per 2·04 of carbon as carbonic oxide.

I have in addition been able to compare the performance of furnaces 52 feet high, smelting the oolitic ores of the Midland counties with the raw coal of Derbyshire, which contains about 35 per cent. of volatile matter. The ore, chiefly a hydrated peroxide, gives off $22\frac{1}{2}$ per cent. of moisture; it yields in the wet state $32\frac{1}{3}$ per cent. of metal, with a consumption of $6\frac{1}{4}$ cwts. of limestone. After making allowance for evaporating the water from the ore, and for extra weight of slag to be fused, the heat absorbed in these Derbyshire furnaces turns out to be about 123,000 calories per 20 of iron; and this being divided by 46 units of coal used, we have $\dfrac{123,000}{46} = 2,674$ calories per unit of fuel burnt.

So far as previous enquiries enable me to judge, the oolitic ore used in Derbyshire, when smelted with coke, gives pretty nearly the same results as those which attend the treatment of Cleveland stone. Under these circumstances I am led to infer that with furnaces of about the same dimensions in each case, say 6,000 cubic feet, there is nearly 25 per cent. more carbon in the form of carbonic acid in furnaces using coke than in furnaces using raw coal.

Admitting the correctness of this inference, there seem but two ways in which to account for this disappearance of carbonic acid in the gases. Either the raw coal, as it is freshly coked in the upper region, produces coke in that soft condition which permits its being easily attacked by the carbonic acid ($CO_2 + C = 2CO$); or the presence of *hydrogen* gives rise to a decomposition of carbonic acid, as already mentioned in the present section ($CO_2 + H_2 = CO + H_2O$).

The large amount of water, shown to be present in the gases of a furnace using raw coal, or at all events a portion of it, must undoubtedly exercise a restraining influence on their deoxidizing power in a part of the reducing zone. Although the hygrometric moisture of the coal, or of the ore, will be mostly expelled in the highest zone of the furnace, *i.e.* before the minerals are sufficiently heated to bring their action on the gases into play, still a considerable portion of water, the result of the destructive distillation of the actual substance of the coal, will not be driven out until the lower and hotter regions of the furnace are reached. In such a case the heated vapour of water will mingle with

SECTION XI.—HYDROGEN IN THE BLAST FURNACE.

the reducing gases, and passing with them through the oxide of iron will act in the manner supposed.

The practical question, which arises from comparing the use of raw coal with coke in the blast furnace, is one of economy; and this is capable of being dealt with in a very few words.

In two furnaces of the same capacity, and blown with air practically of the same temperature (485° C. at Clarence and 475° C. in Scotland), we obtain the following factors:—

Clarence—coke used per 20 of iron 28·92 units containing carbon 26·36 units.
Scotland—raw coal per 20 of iron 42·39 units containing *fixed* carbon 22·64 units.
Clarence—Calories required for reduction, fusion, etc. 103,123.
Scotland—Do.   do.   do.   do.,
    and expelling gases from coal   109,379
Less absorbed in expelling gases   24,560
                                                   —————— 84,819

To compare the duty performed by pure fixed carbon we have:—

Clarence ... $\dfrac{103{,}123}{26·36}$ = 3,912 calories per unit of carbon.

Scotland ... $\dfrac{84{,}819}{22·64}$ = 3,746   ,,   ,,

This shows an inferiority of about $4\tfrac{1}{2}$ per cent. on the side of the Scotch coal, due no doubt to its producing less carbonic acid in the gases than the coke at Clarence.

Viewed in this fashion, the Scotch furnace ought to have done its duty with 40·50 cwts. of raw coal, in order to be on a level with the Clarence furnace.

On the other hand, in coking coal in the ordinary beehive oven, about 10 per cent. of the fixed carbon of the coal is wasted; *i.e.* instead of getting 75 only 65 per cent. of coke is obtained from the best coal. Hence the 26·36 pure carbon used at Clarence really represents 30·40 fixed carbon in the coal; and $\dfrac{103{,}123}{30·40}$ = 3,392 calories per unit of pure carbon.

Of course in such calculations the proper way is to institute the comparison between raw coal and coke obtained from the same raw coal, when applied to the same ore. Admitting however that by using the fuel raw, as in Scotland, there is a loss of about 2 cwts. of coal per ton of iron, this is compensated fourfold by the saving of wages together with interest and wear and tear on coke ovens. There is besides this the gas given off at the blast furnace, which can be much more beneficially applied than at the coke ovens, and if to this is added the expected results from the recent proposal to collect tar and ammonia we should probably be within the mark in setting down 8s. to 10s. on the ton of iron as the difference between using raw coal instead of coke. From this however some allowance must be made, owing to the circumstance that the entire produce of the pit may be converted into coke, whereas only the large pieces can be used raw in the furnace. Probably the best plan therefore would be to use the large coal raw and coke the small, wherever the quality of the latter permits its application in this form.

Having now dwelt at some length on what may be considered the chief functions of a furnace using raw coal, I would conclude my observations on this subject by a few remarks on a subject which some pages back received a casual notice—viz., the collecting of ammonia and tar from the furnace gases.

With the knowledge that both these substances are given off when coal is subjected to destructive distillation, their presence in a blast furnace using raw coal is only what might be expected. Messrs. Bunsen and Playfair in 1846 not only mentioned the fact in their paper on the Alfreton furnace already referred to, but determined the quantity of ammonia in the gases.

At page 127 in my former work, mention was made of a considerable amount of ammoniacal salts having been collected from one of the coke furnaces of my firm. Looking at the fact that the nitrogen of the coal was almost entirely dissipated by the process of coking, I accounted for the appearance of an unusual quantity of ammonia by the decomposition of some of the cyanogen compounds, known to be present in the furnace. The quantity of ammonia collected from different parts of the furnace per 1,000 litres of gas upon the occasion referred to, was as follows:—

Distance above tuyeres	8 ft.	$24\frac{1}{4}$ ft.	60 ft.	$76\frac{1}{2}$ ft.	escaping gases.
Grammes of ammonia	3·28	2·25	2·78	2·36	4·64.

SECTION XI.—HYDROGEN IN THE BLAST FURNACE.

In order to see whether the weights condensed were due to an abnormal condition of things, the experiments were recently repeated at Clarence, upon which occasion the following results were obtained per 1,000 litres of gas, freed from dust, in three different experiments:—

·0075—·0077 and ·0114 grammes of ammonia.

The flue dust separated in the last experiment gave an additional ·0082 grammes per 1,000 litres of gas. Taking the volume of escaping gases at 180,000 cubic feet per ton of pig iron and 156,000 cubic feet per ton of coke this would represent:—

	Per Ton of Pig Iron.		Per Ton of Coke.	
	Lbs. Ammonia	= Sulphate.	Lbs. Ammonia	= Sulphate.
From gas freed from dust	·098	·38	·085	·330
From gas including that from accompanying dust	·218	·846	·190	·738

It would thus appear, as I suspected, that the larger quantities of ammonia collected on the previous occasion must have been due to an unusual condition of the furnace at the time. A probable cause might be decomposed cyanides which, upon some occasion, are found present in the escaping gases, and upon others entirely absent.

On referring to the analysis of the Scotch coal given at p. 315, it will be seen that nitrogen exists in it to the extent of ·94 per cent., equal therefore to ·398 units for each 20 units of pig iron produced. If then all this nitrogen were converted into ammonia, we should have ·483 units per ton of metal; whereas according to analysis the quantity was only ·11 units (v. p. 318). From this we may infer that only about 23 per cent. of the volatile alkali, capable of being generated by the nitrogen present in the coal, is actually formed.

According to a paper recently read before the Society of Chemical Industry by Mr. Walter Weldon, for each ton of coal used in the Gartsherrie furnaces 20 lbs. of sulphate of ammonia is produced, equal therefore to 5·15 lbs. of ammonia; while the ·11 unit mentioned above represents 12·32 lbs. per ton of iron or 5·81 lbs. per ton of coal.

The furnace gas, thus freed not only from the water and tar but also from the furnace fume, is well adapted for the steam boilers and the hot air apparatus.

Some years ago Mr. Carvés described a coke oven which was virtually a retort. The gases were conveyed through condensers in which the tar and ammoniacal water were separated; and the gas, freed from these substances, was used at the oven in order to distil the coal.

In the year 1867 my firm erected 36 of these ovens; but partly owing to the difficulty of keeping them tight, and partly owing to the lower value of the products, they were discontinued. Recently the subject has been revived in the County of Durham by my friends Messrs. Pease and Co., who have applied M. Carvés' principle to the Coppée oven. Hitherto they are well satisfied with their progress; and so far as ammonia and tar are concerned their experience indicates a state of things not inferior to that given by Mr. Weldon as obtained at Gartsherrie.

It has still to be seen whether the coke made in such a form of oven as that described is equal in quality to that obtained by the higher temperature of old-fashioned ovens of the beehive or similar construction.

One disadvantage attends the installation of the apparatus erected by Messrs. Pease—it means the demolition of the present ovens and the erection of a much more expensive plant in their room. From this inconvenience the plan recently patented by Mr. J. Jameson is exempt. He removes the floor of the present oven, and replaces it with one provided with flues. Through these, by means of an aspirator, the gases as they commence to distil from the coal are drawn *per descensum*, and made to pass through condensers where the tar and ammonia are left. The gas freed from these may be reconducted to the interior of the oven, to assist in the distillation of the coal.

Many coals have been tried at the Felling Chemical Works of Messrs. H. L. Pattinson and Co., with very excellent results; for from some of them as much as 12 lbs. of sulphate of ammonia, and 11 gallons, not of tar but of an inflammable oil, have been received per ton of coal.

The value of these products per ton of coal may be taken at:—

12 lbs. of sulphate of ammonia at 2d. per lb.	£0 2	0
11 gallons of oil at 3d. „	0 1	9
Total	£0 3	9

From this sum has to be deducted the expense of condensing and the cost of converting the ammonia into sulphate.

SECTION XI.—HYDROGEN IN THE BLAST FURNACE.

Taking the subject as a whole, there is good reason for believing that we are on the eve of an important change in the application of coal as a combustible; for it seems not improbable, that we may be compelled to use our fuel in the form of gas, with a view of obtaining the valuable products destroyed by the present mode in which it is burnt.

Influenced probably by the quantity and intensity of the heat obtained by the combustion of hydrogen, the idea has occurred to various inventors to inject this gas at the tuyeres of the blast furnace. Many years ago it was proposed by Mr. Mickle to collect the hydrocarbons given off in coking coal, and apply them to the purpose referred to. Subsequently Mr. Dawes in Staffordshire tried for some time the use of "water gas," prepared by passing steam over carbonaceous matter kept at a high temperature in a retort. These trials, and it may be others, having the same object in view, were abandoned; and until recently nothing further, I imagine, has been heard of using hydrogen in any form as a fuel for smelting iron.

The attention which has, from time to time, been bestowed on the use of hydrogen, the general costliness of such experiments, and its general interest in connection with the economy of the blast furnace, have induced me to make some experimental enquiry into the behaviour of mixed gases, resembling in composition the water gas now proposed to be used.

So far as I know, all plans in this direction contemplated employing the gas as a supplement to coke or coal; but it is now expected that all solid fuel in the furnace itself may be dispensed with, and the so-called water-gas employed in its room. It is proposed to prepare this substance, which theoretically is composed of equal volumes of hydrogen and carbonic oxide, by passing very highly heated steam over refuse coke, or over coal. In the latter case the two gases mentioned above are mixed with a further quantity of hydrogen and hydro-carbons, given off by the coal.

Looking at the price of blast furnace coke, and that of such combustible matter as might suffice for the manufacture of water gas, it is quite possible that a given amount of *heating* power might be produced more economically from water-gas than could be obtained from the coke usually employed in smelting iron. As a mere question of heat evolution, it cannot of course be pretended that on burning carbon in order to decompose steam, a larger quantity of heat can be generated by burning the hydrogen and carbonic oxide thus obtained, than that capable of being produced by the carbon itself burnt in the usual way. On the contrary, as is well known, there is an absolute loss attending the manufacture of water-gas even when pure; which it never is in practice, being contaminated more or less with carbonic acid, etc. Another possible advantage in employing the gaseous fuel would be the possession of some virtue which fitted it for reducing oxide of iron. No doubt the laboratory experiments quoted in this section would indicate that hydrogen gas is more energetic in this respect than carbonic oxide; but, on the other hand, the action of the blast furnace, as already described, points pretty conclusively to the fact, that little, if any, hydrogen is directly concerned in reducing the metal, and further that its presence in the furnace is not attended with any economy in the quantity of solid carbon required.

We have in this enquiry first to consider the difference of the conditions under which it is now intended to employ hydrogen and those which attend its use when contributed by the use of raw coal in the furnace. In the first mentioned case it is introduced at the hearth and in the second it is generated chiefly in the reducing, or at all events in the upper zone.

The most recent idea is to pass steam through vessels filled with highly heated fire bricks, resembling in principle the well known regenerators of Messrs. Siemens. By this means it is expected to obtain the water-gas at a temperature of 2,000° C. (3,632° F.), and this it is proposed to burn with air heated to 1,500° C. (2,732° F.)

I will consider immediately the use of pure water-gas when employed under the conditions just named; but in the meantime I would point out what I conceive to be serious difficulties in securing the proposed conditions themselves.

In the first place water-gas prepared in the manner suggested is never pure. My friend, Mr. G. S. Dwight, who is interested in a company who have applied it extensively to certain purposes, to which it is no doubt admirably adapted, has furnished me with an analysis

SECTION XI.—HYDROGEN IN THE BLAST FURNACE.

of the gas prepared in an excellently devised form of apparatus. This specimen, as I shall hereafter show, contained close on 17 per cent. of gases which are worse than useless in the blast furnace.

I was permitted by the courtesy of Mr. Dwight and his friends to examine the operation of producing this gas on a large scale at Essen. At that time the carbonaceous element was small refuse coke. The fire-brick regenerator was heated to as high a temperature as a coke fire blown with a powerful fan could generate. The blast was then discontinued, and steam introduced in an opposite direction; *i.e.*, it passed through the heated regenerator and down through the incandescent coke. This was maintained as long as the steam was supposed to be entirely decomposed.[1] The steam was then shut off, and blast applied as before preparatory to a second admission of steam.

The decomposition of the vapour of water of course was attended with a great reduction of temperature; so much so that the water-gas left the generator, as I was informed, at a temperature of 600° C. (1,112° F.) If instead of the coke which was employed in the apparatus when I examined it, raw coal were used, an additional amount of heat would be required for effecting the volatilization of its gaseous constituents.

Of the practicability of heating any vast volume of air to 1,500° C. (2,732° F.) they who have worked with fire-brick stoves can say something. Hitherto I believe no one has succeeded in maintaining a steady temperature of 1,000° C. (1,832° F.)

In a paper recently published, the writer apparently recognizes the oxidizing influence which burnt water-gas—consisting of carbonic acid and watery vapour—must have on metallic iron. To counteract this he limits the quantity of air employed to that required for burning one-tenth only of his gaseous fuel.

If we examine theoretically the conversion of steam into what may be designated pure water-gas, *i.e.*, a mixture of equal volumes of carbonic oxide and hydrogen (or 28 parts by weight of the former and 2 of the latter), we bring out the following results:—

---

[1] In my own experience, in passing steam through coal to generate a kind of water-gas, undecomposed vapour of water was often present. This of course sent into a blast furnace is eminently undesirable.—I.L.B.

Splitting up of one equivalent of steam, H_2O, into its elements, by means of one equivalent of carbon—

	Calories.
2 H × 34,000 cal. = absorption of	68,000
12 C × 2,400 cal. = generation of	28,800
Net amount of heat absorbed	39,200

Assuming, according to information afforded by Mr. Dwight, that water-gas generated from coke is given off at a temperature of 600° C. =1,112° F., we have in burning this pure water-gas :—

	Calories.
2 of hydrogen × 34,000 calories	68,000
28 of carb. oxide × 2,400 calories	67,200
Heat by combustion of 30 units of cold water-gas	135,200
Heat in water-gas at 600°, 2 H × 3·409 sp. heat × 600° C. = 4,091	
28 CO × ·245 „ × 600° C. = 4,116	
	8,207
Total heat by combustion of 30 units of hot water-gas	143,407

In practical working the 8,207 calories, corresponding to the sensible heat of the generated gas, have to be supplied in addition to the 39,200 calories absorbed by the decomposition of the steam. In all 47,407 calories have therefore to be generated by the combustion, to carbonic acid, of a portion of the fuel, the products of this combustion heating the regenerator and leaving it, as was the case at Essen, at 225° C.

To ascertain how much carbon is needed to obtain this amount of heat, we must proceed as follows :—

		Calories.
Heat afforded by 1 unit of carbon burnt to carb. acid with production of 3·666 CO_2 accompanied by 8·797 of nitrogen 225° C.		8,000
Less carried off by CO_2 3·666 × ·216 sp. heat × 225° =	178	
„ „ 8·797 × ·244 „ × 225° =	483	
		661
Net heat afforded by 1 unit of carbon, burnt to supply heat for generating water-gas		7,339

SECTION XI.—HYDROGEN IN THE BLAST FURNACE. 333

Then $\frac{47,407}{7,339} = 6\cdot46$ units of carbon needed in addition to the 12 units contained in the available gas, to yield 30 units of the latter at 600° C. It follows then that 12 + 6·46 or 18·46 units of carbon yield gas capable, in the cold state, to generate 135,200 calories, or each unit of carbon consumed yields 7,324 units in the form of water-gas.

In the estimate just given no allowance whatever is made for loss by radiation, etc., which, as is well known, must add something to the expenditure of fuel.

In the event of this water-gas being applied as a means of heating only, the useful effect of the carbon consumed in its production will be that used by the full powers of the latter *minus* the waste; which cannot be less than close on 10 per cent., and may be double this amount or more. If however the hydrogen it contains is only partially applicable as a means of heat evolution, as appears to be the case in the blast furnace (*vide* p. 320, *et seq.*, on Scotch furnaces), then out of 135,200 calories a large portion might for this purpose be useless.

In the estimate just given it is assumed that the gas is pure; in other words that all the steam is decomposed, and decomposed in such a way as to ensure the products being all combustible gas. Instead of this, some steam escapes decomposition, and some carbon is converted into carbonic acid—the latter being a direct source of loss and the former very hurtful in the blast furnace itself. Besides these sources of contamination, a little atmospheric air is always to be found in water-gas made in the way already described.

Mr. Dwight has kindly furnished me with an analysis of this gaseous fuel obtained by passing superheated steam over raw coal. Its constituents by weight were:—

	Per Cwt.	Carbon Content.
Oxygen	1·740	—
Carbonic acid	6·372	1·738
Carbonic oxide	70·969	30·415
Hydrogen	7·473	—
Hydro-carbon (CH_4)	4·648	3·486
Nitrogen	8·798	—
	100·	35·639

SECTION XI.—HYDROGEN IN THE BLAST FURNACE.

According to the information supplied to me, one pound of coal employed in the apparatus affords 25 cubic feet of gas, which, if of the composition just given, will weigh 1·029 lbs.

From this it would follow that 100 lbs. of coal may be considered as giving 103 lbs. of the gas; and as the gas itself contains 35·639 per cent. of carbon, we have in the 103 lbs. of gas 36·71 lbs. of carbon obtained from treating 100 lbs. of coal. But if the coal used contains 66 per cent. of carbon, which is probably the least we may expect, then each pound of carbon in the gas has been obtained by an expenditure of $\frac{66}{36\cdot71}$ or 1·8 lbs. of carbon in the coal.

In the two cases we are considering, it appears that for 100 parts of carbon employed the following are the products:—

	Useful Carbon (CO) and Hydro-carbons.	Useless Carbon.	C consumed in production of gas.
Pure water gas	65 CO.	nil.	35
Water gas as made with raw coal	52·9 CO & CH	2·7	44·4

With regard to any idea that either of the two gases just referred to could be suitable forms of combustible for a blast furnace, and that they and the blast could be delivered at the tuyeres at a temperature of 1,000° C. = 1,832° F., which is probably above that to which they could be heated, we may take the following as the quantities of heat which would in such case be generated.

Pure water gas.—Quantity of heat capable of being evolved per unit of carbon consumed in making the gas, of which unit ·35 supplies the heat for decomposing the water, leaving ·65 available in the blast furnace.

Weight of gas afforded per unit of carbon consumed	1·625 units.
Air required for burning 1·625 units of gas	7·455 „
Specific heat of the water-gas	·456
Heat afforded by combustion of gas per unit of C with air at 0° C. = 32° F.	7·324 calories.

Let it further be supposed that in order to preserve the reducing character of the mixed gases, one tenth only of the water-gas was burnt.

SECTION XI.—HYDROGEN IN THE BLAST FURNACE. 335

	Calories.
Heat in 1·625 of gas (equal to 1 of C burnt in generator)— × ·456 sp. heat × 1,000° C.	= 741·0
Heat in 1/10th of air required for complete combustion— $\frac{7\cdot455}{10}$ × ·237 sp. heat × 1,000° C.	= 175·5
Heat of combustion of 1/10th of gas— 1·625 ÷ 10 × 7,324 calories	= 732·4
Total heat per unit of carbon	1648·9

Water gas as obtained from steam and raw coal.—Quantity of heat capable of being evolved per unit of carbon consumed in making the gas, of which unit only ·529 is available in the furnace; weight of gas afforded per unit of carbon consumed, 1·029 unit; air required for burning 1·029 unit of the gas, 5·185 units; specific heat of the gas itself, ·494; number of calories afforded by its combustion, when burnt with air at 0° C. = 32° F., 5,026.

	Calories.
Heat in 1·029 of gas (equal to 1 of C. burnt in generator) × ·494 sp. heat × 1,000° C.	= 508·3
Heat in 1/10th of air required for complete combustion— $\frac{5\cdot185}{10}$ × ·237 sp. heat × 1,000° C.	= 122·9
Heat of combustion of 1/10th of gas— 1·029 ÷ 10 × 5,026 calories	= 502·6
Total heat per unit of carbon...	1133·8

Thus it appears that, when burnt in the way suggested, there is obtained useful heat in the furnace, for each unit of carbon delivered to the generator, to the amount of say 1,648 calories from pure water-gas made with coke, and say 1,133 calories from water-gas as it is actually obtained from raw coal.

Leaving the theoretical gas on one side, and confining ourselves to the practical, we can easily compare this result of 1,133 calories with the quantity of heat evolved by each unit of raw coal in the blast furnace, as described in the present section. There it will be seen we had 110,577 calories afforded by 42·39 of coal; equal therefore to 2,609 calories for each unit of coal consumed.

Of these 110,577 calories 24,560 were absorbed in expelling the volatile constituents contained in the coal; which work in the present case has already been performed in the water-gas generator. This leaves 86,017 for the other work of the furnace. The gases contained 4·43 units of carbon in the form of carbonic acid, equivalent to 24,808 calories evolved in raising this weight of carbon from carbonic oxide to carbonic acid.

	Calories.
If then to make 20 units of pig iron there is required	86,017
Of which the generation of carbonic acid from carbonic oxide gives	24,808
There is left for the combustion of the water-gas to supply	61,209

Now $\frac{61,209}{1,133}$ = 54 units of raw coal used in the water-gas generator.

According to this statement then, there would be required 54 units of coal, burnt with air at 1,000° C., to produce the same quantity of heat from water-gas as 42·39 units of coal burnt with air at 427° C.; showing a loss of about 22 per cent.

But the *quantity* of heat is not the only question which has to be considered; for the intensity is of equal importance. Now the specific heat of the products of combustion of water-gas made from coal, and having the composition already mentioned, is ·268.

Its quantity, per unit of carbon, we found to be—

$$\frac{1·029 + 5·185}{10} = ·6214.$$

The specific heat of the unburnt part of the gas = ·494, its quantity—say 9/10ths of the whole—9/10 × 1·029 = ·9261.

From these data the temperature of this imperfect combustion is thus estimated :—

$$T = \frac{1,133 \text{ cal.} - 64 \text{ cal. latent heat of vap. of water}}{·6214 \times ·268 + ·9261 \times ·494} = 1,715° \text{ C. or } 3,119 \text{ F.}$$

A temperature of 1,715° C. is probably below that at which a furnace could be worked; but this no doubt might be met by a more perfect combustion of the water-gas. Thus it is found by estimation that the temperature produced by burning water-gas made from raw coal with one-third of the air needed for perfect oxidation is 2,495° C. (4,523 F.)

It is however needless further to examine the system in this direction; because, if all the difficulties already pointed out were effectually set aside there is, I am of opinion, no question but that the principle

SECTION XI.—HYDROGEN IN THE BLAST FURNACE.

blast is fundamentally erroneous. Although the minerals in the ordinary furnace are freed from almost all their oxygen in the upper zone it is at the same time a powerfully reducing instrument at its lower extremity. There is no necessity whatever to discuss the reducing power in the upper zone of a furnace attempted to be driven by water-gas; because the difficulty will arise in the zone of fusion where this gaseous fuel meets the blast. The heat produced at the hearth must of course be dependent on the oxidation of the hydrogen and of the carbonic oxide, the products being vapour of water and carbonic acid.

It seems like a work of supererogation to ascertain by experiment what the effect of highly-heated iron must be on vapour of water and carbonic acid, but in the absence of any experience of the conduct of a mixture of these two substances, the following trials were made:—

Ten grammes of pure hematite ore, in grains about the size of mustard seed, were exposed in a porcelain tube, at a temperature of about 1,000° C. (1,832° F.), to a current of a mixture of equal volumes of hydrogen and carbonic oxide.

	Grammes.	
The ore lost in weight oxygen		2·9230
The water collected indicated deoxidation by hydrogen...	1·6681	
The carbonic acid collected indicated deoxidation by carbonic oxide ...	1·1909	
		2·8590
Difference due to experimental error and carbon absorbed by iron		·0640

The spongy iron thus produced was then exposed to a current of watery vapour and carbonic acid, generated by exposing an equal quantity of ore to the water-gas, the iron sponge and ore being heated as before to 1,000° C.

	Grammes.	
The ore lost of its oxygen ...		1·5153
Of this hydrogen removed	·5347	
" carbonic oxide removed	·5268	
	1·0615	
" the spongy iron absorbed ...	·3990	
		1·4605
Difference due to experimental error and to deposited carbon ...		·0548

In this case the gain of oxygen by the metallic iron was close on 39 per cent. of the loss of oxygen experienced by the ore.

v

The same mixture of gases was then tried on calcined Cleveland ore also broken to the size of mustard seed and containing 43·7 per cent. of iron. Six litres were passed in two hours over 10 grammes of ore, the temperature being maintained at 1,000° C. = 1,832° F.; during which time the metal lost 88 per cent. of its original oxygen. The weight of the six litres of water-gas, namely, 4·018 grammes, was increased to 5·712 grammes after passing over the ore.

The oxygen thus separated weighed per 100 of ore ...		16·470
As estimated from the carbonic acid collected, the carbonic oxide removed	8·209	
As estimated from the water collected, the hydrogen removed	8·729	
		16·938
Difference due to experimental error and carbon deposition ...		·468

The gas after passing over the ore would have the following average composition in 100 parts by weight:—

1.—*Oxidized gases.*—Carbonic acid ... 39·5   Water vapour 17·2 = 56·7
2.—*Unaltered gases.*—Carbonic oxide 40·5   Hydrogen ... 2·8 = 43·3
                                                               ———
                                                               100·

Volumetrically 100 volumes would contain—

                                                              Vols.
   1.—Carbonic acid ... 19·1   Vapour of water ... 20·3 = 39·4
   2.—Carbonic oxide ... 30·8   Hydrogen ... ... 29·8 = 60·6
                                                           ———
                                                           100

By weight 1 of C as CO_2 is accompanied by 1·61 of C as CO.
    „   1 of H as H_2O „   1·46 of H as H.
         —   ————
         2   3·07

Showing 1·535 of reducing gases to 1 of C and H as oxidizing gases.

By measure 100 volumes of the reducing (CO and H), are associated with 6·5 of the oxidizing, gases. Over ten grammes of the same ore as that used in the former experiment 6 litres of pure water-gas was passed, equal to 60 litres per 100 of ore; the temperature as before being 1,000° C. = 1,832° F.

SECTION XI.—HYDROGEN IN THE BLAST FURNACE.

Calculated on 100 grammes we have the following numbers:—

Reduced ore, equal to 83·53 grammes, as placed beyond the 100 grammes of calcined ore so that the water-gas, partly converted into water and carbonic acid, passed over it.

	Grammes.
The 100 of fresh calcined ore lost of oxygen	15·04
Accounted for as follows:—	
Hydrogen took up ... ... 5·99	
Carbonic oxide ,, ... 6·22	
Reduced iron ,, ... 3·16	
	15·3
Difference due to experimental error and carbon deposition ...	·33

The escaping gases were also examined with the following results:—

The 60 litres, calculated on 100 grammes of calcined ore, consisted of 30 litres of carbonic oxide, or 37·50 grammes, and 30 litres, or 2·68 grammes, of hydrogen = 40·18 grammes of original gas raised to 57·12 grammes in the former experiment. By the loss of oxygen absorbed by the metallic iron the weight was now reduced to 52·39 grammes, and 100 parts by weight consisted of—

1.—*Oxidized gases.*—Carbonic acid 32·6 + Vapour of water 12·8 = 45·4
2.—*Unaltered gases.*—Carbonic oxide 50·8 + Hydrogen ... 3·8 = 54·6
                                                                                             100·

Volumetrically 100 contain—
    1.—Carbonic acid ... ... 14·3 + Vapour of water 13·7 = 28
    2.—Carbonic oxide ... ... 35·0 + Hydrogen ... 36·8 = 72
                                                                                           100

By weight.—1 of C as CO_2 accompanied by 1·24 of C as CO.
              1 of H as H_2O ,,   1·26 of H as H.
                    2   3·50

Showing 1·75 of reducing gases to 1 of C and H as oxidising gases.

By volume 100 volumes of reducing gases (CO and H) are associated with 39 of oxidising gases.

In the experiment just described the oxygen removed from the calcined ore (by H 5·99 grammes and by CO 6·22 grammes) was 12·21 grammes; while the reduced ore took up only 3·16 grammes or a little under 26 per cent. of that lost by the oxidized ore, both containing the same quantity of metallic iron.

There is not the slightest reason to suppose that the immense surface of highly heated iron which fills the interior of a blast furnace would not amply suffice, so long as it retained the temperature proper to the smelting process, to decompose all the carbonic acid and vapour of water exposed to its influence. Admitting however that only 26 per cent. of the metal was re-oxidized, and taken up, as it would be, by the slag, the cost of the remainder of the iron would be raised to a corresponding extent—not to speak of the physical difficulty of working a blast furnace in which one-fourth of its iron was running over the slag notch. Instead of cast iron we should, under the supposed conditions, have an infusible mass and a cinder, the corrosive action of which no brickwork could withstand.

What the ultimate fate of the furnace itself would be under such a mode of treatment we shall probably never know, for it is not to be supposed that the operation would be continued long enough to supply us with this information.

I met recently with the account of an experiment in which an attempt was made to drive a furnace partly with coke and partly with water-gas. On the circumstances and conditions of this experiment I should like to say a few words; because they seem to me to confirm the general views just laid down. A small furnace was constructed, 21 feet high, in which with blast heated in a Cowper's fire-brick stove a few cwts. only of metal were made per day with an expenditure of about 7 tons of coke per ton of iron. No information is given to account for the singularly defective character of these results. Forty years ago I worked a furnace 20 feet high—of larger diameter it is true; but out of it 7 tons a day were run with an expenditure of only 2 tons of coke per ton of metal, although the blast was under 650° F.

To the furnace now referred to water-gas was then applied—not alone as has been intimated, but with the retention of a considerable quantity of coke, which was introduced into it with the ore and flux. The daily produce, in 24 hours, was then about $6\frac{1}{2}$ tons and the consumption of coke per ton of iron was:—

SECTION XI.—HYDROGEN IN THE BLAST FURNACE.

	Cwts.
Charged as such into the blast furnace	30·34
Consumed in the water-gas generator	25·74
Total coke per ton of iron	56·08

The temperature of the blast is not given; but that of the water-gas is stated to have been 600° C. (1,112° F.) exactly that previously assumed in these pages. The limestone per ton of iron was under 4 cwts.

The description of the experiment does not contain the details necessary for making any critical examination of all the results; but it is stated, that operations had to be suspended owing to the "chilling" of the furnace. This difficulty was ascribed to the quantity of moisture in the ore—the oolitic of Luxemburg—which contained in all 19·50 per cent. of volatile matter. To expel this, probably a couple of cwts. of coke would have sufficed on each ton of iron made.

I am however confirmed by what is said, in my view of the oxidizing tendency which the presence of carbonic acid and vapour of water must confer on the gases of the hearth; for it is stated in the account given of the trial that the mineral used contained 34·50 per cent. of iron, while the pig iron produced only corresponded to a yield of 25·6 per cent. Here then we have a furnace within which and independently of the water-gas nearly as much coke is consumed per ton of metal, as would alone in the ordinary blast furnace suffice for its production, and at the same time actually wasting $25\frac{1}{2}$ per cent. of all the iron in the ore—no doubt by the oxidizing character of the products of the combustion of the water-gas, that combustion being obtained at the expenditure of 25·74 cwts. of coke on the ton of iron.

While admitting that the dimensions of this experimental furnace were not such as to command any large measure of success, to my mind the indications it has furnished, confirm the opinion that for *reducing* iron in a blast furnace water-gas will be found entirely unsuited. That the introduction of a moderate quantity of this form of fuel, in the way practised and abandoned by Mr. Dawes many years ago, may, while wasting a portion of the iron, partly free the remainder from silicon and phosphorus, is not impossible.

SECTION XI.—HYDROGEN IN THE BLAST FURNACE.

When in the discharge of my duty as a judge at the Philadelphia Exhibition in 1876, my attention was drawn to a specimen of iron described as having been smelted with petroleum injected at the tuyeres. No information was given as to the circumstances under which it was produced.

The experiment has been repeated by Mr. E. W. Shippon[1] of Meadville in Pennsylvania, who forced in a stream of this oil previously heated. At the commencement of the trial the furnace was being worked with vegetable fuel; and although the temperature of the hearth previously was satisfactory, the moment this liquid hydrocarbon was introduced, the cooling effect was such that it was no longer possible to effect a separation of the iron from the slag.

[1] Journal of United States Association of Charcoal Workers, Nov., 1882.

SECTION XII.

ON THE PRODUCTION OF MALLEABLE IRON IN LOW HEARTHS FROM PIG IRON.

We have now seen how the metal iron, gradually freed from its associated oxygen during its passage through the blast furnace, arrives at the zone of fusion intensely heated, and there enters into combination with carbon and other substances, the product being cast iron.

In the arts this commodity is one of great importance. It melts readily, takes the form of any mould into which it is poured, and is easily shaped in the lathe and by other well-known appliances. Compared with wrought iron, cast iron is deficient in strength by about one half to three fourths; and above all it is entirely devoid of that property which renders the malleable metal so valuable, namely its capability of being welded or worked, when softened by heat.

To free cast iron from those substances which interfere with malleability, various modes of treatment are adopted; and it is to a consideration of these that the remainder of this work will be devoted, so far as the manufacture of the metal is concerned.

It is proposed to do this in a much more concise manner than has been followed with regard to the smelting of the ore. The latter, from its more complicated character, necessarily involved a more lengthened discussion; and it is moreover a section of the processes to which my own attention has been more especially devoted, particularly during the last dozen years.

The chief object of what follows is to set forth certain principles, more or less apparent in all the processes by which the impurities, as we may regard them, are separated from the iron itself.

When pig iron was first placed in the hands of the bar iron maker, the only means at his disposal was the low hearth, in which he had been accustomed to obtain his product direct from the ore. Of this primitive contrivance there exist several modifications, used for separating the carbon, silicon, etc., from the pig.

At first sight there appears a certain amount of inconsistency in the adaptation of an apparatus to a function entirely opposed to that to which it was originally devoted. A chief office of the Catalan hearth was the reduction or *deoxidation* of the ore; in the Lancashire fire and other analagous appliances the object to be achieved is the *oxidation*, more or less completely, of the carbon and metalloids, found in combination with the metal in the pig.

No doubt the workman endeavoured so to manipulate his fuel and blast, as to mitigate the evil referred to. His endeavour in the ore hearth was directed to maintain, as far as he was able, a reducing action among the charcoal; while in the operation of "fining," as it is termed, the removal of the carbon and other substances required the presence of an oxidizing atmosphere.

We have already seen how large a waste of metal accompanied the use of the Catalan furnace, as compared with the blast furnace, where the loss from this cause is reduced to a minimum. The low hearths employed for obtaining wrought iron from the pig had to contend against the same disadvantage, caused by the free exposure of the porous malleable product to an atmosphere which, from its richness in carbonic acid, could not, at the high temperature of the operation, fail to be one of a highly oxidizing nature. In this particular however it occupied a different position, when compared with its eventual rival the puddling furnace, to that of the Catalan furnace when contrasted with the blast furnace; for in both these modes of obtaining malleable iron from the crude metal the waste is considerable, whereas as has just been stated the loss of iron in the blast furnace is practically *nil*.

When the Lancashire fire is looked at as a means of producing an article of· excellent quality, it is a process which, in spite of its imperfections, deserves some notice; for it has maintained its ground amid all the commercial changes, which have brought a certain embarrassment upon some other branches of the iron trade.

SECTION XII.—IN LOW HEARTHS.

It is a little hazardous to venture on expressing an opinion as to the future of a mode of manufacture which still continues to be followed in Sweden and Russia, for producing the finest brands of iron known in commerce. If the system is expensive, it yet possesses the recommendation of being calculated to remove impurities, by the very waste of iron which accompanies its practice.

It might be considered that any extraordinary precautions in treating such pig-irons as are employed for the high qualities of bar iron in question would be superfluous; since the ore employed for their manufacture is almost free from sulphur and phosphorus, and charcoal, the fuel used in Sweden and Russia for the blast furnace and Lancashire and Walloon fires, is the purest form of solid combustible with which we are acquainted.

Much progress may fairly be claimed as having been made in connecting certain defects in malleable iron with the presence of certain foreign bodies; although there is no doubt something yet to learn before any well defined law can be laid down, which will determine the quality of iron by the exact amount and nature of the impurities found associated with it. The difference in content of phosphorus may easily account for the superior excellence of the three foreign brands enumerated below, when compared with the make of some of our best known British firms. One of the latter, from West Yorkshire, is of its kind much esteemed, and commands a very high price; and the other, although less well known, is remarkable for its purity, bearing in mind the fact that it is made exclusively from Cleveland pig iron, which in itself contains an excessive amount of phosphorus.

	Swedish.		West Yorkshire.	Hopkins, Gilkes, and Co. DANKS.
Carbon	·040	·200	·192	·055
Silicon	nil.	·100	·016	·012
Sulphur	nil.	·025	·010	·025
Phosphorus	·005	·100	·067	·085
Manganese	nil.	·050	·086	·058
Arsenic	nil.	nil.	Copper ·010	·025
Total foreign bodies	·045	·475	·381	·260
Iron	99·955	99·525	99·619	99·740
	100·	100·	100·	100·

346   SECTION XII.—PRODUCTION OF MALLEABLE IRON

A great advantage attending the Swedish mode of making malleable iron is the perfect manner in which scoria as well as the associated metalloids are separated from the product. This will be seen on referring to the analyses in which Swedish bars are shown to contain fully $99\frac{1}{2}$ per cent. of iron (Fe). Some addition to the "foreign bodies" in the British iron ought to be made for the scoria, which however is very variable in its amount.

The following is the composition of the pig metal used for making the malleable iron given in the previous page:—

	Swedish.		West Yorkshire.	Hopkins, Gilkes, and Co., Cleveland
Carbon	4·00 to	4·50	3·686	3·350
Silicon	·20 ,,	·50	1·255	1·550
Sulphur	·01 ,,	·03	·033	·140
Phosphorus	·01 ,,	·15	·565	1·550
Manganese	tr. ,,	1·80	nil.	nil.
Total impurity	4·22	6·98	5·539	6·590

When we recollect that the ore from which Swedish iron is manufactured contains from 15 per cent. to 20 per cent of foreign matter, it is no small thing to say in favour of a process that in two operations or three including the forge, it gives us a product of which 99·95 per cent. may be pure iron.

Forty specimens of scoriæ from the Lancashire hearth gave on analysis an average of:—

	Per Cent.
Silica	9·52
Peroxide of iron	12·97 } = 64·33 iron.
Protoxide ,,	71·05
,, manganese	3·65
Lime, alumina, etc.	2·81
	100·

Partly from my own observations, but chiefly from those of my friend Professor Richard Akerman of Stockholm, I have drawn up the following short account of the work performed in the so-called "Lancashire fire," as imported into Sweden from South Wales, where it was at one time extensively employed in making iron for tin plates.

SECTION XII.—IN LOW HEARTHS.

The fires vary in size, so that the produce ranges from 7 to 15 tons of blooms per 12 shifts, each fire being managed by 2 to 3 men according to size. Taking two of these men as earning 3s. 6d. per shift, which was the average wage when I was in Sweden, the labour only amounts to from 10s. to 12s. per ton. The consumption of charcoal, using air at 212° F. (100° C.), is given at about 1,400 lbs., per ton of blooms. In some of the Swedish works charcoal is obtained at a small outlay. Immense quantities of logs are sawn up into planks, and the cuttings, valueless for other purposes, are burnt into charcoal. In this way the fuel supplied to the iron works occasionally costs as little as 20s. per ton, and in some rare instances even less than this. The waste of pig iron in the fire is stated to be 13 per cent.

These blooms are drawn into bars under the hammer for the finer qualities. For this the labour costs about 12s. per ton, the waste being 10 per cent. and the fuel about 12 cwts. per ton of bars.

The expense incurred for fuel is of course enormous; but this item in the cost of manufacture is high, not from the extraordinary quantity required, but from its high price. The waste of iron is something more than would be incurred in making bars of a similar size from puddled iron. The fuel used in the conversion of the pig to the malleable form is about 5 cwts. more than that required in a modern forge and rolling mill.

Taking pig iron at 60s. per ton, at which price or even less it is sometimes made in Sweden, we have the following as the cost of metal, labour, and fuel required for producing wrought iron in these old-fashioned Swedish iron works:—

	£	s.	d.
26 cwts. of pig iron at 60s. per ton.	3	18	0
25 to 30 cwts. of charcoal at 20s., say 27½ cwts....	1	7	6
Labour	1	3	0
	£6	8	6

To this cost, which is a very favourable one, a considerable sum for other expenses has of course to be added. When however the selling price of the commodity, which is often as high as £10 or £12 per ton, is taken into the account, it is easy to perceive that for the finer brands there is no reason to abandon the manufacture of Swedish bars, nor indeed to substitute other more modern systems for that in present use.

In the statements of cost and waste it must be borne in mind that the Swedish bar is a *finished* product, whereas to iron made in a puddling furnace has to be added the expense and loss of metal incurred in re-rolling.

The total quantity of hearth-refined malleable iron produced in Sweden is about 240,000 tons, and in Russia about 300,000 tons, of which latter a considerable proportion is obtained by means of the puddling process. The actual quantity therefore of the finer qualities of bars made in the old fires, is but a very insignificant proportion of the entire make of the world.

Until very recently the same process as that just described was extensively followed in Great Britain for the manufacture of "tin-plate bars." The pig iron employed was of good quality smelted and refined with coke. The "metal" (refined pig) was placed in the hearth and kept covered with charcoal and blast applied, which was continued until the fluid metal was converted into a solid mass of malleable iron. This was drawn from the fire and hammered into a "lump." A second heat applied to the lump enabled the workman to produce a "bloom" under the hammer which, after being reheated, was rolled out into bars ready for the sheet mill. Mild steel however has now almost, if not entirely, superseded iron in the tin plate works, so that this connecting link with byegone days in iron making is fast disappearing in the United Kingdom, if indeed it has not already so disappeared.

The antiquated method of producing malleable iron which we are considering, has a certain technical as well as a historical value. Having, by the favour of Messrs. Knights of Kidderminster, had access to the books of their ancestors, beginning 165 years ago, I cannot refrain from making a few extracts from documents which, perhaps, will not fail to interest those whose furnaces make in one week as much pig iron as the Hales furnace at the Stour works did in the whole of 1727.

SECTION XII.—IN LOW HEARTHS.

In that year 477 tons 6 cwts. 1 qr. 5 lbs. of "raw iron" was produced at the following cost:—

	£	s.	d.	
Rent of furnace	0	2	10½	per ton.
Clerk's salary	0	4	2	
Travelling expenses, etc.	0	1	0	
Wages and common charges	0	8	0¼	
Ironstone, 947 blooms, costing 12s. per *bloom*, consumption 1·98 per ton pig	1	3	6	
Charcoal, costing 42s. 8d. per *load*, 2·44 per ton pig	5	4	5¾	
Utensils	0	0	7¾	
				7 4 8¼
Interest on capital				0 9 6¾
				£7 14 3
The selling price appears to have been about				8 11 6

The firm possessed, at that time, two forges at Whittington and Cookley where the chief part of the produce of their furnace, along with other pig iron was made into bars. In that year the production of the two establishments was about 621¼ tons which was sold at £20 12s. 3d., leaving a profit of £1 16s. 6d. per ton.

In the year 1740, the Hales furnace ran 780¾ tons of pig iron, the cost of which works out as follows:—

	£	s.	d.
Rent of furnace, 1s. 8¼d.; clerk's salary, 2s.; travelling expenses, 5½d.	0	4	1¾
Wages and common charges	0	9	7¾
Ironstone, 1,822 *blooms*, 13s. 6d. per *bloom* (2·33 blooms per ton pig)	1	11	6
Charcoal, 1,658 loads, 33s. 6d. per load (2·08 loads per ton pig)	3	9	10½
	5	15	2
Interest	0	6	4
	£6	1	6
Selling price for this year was	6	9	4

At five forges the make of hammered bars was 2,002 tons 16 cwts. 1 qr. 28 lbs. or an average of 400 tons at each establishment. The cost per ton delivered at works, as nearly as I can make out from the accounts seems to have been:—

		£	s.	d.
Pig iron (average price, £6 4s. 9d. per ton)	23·60 cwts.	7	7	3
Charcoal (,, 32s. per load)	1·36 loads	2	3	6
Pit coal for drawing out iron (no weight given)		0	4	10
Wages to finers, 9s. 1½d.; wages to hammermen, 7s. 2½d....		0	16	4
Wages: carpenter, 1s.; smith, 6d.; stock taker, 9¼d.; clerk, 1s. 5d.		0	3	8¼
Common charges—probably including some labour	...	0	6	6
Coals to workmen, 10d.; utensils, 6d.		0	1	4
Travelling expenses, 9¼d.; rent, 4s. 10¼d.		0	5	7½
		11	9	0¾
Interest on stock charges		0	8	10¼
		£11	17	11

A large proportion of the entire make, viz. 1,603 tons, appears to have been converted into nail-rods at some neighbouring slitting mills, for which about £15 per ton was paid when the forge owner stood the waste, and £31 when this was at the charge of the slitting mill proprietor.

It is a little curious to compare the mercantile transactions of what, at that day, must have been a large concern, for as we saw in Section II. the entire make of pig for the year in question was only 17,350 tons.

In the books of this firm for 1740 there appear the names of 64 customers. The largest quantity received by any one firm during the year was 187 tons and the average of the whole was about 28 tons. The profit realized in the forges was 42s. per ton.

As a matter of strict comparison the figures illustrating the manufacture 142 years ago have little value, owing to the change in the value of money and to the great increase in the cost of labour.

The waste, 23·60 cwts. of pig per ton of bars, is low; but this is probably due to pig iron being charged by a different ton to that which was adopted in the sale of finished iron. I may add that so far as my examination of these old books goes I have not found anything which explains what is meant by a *bloom* of ironstone or a *load* of charcoal.

SECTION XIII.

ON THE REFINERY AND THE PUDDLING FURNACE.

The carbon and silicon which enter into the composition of pig iron confer upon it fluidity at high temperatures; but the silica resulting from the oxidation of the silicon during the process of puddling has a great tendency to corrode the furnace in which this operation is performed.

I am unacquainted with any minute details of the difficulties, chemical or otherwise, which no doubt beset Cort in his first attempts to puddle iron on sand bottoms. The formation of a fusible silicate of iron, generated by the action of the fire on the metal itself, would speedily melt the masonry or sand fettling with which it came in contact;—an inconvenience to which the floor of oxide of iron suggested by S. B. Rogers would also be liable, although to a much lesser extent.

Mr. Edward Williams informs me that the use of the "sand bottom" had been generally abandoned before he was old enough to remember much of its peculiarities. His impression is that it did its work badly; because it was necessary to avoid the accumulation of fluid cinder, for the exit of which there was ample provision. This comparative absence of cinder, rich in oxide of iron, which in puddling, as in refining, is the agent which effects the removal of the phosphorus, could not fail to damage the quality of the product. To some extent I apprehend this evil would be counteracted by previously passing the pig iron through the refinery, and by the time required in puddling a "heat" which occupied *three hours*.[1]

The improvement suggested by S. B. Rogers for puddling iron was scarcely inferior in point of importance to the original idea of Cort's. By it, the iron floor covered with oxide of iron, already mentioned, the process was greatly shortened and the phosphorus more effectually

[1] Iron Metallurgy, 1858, by S. B. Rogers, p. 237.

removed than in Cort's original furnace. The value of Rogers' invention, in respect to the removal of phosphorus, seems not to have been suspected by himself; for in his work on Iron Metallurgy he rarely alludes to phosphorus, but confines himself to the importance of ridding the metal of *sulphur*. This, however, signifies nothing in estimating the value of the invention, for which its author, like his predecessor Cort, never received any reward from the fruits of their labours.

Anterior to the use of the iron bottom in puddling iron the forge and mill cinders were, owing to their content of phosphorus, of little or no value. By Rogers' improved mode of puddling, pig iron containing an increased amount of this substance could be employed in the manufacture of malleable iron, at all events for certain purposes. This change of conditions enabled the iron masters of South Wales to employ profitably the large accumulations of "cinder" remembered by many of the older members of the trade.

It would soon be discovered that the exposure of pig iron to the action of the blast, in one of the old fires, was advantageous, because by this means a large proportion of the silicon was separated, although the operation was arrested before any very great quantity of the carbon was burnt. This preliminary treatment or refining of crude iron, in the so-called "running-out-fire" before it was rendered malleable by puddling, continued in general use for fifty years after Cort first introduced the latter process, which speedily superseded almost entirely the Lancashire and analagous fires.

The nature of the smelting furnace enables us to follow with a considerable amount of precision the progress of the reduction of the ore. This is effected, as we have already seen, by an examination of the composition of the gases, as they ascend and are brought in contact with the minerals under treatment. This mode of procedure is however not practicable in the refinery, the action of which it is now proposed to consider. Owing to the very small dimensions of the latter, it would be impossible to collect specimens of gas, which would correctly set forth the precise nature of the combustion produced by the action of the blast on the coke.

It is, however, easy to calculate the quantity of air delivered to the refinery, based upon the displacement which takes place in the

SECTION XIII.—REFINERY AND PUDDLING FURNACE. 353

blowing cylinder. The difference between the real and the estimated quantity of blast upon this mode of estimate, has been ascertained, by analysing the furnace gases at the Clarence works, to be about 11 per cent. This assumption is based on the air admitted being converted into carbonic oxide. Returns obligingly furnished to me from the Bowling and Monkbridge works would indicate, after making a similar allowance of 11 per cent. that the result shows a considerable waste of air, even supposing all the carbon were burnt to the state of carbonic acid. I have made out two accounts of refinery workings so as to show the amount of oxygen needed to burn the carbon of the pig iron and coke to carbonic acid, and to oxidise the iron and metalloids.

		Lbs.	Oxygen. Lbs.
To oxidize the iron wasted to state of protoxide	...	116·5 =	33·3
Carbon of the pig iron to CO_2	...	22·4	59·7
Silicon ,, SiO_2	...	67·2	76·8
Phosphorus ,, P_2O_5	...	17·9	23·1
			192·9

The coke used represented in point of quantity, in example No. 1, 430 lbs. of carbon, and in example No. 2, 490 lbs. of carbon per ton of refined metal.

	No. 1. Lbs.	No. 2. Lbs.
Oxygen to burn carbon in coke to CO_2	1,146·7	1,306·7
Oxygen to oxidize iron, etc. as above, considered same in both cases	192·9	192·9
	1,339·6	1,499·6

The blast consumed as estimated by displacement was 94,000 cubic feet for No. 1, and 107,100 cubic feet for No. 2.

	Lbs.	Lbs.
This means at ordinary temperature and pressure (60° F. and 30 inches mercury) of oxygen ...	1,757	2,003
Less 11 per cent. difference between displacement and actual effect	193	220
	1,564	1,783
Oxygen required as per statement given above ...	1,339·6	1,499·6
Excess if all C were burnt to CO_2	224·4	283·4

Although there appears on the face of these statements sufficient atmospheric oxygen to burn all the carbon to carbonic acid, the existence of a considerable volume of flame at certain stages of the process

W

is indicative of the generation of some carbonic oxide. Nevertheless, looking at the question as a whole, and having regard to the process itself we are safe in regarding the action of the refinery on the impurities of the iron as being one of an oxidizing character. This oxidizing tendency of the atmosphere of the refinery is evinced not only by the oxidation of the metalloids but also by the fact that a portion of the iron in the cinder is actually peroxidised. As is well known, the mode of conducting the operation is to make use of a certain quantity of cinder, either obtained from a previous charge or derived from the forge and mill. The following particulars of refining Cleveland iron were furnished by the kindness of the Weardale Iron Company, through their mill manager Mr. Wm. Hutchinson:—

	Composition of Pig Iron.	Composition of Refined.
Iron by difference	92·50	96·54
Combined carbon	tr.	1·96
Graphitic „	3·12	·54
Manganese	tr.	nil.
Silicon	2·80	·12
Sulphur	·11	tr.
Phosphorus	1·47	·84
	100	100

Cinder.	Composition before Operation.	Composition after Operation.
Peroxide of Iron	8·29 } = Fe 52·42	2·57 } = Fe 46·78
Protoxide „	59·95	57·85
„ manganese	5·71	3·90
Silica	17·40	26·41
Lime	1·74	2·20
Magnesia	tr.	·24
Alumina	2·98	2·47
Sulphur	0·39	·05
Phosphoric acid	3·12	4·14
	99·58	99·83

Pig iron used per ton of refined iron 22·06 cwts.

	Lbs. per ton of metal.
Cinder added to refinery	375
„ received from refinery	775

SECTION XIII.—REFINERY AND PUDDLING FURNACE. 355

The changes which an inspection of these figures indicate are as follows:—A reduction of about 20 per cent. in the quantity of carbon, and a conversion of the greater portion of the remainder from the condition of graphitic to that of combined carbon. The small quantities of manganese and sulphur have almost entirely disappeared; while the silicon has been eliminated to the extent of 96 per cent., and the phosphorus to that of 43 per cent. of their original quantities.

It has to be remarked however that there is considerable irregularity in the composition of the refined metal, as regards the last two mentioned elements, which it is the more immediate object of the process to separate. Seven casts made at Tudhoe from Cleveland pig gave the following numbers:—

	Per Cent.	Per Cent.	Per Cent.	Per Cent.	Per Cent	Per Cent.	Per Cent.	Average. Per Cent
Silicon in the refined metal	·023	·013	·046	·093	·240	·140	·090	·092
Phosphorus „	·424	·412	·445	·674	·590	·470	·450	·495

In these examples the figures indicate an average removal of about 96 per cent. of the silicon and 66 of the phosphorus, the irregularities being possibly due to the quantity or quality of the cinder employed.

The following table contains the composition of four casts of iron refined at the Bowling works, and analysed at the Clarence laboratory.

	PIG IRON CONTAINED.				REFINED IRON CONTAINED.			
	Per Cent. Carbon.	Per Cent. Silicon.	Per Cent. Sulphur.	Per Cent. Phosphorus.	Per Cent. Carbon.	Per Cent. Silicon.	Per Cent. Sulphur.	Per Cent. Phosphorus.
	3·72	1·27	·029	·719	3·49	·114	·025	·370
	3·70	1·26	·042	·744	3·33	·128	·023	·358
	3·81	1·22	·033	·544	3·41	·128	·023	·303
	3·686	1·255	·033	·565	3·342	·130	·025	·490
Av.	3·729	1·250	·034	·643	3·412	·125	·024	·380

The average percentage removed, calculated on the original quantities, is carbon 8·5 per cent., silicon 90 per cent., sulphur 29·4 per cent., phosphorus 40·9 per cent. Of the four metalloids just enumerated carbon is the only one which may be regarded as indispensable, and also harmless; because we can in the subsequent stage of puddling almost completely rid the iron of its presence, as soon as it is no longer required to impart fusibility to the metal. Within certain limits therefore the more carbon retained in the refined iron the better, because the longer will the puddlers' charge possess the property of remaining liquid—a condition of things which is most advantageous for freeing the metal of its silicon, sulphur and phosphorus.

In the case of refined metal we commence with a material comparatively low in these three elements, as compared with the pig iron in its unrefined state. It is therefore more quickly reduced to a given point of purity in respect to the same, than would happen with pig iron, containing much larger percentages of them. The Bowling puddler, instead of receiving refined metal containing 2 to $2\frac{1}{4}$ per cent. of carbon and ·6 to ·8 per cent. of phosphorus, as happens to the puddlers operating with Cleveland iron, has to deal with his charge with half the above quantity of phosphorus and one half more carbon. He has therefore to contend with an evil of a much less magnitude, and under much better conditions for dealing with it, than falls to the lot of the puddlers in Cleveland.

In refining good grey iron, a fraction above 22 cwts. of pig is consumed for each ton of the product; a result which, having regard to the actual quantity of metal in both, shows an actual loss of about 5 per cent. of iron. In reality, the waste is perhaps a little more; as the cinder brought from other departments often contains fragments of malleable iron, which of course *pro tanto* improve the apparent yield.

In the manufacture of malleable iron, each step in the process leads to the production of slags, in which the iron is accompanied by all the silicon, phosphorus and sulphur which have been separated from the metal employed in the process. Where the object is the manufacture of a low class of iron, the whole of the slag is returned to the blast furnaces. This slag is of such an amount, that every ton of such iron delivered to the consumer contains about 6 cwts. which has been derived from the refuse supplied by the forges and mills.

The silica need not be regarded as being hurtful in the blast furnace from its excessive quantity; but the very fusible character of the slags prevents their being so thoroughly acted on by the reducing gases in smelting, as is desirable. When therefore they are employed in the proportion referred to, the resulting pig iron is perfectly white, and is besides deficient in carbon, containing only $2\frac{1}{4}$ to $2\frac{1}{2}$ per cent. The temperature of the hearth is lowered, by the arrival within its precincts of fused unreduced oxide of iron, which leads to a considerable absorption of sulphur by the metal, amounting in some cases to as much as ·75 per cent. So far as the silicon itself is concerned, no inconvenience arises from its presence; for such white iron as that in question contains as a rule under 1 per cent. of this element, or much less than the usual run of grey pig made without any admixture of cinder. The cause of this is easy to understand: the furnaces making cinder iron are worked below the temperature required for a plentiful reduction of silicon, and absorption by the iron. The sulphur is of course a very undesirable ingredient in iron, but it is much less objectionable in the raw materials than phosphorus, because a considerable proportion of the former is carried off in the blast furnace cinder, which is not the case to any extent with the latter. The phosphorus would therefore go on increasing in the pig, and would then be again returned to the forge, were its accumulation not kept down by the use of hematite ores, comparatively free from its presence.

I endeavoured to estimate the quantity of phosphorus and sulphur entering the furnace in order to compare the quantity contained in the pig iron and slag:—

	Phosphorus. Per Cent.	Sulphur. Per Cent.
Calcined ironstone contained	·522	1·052
Limestone	·011	·059
Coke	·265[1]	1·580

One hundred parts of pig iron required:—

	Phosphorus. Per Cent.
240 of calcined ironstone, containing	1·253
60 of limestone	·007
120 of coke	·318
	1·578

		Per Cent.
100 of pig contained of phosphorus		1·441
140 of slag „ „ ·105 % =	·147	
		1·588

It would thus appear that 86 per cent. of the entire phosphorus has been absorbed in the iron, and 14 per cent. has been taken up by the slag.

---

[1] This is an unusually high percentage of phosphorus for coke and is probably overstated.

The before-named quantities of materials contained of sulphur:—

		Per Cent.
240 of calcined ironstone	... ...	2·525
60 of limestone	... ...	·035
120 of coke	... ...	1·896
		4·456

		Per Cent.		Per Cent. of Whole.
100 of pig contained of sulphur	...	·093	=	2·09
140 of slag „ at 2·63 %		3·682	=	82·64
Carried off in the gases and in the fume		·681	=	15·27
		4·456	=	100·00

Notwithstanding the use of purer ore, the pig iron obtained when so much cinder is employed in the furnace, being low in carbon and high in both sulphur and phosphorus, is but of indifferent quality. Its lowness in carbon and silicon renders it difficult of fusion, and hence, when treated in the refinery, the loss is fully 50 per cent. more than that which accompanies the use of good grey iron. Imperfect fusion being inconsistent with the removal of phosphorus, etc., this removal can only be conducted under circumstances which necessarily lead to an increased burning away of the metal itself; and even then the elimination of the hurtful ingredients is less perfect than it should be. Any attempt at a more complete purification of the refined metal by a mere prolongation of the process only increases the waste without thereby improving the quality of the product. More iron passes into the cinder, while the semi-fluid metal retains its phosphorus with increased tenacity. Such a state of things is shown in the following analysis of a charge of Clarence iron, overblown in the refinery, and that of its accompanying cinder:—

Refined Iron.		Per Cent.	Cinder.		Per Cent.	
Carbon	... ...	2·07				
Silicon	... ...	·36	Silica	... ...	15·47	
Sulphur	... ...	·06	Sulphur	... ...	·68	
Phosphorus	... ...	·82	Phosphoric acid	...	4·11	
Calcium, etc.	...	·28	Lime, etc.	... ...	1·95	
Manganese	... ...	·08	Protoxide manganese		·33	
Iron	... ...	97·27	Do. iron	...	76·81	= iron 59·74
		100·94			99·35	

SECTION XIII.—REFINERY AND PUDDLING FURNACE. 359

In order to follow out the behaviour of the metalloids while a charge was being refined, Mr. Evans of the Bowling works kindly sent me the following samples, prepared at my request, which were examined in the Clarence laboratory.[1]

	Carbon. Per Cent.	Silicon. Per Cent.	Sulphur. Per Cent.	Phosphorus. Per Cent.
Original pig iron contained ...	3·686	1·255	·033	·565
After fusion in the refinery ...	3·510	·575	·034	·557
10 minutes after complete fusion	3·707	·478	·038	·537
20 ,, ,,	3·644	·273	·032	·530
28 ,, ,,	3·544	·154	·025	·509
Refined metal	3·342	·130	·025	·490

By the favour of the Bowling Iron Company, I have also received some analyses, made by Mr. J. W. Westmoreland, of the cinders produced at that well known establishment. Among these are two specimens from the refineries.

	Date of Production—Average of One Day's Work.	
	Dec., 1879.	Mar., 1880.
Silica	31·05	30·31
Alumina	4·37	5·73
Lime	1·56	1·55
Magnesia	·85	·73
Peroxide of iron ...	10·08 ⎱ = iron 41·47	3·60 ⎱ = iron 42·44
Protoxide ,, ...	43·90 ⎰	51·33 ⎰
,, manganese ...	5·90	4·96
Phosphoric acid ...	2·79	2·56
Sulphur	·032	·04
	100·532	100·81

The composition of these slags, containing as they do so large a percentage of a powerful acid such as silica, explains the reason of a less quantity of phosphorus being absorbed by the bases than happens, as will be shown, when the silica is in a less preponderating quantity.

As we shall hereafter see, experience with the Bessemer converter proves that atmospheric air at high temperatures rapidly desiliconizes pig iron; and hence it will be remarked that during the mere act of

---

[1] These will be shown diagrammatically in a future Section.

fusion, in the analyses just given, a quantity of silicon, approaching 60 per cent. of the original quantity, has disappeared from the metal. The same observation applies to the carbon ; but, owing no doubt to the superior affinity of silicon at this particular temperature for oxygen, the former was but slightly affected during the period in which the charge was being melted. On the other hand, neither the sulphur nor phosphorus manifest any marked tendency to separate themselves from the iron during the act of fusion, and such portions as pass into the cinder doubtless do so from their affinity to the oxide of iron contained in the slag. This last named substance gradually increases in amount in the scoriæ, as the metal of the charge is oxidized by the atmospheric air; and then the iron-oxide, agitated with the cinder by the blast, has its phosphorus acidified. In this condition the phosphorus is absorbed by the slag, the iron of which acts as a vehicle for acidifying and removing such remaining portions of the metalloids as are separated by the process of refining.

The refined metal is now transferred to the puddling furnaces, where it is agitated, by means of the puddler's tool, with oxide of iron, derived partly from the floor and sides of the furnace and partly from the oxidation of the iron itself. It has been freed in the refinery from nearly half of the phosphorus it contained as pig iron, and from nine tenths of the silicon which in the form of silica constitutes so formidable an obstacle to the combination of phosphoric acid with the oxide of iron of the cinder. The absence too of an excessive quantity of silica proves such a protection to the structure of the furnace that the tendency, when using refined iron, is for the floor to gather; so that at the end of the week a certain portion of the accumulation has to be removed by mechanical means.

The following statement illustrates the gradual separation of the four metalloids, from high class pig to malleable iron, as to which it should be remarked a certain portion of carbon in this case is purposely retained in the final product:—

	Carbon. Per Cent.	Silicon. Per Cent.	Sulphur. Per Cent.	Phosphorus. Per Cent.
Cold blast iron, Bowling	3·656	1·255	·033	·565
,, as refined metal	3·342	·130	·025	·490
Malleable iron from refined metal	·226	·109	·012	·064

SECTION XIII.—REFINERY AND PUDDLING FURNACE. 361

In speaking of the process of refining, the superior affinity of silica for the bases in the cinder produced was referred to, as the probable cause of so small a quantity of the phosphorus being removed from the iron. This removal can only be effected by the acidification of this substance, and its subsequent combination with oxide of iron or other base in the cinder itself. In the case of the refinery, owing to the presence of the silicon in the pig the bases in the slag are speedily saturated with acid, chiefly silicic, so that there is no room as it were for an acid of a less energetic character. Such a state of things does not exist in puddling. At Bowling the silicon has now been to a great measure removed, and the cinder in consequence contains a much less proportion of silica; and hence we see that the phosphorus has been greatly reduced in quantity by this second operation.

The Bowling puddling cinder, taken over a whole day from six furnaces, was found to be composed of—

	Per Cent.	
Silica	12·75	
Alumina	2·35	
Lime	·18	
Magnesia	·18	
Peroxide of iron	14·39	} = iron 62·25
Protoxide ,, ...	67·09	
,, manganese	·55	
Phosphoric acid	3·12	
Sulphur	·05	
	100·66	

The superiority of such a cinder for removing phosphorus as that just described, when compared with those from the refinery, is sufficiently apparent. Not only have we a great diminution in the amount of the rival acid (silicic) which contends for the basic substances with the phosphoric, but we have also a great increase in the bases themselves, and notably in the oxide of iron.

The loss in puddling refined metal of the character in question is $7\frac{1}{2}$ per cent.; and we have seen (p. 355) that the loss in refining is about 9 per cent. Hence, confining our attention to the actual metallic iron which has been scorified during the two operations of refining and puddling, the waste, reckoned upon the total amount of pure iron, may be taken at 9 per cent., or reckoned upon the pig iron used, at about $15\frac{1}{2}$ per cent.

The system of producing malleable iron just described is virtually a combination of the old low hearth with the puddling furnace as designed by Cort. In both stages of the process the oxide of iron required for purging the metal of the impurities it had contracted in the blast furnace, is obtained by the expensive method of wasting metallic iron; of which about one-tenth (9 per cent.) was returned to the condition of oxide in order to purify the remainder. Such was about the extent of the loss in using good grey iron, but in the case of the lower grade of metal already referred to in these pages the proportion of actual iron re-oxidized in the refinery and puddling furnace would amount to considerably more than this.

This and other expenses connected with the refinery led the manufacturer to endeavour to dispense with its use; for in one way or other it had the effect of adding 20 to 30 per cent. to the cost of the pig iron. There was no difficulty attending a replacement of refined by crude iron in the puddling furnace, to a moderate extent, but when this was exceeded the wear on the furnace became a source of inconvenience, which it was attempted to meet by the application of fire-clay, limestone, and other refractory materials. The difficulty arising from puddling pig iron was further augmented by the application of hot air to the blast furnace; since thereby, according to general testimony, the reduction of silicon and the amount consequently absorbed by the pig is increased.

In later years iron ores have been largely employed, as a "fettling" for the puddling furnace. At first the ore was used together with limestone or fireclay, but Mr. E. Williams, alive to the importance of getting rid of phosphorus by means of oxide of iron, was among the very first to employ it exclusively as a mode of protecting the sides and bottom of the furnace. For this purpose choice was made of those which were as free as possible from earthy ingredients, and contained but a small amount of phosphorus. In districts such as South Wales, red ore was used in considerable quantities for the purpose in question, without in reality adding materially to the cost of production. Such ore as that

SECTION XIII.—REFINERY AND PUDDLING FURNACE.

referred to was already extensively employed in the blast furnace; and it was found that to use it as a lining for the puddling furnace before it was smelted did not greatly affect its value as a mere ore of iron, *i.e.*, the cinder formed of ore and waste of pig iron did not augment the quantity of phosphorus beyond what it would have been had the ore been consumed direct in the blast furnace. So successful were the South Wales ironmasters in this modified form of puddling, that the waste in the puddling furnace, when using pig, of somewhat improved quality it is true, did not greatly exceed that which took place in puddling refined metal, the latter amounting to about 9 per cent.

Returns in my possession, extending over six months, contain the following facts, which, if not absolutely correct, may be regarded as being substantially true:—

Weekly Average Make per Furnace.				Metal used per Ton of Puddled Bar.									Coal per Ton of Puddled Bars.		
				Refined.			Pig.			Total.					
T.	C.	Q.	L.	C.	Q.	L.	C.	Q.	L.	C.	Q.	L.	C.	Q.	L.
16	11	1	24	10	1	24	11	3	6	22	1	2	13	2	26
17	6	1	24	2	3	12	19	1	19	22	1	1	14	0	23

The labour in puddling pig iron was more severe, and was consequently rather more highly paid than that in puddling refined metal; but in other respects these figures, so far as they go, prove a great advantage in point of economy by omitting the operation of refining. The make per furnace, it will be perceived, is larger, while the waste is no more, in the case where a much smaller quantity of refined metal was used.

There still remains the question of the extent to which the quality of the product might be affected by the suppression of the refinery. Fair grounds have been given for assuming that the chemical purity of the malleable iron will be promoted by removing the silicon and phosphorus from the process as early as possible, and in as large a quantity as possible. It will however be proved immediately that, by the use of increased quantity of oxide of iron, it is perfectly feasible to begin with a pig iron containing twice as much of these two metalloids as in such iron as that of Bowling, and yet to obtain malleable iron with only half as much of either as is usually present in West Yorkshire brands. We cannot however assume—remembering all the properties looked for in wrought iron of the highest class, such as malleability, etc.—that purity of composition is the only requirement. The interposition of non-metallic matter, such as cinder, in a mass of malleable iron intended for certain purposes, may be far more hurtful in its consequences than a much larger excess even of phosphorus. A proper attribute of a railway bar, for example, is as much hardness as is consistent with strength, combined with as great an amount of homogeneity of structure as can be obtained. As much as ·4 per cent. of phosphorus is quite compatible with the first of these conditions; but probably ·04 per cent. of intercalated cinder in the iron might be fatal to its longevity, by preventing perfect welding and cohesion of the different bars which form the "pile." It is perhaps not unreasonable to conceive that, where the relative quantity of cinder to iron is largely increased, as happens from the increased use of oxide in the puddling furnace, the danger just referred to may be greatly intensified.

There is nothing speculative in the assertion that iron rails, made before the complete discontinuance of refining, were generally speaking longer lived than those of more recent manufacture. No doubt in later days the permanent way of our main lines has been much more severely taxed than was formerly the case. The engines are more ponderous, the traffic is heavier, and the speed greater; but the experience of the North Eastern railway at all events indicates that *ceteris paribus* rails of iron have occasionally been made equal in point of durability to some of steel. Whether this be due to their having been made from refined metal, or whether indeed they were so made, we have unfortunately little means of proving.

The rapid growth of the Cleveland iron trade which furnished metal unusually rich in silicon and phosphorus, rendered additional care in the forge absolutely necessary. For best purposes grey iron was preferred, in which the large percentage of carbon and silicon maintained the bath of metal long enough liquid, to enable a greatly increased dose of oxide fettling to remove the phosphorus. The cost of the ore used for this purpose was met to some extent by the improved yield obtained by the puddlers; for, the oxidation of the metalloids being performed at the expense of the fettling, the waste of iron is not more than that incurred in the mere preliminary processes of refining, viz., between $7\frac{1}{2}$ and 10 per cent. on the pig iron used, and it is often even lower than this.

SECTION XIII.—REFINERY AND PUDDLING FURNACE. 365

The amount of oxide of iron employed in the manner just referred to often reached 5 cwts. per ton of puddled bars, and in some cases as much as 8 cwts. is used. The North Eastern Railway Company are accustomed to analyse the rails received from their contractors; and the following ten examples exhibit the extremes of the phosphorus content in iron rails, as delivered indiscriminately by manufacturers using exclusively Cleveland iron:—

	Carbon.	Silicon.	Sulphur.	Phosphorus.	Manganese.
	·08	·16	·05	·17	·10
	·03	·17	·05	·24	·30
	·08	·20	·05	·26	·14
	·08	·16	·04	·28	·17
	·08	·15	·04	·32	·21
	·08	·12	·05	·33	·08
	·08	·13	·02	·37	·19
	·08	·13	·03	·39	·18
	·08	·19	·04	·43	·15
	·08	·18	·04	·45	·08
Average content in rails	·075	·159	·041	·324	·160
,, ,, in pig iron assumed to be	3·50	1·75	·15	1·55	·50
The percentage removed of the original quantities averages	78·5	90·9	72·6	79·1	68·0

The severity of the labour in puddling, and the consequent liability to imperfect work, as evidenced in the figures just quoted, naturally turned men's minds to the question of superseding human exertion by mechanical appliances. The Dowlais Iron Co., under the superintendence of the late Mr. Wm. Menelaus, led the van in this branch of experimental metallurgy by testing the merits of a revolving furnace designed by Mr. Tooth. A certain amount of success attended these costly trials; sufficient to induce others to listen favourably to a paper read by Mr. Samuel Danks before the Iron and Steel Institute, in 1871, on mechanical puddling. The furnace proposed by this gentleman differed very slightly from those previously used at Dowlais, except, as Mr. Menelaus observed, in the matter of success.

Three commissioners were appointed to proceed to America, to report on the manner in which the furnace of Mr. Danks was performing its duty in the Western World; and the results of their observations there were fully set forth in an able paper by Mr. G. J. Snelus, himself a member of the commission.

It has to be remarked that one condition of the success of this furnace was a plentiful use of oxide fettling, amounting to as much as 30 to 75 per cent. of the pig employed. This was deemed necessary for the protection of the iron work of the rotating barrel. A considerable proportion of the fettling was introduced in a liquid state, along with the molten pig iron. When the apparatus was made to revolve, an intimate and immediate admixture of the two fluids was effected, no doubt more perfect in its nature than that which happens with manual labour. Not only has the removal of the metalloids, by the Danks' furnace in this country, been more thoroughly performed than in hand puddling, but the reduction of iron from the fettling has been as much increased, as to enable the rotating furnace to produce a ton of puddled iron from 18·5 cwts. of pig iron employed, inclusive of the waste incurred in remelting the metal in a cupola.

The various analyses given by Mr. Snelus in the paper referred to, are so instructive that they deserve careful study by those interested in the subject. It is however needless to repeat them here, because, since the time at which they were performed, the North Eastern Railway Company have received 2,000 tons of rails made in rotating furnaces by Messrs. Hopkins, Gilkes, & Co., of Middlesbrough. From the laboratory book of the Railway Company I have extracted 81 consecutive analyses, which show a much more perfect separation of phosphorus than those set forth in the paper of Mr. Snelus; and these have been tabulated as follows, the phosphorus only being absolutely correct:—

		Phosphorus. Per Cent.	Silicon. Per Cent.	Sulphur. Per Cent.
7 specimens contained	... ...	·071	·072	·020
7 ,, ,,	... ...	·073	·093	·093
19 ,, ,,	... ...	·145	·109	·030
21 ,, ,,	... ...	·147	·105	·036
12 ,, ,,	... ...	·244	·120	·029
5 ,, ,,	... ...	·251	·127	·028
6 ,, ,,	... ...	·326	·138	·035
3 ,, ,,	... ...	·346	·123	·039
1 ,, ,,	... ...	·402	·126	·055
81 average content in rails	... ...	·170	·109	·032

SECTION XIII.—REFINERY AND PUDDLING FURNACE.

	Phosphorus. Per Cent.	Silicon. Per Cent.	Sulphur. Per Cent.
Average content in the pig iron, assumed as before	1·55	1·75	·15
Average removal, reckoned on original quantities	89·0	93·7	78·6
Average by hand puddling, as given above	79·1	90·9	72·6

As an indication of what has been achieved in freeing iron from foreign matter two sets of analyses are inserted, one, of the pig used, and the other of the malleable iron produced. In both, the article produced was the same, viz. rivet iron; but in one case obtained from Low Moor pig, by means of refining and hand puddling, and in the other obtained from Middlesbrough unrefined pig iron, by means of the revolving furnace at Messrs. Hopkins, Gilkes, & Co.'s works:—

	Silicon. Per Cent.	Sulphur. Per Cent.	Phosphorus. Per Cent.
Low Moor cold blast pig iron gave	1·380	·075	·620
Middlesbrough pig gave	1·750	·100	1·500

	RIVET IRON FROM	
	Low Moor Iron Co. Per Cent.	Middlesbrough Pig. Per Cent.
Carbon	·191	·055
Silicon	·016	·012
Sulphur	·010	·025
Phosphorus	·067	·085
Manganese	·086	·058
Copper	·010	·025
Slag, consisting of silica, oxide of iron, and phosphoric acid	·262	·417
	·642	·677
Iron	99·300	99·400
	99·942	100·077

The strength, as certified by Kirkaldy, was as follows:—

	Dia. Inch	Area Square Inch.	Ultimate Stress.		Fractured Area.		Ultimate Stress.	Extension = per Cent.
			Total Tons.	Per S. In. Tons.	Dia. Ins.	Area. Sq. Ins.	Per Sq. In. on Fractured Area. Tons.	
Low Moor ...	·725	·4128	9·4	22·771	·52	·2134	44·25	29·68
Middlesbrough ..	·7225	·4096	9·5	23·227	·515	·2075	45·783	29·68

In both these specimens of iron there is a notable quantity of slag, but that made in the revolving furnace is the larger of the two. It does not follow however that this is due solely to the excessive quantity of oxide of iron used in the puddling furnace. The practice at Low Moor is to work small heats, under 4 cwts. ; and from these, by repeated hammering, the cinder can be more effectually expelled than from masses weighing half a ton, which is the weight of a bloom obtained from the revolving furnace. The presence of this slag does not seriously impair the tensile strength of malleable iron, but it effects the quality of the metal as to malleability, as well as in other attributes acquired for such high-class products as that of Low Moor.

The origin of this evil is easily accounted for. During the process of puddling, the grains of iron, as they form under the action of the workman's tool, adhere together, forming a sponge-like mass. Each particle is necessarily surrounded by a coating of cinder; besides which the fused slag may, and often does, occupy a cavity in the puddler's ball, of greater or less dimensions. The action of the hammer expels by far the greater portion of the objectionable matter; for when a well hammered slab, 2 or 3 inches in thickness, is broken, the fracture to the naked eye generally presents a continuous bright surface, mostly quite crystalline in its character. On the other hand, in the very heart of the mass black blotches are occasionally perceptible, consisting of imprisoned cinder so enveloped in solid iron that, when the slab comes to be rolled, this cinder will spread out into a thin sheet, and prevent union between the metallic surfaces so completely that there is often no difficulty in separating the two parts by means of a cold chisel.

Something of the same effect may be produced by roughness on the outside of the slabs, etc. These fill up with cinder; and, when they come to be piled in the balling furnace, this cinder may get sealed up in the boiler-plate slab, or billet, or pile. In such a case however the cinder, being melted when it is brought to the hammer or rolls, is mostly expelled before the surfaces by which it is enclosed are united.

Not unfrequently the interposed matter appears from its colour to be derived from pieces of fire brick, probably dropping from the roof

SECTION XIII.—REFINERY AND PUDDLING FURNACE. 369

of the furnace; and if it or other substances get enveloped in a puddled ball, before they have time to escape, it is certain to produce the inconvenience under consideration.

The following analyses give the composition of imprisoned matter referred to, as taken out of a boiler plate made by a well-known West Yorkshire house:—

	Cinder from Puddled Lump.	Cinder from Boiler Plate.
Silica	12·860	1·553
Alumina	1·340	—
Protoxide of iron	55·713	85·875
Peroxide ,,	26·636	11·505
Protoxide of manganese	·511	·324
Phosphoric acid	3·277	·517
Sulphur	·124	·028
	100·461	99·802

It is worthy of observation that in the case of boiler plates the want of soundness in the welding is often accompanied by the formation of a hollow cavity filled with gas. This possibly may arise from carbonic oxide passing through the pores of the iron, acting on the oxide of iron in the cinder, and remaining there as carbonic acid. The analysis itself indicates a greatly reduced quantity of peroxide of iron in the boiler plate, as compared with the puddled lump. Whether this is due to such an action as that referred to I am unable to say.

By extraordinary care, the loss arising from unsoundness is, in the case of the best Yorkshire plates, reduced to a minimum. Nevertheless the evil results, as regards the boilers of sea-going steamers, etc., would be infinitely greater than they now are, were not the most minute precautions taken in the examination and rejection of any work of a suspicious character.

If the evil complained of sometimes makes its appearance, in spite of great precaution to prevent its occurrence, it is never, as may be easily imagined, entirely absent, where puddled iron, instead of being hammered, is only reduced in size under a squeezer, to fit it for the rolls. Such puddled bars, on fracture, exhibit numerous streaks of scoriæ, which have solidified during the act of rolling. When these

x

are reheated, the cinder melts, and bursts through the iron walls of the cavities enclosing it with explosive violence, under the pressure of the rolls. This however does not suffice for its entire expulsion; and in consequence the merchantable articles indicate too clearly the existence of a serious defect in the mode of manufacture.

If a rolled bar of common iron, made from piled puddled bars, be immersed in weak acid, it is often dissolved into such a form as resembles a bundle of rough coarse wires laid alonside of each other; and the same thing happens when a rivet in a ship's bottom is exposed to the corrosive action of bilge water.

The altered state of public opinion, respecting the future of puddled iron, has greatly diminished the interest felt in questions connected with its production. The process however, from having been practised for nearly a century, and from its ability to separate so large a proportion of the phosphorus contained in certain makes of pig iron, deserves some notice. Such is particularly the case when the examination is connected with improvements, either in efficiency or economy, such as may possibly still enable puddling to hold its place, at all events for some time, in spite of the present revolutionary tendency of events in the iron trade.

The more perfect way as compared with hand puddling, in which the mechanical furnace, as designed by Mr. Danks, performs its duty, has been sufficiently dwelt upon. As regards its economy it would be difficult to speak with the same amount of confidence. The experience of my friend Mr. Robert Heath has enabled him to report favourably even in this respect; but so far as the experience of the North of England is concerned the advantages of the rotating furnace in question must, in my opinion, be confined to improvement in the quality of the product and must not be extended to greater economy.

Probably no firm has done more in the development of this instrument than Messrs. Graff, Bennett, & Co., of Pittsburg; and no one has more carefully studied its use than their manager, Mr. John Williams. In very terse language he describes it thus: "As a worker of metal it is without an equal; as a melter it is inferior to many; as to endurance it is the shortest lived of any; and as to convenience of repairs it is one of the most difficult."[1]

Transactions Iron and Steel Inst., 1877.

SECTION XIII.—REFINERY AND PUDDLING FURNACE.

In the old furnace, worked by manual labour, the puddler and his "underhand" turned out in each shift about 24 cwts. of puddled iron, say under 2½ tons per 24 hours per furnace with two sets of men. In Wales, in the days when refined metal was used, about 3½ tons represented the daily product. In times of average prices, the men's wages would amount to about 20s. for the 24 hours at each furnace. Now it is evident that, with any improved process, there is not a large margin for economy, unless the output is sufficient to effect great savings in the general expenses. I believe that in the Danks furnace as much as 8 tons of puddled iron was turned out in the 24 hours; which, if performed by hand, at 8s. per ton, gave £3 4s. 0d. for labour. The increased expenditure however for repairs and other incidentals, according to the information afforded to me, indicate too clearly that the application of mechanical puddling, in the North of England at least, has not been attended with any economy, even after crediting it with the superior yield from the metal used.

So far as the experience of nearly two years enables us to speak, the prospects of success are very much brighter in connection with two revolving furnaces at Creusot, than with any of their predecessors. One serious source of difficulty with previous inventions of this class has been the maintenance of the lining. To mitigate this, the revolving barrel in which the iron is puddled is at Creusot so constructed, that there is a space through which a current of water is kept constantly running.* In this way the coating of cinder is much more easily retained in its position and the costly necessity of previously fusing this material is entirely avoided.

The Creusot furnaces are on a much larger scale than those hitherto used; for they operate on charges of a ton each, of which one per hour is turned out, or, to cover stoppages, say 20 tons for each furnace per 24 hours. The actual time required for converting the pig, which is introduced in a melted state, I took down upon the occasion of my visit as being 30 minutes. The fettling did not amount to above 30 per cent. of the weight of the pig used, and consisted of 100 parts of ore, 100 of roll scale, and 40 of scrap iron. The loss was 3 per

* Mr. Crampton was the first to put the water jacket in operation in this country, but I believe that Mr. Wm. Sellers of Philadelphia also claims priority in suggesting its use.

cent. on the pig used, which contained ·07 of phosphorus; and this quantity was reduced to the infinitesimally low proportion of 005 per cent. in the puddled iron.

Besides the question of quality of product, that of economy demands the attention of the manufacturer, hence the following tables have been constructed to show the relative position in respect to cost of different modes of puddling. The pig iron in all cases is taken at the same price; and, as the fettling is an important element, its value and that of the pig iron are included in the first column.

	Creusot.	Mechanical Rabble.	Hand Puddling.	Hand Puddling, France.	Danks.
Cost of pig and fettling	100	114·4	118·4	119·7	123·6
,, coals, by weight	100	163·4	240·3	225·9	192·3
,, wages, as money	100	125·	117·3	113·	123·

The high cost of pig and fettling charged against the Danks furnace is mainly due to the serious expense attending the melting of the latter. The consumption of fuel in the Creusot furnace would be much below its present figure, were it not for a second furnace being kept heated, containing a stock of melted iron always in readiness for the puddlers. In the case of the Danks furnace, less than one-half the weight of coal consumed was required in the actual operation of puddling. Each 100 parts of coal were applied as follows :—

Melting pig in cupola	12·5
Melting cinder	40·0
Puddling furnace	47·5

In respect to the labour, my data do not enable me to speak very confidently as regards Creusot; at the same time there is every reason to consider that the basis of my calculation is pretty near the truth.

Taking the information as a whole, comprising as it does the three most important items of cost in puddling, we seem justified in supposing that, if the art is not destined to be so rapidly extinguished as was at one time imagined, the task of effecting it by mechanical means has been satisfactorily solved, at all events at Creusot. Such appears to be the opinion of one very competent judge; for Mr. James Kitson, Junior, has recently erected a similar furnace at the Monkbridge works in Leeds, and has expressed himself entirely satisfied with the results so far as economy is concerned.

SECTION XIII.—REFINERY AND PUDDLING FURNACE. 373

The iron made in the revolving furnace at Monkbridge is as a rule of very high quality, but unfortunately it sometimes happens that it exhibits signs of redshortness. This defect was not so serious as to have interfered with its use for ordinary purposes, nor would it have prevented its being employed for Siemens-Martin steel. For plates of the class sent out by the Monkbridge Company for locomotive and marine boilers it was inadmissible. Mr. Kitson having requested my advice on the subject, steps were taken to examine the composition of the best British makes of malleable iron known to be free from redshortness.

In this view the following analyses were made in the Clarence laboratory:—

	Carbon. Per Cent.	Silicon. Per Cent.	Sulphur. Per Cent.	Phosphorus. Per Cent.	Copper. Per Cent.
Bowling 1¼ square bar	·015	·157	·005	·220	—
Low Moor rivet iron	·200	not given	trace	·115	—
Monkbridge best plate, puddled by hand	·018	·151	·004	·097	·013
Monkbridge best plate, rotary furnace, not redshort	·016	·118	·015	·072	·011
The sample of Monkbridge plate, also made in rotary furnace, being redshort	·015	·121	·019	·138	·016

There is obviously nothing in the fixed impurities which tends to throw any light on the cause of the defective quality complained of, and attention was therefore directed to ascertaining whether oxygen gas in some form might not be concerned in the matter.

Pure hydrogen gas was passed during two hours over coarse borings heated to a temperature about the melting point of cast iron. The oxygen ascertained by conversion of hydrogen into water was:—

	Per Cent.
Best Yorkshire hand-puddled	·349
Plate free from redshortness	·674
Redshort sample	·713

Nothing was gained from this experiment, for it would appear as if there were only a slight excess of oxygen in the redshort specimen as compared with the other.

A recent plan of ascertaining the presence of oxygen devised by M. Tucker was then tried. It consists in fusing a weighed sample of the iron in a carbon crucible. The increase of weight due by carbonization is obtained, and, after allowing for this, the loss is considered as oxygen. By this means the following results were obtained:—

	Per Cent.
Best Yorkshire hand puddled	·750
Made in rotary furnace, not redshort	·704
,, ,, redshort	1·384

It should be noted that the oxygen got by the hydrogen treatment is possibly nothing more than might be due to oxide of iron and it certainly is a question whether the action of the hydrogen gas would affect the entire substance of the metal. Similar experiments had shown that while borings were nearly constant in weight after two hours exposure small bars only decreased gradually till after 20 hours treatment. Against this inconvenience the carbon experiment probably will give too high a figure for oxygen, as the bases of the enclosed cinder—other than iron oxides—and part of the silica will not remain united with the metal and therefore be accounted for as oxygen. Both sets of experiments were however made under exactly similar conditions, and by the latter certainly the redshort iron shows a great excess of oxygen. The fact is given without any intention of considering the question as finally settled.

The present is a fitting opportunity to compare the economy of producing malleable iron from pig, rich in phosphorus, with that of making blooms by the direct process from ores used in the smelting of such cast iron. For this purpose we will assume, that which often happened, viz., that iron obtained in a revolving furnace was as free from phosphorus as that mentioned in Section III. got by the direct process.

Now allowing as much as 12 cwts. of fettling per ton of iron one-half of which is cinder we have in reality only half this weight of fresh material to deal with, which enables the Danks or similar furnace to give as great a weight of puddled iron as they use of pig. We have therefore—

	Cwts. of Iron (Fe).
20 cwts. of pig iron equal to	18·6
6 ,, fettling containing 60 per cent. = ...	3·6
	22·2

So that in reality the total loss of actual iron is under 10 per cent. instead of 35 per cent. as shown in Section VII. on the direct process at page 41. If indeed the iron in the full weight of fettling were taken into the account, the waste of metal would be very little more than one half of 35 per cent.

On discussing the subject of the revolving furnace with Mr. Williams, he quite agrees with me in the high character of the product it is capable of turning out, as indeed has been shown by the analysis given in the present section. According to the opinion of this experienced authority, this form of apparatus did not receive at the hands of the skilled men of the trade, that attention its merits deserved. At Dowlais, efforts were exclusively directed to prevent corrosion of the fettling; but Mr. Danks, some years afterwards, propounded the sound doctrine that it was not by endeavouring to maintain the lining intact that real success was to be secured. The proper course to follow was to use an ore as free from earthy impurity and phosphorus as possible, which, wearing away, would add to the oxide of iron in the bath, and thus assist in purging the pig iron from those substances which deteriorated the quality of the product. As Mr. Williams however observes Mr. Danks appeared "a day after the fair." Steel began to take the place of iron and manufacturers found little inducement to attempt the improvement of a process, which, according to many, is not destined to be long-lived in the iron trade of the world. But for this, he thinks that the revolving puddling furnace would have superseded that of Rogers, both in the matter of quantity and quality of the iron produced.

So far as the information afforded by Ritter v. Tunner enables one to form an opinion, the balance against the direct process for labour and coal as compared with the blast and puddling furnaces taken together renders it, in competition with the former, to say the least of it, all but hopeless, charging both operations with ore and fuel at the same price.

The process of puddling being one of a purely chemical character, it is not surprising that means should have been devised to promote its efficiency by additions of substances known under the general designation of chemicals. I do not propose to examine this subject at length, because there is no method hitherto suggested which has come into general use. Suppose for the moment however that there is to be found, in nature or elsewhere, matter which is capable of separating the hurtful ingredients from pig iron, more expeditiously than is done by oxide of iron. If such a substitute be expensive, but from its energetic action need only be used in small quantities, we are met by the difficulty of having to bring a few pounds of it—indeed in one process only a few grains—into intimate contact with 4 or 5 cwts. of iron; without which contact no substantial good can be effected. If a large quantity of an inexpensive substance would suffice, it would seem difficult to obtain a cheaper material than that now in use, viz., a mixture of ore with the cinder produced in the works. Such a material is not only capable of effecting its intended purpose in a very efficient manner, but it possesses the advantage of being available in the blast furnace; which cannot be said of any other substance, with the exception of lime, and to a limited extent of manganese, applicable for "physic" in puddling.

In the next section the more recent methods of eliminating those foreign substances which, so to speak, are dissolved by iron during the operation of smelting will be considered. It will then be shown that their separation from the metal itself can be as completely effected by processes of a much more economical nature than that of puddling. Under these circumstances it may well be asked, what is it that causes Cort's invention still to retain so conspicuous a position in the manufacture of iron as it does. The difficulty which has hitherto been a barrier to the more rapid substitution of ingot iron for welded iron is partly due to some occasional irregularities in the quality of the former and partly owing to the unwillingness and want of experience in the workmen to deal with a new material. Time before long will remove most of these obstacles and every year will, no doubt, see a larger proportion of steel or ingot iron taking the place of puddled or welded iron. Strength and durability of course are very important factors in determining the selection of materials, but so long as these are not accompanied by economy in the cost of production the cheaper article will continue for a long time to hold at least a portion of its former ground. Hitherto plates used for boilers and ships can be manufactured more cheaply from puddled iron than from ingot iron or steel and in consequence the former is still largely employed. In all pro-

SECTION XIII.—REFINERY AND PUDDLING FURNACE.

bability however this difference of cost is a mere question of time, and then a puddling furnace may be as rare an object as a Catalan fire or a blast furnace driven with cold air is at the present time.

The statistics of the iron trade are not sufficiently complete to enable us to institute a very exact comparison between the position of the manufacture of puddled iron of the present time and that of previous years. Roughly however the figures given below exhibit the nature of the change which has been going on during the last ten years:—

	1873.	1882.
Pig iron made in the United Kingdom	6,566,171	8,493,287
„ exported	1,142,065	1,758,152
Consumed or stocked for home consumption...	5,424,106	6,735,135

Neglecting any differences in stocks on hand there appears to have been an increase of about 20 per cent. in the pig iron worked up in the year 1882 as compared with 1873.

According to the same authority—Reports of the British Iron Association—the number of puddling furnaces at work was as follows:—

1873—7,264 furnaces. 1882—4,369 furnaces.[1]

Thus while the general consumption of pig during the ten years in question has been increased about 20 per cent., it appears possible that the make of puddled iron in the United Kingdom has decreased in quantity.

The puddling furnace cannot however be taken leave of without remembering what immense service it has rendered to the manufacture of iron and to the spread of civilisation. Something must be said in its favour as a mere means of freeing the metal of its associated impurities. It is not a little singular that these substances can be so completely separated by puddling that we can have a product in which their united amount often does not exceed one quarter per cent. of the total weight of the bar.

Now that rails can be made of steel, even from a comparatively expensive hematite ore brought from Spain or raised in England, more

---

[1] The accuracy of this return is questionable.

economically than a good quality of rail can be produced in iron from the cheaper ironstone of Cleveland or elsewhere, steel, to the entire exclusion of iron, must be henceforward looked upon as the proper material for railroads. Iron, in the matter of wear, exhibited very great irregularity, some rails showing signs of distress within a year or two of being laid down, while others afforded very satisfactory results. This uncertainty of quality has perhaps been the means of procuring for iron a worse reputation than its merits as a whole deserved.

In illustration of this assertion I would instance the experience on the main line of the North-Eastern Railway on certain sections of its system, which may be taken as fair samples of the others. On that extending between Newcastle and Berwick, 66·8 miles of double way, the iron rails laid down in 1847 weighed 65 lbs. per yard. Renewals commenced in 1855 and terminated in 1867. In these the weight was increased to 82 lbs. per yard.

The maximum duration of the 65 lbs. rails was 21 years and the minimum 8 years, the average being 12·8 years.

A second relaying with the 82 lbs. rails was commenced in 1864, and the average age of those taken out has been 14·18 years. The iron rails still in use on the Newcastle and Berwick section is 37·30 miles of single road, having a present (June, 1883) average age of 14·87 years.

The Newcastle and Darlington section has a length of 35·15 miles of double way. Rails of 65 lbs. per yard, laid down in 1844, began to be renewed in 1853 and others of 82 and 83 lbs. were laid in their place. The longest life of those of 65 lbs. was 19 years and the shortest 9 years the average of the whole being 11·47 years. The average life of the heavier rails (82 and 83 lbs.) was only 9·38 years, but the traffic became much heavier after they were brought into use. There only remains 3·32 miles of single way of iron rails on this section, the average present life of which is 8·40 years.

The Newcastle and Carlisle section comprises 59·4 miles of double way. It was laid with Losh's patent "fish bellied rails," weighing only 42 lbs. per yard, partly rolled at Walker-on-Tyne but chiefly in South Wales. The maximum duration of these was 26 years and the minimum 13 years, the average being 17·25 years.

SECTION XIII.—REFINERY AND PUDDLING FURNACE.

The average life of 83 lbs. rails, commenced to be laid down on this section in 1849 was 14·21 years. There still remains (June, 1883) 69·2 miles (single way) of iron rails on this section, the average age of which is 10·05 years.

The statements just submitted do not afford any proper criterion of the resisting powers of iron rails; for this can only be determined by the comparative weights of the engines, the amount of traffic and the speed of the trains which have passed over them. According to Mr. R. Price Williams the average life of an iron rail, on the most heavily worked portions of the railways in the United Kingdom, may roundly be taken at about $17\frac{1}{2}$ millions of tons. Quoting the experience of Mr. Webb of the London and North-Western Railway, Mr. Williams proceeds to estimate that the average life of a Bessemer steel rail will be 161 millions of tons, or approximately about nine times that of an iron rail. This was an opinion expressed about the year 1878, but it is to be apprehended that actual experience, in which even now (1883) we are greatly deficient, will not be found to corroborate this opinion of the relative durability of the two kinds of rails.

On the North-Eastern system comprising 1,508 miles of main line of railway equal to 2,499 miles of single line, steel rails were only introduced, at first very sparingly, about the year 1862 and it was not until 1874 that they were largely used. Since the year 1877 no iron rails have been purchased by this company. At the present moment 1,303 single miles of their way are laid with steel and the quantity renewed as not fit for further use, is too insignificant to afford any data upon which to form any opinion of their durability. The only accurate data in my possession is that afforded by 1,216 yards in the Shildon tunnel, on the Stockton and Darlington Section, having a gradient of 1 in 250 against the load. In June and July, 1865, this length was laid with double-headed steel rails weighing 77 lbs. per yard, of which the wear was ascertained from time to time.

By March, 1869, the average loss of weight on twenty-four rails was found to be 8 lbs. per yard, equal to 10·39 per cent., an amount of abrasion which necessitated their being turned.

In June, 1872, an average of twenty-three rails showed a further wear of 9 lbs. per yard equal to 11·69 per cent., making 22·08 per cent.

during the seven years they had been in use, equal therefore to 3·15 per cent. per annum. At this period the whole were taken up as being no longer serviceable.

The total net traffic which passed through the tunnel was 31,149,473 tons, or including the weight of the trucks, etc., say 46 million tons. Roughly we may consider the return load would be that of the empty trucks, etc., say $15\frac{1}{4}$ million tons. Taking the entire weight which passed over the two lines of way we have $61\frac{1}{4}$ millions, which would give 30·62 million tons as the life of a rail in this particular tunnel. I am not prepared to say that the wear of a rail in such a situation is the same as that in the open air; but I do know that one-ninth of the period during which they were in use, viz., $\frac{7}{9}$ year, or less than ten months, did not express the life of an iron rail in this locality, which it ought to do according to the formula laid down by Mr. Price Williams.

So far as the experience of the officials of the North-Eastern Railway enables them to judge at present, they will be well satisfied if the steel rails now in use have double the life of those of iron of the same weight per yard (82 lbs.) which preceded them, *i.e.*, instead of remaining serviceable for ten to twelve-and-a-half years they are found in use at the end of twenty-two years, calculated on the present traffic.

Whatever the power of steel rails to resist abrasion may be, all experience points to their life being much more uniform than was the case where iron was the material employed. An iron rail however became unserviceable long before it had lost much of its weight. Instead of actual wear, the head of the rail became crushed, its comparative toughness apparently resisting abrasion. Thus while the average loss of weight in iron rails would not exceed $7\frac{1}{2}$ per cent. at the period of removal, the loss in 82 lbs. steel rails appears to be 1 lb. per yard per annum—say $1\frac{1}{4}$ per cent. for each year of actual service.

Before economy in the manufacture of steel rails had reached its present position, the North-Eastern Railway Company tried iron, in the form of a solid bloom, obtained from the pig of their own district in the revolving furnace. The puddled iron was treated in machinery constructed by Sir W. G. Armstrong & Co., and the rails were case-hardened by Dodds' process. The change in the cost of producing steel rails however put an end to further trials in this direction.

SECTION XIV.

ON MORE RECENT METHODS OF SEPARATING THE SUBSTANCES TAKEN UP BY IRON DURING ITS PASSAGE THROUGH THE BLAST FURNACE.

IN a former portion of this work the advantages were set forth of the circuitous mode of treating the ores of iron by the aid of the blast furnace, even when the metal in its malleable state was the ultimate object in view. When iron is required in this last mentioned form, the manufacturer has hitherto been unable to obtain it in large masses, by the ordinary method of simple fusion, as pursued by smelters of other metals in common use, such as gold, silver, copper, lead, or zinc. The melting point of wrought iron is so high, that it is only within the last quarter of a century that we have been able to bring any quantity, beyond a few pounds, to the fluid state at one time, by means of heat. Under these circumstances the iron maker had to avail himself of that other valuable property possessed by this metal, namely welding; by the aid of which his granules were built up into the shape of the "puddler's ball" by the manual exertion of the workman. This required extensive shops containing squeezers, hammers, puddle-rolls, etc., by means of which, and of a considerable expenditure of metal, fuel, and labour, the puddled iron was welded together. The product thus obtained is liable to unsoundness, from the difficulty of separating the adhering cinder, which, as is well known, gets sealed up in the interior of the mass.

In 1856 Henry Bessemer announced that he was able to obtain fluid malleable iron by blowing air through the molten crude metal employed in its production, the heat being obtained by the combustion of those substances which had united themselves with the iron in the blast furnace. In other words the very heat, as it were, which had effected the reduction of the metalloids, in the smelting process, was rendered available in the Bessemer converter. The surprise such a statement was calculated to excite was a good deal subdued by the small amount of faith with which it was received.

No doubt it did seem at first sight incredible that such an effect as that represented could be obtained in any form of apparatus; seeing that, in the old fashion of making malleable iron by puddling, the metalloids contained in the pig were equally oxidised, with the same evolution of heat as that generated in the converter. The result with the latter was the more surprising, inasmuch as in that case there is an enormous *excess* of heat, which escapes in the form of intensely brilliant flame from the mouth of the vessel; while in puddling the metal a quantity of coal has to be burnt, most wastefully it is true, equal to half that required in the smelting furnace.

In forming a judgment on the subject the difference of the conditions was entirely overlooked. In the puddling furnace the internal heat was produced by the comparatively slow combustion of the carbon and slow oxidation of the metalloids; while the heat was applied externally to the raw material in an apparatus from which there was a great loss by radiation, convection, etc. In the converter, on the other hand, an immense volume of air was *rapidly* poured through the metal, generating the heat in the very heart of that mass the temperature of which it was desired to raise. The operation was also performed under circumstances where the loss by radiation, having regard to the quantity under treatment, was infinitely less than in an ordinary furnace.

For some years after the first brilliant exhibitions of a Bessemer converter at work, those interested in its success encountered great disappointment. The fact that atmospheric air, when forced into molten cast iron, would burn off carbon, silicon, and sulphur, and would partially remove the phosphorus, was well known to every one who had any experience in the violent action set up towards the close of the old process of refining. No one suspected however, until Bessemer proved it, that by forcing the air upwards through the mass of liquid iron in sufficient quantity, and with sufficient velocity, so intense a temperature would accompany the process. Unfortunately the absence of a sufficient quantity of oxide of iron in the cinder causes the whole of the phosphorus, originally contained in the pig, to accumulate in the decarburized metal.

The presence of this phosphorus was a fatal blow at first to Bessemer's hopes; and the remarkable thing is that a metal containing above a certain quantity of it, made in the converter will be entirely useless, although it contains a smaller quantity of phosphorus and other non-metallic matter than a piece of malleable iron of good merchantable quality, produced by means of the puddling furnace.

Ultimately success was achieved by employing Swedish iron of great purity; and in the end certain English makes were found available, provided they did not contain above one part in a thousand of phosphorus, and provided they either contained manganese, or had this metal added, as was proposed by R. F. Mushet when he patented the addition of spiegel iron to the contents of the converter. By the aid of this valuable and important discovery "the red shortness" which rendered Bessemer metal unmanageable in the subsequent stages of manufacture was entirely removed.

The fact that manganese was often present in ores from which the finest descriptions of malleable iron had been made in Sweden and Russia, probably suggested its early use, by Hassenfratz and others, in the manufacture of steel as well as of bar iron. Forty years ago my friend Josiah Marshall Heath rendered great service to the steel trade of this country by the addition of manganese to crucible steel.

Bessemer steel as now manufactured can be obtained of various qualities—it may have the carbon so reduced in quantity as to resemble in this respect many well known brands of malleable iron, or it may be run in the form of steel of any degree of hardness, by a mere increase in the quantity of carbon it contains. As the spiegel iron which is added immediately before running the steel into the ingot mould, may bring with it more carbon than is required for certain purposes, the alloy known as ferro-manganese is in those cases used in place of the spiegel. This substance was, I think, first made commercially by Mr. Henderson in a reverberatory furnace; but since then M. Gautier, formerly of Terrenoire, succeeded in obtaining it from the blast furnace. Owing to the high temperature required, as much as three tons of coke are consumed for each ton of ferro-manganese produced, which may be made to contain as much as 80 per cent. of manganese; but owing to the difficulty of reducing this metal a considerable quantity runs off as oxide in the slag.

In recent years the progress of this branch of metallurgic science has advanced with unexampled rapidity. By means of Bessemer's own mechanical appliances, added to the use of hydraulic machinery, and of rolling mills of unprecedented power and capacity, the manufacture of steel, particularly in the form of rails, has been so economised as to defy the competition of iron, even when produced from the cheapest known ores. This in part is due to improved machinery, partly applicable no doubt to the manufacture of iron; but it chiefly arises from the greater simplicity of the process itself. To make an iron rail in the old fashion, the metal is puddled in small heats, hammered, and then rolled into puddled bars often not exceeding 120 lbs. in weight. These pass through many hands, have to be cut up, and even to the extent of 25 to 33 per cent. have to be re-rolled into slabs, before they form the pile for the rail itself; the iron being necessarily allowed to cool between each operation.

For steel 7 or 8 tons of metal, brought hot from the blast furnace, is run into the converter. In less than half-an-hour the whole is in the form of steel ingots large enough for three of the heaviest and longest rails in use, instead of one, as in the case of the iron pile. The heat which has served for effecting its conversion into steel would, if properly used, suffice for its manipulation by the rail mill; at all events the ingot, being thrust hot from the mould into the re-heating furnace, needs in actual practice very little additional fuel for finishing the work.

When the two modes of dealing with pig iron are reduced to actual figures, the advantage of Bessemer's pneumatic process over that of puddling leaves no doubt as to the impossibility of the latter ever being able to hold its own against this invention, improved as the latter has been by the experience of later years. In support of this opinion we have only to compare the waste of metal, the quantity of fuel consumed, and the expense of labour for the two systems. These are set forth with a sufficient approach to correctness in the two sub-

SECTION XIV.—SUBSTANCES TAKEN UP IN BLAST FURNACE.

joined columns of figures. Taking steel as unity, the comparative consumption for iron rails is stated in the column under that head, including in both cases the fuel and labour at the blast furnace:—

		Steel Rails.	Iron Rails.
Pig iron employed	... ...	100	105
Coal „	... ...	100	150
Labour „	... ...	100	190

Against the statement just made it must be recollected that the iron rail manufacturer is able to employ a much cheaper quality of pig iron than can be used in the Bessemer converter; unless indeed the so-called Basic process is employed for ridding the cheaper iron of its phosphorus, an operation which will be attended with some expense, from which non-phosphoric iron is exempt.

In speaking therefore of competition between iron and steel, in the comparison just referred to, the observation is for the present confined to that of the facility and cost of extracting, from certain well known deposits, supposed equally accessible, those ores which, without further preparation, are fit for making steel, and those which, from the presence of phosphorus, are not suitable for steel, but can be employed in the manufacture of what is known as malleable or wrought iron.

One of the objects, however, of the present section is to consider the prospect of our being able to enlarge the choice of our raw material; without which the extended use of steel might before long receive a very serious check. This enlarged area of selection will be considered at a later period.

If to simplicity and economy of manufacture we add the admitted superiority of strength possessed by steel, and the absence of the imperfect welding which frequently attends the use of the puddling furnace, the importance of the question becomes obvious. To Great Britain, whose position as an iron making nation has been so largely dependent on her possession within her own shores of the requisite supplies of ore, the question has an especial interest. This position, it is clear, will be materially affected if, instead of feeding our furnaces with mineral obtained in the country itself, we have to derive our supplies, or a large portion thereof, from quarters to which all the world has as ready access as ourselves; leaving any difference in the cost of transport to tell for or against us, as the case may be.

By comparing the change which has taken place during recent years in the relative quantities of ore of different qualities smelted in the iron works of Great Britain, it will be easy to appreciate the nature of the alteration which is taking place in this truly national branch of industry.

	Iron Ore suitable for Steel.[1]			Clay Ironstone, &c. Unsuitable for Steel.	Grand Total.
	British.	Imported.	Total.		
1871	2,233,751	524,175	2,757,926	14,101,137	16,859,063
1880	2,757,887	3,060,331	5,818,218	15,268,522	21,086,740
1881	2,805,472	2,803,198	5,608,670	14,640,593	20,249,263
1882[2]	2,807,372	3,282,496	6,089,868	14,638,694	20,728,562

In order to ascertain the real progress of the manufacture of pig iron comparatively free from phosphorus, and in consequence suitable for steel purposes, consideration must be given to the relative richness of the two classes of ore. For the present purpose the purer descriptions are regarded as yielding 52 per cent. of metal against 30 per cent. afforded by the other.

	Pig Iron obtainable from.		Increase over 1871.	
	Steel Ores.	Ordinary Ores.	Iron from Steel Ores.	From Ordinary Ores.
1871	1,434,121	4,230,341	—	—
1880	3,025,473	4,580,456	111 %	8¼ %
1882	3,166,731	4,391,608	120¾	3¾

It will further be perceived, from the former table, that while the increase of the steel ore raised in Great Britain during ten years amounts to only 25½ per cent., that on the importation of foreign ores is no less than 434¾ per cent.

For some months previous to the autumn of 1879, the demand upon the two kinds of pig iron was so nearly balanced, that the difference in their market values was a near approximation to the difference of cost. At that period the lowness of price greatly stimulated the demand for steel rails; which was followed temporarily by a considerable and disproportionate rise in the selling rates of hematite pig iron, as compared with other descriptions of the crude metal. The following figures exhibit the extraordinary change, in both directions, which a few months wrought in the values both of raw materials and finished products.

---

[1] Mineral Statistics, by Robert Hunt, F.R.S. Imported ore includes burnt cupreous pyrites.

[2] Report of Iron Trade Association.

SECTION XIV.—SUBSTANCES TAKEN UP IN BLAST FURNACE.

	1879.	1880.	
	July.	Jan.	June.
	£ s. d.	£ s. d.	£ s. d.
Selling price of Cleveland No. 3 pig at works	1 12 6	3 1 6	1 16 0
,, Hematite pig ,,	2 7 6	6 2 6	2 12 6
,, Steel rails ,,	4 10 0	9 10 0	5 7 6
Cost of raw materials—			
Red ore delivered at hematite furnaces, in Cumberland and Lancashire, 52 per cent. of iron	0 12 6	1 8 6	0 17 0
Coke delivered at hematite furnaces	0 17 0	1 6 0	0 18 6
,, ,, ,,	0 10 6	0 19 0	0 12 0
Ironstone delivered at Cleveland furnaces, 31 per cent. of iron	0 4 2	0 4 9	0 4 4

Thus, at the beginning of 1880, we seemed to be approaching the position of having a cheaper system of manipulation, affording for most uses a superior article, but threatened, as is illustrated by the altered values, with a deficiency of the necessary raw material. This is the fate which will, in the absence of new discoveries of pure ore, unquestionably overtake us, and indeed the rest of the world as well, unless we can extend the area of our supplies, by some mode of improving the iron hitherto found useless for steel making purposes. As we have seen, the obstacle to the use of certain varieties of pig iron for steel making is the presence of too large a quantity of phosphorus. Sulphur no doubt is also highly objectionable, but it is usually present in very small quantities. It is however impossible to consider the elimination of either of these substances without studying the behaviour of the other two metalloids always found in pig iron, viz., carbon and silicon.

Reference has been made at some length, in a previous section, to certain modes of ridding Cleveland pig iron of the four metalloids just referred to. These are summarised in the table given below. In the carbon column, the separation of this element is dealt with as being complete at every stage, except that of refining; because in the others, although rarely entirely absent, its presence is no longer either necessary or hurtful. I the other columns an average effect is represented. Following these are figures setting forth the change produced by blowing the purer Bessemer pig in the converter.

It is, perhaps, superfluous to mention, that with the great variations in the composition of the products of the different processes, it is impossible to present the information intended in a very well defined form. The table must therefore be considered as a comparative one, constructed from different sources in my possession.

	Carbon.	Silicon.	Sulphur.	Phosphorus.
Cleveland pig contains, say	3·50	1·75	·15	1·55
,, when refined ,,	2·50	·091	·11	·50 to ·75
,, when refined and then puddled may contain	Nil	·19	·01	·260
,, hand puddled with oxide of iron	Nil	·159	·041	·324
,, mechanically puddled with excess of oxide of iron	Nil	·109	·032	·10 to ·170
Bessemer pig contains	4·10	2·50	·060	·050
,, when blown in converter contains	·10	·03	·04	·060

The apparent increase of phosphorus in the last line is due to the diminished weight of the product, which still contains the whole quantity of this element originally charged into the converter.

It may be convenient at this period to consult a series of diagrams which have been constructed, for the purpose of exhibiting the times required for expelling the four metalloids, as well as the extent to which such expulsion is usually carried. In all cases excepting in that of the Bessemer Acid process, so named by Prof. von Tunner to distinguish it from the Basic, it will be considered that the iron subjected to treatment is that of Cleveland, containing, say, carbon 3·50 per cent., silicon 1·75 per cent., sulphur ·15 per cent., phosphorus 1·55 per cent. The sulphur in all cases is omitted from the diagrams, partly owing to its smallness in quantity and the uncertainty as to the extent of its separation, and partly to avoid complication in the drawings. In every instance it will be perceived that it is the silicon which disappears most speedily, next follows the phosphorus, except in the case of blowing in the Bessemer converter, whether Acid or Basic; and lastly comes the carbon, which holds fast the longest by the iron, except in the case of the so-called Basic treatment as it is applied to the Bessemer process.[1]

The real difficulty, which besets the question of an indiscriminate use of pig iron for steel purposes, being that of phosphorus, it will be

---

[1] In figure No. 6—Hand-Puddling—it would appear that the phosphorus is the first to be affected, but this is believed to be an exceptional case.

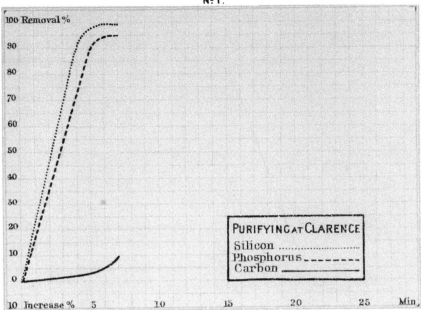

Nº 1.

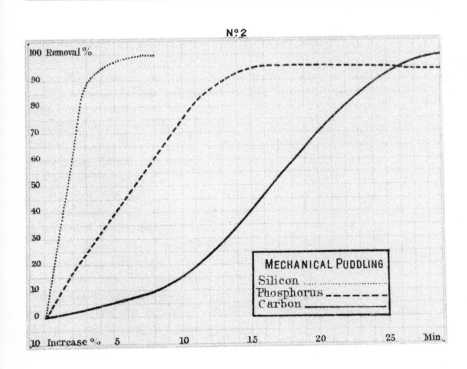

Nº 2

Plate 8.

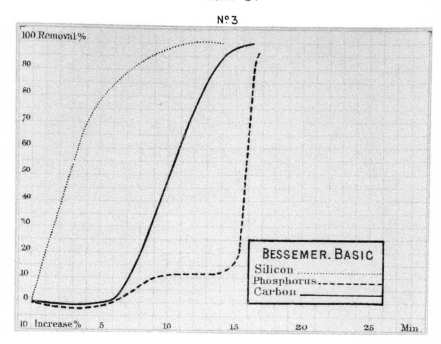

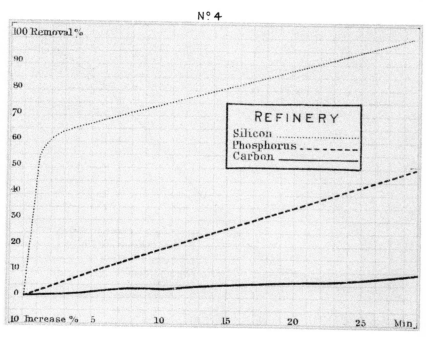

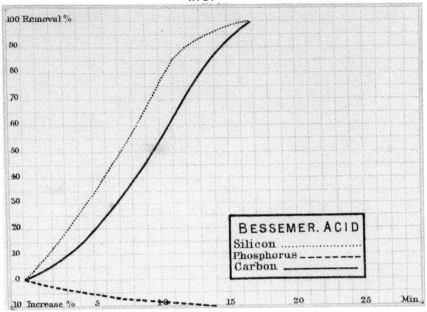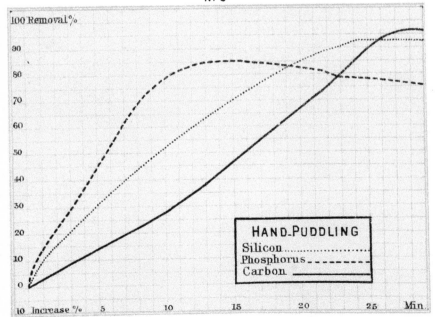

convenient to address ourselves more particularly to the conditions which influence the behaviour of that element in the different modes of treatment just enumerated. In these different processes, we may regard the percentage of phosphorus expelled from Cleveland pig, originally containing from 1·5 to 1·75 per cent., to be as follows:—

	Per Cent.
In the Bessemer converter—Acid process—practically ...	Nil
,, Refinery	50
,, Puddling furnace, ordinary treatment	80
,, ,, (Danks') with excess of ore ...	90

Now the conditions which produce such different results, as regards the phosphorus, are:—

1.—The duration of the operation;
2.—The quality of the atmosphere—reducing or oxidising—in which it is carried on;
3.—The composition of the slag;
4.—The temperature at which the process is conducted.

It would be superfluous to dwell at any length on the first of these four conditions, because there is not a very marked difference between the length of time required for each; but the main reason for passing quickly from it is that no length of time, as will be shortly seen, suffices for expelling the phosphorus in the converter as originally worked, while in the other processes, the previous disappearance of the carbon and silicon, by depriving the metal of its fluidity, would prevent any further diminution of the element (phosphorus) we are specially seeking to expel.

We may now pass on to the second of the conditions referred to, viz., the nature of the gases which permeate or surround the metal, as it is exposed to the different modes of treatment we are considering.

The composition of the atmosphere, in the processes of refining and puddling, has already been shown to be one of a strongly oxidising character. As regards that of the converter, we possess some very valuable information, for which we are indebted to Mr. Snelus. Mr. Tamm has also published a series of analyses of the gases given off during the blowing of a Bessemer charge.

According to the first of these authorities, the following was the composition by volume of the gases taken at various periods of a "blow."

Time after starting	2 min.	4 min.	6 min.	10 min.	12 min.	14 min.	18 min.
Free oxygen	·92	—	—	—	—	—	—
Carbonic acid	10·71	8·59	8·20	3·58	2·30	1·34	—
Carbonic oxide	—	3·95	4·52	19·59	29·30	31·11	—
Hydrogen	—	·88	2·00	2·00	2·16	2·00	—
Nitrogen	88·37	86·58	85·28	74·83	66·24	65·55	—
...	100·	100·	100·	100·	100·	100·	blow completed.

The rate of absorption of the oxygen by the silicon, etc., is easily calculated by taking the quantity originally accompanying the nitrogen in the gases, and deducting from it the quantity still remaining. The volumes of atmospheric oxygen corresponding with the nitrogen, regarding the blast as dry, were as follows:—

Time after starting	2 min.	4 min.	6 min.	10 min.	12 min.	14 min.
Volumes of oxygen blown into converter	23·21	22·74	22·39	19·65	17·39	17·21
„ in gases	11·63	10·57	10·46	13·38	16·95	16·89
Difference, vols. absorbed	11·58	12·17	11·93	6·27	·44	·32

It will thus be observed that the gaseous current passing through the converter is of a very varying character, possessing at the commencement, from the carbonic acid it contains a powerfully oxidising nature, after which it is gradually converted from the excess of carbonic oxide into one having the very opposite tendency. In illustration of this, I have divided the analyses as given by Mr. Snelus and Mr. Tamm into three periods, corresponding with three different stages of the blow; and have estimated the percentage of oxidising and reducing gases to be as follows:—

	Oxidising Gas. Vols.	Reducing Gas. Vols.		Total. Vols.
Mr. Snelus' analyses, 1st period	80·8	19·2	=	100
„ 2nd period	29·5	70·5	=	100
„ 3rd period	5·3	94·7	=	100
Average	38·5	61·5	=	100

SECTION XIV.—SUBSTANCES TAKEN UP IN BLAST FURNACE. 391

	Oxidising Gas. Vols.	Reducing Gas. Vols		Total. Vols.
Mr. Tamm's analyses, 1st period ...	93·0	7·0	=	100
,, 2nd period ...	22·5	77·5	=	100
,, 3rd period ...	16·4	83·6	=	100
Average	44·0	56·0	=	100

From the quantity of oxygen absorbed and the smallness of the amount of carbon in the gases, it is obvious that it is the silicon, not the carbon which is at first chiefly oxidised—an inference which is confirmed by the analyses of the metal itself. Omitting fractions, Mr. Snelus estimates the relative amounts of silicon and carbon oxidised to be as follows :—

Time after blowing	2 min.	4 min.	6 min.	10 min.	12 min.	14 min.
Silicon	73	70	69	40	4	3
Carbon	27	30	31	60	96	97

The alteration in the composition of the gases is an indication that at the outset a large proportion of the oxygen is retained by the bath of metal, *i.e.*, is taken up by the silicon, by the manganese, if any, and by the iron itself. The action on the silicon falls off rapidly towards the end of the operation, owing to its exhaustion from the iron; the manganese follows; but the iron would continue to be oxidised, were the operation unduly prolonged, as long as there is heat enough to enable the oxygen to act on it. The correctness of these views is manifested in a series of analyses by Prof. Kupelwieser, of samples from the blowing of the manganiferous pig iron of Styria. The last column contains the particulars before adding the spiegel at the end of the blow.

	Pig Iron.	1st Period. Metal after Slagging.	2nd Period. End of Ebullition.	3rd Period. End of Blow.
Graphite	3·180	—	—	—
Combined carbon ...	·750	2·465	·949	·087
Silicon	1·960	·443	·112	·028
Sulphur	·018	trace	trace	trace
Phosphorus	·040	·040	·045	·045
Copper	·085	·091	·095	·120
Manganese	3·460	1·645	·429	·113
	9·493	4·684	1·630	·393
Iron	90·507	95·316	98·370	99·607
	100·	100·	100·	100·

Corresponding Slags.	Blast Furnace.	1st Period Converter.	2nd Period Converter.	3rd Period Converter.
Silica	40·95	46·78	51·75	46·75
Alumina	8·70	4·65	2·98	2·80
Lime	30·35	2·98	1·76	1·19
Magnesia	16·32	1·53	·45	·52
Potash and soda ...	·32	trace	trace	trace
Sulphur	·34	·04	trace	trace
Phosphorus	·01	·03	·02	·01
Protoxide of manganese	2·18	37·00	37·90	32·23
,, iron	·60	6·78	5·50	16·86
	99·77	99·79	100·36	100·36

As regards the matter of atmosphere in the converter, the composition, according to Mr. Snelus' analyses, four minutes before the end of a Bessemer blow, was as indicated below. Above this is placed the mean of two recent examinations of the gas at the tuyeres of one of the Clarence furnaces; by which it will be seen that the reducing power in the converter is somewhat superior to that of the hearth of a blast furnace.

	Carb. Acid. per 100 vol.	Carb. Oxide. per 100 vol.	Hydrogen. per 100 vol.	Nitrogen. per 100 vol.
Clarence gas above tuyeres...	1·90	39·18	1·97	56·95 = 100
Bessemer gas four min. before end of the blow ...	1·34	31·11	2·00	65·55 = 100[1]

Notwithstanding the energetic reducing power of the gases, even towards the close of the Bessemer process, as exhibited by the above figures, the iron appears to continue to suffer oxidation to the last. It will however be remembered that, even in the case of the blast furnace itself, *perfect* reduction of all the oxide of iron is never accomplished, and this was ascribed to the dissociation of carbonic oxide, setting up a condition of static equilibrium, in which a sub-oxide of iron was a necessary accompaniment. There is perhaps no reason for doubting the splitting up of carbonic oxide in the converter, it seems therefore very possible that the oxidation of iron may proceed from the same cause. Of course, as the carbon in the pig iron is removed,

---

[1] The reducing power is determined by the relation of carbonic oxide and hydrogen to carbonic acid. Bessemer contains 33·11 CO + H to 1·34 CO_2, then $\frac{33·11}{1·34}$ = 24·7; Clarence blast furnace gas contains 41·15 CO + H to 1·90 CO_2, then $\frac{41·15}{1·90}$ = 21·6. It must be observed however that at the point of entry the blast possesses an oxidising character, and that this continues until it becomes saturated with carbon derived from the bath of iron.

SECTION XIV.—SUBSTANCES TAKEN UP IN BLAST FURNACE. 393

carbon gases cease being generated, and we have atmospheric oxygen passing into the bath of iron. It is then, I presume, that the phenomenon known as overblowing commences. The oxygen, having no other base to seize on, concentrates its action on the iron, which it rapidly oxidises; and partly escapes in the form of the brown smoke so characteristic of the termination of the operation, particularly when it is continued very long, as occasionally happens in dealing with certain qualities of pig iron.

This comparison of the Bessemer converter with the hearth of the blast furnace leads us naturally to a consideration of the slags produced in each case.

The ability of silica at high temperatures to expel even the most powerful acids, such as sulphuric, is well known. Glass makers avail themselves of this property; for instead of using pure soda they employ that alkali in its cheaper form of sulphate, leaving the sulphuric acid to be driven off by the silica in the glass. Now it happens that the slags produced at blast furnaces making Bessemer pig contain about 41 per cent. of silica; and the slag from the converter at the end of the blow just quoted contained 46·75 per cent. of this substance, and it often greatly exceeds this. Hence even admitting that the gases flowing through the converter are possessed of as intense a reducing power as that of the blast furnace, which is perhaps not always the case, this gas being accompanied by a more silicious slag than is obtained in smelting pig, forbids all possibility of phosphorus being acidified and separated in the Bessemer process, as that process is described by the inventor himself, and now often distinguished by the prefix of acid. This inference is of course based on the experience taught us by the blast furnace, when practically under almost precisely similar circumstances as those attending the converter, all the phosphorus is retained by the metal.

It often happens that the silica is found in Bessemer slags of a quantity greatly exceeding that mentioned in connetion with blowing Styrian iron. This of course adds to the difficulty of acidifying and removing the phosphorus. The following are three such examples, which represent the composition, as determined in the Clarence laboratory, of three specimens from a Sheffield manufactory, using the ordinary West of England hematite iron. The fourth specimen, kindly sent me by Mr. Menelaus, is one from the Dowlais works, and was analysed in the laboratory of that establishment. Alongside of it, Nos. 5 and 6 are analyses of slags from a Cleveland and a West Yorkshire furnace.

	Bessemer.				Blast Furnace.	
	1.	2.	3.	4.	Cleveland. 5.	Bowling. 6.
Silica	81·32	78·12	80·84	64·10	30·84	37·30
Alumina	3·62	1·79	3·40	10·48	25·71	17·58
Lime	1·42	4·15	1·32	—	34·03	30·96
Magnesia	—	—	—	—	6·92	7·96
Phosphorus	trace	trace	trace	·02	·15	trace
Protoxide of iron ...	7·65	9·19	8·36	18·90	·23	·58
Peroxide ,, ...	—	—	—	4·60	Sulphur 1·82	2·06
Protoxide of manganese	5·85	6·81	6·04	1·25	·26	3·38
Potash and soda ...	—	—	—	—	1·30	—
	99·86	100·06	99·96	99·35	101·26	99·82

I believe M. Grüner, of Paris, was one of the earliest, if indeed he was not the first, to point out that it was the excessive amount of silica in the Bessemer slags which prevented the basic matter they contain from combining with phosphoric acid.

So far as the slags are concerned, the essential difference between those formed when phosphorus is separated—viz., in refining and puddling—and those formed in the Bessemer process, is in the content of silica and oxide of iron. In the converter the quantity of silicic acid is usually much larger, and that of oxide of iron much smaller, than in the refining and puddling furnace.

For the purpose of contrasting the differences in question the following analyses are appended:—

Refinery Cinders from Running out Fire:—

	Clarence Iron.	Bowling.	Dowlais.
Silica	26·41	31·05	33·33
Alumina	2·47	4·37	5·75
Lime	2·20	1·56	1·19
Magnesia	·24	·85	·50
Protoxide of iron	57·85	43·90	55·11
Peroxide of iron	2·57	10·08	—
Protoxide of manganese	3·90	5·90	2·71
Sulphur	·05	·032	·17
Phosphoric acid	4·14	2·79	2·26
	99·83	100·532	101·02
Iron present	42·28	41·19	42·86

SECTION XIV.—SUBSTANCES TAKEN UP IN BLAST FURNACE.

It must be observed that the composition of the final cinder does not represent that of the cinder always added during the process. Thus the content of silica in the cinder added was only 16 per cent. in the case of the Clarence example, so that at the period when the phosphorus was carried off into the slag the content of silica in the latter would probably be much less than 26·41 per cent.

Puddling Furnace Cinder:—

	Hand Puddling.	Danks' Furnace.
Silica	16·53	14·17
Alumina	1·04	1·76
Lime	·70	·25
Magnesia	—	·42
Protoxide of iron	66·23	59·14
Peroxide of iron	—	20·94
Peroxide of manganese	4·90	1·21
Phosphoric acid	3·80	1·20
Sulphur	2·48	·33
Iron combined with sulphur	4·32	—
	100·	99·42
Iron present	55·83	60·65

The small amount of silica present in both these slags and their extreme richness in oxide of iron, render them well fitted for removing the metalloids associated with iron in the crude metal. Nevertheless in ordinary practice it is by no means an uncommon thing to find, with iron made from Cleveland pig metal, 25 per cent. of the original phosphorus left in the bar. For superior purposes the content of phosphorus is often reduced to as low as 10 per cent. of its original quantity; but then as much as 40 per cent. of the weight of the pig of very rich fettling is used, containing above 60 per cent. of iron in the form of peroxide. In the Danks furnace Messrs. Hopkins, Gilkes, and Co. employed sometimes as much as 80 per cent. of fettling, consisting of rich oxides, which enabled them to bring down the phosphorus to about ·10 per cent. in the finished bar, say from 1·5 to 1·7 per cent. in the pig.

Having regard then to the value of oxide of iron as a dephosphorizer, I endeavoured in 1875 to remedy its absence in the Bessemer converter by the addition of fused oxide of iron to a charge of metal before commencing the blow. The action however was so violent that much of the iron was projected out of the vessel. A second attempt was made in the same direction by continuing the blast until a large quantity of the iron itself, nearly 25 per cent., was oxidised. The fused oxide however acted so strongly on the silicious lining of the converter, that I did not continue the experiment beyond a point where the silica in the cinder had fallen to about 46 per cent. It had the following composition :—

Silica	45·38	
Alumina	·51	...
Lime	1·40	...
Protoxide of manganese ...	2·92	... ...
Protoxide of iron	47·19 ⎫	= 38·70 of metallic iron.
Peroxide of iron	2·86 ⎭	
	100·26	

	Carbon.	Silicon.	Sulphur.	Phosphorus.
The iron (Cleveland), as it flowed into the converter contained ...	3·13	1·87	·12	1·33
The blown metal, which was perfectly liquid, contained	nil	·32	·05	1·66

Up to the present point of our examination of the conditions which affect the separation of phosphorus from pig iron, no account has been taken of the temperature at which the processes are carried on. It has been unnecessary to do so because, irrespective of intensity of heat, the difference between the composition of the slags and of the gases which permeate the metal may be the cause of the refining and puddling processes removing a large proportion of the phosphorus, while in the Bessemer converter this element remains almost entirely unaffected. The similarity of results between the blast furnace and the converter, in respect to the behaviour of phosphorus, I am supposing as possibly due to certain conditions being common to the two processes, viz., composition of gases and constitution of slags. Still I would have it borne in mind that in the three processes in question we find that the quantity of phosphorus passing into the cinder does correspond inversely with the temperature.

Percentage of Phosphorus separated :—

In the Bessemer Process.	In Refining.	In Puddling.
Practically nil	50	90

SECTION XIV.—SUBSTANCES TAKEN UP IN BLAST FURNACE.

In the year 1877[1] I communicated to a meeting of the Iron and Steel Institute an opinion I had been led to entertain, as to the possible influence of temperature in modifying the removal of phosphorus from iron by oxide of iron. I am aware that it is believed by many that instead of temperature being a factor of any value in the operation, it is the differences in the amount of silica, contained in the bath of oxide of iron by which it is sought to acidify the phosphorus, which alone influence the separation of this substance.

The power of silica to neutralise oxide of iron, and prevent the absorption of phosphoric acid, cannot be doubted, as I shall presently demonstrate; but I am not disposed to acquiesce in the opinion that temperature may be wholly disregarded.

Mr. Pourcel, in an admirable paper read before the Institute in 1879, appears to agree with me in setting down temperature as an element which does materially affect the interchange in question.

On the other hand, according to my distinguished friend Professor Grüner, when silica is present to the extent of 30 per cent. in the slag, phosphoric acid is unable to combine with oxide of iron; and I have gathered from the general tenor of his remarks that he attaches little if any importance to differences of temperature, as affecting the separation of phosphorus.

Phosphoric acid we know is volatile at high temperatures; and taking advantage of this the following experiments were performed in the Clarence Laboratory, to ascertain the power of silica to expel this acid. A quantity of phosphate of iron ($Fe\ O$, $P_2\ O_5$ containing $66.77\ P_2\ O_5$) was prepared, mixed with half its weight of silica, and kept in a state of semi-fusion at a bright red heat for five hours; at the end of that time 4·2 per cent. of the phosphorus it had contained had disappeared. The fused mass was then cooled, finely pounded, and exposed for two hours to a temperature sufficient to melt malleable iron. The loss of phosphorus at the end of this time, estimated on its original quantity, was equal to 17·6 per cent.

The slowness and the limited extent to which this evaporation takes place would, under any circumstances, prevent this property of phosphoric acid being practically available in any of the processes we

[1] Journal of Iron and Steel Institute.

are considering; besides the presence of metallic iron would deoxidise the acid and thus permit its reabsorption by a second portion of the metal.

The diagrams given in a previous page, having been constructed to exhibit the percentages of the original amounts of carbon, silicon, and phosphorus, which are removed at different periods of the various operations, do not set forth the actual quantities of each metalloid separated at any particular time. A table in consequence is given below which contains the weights of each of the three substances which were oxidised at the end of seven minutes from the commencement. Seven minutes was the period selected because in one case, that of the exposure of the melted iron with oxide of iron, all the silicon had then disappeared from the iron.

The pig iron employed was Cleveland and hematite, and may be considered as having the following composition:—

	Silicon.	Carbon.	Phosphorus.
Cleveland	1·75	3·50	1·55
Hematite	1·75	3·50	·05

The units of metalloids removed and remaining were as follows:—

	Silicon Removed.	Carbon Removed.	Silicon and Carbon Remaining.	Phosphorus. Removed.	Remaining.
Purifying process, Cleveland pig[1]	1·75	·35	3·15	1·47	·08
Mechanical puddling "	1·68	·28	3·29	·85	·70
Basic Bessemer process "	1·54	·31	3·40	1·50	·05
Refinery "	1·22	·15	3·88	·20	1·35
Acid Bessemer process, hematite pig	·91	1·12	3·22	·00	·06
Hand puddling, Cleveland pig	·70	·78	3·77	·96	·59

As has been already intimated in these pages, an essential condition for effecting the acidification and removal of phosphorus is that the iron containing it should preserve the liquid state. Now the fusibility of iron being aided by the presence of silicon or carbon, or by both together, the approach to uniformity in the united quantity of these two metalloids, still remaining combined with the metal, may be accepted as an evidence of the mass still remaining fluid at the expiry of the seven minutes.

[1] This process will be afterwards described.

SECTION XIV.—SUBSTANCES TAKEN UP IN BLAST FURNACE. 399

The manner in which silicon interferes with, or entirely prevents the removal of phosphorus may be supposed to be as follows:—

In the Bessemer converter we know that silicon is the first of the metalloids which undergoes oxidation. Let us imagine that it were possible for silicon and phosphorus, in the case of hematite iron, to be almost simultaneously acidified, as indeed it will be seen in the diagrams is the case in some other of the processes; then we should have each unit of silicon giving 2·14 units of silica, while each unit of phosphorus afforded 2·29 units of phosphoric acid. Hematite iron however contains generally fifty times as much silicon as it does of phosphorus; so that the generation of 2·29 units of phosphoric acid would be accompanied by the formation of (2·14 × 50) 107· of silicic acid.

Concurrent with this acidification of Si and P, manganese, or in its absence, Fe may be oxidised in sufficient quantity to combine with the acids (SiO_2 and P_2O_5) thus formed. The composition of the resulting salts is probably determined by the temperature of the bath of iron, and they may be regarded as the most stable compounds under the existing conditions—probably $(FeO)_3 P_2O_5$ and $(FeO)_2 SiO_2$—the latter containing 29·4 per cent. of SiO_2, and corresponding nearly with refinery cinder. Such a cinder may be regarded as the result of the rëaction between pig iron and air without much admixture of foreign matter. This primary slag may be looked upon as a neutral silico-phosphate with a large preponderance of silica in relation to phosphoric acid. This silico-phosphate of iron is exposed, in the heated converter, to the action of an excess of fused silica, derived from the sides of the vessel itself. The result of the experiment, already described, entitles us to infer that a portion of the oxidised phosphorus may probably be volatilized. In this event it is pretty certain that the phosphorus thus vapourised would be retransferred to the metallic portion of the bath.

In practise therefore not even incipient oxidation or vapourization of phosphorus is perceptible: there is a great excess of SiO_2 in the slag, from the very beginning of the oxidation and throughout the entire "blow," above that contained in the neutral silicate (see analysis, p. 392). The liberated oxidised phosphorus is consequently rëabsorbed, in other words, none is finally separated.

This supposed check to the formation of phosphate of iron just described would of course cease within the bath the moment all the silicon was removed. We may then imagine a moderate amount of iron (and the analyses of the Bessemer slags prove that it never exceeds a moderate amount) to be oxidised, and that the oxide of iron so formed might acidify phosphorus with the formation of phosphate of iron. The phosphate thus formed would, however, rise to the surface and coming in contact with a highly siliceous cinder, containing sometimes as much as 60 and up to 80 per cent. of silica, would probably suffer instantaneous decomposition. Phosphoric acid would be set free, and the phosphorus would be immediately seized by and returned to the metallic iron, from which it had been just previously separated.

M. Grüner assigns 30 per cent. of silica in the slags, as the quantity sufficing to prevent the removal of phosphorus, or, as I have just put it, in some cases, to ensure its return, if ever removed, to the iron. On referring to my own experiments I must admit that in no case where there was any notable separation of phosphorus did the silica in the slag exceed the proportion assigned by M. Grüner as the limit.

Whatever may be the truth as regards phosphorus, it is highly probable that as regards the other metalloids temperature cannot be disregarded. For the purpose of examination of this subject, reference may be made to the diagrams Nos. 1 and 5, which refer to the purifying and Bessemer (acid) processes respectively. It may be that the silicon is more rapidly removed in No. 1—the purifying process— by intimate and immediate contact with a large quantity of oxygen, although combined with iron as it is, in the oxide used in the operation. This is sufficient for the oxidation of the whole of the silicon, while in the Bessemer converter the oxidation can only proceed as the oxygen is supplied. That state of things however which has promoted the separation of silicon and has permitted the separation of phosphorus, has operated in a contrary direction with carbon, for while ·35 units only have been driven off in the purifying process 1·12 units of this substance have disappeared in the converter. No other explanation occurs to me than that carbon is acted on more readily at the higher temperature of the converter than it is at the more moderate one in which the purifying experiments were conducted.

A comparison between the process of refining and that of puddling, (diagram No. 2) is not calculated to throw much light on the question.

SECTION XIV.—SUBSTANCES TAKEN UP IN BLAST FURNACE. 401

The occasional existence of peroxide of iron in the cinder of the puddling furnace would indicate that this operation is, as might be expected, of a more strongly oxidising tendency than that of refining. The temperature of the refinery no doubt exceeds that of the puddling furnace, but it is probably rather the excessive quantity of silica in the cinder of the former against the highly basic character of the latter, than any difference in the temperature which causes the iron to retain so much larger a proportion of its original phosphorus than is found in iron puddled from pig iron direct.

Notwithstanding the uncertainty which attaches to any explanation of the phenomena as they have just been described, it seemed to offer a field for further enquiry. For if a reduction in the temperature at which the iron was maintained does not expedite the removal of the phosphorus, it may at least retard the oxidation of the carbon. This would retain the metal in a liquid state and thus afford a better opportunity of dealing with the phosphorus.

In order to pursue this subject, means were taken to submit pig iron to the action of oxide of iron at a much lower temperature than that at which either puddling or refining are usually carried on. To ensure a proper measure of success, it is absolutely indispensable that rapid contact is effected, because, failing this, the metal solidifies before the dephosphorizing is completed. The excellence of the results obtained in this so-called Purifying Process, during many trials made in connection with this subject, are to be attributed to the precautions which have been taken to secure this necessary condition.

A revolving puddling furnace, water jacketed, had a longitudinal bridge formed on the interior cylinder, so that when the lining was in its place, this prominence constituted a kind of shelf. By reversing the engine, so as to give the furnace barrel a half turn, a cascade of iron and cinder fell continuously over the shelf, by which an intimate and rapid mixture of the two fluids was secured.

After subjecting ordinary Cleveland iron to this treatment for five or six minutes, the whole was run out; and on some occasions—

Per Cent.
Carbon was found only reduced to the extent of 5 of its original quantity.
While silicon had disappeared to the extent of 99 ,, ,,
And phosphorus ,, ,, 95 ,, ,,

Ten cwts. of iron were treated on each operation, with about 4 cwts. of liquid oxide. The furnace was kept up to a good heat, but owing to the weight of material employed the temperature was inferior to that which obtains even in a puddling furnace. No. 1 in the series of diagrams, page 388, gives the particulars of this mode of removing silicon and phosphorus, which I have already designated as a process of Purification. For the sake of greater clearness, I subjoin in figures, taken from the diagrams, the position of matters in each process after six minutes had elapsed, beginning with the metal (Cleveland) in a state of fusion:—

Percentage removed in six minutes of	Carbon. Per Cent.	Silicon. Per Cent.	Phosphorus. Per Cent.
Bessemer converter—Basic	1	88	Nil.
Refinery ...	3	68	13
Hand puddling ...	17	36	52
Mechanical puddling with excess of oxide	8	97	47
Purifying with about half the quantity of oxide used in mechanical puddling	5	99	95

The circumstance of the mechanical puddling removing the silicon as quickly as the purifying process, while the phosphorus is only eliminated to half the extent, would indicate some difference beyond that of rapid admixture, inasmuch as there is a considerable resemblance in the mechanical appliances employed in both. While suggesting that it may be due to differences of temperature, I am bound however to admit that more extensive experience is desirable before speaking with confidence on this branch of the subject.

In reference to the oxide employed in the purifying process, it may be remarked that, as a rule, it contained about 20 per cent. of silica, which is more than that commonly employed in the fettling used in the puddling furnace.

In the process of puddling, as it is usually conducted, an increase of heat is applied to the ball just before it is withdrawn from the furnace. At this period all the granules of iron are of course coated with the cinder, more or less enriched with the phosphorus which has been previously drawn from the iron. If moderate temperatures are best adapted for enabling the oxide of iron in the cinder to acidify, and thus absorb the phosphorus, it seemed reasonable to infer that by exposing the two substances to an intense heat the action should be reversed; for we know that iron at high temperatures can rob even phosphate of lime of its phosphorus.

SECTION XIV.—SUBSTANCES TAKEN UP IN BLAST FURNACE. 403

Some Cleveland pig iron, containing 1·516 per cent. of phosphorus was puddled at a lower temperature than commonly obtains in a Danks furnace. The iron, when ready for balling, contained ·122 per cent. of phosphorus, but when the balls were cleansed from adhering cinder by fusion in alkaline carbonates, the phosphorus was reduced to ·068 per cent. After heating this granular iron with the cinder accompanying it to a welding point, the bloom similarly cleaned contained ·145 of phosphorus. In like manner other granules of iron, also containing ·086 of phosphorus, were exposed to the highest heat of a reverberatory furnace, in contact with some of the cinder with which they had been puddled. In two hours and ten minutes the phosphorus combined with the iron had risen to ·255 per cent.

These experiments appear to indicate that at the lower temperature oxide of iron acidifies and combines with the phosphorus existing in pig iron; and that the phosphorus, after the carbon has disappeared, is retained in its new form of combination. When however the temperature was raised, the metallic iron recovered a portion of the phosphorus, which it had parted with when the heat was of a less intense character.

It was soon remarked, during the progress of my trials, which extended to the treatment of a few hundred tons of metal, that silica greatly impeded the object I had in view. Calcined Cleveland stone, which contains about 11 or 12 per cent. of lime and magnesia, and about 12 per cent. of silica, was then tried both alone and with lime superadded, with such results as led me to consider that, in point of economy and efficiency, the oxide as it exists in the Cleveland ore would be found to satisfy all the requirements of the case.

It sufficed to heat the ironstone, or other form of oxide, red hot, and then to run the metal from the blast furnace upon it: the silica, furnished by the oxidation of the silicon in the iron, rapidly fused the oxide of iron, which at once commenced to act on the phosphorus in the manner already described.

The late Mr. John Price, while Superintendent of the Government iron works at Woolwich, practised this system of freeing Cleveland pig iron from its associated phosphorus at that establishment. The molten metal was simply agitated with melted oxide in an ordinary puddling furnace and then run out into moulds. The purified iron was then melted in a Price's retort-furnace and spiegel or ferro-manganese was added in the way usually practised in the open hearth ore-process. The steel so made was used to some extent for castings or forged into axles and other objects required at Woolwich. A number of ingots was run from which perfectly sound rails were rolled for the North-Eastern Railway. By this imperfect mode of manufacture the phosphorus was reduced to under 2 per cent. in the steel.

Soon after I communicated the results of these experiments to the Iron and Steel Institute,[1] Mr. Krupp, of Essen, who had been engaged in a similar line of enquiry, commenced to practice the process on a large scale, and my late friend Alex. L. Holley has done the same in the United States. In addition to oxide of iron, it was stated that it had been found advantageous to employ a certain quantity of oxide of manganese. Mr. Holley has described at length the result of his observations.[2] According to information received by this gentleman upon this occasion, something like 17,000 tons of pig iron had been so treated at Essen, where the product was partly used for the manufacture of steel tyres and axles for railway purposes.

The iron operated on contained only about half the phosphorus usually found in Cleveland pig; and the following shows the composition of the metal before and after the process of purification:—

	Original.	After 4 mins.	After 5½ mins.	After 7 mins.
Carbon	3·32	3·27	3·27	3·32
Silicon	0·39	·02	·01	·023
Sulphur	·09	·024	·026	·029
Phosphorus	·74	·16	·146	·106
Manganese	2·32	·038	·116	·058

The cinder contained as follows:—

Silica	13·0
Lime	·7
Alumina	11·6
Oxide of manganese	16·6
,, iron	51·0
Phosphoric acid	6·0
Sulphuric ,,	·2
	99·1 per cent.

[1] Journal of Iron and Steel Institute, 1877.
[2] Trans. American Institute Mining Engineers, 1880.

SECTION XIV.—SUBSTANCES TAKEN UP IN BLAST FURNACE.

The process was carried on at Essen by treating the metal in a Pernot furnace, the revolving motion of which renewed the surfaces of iron and cinder; but in my judgment this appears a less perfect manner of securing a rapid admixture between the two substances, than the plan adopted in my own experiments.

The immediate object of my trials was to obtain a material which, coming from phosphoretic iron, might yet be serviceable for steel making. At the same time it was obvious that such crude iron, purged much more completely of its silicon, sulphur, and phosphorus than by any process of ordinary refining with which I am acquainted, and still retaining nearly its original amount of carbon, ought to be valuable as a material for the manufacture of malleable iron. I would submit, as an instance for comparison, the composition of two samples, which may equally be regarded as refined metal, as under:—

	Carbon.	Silicon.	Sulphur.	Phosphorus.
Purified hot blast Cleveland iron	3·251	0·22	0·12	0·089
West Yorkshire cold blast refined	3·412	0·125	0·024	0·380

In former times it was considered necessary to avoid not alone impurities, in the selection of pig iron for forge purposes; but it had to be chosen from some imaginary quality in the actual iron, which in some mysterious manner fitted it especially for its intended use. This idea indeed still lingers in the minds of some manufacturers of the present day; but it remains yet to be proved that perfectly good malleable iron may not, by the same careful treatment in the forge, be made from metal however obtained, provided only the composition of the puddlers' charge resembles the refined cold blast pig of West Yorkshire.

The Bessemer process sprung into existence towards the close of the reign of empiricism. Its original failure and subsequent success drove the manufacturers to seek for scientific aid in order to ascertain the reason of both; and now pig iron of suitable quality for this purpose is chosen by analysis, with the utmost certainty as to the results which metal of a given composition will afford. So long as this composition is retained, no one enquires from what kind of ore the metal is produced.

Under the competition which iron, purified in the manner just referred to, has to meet, the margin of economy is not a large one.

This observation however, is limited to a comparison of the actual cost involved in the smelting of such ores as those of the Cleveland hills, and of the purer hematites of the north-west of England, or those of Spain and elsewhere. Of course this margin is liable to fluctuations of a very great magnitude, when the demand for the finer kinds of iron rises, as it has done recently, in excess of the supply. In Mr. Holley's report of the practice at Essen, he sets down the loss at 2·5 per cent., which coincides pretty nearly with my own practice; and this with the cost for coal, repairs, ore, labour, etc., will bring the total expense up to about 6s. or 7s. per ton. M. Peterson of Eschweiler gives the following as his estimate:—18 to 30 of ore per 100 of pig, according to the quantity of phosphorus contained in the iron; coal about 2 cwts. per ton of crude iron. Assuming Cleveland pig iron to be bought at 15s. per ton below the price of the purer qualities made from hematite ore, it would appear that in the absence of any better method of treating phosphoric iron such a process as that just described might be employed with advantage.

In the experiment with the converter described in page 395 and undertaken with a view to reduce the acid composition of the slag, oxide of iron was the base employed. Failure attended my researches in this direction, because the basic addition corroded the silicious lining so rapidly, as to maintain in the cinder an excessive amount (above 45 per cent.) of silica. The want of success which had attended the attempts of Dr. Siemens and others to use other linings than silica deterred me from prosecuting any expensive trials as to a better mode of protecting the vessel itself.

Mr. G. J. Snelus had in the year 1872 met with a certain amount of success in lining a small converter with lime; and during his experiments he undoubtedly had observed that the neutralising of the silicic acid by this earth exercised a marked effect, in reducing the quantity of phosphorus as contained in the pig iron. Either this success, commercially speaking, was not sufficiently marked, or other circumstances deterred Mr. Snelus from prosecuting his work with the spirit which a further inquiry has led many competent judges to believe it deserved. Messrs. Thomas and Gilchrist, in the years 1878 and 1879, read before the Iron and Steel Institute papers descriptive of their work in the use of lime in the converter. They also use a vessel lined with lime, as first suggested by Mr. Snelus, or with a mixture of magnesia and lime; but, inasmuch as something like 20 parts of quick lime have been found necessary for the treatment of 100 of Cleveland pig iron, it is clear it would not do to have to depend on the lining for furnishing so large an amount of the calcareous base; even were it capable of affording it which is perhaps doubtful. This has led these gentlemen to add lime to the charge, for the avowed purpose of neutralizing the silica formed during the blow—an object which is not so conspicuous in Mr. Snelus' original publication bearing on the subject.[1]

The use of lime in the converter, suggested in the manner just described, has led to a large quantity of phosphoric pig iron being now used for the manufacture of steel rails. Indeed as regards the quality of the product, so far as chemical composition and the ordinary mechanical tests enable us to judge, there is no reason whatever for supposing that it differs from the steel made from hematite iron. According to information supplied to me by Mr. Percy Gilchrist the works now using the Basic process produced in the half-year ending 31st March, 1883:—

Works in Germany	...	...	...	...	152,479
,, England	...	...	...	...	57,911
,, Austria	...	...	...	...	37,476
,, other countries	...	...	...	33,034
Total Tons	...	...	...	280,900

The extent to which this Basic process will take the place of its acid predecessor must of course depend on the difference in price between the two kinds of pig-iron—the greater this difference, the greater will be the margin to cover the larger waste and additional expenses connected with the Basic mode of treatment. At Middlesbrough this difference of cost of production of the crude metal is about 10s. to 15s. per ton, whereas in Westphalia and the East of France it is sometimes almost double this amount. Accordingly in certain localities this new process has been adopted much more rapidly than in the United Kingdom.

In speaking of the development of this recently introduced branch of industry, it would be unpardonable to omit mentioning the claims of Mr. E. Windsor Richards in its promotion. For half-a-dozen years

[1] Mr. Snelus and Mr. Thomas received Bessemer Gold Medals for their inventions in 1883 from the Iron and Steel Institute.

after Mr. Snelus' first specification, no notice was taken by the steelmakers of this gentleman's invention. This however may be excused by the circumstance of its value, as a means of removing phosphorus, not being very clearly stated. Fortunately when Messrs. Thomas and Gilchrist brought their ideas before the public, Messrs. Bolckow, Vaughan, & Co. had just erected a very large steel rail mill on the banks of the Tees, and it was fortunate for the inventors, that this work was being carried out under the direction of Mr. Richards. No one who is acquainted with the difficulties which were encountered in the practical application of what may appear a simple matter, will deny that to this gentleman and to the enterprise and boldness of his Directors merit is due not inferior to that of the originators of the process itself.

Messrs. Thomas and Gilchrist appear to fix 20 per cent. of silica in the slag, as the limit to which this substance can be permitted to attend without prejudicially interfering with the removal of phosphorus. These authorities, and also Mr. Snelus, insist on the formation of a highly basic slag, so as to present a strong base with which the phosphoric acid may unite *at the moment of its formation*. Messrs. Thomas and Gilchrist further give us to understand that a sufficiency of lime prevents the necessity of "over-blowing;" which may be presumed to mean that it avoids the evil of wasting metallic iron, and the consequent admission of oxide of iron into the cinder.

We are indebted to Mr. E. W. Richards of the Eston Works, and to M. M. Massenez and Pink of Hörde, for much valuable information in the practise of this now so-called Basic process. From it we learn the nature of the gradual changes experienced by iron, in the presence of lime, while throwing off the manganese and the four metalloids with which it is associated while in the form of pig.

The following table, and an appropriate diagram, were communicated by Mr. Richards to the Iron and Steel Institute at Liverpool in 1879, in connection with his experience in removing phosphorus by the Basic treatment.

Fluid metal—

Minutes after commencement of blow	0	3	6	9	12	$14\frac{1}{2}$	$16\frac{1}{2}$
Carbon	3·5	3·6	3·40	2·40	·09	·075	·00
Silicon	1·7	·8	·28	·05	·01	·00	·00
Sulphur	·05	·05	·05	·05	·05	·05	·05
Phosphorus	1·50	1·60	1·63	1·43	1·42	1·20	·08

SECTION XIV.—SUBSTANCES TAKEN UP IN BLAST FURNACE.

Corresponding slags—

	3	6	9	12	14½	16½	16·35
Minutes after commencement of blow							
Silica	32·60	42·60	36·00	35·60	33·00	15·60	16·60
Phosphoric acid ...	·60	·15	1·60	2·61	5·66	15·06	16·03
	33·20	42·75	37·60	38·21	38·66	30·66	32·63
Oxide of iron... ...	7·26	2·57	5·91	6·17	7·89	13·43	14·59
Difference — Lime, magnesia, alumina, and oxide of manganese	59·54	54·68	56·49	55·62	53·45	55·91	52·78
	100·00	100·00	100·00	100·00	100·00	100·00	100·00

Two heats, blown from Cleveland iron in my presence, at Eston, gave before the addition of spiegel the following results:—

	Per Cent.	Per Cent.
Carbon	·036	·068
Silicon	Nil.	Nil.
Sulphur	·065	—
Phosphorus	·150	·064
Manganese	·125	—

Corresponding Slags—

Silica	21·10	17·79
Phosphoric acid	10·36	13·83
	31·46	31·62
Protoxide of iron ...	13·89 } = Iron 12·00	13·59 } = Iron 11·96
Peroxide ,, ...	1·71 }	2·00 }
Protoxide manganese ...	12·49 ⎫	6·24 ⎫
Lime	29·44 ⎬ 52·94	34·73 ⎬ 52·79
Magnesia	7·38 ⎪	9·56 ⎪
Alumina	3·63 ⎭	2·26 ⎭
	100·00	100·00

The Diagram No. 3 in the series, given at page 388, is constructed upon the same principle as that adopted in this work for the other processes of treating pig iron, and it shows in a blow of Basic Bessemer steel the percentage rate at which the metalloids are removed from the iron.

The analyses of the metals and slags which have been blown under the Basic treatment leave no doubt that the phosphorus originally in the pig is acidified, and transferred to the slag as a phosphate. Up to this time however those who have examined the subject most frequently are not agreed as to whether the phosphoric acid is combined with lime, or with oxide of iron.

Experiments by M. Pourcel, described by him in a very able paper,[1] prove apparently beyond all doubt that at high temperatures cast iron decomposes phosphate of lime. It is somewhat improbable therefore that, were the phosphorus in pig iron acidified in a converter, it would combine with lime in preference to oxide of iron, if in the presence of the latter base. On the other hand, it is equally difficult to see how, in the absence of oxide of iron, acidification of the phosphorus can be accomplished. The lime cannot be regarded as a possible source of the oxygen required, and it is certain that air blown through melted cast iron will oxidise the metal before the phosphorus is sensibly affected. Phosphate of iron may no doubt be formed in the first instance, which if decomposed by lime would give phosphate of lime and oxide of iron. This oxide of iron if reduced by carbon or silicon might then pass as metal into the steel.

In respect to the source of the oxygen required for acidification, there is perhaps little, if indeed any, difference of opinion; but while M. Pourcel gives good reasons for believing that the phosphoric acid is combined with iron in the cinder, Mr. Stead,[1] who has had ample opportunity of examining the subject at the works of Messrs. Bolckow, Vaughan, & Co., arrives at a different conclusion; and certainly his opinion is one we cannot disregard.

It may appear of little moment by what precise mode of action the phosphorus is transferred to the cinder, so long as it is effectually removed from the iron. There is connected with the question however one point of some economic importance; for, if lime could at once lay hold of phosphoric acid, however formed in the process, overblowing and the waste of iron consequent thereon might possibly be dispensed with. If, on the other hand, phosphate of iron must first be generated, it seems probable that an additional waste of metal, as compared with the ordinary Bessemer blow, is unavoidable.

[1] Journal Iron and Steel Institute, 1879, No. 2.

SECTION XIV.—SUBSTANCES TAKEN UP IN BLAST FURNACE. 411

For, on referring to the analyses (page 409) of iron and its corresponding slags, taken at various stages of the operation, it will be observed that the iron has sensibly increased in the slag, before the phosphorus has greatly diminished in quantity in the metal under treatment. It therefore seems highly probable that if the phosphoric acid in the cinder is combined with lime, it has changed places with silica previously combined with that earth; the result being a conversion of phosphate of iron and silicate of lime to phosphate of lime and silicate of iron.

The large quantity of a somewhat refractory cinder has undoubtedly presented, as has been already intimated, an inconvenience in the manipulation of this new process. I have been unable to ascertain the proportion which the cinder actually bears to the iron operated on; but an approximate estimate can be made, based on the changes expressed by the iron itself and by the composition of the slags. Adopting these data as a guide in the computation I estimate that there will be present in the converter in which ten tons of metal are being blown something like $2\frac{1}{2}$ to 3 tons of slag.

This large amount of additional matter as compared with the acid process must undoubtedly continue to be a source of some inconvenience owing to its tendency to choke the apparatus. In the ordinary Bessemer process, every constituent of the scoriæ may be regarded as having afforded by its previous oxidation a great amount of heat. In the Basic process, on the contrary, the fusion of the lime and magnesia demands an intense temperature without contributing in any way to its generation. In the matter of volume too, this cinder may present some difficulty; for, swollen as it is with gaseous matter, it will probably occupy a much greater space in the converter than the steel itself; but the difficulty can hardly be regarded as one of serious magnitude in a process where the chief object, viz., the removal of the phosphorus, has been so completely achieved.

The information contained in the analyses of the cinder already given enables us to form some idea of the actual quantity of iron lost during the Basic process, over and above that combined with the cinder as oxide of iron.

In the computation showing the weight of slag formed during the operation, the quantity of metalloids separated was considered as being

7·38 parts per 100 of iron treated. The actual loss in weight however was 17 per cent.; and the difference between this figure and that for the metalloids removed (9·62) could only represent loss in iron, which is therefore 9·62 per cent. in all. Of this no doubt a portion was represented by granules of metal left in the cinder, for on one occasion a specimen of the latter was found in the Clarence Laboratory to contain 11·4 per cent. of its weight, which was separated by means of a magnet.

Now if we assume that 100 parts of pig gave 23 of cinder, and that the latter contained 14·80 per cent. of oxide of iron, we have 2·64 parts of iron accounted for in this way. The actual loss of iron, however, has been shown to be 9·62 parts, which leaves 6·98 unaccounted for.

The estimated loss in blowing ordinary Bessemer iron at Neuberg, already referred to, amounted to 10·936 per cent., and consisted of—

Carbon.	Silicon.	Sulphur.	Phosphorus.	Manganese.	Iron.	
3·930	1·932	·018	Nil.	3·327	1·729	= 10·936

This calculation, which corresponds within a mere trifle to the actual loss, indicates that little or no iron has disappeared which cannot be accounted for. Again, in treating ordinary Bessemer iron in Great Britain, where the loss is said to be 12 per cent., there is no such difference between the iron in the steel and cinder and that delivered to the converters. Mr. Stead regards the quantity of iron actually evaporated in the Basic process as being exceedingly small. If so, then a large portion of the additional loss of metal must be due to the pasty consistence of the slag, in consequence of which a much less perfect separation of the mechanically suspended globules of steel takes place, than from the fluid cinder when blowing hematite iron. These globules of metal, being irregularly disseminated through the slag, their quantity is not easily ascertained.

Looking at the larger quantity of slag which has to be melted, there can be no reason for supposing that there is not a larger quantity of heat required in the Basic system than in a converter worked in the ordinary Bessemer fashion. Admitting this, it will be demonstrated, when we come to estimate the quantity of heat evolved in the two processes, that possibly the burning of a quantity of iron may be indispensable, when a large quantity of lime is added to the

SECTION XIV.—SUBSTANCES TAKEN UP IN BLAST FURNACE. 413

charge. If the oxide of iron resulting from this combustion is not in the slag, it must have been vapourized, an action of itself demanding the expenditure of further heat.

The overblowing supposed as being indispensable to the success of the Basic process induced me to observe the effect of exposing borings of different rails heated to a full red to a current of hydrogen gas. This was done in order to discover to what extent the steel had absorbed oxygen gas.

An iron rail of the ordinary quality lost in 2 hours ·149 per cent., and in 4 hours ·190 per cent. A steel rail made by the Basic process before adding the spiegel lost in $3\frac{3}{4}$ hours ·248 per cent., and in $8\frac{3}{4}$ hours ·284 per cent. After the addition of the spiegel, which is known to remove oxygen, the loss in $3\frac{3}{4}$ hours by exposure to hydrogen was ·079 per cent., and in $8\frac{3}{4}$ hours ·105 per cent.

These two experiments point to the conclusion that the iron rails, in which, from the presence of cinder, the most oxygen might be expected, in reality contained somewhat less of this element, liable to be affected by hydrogen than steel obtained by the Basic treatment.

Phosphorus, constituting the dangerous element, it is of importance that the separation of this substance should be so effectual by the Basic mode of treatment, that it does not exceed in quantity that found in rails made by the acid process.

For the purpose of illustrating how completely this is achieved two tables of analyses of each kind of rail have been taken from the laboratory records of the North-Eastern Railway Company.

CONSECUTIVE ANALYSES MADE IN NORTH-EASTERN RAILWAY COMPANY'S LABORATORY BY Mr. ROUTLEDGE OF 20 STEEL RAILS MADE FROM HEMATITE IRON. CONTAINED PER 100 PARTS.

No.	Carbon.	Silicon.	Sulphur.	Phosphorus.	Manganese.	Iron.
1	·47	·14	·08	·06	1·59	97·66
2	·50	·13	·12	·06	1·31	97·88
3	·45	·10	·15	·06	1·31	97·93
4	·52	·09	·10	·06	·95	98·28
5	·45	·09	·10	·06	1·31	97·99
6	·43	·11	·11	·05	1·04	98·26
7	·42	·12	·12	·05	1·22	98·07
8	·41	·14	·15	·05	1·46	97·79
9	·44	·16	·13	·05	1·70	97·52
10	·44	·06	·13	·05	·97	98·35
11	·45	·06	·14	·05	·92	98·38
12	·52	·08	·17	·05	1·04	98·14
13	·52	·12	·15	·05	1·01	98·15
14	·50	·10	·13	·05	1·08	98·14
15	·45	·10	·11	·05	1·20	98·09
16	·43	·05	·09	·05	1·12	98·26
17	·39	·08	·09	·05	1·04	98·35
18	·39	·10	·14	·05	1·13	98·19
19	·35	·08	·12	·05	1·01	98·39
20	·51	·10	·08	·04	1·16	98·11
	·452	·105	·121	·052	1·178	98·092

In the finished steel there are three elements which the manufacturer seeks to get rid of, viz. the silicon, sulphur and phosphorus. The united percentage of these in the rails made from hematite pig is ·278 whereas in the Basic steel, although the iron used in its manufacture may contain 2 per cent. of phosphorus, this substance is practically reduced as low in quantity as in the hematite steel.

SECTION XIV.—SUBSTANCES TAKEN UP IN BLAST FURNACE.

CONSECUTIVE ANALYSES MADE IN NORTH-EASTERN RAILWAY COMPANY'S LABORATORY BY Mr. ROUTLEDGE OF 20 STEEL RAILS MADE FROM CLEVELAND IRON BY THE BASIC PROCESS.

	Carbon.	Silicon.	Sulphur.	Phosphorus.	Manganese.	Iron.
No. 1 ...	·38	·07	·03	·06	1·15	98·31
,, 2 ...	·47	·04	·05	·06	1·16	98·22
,, 3 ...	·47	·07	·03	·06	1·17	98·30
,, 4 ...	·46	·03	·12	·06	1·31	98·02
,, 5 ...	·48	·05	·10	·06	1·45	97·86
,, 6 ...	·46	·04	·11	·06	1·24	98·09
,, 7 ...	·48	·06	·11	·06	1·07	98·22
,, 8 ...	·46	·09	·11	·06	1·19	98·09
,, 9 ...	·37	·09	·07	·05	1·22	98·20
,, 10 ...	·40	·05	·10	·05	1·09	98·31
,, 11 ...	·46	·03	·11	·05	1·28	98·07
,, 12 ...	·46	·04	·07	·05	1·18	98·19
,, 13 ...	·46	·06	·14	·05	1·29	98·00
,, 14 ...	·48	·06	·11	·05	1·40	97·90
,, 15 ...	·44	·06	·11	·05	1·46	97·88
,, 16 ...	·47	·06	·09	·05	1·46	97·89
,, 17 ...	·48	·07	·10	·05	·90	98·40
,, 18 ...	·46	·12	·12	·05	1·07	98·18
,, 19 ...	·44	·10	·12	·05	·82	98·47
,, 20 ...	·42	·02	·10	·05	1·11	98·30
	·450	·065	·095	·054	1·201	98·135

The sum of the silicon, sulphur, and phosphorus is only ·214 per cent., or nearly 30 per cent. less than that which appears in the previous table.

In selecting pig for the Bessemer converter, manufacturers hitherto have had two objects in view—a sufficiently low percentage in phosphorus (under ·1 per cent.), as not to impair the strength of the steel, and a sufficiently high percentage of silicon as to afford the necessary heat required during the blow.

In the Basic treatment an excessive amount of silicon is accompanied with considerable inconvenience; for the higher it is in quantity the greater is the proportion of lime required to neutralise it. This however is by no means the most serious difficulty entailed by the excessive amount of lime required in order to secure the removal of the phosphorus. The slag, which may be regarded as a silico-phosphate of lime and iron, contains such an excess of basic matter as to be very infusible, and in consequence interferes with the current work of the converter by incrustations which gather on the sides and particularly on the throat of the vessel.

The idea, which seems to have first suggested itself to the manager of the Hörde works of replacing the silicon by phosphorus, as a heat-evolving medium, has been attempted in practise apparently with great success. Mr. Massenez of that establishment has announced that by using a pig iron containing only $2\frac{1}{2}$ per cent. of carbon and 2 per cent. of phosphorus, he can produce steel having only ·03 to ·06 of the latter, provided the silicon in the pig does not exceed ·5 per cent.

These considerations have now led to the frequent use of white or of hard grey iron; since these, being smelted at a lower temperature than "richer" grey metal, contain a much smaller quantity of silicon. Although it has been occasionally stated that the reducing of the percentage of silicon by a lowering of the temperature in the blast-furnace is accompanied by a corresponding reduction in the phosphorus, such has not been my own experience. This is of importance for the reason that the deficiency of the heat evolved in the converter, due to suppressing, not only the silicon but some of the carbon, has to be made up, as described above, by heat due to the oxidation of phosphorus.

Very recently some white iron was made at the Clarence works; and the following figures prove that there has been no diminution in the content of phosphorus:—

	Carbon	Silicon.	Sulphur.	Phosphorus.
Clarence grey iron, No. 4	3·294	1·531	·072	1·480
,, ,,	3·606	1·353	·023	1·676
White iron, made 9th June, 1880	2·918	·578	·490	1·670
,, 10th ,,	3·005	·630	·401	1·660
,, 11th ,,	3·091	·700	·383	1·652

SECTION XIV.—SUBSTANCES TAKEN UP IN BLAST FURNACE. 417

The following are analyses made in the Clarence laboratory to ascertain the extent to which phosphorus had been removed from a sample of white iron blown at Hörde. The operation was performed in the presence of my son, Mr. T. Hugh Bell, and Mr. Thompson the manager of the Clarence works. Through the courtesy of my friend M. Massenez, the necessary samples were brought away by these gentlemen. They contained:—

	White Iron obtained by re-melting Grey and White Pig.	Steel Bar after addition of Spiegel.
Carbon	2·923	·171
Silicon	·710	·016
Sulphur	·307	·085
Phosphorus	1·649	·071
Manganese	·574	·732

Of course what has to be dreaded in white pig iron is the excess of sulphur, which a metal of this class generally absorbs in the blast furnace. Now in the present case, although this substance is found in the pig to the extent of ·307 per cent., it has been removed in this particular steel to the extent of 73 per cent. of its original quantity: and the proportion compares favourably with steel made from ordinary hematite pig.

The following three sets of analyses, made on the pig and on the metal before the addition of the spiegel, apply to three other charges blown on the Basic system:—

	Experiment No. 1.		Experiment No. 2.		Experiment No. 3.	
	Pig Iron.	Blown Metal.	Pig Iron.	Blown Metal.	Pig Iron.	Blown Metal.
Carbon	2·58	·040	2·82	·081	2·73	·07
Silicon	1·08	·005	·45	—	·72	·004
Sulphur	·22	·140	·16	·09	·25	·12
Phosphorus	1·04	·045	·96	·05	1·04	·06
Manganese	1·35	·370	1·04	·029	1·27	·19

The average composition of these three charges therefore appears to be:—

	Pig Iron.	Blown Metal.	Average Percentage. Removed.[1]
Carbon	2·710	·063	97·68
Silicon	·750	·003	99·60
Sulphur	·210	·116	44·76
Phosphorus	1·013	·052	94·86
Manganese	1·220	·196	83·93
Total	5·903	·430	92·72

[1] These percentages and some of the subsequent figures may perhaps not be strictly correct, because the charges are treated as being all pig; whereas about 22 per cent. of "scraps" were used, of which no composition is given.

A A

The times required for blowing were as under:—

	Exp. 1. Min. Sec.	Exp. 2. Min. Sec.	Exp. 3. Min. Sec.	Average. Min. Sec.
Time previous to overblow	11 45	14 0	12 0	12 35
Duration of overblow	1 55	2 5	2 0	2 0
	13 40	16 5	14 0	14 35

In order to ascertain the quantity of heat developed during a blow in the converter, the loss of each substance by combustion, and the extent, as near as possible, to which oxidation is carried, must be ascertained.

The heat evolved by the carbon depends of course on the relative quantities of carbonic oxide and carbonic acid generated during its combustion. These vary at different periods of the blow; but, guided by the researches of Snelus and Tamm, I have assumed that the carbon is burnt off, at the rate and in the manner set forth in the table given below, in which the combustion is divided into seven periods of two minutes each:—

Periods—Number of Period.	1	2	3	4	5	6	7	
Carbon burnt	5%	12·5%	12%	13·8%	15·5%	19·8%	21·6%	=100
Percentage of CO_2 vols.	100	68	64	38	15	7	4	
„ CO „	—	32	36	62	85	93	96	
Calories (C) per unit of carbon	8,000	6,208	5,984	4,528	3,240	2,792	2,624	

From these figures 4,144 calories appear to be the heat equivalent of each unit of carbon burnt; which may therefore be taken at 4,150.

The iron, which leaves the converter in the form of dense brown fumes during the overblow, I have calculated as escaping in the form of magnetic oxide, *i.e.* one equivalent each of Fe O and $Fe_2 O_3$. In the slag the iron is chiefly found as protoxide with varying quantities of peroxide. In an elaborate paper by Mr. Jos. Massenez[1] of Hörde, 100 parts of slag are stated to have contained, at different periods of the blow, the following quantities of the two oxides of iron:—

Min. after commencement of blow.	2	4½	6¼	9	10⅔	11½	11¾
Protoxide of iron	4·21	4·84	6·77	4·13	5·97	11·58	9·42
Peroxide of iron	·14	·50	1·62	2·00	2·57	3·52	1·61
Equal to Fe	3·38	4·11	6·41	4·51	6·28	11·46	8·44

[1] Dephosphorizing in the converter by J. Massenez, Hörde. Trans. Iron and Steel Inst., 1880, p. 475.

SECTION XIV.—SUBSTANCES TAKEN UP IN BLAST FURNACE. 419

Another sample contained :—

Min. after com. of overblow.	2	4	6	8	$9\frac{1}{4}$	$10\frac{3}{4}$	$11\frac{3}{4}$	$11\frac{1}{2}\frac{1}{2}$	$12\frac{1}{12}$
FeO	4·21	3·69	3·24	2·90	5·42	4·45	11·21	12·27	8·58
Fe_2O_3	·80	·60	1·18	1·81	2·54	·57	2·95	4·94	3·81
Equal to Fe	3·83	3·30	3·35	3·52	6·00	4·85	10·14	13·01	9·34

To avoid complication, the heat-equivalent of the iron is taken uniformly at 1,582 calories.

Mr. Massenez gives the average composition of the slags from four charges of steel at the end of the blow, as follows :—

Silica	11·46
Phosphoric acid	9·63
Total acids	21·09
Alumina	2·50
Lime	49·57
Magnesia	7·69
Sulphide of calcium	1·72
Protoxide of iron	8·87 = 6·89 ⎫ 8·62 Fe.
Peroxide of iron	2·48 = 1·73 ⎭
Protoxide of manganese	6·33
	100·25

The weight of slag was equal to 19·3 per cent. of the charge used, or 23·6 per cent. on the steel produced : equal therefore to 4·72 cwts. per ton of steel.

The actual weight of slag from five blows was 3,795 kilogrammes, which, according to the above composition, would contain 2,173 kilogrammes of lime and magnesia. The calcareous matter used is given at 3,400 kilogrammes so that the caustic lime, as usually happens, must have contained a considerable quantity of carbonic acid.

The composition of the materials used was as follows :—

	C.	Si.	S.	P.	Mn.	Fe.	
Pig iron, $\frac{3}{4}$ white, $\frac{1}{4}$ grey, average	2·83	0·66	·29	1·28	·52	94·42	= 100
Steel	·234	—	·073	·022	·522	99·149	= 100
Spiegel	4·01	·67	—	·21	11·25	83·86[1]	= 100
Steel produced	·234	—	·073	·022	·522	99·149[1]	= 100

[1] Includes a little copper.

Produce of steel 81·52 per cent., or 100 parts of steel equal to 122·7, or, say 123 of the mixture.

The following table exhibits the weight of the substances which are expelled during the blow:—

			C.	Si.	S.	P.	Mn.	Fe.
Pig iron	...	100	2·830	0·660	·290	1·280	·520	94·420
Steel scrap	...	13	·030	—	·009	·003	·068	12·890
Spiegel	...	10	·401	·067	—	·021	1·125	8·386
		123	3·261	·727	·299	1·304	1·713	115·696
Steel	...	100 contains	·234	—	·073	·022	·522	99·149
Composition of the 23 parts which have disappeared =			3·027	·727	·226	1·282	1·191	16·547

The heat evolved is thus estimated:—

					Calories.		Calories.
Carbon	...	...	...	3·027 ×	4,150	=	12,562
Silicon	...	...	...	·727 ×	7,830	=	5,692·4
Sulphur[1]	...	...	...	·226 ×	Nil.	=	Nil.
Phosphorus	...	...	...	1·282 ×	5,868	=	7,522·7
Manganese	...	...	...	1·191 ×	1,724	=	2,053·2
Iron	...	...	...	16·547 ×	1,582	=	26,177·3
				23·			54,007·6

Since the period at which the sample was taken the pig employed is white and contains per cent.:—C 2·75 to 3; Si ·01 to ·03; S ·01; P 2·5 to 3; Mn 2·5.

The loss in blowing is 16 per cent. *i.e.* 120 parts of it and spiegel are required per ton of steel.

The composition of the slag is:—

Silica	...	...	...	...	...	5·03
Phosphoric acid	...	...	...	...	...	19·46
Total acids	...	...	...	...		24·49
Alumina	...	...	...	...	...	2·50
Lime	...	...	...	...	...	49·27
Magnesia	...	...	...	...	...	5·92
Sulphide of calcium	...	...	...	...		2·13
Protoxide of iron	...	...	...	...	...	10·15
Peroxide of iron	...	...	...	...	...	1·85
Protoxide of Manganese	...	...	...	...		4·36
						100·67

[1] The chief part of the sulphur being found in the slag as sulphide of calcium no credit is given for any heat by its removal.

SECTION XIV.—SUBSTANCES TAKEN UP IN BLAST FURNACE.

Another manufacturer gave me the following as an approximate consumption of material per 100 of steel:—

Composition of materials:—

	C.	Si.	S.	P.	Mn.	Fe.	
Basic pig iron	3·50	1·50	·12	1·75	1·00	92·13	= 100
Hematite pig	4·00	2·50	—	·05	—	93·45	= 100
Spiegel	4·00	1·00	—	—	20·00	75·00	= 100

Elements in 120 of material:—

			C.	Si.	S.	P.	Mn.	Fe.
Basic pig iron	110	=	3·850	1·650	·132	1·925	1·100	101·343
Hematite pig	5	=	·200	·125	—	·003	—	4·672
Spiegel	5	=	·200	·050	—	—	1·000	3·750
	120	=	4·250	1·825	·132	1·928	2·100	109·765
Steel	100	=	·450	·010	·060	·055	1·200	98·225
Substances expelled	20	=	3·800	1·815	·072	1·873	·900	11·540

Heat evolved:—

		Calories.		Calories.
Carbon	3·800 ×	4,150	=	15,770·0
Silicon	1·815 ×	7,830	=	14,211·4
Sulphur	·072 ×	Nil.	=	Nil.
Phosphorus	1·873 ×	5,868	=	10,990·7
Manganese	·900 ×	1,724	=	1,551·6
Iron	11·540 ×	1,582	=	18,256·2
	20·			60,779·9

The loss in this case was named as being the outside, 117 of pig, etc., being considered as a practicable amount, for the production of 100 of steel. This alteration would reduce the heat evolved to about 51,600 calories.

To contrast the heat required in the Basic process, as compared with the Acid process, we will assume in the latter an average consumption of 112·5 units of pig to be needed per 100 of ingots: the pig consisting of 105 units of hematite pig and 7½ of spiegel.

The composition of the materials employed and steel made may be taken as follows:—

	C.	Si.	S.	P.	Mn.	Fe.	
Hematite pig	4·000	2·500	·025	·060	0·500	92·915	= 100
Spiegel	4·500	·750	—	—	18·000	76·750	= 100
Steel	·450	·100	·020	·060	1·110	98·260	= 100

Hence the elements in 112·5 units of materials would consist of:—

			C.	Si.	S.	P.	Mn.	Fe.
Pig iron ...	... 105·	=	4·200	2·625	·026	·063	·525	97·561
Spiegel ...	... 7·5	=	·337	·056	—	—	1·350	5·756
	112·5	=	4·538	2·681	·026	·063	1·875	103·317
Steel ...	... 100·	=	·450	·100	·020	·060	1·110	98·260
Disappeared during blow ...	12·5	=	4·088	2·581	·006	·003	·765	5·057

The heat evolved is computed to be as follows:—

					Calories.		Calories.
Carbon	...	...	...	4·088	× 4,150	=	16,965·2
Silicon	...	...	...	2·581	× 7,830	=	20,209·2
Sulphur	...	...	...	·006	× Nil.	=	Nil.
Phosphorus	...	...	...	·003	× 5,868	=	17·6
Manganese	...	...	...	·765	× 1,724	=	1,318·8
Iron	...	...	...	5·057	× 1,582	=	8,000·1
				12·500			46,510·9

By estimate the slag from the oxidized materials should weigh about 11·52 units, equal therefore to 2·30 cwts. per ton of steel ingots.

Such calculations as those just given, where applied to the blast furnace, give results, as to the sum of the heat evolved and appropriated, which correspond, as has been demonstrated at the proper place, very closely with each other. In the case of intense temperatures, like those which obtain in the Bessemer converter, the problem is a more difficult one. On the assumption that the liquid steel has a temperature of 2,000° C. (3,632° F.) and the melted pig iron 1,200° C. (2,192° F.), and that the gases are given off at an average temperature of 1,600° C. (2,912° F.), we obtain, for the heat absorbed, the following results per 100 units of steel in the case of the Hörde blow, p. 420:—

									Calories.
100	units of	steel	×	2,000	×	·25	Sp. heat	=	50,000
23·6	„	slag	×	2,000	×	·22	„	=	10,384
3·44	„	CO_2	×	1,600	×	·216	„	=	1,188 ⎫
4·87	„	CO	×	1,600	×	·245	„	=	1,908 ⎬ 21,555
48·00	„	N	×	1,600	×	·243	„	=	18,459 ⎭
									81,939

SECTION XIV.—SUBSTANCES TAKEN UP IN BLAST FURNACE. 423

To supply this heat we have:—

1st.—That evolved as calculated from the results given by Mr. Massenez, viz.		54,007
2nd.—That brought in by 123 of melted pig iron 123 × 1,200° × ·22 sp. heat	32,472	
3rd.—That brought in by 61·84 blast × 37·5° C. × ·237 sp. heat	549	33,021
		87,028
Sum of heat estimated to be absorbed		81,939
Difference		5,089

Although the two sides of the amount agree pretty closely, I am not able to claim absolute correctness for the statement; because we are not in possession of very exact data as to the specific heat of bodies at such high temperatures as those with which we are dealing. One thing however seems pretty certain, viz., that the heat evolved by substances other than iron would of themselves not suffice, in the Basic treatment, to raise the temperature of the bath to the required point; for it will be perceived that, in this particular case, no less than 26,177 calories, or nearly one-half of the whole are due to the oxidation of the iron. Under no circumstances can iron be regarded as an economical source of fuel, as we may see by comparing the calories developed in burning off that portion of the constituents of pig iron which disappears in the converter. Of all these constituents silicon is the best, and iron the worst. Adopting the former as a standard, and calling it unity, we have the following heat-values for the other bodies:—

Silicon.	Phosphorus.	Carbon.[1]	Manganese.	Iron.	Sulphur.
100	75	60	22	20	31

To obtain the effectual heat produced by burning these substances, a deduction must be made to cover that carried off by the nitrogen of the blast and by the gaseous products of combustion. Thus corrected the figures stand thus:—

Silicon.	Phosphorus.	Carbon.	Manganese.	Iron.	Sulphur.
100	66	14	21	17	12

[1] As burnt in the converter to CO and CO_2.

A reference to the appropriation of the heat evolved in the operation will show that out of 81,939 calories no less than 21,555 calories are carried off in the gases, which rush with characteristic violence from the converter.

The escape of heat, from this cause, is of course unavoidable; and to supply the waste, as well as to maintain a high temperature in the bath of metal, the expenditure of a considerable quantity of combustible is indispensable. As a source of heat phosphorus stands next to silicon, and by returning the cinder from the Basic process to the blast furnace pig containing as much as 3 or 4 per cent. of phosphorus has been obtained for use in the converter. If however oxide of iron is needed for acidifying this element it is probable, irrespective of any question appertaining to heat, that any addition to its quantity means an extended period of overblow to rid the steel of its presence.

In regard to the temperature at which the blow is effected, Professor Åkerman states,[1] that the greater the heat, the closer and more homogeneous, as a rule, the metal will be: besides which the danger of its becoming red-short is proportionately dimished. He adds it is not improbable that the brittleness, which at times characterizes Bessemer metal, stands in some relation to a deficiency of heat during the process. Lastly, he considers that a good heat in the contents of the converter prevents waste, by avoiding the formation of scull, and by preventing so much steel being scattered by the blast as takes place when the metal becomes sluggish through cooling.

The Basic treatment of phosphoriferous pig in the converter has now been before the world for a sufficiently long time to enable us to judge of its future. In the year 1880, upon the occasion of a discussion on a paper read by M. Jos. Massenez at Düsseldorf,[2] Ritter v. Tunner spoke of the economical view of the question being the weak side of the Basic process. According to the data he had collected, the cost of the Basic process exceeds that of the Acid process by 12s. 6d. to 16s. 6d. per ton. In a Report prepared by the same learned authority for the Iron Masters Union of Styria and Carinthia,[3] it is stated that the

[1] On the generation of heat during the Bessemer process. Trans. Iron and Steel Inst., 1872, p. 110.
[2] Zeitschrift der Berg und Hütten Kunde, 1880, Nos. 5 and 6.
[3] Trans. Iron and Steel Inst., 1880, p. 296.

SECTION XIV.—SUBSTANCES TAKEN UP IN BLAST FURNACE. 425

cost of Bessemer hematite iron in Westphalia is 66s. per ton, while that of phosphoric pig is only 45s., showing therefore a difference of 21s. per ton in the cost of the crude iron. At the same time it is stated, with pig iron at these prices, that the cost of a ton of ingots made by the Basic process cost 15s. 8·28d. less than ingots made by the Acid process.

The basis upon which this last estimate is framed supposes the iron in the Acid process to be used direct from the blast furnace, while that for the Basic process is remelted. Exclusive of waste of iron, this probably means a difference of 2s. 6d. to 3s. per ton. Taking it at 2s. 8·22d. we have the extra cost in the case of the Basic process (both using the metal direct) standing at 13s. per ton of ingots.

In a paper by Messrs Thomas and Gilchrist, read before the Society of Arts, about two years subsequent to the investigations of Ritter v. Tunner, viz., in April, 1882, the cost for extra labour, Basic linings, and lime added in the converter is given at 6s. 6d. to 7s. on the ton of steel made. Although the waste of iron exceeds in weight that incurred in the Acid process, the actual expense for this item is something less on the Basic process than in the other, arising from the lower cost per ton of the pig iron employed.

As already stated, the chemical composition as well as the power of resisting the blow of a falling weight, in the case of steel rails made by the Basic process, prepares us for the expectation that they are, in every respect, equal in quality to those produced from the best hematite pig iron by the Acid process.

The North-Eastern Railway Company have during the last four years received upwards of 37,000 tons of rails made by this method at the Eston Works of Messrs. Bölckow, Vaughan, & Co. Both in places of ordinary wear, and on curves where faulty rails soon give way, the deliveries made by this firm have, so far, fully realized the hopes entertained of their quality.

More than once in the present work reference has been made to the very valuable service rendered to the metallurgy of iron by the admirable furnace of Dr. now Sir William Siemens and his brother. That heat which in an ordinary reverberatory furnace escapes into the chimney and, so far as its own work is concerned is so much loss, is returned to the hearth by means of the atmospheric air and gaseous fuel employed as the source of heat. This economy of combustible although important, is less so than the intensity of the heat, commanded by the general arrangement of the apparatus; for we are, by its help, enabled to fuse steel and even wrought iron with the utmost ease.

The idea of melting wrought and pig iron in such proportions as to afford steel is a very old one, first suggested I believe by Réaumur above 150 years ago. In recent times comparatively, viz., 40 years ago, I assisted the late Josiah Marshall Heath in making crucible steel at the Walker Iron Works by this method. This process would doubtless have been a commercial success, if we had been acquainted with the resources of the Siemens furnace. The application of this invention to such a purpose as that in question is so obvious, that its aid was speedily brought into requisition in what is now generally known as the Siemens-Martin or open-hearth process.

After the meeting of the Iron and Steel Institute held at Vienna in 1882, its members were permitted by the Directors of the Southern Railway of Austria to examine their establishment at Gratz, conducted exclusively on the Siemens-Martin system under the very excellent management of M. Prochaska.

The works have a certain historical interest, for they were among the earliest, if indeed they were not actually the first, put down in Austria, for the manufacture of steel rails. In 1864, when I visited the place, they were already making the rails they required of Bessemer steel, under the management of Mr. Hall, a native of Newcastle-on-Tyne. A pile was made of the worn-out Vignoles iron rails, and a hammered steel ingot served to form the head of the new rail.

In the meantime the Siemens-Martin process was being developed, and as the native iron used in the manufacture of the iron rails, as well as the pig iron of Styria and Carinthia, were sufficiently free from phosphorus to permit their use in the manufacture of steel, the Bessemer plant was discontinued, and the open-hearth mode of manufacture introduced in its room.

The fuel used in the gas producers is the brown coal of Leoben. No blowing machinery being required, the boiler fuel consumed in the

SECTION XIV.—SUBSTANCES TAKEN UP IN BLAST FURNACE. 427

Bessemer process is not needed in that of the open-hearth. Making allowance for this something like 5 or 6 cwts. more coal is burnt in the latter operation than in the former; which, together with additional labour, places it at a disadvantage, in these items, of about 10s. per ton of ingots as compared with the converter. The open-hearth process, as conducted at Gratz and elsewhere, has on the other hand the advantage of the waste of iron being only about one-third of that incurred in blowing pig iron. This, together with the relative local values of old iron rails and pig iron, afford so great a margin in favour of the Siemens-Martin mode of manufacture that the Southern Railway Company found it greatly in their interests to remove their Bessemer plant and to erect Siemens' furnaces in its room.

A similar plan of using old iron rails of British make was adopted in South Wales and probably elsewhere. In this case however the old material contained so much phosphorus that when mixed with hematite pig the resulting rails often contained ·2 per cent. or even more of this substance. It is well known that finished rails with this quantity of phosphorus are apt to be brittle. Whatever the cause may be, the use of old iron rails of native manufacture, as a material for producing steel rails, has, I believe, been entirely discontinued in the United Kingdom.

An ordinary charge consisted of 60 per cent. of old Welsh rails and 40 per cent. of hematite pig. The former contained ·33 to ·40 per cent. of phosphorus and the latter ·07. The average of this mixture was therefore about ·268 per cent. phosphorus; and the rails were given me as containing actually from ·20 to ·26 per cent. of this substance and ·20 of carbon. If to these ·10 for silicon and ·075 for sulphur are added we have a total of ·615 per cent. for the four metalloids as against ·730 per cent. and ·664 per cent. given at pages 414 and 415 for Bessemer steel. The manganese in the steel made from old iron rails and hematite pig was only ·547 per cent. against 1·178 and 1·201 in the Bessemer steel.

With regard to the strength of a rail containing such a quantity of phosphorus as that just named, a common test to which an iron rail is exposed is that of a weight of 1,800 lbs. falling 4 feet, but a rail made from Cleveland pig having the composition given below was subjected to three blows from a weight of 2,254 lbs. with a fall of 2 feet, 3 feet, and 4 feet respectively, without being broken. The total deflection was ⅜ inch. The rail contained:—

Carbon	·175
Silicon	trace
Sulphur...	trace
Phosphorus	·216
	·391
Manganese	·150
Iron by difference	99·459
	100·000

Reference has already been made in the present section to the fact that phosphorus is always to be found in an iron rail to an extent which would be fatal to the use of a steel rail. Of course the difference of the conditions between the two cases must not be lost sight of. As may be seen at pages 414 and 415, Bessemer steel rails contain of manganese and the four metalloids (C, Si, S and P) nearly 2 per cent. of their weight; whereas an iron rail rarely contains one-half this quantity, and of this a certain proportion is to be found only mechanically blended with the iron, in the form of cinder, and not therefore affecting its quality in the same way as if it were chemically combined. The following analyses exhibit the composition of some iron rails made from Cleveland iron, and received by the North-Eastern Railway Company:—

Carbon.	Silicon.	Sulphur.	Phosphorus.	Manganese.	Iron.
·10	·16	·07	·67	·07	98·93
·08	·18	·04	·51	—	99·19
·07	·22	·08	·45	·11	99·07
·07	·16	·05	·43	—	99·29
·17	·20	·03	·42	·05	99·13
·06	·15	·05	·40	·08	99·26
·08	·18	·04	·40	—	99·30
·10	·20	·04	·31	—	99·35
·06	·17	·03	·29	—	99·45
·06	·19	·05	·28	—	99·42
·08	·10	·02	·23	—	99·57
Average 100 = ·084	·174	·045	·400	·028	99·269

SECTION XIV.—SUBSTANCES TAKEN UP IN BLAST FURNACE. 429

These considerations render it not improbable that in the future the Siemens furnace may form a much more important adjunct to every large steel works for remelting than has been the case hitherto. Up to this time, no quantity of old steel rails worthy of notice has been returned to the hands of the manufacturer; but in a few years there will be a great change in this respect, because then the railway companies will be lifting annually a great weight of worn-out material from their permanent way. It is very possible by that time that the old rails thus thrown on to the market may, to some extent, be simply reheated in an ordinary reverberatory furnace and rolled into steel, to be used for purposes for which malleable iron is now frequently employed. When the market is more plentifully supplied with old steel rails than has hitherto obtained, it may well happen that at the price they will command this old material will be remelted in the Siemens furnace to be rerolled into railway bars. By that time also it is pretty certain that the use of mild steel or "ingot iron" will have largely superseded the use of "welded iron;" and the demand for the former will be conveniently supplied by fusing old steel rails with a small quantity of ore on the plan first practised with pig iron by Sir W. Siemens, at Landore.

It may be interesting to consider the effect on the market between returning iron rails having an average life of ten years and steel rails having a life of twenty years.

If we assume new rails to be exclusively manufactured from old rails, of which one-third for tops and bottoms of the pile consists of No. 2 bars, about $22\frac{1}{2}$ to 23 cwts. would be consumed for each ton of new rails produced. If in addition to this the iron rail lost $7\frac{1}{2}$ per cent. of its weight while in use, we may consider each ton of new rail to be represented by $24\frac{1}{4}$ to $24\frac{3}{4}$ cwts. of rail as originally laid down —or an approximate waste of $18\frac{1}{2}$ per cent., spread over ten years, equal therefore to 1·85 per cent. per annum.

In the case of a steel rail, we may regard the remelting to be accompanied by a loss of 3 per cent., and the subsequent reheating by a further loss of 5 per cent. To this has to be added the waste from abrasion, which we have assumed to amount to 20 per cent. With these factors as a basis of computation, I estimate the loss upon the original weight of rail laid down at about 25 per cent., equivalent, over twenty years, to $1\frac{1}{4}$ per cent. per annum.

There is however this difference to the owner of a rolling-mill, that whereas, in the case of the iron rails, he would be called upon to re-manufacture 10 per cent. of the rails in each year, the steel-mill owner will only be required to work up 5 per cent. of the rails in use during the same period.

My friend Mr. T. E. Harrison, the experienced Engineer in Chief of the North Eastern Railway, has estimated roughly for me that the weight of rails laid down in the United Kingdom is 7,731,000 tons, this means 773,100 tons to reroll per annum were iron rails exclusively in use, as against 386,550 tons were our lines laid with steel alone.

In this computation it is assumed that old rails in each case are employed exclusively for the manufacture of new rails. This, particularly in the case of iron rails, would not happen; but it is unnecessary for our present purpose to consider the question with greater minuteness.

If we examine the composition of the materials employed in the Siemens-Martin mode of manufacturing steel rails we shall see that something more is accomplished than mere fusion, and that all the metalloids have been more or less reduced in point of quantity. Approximately the extent to which this is done is as follows:—

	C.	Si.	S.	P.
60 parts of iron rails contain...	·050	·104	·270	·240
40 ,, hematite pig ,, ...	1·600	·100	·001	·006
	1·650	·204	·271	·246
100 ,, steel rails ,, ...	·230	·100	·075	·230
Leaving to be oxidised in the furnace ..	1·420	·104	·196	·016 = total 1·736

The conditions attending the treatment of the molten mass in the open hearth render the oxidation of the metalloids a very slow process. The metal is lying for the most part undisturbed in the furnace, hence the combustion takes place on the surface alone, which prevents more than two charges being completed in the 24 hours, giving from 16 to 20 tons of ingots, according to the size of the furnace.

The existence in pig iron, of 6 to 7 per cent. of powerful reducing agents, like carbon and silicon, naturally suggests the idea of employing them to reduce oxide of iron, instead of being uselessly consumed as they are in the Siemens-Martin process. This sound principle, which was the subject of a patent granted to Samuel Lucas in 1791

SECTION XIV.—SUBSTANCES TAKEN UP IN BLAST FURNACE.

for making crucible steel, has been most successfully introduced and brought to great perfection, at Sir W. Siemens' works at Landore, near Swansea. The pig iron is melted in one of his regenerative furnaces; and, when at a full heat, iron ore as free from phosphorus and sulphur as possible is thrown into the bath of metal. Violent ebullition ensues; the carbon escapes as carbonic oxide, and the silicon as silica passes into the slag, together with a part of the sulphur and it may be a minute quantity of the phosphorus.

The result of the reäction of the metalloids on the ore is such that it is not uncommon to obtain a weight of steel nearly equal to that of the pig iron used. In one case, of which I have the particulars, the following numbers represent the nature of the action as it takes place in the furnace:—

	Tons.
Actual weight of iron (Fe) in the pig iron, including spiegel, ferromanganese, etc., introduced into the furnace	331·92
Actual weight of iron (Fe) estimated as existing in the steel ...	343·53
Iron reduced from the ore	11·61

	Tons.
The ore used was 58·14 tons containing 55 per cent. of iron ...	31·97
Iron in the pig, etc., as above	331·92
Total	363·89

The actual weight of iron (Fe) delivered to the process in the pig iron, spiegel, scrap, and ore together being 363·89 tons, and that contained in the steel being 343·53 tons; the loss of iron (Fe) (20·36 tons) is equal to 5·53 per cent. The pig iron, spiegel, and scrap weighed 355·7 tons, so that reckoned upon the steel received (349·2 tons) the waste amounts to about 2 per cent.

The slag obtained is highly siliceous, which forbids any appreciable amount of phosphorus being removed from the iron during the process. The following is an analysis of a specimen:—

Silica	69·22
Protoxide of iron	20·57 ⎫ = iron 18·40
Peroxide ,, ...	3·43 ⎭
Protoxide of manganese	3·06
Lime	2·30
Phosphorus	trace.
	98·58

I have always been impressed with the advantages in point of economy of which this so-called ore process of Landore seems to be susceptible, and the statistics of the trade would indicate its growing importance. Contrasting its production with that of the Bessemer process, we have the following figures :—

Years.	Make of Bessemer Steel.	Make of Open Hearth Steel.[1]
1876	400,000	128,000
1877	508,400	137,000
1878	633,733	175,500
1879	519,718	175,000
1880	739,910	251,000
1881	1,023,740	338,000
1882	1,235,785	436,000
Increase of 1882 over 1876	209 per cent.	240 per cent.

So far as my information enables me to judge, 7s. 6d. per ton will cover in Great Britain, the expense of the additional labour and fuel in the open-hearth process as compared with the Bessemer, as both are carried on at the present moment. In the matter of iron materials employed, taking the average of hematite pig iron and spiegel as used in the usual proportions at 54s. and ore at 16s., we have the following as the cost per 100 tons of steel :—

	£ s. d.	£ s. d.	
112 of pig and spiegel at 54s.	302 8 0	= 3 0 5·76	per ton of steel.
102½ „ „ 54s.	276 15 0		
Add 17 of ore „ 16s.	13 12 0	2 18 0·84	„
Total	£290 7 0	0 2 4·92	in favour of the ore process.

This calculation shows a difference (7s. 6d. — 2s. 5·76d.) of about 5s. in favour of the Bessemer process, but the estimate is based on the Siemens furnace making two charges in the 24 hours. Reference has been already made to the tranquil state of the bath of metal lying on the hearth, being not very favourable for the oxidation of the metalloids. One mode of endeavouring to meet this difficulty consists in blowing steam through the iron by means of a pipe plunged below the surface. I was afforded, by the courtesy of the owners of the Horse

---

[1] Report of the Iron Trade Association, 1881 and 1882. The figures include the Siemens-Martin steel, but in it ore is now generally added.

SECTION XIV.—SUBSTANCES TAKEN UP IN BLAST FURNACE. 433

Nail Works in London, an opportunity of examining the operation. The pig iron treated contained 4·53 per cent. of carbon and 1·02 of silicon. By the time the iron was melted, and after some ore had been added, the carbon stood at 2·20 and the silicon at ·037.

The following table shows the rate of the disappearance of these elements, after commencing to inject steam into the bath, and after a further quantity of ore has been added:—

	Metal contained.	
	Carbon.	Silicon.
After 13 minutes blowing in steam	2·02	·037
28 ,,	2·19	·007
43 ,,	2·11	·007
58 ,,	1·97	·007
73 ,,	1·76	·007
88 ,,	1·66	·007
Finished steel contained after addition of ferromanganese	·065	·021

A certain amount of chemical action is expected to be exercised by the decomposition of the steam, but against this has to be set the cooling effect its introduction is sure to produce, and in point of fact I did not learn that the operation as a whole was materially shortened.

A more promising modification of the Siemens furnace for the production of steel is that known as the Pernot system. In it the hearth is set on the slope and is made to revolve; by means of this movement fresh surfaces are continually exposed to the action of the ore and the atmospheric oxygen. I believe that it is to the enterprise of the managers of St. Chamond near St. Etienne that we are indebted for what progress has been made in this new development of the open-hearth process. At this establishment furnaces capable of holding 8 tons, and others equal to 25 tons, are employed; and appeared on the occasion of my visit to be working very satisfactorily. Local circumstances formed the reason assigned for using pig iron and scrap steel without any admixture of ore; so that it was at that time an adaptation of the Pernot furnace to the Siemens-Martin process which was in progress. My late friend, Alexander L. Holley, carefully examined the work as it was being performed at the St. Chamond works, and in a paper he describes its action.[1]

The average duty in 1877 over four weeks was 3·27 charges in twenty-four hours instead of two charges as in the case of the simple

[1] Trans. American Institute of Mining Engineers, 1879, Vol. VII., p. 241.

regenerative furnace of Siemens. In place of using 12 cwts. of coal per ton of ingots, 8 to 9 cwts. sufficed. In the year 1879 the average number of charges was 4·09 per twenty-four hours and the coal was less than 8 cwts. to the ton of ingots; in every case the iron and scrap steel being charged cold into the preheating chamber.

Opinions are very much divided respecting the quality of the product of the open-hearth process, as compared with that of the Bessemer converter. It is alleged that the lengthened period over which the former process is extended is greatly in favour of commanding an uniformity of composition in the steel entirely beyond the reach of the latter. I have endeavoured to ascertain the grounds upon which this claim rests, by a reference to the analyses made in the North-Eastern Railway and Clarence laboratories; these are given below:—

Open-hearth Steel.	Carbon.	Silicon.	Sulphur.	Phosphorus.	Manganese.	Iron by diff.
Railway axle ...	·46	·08	·04	·07	1·11	98·24
,, ...	·52	·08	·10	·07	·83	98·40
Boiler plate ...	·26	·05	·00	·06	·47	99·16
Steel bar ...	·26	·02	·06	·05	·79	98·82
Boiler plate ...	·205	·011	·081	·049	·569	99·085
,, ...	·190	·013	·076	·044	·326	99·351
,, ...	·226	·016	·083	·042	·610	99·023
,, ...	·204	·035	·050	·049	·417	99·245
,, ...	·167	·031	·066	·053	·235	99·448
,, ...	·160	trace	·050	·040	·320	99·430
Average ...	·265	·033	·061	·053	·567	99·021

As a matter of comparison, the analyses of some of the best West Yorkshire brands of malleable iron are added:—

Malleable Iron.	Carbon.	Silicon.	Sulphur.	Phosphorus.	Manganese.	Iron.
Railway wheel tyre ...	·10	·17	·01	·09	·08	99·55
Angle iron	·11	·14	·02	·09	·06	99·58
Railway axle	·12	·15	·01	·09	·00	99·63
,, ...	·14	·25	·02	·09	·01	99·49
,, ...	·13	·14	·00	·16	·00	99·57
,, ...	·13	·14	·00	·10	·00	99·63
Boiler plate	·08	·12	·02	·08	·03	99·67
Crank axle	·05	·10	·01	·03	·06	99·75
Bar iron	·015	·157	·005	·220	—	99·603
Boiler plate	·044	·121	·008	·195	—	99·632
Average	·092	·149	·010	·115	·024	99·610

SECTION XIV.—SUBSTANCES TAKEN UP IN BLAST FURNACE.

The composition of steel and iron differs too much to enable us to draw any satisfactory comparison between the two, so as to infer therefrom any superiority of quality. Thus steel from its nature contains more carbon, and from circumstances connected with its manufacture is richer in manganese. Of the three elements (Si, S, and P), which are known to weaken both forms of the metal, steel in the examples given enjoys some advantage. The averages stand as follows:—

	Open-hearth Steel.	Iron.	Bessemer Steel.[1]	
Silicon, sulphur and phosphorus...	·147	·274	·278	·214
Carbon	·265	·092	·452	·450
Manganese	·567	·024	1·178	1·201
Iron	99·021	99·610	98·092	98·135
	100·	100·	100·	100·

Again it may be pointed out that in the puddling process, of these hurtful ingredients which are sought to be expelled, all but ·274 per cent. of the total weight has disappeared in the case of the iron. In this respect the open-hearth steel gives more favourable results than the Bessemer steel, and this with a smaller proportion of manganese. Whether this difference in the content of the hurtful ingredients suffices to account for the alleged superiority of open-hearth steel, or whether it is due to some difference in the quantity of gas occluded by the steel in the open-hearth, I am unable to say.

In the few sentences respecting the comparative excellence of Bessemer and open-hearth steel, I am not venturing upon any opinion of my own, my intention being confined to an endeavour to account for that which is asserted by others, who have had more ample opportunity than myself of judging of the respective merits of both. In cases where the steel has to be exposed to great strain, either in preparing it for its future application or when so applied, engineers not infrequently stipulate for open-hearth steel being employed. At the same time I have personally met with innumerable instances in which Bessemer steel, in both the respects referred to, left nothing to be desired.

Until the time of Sir Henry Bessemer's invention, steel was known in commerce in comparatively very limited quantities; and a short time

---

[1] Examples given, pp. 414 and 415.

anterior to that period its use was chiefly confined to those purposes, such as engineers' tools and cutlery, for which high prices could be paid without inconvenience to the consumer. Although pig iron contained far more than the proportion of carbon required to constitute steel, it was found necessary to incur the loss due to burning off the whole of this element, in order to free the product from those other substances which affected prejudicially the quality of an article required for the purposes above mentioned. This was so essential to success that not only were the purest ores selected for the purpose of manufacturing the pig iron, but the malleable iron was obtained in the low charcoal hearths mentioned in Section XII., so as to eliminate the silicon, sulphur, and phosphorus as completely as possible from the bar. The malleable iron thus obtained had a portion of carbon returned to it by the well-known process of cementation, and the resulting steel was melted in crucibles. Little is known of the early history of the cementation process and nothing of the date of its introduction. It still continues to be the source of the finer kinds of steel, although it is stated that a good deal of the crucible steel made at the present day, instead of being Swedish or Russian iron converted by cementation, consists chiefly of Bessemer steel melted in crucibles.

It is believed that the most ancient method known for making steel is an invention of Hindoo origin. In it fragments of wrought iron, made in the rudest way direct from the ore, by means of charcoal, was melted in pots in the presence of vegetable matter. The resulting steel is known in the market as "wootz," and containing as it does nearly 1·75 per cent. of carbon, possesses great hardness.

On referring to the diagrams given at page 388, it will be seen that when pig iron is exposed to the action of the refinery, or of the puddling furnace, silicon and phosphorus are separated sooner than the carbon. It is true that the two first named elements are never entirely removed by either of the processes referred to: but, inasmuch as pig iron, in passing to the condition of malleable iron, must pass through what may be considered an intermediate condition, viz.: that of steel, the process only requires to be stopped at that point to obtain this form of the metal.

Many years ago I witnessed the manufacture of steel in Rhenish Prussia as performed in a small refining fire. The hearth was about

SECTION XIV.—SUBSTANCES TAKEN UP IN BLAST FURNACE. 437

2 feet square, the fuel used was a mixture of charcoal with a little coke, and the material treated was the pig iron and spiegel-eisen of the country. To this was added a portion of slag, obtained from previous operations, together with a certain quantity of iron ore. The iron as it melted, meeting with the blast and fused cinder, became decarburized, and acquired a pasty state; when fresh pig was added according to the judgment of the workman. As may be imagined, the process was somewhat uncertain in its results, and depended greatly on the skill of the finer. The operation lasted about eight hours, and the mass of steel was withdrawn in the form of a flat cake, which weighed about 6 cwts. It was flattened under a hammer, and broken into pieces of about 1 cwt. each. These were heated by the spare heat of the succeeding charge, and drawn out under the hammer into the form of bars. The quantity produced per week from each fire was about 3 tons, with a consumption of something like 20 to 22 cwts. of charcoal for each ton. For every 100 parts of steel bars obtained, about 140 of pig iron was consumed: it was therefore, it will be perceived, a wasteful process both in the matter of fuel and metal. The product was sorted according to quality, the proportions averaging about two-thirds of high quality, and the remainder of an inferior description. The very rudeness of the process, giving rise to a large production of cinder, tended to remove the impurities from the pig employed, which moreover was originally of excellent quality. This and the presence of the manganese in the spiegel enabled the German manufacturers of Siegerland to furnish a very high class quality of steel in the primitive appliances referred to.

We have thus seen how the ancient hearth, which in succession had been employed for obtaining malleable iron from the ore and malleable iron from pig iron, was also used for producing steel from pig iron. The puddling furnace has, in its turn, been enlisted in the same sequence of services. Invented originally by Cort for the conversion of pig into malleable iron, it was sought by Clay some 50 to 60 years afterwards to employ it as a means of avoiding the use of the blast furnace by reducing the ore at once into wrought iron. This was, like all the direct processes hitherto tried, a commercial failure. It was otherwise with an attempt to apply the same principle to the puddling furnace, which, for many years, had been in successful operation with the finery as shortly described above, the idea being simply to stop the fining process before complete (or rather all but complete) decarburization had taken place. Although this mode of producing steel had been tried in Austria and Germany so early as 1835, I have no recollection of having ever heard of it before the year of the first International Exhibition in London (1851). The practise, after the iron became fluid in the puddling furnace, was to lower the damper, which enabled the workman thoroughly to incorporate the iron and cinder, rendered less liquid by the cooling, so as thereby to remove the silicon and phosphorus more effectually than was usually accomplished by the ordinary method of puddling. It seems to me possible that this reduction of the temperature of the furnace might also act as a means of retaining the carbon, and therefore of preserving the fluidity of the metal in the manner mentioned when describing the process of purifying pig iron (p. 401).

The late Mr. Parry of Ebbw Vale gives the composition of puddled steel and that of the grey pig iron used in the manufacture. They are as follows:—

	Pig Iron.	Puddled Steel.
Carbon	2·680	·501
Silicon	2·212	·106
Sulphur	·125	·002
Phosphorus	·426	·096
Manganese	1·230	·144
Iron by difference	93·327	99·151
	100·	100·

The average composition of the cinder as given by Mr. Parry was as follows:—

Silica	24·75	
Alumina	4·55	
Lime	·30	
Protoxide of iron	60·95	= Fe 47·40
,, manganese	9·45	= Mn 7·35
	100·	

The loss on the pig iron used was about 10 per cent., and the coal required in the process was approximately 30 cwts. per ton of steel; the produce of each furnace being about 1¼ ton in the 12 hours.

The quantity of puddled steel made was, at one time, not inconsiderable; and but for the introduction of the Bessemer process might have become a very important branch of industry. Such steel as that of the composition just named, if melted in a Siemens furnace, would have afforded an excellent material for the top of a rail, and in former years might have been profitably employed for this purpose.

Reference was made in Section II. (p. 10), to the same furnace having been employed in comparatively recent times, as a means of obtaining wrought iron, pig iron, and also steel. For this threefold purpose the furnaces probably occupied an intermediate position in point of size between the low hearth and the Stückofen or Blauofen. Dr. Percy, in his admirable work of reference,[1] mentions that the process of reduction can be so conducted in the low hearth, the most ancient form of furnace known to us, as to produce *fer aciereux* or steely iron. By a still further modification of the operation, more carbon was made to unite with the iron and steel was then obtained. The quality, it is true, was irregular, but in the absence of better methods it is not improbable that the Catalan or analogous fire, may have been used in very remote ages for the production of the material for tools required in the construction of the oldest monuments of stone with which we are acquainted.

POSTSCRIPT.

In these pages mention was made of the fact that the heat in an ingot of steel, as drawn from the mould into which it had been poured from the converter, would, if equally distributed through the mass, enable it to be converted into a rail. Some few years ago Mr. Alexander Wilson, of the Dronfield Steel Works, near Sheffield, informed me that he had succeeded, by transferring the hot ingot quickly into a furnace, in delivering it to the mill at an expenditure of only half a cwt. of coal per ton of rails made: indeed he had, if I remember rightly, rolled an ingot into a rail direct from the ingot mould. The consumption of coal being reduced to so small an amount, it was found more profitable to continue to use the furnace for reheating, after which the rail was rolled off without further preparation.

[1] Iron and Steel, p. 764.

At the Vienna meeting of the Iron and Steel Institute, held in 1882, Mr. John Gjers describes a plan of his own for avoiding the reheating of the ingots. A series of holes, or "soaking pits" as he calls them, sunk in the ground, are lined with fire brick. Into these the hot ingots are introduced as rapidly as they can be got from the ingot moulds and covered with an iron plate. The brickwork is speedily raised to a good red heat by the hot steel, so that the ingots after half an hour's exposure in the pits, become equally hot throughout and are then taken direct to the rolls without being passed into any furnace. Owing to the state of the machinery at the Darlington and West Cumberland Works, where the plan has been in operation for some time, it has not been found convenient to finish the cogged ingot into a rail without submitting it to a wash heat; but it is confidently expected to accomplish this when the rolls and other appliances are provided of suitable strength.

SECTION XV.

STATISTICAL.

The fourteen preceding Sections in these pages have dealt exclusively with some of the theoretical and practical questions, which chiefly interest those who are actually engaged in the manufacture of iron.

It is proposed in the remaining divisions to consider certain facts and figures more calculated to interest the trader and economist than the manufacturer, yet of sufficient importance to be included in a work of a more purely technical character.

As stated, in the opening remarks of this volume, it was a desire expressed by certain friends to receive some report on the manufacture of iron in foreign countries which led to its compilation. It was considered by them that the discharge of my official duties at the International Exhibitions of Paris and Philadelphia would have placed ample materials at my command for the preparation of such a report as that referred to. It is however open to question whether such displays as those just mentioned afford the true means of arriving at very correct conclusions respecting the condition of this branch of industry in the different exhibiting nations. The necessary information for such a result can only be obtained by a more minute examination than that afforded by a mere inspection of samples of workmanship exposed to view upon such occasions. Although the work when specially prepared for the purpose may not represent an average make of the exhibitor, the collection of specimens as were to be found at Paris and Philadelphia are most useful to those who are desirous of studying the position of a manufacture like that of iron. Articles of the metal as well as of steel were shown in both cities in which mere magnitude proved the possession of the mechanical means of dealing with such masses as are only to be found in establishments of the highest importance and development. Various modes of testing the quality are adopted which, from the expense and labour attending them, are only practised when it is certain that the results will be carefully inspected by a numerous class of competent judges. Besides these modes of affording instruction to those interested in acquiring it, questions bearing on the purely scientific aspects of this branch of metallurgy were, on both occasions, largely illustrated by a well-chosen series of examples.

So far as the word "international" is concerned, as applied to an exhibition of iron, in its various forms, the title is perhaps somewhat exaggerated. When prominence in manufacturing skill has to be manifested by the display of an armour-plate of 40 tons, or a cast-steel ingot of 100 tons, as was done at Paris, firms at a distance naturally shrink from encountering the expense of sending such ponderous masses far away from home. Under such circumstances the iron section, in point of extent and completeness, is chiefly an exhibition of the products of the country in which the gathering is held. In my Report to Her Majesty's Government on the International Exhibition at Philadelphia, I remarked on the poverty which was so conspicuous in the number and character of the specimens of iron sent from Great Britain, notwithstanding the fact of its being the largest producer of the metal in the world. In like manner, in an excellent report to the Government of the United States on the Paris Exhibition, my friend the Hon. Dan. J. Morrell laments the absence of American products from the vast buildings in the Champ de Mars.

In one respect it must be admitted that the French stood out, if not foremost, certainly very conspicuously in the world as manufacturers of iron; viz., in the scientific completeness with which they presented the information intended for the use of visitors. This completeness was not limited to the mere objects of manufacture themselves, but embraced the locality, the natural conditions and the method of extraction of the raw materials, models and drawings of the machinery employed, as well as minute descriptions of the processes, and detailed accounts of the condition of the workmen engaged in conducting them. No doubt in these respects this nation was closely followed at Paris, by the ironmasters of Sweden and Belgium, and by the establishments of the Austrian Government; but on the whole I question whether anyone will dispute the place of honour, at all events on that occasion, with the French.

Descriptions of what was to be seen in the way of iron metallurgy in Philadelphia in 1876, and in Paris in 1878, have been so amply supplied by my friends Messrs. Åkerman, Morrell, Tunner, and Wedding, that I do not propose to extend my labours by referring in detail to my own lengthened experience at either of these exhibitions, but rather to make that experience useful in the preparation of what follows in the succeeding Sections of this work.

The progress which Great Britain has made in the means of economising the production of iron and steel has made itself apparent in the enormous increase of its annual make. One hundred and forty years ago, *i.e.*, in 1740, the make of pig-iron in the United Kingdom, as we have seen, was under 20,000 tons—in 1880 it was $7\frac{1}{2}$ million tons. With the means at the disposal of the iron manufacturers in 1740 it would have been physically impossible to have achieved this result. To effect what has been done charcoal has had to give place to coal, a change first successfully accomplished by Abraham Darby; James Watt had to place the steam-engine at our disposal, Cort had to teach the puddling process, Neilson the use of hot air in the blast furnace, and Bessemer the pneumatic mode of making steel.

These discoveries, the offspring of British invention, were rapidly adopted abroad; and if the United Kingdom is still far ahead in the quantity of iron produced, this has not prevented alarm being felt lest we are at length being overtaken in the race. As an instance, it was recently stated in the public prints, that Belgium was outstripping us in economy of manufacture, because Cleveland pig iron was being converted in Belgian works into wrought iron girders, and then returned to London and sold at a cheaper rate than the rolling mills in England could supply them. Now Cleveland pig iron, it is true, is imported into Belgium from Middlesbrough for *foundry* purposes; but the girders sent to England are made of an entirely different description of pig iron, smelted on the spot in Belgian furnaces. Nevertheless there is no doubt that in late years our foreign trade, relatively speaking, has been greatly encroached upon. This will best be seen by an examination of the figures contained in the subjoined table,[1] which shows roughly the proportion of pig iron made in Great Britain and elsewhere, the figures representing thousands of tons:—

	1871.	1872.	1873.	1874.	1875.	1876.	1877.	1878.	1879.	1880.	1881.	1882.
Great Britain ..	6,627	6,741	6,566	5,991	6,365	6,555	6,608	6,300	6,009	7,721	8,377	8,493
Other countries ..	5,399	7,165	7,604	7,060	6,776	6,374	6,822	7,255	7,759	9,764	10,589	11,582[2]
Total	12,026	13,906	14,170	13,051	13,141	12,929	13,430	13,555	13,768	17,485	18,966	20,075
Percentage of whole made in Gt. Britain	$55\frac{1}{4}$	$48\frac{1}{2}$	$46\frac{1}{3}$	46	$48\frac{1}{2}$	$50\frac{3}{4}$	$49\frac{1}{4}$	$46\frac{1}{2}$	$44\frac{1}{3}$	$44\frac{1}{4}$	$44\frac{1}{4}$	$42\frac{1}{3}$
Do. in other countries	$44\frac{3}{4}$	$51\frac{1}{2}$	$53\frac{2}{3}$	54	$51\frac{1}{2}$	$49\frac{1}{4}$	$50\frac{3}{4}$	$53\frac{1}{2}$	$55\frac{2}{3}$	$55\frac{3}{4}$	$55\frac{3}{4}$	$57\frac{2}{3}$
	100	100	100	100	100	100	100	100	100	100	100	100

Thus it will be seen that twelve years ago (1871) we produced about (6,627,000—5,399,000) 1,228,000 tons more than all the rest of the world put together, and we end in 1882 by producing (11,582,000—8,493,000) 3,089,000 tons less than the joint produce of other nations.

Approximately our exports of all kinds of iron have been as follows (in thousands of tons):—

	1871.	1872.	1873.	1874.	1875.	1876.	1877.	1878.	1879.	1880.	1881.	1882.
As pig	1,061	1,331	1,142	774	947	910	881	924	1,223	1,632	1,482	1,758
Other kinds reduced to pig[3]	2,635	2,563	2,268	2,141	1,888	1,642	1,831	1,715	2,075	2,693	2,922	3,240
	3,696	3,894	3,410	2,915	2,835	2,552	2,712	2,639	3,298	4,325	4,404	4,998
Left for home consumption[4]	2,931	2,847	3,156	3,076	3,530	4,003	3,896	3,661	2,711	3,196	4,173	3,495
Total make of United Kingdom	6,627	6,741	6,566	5,991	6,365	6,555	6,608	6,300	6,009	7,521	8,577	8,493

[1] Report of the British Iron Trade Association, 1882, page 138. Corrected, when necessary, approximately to tons of 2,240 lbs. avoirdupois.

[2] In the figures setting forth the quantities produced in "other countries" certain corrections have had to be made in the numbers given in the Report of the Iron Trade Association. It is not until 1881 that the production of the German Empire is given as a whole when there is a sudden increase of about one million tons in the column (p. 138) previously given as Prussia. On referring to Pechar's "Coal and Iron in all Countries of the World" I find that the quantities exceed those in the Report referred to, beginning with an excess of 194,000 tons in 1871 and ending with 477,000 tons in 1876. Pechar ends his account with 1876, so that between that year and the end of 1880 I am without any data. I have therefore divided equally the difference among the four years which intervene. In Austro-Hungary there is a blank for 1878 which is filled up with 400,000 tons. A similar omission for 1882 occurs for Sweden and Austro-Hungary together for which 1,000,000 tons have been inserted.

[3] Obtained by adding 25 per cent. for waste to the weights exported as given in Report of Iron Trade Association.

[4] These numbers disregard any alterations in the quantity of stocks of iron on hand.

SECTION XV.—STATISTICAL. 445

The same statistical tables of the British iron trade, upon which the last series of calculations have been based, give the following as the exports of pig iron alone from different countries during the ten years ending 1880 (in thousands of tons) :—

	1871.	1872.	1873.	1874.	1875.	1876.	1877.	1878.	1879.	1880.[1]
From Great Britain as above	1,061	1,331	1,142	774	947	910	881	924	1,223	1,632
From other countries ..	217	257	305	355	455	389	451	469	517	512
Total	1,278	1,588	1,447	1,129	1,402	1,299	1,332	1,393	1,740	2,144

Considered as percentages, we have the following numbers:—

	1871.	1872.	1873.	1874.	1875.	1876.	1877.	1878.	1879.	1880.
Great Britain ..	$83\frac{1}{4}$	$83\frac{3}{4}$	79	$68\frac{1}{2}$	$67\frac{1}{2}$	70	66	$66\frac{1}{4}$	70	76
Other countries ..	$16\frac{3}{4}$	$16\frac{1}{4}$	21	$31\frac{1}{2}$	$32\frac{1}{2}$	30	34	$33\frac{3}{4}$	30	24
	100	100	100	100	100	100	100	100[1]	100[1]	100[2]

These tables clearly show that, although Great Britain in the transactions of the ten years has largely increased her exports of pig iron, yet foreign countries, chiefly Germany it may be observed, have sent abroad more than their former share of the additional quantities required by importing nations. Thus while the increase of Great Britain amounts to only about 56 per cent. on the exports of 1871, that of other countries is no less than 136 per cent. Of course due allowance must be made for the circumstance that the percentage addition, in the case of other nations, is on an initially small quantity; the actual increases being 571,000 tons for Great Britain against 295,000 tons collectively for other countries.

Practically the increased production of iron, during the ten years under consideration, has arisen in four countries as follows (in thousands of tons) :—

	Great Britain.	United States, Net Tons 2,000 lbs.	Prussia.	France.		TOTAL.
Make, 1880 ...	7,721	4,295	1,950	1,733	...	15,699
,, 1871 ...	6,627	1,911	1,297	859	...	10,694
Increase ...	1,094	2,384	653	874		5,005
Rate of increase per cent. ...	$16\frac{1}{2}$	$124\frac{3}{4}$	$50\frac{1}{3}$	102	Average...	$46\frac{3}{4}$

[1] The last three years contain some blanks which have been filled up with assumed numbers.

[2] The last three years are given partly by estimate owing to no returns being inserted for some of the countries.

By far the most remarkable among these increases is that of the United States; and of this more than one-half, or a total of 1,225,000 tons, took place in the last year of the period referred to.

The make of pig iron in the United States has further been maintained during the two succeeding years, the weight produced being:—

	Net Tons.
For 1881	4,641,676
For 1882	5,178,121

Thus while in 1871 the make of the American Union was only 26 per cent. of that of Great Britain, it has grown for the two years just recorded to 52½ per cent. of that of this country, an increase unparalleled in the history of the iron trade.

With the figures just given before us, it seems quite idle to shut our eyes to the fact that the supremacy we have long held as an iron making nation is threatened with a permanent abatement, at all events in point of its relative extent. No one who has examined the ironworks in Europe or America can deny that they are managed by men who in science and skill are our equals. The question therefore resolves itself into one of the relative natural advantages possessed by Great Britain and her competitors; that is to say into the relative abundance and facility of extraction of the necessary minerals, and the relative ease with which they can be brought together and the product conveyed to the market.

In endeavouring to define the circumstances which have affected the development of the iron trade in foreign countries, it is impossible entirely to ignore the consequences of foreign legislation. In considering the effect of protective duties, I am not going to regard it as within my province to question the propriety of a country adopting any line of policy which it may consider best calculated to promote its own advancement. Thirty-six years ago we embraced, without consultation with other powers, what is known as the Principles of Free Trade. We did this in our own interests, believing, it is true, that we were at the same time promoting those of the world at large. We continue in this course; because we rely on the soundness of the policy itself; and we must leave those with whom we deal to be their own judges of what suits them best, whether we agree with them or not. The question therefore is only referred to in these pages, because any circumstance which favours the extension of the iron trade in a foreign country necessarily interferes with our transactions with that country in this particular industry. We have moreover to enquire whether such extension is of a nature to lead us to expect future competition, in supplying those parts of the earth which, from a variety of circumstances, cannot, at all events for many years to come, or in some cases probably for ever, supply themselves cheaply with home made iron. To determine this, it seems absolutely indispensable that we should distinguish between those circumstances, such as the cost of labour and protective duties, which are more or less of an artificial character, and those, such as the accessibility of coal or ore, which are of a purely natural kind.

The matters involved in such an enquiry as that just referred to, possess an interest to the British public at large, as well as to the iron manufacturers of these islands. The great change which has taken place in late years in the price of labour in the United Kingdom, renders it desirable that we should consider whether a similar alteration obtains among those nations we have to meet as competitors in the markets of the world; and if so, what is its cause, its nature, and its extent. This important question will receive some attention in the section which follows.

An opinion is not unfrequently expressed that the duties levied abroad on foreign iron have greatly restricted the exportation of this metal from Great Britain to such countries. There is no doubt that this is perfectly true: as an example, except in cases of extraordinary emergencies, the trade in iron between this country and the United States may be said to have been almost annihilated; and this in the first instance by the prohibitive nature of the American tariff. It is however quite another thing to conclude that if all duties were abolished all, or indeed many, foreign iron works would be closed; and that on Great Britain would then rest the responsibility of supplying almost the whole world with iron. The cheaply wrought Black Band ore of Scotland, often brought out of the same pit which furnished the coal for smelting it no doubt for some years, when the total make was not one-fourth of what it now is, placed this country in a position which defied at the time all competition in point of cheapness of production. Since that period however, in seeking to produce iron economically, we are dependent on the lias formation for metal of ordinary quality, and for other kinds on the rich deposits of hematites in Lancashire and Cumberland. For the treatment of both these descriptions of mineral, the fuel has to be conveyed to the ore, or *vice versa,* for greater or shorter distances. It must be remembered however that our cheaply-worked ironstones of North Yorkshire, Lincolnshire, and Northamptonshire have now their counterparts near the Moselle, in Alsace, and in Luxemburg; and the rich ores of the West Coast of England are rarely quite so favourably situated for cheap extraction as those of Bilbao, Lake Superior, and some in Pennsylvania, or the immense tracts of brown hematite, and the so-called "fossiliferous ore," found in some of the southern states of America.

In order to occupy a position of any importance as an iron-exporting country, the possession of a plentiful and cheap supply of coal is of course indispensable. In this respect Great Britain has nothing to fear, inasmuch as for many years she has raised nearly as much as all the rest of the world put together; and out of her superabundance there is exported for foreign use a weight equal to that worked either in France or Belgium, which after Great Britain and the United States are two of the principal iron-making countries of the world.

Great however as is our wealth in fossil fuel, we are far outstripped by North America whose known coal-fields extend over an area of 24 times as great as those of England and Scotland put together.

Germany also has nothing to fear at present for want of coal. The Ruhr district alone, which in 1840 produced short of one million tons, has now an annual output closely approaching twenty millions of tons. The area of coal-bearing strata on that river is said to exceed 1,000 square miles, which is about one-eighth of that of the entire United Kingdom.

The production of coal as given in the Iron Trade Report for 1882 was as follows:—

	Tons.
In Great Britain...	156,499,000
United States...	72,000,000
Germany	53,000,000
France...	20,803,000
Belgium	17,485,000

Of these five nations, Great Britain and the United States are the only countries which import no coal. The imports and exports for the year 1882 was as follows:—

	Received. Tons.	Exported. Tons.
United Kingdom...	Nil.	20,958,000
United States ...	Nil.[1]	650,000
Germany ...	7,631,000	2,090,000
France ...	10,293,000	587,000
Belgium ...	1,058,000	5,853,000

Of these France alone stands in the position of having its iron trade susceptible of being impeded for want of coal; for, although Germany appears as having imported above $7\frac{1}{2}$ million tons, no doubt to places remote from her coal-fields, she appears to have exported above two million tons. In the iron works of France, including the Bessemer and Siemens-Martin steel works the coal consumed is probably about six million tons; but it has been just proved that she is indebted for a very much larger quantity even than this to localities outside her own dominions.

In the matter of iron ore the United Kingdom will also bear comparison with any part of the world.

After the discovery of the Bessemer and open-hearth processes, the ores of iron soon came to be practically divided under two heads— viz., those which afforded metal fit for the steel manufacturer, and those which, from the presence of phosphorus, were unfit for his purposes. It is true this division held more or less good under the old method of making steel by cementation; but the presence of phosphorus in small quantities was then less objectionable, owing to its partial removal during the conversion of the pig into malleable iron, a process which was a preliminary step in the manufacture of steel.

Clay ironstone, so called from the presence of alumina, constitutes by far the largest proportion of the ores of iron; but I am unacquainted with a single case where this mineral is sufficiently free from phosphorus to be capable of treatment in the Bessemer converter, or in the original mode of making steel in the Siemens furnace.

---

[1] It is doubtful whether this return is quite correct as vessels occasionally take out small quantities of coal to the United States. In 1876 the importations were, according to Pechar, 407,000 tons.

On the other hand, carbonate of iron, or spathose ore, is, I believe, invariably suitable for the steel-maker's operations; but it is of too rare occurrence to form an important element in considering the question. The oxides of iron, as found in the magnetites, and in the hematites, red and brown, constitute by far the most important source of metal possessing the purity required for the production of steel. It often happens however that, even in these varieties of ore, phosphorus occurs in such proportions as to render them unfitted for the purpose.

The only important deposits in Great Britain of native oxide of iron fit for steel-making, as it is now commonly practised, are those in Cumberland and Lancashire, which frequently are, for this class of mineral, among the most economically worked in the world.

So long as steel was only required in moderate quantities, and was all produced, chiefly from foreign iron, by the process of cementation, the mines of the two counties just named amply sufficed for the demands made on their resources. These demands were confined to supplying the means of diluting the phosphorus found in the forge cinders, when the same were sent to the blast furnace, or of making for certain purposes, pig iron of a higher quality than that obtained from the usual run of clay ironstones.

The necessities of the Bessemer steel-makers, stimulated by ameliorations in the manufacture, soon exceeded the powers of the Lancashire and Cumberland mines which with some smaller sources of supply may be regarded as capable of furnishing close on to three million tons per annum of ore, equal to a little above one and a half million tons of pig iron. The deficiency has been made up by considerable importations, chiefly from Spain; since ore can be obtained from Bilbao and delivered at blast furnaces on certain coal-fields in Wales, and in the North of England, at a cost of 7s. to 10s. for carriage, and at the present moment (July, 1883) is being conveyed for 5s. 6d. to 6s. 6d. per ton.

So far as present appearances of the Spanish mines enable us to judge, there is every reason for believing that the steel works of the world can count for many years to come on a very large supply of suitable ore, for I am informed that in the year 1881 the mines of Bilbao alone have furnished nearly two and three-quarter million tons.

SECTION XV.—STATISTICAL.

Having regard to the position of Great Britain, and the situation of her coal-fields in relation to the sea and to the markets generally, there is thus no nation more favourably situated for smelting Bessemer pig, whether from native or from foreign mineral.

There are many persons, myself among the number, who think that steel, or perhaps more properly ingot iron, may ultimately supersede in in a great measure, if not entirely, the use of malleable iron as produced from the puddling furnace. At one time the chief barrier to this possible change appeared to be the want of metal sufficiently free from phosphorus to supply the greatly increased demand, which would be necessitated by the extended use of such a quality of iron.

From this difficulty the Basic process has undoubtedly liberated us. It may be questioned, as it is by many, whether circumstances connected with the position of the minerals, and that of the markets may not in many cases enable the hematite pig iron to compete on more than equal grounds with the Basic process, as applied to the cheaper pig iron obtained from the lias-measures of North Yorkshire, Northamptonshire, or Lincolnshire. There is however no doubt that the question of an inexhaustible supply of raw material being now at our command is entirely settled by the Basic process in our favour; and that, so far as economical production of steel and economical transport to a neutral country are concerned, the British manufacturers are in a position not inferior to that of any other nation. At the same time, as will be hereafter shown, we do not by any means stand alone in the possession of what may now be regarded as the most abundant and cheapest source, whence to supply the blast furnaces with the needful ore.

To avoid the use of unnecessary figures the output of ore for three years beginning 1870 in the five countries selected for examples are given below (000 omitted):—[1]

	United Kingdom.	United States.	Germany.	France.	Belgium.
1870	14,370	3,210	3,839	2,899	654
1875	15,821	4,500	4,730	2,540	365
1881	17,446	7,974	7,573	3,500	200

The importation of iron ore by iron-making countries has of late

[1] Authorities—Robt. Hunt, F.R.S., Mining Record Office; Iron Trade Association; and Joh. Pechar.

years assumed proportions of great magnitude. Twenty years ago such transactions were all but unknown, and so late as 1870 the quantities so dealt with were insignificant.

Iron ore exported and imported by the five principal iron making countries, and by Italy, for the three years named, in thousands of tons, was :—

		Imported.	Exported.
Great Britain...1870	... ...	400[1]	Nil.
1880	... ...	3,060[1]	95
1881	... ...	2,803[1]	68
United States...1870	... ...	17	Nil.
1880	... ...	493	Nil.
1881	... ...	782	Nil.
Germany ...1870	... ...	300	84
1880	... ...	607	1,263[2]
1881	... ...	615	1,443[2]
France... ...1870	... ...	489	145
1880	... ...	1,168	114
1881	... ...	1,287	88
Belgium ...1870	... ...	568	179
1880	... ...	921	347
1881	... ...	1,169	366
Italy 1870	... ...	—	40
1880	... ...	—	297
1881	... ...	—	330

In respect to the five iron making countries, with the exception of Belgium, this large importation cannot be said to arise from any inability in the domestic mines to compete with those abroad in the matter of cheapness. The change, as already intimated, has been almost exclusively due to the demand for iron sufficiently free from phosphorus to afford material for the Bessemer and Siemens-Martin steels, the make of which has so greatly increased in the last ten years. The nature of

---

[1] Includes purple ore, *i.e.* the oxide of iron obtained after treating cupreous pyrites in the wet way.

[2] This large increase of exports from Germany is probably due to the annexation of Alsace-Lorraine and the Duchy of Luxemburg being included in the German Zoll Verein.

this increase will be best understood by a reference to the increase in Bessemer steel between 1870 and 1881; the weight, in thousands of tons, being:—

	Great Britain.	United States.	Germany.	France.	Belgium	Total.
1870 ...	215	40	125	83	6	469
1880 ...	1,044	1,074	686	384	95	3,283
1881 ...	1,673	1,696	993	454	170	4,986

The Mining Records for the year 1865 do not distinguish foreign ores imported into the United Kingdom from those which have arrived coastwise from British mines. From the information given it would not appear that the quantity received from foreign countries can, for that year, have amounted to 10,000 tons. In the year 1867 as much as 86,568 tons of foreign ores were imported which, as we have seen, grew to 400,000 tons in 1870 and to above three millions in 1880.

So long as Great Britain could supply her own wants from more cheaply wrought ores for Bessemer iron, than other nations could obtain similar ores from native mines or by way of importation, the British manufacturer enjoyed corresponding advantages as compared with his foreign competitors. Now however that there is imported into the United Kingdom, a quantity of ore more than equivalent to furnish pig iron for all the steel made within its own boundary, it is clear that it no longer enjoys the favourable position, comparatively speaking, which it formerly occupied. The British and foreign steel maker pay the same price for their ore at the port of shipment, and any difference, on the one side or the other, is confined to the respective costs of conveyance from such port to the place of consumption.

Before leaving the subject of iron ore, I would wish not only to point out the extraordinary increase in the production of iron furnished by the mines of the United Kingdom in the last fifty years, but also to direct attention to the extraordinary change in the sources of that production. For this purpose the makes of pig iron for the years 1830, 1860, 1870, and 1880 are compared, and to these 1882 is added. I am not acquainted with any trustworthy records of the ore raised in 1830, and those for 1882 not being issued, my comparison in respect of ore must necessarily be confined to the years 1860, 1870, and 1880.

MAKE OF PIG IRON IN GREAT BRITAIN.

	1830.	1860.	1870.	1880.	1882.
Cleveland, including Durham and Northumberland	5,327	658,679	1,627,557	2,416,418	2,688,650
Northamptonshire...	Nil.	7,595	43,166	178,714	192,115
Lincolnshire ...	Nil.	Nil.	31,690	207,704	201,561
Derbyshire and Notts	17,999	125,850	179,772	366,792	445,735
	23,326	792,124	1,882,185	3,169,628	3,528,061
Cumberland and Lancashire	Nil.	169,200	677,906	1,541,227	1,783,920
Scotland	37,500	937,000	1,206,000	1,049,000	1,126,000
South Wales ...	277,643	969,025	979,193	889,738	883,305
South Staffordshire	212,604	469,500	588,540	384,556	317,117
North Staffordshire	Nil.	146,950	303,378	225,023	398,443
West and South Yorkshire	28,926	98,100	77,717	306,560	279,253
Shropshire	73,418	145,200	112,300	88,338	80,475
Gloucester, Wiltshire and Somersetshire	—	50,293	93,601	37,351	48,000
North Wales ...	25,000	49,360	42,695	57,812	48,713
	678,417	3,826,752	5,963,515	7,749,233	8,493,287

In a French work[1] written in 1837 the authors mention that in 1740 the make of iron in England was 17,350 tons obtained from 59 furnaces; equal therefore to 294·11 tons each per annum, or 5·13 tons per week. They state that the ores chiefly used were brown and red hematites. Earthy ores, "mines terreuses," were also smelted; but it does not appear, according to the authority just quoted, that the clay ironstones of the coal-measures, "minerais de fer carbonaté des houillères," were known in the beginning of the last century. In opposition to this condition of things the authors of the "Voyage Métallurgique," when they wrote, attribute that marked superiority of Britain as an iron-producing country over all other European nations[2] to the possession of the clay ironstone referred to.

[1] "Voyage Métallurgique" par MM. Dufrenoy, Élie de Beaumont, Coste et Perdonnet. Vol. I., p. 221.

[2] Idem. Vol. I., p. 2.

According to the figures contained in the Mining Records of Great Britain, it would appear that in the year 1880 the iron ores raised may be divided under the following heads:—

		Tons.
Iron ores from the coal-measures	... ...	5,397,477
,, from hematite veins, etc.	... ...	3,309,074
,, from the the lias-measures	... ...	9,319,498
		18,026,049
Imported—Burnt ore from cupreous pyrites	427,730	
,, Hematite	2,632,601	
		3,060,331
		21,086,380

Approximately the pig iron capable of being produced by these ores would be as follows:—

		Tons.	Per Cent.
From the clay ironstone of the coal-measures ... ... 30 per cent.	=	1,619,243	= 20·59
From hematites of all kinds ... 50 ,,	=	1,654,537	= 21·04
		3,273,780	
From the lias-measures, average yield taken at ... ... 32½ ,,	=	3,028,836	= 38·52
From the imported ore ... 51 ,,	=	1,560,768	= 19 85
		7,863,384	100·

The actual make of pig iron in 1880, according to the Mining Records, was only 7,749,233 tons, of which a certain proportion would however be smelted from mill and forge cinders.

The difference between the two sets of figures may chiefly arise from the yield of the different kinds of ores being taken a little too high; or from an accumulation of stocks of ore, all that which was raised or imported not having been used.

In order to exhibit the position of each iron-producing centre of the United Kingdom, tables have been constructed for the years 1860, 1870, and 1880, showing the ore raised and the pig iron produced, from which an estimate is made of the quantity of metal smelted in other districts than those where the ore is obtained.

For the Year 1860.

	Per Cent. of Iron.	Ore Raised. Tons.		Equivalent Pig Iron. Tons.	Pig Iron Made. Tons.	Equivalent to Pig Iron from	
						Ore brought from elsewhere. Tons.	Ore sent elsewhere. Tons.
Cleveland, including Northumberland and Durham, ore taken at	31	1,483,819	=	459,983	658,679	198,696	—
Northamptonshire and Oxfordshire	37	101,497	=	37,553	7,595	—	29,958
Lincolnshire	33·3	16.892	=	5,631	Nil.	—	5,631
Derbyshire and Notts.	33·3	375,500	=	125,166	125,850	684	—
Cumberland and Lancashire	53	989,610	=	524,493	169,200	—	355,293
Scotland	32	2,150,000	=	688,000	937,000	249,000	—
South Wales	30	630,705	=	189,211	969,025	779,814	—
South Staffordshire and Warwick	35	805,200	=	281,820	469,500	187,680	—
North Staffordshire	32	738.229	=	236,233	146,950	—	89,283
West and South Yorkshire	33·3	255,700	=	85,233	98,100	12,867	—
Shropshire	33·3	165,500	=	55.166	145,200	90,034	—
Gloucester, Wilts., Hampshire, and Somersetshire	40	196,892	=	78,756	50,293	—	28,463
North Wales	33·3	85.097	=	28,366	49,360	20,994	—
Ireland, etc.	33·3	29,571	=	9,857	—	—	9,857
		8,024,212		2,805,468	3,826.752	1,539,769	518,485

No foreign iron ore imported, according to "Mining Records," in the year 1860.

For the Year 1870.

	Per Cent. of Iron.	Ore Raised. Tons.		Equivalent Pig Iron. Tons.	Pig Iron Made. Tons.	Equivalent to Pig Iron from	
						Ore brought from elsewhere. Tons.	Ore sent elsewhere. Tons.
Cleveland, including Northumberland and Durham, ore taken at	31	4,298,220	=	1,332,448	1,627,557	295,109	—
Northampton and Oxfordshire	37	800,051	=	296,018	43,166	—	252,852
Lincolnshire	33·3	248,329	=	82,776	31,690	—	51,086
Derbyshire and Notts.	33·3	384,865	=	128,288	179,772	51,484	—
Cumberland and Lancashire	53	2,093,241	=	1,109,417	677,906	—	431,511
Scotland	32	3,500,000	=	1,120,000	1,206,000	86,000	—
South Wales	30	560,055	=	168,016	979,193	811.177	—
South Staffordshire	35	467,500	=	163,625	588,540	424,915	—
North Staffordshire	32	910,134	=	291,242	303,378	12,136	—
West and South Yorkshire	33·3	307,717	=	102,572	77,717	—	24,855
Shropshire	33·3	337·627	=	112,542	112,300	—	242
Gloucester, Wilts., Hampshire, and Somersetshire	40	304,665	=	121,866	93,601	—	28,265
North Wales	33·3	59,240	=	19,746	42,695	22,949	—
Ireland, etc	33·3	99,010	=	33·002	Nil.	—	33,002
		14,370,654		5,081.558	5,963,515	1,703.770	821,813

No iron ore imported, according to "Mining Records," in the year 1870.

For the Year 1880.

	Per Cent. of Iron.	Ore Raised. Tons.	Equivalent Pig Iron. Tons.	Pig Iron Made. Tons.	Equivalent to Pig Iron from Ore brought from elsewhere. Tons.	Ore sent elsewhere. Tons.
Cleveland, including Northumberland and Durham. ore taken at	31	6,528,011 =	2,023,683	2,416,418	392,735	—
Northampton, Leicester and Oxford	37	1,610,760 =	595,981	178,714	—	417,267
Lincolnshire	33·3	1,154,584 =	384,861	207,704	—	177,157
Derbyshire and Notts.	33·3	152,512 =	50,837	366,792	315,955	—
Cumberland and Lancashire	53	2,759,407 =	1,462,485	1,541,227	78,742	—
Scotland	32	2,664,483 =	852,634	1,049,000	196,366	—
South Wales	30	343,927 =	103,178	889,738	786,560	—
South Staffordshire	35	399,745 =	139,910	384,556	244,646	—
North Staffordshire	32	1,398,693 =	447,581	225,023	—	222,558
West and South Yorkshire	33·3	286,698 =	95,566	306,560	210,994	—
Shropshire	33·3	226,721 =	75,573	88,338	12,765	—
Gloucester, Wilts., Hampshire, and Somersetshire	40	189,854 =	75,941	37,351	—	38,590
North Wales	33·3	43,016 =	14,338	57,812	43,474	—
Ireland, etc.	33·3	267,849 =	89,283	Nil.	—	89,283
		18,026,260	6,411,851			
Imported	50	3,060,331	1,530,165	—	—	1,530,165
		21,086,591	7,942,016	7,749,233	2,282,237	2,475,020

What is particularly wished to be pointed out is the great change which fifty years have produced in the character of the minerals employed in the iron works of Great Britain. In 1830 there were only 5,327 tons of iron furnished by that class of ore now so extensively mined in North Yorkshire, Northamptonshire, Lincolnshire, etc. Against this we have in 1880 no less than 3,028,836 tons of pig metal were derived from the lias-measures of these counties.

The possession of a bed of ironstone varying from four to fourteen feet in thickness, and occurring in several large fields in a line extending from the mouth of the Tees to the British Channel, is of course an enormous advantage to the United Kingdom in an iron-producing point of view. We do not however, as has been intimated in the present section, stand alone in this enviable position as mineral owners.

A similar stretch of country, geologically speaking, takes its rise in Germany, and passing through the Duchy of Luxemburg terminates in France. At Luxemburg, Hayange, and in the neighbourhood of Nancy the beds of ironstone, found in the measures referred to, are extensively worked. The importance of the deposit will receive further notice when a comparison is attempted to be drawn between the iron-making minerals of the United Kingdom and those of other nations.

According to the Mining Records no pig iron was produced in Lancashire or Cumberland in 1830. Practically this is so, although a small quantity of charcoal iron was smelted from the hematite of the district anterior to the year in question—a process which is still continued. Whatever ore was raised there—and it amounted to 579,924 tons in 1854—was used almost entirely in the furnaces of South Wales and South Staffordshire. In 1830 probably not one-half this quantity would be raised; whereas in 1880 there were mined in Cumberland and Lancashire 2,759,407 tons and there were imported 3,060,331 tons, making together nearly 5,750,000 tons of hematites, mainly used for Bessemer and open-hearth steel. Of clay ironstone from the coal-measures, which, when the authors of the "Voyage Métallurgique" wrote, was the source of our supremacy as iron makers, we raised in 1880 5,397,477 tons, equal perhaps to 1,800,000 tons of pig iron, or less than one-fourth of the make of that year.

Since the work just referred to was written, great changes have taken place in the comparative importance of the clay-ironstone of the coal-measures. South Wales still occupies a prominent place as an iron-making centre, but its blast furnaces are almost exclusively supplied with imported ores for the production of Bessemer steel. South Staffordshire has declined nearly 20 per cent. in its production of pig iron between 1860 and 1880, and of the reduced quantity a part is obtained from Northamptonshire ore. The Derbyshire furnaces are similarly supplied with ironstone, the quantity obtained from the neighbouring collieries being insignificant. The only important iron-making locality which maintains its ancient position in connection with the output of clay ironstone, is Scotland, which in 1880 raised 2,659,317 tons. Of this quantity very little more than one-half, 1,435,647 tons, was the famous Black Band which formerly was almost exclusively used in the Scotch furnaces. The remainder, owing to the partial exhaustion of the Black Band mines, was ordinary clay band ironstone.

We have just seen how a great revolution materially altered the complexion of the pig iron trade. Within the last ten years the manufacture of malleable iron received a check which by many was regarded as its death-blow; and indeed this prophecy has been actually realized so far as the works in South Wales are concerned. It became evident about the year 1874 or 1875 that steel rails were being made at such a price and of such a quality that iron as a material for rails would speedily become a thing of the past. This prediction has been fulfilled in the district of which Middlesbrough is regarded as the centre, and the make of malleable iron in consequence fell in 1879 to 47 per cent. of what it was in 1873. Fortunately in the last three years the demand for iron for shipbuilding purposes has absorbed all the iron formerly required for rails, as may be seen by comparing the year of the largest make of this article, viz., 1873, with 1882 :—

	Production in Thousands of Tons. 1873.	1882.
Rails	374	7
Plates	191	498
Angle bars	51	150
	616	655

Mr. Edward Williams has prepared a statement of the make in the neighbourhood referred to, and by his permission I give it a place here.[1] The make in thousands of tons was distributed as follows :—

	Total Tons.	Rails. Per Cent.	Plates. Per Cent.	Angles. Per Cent.	Bars. Per Cent.	
1872	702	49	29	10	12	= 100
1873	707	53	27	7	13	= 100
1874	671	45	31	9	15	= 100
1875	646	44	31	7	18	= 100
1876	484	26	41	12	21	= 100
1877	455	9	54	17	20	= 100
1878	485	5	55	21	19	= 100
1879	333	2	60	17	21	= 100
1880	584	5	63	18	14	= 100
1881	667	3	67	18	12	= 100
1882	726	1	68	21	10	= 100

[1] The table is not quite correct as one or two works made no return.

According to these returns not only are there no signs *at present* of decadence in the malleable iron trade of the centres connected with Cleveland, but there is an increase in 1882 over any previous year of its history.

No authentic record having been kept of the malleable iron trade of the entire kingdom until very recently, we are unable to judge of the comparative position of Cleveland by a reference to the total production. Enough however is known to justify the assertion that the favourable turn which has manifested itself in the more northern counties, is merely the result of the decay of the trade elsewhere. Mr. Robert Hunt, in the records so often referred to, gives the number of puddling furnaces at work; and the change which has taken place in South Wales and South Staffordshire speaks for itself :—

	South Wales	South Staffordshire.	Total.
1872.—Puddling furnaces at work	1,251	2.155	3,406
1880.— ,, ,,	517	1.625	2,142
Decrease	734	530	1,264
Percentage of decrease	58·6	24·6	37·1

Mr. Jeans gives 2,681,000 and 2,841,000 tons as the make of malleable iron in the United Kingdom for the years 1881 and 1882 respectively. In 1873 the make of Bessemer and open-hearth steel would not much exceed 500,000 tons. In 1882 this has risen to more than 2,100,000 tons; and how far or how rapidly this substitution of steel for iron has yet to advance it would be difficult to predict. Iron rails have been displaced by those of steel, and the puddling furnaces, thus laid idle, have found employment in furnishing plates for the shipbuilders. But whereas in 1877 the tonnage of vessels built of steel was 1,118, in 1881 it had risen to 71,533. Has the puddler before long to see his occupation in connection with shipbuilding follow the example of the rail trade? If so, where has he to look for relief? No one questions the superiority of steel over iron for naval architecture. The only barrier to its exclusive use is one of price, which is due simply to considerations of a mechanical order in the rolling mill itself. So far as the mere conversion of the pig iron into a malleable product is concerned, the puddling furnace is doomed, at all events for the production of a material

SECTION XV.—STATISTICAL. 461

for the shipbuilder. And many other, we may say most other trades, for which puddled iron is now used, in the course of time, will be supplied from the Bessemer converter or from the open-hearth furnace.

The consumption of limestone as a flux is not a serious item in the matter of quantity, varying as it does from 5 to 15 cwts. per ton of iron, when of moderate purity. It is moreover a mineral of very inexpensive extraction, from the great thickness of the strata which are usually accessible by mere quarrying. Variations of cost at the smelting works are chiefly due to distances of carriage; but usually these are not very great on account of the vast calcareous beds, which are found equally among the older and newer geological formations. The cretaceous, oolitic, magnesian, carboniferous, and silurian limestones, are thus all contributors to the flux needed by the iron smelter. It often happens that the limestone of the magnesian formation is as free from magnesia as any other; but frequently this is not so, and then the rock sometimes containing 30 per cent. and more of magnesian carbonate is used. The consequent inconvenience is the introduction of a quantity of matter, which having little or no value in the process occasions a decrease in the production of the furnace, and a somewhat greater consumption of fuel; not only on account of the increased weight of inert matter to be fused, but probably also on account of the solvent action of the additional carbonic acid on the coke.

It has been observed on a previous page that the abolition of protective duties would not necessarily be accompanied by a very large extension of our dealings with those countries which seek to exclude us from their markets. It may quite well happen that the importing country can manufacture a commodity in competition on equal terms with the country which supplies it; and yet, the demand at certain periods being beyond the powers of the former, the latter is called upon to make up the difference, which it is able to do at the higher prices then ruling.

Germany may be taken as an example for illustration.

Beginning with the year 1860 the duty that year was 20s. 5d. per ton on pig iron imported. Taking the selling price at Middlesbrough at that time, viz. about 45s., and adding the carriage and duty, a ton of Cleveland iron delivered in the Westphalian Works would cost about

80s. The entire make of Prussia during 1860 was 395,000 tons, while that imported would probably be about 100,000 tons.[1] I cannot speak positively as to the net cost of pig iron in Germany at this period; but that the difference between the cost and the selling price was not inconsiderable may be inferred from the fact that by the end of 1864 the Prussian make had risen to 705,000 tons. It has to be observed that the average price of Cleveland iron, over the five years under consideration, was 50s. at Middlesbrough; and that its value had now begun greatly to influence Continental prices.

In 1865 the duty was reduced from 20s. 5d. to 15s. 3d. yet without stimulating the importations, which remained stationary until the end of 1868. In that year the price of pig iron at Middlesbrough had fallen to 43s.; but in the meantime the product of Prussia had risen to 1,053,000 tons, although Middlesbrough iron could now be landed on the Rhine at fully 7s. per ton lower than in 1860.

In 1868 the import duty experienced a further reduction of 5s. 1d., leaving it at 10s. 2d. per ton, at the same time that Middlesbrough iron was 4s. per ton cheaper than when the duty was reduced from 20s. 5d. to 15s. 3d. Notwithstanding the increased competition arising from this combination of circumstances the Prussian make reached 1,180,000 tons by the end of 1869.

In the following year (1870) the duty was lowered to 5s., the price of iron in Middlesbrough being only 2s. higher than it was in 1860: and at this price or thereabouts it remained till the close of 1871, by which time the production in Prussia had risen to close on 1,300,000 tons. From 1873 to 1876 pig iron was admitted duty free.

All this movement indicates that the reduction of duty from 20s. 5d. in 1860 to 5s. in 1870 had not prevented the make in the Prussian dominions from advancing from 395,000 to 1,300,000 tons, or threefold.

It has to be remarked that the make was increased by 168,000 tons in the three years ending 1870, although the average price of pig iron was no more than it was in 1860, when it was considered that the home pig iron manufacturers required a protective duty of 20s. 5d. per ton.

---

[1] Scrap iron being included in the official returns, the actual weight of pig iron can only be given approximately. Holland is included, most of its imports being *in transitu.*

In 1872 the production was close on one and a half millions of tons; and this was increased during 1873 by 120,000 tons, although during this year the duty was entirely abolished. This latter circumstance however had no influence on the action of the Prussian iron trade, probably owing to the great rise in prices generally throughout the world; Middlesbrough pig iron having in 1873 touched £6 and Scotch £6 15s. per ton.

All experience justifies the conclusion that, whenever profits in any particular branch of industry attain a position superior to that in other manufactures, capital is attracted thither; and apparently it matters little by what means this position is arrived at.

It is therefore not surprising that the Prussian nation, encouraged by the price of pig iron artificially raised by duties levied between 1860 and 1872 and afterwards by the commercial excitement of 1872, and following year, should have added to their powers of production in the manner described.

Almost contemporaneously with the abolition of duty in Prussia, pig iron reached its highest point in Great Britain; from which it steadily declined until in 1878, it was selling as low as 32s. at Middlesbrough, and all Europe felt this change, as we did. From the year 1830 up to 1874, it very rarely happened that the make of any one year was not in excess of that of the twelve months preceding; and so far as the means I possess enable me to judge the falling off did not at any time extend beyond one year. But both in 1874 and 1879 the production of pig iron in the United Kingdom was about three-quarters of a million below that of 1872; and the average make of the entire six years ending in 1879 was 440,000 tons under the average of the three years terminating with 1873. No doubt the Prussian markets were affected by the large importations from Great Britain; which averaged 523,000 tons per annum, for the six years ending 1879, instead of 796,000 tons (Holland included), which was the average of the two years anterior to 1874. Prussia in the meantime had raised her powers of production of pig iron to within a trifle of two millions of tons per annum; or nearly one-half more than it stood at in 1871 and 1872, the years of high prices. In other words, her increase of make was about 600,000 tons, not far from the same quantity she was receiving in the way of importation. It is not extraordinary, under such circumstances, that the Prussian ironmasters should feel oppressed by the competition which had arisen *among themselves,* and which in point of fact was, it seems to me, far more to blame than that offered by Great Britain.

Another difficulty, to which the iron manufacture is exposed, possibly intervened to aggravate the distress felt in Germany by those engaged in the trade. In the prosperous times wages had naturally risen; but the enhanced rates continued to be paid, long after the ironmaster could afford to pay them. This was certainly the case in the United Kingdom, and it was pre-eminently so in the United States. The imposition of a duty of 25s. per ton on pig iron, and double this on malleable iron, was followed by a corresponding rise in value of these commodities in America. The high prices of 1872 and 1873 in Great Britain were immediately felt on the other side of the Atlantic; and in consequence the cost of producing pig iron rose from 47s. 9d. to 98s. 7d. or 108 per cent., at which, owing to the increased prices of materials and of labour, it continued long after many of the iron manufacturers were known to be losing money on every ton of iron they sold.

In Prussia the increase in the pay of the men was, as will be shown hereafter, very much less than in the United Kingdom or in America; but still the trade was an unprofitable one, and those who pursued it sought relief by weighting the imported iron with a 10s. duty, which was imposed in 1879. It remains to be seen whether in the course of years and in times of good demand, this artificial augmentation of price will not be followed by the same results which have marked the history of the German iron trade between 1860 and 1873.

Belgium, already referred to as a country supposed to be able to compete on British soil with British iron makers, in reality has not greatly altered her position during the last ten years, until 1880, when the make of iron showed an increase of 8 per cent. as compared with 1870.

In the year 1865 this kingdom raised 1,018,000 tons of iron ore; but in 1876 this had fallen to 269,000 tons, of which 166,000 were exported, chiefly to France. In 1881 the product was only 200,000, so that practically Belgium is entirely dependent on imported ore for her blast furnaces, the amount imported in 1881 being 1,169,206[1] tons, as against 301,000 in 1866 and 1,206,717 tons in 1882.

[1] Part of this was probably in transit to Germany.

Up to the year 1866 the duty on pig iron imported into Belgium was 19s. 7d., and on bar iron 39s. 2d. per ton: in that year they were reduced to 4s. 2d. and 8s. respectively. This alteration produced little effect on the quantity of pig received into that kingdom. In 1865 24,864 tons were imported at the high duty, and two years after, under the low duty, the quantity had only risen to about 42,549. The import gradually increased up to 1872, when it reached 137,000 tons. In the meantime however the home production had risen from 470,767 tons in 1865 to 655,565 tons in 1872—proving conclusively that the lowering of duty had not impeded the manufacture of pig iron.

Since 1872 the iron trade in Belgium has experienced the vicissitudes which befel the same trade all over the world; but it may be questioned whether, in spite of its small protective duties, and of its being nearer to, and therefore more accessible from, the United Kingdom than Germany and most parts of France, the home manufacture of this enterprising little kingdom has not stood its ground as well as others protected by much higher tariffs.

The following tables exhibit the position of the Belgian iron trade, from 1870 to 1880, in thousands of tons:—

	1870.	1872.	1873.	Average of 4 Years ending 1878.	1880.
Pig iron made	565	655	607	497	610
,, imported	82	137	145	183	206
	647	792	752	680	816
,, exported[1]	10	49	27	11	41
Consumed at home	637	743	725	669	775
Malleable iron made	491	502	480	404	450
Steel made	6	15	21	81	95
Steel and iron	497	517	501	485	545

The various articles of wrought iron or steel imported into Belgium are insignificant in amount, not exceeding 15,000 tons per annum. The cheap rate however at which ore and pig iron are imported, aided

[1] Possibly chiefly in transit to Germany, and is, therefore, deducted from the imports.

by cheap labour, has enabled that country to carry on an export trade, fully equal to a consumption of one-third of the pig iron it has made and imported:—

EXPORTS FROM BELGIUM.

	1877. Tons.	1878. Tons.	1879. Tons.	1880. Tons.
Steel (chiefly rails)	18,237	26,956	46,657	46,577
Iron rails	42,599	34,273	29,596	28,124
Plates	16,223	23,198	23,301	32,302
Wire	2,036	2,898	3,393	4,560
Section irons	113,150	132,769	154,495	162,339
Nails	10,296	9,096	8,308	10,871
Unclassed	13,543	17,610	20,327	20,247
Castings	1,999	10,551	12,862	14,528
	218,083	257,351	298,939	319,548

During the three years 1875, 1876, and 1877, England received of the above articles 33,792, 36,752, and 52,662 tons respectively.

52,000 tons is unquestionably a large quantity of malleable iron to be received by England from Belgium; but Belgium is the last place we ought to regard with any ill-will in reference to such a transaction, seeing that at least three times this weight of British pig iron has been used up by the Belgian manufacturers.

In 1881 and 1882 the weight of iron and steel received from Belgium was 50,100 and 43,800 tons respectively, while from Germany, which demands protection against British iron, 75,000 tons in the year 1881 and 72,000 tons in 1882 were landed at British ports.

With France, between which and ourselves extensive commercial relations exist, we do but little in iron, owing to the high rate of import duties levied on all forms of the metal. These rates are as follows:—

Pig Iron.	Bar Iron.	Iron Rails.	Iron Plates.	Steel Rails.	Steel Plates.
12s.	40s. 10d.	40s. 10d.	57s. 1d.	48s. 9d.	73s. 4d.

With the exception of pig iron, our transactions may be said to be absolutely *nil*. In 1860 the duty on this article was 39s. 2d., from which it was reduced to 32s. in 1864, and finally in the following year to 12s. where it now remains.

The following shows the make and importations since 1866, in thousands of tons:—

	1866.	1867.	1868.	1869.	1870.	1871.	1872.	1873.	1874.	1875.	1876.	1877.	1878.	1879.	1880.
Make	992	931	934	1,018	923	859	1,217	1,366	1,423	1,416	1,453	1,522	1,508	1,314	1,733
Add import	143	155	107	127	83	77	122	125	122	202	184	212	166	153	167
	1,135	1,086	1,041	1,145	1,006	936	1,339	1,491	1,545	1,618	1,637	1,734	1,674	1,467	1,900
Deduct exported	23	18	21	22	16	14	36	46	51	48	52	49	50[1]	50[1]	50[1]
Total used in France	1,112	1,068	1,020	1,123	990	922	1,303	1,445	1,494	1,570	1,585	1,685	1,624	1,417	1,850

From these figures it would appear that at least 90 per cent. of all the iron used in France is French make.

As regards ore, the quantity raised in France during the twelve years ending 1879 varied from 2,099,000 tons to 3,790,000 tons; the average being 2,770,000.

In 1869, the first of these years, France imported 592,000 tons of ore; in 1880 she brought in 1,168,000, equal to 38 per cent. of her consumption during the three years ending 1880.

Russia produces a small quantity of iron, chiefly smelted with charcoal and discourages all importations by imposing prohibitive duties as follows:—

Pig Iron.	Bar Iron.	Iron and Steel Rails.	Iron and Steel Plates.
11s. 8d.	78s. 9d.	98s. 4d.	108s. 4d.

As already intimated the United States of America, Germany, France, and Belgium, from their mineral wealth and geographical position, constitute our most formidable rivals in the manufacture of iron. With the exception of Belgium efforts more or less vigorous have been, and are still being made, to exclude the produce of Great Britain from entering into competition with that of home origin. Under these circumstances it is perhaps worth speculating on the prospects of the non-producing countries being able to absorb any portion of British made iron, for which future legislation abroad may compel us to seek other markets. For this purpose a table for ten years has been compiled from the statistics in the Report of the British Iron Trade Association.[2] In it have been inserted the quantities of iron imported by the chief iron producing countries to which Holland has been

---

[1] Assumed.

[2] The weights included all descriptions without reference to their being stee wrought or cast iron.

added, because a large portion received by it is merely *in transitu*, chiefly to Germany. It is almost needless to say that nearly the whole of the imports in question is derived from Great Britain.

The following are the quantities in thousands of tons:—

	1871.	1872.	1873.	1874.	1875.	1876.	1877.	1878.	1879.	1880.
United States	927	970	488	283	206	158	167	157	717	1,355
Germany	306	423	381	216	299	298	284	289	259	269
Holland	290	389	397	237	263	267	224	259	239	210
Belgium	—	163	175	110	109	115	98	90	83	116
France	84	108	110	84	104	112	123	112	101	117
Russia	110	137	243	199	170	132	100	86	211	204
	1,717	2,190	1,794	1,129	1,151	1,082	996	993	1,610	2,271
Other countries	1,452	1,192	1,163	1,358	1,307	1,142	1,350	1,303	1,273	1,516
	3,169	3,382	2,957	2,487	2,458	2,224	2,346	2,296	2,883	3,787

In a communication to the Iron and Steel Institute in the year 1875, I ventured, after my examination of the American works, on the prediction that we must prepare to look upon our vast dealings in iron with the United States as a thing of the past. A glance at the first line of figures up to the year 1878 may be taken as an evidence of the correctness of this view of the case. A different condition of things obtains in 1879 and the following years, but this is one upon which no estimate of the future can be grounded. There are, no doubt, many in the United States who dissent strongly from the policy of protective duties, but their opinions are not those which as yet are permitted to guide the conduct of affairs. High duties still continue to be levied and the large importations perceptible in the last two years are solely attributable to a sudden demand for materials required for a large and unexpected expansion of railway communication. This demand still continues in a modified form but we have only to look to the extraordinary and unprecedented efforts made by the iron trade of the country to satisfy ourselves that this demand is only temporary in its character.

To judge of the requirements and position generally of the United States in respect to pig iron the following figures, in thousands of net tons of 2,000 lbs. are given:—

	1871.	1872.	1873.	1874.	1875.	1876.	1877.	1878.	1879.	1880.	1881.	1882.
Home make	1,912	2,854	2,868	2,689	2,266	2,093	2,314	2,577	3,070	4,295	4,641	5,178
Imported	199	277	241	103	59	83	66	74	340	784	520	604
	2,111	3,131	3,109	2,792	2,325	2,176	2,380	2,651	3,410	5,079	5,161	5,782

The history of the iron trade of the United States ought to be a lesson to any one who ventures on looking far into the future of this branch of industry. In the year 1871 the American Iron Trade Association estimated the annual producing power of the States in pig iron, to be equal to two and a half millions of tons, and the actual make for that year was close on two millions. The extravagant prices of 1872 and 1873 in Great Britain made itself felt all over the world; and this led to such an increase in the number of furnaces on the other side of the Atlantic, that by the end of the latter year the American ironmasters stated their ability to turn out 5,439,000 tons, being an increase of 171 per cent. in two years.[1] From some cause or another a very large proportion of the augmented capacity remained unutilized, while the importations, as the table shows, manifested a marked increase; but the whole quantity of pig iron made and imported was under 60 per cent. of the alleged powers of home production.

It is unnecessary to pursue at greater length the history of the wonderful development of the iron trade in America, for the figures just given speak for themselves. It may be mentioned however that besides pig, there was imported into the United States a considerable weight of rails and other descriptions of iron. Thus in the four years ending 1882 there was received of all kinds 721,126, 1,370,428, 1,175,259, and 1,192,683 tons respectively. When to these large figures, those representing the make of the States themselves are added, we have a weight which may be regarded as being required for home consumption, far in excess of anything hitherto devoted to a similar purpose by ourselves. At page 444 the quantity estimated as having been used in the United Kingdom in 1882 was 3,495,000 tons; whereas the consumption for domestic use in the United States would not, making allowance for the waste in conversion, be far short of 6,660,000 tons.

The table, exhibiting the proportions of iron taken from us by iron-producing countries and by those nations who, practically speaking,

[1] The exports being insignificant no notice is taken of them.

make no iron of their own, tells us that about 60 per cent. of all our exports is received by the former. If the time should arrive when our present competitors manufacture sufficient for their own wants, we should, according to the present figures, have this 60 per cent. of our exports, equal to about one-fourth of our production, thrown on our hands, less that further quantity which the non-producing countries may in the meantime be able to absorb. Such a change as that referred to is not likely to occur suddenly or in the immediate future. At the same time we cannot shut our eyes to the great fluctuations which have taken place during recent years in our foreign trade. Our exports in 1872 to iron-producing nations (*vide* Table, page 468) were very nearly what they were in 1880, but in the meantime, *i.e.* during the five years ending 1878, they declined fully one-half. They fell in two of the years of this period from about two million to less than one million tons.

It is highly improbable that dealings in iron will entirely cease between Great Britain and even those nations best able to supply their own requirements; but whatever the loss of trade with such countries may prove to be, it would be unwise to count on the remainder of the world being able, for some time to come, to relieve Great Britain of any great amount of the surplus, arising from any large alteration in our foreign commerce.

As pertinent to this question I have extracted from a Table, compiled by Mr. Jeans,[1] the quantity of iron of all kinds consumed per head in 1881 by seven of the chief iron-producing countries, and the following is the result of my examination:—

Iron Producing Countries.	Population.	Tons Consumed.	Lbs. per head Consumed.
United Kingdom	35,968,000	4,618,932	287·53
United States of America	50,152,866	6,065,919	270·92
France	37,672,048	2,508,706	149·16
Germany	45,194,177	2,488,957	123·36
Belgium	5,519,844	587,000	238·20
Russia	88,000,000	965,000	24·56
Austrian territories	37,741,434	625,000	37·09
Sweden and Norway	6,391,098	220,000	77·07
	306,639,467	18,079,514	132·07

[1] British Iron Trade Association, 1882, page 152.

SECTION XV.—STATISTICAL.

Iron Producing Countries.	Population.	Tons Consumed.	Lbs. per head Consumed.
Other countries of Europe, none of which can be regarded as iron-producing	91,694,283	978,449	23·90
Total for Europe and United States	398,333,750	19,057,963	107·16
British possessions exclusive of India	11,465,079	621,483	121·40
	409,798,829	19,679,446	107·57
Do. in India... ...	446,759,606	481,951	2·40
Egypt	5,517,000	18,614	7·55
South America and Islands	45,449,357	274,353	13·50
Asia exclusive of British possessions	517,161,778	113,382	·49
	1,424,686,570	20,567,746	32·33

Thus it will be seen that something short of 410 millions of the inhabitants of the earth consume more than nineteen-twentieths of all the iron produced, leaving the remaining twentieth to satisfy the necessities of the other inhabitants, amounting to about 1,014 million persons. This gives an annual consumption of 107·57 lbs. per head for the 410 millions, against 1·96 lbs. for the other 1,014 millions. If the spread of civilization were such as to raise the consumption among these 1,014 millions of persons to a little above 6 lbs., from a little below 2 lbs. per annum, as it now stands, the two million tons of iron, at present exported from the United Kingdom, would find a market elsewhere, in case it was no longer required by our rivals in the production of the metal. It is to be apprehended however that the power of increasing the make of iron among civilized nations will advance more rapidly than the means of consuming it will grow among those who, like the 517 million Asiatics, pass their lives without needing above ·49 lbs. per individual per annum.

There is another source of relief to the manufacturer, which promises a more ample area of consumption than the advancing civilization of semi-barbarous countries, viz., new fields of enterprize among those nations which are already large consumers. No industry forms a better illustration of my meaning than that of shipbuilding. I remember the arrival of John Coutts on the Tyne in the year 1840, when he began the construction of iron vessels in the yard now occupied by Sir W. G. Armstrong, Mitchell, & Co. But, according to the

Report of the British Iron Trade Association, so slow was the progress of this new art that ten years afterwards, viz. in 1850, the total tonnage of iron vessels built in the United Kingdom was only 12,800 tons, while that of timber vessels was over 120,000 tons.

According to the same authority, there were built in subsequent years as follows:—

	Timber.	Iron.
1860	147,000	64,679
1870	161,000	255,000

Wooden shipbuilding is now fast disappearing, while that of iron is extending beyond the wildest expectations. In the year 1880 the tonnage of iron and steel vessels built was 796,221 tons, and in 1881 it reached 1,013,208 tons. For the amount of shipping built in the latter year, including the machinery for that portion which was to be propelled by steam power, Mr. Jeans estimates that not less than 800,000 tons of iron were consumed.

I would conclude these few remarks on iron shipbuilding by contrasting the small amount of success which attended Coutts' efforts with the present position of the Tyne, on the banks of which he was the pioneer in the art. In the year 1881 the iron vessels launched from the yards on this river alone reached 177,165 tons; or 84 per cent. of the entire tonnage of the United Kingdom, wood and iron, constructed 21 years before, viz. in 1860.

Mention was incidentally made, a few pages back, of the resources possessed by the United States in the minerals required for the production of iron. This may be supplemented by the assertion that so enormous are those resources that, taking the country as a whole, no conceivable demand can have to face the possibility of being, even within many centuries, followed by exhaustion. Such at least is the impression left on my mind as the result of repeated enquiries in all the large coal-fields on this side of the Rocky Mountains, and of an examination of all the chief iron ore districts, from the shores of Lake Superior to the Southern States of Tennessee and Alabama.

Under such circumstances the question may well be asked as to the probability of the United States, after satisfying their domestic wants, being able to administer to those of the world at large. Now so far as the possession of mineral wealth is concerned, there is nothing to prevent the production there of any imaginable quantity of pig iron, and its conversion into the usual forms known in commerce. The accomplishment of such a task would however demand a large addition to the labouring classes of America; an addition which not even its rapid increase of population would be able, for some years, to meet.

An important factor in the solution of such a problem ought to be the cost of production. Wages already high, stimulated by high prices and very large profits, produced, as will be pointed out in a section devoted to the iron trade of America, a very large inflation in the value of labour. This impediment to the meeting of competition in the open market is more or less of an artificial character, and probably in time will be partly removed. The geographical position of the ore and coal, and of the markets themselves, constitute on the other hand obstacles of a more insurmountable description. The distances over which ore is conveyed are sometimes very great; as an example the produce of the Lake Superior Mines is carried to Pittsburg, involving carriage of 790 miles. The cost of transport on the minerals consumed for each ton of pig iron I have calculated[1] to average 10s. 9d. at the eight chief seats of the iron trade in Great Britain; whereas in the United States the mean charge at fourteen of the large centres is 25s. 8d. No doubt a much more favourable condition of things obtains in the Southern States, where the fuel and ore lie in immediate proximity. For the purposes of foreign trade the Northern States are under a great disadvantage, as compared with some of the principal centres of the iron manufacture in Great Britain. It costs the Scotch or the Welsh ironmaster about 3s. to put a ton of his produce on board a ship, and at Middlesbrough vessels are loaded at the wharves attached to many of the furnaces at a charge of less than sixpence. Against these favourable terms the American manufacturers in many of the iron districts are rarely less than 100 miles and often far more than 100 miles distant from the shipping port. But this and the position of the Southern States will be most conveniently considered in the Section on the United States.

If however the future action of the American protectionist manufacturer has to be judged by his prototype in Germany it is difficult to predict the precise line of conduct he may adopt. The

[1] Report to Her Majesty's Government on Iron Manufacture of the United States compared with that of Great Britain.

same ore is shipped at Bilbao for Westphalia and for Middlesbrough. The sea freight is at least as favourable to the Tees as to Rotterdam, but, when in the Tees, the ore is virtually at the furnaces, whereas the Westphalian manufacturer has to pay 5s. to 7s. per ton before it arrives at the smelting works. In spite of the higher wages in every department, I am within the mark when I say that with pig iron at the same price, the mere act of conversion in most cases is performed in England fully as economically, if not more so, than it is in the Rhenish provinces. Notwithstanding an increase of cost on the pig of 10s. to 14s. due to the inland carriage, and a second charge of 8s. to 10s. per ton for bringing the steel rails back to the port, the English makers have been, on several occasions, undersold, at a time when scarcely the cost of production was being realized in England. The payments mentioned above, for inland carriage, involve a charge of 20s. to 25s. per ton of rails, from all, or a greater part of which, many British makers are exempt. The German manufacturer, however, seems to prefer a loss so incurred, to stopping his works.

I would point out that the quantity of iron exported from Germany is very considerable. It was above a million tons in 1882; and although the observations made above are not applicable to the same extent to all this quantity, yet the orders for which the Westphalian makers competed successfully were of a very important character.

Is it not true then that German capitalists, attracted by the prices artificially raised by the import duties, have extended the powers of production to a dangerous point? From this the only escape has been the exportation of a portion of their produce at a loss, which loss is recouped by the better prices paid by the inland consumer.

The object of the present section is not to render a complete statistical account of the make of iron and its destination. This has been well done by the British Iron Trade Association in its yearly publications, which merit the attention of all those interested in the subject. My attention was intended to be confined to the use of such figures as would exhibit the relative positions of the different iron-producing nations, accompanied by some information of a general kind to indicate the application or destiny of their products.

SECTION XVI.

BRITISH LABOUR COMPARED WITH THAT OF THE CONTINENT OF EUROPE.

OF the sums which constitute the cost of British-made iron, those which are not directly effected by the money-value of British labour form comparatively a small part. The item least dependent on the wages market is the not unimportant one of royalty; but even there we have indications which may lead us to suppose that the owner of the soil does not remain altogether uninfluenced by the upward tendency of the earnings of those who extract the mineral from beneath the surface.

Be this last-named conjecture as it may, it could be shown that the labour engaged in mining, conveying the raw materials to the furnace, and smelting the ore, forms about 80 per cent. of the cost of pig iron. When we come to examine the details of the expense of producing malleable iron or steel, even in their simplest forms, it will be found that wages make up about 90 per cent. of the whole. Under such circumstances the actual cost of a given amount of labour is an all-important factor in examining the subject of the present section.

The lowest wage earned by the working class in Europe is usually to be found in those parts which are purely agricultural. The personal expenses of those who devote themselves to the cultivation of the soil, or who follow a pastoral life, are usually less than the outgoings of those occupied in other industrial pursuits: and when this most ancient occupation is followed in a warm climate, lightness of clothing, and a diet chiefly vegetable, assist in keeping down the cost of living.

The sudden introduction on a large scale of a manufacturing industry into an agricultural district, tends to disturb the value of labour which may have prevailed for many years. On the other hand mining, or any employment not requiring long training, may find a footing to a moderate extent without greatly affecting the normal rate of wages of a district. Such at least has been the experience of the lead-mining districts of the North of England.

In illustration of what has been advanced may be quoted the daily pay of the labourers on an Indian tea plantation, who, as well as the workmen at a small blast furnace adjacent, receive only five pence a day. In this case the furnacemen's wages, although higher perhaps than those of the ordinary agricultural labourer, do not exceed those of the tea cultivator.

On the shores of the Mediterranean, between Malaga and Gibraltar, the rate of agricultural wages was given me in 1872 as equal to 1s. 0½d. per day; but the labourer was able to keep himself on 4d., his daily nourishment consisting of about 2 lbs. of bread and one ounce of olive oil, with a little salt, boiled into the national dish *gaspacho*.

The workmen at a charcoal iron furnace, situate between these two towns, were being paid as follows :—Iron ore miners, 1s. 3d. ; furnace keepers, 2s. ; slag men, 1s. 5½d. for one day's work.

In the event of a new industry being established on a small scale, the wages earned by the workmen continue to be regulated more or less by the rates previously current in the country. Thus in another part of Spain I visited furnaces making 280 tons of iron per week, where the head keeper was paid 4s. 4d. for his day's work, and assistants and slagmen from 2s. 2d. to 2s. 10d. The blast engineman for three furnaces had only 2s. 10d., and his fireman 1s. 10d. One-third of the workpeople were women, who received 1s. 5d. Partly owing to less perfect arrangements than are found at the latest erected works in Great Britain, and partly no doubt owing to the inferior living, there were nearly one-half more people employed at the establishment in question, for very little more than half the iron produced, as compared with the best Cleveland furnaces. The average earnings of the workpeople, all told, were under 2s. per day, or less than one-half the rate paid in England; notwithstanding which, for the reasons already assigned, the cost for labour per ton of iron was about the same as that incurred in the district with which the comparison has been made.

Old as is the iron trade in Sweden, its relative extent, and that of manufactures in general, as compared with agriculture, is not such as to have greatly affected the price of labour, as determined by the rates paid for cultivating the soil. The same may be said of the Austrian dominions.

	Austria, 1865.		Sweden, 1866.
	s. d.		s. d.
Coal miners	2 0	(underground, working 12 hours).	—
Iron ore miners	2 0	(working 12 hours)	1 7 (working 12 hours).
Blast furnace keepers			2 0
Rollers			2 6 to 3s.
Joiners			1 7
Blacksmiths			2 3 (working 10 hours).

It is some years since these rates were obtained in the two countries just mentioned; but a Government Report[1] prepared for the Paris Exhibition gives 1s. 3d. to 2s. 6d. as the earnings in 1878 of an Austrian coal hewer.

Having thus shown in general terms that on the first appearance of manufacturing enterprises in an agricultural country they have their rate of wages determined in a great measure by the value of agricultural labour, we will now proceed to review the circumstances attending a more advanced state of development as applied to the iron trade.

The cost of food is an all-important element in considering the question before us, for it is to the labourer what wages are to the ironmaster. In both cases we have to deal with the chief expense in producing that article which the workman has to sell and the manufacturer has to buy, viz. labour.

Some remarkable changes have taken place in later years in the nominal value of objects of daily consumption. Often however, in cases where there has been an increase in apparent value, the alteration is due to the greater abundance of the precious metals; and on the other hand, where prices have fallen, the difference is generally the result of improved methods of production or in the conveyance of material. At other times these two opposing forces, so to speak, may have a tendency to counterbalance each other in the effect upon the commodity.

The cost of the food previously mentioned as consumed by an agricultural labourer in Spain at the present day, appears to correspond

[1] Die Mineralkohlen Oesterreichs, 1878, p. 66.

in amount with that of his predecessor in England 200 years ago; for, according to Sir William Petty, a countryman's wage was then fourpence per day with his food, or eightpence if he found himself in victuals.

Macaulay, in his History of England, quotes beer and butcher meat being much cheaper in 1685 than it was when he wrote, about thirty years ago. Yet the price of meat, 3d. to 3¼d. per lb., was such that thousands of families, he says, scarcely knew the taste of it. On the other hand, according to this author, the produce of tropical countries and the produce of the mines was positively dearer than at present. Under such a condition of things the labourer in 1685 would have to pay more for such absolute necessaries as sugar, salt, coals, candles, soap, shoes, stockings, and generally all articles of bedding. At the same time Macaulay gives it as his opinion that blankets and garments in the older time were not only more costly, but less serviceable than they are in our own.

There is no more important item of household consumption of the working classes than flour; yet Macaulay mentions that during the last twelve years of the reign of Charles II. wheat was 50s. per quarter. Bread, he says, such as now is given to the inmates of a workhouse, was then seldom seen even on the trencher of a yeoman or a shop-keeper. The great majority of the nation lived almost entirely on rye, barley, and oats. The average price of wheat during the last five years ending 1879 has fluctuated between 41s. 7d. and 54s. 7d., the average being 48s. 1d., or 4 per cent. lower than it was two centuries ago, irrespective of the greater value of money in former times. This therefore is a case where improved agriculture and ameliorations in transport have countervailed and much more than countervailed the enormous depreciation in the circulating medium.

With the facts and figures just enumerated before us, and with the knowledge that the lowest wages in the agricultural districts contiguous to our great centres of industry are at least four times as high as they were two hundred years ago, no one can deny an immense improvement in the condition of the labouring population of our time, as compared with that of their predecessors. As has been shown above, the purchasing power of money in the reign of Charles I. was much greater than it is at the present time; but so far as wheat, which forms now so important an ingredient in the workman's food, is concerned, as much can be bought for a shilling to-day as could be obtained for that coin, two hundred years ago.

Labouring men at present live in better houses, eat better food, have better clothing, and are able to command enjoyments utterly beyond the reach of their forefathers. Their employers on the other hand receive a larger amount of labour, afforded with greater ease by those who produce it than the lesser quantity by the badly paid, badly housed, badly fed, and badly clothed workmen of two centuries ago. Indeed, for an illustration of the connection of good wages with good work, we do not require to go back two hundred years; for our own times afford us the means of comparing the condition and value of well paid men with those who are the reverse.

It matters little however, in a comprehensive view of the labour question, how well a man is paid, unless some consideration is given to the purchasing power of the money he receives.

This purchasing power of the workman's earnings is evidenced of course by the prices paid for the usual articles of domestic consumption; and any advantages in this respect would necessarily permit a workman to dispose of his labour more cheaply than one less favourably situated could afford to do. There is no doubt that forty years ago the foreign workman was able to command the necessaries of life on much easier terms than were our own countrymen. I have myself remarked this at the time, when on visits to Continental iron works; but it has also come within my observation that since 1840 a remarkable alteration in this respect has taken place, many articles being now supplied in Great Britain at much cheaper rates than were formerly paid, while in France, Germany and elsewhere the same commodities have greatly risen in price. The adoption of Free Trade in this country, and the facilities afforded to transport both on land and water by the introduction of steam as a propelling power, have placed the United Kingdom in economical correspondence with foreign seaports.

The actual weight of weekly food, required for a family of two adults and four or five children, does not amount to 50 lbs. Under such conditions, a very few pence per week will represent the extent of the disadvantage of the British workman, as compared with the foreigner, supposing the former were exclusively fed on produce grown near the home of the latter

The Continental farmer has thus found in the British consumer a customer for produce rarely if ever sent from his own neighbourhood in former days. Such a development of our relations with foreign nations has of course tended to equalise prices here and abroad—the increased demand on the Continent has raised the prices there, while the increased supply with us has had an opposite effect—greatly to the benefit of the producer abroad and to the consumer in the British Isles.

As examples of this, I would mention that at the time of a journey undertaken in 1847, butcher meat was being sold in France and Germany as low as 4d. per lb. I made no memorandum of the prices charged in the North of England at that time; but it would probably be one-half more at least than the figure just named.

Twenty years after this, viz. in 1866 and 1867, I find the following quotations in my note books, for animal food:—

	Per Lb.
Sweden	6 to 7
France, St. Etienne	$6\frac{1}{4}$
" Bouches du Rhône	$6\frac{1}{4}$ „ $7\frac{1}{2}$
" Central	$5\frac{1}{4}$ „ $5\frac{3}{4}$
" Eastern	$6\frac{3}{4}$ „ $7\frac{1}{4}$
Belgium	$7\frac{3}{4}$ „ $8\frac{1}{4}$
Austria	6

These prices show an increase of about 75 per cent. as compared with 1847, and were only about 10 per cent. cheaper than the rates prevailing at the same time in the mining districts of Durham and North Yorkshire.

As a further instance of the levelling tendency in prices in recent years, may be quoted the fact that Continental Europe as well as ourselves is now drawing large supplies of provisions from the United States. American cheese and American bacon are now to be found in many parts abroad; with this difference however that they are usually saddled with an import duty, from which legislation has fortunately relieved the British subject. The advantage, from a consumer's point of view, of the legislation for which we are mainly indebted to Villiers, Cobden, and Bright, is conspicuously seen in comparing the prices of wheat, given a few pages back, with those which obtained before the change in our Corn Laws just referred to.

SECTION XVI.—THAT OF THE CONTINENT OF EUROPE.

The following are decennial averages for England and Wales, beginning with that ending in 1810, taken from the statements prepared by the Receiver of Corn Returns:[1]—

10 Years ending	1810.			1820.			1830.			1840.			6 Years ending 1846.		
	£	s.	d.	£	s.	d.	£	s.	d.	£	s.	d.	£	s.	d.
Average per quarter	4	4	0	4	7	6	2	19	6	2	16	11	2	14	9
Highest price ...	5	19	6	6	6	6	3	8	6	3	10	8	3	4	4

The average price of wheat during the eight years ending 1860 was £2 17s. 8d. per quarter; this in the succeeding eight years fell to £2 12s. 2d., then to £2 12s., and in the years 1876–1879 it was only £2 9s. On the other hand, in the year 1852–53 we grew 11·57 millions of quarters, and imported 5·9 millions: in the year 1878–79 we produced 12·64 millions, and imported 14·43 millions. It has also to be remarked that our home-grown wheat and that supplied by other countries in 1852–53 sufficed to keep the price at £2 4s. 7d. per quarter; but in 1853–54, owing to a bad harvest, our home supply fell to 10·64 millions; and the market being unprepared, having only imported 6·09 millions, or practically the same as the previous year, the price rose to £3 12s. 11d. The Crimean war then broke out, and interrupted our relations with Russia; our imports of wheat fell off in 1854–55 and 1855–56 to 2·98 and 3·26 millions respectively, and the prices consequently averaged £3 12s. during those two years.[2] When it is remembered that this change would inflict an extra outlay of about 2s. per week on a family of two adults and four children, and that in cases where the income did not exceed 18s. or even £1 per week, it is easy to conceive the inconvenience, not to say actual privation, which would flow from such a state of things.

It is quite true that with free importations of corn, we have sometimes had very high prices. Thus in 1866–67, 1867–68, and 1873–74, wheat averaged £3 0s. 4d., £3 8s. 4d., and £3 1s. 3d. respectively; but in each of these years the harvest in the United Kingdom was very bad, and the importations were not large enough to supply the deficiency. The pinching effect of these high prices is manifested in the quantity consumed, which, instead of being 5·67 bushels as it was in 1876–1879, fell to 4·80–4·63 and 5·17 bushels per head of the population in the three years just mentioned; and as the short

[1] Encyclopædia Britannica, Vol. 7, p. 398, Ed. 1860.
[2] Paper by Lawes and Gilbert.

allowance would fall exclusively on the poorer orders, it is easy to comprehend the suffering which would be their lot. Compare however a sufficient number of years, and then the advantage of the change is clear enough. The average price of wheat over the 25 years ending 1843 was 59s. 2d. per quarter; whereas in the same period beginning 1848 after the repeal of the Corn Laws the price was only 51s. 6d.

The upward tendency already mentioned in prices of food abroad has continued from 1867 down to the present time; for which statement, I will here insert figures gathered from the writings of other authors.

M. Chatelant, of the Statistical Bureau in Berne, shows by quotations of the actual prices of provisions in Switzerland that between 1840 and 1850 there was an increase of about 75, and in some articles of 100 per cent. There was a further rise between 1861 and 1872 of from 30 to 40 per cent.; and at the same time clothing, house-rent, and in short all the necessaries of life, had experienced a corresponding change in value.

I had placed in my hands by M. Vuillemin, the Director of the large collieries of Aniche in the North of France, a printed table showing the prices of articles of domestic use in the Pas de Calais, given in francs, and in periods of ten years:—

10 Years ending	1830.	1840.	1850.	1860.	1870.	1870 to 1878.	Increase per cent. from 1830 to 1878.
Butcher meat, per kilo.	0·62	0·75	1·01	1·06	1·34	1·66	157
Butter, per kilo.	1·50	1·55	1·74	1·86	2·57	3·04	102
Potatoes, per hectolitre	2·33	3·12	4·34	4·25	4·06	4·82	107
Shoes, per pair	2·45	2·17	2·50	3·24	4·68	6·53	166
Coarse blue cloth per metre	4·16	8·05	7·56	8·20	7·68	7·12	71

According to Dr. Edw. Young,[1] Chief of the United States Bureau of Statistics, who made labour and cost of living in Europe and America the subject of extensive enquiry, the advance in the cost of provisions in France in 1855, as compared with the ten years ending 1833, was as follows:—

	Butcher Meat. Per Cent.	Fowls. Per Cent.	Butter. Per Cent.	Eggs. Per Cent.	Potatoes. Per Cent.
Increase of price	34 to 36	35 to 51	29	25	103

[1] "Labour in Europe and America."—*Young.*

SECTION XVI.—THAT OF THE CONTINENT OF EUROPE.

This writer further states that beef, which was selling for 4½d. at Lyons in 1855, had risen in 1875 to 6¼d. He also gives some comparative tables of prices in different countries of Europe. His quotations unfortunately do not apply to the same year for the various localities, being for 1871 in some places, and 1872 in others. Wheat however in England and Wales only exhibited a difference of 4d. per quarter between the two years in question. We may therefore infer that there would not be any material change in other articles of food.

PRICES OF PROVISIONS AS GIVEN BY DR. YOUNG.

	England, Bradford, 1872.			Germany, Düsseldorf, 1872.			France, St. Etienne, 1871.			Belgium, Charleroi, 1872.			Austria, Trieste, 1872.			
	£	s.	d.	£	s.	d.	£	s.	d.	£	s.	d.	£	s.	d.	
Wheat flour	1	13	0	1	13	0	1	13	0	1	13	10	2	8	8	per barrel, 280 lbs.
Beef	0	0	9¼	0	0	8½	0	0	9	0	0	8¾	0	0	6	per lb.
Mutton	0	0	8	0	0	8½	—			0	0	9	0	1	0	,,
Lard	0	0	8	0	0	8½	—			—			0	1	0	,,
Cheese	0	0	9	0	0	7	0	0	7	—			0	0	5	,,
Tea	0	2	6	0	2	6	0	3	4	0	3	2	0	3	4	,,
Roasted coffee	0	1	4	0	1	3	0	1	6	0	1	1½	0	1	8	,,
Brown sugar	0	0	4	0	0	6	0	0	6½	0	0	7½	0	0	6	,,

Coming down to later years, and quoting from the result of my own enquiries, in October, 1878, I find the following information :—

	GERMANY.								GREAT BRITAIN.	
Town	Saarbrück.		Mühlofen.		Ruhrort.		Hörde.		Middlesbrough.	
	s.	d.	s.	d.	s.	d.	s.	d.	s.	d.
Wheat flour, per stone of 14 lbs., seconds	2	6¼	2	1	3	0	1	9½	1	11	—
Bread, per lb.	0	1½	0	1½	0	1⅓	—		—	
Potatoes, per st.	0	5	0	3½	0	3	0	5	0	8	—
Beef	0	7	0	7¼	—		0	7¼	0	6 to 10	according to quality.
Mutton	0	9	0	7¼	—		—		0	6 to 10	,,
Bacon	—		—		—		—		0	5½ to 6	,,
Lard	—		—		—		—		0	7	—
Cheese	—		—		—		—		0	7	—

Admitting differences of climate, fertility of soil, etc., to afford certain advantages to the inhabitants of a Continental locality, in obtaining the necessaries of life, it is clear that such advantages must be limited, as far as the British consumer is concerned, by the cost of transporting agricultural produce thence to the markets of Great Britain; and this, on about 10 lbs. per head, amounts to an insignificant sum. Practically the general result of my investigations has led me to think that this is in reality the case; and that any difference against the British workman, in the cost of certain articles of food, is probably compensated for by the superior facility with which his insular home, as compared with Continental Europe, is supplied with other provisions from the United States.

As a set-off against the supposed equality of position just spoken of, it has often been asserted that the foreign workman lives less luxuriously than do the inhabitants of Great Britain of the same class. In the matter of quality of food, our own workmen are perhaps more fastidious than are their fellows on the Continent; for while the latter are satisfied with animal food of a second quality, the former, certainly in times of moderate prosperity, eat the best of its kind. There may be some economy in this, for no one knows better than does the hard worked man about our blast furnaces and rolling mills that his occupation demands a generous diet; without which he could not face the severe strain on his physical powers.

Many enquiries have been directed towards ascertaining the actual cost of living in different countries. Without claiming for any such statements absolute accuracy, I here insert lists of various localities, supposing it probable that the errors on the whole are not greater on one side than on the other.

In the year 1879 a commission composed of German iron masters met to consider the question of further protection being granted to their trade, by an increase in duties levied on iron imported into Germany. Upon that occasion they took evidence on the weekly expenses of maintenance incurred by men occupied at iron works.

Case No. I. was that of a man engaged in unloading minerals at blast furnaces (generally done by shovel work), with a wife and six children, two to twelve years of age.

Case No. II.—A man mixing ores and limestone (shovel work), with wife and two children.

Case No. III.—A man unloading castings, with wife and four children, two to eighteen years of age.

	Case No. I. £ s. d.	No. II. s. d.	No. III. £ s. d.
Black bread (Schwarz Brod)	0 2 7	0 9	} 0 3 0
White bread	0 0 8	1 0	
Flour	0 1 5	1 3	
Butter	0 1 7	1 4	0 1 9
Meat, bacon, and lard ...	0 3 5	2 6	0 3 2
Vegetables	0 2 6	2 10	0 4 1
Coffee and chicory	0 0 11	1 2	0 0 9
Milk	0 0 7	0 9	0 0 7
Groceries, etc....	0 0 8½	0 11½	0 1 1
Total for provisions ...	0 14 4½	12 6½	0 14 5
Clothing	0 2 9¾	2 5½	0 2 9½
Washing materials	0 0 3	0 4¼	0 0 5
Tobacco	0 0 3¾	—	0 0 3¾
School	0 0 4½	—	0 0 2¾
House rent	0 2 6	2 0	0 1 10½
Firing and lighting	0 1 8	1 6¾	0 1 7½
Taxes	0 0 2½	0 2½	0 0 7¼
	£1 2 6	£19 1½	£1 2 3¼

Dr. Young, in his work already quoted, gives the following as the cost of living in various parts of Germany, in the year 1872. Each family consisted of two adults and of the number of children named at the head of each column:—

Locality.	Barmen.		Essen.		Aix la Chapelle.	Berlin.	Stuttgart.	
No. of children ..	Two.	Four.	Three.	Three.	Two.	Three.	Three.	Five.
	£ s. d.	£ s. d.	£ s. d.	£ s. d.	£ s. d.	£ s. d.	£ s. d.	£ s. d.
Provisions	0 17 2¼	0 18 9	0 15 5	0 17 4	0 16 10	0 15 8	0 16 3½	0 16 5
Clothing	0 1 2½	0 5 2	0 4 0	0 4 0	0 1 9	0 2 3½	0 2 10½	0 1 3½
Tobacco, beer, etc...	0 2 0	0 0 6	0 1 5	0 1 4½	0 2 6	0 2 3½	0 1 0½	0 5 0
School and church ..	0 1 7	0 0 6	0 0 5	0 0 5	—	—	0 1 8	—
Rent, fuel, and light	0 3 4	0 3 5½	0 3 7½	0 4 5	0 5 4	0 7 0	0 9 7½	0 4 10
Sundries, including taxes	0 0 2	0 0 1½	0 1 4½	0 1 4	—	0 0 10½	0 0 4½	0 1 8
	£1 5 5¾	1 8 6	1 6 3	1 8 10½	1 6 5	1 8 1½	1 11 10½	1 9 2½
Weekly earnings ..	£1 7 0	1 10 0	1 7 6	1 10 6	2 0 3	1 7 0	1 13 4	1 13 4

In the case of the Barmen operatives the wages include the earnings of the wife. At Aix la Chapelle a son assisted the father. At

Essen and Stuttgart the head of the family alone contributed to its support. The occupations in these instances are not given; but they all go to prove that with an income of 27s. to 30s. per week the expenses are usually within a trifle of the earnings.

A case is quoted by Dr. Young of a family in Belgium, consisting of two adults and four children, whose weekly expenditure seems to be on a miserable scale, and was as follows:—

	s.	d.
Provisions...	10	0
Clothing and bedding	4	0
Rent	0	7
Fuel, lighting, etc.	0	7
Total	15	2

I obtained from four different classes of men working for my own firm a statement of their weekly expenditure. From these I have selected four cases of labourers, not miners, each with a wife and four children. These, it may be remarked, are very much of the same class as the three cases of German workmen given above. No. I. and II. were engaged at a colliery and III. and IV. at an ironstone pit:—

	No. I.		No. II.		No. III.			No. IV.		
	s.	d.	s.	d.	£	s.	d.	£	s.	d.
Flour and potatoes	5	5	6	1½	0	4	11½	0	7	0½
Groceries, milk, soap, etc. ...	1	8	2	11	0	2	5½	0	3	2
Meat, fish, butter, etc. ...	2	9½	2	0	0	5	6	0	2	11
Total provisions and groceries	9	10½	11	0½	0	12	11	0	13	1¼
Clothing	1	6	3	0	0	4	0	0	1	3
Tobacco and beer	0	6	—		—			—		
School	0	8	1	0	0	0	8	0	1	0
House rent and coal	3	9	4	6	0	3	6	0	3	6
Lighting	0	3	0	3	0	0	3	0	0	8
Doctor, relief fund, and union levies	0	9	—		0	0	9	0	1	0
Sundries	0	8½	—		0	0	6½	0	0	2¼
	18	0	19	9½	1	2	7½	1	0	9

The prices on which these estimates were based were:—

SECTION XVI.—THAT OF THE CONTINENT OF EUROPE. 487

Article.	Quantity. Lbs. Oz.	Price. s. d. s. d.	
Flour	about 25 0	2 1 to 2 6	per stone of 14 lbs.
Barley, &c.	,, 1 0	— —	
Potatoes	,, 14 0	0 10 to 1 0	,,
Animal food	2 5	— —	
Beef	—	0 7 to 1 0	per pound.
Bacon	—	0 5 to 0 7	,,
Lard	—	0 7 to 0 9	,,
Tea	0 4	2 6 —	,,
Coffee	0 6	— —	
Sugar	3 0	0 3 to 0 3½	,,
Coals	—	8 6	per ton.
Total weight	45 15		

It will thus be seen from the two sets of figures given, setting forth the cost of living in Germany and in England, that the difference between the two is insignificant for men engaged in similar occupations.

According to a return in my possession, the following shows the weekly expenditure of men engaged in labour appertaining to the production of pig iron, in the North of England; the family in each case consisting of two adults and four children:—

	Coal Miners.			Ironstone Miners.		Limestone Quarrymen.	Blast Furnace.
	£ s. d.	£ s. d.	£ s. d.	£ s. d.	£ s. d.	£ s. d.	£ s. d.
Flour and potatoes	0 5 4½	0 7 5	0 5 4	0 4 11½	0 6 6½	0 5 0	0 4 0½
Groceries and milk	0 3 4	0 3 3	0 4 6	0 2 9	0 3 0½	0 2 6	0 2 7
Meat, fish, butter, etc.	0 6 6	0 8 1	0 7 10	0 8 3	0 3 8	0 5 6	0 8 6½
Total provisions	0 15 2½	0 18 9	0 17 8	0 15 11½	0 13 3	0 13 0	0 15 2
Clothing	0 3 3	0 7 0	0 6 0	0 5 0	0 2 3	0 5 0	0 6 0
Beer and tobacco	0 1 6	—	0 0 7	—	0 0 3	0 1 0	—
School	0 0 11	0 1 0	0 0 3	0 0 8	0 0 6	0 1 0	0 1 0
Rent and fuel	free.	free.	free.	0 4 6	0 4 10	0 5 0	0 0 6
Lighting	0 0 4	0 1 0	0 0 5½	0 0 8½	0 1 2	0 1 0	0 4 8
Sundries	0 1 8½	—	0 0 6½	0 0 7½	0 0 8½	0 0 1	0 0 6
Maintenance	1 2 11	1 7 9	1 5 6	1 7 5½	1 2 11½	1 6 1	1 7 10
Doctor, Relief Fund, and Union	0 1 10	0 1 0	0 0 9	0 0 9	0 0 9	0 1 0	0 0 3
	£1 4 9	1 8 9	1 6 3	1 8 2½	1 3 8½	1 7 1	1 8 1

Among the difficulties the labouring classes have to contend with is the occasional high price of provisions. A return was made to a large mineral owner in the Cleveland district, from a man on whom he could rely, by which it appeared that, in the years 1873 and 1879, on six articles of ordinary consumption there was a difference of 50 per cent. in the matter of cost; the same quantities being assumed in each case. The figures were as below:—

Article.	Quantity.	1873.				1879.			
		Price.		Cost.		Price.		Cost.	
		s.	d.	s.	d.	s.	d.	s.	d.
1.—Flour	2 stones	2	6	5	0	1	9	3	6
2.—Bacon	4 lbs.	0	8	2	8	0	5	1	8
3.—Cheese	1 ,,	0	10	0	10	0	6	0	6
4.—Lard	1 ,,	0	8	0	8	0	6	0	6
5.—Tea	½ ,,	3	4	1	8	2	0	1	0
6.—Sugar	4 ,,	0	4	1	4	0	3	1	0
Cost per week				12	2			8	2

Fortunately such differences as are perceptible in the figures just given, are of much less frequent occurrence than when heavy import duties added greatly to the cost of the necessaries of life.

The fall in prices recently observable in bacon, cheese and lard is no doubt due to the large importations from America; and in the case of tea and sugar the decline in price is caused by the changes in duty. The British consumer is therefore greatly indebted to the policy of Free Trade, introduced about thirty-six years ago; its tendency being to afford a constant and plentiful supply of provisions at moderate rates, and thus avoid the privations which must obviously beset a household which, with an income of 20s. a week or even less, finds itself suddenly called upon to expend an additional 4s. on the necessaries of life.

Besides the power of thus obtaining the raw material of food at a reduced cost, great improvements in the machinery for grinding corn have enabled the miller to carry on his business on a much more economical footing than was possible long after the beginning of the present century. Flour was formerly made almost exclusively in small mills, driven by wind or by water, and in such insignificant quantities as to render the application of expensive machinery impracticable. The present large importations of corn from abroad, together with the easy concentration of home-grown produce in one locality by means of

SECTION XVI.—THAT OF THE CONTINENT OF EUROPE.

railways, have led to the creation of enormous establishments; and in consequence, as I am informed by Mr. Appleton of the Stockton flour mills, labour is not only greatly economised, but an improved yield and quality of flour are obtained. The mechanical engineer, in point of fact has done for the loaf of bread that which he has achieved for cotton and for wool; and as a consequence we enjoy advantages in food and clothing unknown to our ancestors.

The price paid by workmen for board and lodging perhaps affords a ready criterion, by which we may judge of the relative cost of living on the Continent and in Great Britain. At a large iron work in Central France, the charge in 1867 was 13s. 2½d to 16s. per week, which included animal food once a day, selling at that time at 7d. per lb. This was quite as much as was being paid in the North of England at the period in question.

In the matter of clothing I obtained many quotations; but it is useless to enumerate these, owing to the great differences in quality, etc.

The markets now open to the British manufacturer for his raw material, and his unsurpassed machinery, forbid the idea that the foreign workman can be better or more cheaply clothed than are our own men; and this is confirmed by the statements of expenditure under this head contained in the tables already given in these pages.

It would also be difficult to institute any exact comparison in the item of house rent, the extent of accommodation varying so much in different localities. In the counties of Northumberland and Durham our miners are infinitely better lodged than they were thirty years ago. It is the custom to provide the colliery population in the North of England with dwellings rent free, and any concession in the way of superior accommodation is so much money out of the pocket of the coal owner. Nevertheless it would probably be far within the mark to say that our miners' houses now cost fully double what they did anterior to the period above mentioned. So far as my own observation goes, I should say that the domestic life of our own workpeople is, to say the least of it as replete with comfort as that of men in a similar position in most places abroad.

The foregoing examples of the cost of living are not given under the belief that they represent either on the Continent or in the United Kingdom, the *average* expenses of all families. All that is wished to be inferred from them and from numerous other cases in my possession, is that the cost of living may now not materially differ in the different countries.

There is little doubt that there are many workmen, who, receiving large wages in the rolling mills and iron shipbuilding yards in Great Britain, spend much more on their households than any of the sums mentioned above. One thing is quite sure, viz. that the style of living is of a much more nourishing character in this country than it was anterior to the repeal of the corn laws.

The Right Hon. John Bright, whose name has been already mentioned in connection with the abrogation of the laws referred to, in a speech delivered a couple of years ago, described the effect of the change in the habits of our labouring population. He said, "I suppose if, at this moment, all the workmen and their families in the three kingdoms could be put into the scales, they would weigh some thousands of tons more than they would have done thirty years ago."

In the previous Section reference was made to the expectation of an increased consumption of iron among the nations of the earth, by many of which it was exceedingly small. The gentleman, whose opinion I have just quoted, observed in a letter, "As to the consumption of manufactures it is now small when compared with the wants of the populations of the various countries of the globe. The great armies of Europe and the exactions of Government tend to limit the enjoyments of the people and thus consume the proceeds of their industry. The time will come, I do not doubt, when nations will be wiser."

There is one item in the tables of expenditure already set forth, which has often formed a theme for observation and discussion, particularly in recent years; viz. the consumption of fermented and spirituous drinks. In several of the instances given this is conspicuous by its absence which probably arises from those who furnished the details being very steady men, and hence selected as being most capable of affording correct information.

Dr. Young in his work devotes 30 closely printed pages to the consideration of the condition of the working classes of Great Britain. He, like many other writers, connects "the evils which afflict the British workmen and their families with the excessive use of spirits and beer." The opinions of this writer are the more worthy of respect because, as a citizen of the United States, he admits that his countrymen, next to the inhabitants of Great Britain, are the most conspicuous for their large consumption of intoxicating beverages.

This national reproach is supported by quotations from various British reports, both statistical and otherwise, not only of the amounts consumed but of the effect produced on the manners and morals of the people. It has to be remarked, however, in reference to this unenviable notoriety, that, in the hope of mitigating the evil, Great Britain has perhaps taken more pains than any other nation to reveal to the world the consequences, as well as the extent, of this sore in our social system. It is therefore possible that the charge of much greater intemperance laid at our door may, to some extent, be the result of more ample confession, rather than of a very much more extensive prevalence of the vice itself.

Dr. Young is almost entirely silent in respect to the drinking customs of other nations. He alludes to the use of the deleterious spirit known as absinthe in France; and, on the authority of the Consul of the United States in Belgium, he describes the working men of that country as being afflicted with a "terrible misfortune, who not only drink beer but gin and rum, the latter being so cheap that thousands of labourers go reeling home daily from their toil."

I cannot say that the impressions produced on my mind by frequent visits to Belgium correspond with those of the American Consul; and his assertion in respect to the Belgian work people is, I think, inapplicable to that class in Great Britain. No doubt there are among the latter many who daily consume beer or spirits, or both, to an extent which proves injurious to their health and usefulness. As a rule, however, any approach to such a state of things as that just mentioned is reserved for the weekly, or the fortnightly, pay days. Apart from these occasions, there is in the demeanour of the people, outside of the public house, little proof of great addiction to drink at least among the miners and iron workers in the North of England. It would appear, however, that there is too intimate a connection between hard drinking and high wages; for the government returns show that in the year 1872, when the earnings of workmen were high, the actual money spent in beer, wine, and spirits exceeded that of 1871 by no less a sum than twelve million pounds sterling.

This unfortunate result of increased prosperity is however not confined to the British people. Between 1874 and 1878 I paid many visits to the iron works and mines in Europe and North America; and the common answer to my repeated inquiries was that the workmen, generally speaking, had not really profited by the increased value of their labour in the good times of 1872 and 1873. In all cases the reason assigned was the devotion of a considerable portion of their additional earnings to drink. In no quarter was this more strongly insisted upon than in the establishments I had an opportunity of examining in the United States.

According to an article which appeared in the *Times* newspaper, quoted in Cunningham's "Conditions of Social Well-being," the inhabitants of the United Kingdom drank in various forms in 1872 the equivalent of 72,000,000 gallons of pure alcohol against 65,000,000 consumed in 1871. Taking the population at 31,513,000 for 1871 and 31,841,000 for 1872, we have a consumption of pure alcohol per head of 2·04 gallons for 1871 against 2·28 gallons for 1872. This increase is equal to 11¾ per cent. and according to the *Times* "was probably due to the increased wages of labour, which have allowed the working man to indulge himself in more luxuries."

If the consumption of beer alone had to be accepted as the standard of comparison, the United Kingdom occupies an unenviable position. According to a writer in the *Journal of Applied Science* in January 1881 the following was the rate of consumption of this beverage among European nations:—

Belgium	33 Gallons.
United Kingdom	32 ,,
Germany	22 ,,
Denmark	12 ,,
Holland	9 ,,
Norway	8 ,,
Austria	7½ ,,
Switzerland	6 ,,
Sweden	5 ,,
France	4¾ ,,
Russia	¾ ,,

It would be very instructive, if we possessed the necessary data, to compare the actual quantity of pure alcohol consumed by different

SECTION XVI.—THAT OF THE CONTINENT OF EUROPE.

nations. I have consulted M. Maurice Block's *Statistique de la France comparée avec les divers pays*, from which I reduced the various beverages to a common standard of pure alcohol. I have not however been able to ascertain the quantity of distilled spirits consumed in foreign countries, and indeed there are so many difficulties in making the calculation, owing to the different and unspecified strengths of the beverages, that any estimate could only be regarded as mere guess at the truth. So far as my figures enable me to judge I am not satisfied that as a nation we really drink much more than some of our neighbours. This however is small consolation, for the fact remains that among our workpeople the habit of drinking is far too prevalent, and we must therefore hope, with the spread of education and the offer of more rational amusement, that we are on the way to some improvement.

In the meantime the facts below are not encouraging:—

Average annual consumption, 10 years ending ..	1850.	1860.	1870.	1879. 9 years.
A.—Spirits per head of population, gals.	·942	1·030	·946	1·201
B.—Wine ,, ,,	·224	·234	·430	·517
A and B.—Equiv. in spirits (5 B = 1 A) ,, [1]	·986	1·076	1·032	1·307
Malt per head of population, bush. ...	1·392	1·458	1·728	1·888

Many writers on the subject seem to be agreed that the multiplication of public houses offers a temptation to drinking which we should do well to curtail, and many among them go the length of asserting that their entire suppression would be a public benefit.

My own personal experience in this direction does not enable me to give an unqualified assent to this opinion, particularly as regards the latter portion of it. Too much rigour in respect to such houses of entertainment is apt to lead, and to my own knowledge does lead, to the frequent introduction of liquor into the houses of the workmen; the effect of which on the women and children I believe to be fraught with most dangerous consequences. The power of prohibiting the sale of intoxicating drinks does not obtain in the United Kingdom; but in certain States in the American Union it is otherwise, and I visited one large railway engineering establishment where, at the request of the company and of the inhabitants, the system had been introduced. The consequence was that the men, after receiving their pay, resorted to a town some twelve miles off, and situate outside

[1] This is a mere approximation.

the prohibited area. Not only was drinking there indulged in to a hurtful extent in the public houses of entertainment, but the men brought back with them large supplies of strong spirits, for future use in the family circle.

It would however be an injustice to our workmen as a class to suppose that, when circumstances permit it, considerable efforts are not made to lay by a portion of the increased earnings in good times. This will be apparent from the figures given me by the managers of the Savings Bank of Newcastle-on-Tyne, where the deposits rose considerably in amount when the staple industries of the neighbourhood were prosperous, and only began to fall off when these became depressed:—

Newcastle Savings Bank.	Increase over previous Year.		Decrease over previous Year.	
	Depositors.	Amount. £	Depositors.	Amount. £
1870 ...	508	14,000	... —	—
1871 ...	286	8,000	... —	—
1872 ...	962	33,000	... —	—
1873 ...	1,239	48,000	... —	—
1874 ...	923	52,000	... —	—
1875 ...	342	16,000	... —	—
1876 ...	—	—	... 368	13,000
1877 ...	—	—	... 363	11,000
1878 ...	—	—	... 433	13.000

It would of course be impossible to form any estimate, of even approximate correctness, as to the extent of private savings and investments effected by our labouring population. Inasmuch, however, as the increase in the amount of deposits in the various Savings Banks in the United Kingdom has averaged for some few years back about 2 millions sterling per annum, it is only reasonable to suppose that the workmen have contributed a fair share of this—such at least, I believe, is the case in Newcastle-on-Tyne. Again it is generally understood that large sums are remitted home by the large Irish population engaged in the mining and manufacturing establishments of Northumberland, Durham, and North Yorkshire; and this practice probably obtains wherever this industrious portion of our labouring population is largely employed.

A very considerable portion of the labour about the ironworks in Cleveland is performed by Irishmen; and public opinion is unanimous that no people are more able, or more willing, to undertake hard work than they are. If different results are perceptible in Ireland, there must be some defect in the social circumstances by which they are there surrounded, which does not obtain in England. It is a matter of deep regret that occasionally these good qualities are marred by indulgence in drink, which prevents such men from rising to that position in the community to which their industry is capable of leading them.

Having thus, I hope, conclusively shown that the labouring man in Great Britain is able to command the necessaries of life on terms almost, if not quite, as favourable as on the Continent of Europe, it remains for us to consider whether this food, when applied to the living body, has the same sustaining effect here as abroad; and whether the results it affords to the employer and to the workmen differ in different countries.

In such a comparison, much more is required in many instances than a mere quotation of rates of payment, or even than the actual cost of a given amount of work. Both may be, and often are, materially affected by the convenience of the arrangements provided for the use of the labourer; while the difference in the materials under treatment, and a great variety of other circumstances, disturb any calculation intended for the purpose of such an examination as that proposed.

Under such circumstances, the only course to be adopted is to consider the question in a very general way, and as far as is practicable to compare cases which resemble each other as closely as possible.

The effect of Free Trade measures combined with greatly improved modes of transport has been, as we have seen, to reduce the cost of living with ourselves, while these very circumstances have had, for reasons already given, an opposite effect abroad. Under such influences it would not have been surprising if, while the price of labour necessarily rose abroad, it should have declined in the United Kingdom. Such however has not been the fact. It is true wages are notably higher on the Continent of Europe than they were in former times; but the increase is still more marked in the case of the workmen in Great Britain.

Before entering upon the question of labour in branches of industry most germane to the present work, I would offer a few brief remarks on the relative cost of agricultural labour in the United Kingdom and on the Continent of Europe, so as to establish a common standard of comparison.

Fifteen years ago, upon the occasion of a journey through France, I found agricultural labourers paid 2s. to 2s. 4½d. per diem, or 1s. 7½d. if fed by the farmer. The same class now receives 2s. 3d. to 2s. 6d. per diem, or an increase of about 11 per cent. In Alsace and Lorraine, in 1879, farm servants were getting 2s. 2d. to 2s. 7d. a day; and according to Dr. Young the best paid men got 2s. in 1858. In some parts of Germany, however, the earnings are still miserably low. They were partly paid in money and partly in provisions, bringing up the total to from 1s. 3d. to 1s. 6d. per day. The existence of these poor people must be a hard one, for wheat was valued to them at £2 per quarter, wheaten flour 1s. 11d. per stone, butter 1s. 2½d. per lb., and potatoes 3½d. per stone. Dr. Young mentions 13 hours a day as being spent in the fields on the Continent, this including meal times.

We still hear in certain parts of England, viz., those remote from manufacturing centres, of very low wages being paid for agricultural labour. Indeed in Cheshire, at no great distance from the mills of Lancashire, 12s. was, a few years ago, the rate received by farm labourers; but in 1879 this had risen to 16s., equal to an increase of 33 per cent. In Cleveland, agricultural servants were paid 14s. per week, anterior to the establishment of ironworks on the large scale which now distinguishes that neighbourhood. Since that time £1 has been paid, and now (1883), when trade is by no means prosperous, the rate usually given is 17s. or 18s. The older farmers there declare that there is no improvement, indeed the reverse, in the amount of work performed by their labourers now-a-days, as compared with former times, notwithstanding their improved means of doing the work. Such would also appear to be the opinion of Mr. Clare Sewell Read as to the county of Norfolk; for he states that on comparing his farm accounts with those of his father, he found that in spite of improved machinery he was paying fully 30 per cent. more wages per acre than had been done 50 years before.[1]

[1] Journal of Statistical Society, June 1880, p. 337.

Speaking from general impressions as a manufacturer, I would say that this is opposed to my own experience. I have seen the common labourers in ironworks paid 14s., sometimes even as low as 13s. per week. Such men are now receiving 17s. to 18s.; and being better fed, I do not doubt, certainly in some cases, that they do more work, simply because they are better able to do it, and probably with less fatigue than under the lower scale.

That a diet, which suffices to keep a man in excellent health with moderate physical exertion, may fail when his energies are more severely taxed, is consistent with well understood natural laws; and the extent to which this happens, is constantly recognised by those who have any experience in the employment of Irish labour. Young healthy men come over from the sister country to England, and being in the vast majority of cases naturally industrious, become, as soon as their improved style of living permits it, very hard working men.

I am greatly inclined to agree with Sir Thomas Brassey, M.P.,[1] who, speaking from his own experience and that of his father, observes that good wages do not necessarily involve a high price of labour; for, unless the workmen be well fed, it is hopeless to expect that his powers can be fully developed.

In connection with that gradual and general improvement in the earnings of the workmen throughout Europe, it may be urged that the great gold discoveries already referred to have tended to depreciate the value of the circulating medium. In other words, gold being more plentiful than formerly, more of it would have to be given for an article in which the relations of supply and demand had not altered. In spite however of the depreciation of the precious metal, the British workman is able to buy more with his money to-day than he could do forty years ago, while his labour now commands a much higher price.

With regard to the numerous class of men known at ironworks under the general name of mechanics—comprising blacksmiths, bricklayers and masons, carpenters and joiners, fitters and men working at lathes, etc.—it is almost impossible to state in figures the exact nature of the change which has taken place in their wages. Different men engaged in the same employment, and in the same establishment, may

---

[1] " Lectures on the Labour Question," p. 13.—*Thomas Brassey, M.P.*

be paid at very different rates, according to their skill and ability. Looking over the figures as they stand in my notes, it looks as if between 1850 and 1870 there had been an average rise of 10 or 15 per cent. Since the latter year, in reduced hours of labour and in advance of wages, the men received up to 1873 an advance of 33 per cent., since which time reductions have been made; so that, as compared with 1850, we shall probably be near the mark in assuming that the men engaged in erecting and keeping in repair the machinery and buildings of our ironworks are better paid by 30 per cent. or more than they were thirty years ago.

To enable me to speak with greater confidence, respecting the general change which has taken place in mechanics' wages, the manager of one of the largest locomotive factories in the North of England has kindly extracted from the books of the concern the average wages actually paid since the year 1850. From these returns the percentage of increase on the rates of that year have been calculated for 1874, when all wages rose to an unprecedented extent; and also for the beginning of 1881, when there is a large falling off as compared with 1874. The wages of all trades gradually rose until in 1874 as compared with 1850, they reached the increase below of:—

	1874. Increase as compared with 1850.	1881. Increase as compared with 1850.
Fitters and machine-men	31·99 per cent.	18·58 per cent.
Black-smiths	41·08 ,,	23·24 ,,
Strikers to black-smiths	30·71 ,,	21·95 ,,
Joiners	50·94 ,,	39·62 ,,
Brick-layers	50·94 ,,	40·12 ,,
Average increase of the five classes	41·73 ,,	28·70 ,,

On January 1st, 1872, the whole of the men in question commenced the nine hours system, *i.e.*, instead of working as hitherto 60 hours in each week they restricted themselves to 54 hours; so that an additional 10 per cent. must be added to the figures just given, as indicating the rise in wages since 1850. So far as the cost of a given amount of labour is concerned, the gentleman to whom I am indebted for the above particulars, considers that at the present time it is in-

SECTION XVI.—THAT OF THE CONTINENT OF EUROPE.

creased fully as much as the calculation, including the nine hours restriction, would indicate—say from 30 per cent. to 50 per cent. In this opinion, it may be added, he is supported by almost every one whom I have consulted on the subject.

In the absence of some common standard to which the price of labour can be referred, it is of course very difficult to compare one workman with another, even of the same nation; and this difficulty is considerably enhanced when we attempt the task as between Englishmen and foreigners. Looking over the quotations in my possession, I have arrived at a general impression that 22 to 25 per cent. higher rates are paid in this country for a given amount of work than it costs abroad; and I found there a common way of expressing the difference was to say that the same labour which cost shillings in England could be done for francs in France, which means a balance against us of 27 per cent. If these suppositions be correct, as applied to the building of a piece of mechanism like a locomotive engine, the difference becomes one of a serious character. Such a machine, if of large dimensions, may cost at present £2,400, of which probably one-third is wages paid by the builder. The materials used in its construction by the best houses are of high quality in which we enjoy little or no advantage over foreign nations in respect to price, and when we come to apply a saving of say 25 per cent. on the wages, in favour of the Continental manufacturer, it means that a locomotive costing £2,400 can be supplied for £2,200 from France or Germany. It is needless to say that such a margin, other things being equal, is quite enough to withdraw custom from our makers.

On the Continent of Europe, as well as in the United States, according to information obtained on the spot, men of other occupations, such as low-priced carpenters, are permitted, if they show any aptitude for the work, to take charge of certain tools in engineering establishments. The object in hand is centred in the lathe, or placed on the bed of the machine, with a little assistance from a skilled mechanic, after which it is superintended by the cheaper labourer.

Without however including any advantage of the nature suggested, the actual wages paid in recent years on the Continent to the different classes of mechanics engaged in repairs at ironworks are here inserted :—

		Fitters and Machine-men. s. d.	Blacksmiths. s. d.	Strikers. s. d.	Joiners and Carpenters. s. d.	Bricklayers s. d.
Germany,	1870 ...	—	2 2	1 6	2 6	2 6
,,	1878 ...	—	—	—	—	3/0 to 3/6
,,	1879 ...	—	3 3	2 6	—	—
,,	,, ...	3/0 to 3/6	3 4	2 6	—	3 2
,,	,, ...	—	3 6	2 6	—	—
France,	1878 ...	—	—	—	—	3/2 to 3/6¾
,,	,, ...	—	—	—	—	3 6¾
,,	,, ...	3 6¾	—	—	3 11½	3 6¾
,,	,, ...	3/6¾ to 4/6	3 9	—	3 6¾	2 9½
Belgium,	1878 ...	—	3/2 to 3/11½	3 2	3/2 to 3/11½	3 2

The average wage paid in the repairing shop of a large Belgian iron works amounted to 2s. 11d., ranging from 2s. 4½d. to 3s. 6¾d.

From these figures we may infer the following averages for 1878 and 1879:—

	s. d.		s. d.	
Fitters and machine-men	3 7	against	4 6¼	paid in the North of England.
Blacksmiths	3 5½	,,	4 9	,, ,,
Strikers to blacksmiths ...	2 6	,,	3 1½	,, ,,
Joiners	3 9	,,	4 7½	,, ,,
Bricklayers	3 4½	,,	4 7	,, ,,

Since in Germany and France 10 hours, and in Belgium 10½ hours, form the working day, instead of 9 as with us, we may assume the average pay of mechanics of the four classes enumerated above to be about 30 per cent. cheaper than it is in the United Kingdom. The relative efficiency of labour in an engine building shop with us and abroad would, as has been already stated, be difficult to estimate; but I learnt that for 3s. 2d. to 3s. 6¾d. a French bricklayer would place 1,200 to 1,500 bricks for a day's work, and in Germany 1,000 to 1,500 bricks were laid in a day at a cost of 3s. to 3s. 6d. for the bricklayers. In both cases there was little or no difference in size of the foreign bricks as compared with those used in England. This I believe is as much as is done by the best of our men, who often receive 5s. 6d. for their day's work; and if these figures can be assumed as indicating the relations between British and foreign labour, the convenient way of expressing the difference in shillings instead of francs is considerably below the mark.

On discussing the comparative cost of labour between the Rhenish provinces and England, a most intelligent director of a foreign work informed me in 1878 that, in his opinion, Great Britain possessed the advantage in the cost of such articles as required a small amount of labour to be expended on their production; but the moment wages, particularly for skilled labour, constituted a more important element in the cost, Great Britain was unable to compete with Germany. According to this gentleman, pig iron fell within the scope of the first of these conditions, and steam engines within that of the second. In the earlier days of the Bessemer process, his firm had obtained a blowing engine from England; which performed its work so admirably that, when requiring a second, they were unwilling to incur the risk involved in employing a German builder, who was without experience in machinery of this description. But in spite of the English engineer having cheaper iron at his command, the price of his German competitor was fully 30 per cent. lower; and, in consequence, the second engine was made at home, and it has done its duty to his entire satisfaction.

The extra cost of labour, however, tells much more severely against Great Britain, when the calculation is applied to articles in which workmanship, instead of forming 30 per cent. of the price of the machine, amounts to double this. Such is the case with regard to spinning machinery, like self-acting mules, etc. I have it from a most trustworthy source, my informant being himself a large manufacturer of this class of tools, that upon a machine which is sold for £230 the nine hours' movement has increased the cost of labour by nearly £14. If this sum be added to the additional wages stated by competent authorities as being already paid to our mechanics, it will be seen that nothing but extraordinary skill, which however may be easily imitated by the foreigner, can secure for our manufacturers a market for their goods beyond the seas.

There is another branch of industry that deserves consideration in this comparison of foreign with British labour, viz., iron shipbuilding. It will hereafter be shown that, although we may in most cases be able to produce pig iron more cheaply than our Continental neighbours, yet when this pig iron is passed on to the malleable iron works, cheaper labour may enable the foreign manufacturer to produce ship plates at as low a price as can be done in any place in the United Kingdom.

Now the tonnage constructed last year in the English and Scotch ship-building yards was such that the wages paid would probably not be far short of £1,500,000. The rates paid range so that the chief men engaged on the iron work in one establishment averaged from 8s. 9d. to 12s. 10½d. per day, over 313 days in 1880.

According to returns in my possession, from two or three English yards and one on the Continent, the average earnings of the entire establishments, including boys, appear to be such that the English hands receive almost exactly double the money paid to the foreign workman. In addition to this the hours of labour abroad are 60 to 62 per week, against 54 in England.

Having received information which thus seemed to place the English ship-builder at such a great disadvantage with his Continental competitor, I have extended my enquiries relating to so important a section of our manufactures, one moreover which has a very close connection with the iron trade of our country. From the owner of a Continental ship yard I obtained the following information respecting the rates of wages paid at his establishment:—

	s.	d.	s.	d.
Joiners			3	2
Carpenters			3	11½
Blacksmiths	3	2 to 3	11¼
Platers			3	2
Riveters			3	2
Fitters	3	2 to 3	11¼

The average amount paid to all employed in the yard (working ten hours per day) is 2s. 11¾d.

So far as information obtained in England enables me to judge, the following table shows the relations of the wages paid in ship-building yards here and abroad:—

	England.				Germany.			
	s.	d.	s.	d.	s.	d.	s.	d.
Pattern makers	29	10	to 32	0	19	8½	to 23	4
Iron founders	30	9	to 32	2⅓	13	3	to 20	1
Turners	29	2½	to 30	9	14	2	to 27	4
Planers	25	9	to 27	8	19	11	to 22	4
Drillers and borers	23	0	to 25	1	15	11	to 22	2
Slotters	27	8	to 28	3½	20	0	to 28	0
Platers	33	8	to 37	0½	17	10	to 33	8
Riveters	32	2	to 33	9½	14	11	to 24	1
Holders-up	24	9	to 27	0	14	2	to 16	4
Average	28	6½	30	5	16	1	24	2

SECTION XVI.—THAT OF THE CONTINENT OF EUROPE.

The average of the English wage, according to these figures is 29s. 6d., compared with 20s. 1½d. for Germany; which makes a difference against the former of nearly one-half. Besides this the Germans work longer hours, which increases the difference by a further 15 per cent.

This at best is a rough way of stating the amount, because it assumes that there is the same number of men in each branch, which is contrary to fact. A still greater cause of disturbance arises from the fact that the above comparisons are all based on day's work; but in English yards about two-thirds of the men are engaged by the piece. If a vessel of 3,000 tons were built on the time system, and German labour were as much cheaper as the above figures would indicate, it would mean a greater cost of more than £2,000 as against the English builder.

Inasmuch however as the greater portion of the work on an iron vessel is performed by contract, no exact comparison can be drawn from the figures given above. It would appear as nearly as may be that one-third of the men working at an iron ship are paid by the day, and two-thirds by the piece. According to a return with which I have been favoured, the former average close on 4s., and the latter nearly 7s. per day. In each case boys are included in the estimate, the average of the entire yard, men and boys, being nearly 6s. per day. This, it will be perceived, is almost exactly double the rate of the foreign yard just mentioned, so that, unless the English artisan turns out double the work of his foreign competitor, the English ship-builder is placed at a disadvantage.

The large amount of wages paid for ship-building renders the business one of national importance; and any interruption to its progress, still more any diminution due to foreign competition, possesses great national interest. In the foregoing paragraph I have contented myself with pointing out, to those more immediately concerned, the difficulty which British ship-builders may have to encounter in competing with foreign houses, so far as the mere construction of a vessel is concerned.

The general subject of expensive labour incurred in the construction of iron vessels, will be apparent at a later period of the present section, when the wages paid in the plate-mills are being considered.

It is often alleged that the law of supply and demand will always regulate the price of a commodity. In regard to labour, this in the long run may be true, but the adjustment is often deferred by the action of those concerned. In isolated cases, however, we have examples of the price of labour never accommodating itself to the ability of the manufacturer to afford to pay it. As an example of a large amount of a particular class of labour being thrown on the market, and the product of this labour selling at unremunerative prices without wages experiencing any reduction, I would quote the chemical trade on the Tyne. It is notorious that for some years past this important branch of manufacture has been carried on at a loss; and although the state of things has been recently intensified by the quantity of soda made by the ammonia process, the owners of the chemical works there were, I believe, losing money before they were being undersold by establishments using this recent invention. Of the reality of the losses there is no doubt, for thirteen establishments in the district have been closed and sold as old material. The discontinuance of these works meant a reduction of 24 per cent. of the make of the neighbourhood.

Between the years 1857 and 1880, or say in a space of 25 years, the products of the works in question, viz. soda ash, refined alkali, crystals of soda, bi-carbonate of soda and bleaching powder, had fallen in value from 35 to 60 per cent., an average decline therefore of about 40 per cent. The price of the raw materials kept pace to some extent with this depreciation in value, but it was otherwise with labour, for I found that in a work paying £564 per week in 1857 their pay bill was £860 in 1881. Making allowance for the increased number of men, it was ascertained that for 20s. earned per head in the former year, 27s. 6d. was paid in the latter. This rise in wages is equal therefore to $37\frac{1}{2}$ per cent. during the 25 years.

Attempts were made from time to time by some firms to reduce the wages, in order to avoid the serious losses which had been encountered in carrying on the works. None of these recent attempts met with any success, doubtless for the simple reason that the general state of wages in the district had advanced, and the men who had been employed in the soda works sought for, and obtained other employment for which their previous life had, to some extent, prepared them.

SECTION XVI.—THAT OF THE CONTINENT OF EUROPE. 505

Thus it may quite well happen that an important branch of industry may, from a change of circumstances, fall into a state of great difficulty, notwithstanding that a considerable demand may still exist for its products. In the case under consideration, that of soda, this alkali can undoubtedly be manufactured more economically by Solvay's ammonia process than by the Leblanc process. By the latter, however, no bleaching powder, a bye-product of the Leblanc system, is made. To obtain this article, a considerable quantity of soda continues, and must continue, to be produced on the old plan, but the soda itself is now (1882) selling at so low a price as to entail the consequences just referred to.

COAL.

Passing from what may be regarded as the more general aspects of the labour question in Europe, I propose now considering the gradual changes which have taken place in the mining and other labour connected with the manufacture of iron. For the purposes of this enquiry the raw materials claim the first place, and of these we will commence with coal.

To make a proper comparison between different localities, as to the cost and efficiency of labour employed in winning coal, it is of course essential that the conditions should be the same. In point of fact, however, this rarely happens: differences in thickness and hardness of the seam, and the greater or less freedom from faults in the strata, or troubles as they are commonly called in Northumberland and Durham, are the chief causes which derange calculations bearing on this question.

The actual quantity of coal cut for a day's work is therefore of itself not necessarily a trustworthy guide in the enquiry; and in addition, while on the Continent the hewers may be engaged ten hours in the pits, five or six hours only is spent at the face of the coal in the Durham collieries, the men remaining seven hours only in the pit.

Besides the natural causes just referred to, the coal hewers sometimes interpose one of an artificial character, viz., that of limiting the output. In the years 1870 and 1871 the men in one case cut on an average nearly $4\frac{1}{4}$ tons of coal for a day's work, the pit working above 300 days in the year; for which they received about 4s. 6d. after paying for lights and powder, besides having house rent and firing without any charge. In 1872 trade began to improve, and the yield fell to a trifle above $3\frac{3}{4}$ tons, for which 5s. $7\frac{1}{2}$d. was paid; and in 1873 and 1874 the average per hewer was only $3\frac{1}{2}$ tons, by which he earned nearly 7s.

Since that time as much as 92 cwts. per shift have been averaged in the same colliery, so that taking $4\frac{1}{2}$ tons as the possible output per man, each hewer sacrificed 1s. per day during the last two years of restricted work, equal to about 15 per cent. on his earnings.

This production, thus reduced by the men, diminished the output in 1873 as compared with 1872 by 755,000 tons; while merely for the iron trade dependent on the Northumberland and Durham coals above 250,000 tons more were needed than in the previous year. There was therefore from this cause alone a deficiency of above one million tons. This circumstance, along with increased activity in all branches of business, led to new pits being sunk, and high wages attracted fresh men to the district. In 1875 and 1876 the produce of the two counties reached 32 million tons. Prices then began to fall, and with them wages declined fully two shillings a day to the hewers; and even this reduced rate was only earned by reverting to their old scale of output.

The effect produced in the coal and iron districts of Durham and North Yorkshire will not soon be forgotten, although arising from a deficiency which probably did not amount to 10 per cent. on the total produce. The iron markets, however, were in so excited a state that anthracite coal was brought in some instances by rail from Wales, to keep the furnaces moving, and coke reached the unprecedented price of 40s. per ton at the pits. Under no conceivable circumstances was it to be expected that wages should or ought to remain uninfluenced by such a state of things; but the unfortunate part of the transaction is that the men, instead of selling as much of their work as possible when it commanded so high a price, sought to influence the labour market by the means referred to. Something considerably short of their old rate of working would yet have enabled the coal owners to make good the quantity in deficit. In the then state of the trade the whole could have been disposed of at fair prices. The competition arising from a needless number of new pits and from a plethora of men has

SECTION XVI.—THAT OF THE CONTINENT OF EUROPE.

been followed by the usual consequences of an attempt to interfere with the principles of political economy. There was worked in the North-East of England under 30 million tons of coal in 1873, a time when we required 32 millions, and could have got this quantity with our then appliances in this corner of the island alone. We can now (1882) work, not only 32 millions, but probably 34 millions; unfortunately however we do not need all we are able to work.

I would not have it supposed that the workmen are alone responsible for the present evils of over-production. Viewed by subsequent events, a good deal of imprudence can be laid at the door of the colliery owners. Misled by the extravagant price of coal, imprudent speculations were entered into, many of which ended disastrously for the adventurers. It may however be claimed for the masters, that they endeavoured to obtain as large a produce from their mines as possible; in which, in some instances, they were certainly not seconded by their men. Cases are known in which fresh pits have been sunk by mine owners, to make good the deficiency accruing from short workings in pits already existing.

The following statement shows the average weight of coal worked by each hewer during 14 years in a South Durham colliery. The net daily earnings given are after paying for oil and powder. Besides this the miner has free house and firing:—

		Weight of Coal. Cwts. per Day.	Net Earnings. Per Day. s. d.
6 years, ending 1869	... ...	80·39	... 4 1·27
2 ,, ,, 1871	... ...	83·87	... 4 5·67
1 ,, ,, 1872	... ...	76·03	... 5 7·40
2 ,, ,, 1874	... ...	70·80	.. 6 10·65
3 { There were several changes in wages in each year during these 5 years.	1875	70·14	... 5 9·13
	1876	78·64	... 5 10·16
	1878	90·00	... 5 0·54
	1879	74·63	... 4 4·42
	1880	91·96	... 4 3·73

14

The coal raised per man employed in the pits, all told, in 1872 was 90 per cent. of that got in 1874, and only 80 per cent. of that of 1870. The same seam was worked during the 14 years, and the pits were actively engaged for 300 to 312 days in the several years.

Taking every one engaged above and below ground in the Inspection District of the South Durham coal mines, the following shows the produce in tons per man employed for the ten years ending 1882:—

	1873.	1874.	1875.	1876.	1877.	1878.	1879.	1880.	1881.	1882.
Tons ...	335	330	389	341	351	356	350	401	398	396

It would thus appear, comparing 1873 with 1880, that the output at the whole of the mines, according to the Government return, shows a deficiency in the former year of 20 per cent.

By the kindness of a friend largely engaged in the Durham coal trade, I have ascertained the following to be the changes which have taken place in the price of labour in that trade between 1850 and 1881.

In the former year the hewers in a particular seam were eight hours "at the face" in the whole coal, and six hours in the broken. During this time they received 3s. 10½d., with free house and fire. In 1881 in the same seam only five-and-a-half hours were spent "at the face" in the whole coal, and four-and-three-quarter hours in the broken coal; and in this shorter period the money earned was 4s. 5¼d., with house-rent and firing as before. The men therefore worked fully a fourth less time, and nevertheless were able to earn about 15 per cent. more money.

According to an article which appeared in the *Newcastle Chronicle* in 1881, the boys in the Rainton and Pittington Collieries received in 1847 10d. per day, and the best paid men 20s. per week. At Shotton and Haswell Pits the average earnings over several months only amounted to 3s. 3d. per day, and at Hartley Colliery 3s. 4d.

An old friend of my own, one of the most experienced mining engineers in the North of England, reminds me that 50 years ago it was no uncommon thing for a hewer to remain 13 or 14 hours in the pit. For this, it may be presumed, he would receive a rate of pay not exceeding that mentioned in the *Newcastle Chronicle*. According to the recollection of my informant, a hewer worked no more coal than his successor does now, but his exertions were spread over a much longer time. Probably the pay at that time did not enable a man to be so fed as to be capable of the more violent exertion now put forth by the better paid miner of the present day.

No heed is taken of any increase in the number of collieries in any of the cases above referred to. Any increase in the powers of produc-

SECTION XVI.—THAT OF THE CONTINENT OF EUROPE. 509

tion, stimulated by the high prices, would probably not be *generally* employed before 1875 although there was an increase in 1873 and 1874. According to the returns of the Mining Record Office, the number of collieries working coal in the Northern counties in 1871 was 304. This was increased in the four following years as follows:—

	1872.	1873.	1874.	1875.
Number of collieries	357	382	387	427

It has to be observed in reference to the restricted output of the English miners, that it not infrequently happens, owing to want of demand for coal, particularly at collieries dependent on the shipping trade, that restriction emanates from the coal owners and not from the coal hewers. In the years 1872 and 1873, however, such an event would be of rare occurrence. Applying actual figures to the case before us, it would appear, according to an account published by the Durham coal trade, that the percentage of shifts lost by the men lying idle was as follows:—

1873.	1874.	1875.	1876.	1877.	1878.	1879.
16·00	13·81	11·72	9·97	8·11	6·58	6·12

There is no country in the world except Great Britain which, having regard to its area and population, produces so much coal as Belgium; and of the total quantity raised in this kingdom, say $14\frac{1}{4}$ to $15\frac{3}{4}$ millions of tons per annum, $10\frac{1}{2}$ to $11\frac{1}{2}$ millions are from the mines of the province of Hainaut. In respect to this portion of the output, I have extracted the following figures from the reports of M. Emile Laguesse, Chief Engineer and Mines Director to the Governor of the province. These extend over the last ten years ending 1879. The table gives the yearly output per workman engaged, the average weekly earnings of the entire staff, and the selling prices of coal during the different years:—

	1870.	1871.	1872.	1873.	1874.	1875.	1876.	1877.	1878.	1879.
	Tons.	Tons.	Tons.	Tons.	Tons.	Tons.	Tons.	Tons.	Tons.	Tons.
Output	149	144	157	146	133	136	133	137	150	155
	s. d.	s. d.	s. d.	s. d.	s. d.	s. d.	s. d.	s. d.	s. d.	s. d.
Wage per week	14 10½	12 11	16 0	21 5	18 2	17 11¼	15 9	12 8	12 8½	12 3¼
Price of coal	8 9	9 1	10 9¾	17 4	13 2½	12 6	10 11½	8 10¾	8 0½	7 6½

During the years 1872 and 1873, the output per man engaged in the Belgian collieries was more than maintained, as compared with 1870 and 1871, the increased produce amounting to 2 per cent. on the former years. The average earnings of all hands engaged in the coal mines of Belgium were 2s. 3½d. per day for 1870 and 1871, and 3s. 1½d. for 1872 and 1873, or an increase of 36 per cent.

Indeed it is a remarkable thing that on the Continent, so far I can judge, the very opposite line of conduct to that of restriction appears to be adopted. There the workmen follow the example of the coal owners by endeavouring, to use a homely phrase, to make hay when the sun shines. At the collieries of Aniche in France, according to a published statement, instead of allowing the output to *decline* 17 to 20 per cent. in the years of high prices, 1872 and 1873, there was an actual increase of 10 per cent. in the rate of production of the men engaged at the mine. The increase in the men's money earnings appears to have been about 15 per cent.

All the authorities I have consulted agree in the opinion just mentioned, that there has been a marked addition to the earnings of colliers on the Continent. Quoting from a publication already referred to, the following table contains the average income and coal worked per man in the collieries in the Départment du Nord et du Pas de Calais:—

Year	1860.	1869.	1872.	1873.	1874.	1875.	1876.
	s. d.	s. d.	s. d.	s. d.	s. d.	s. d.	s. d.
Wages per week	10 7½	12 8¾	15 7¼	17 2	16 4¼	15 4¾	16 1¼
	Cwts.	Cwts.	Cwts.	Cwts.	Cwts.	Cwts.	Cwts.
Quantity worked per day	41·53	61·20	68·82	67·6	60	60	58·46

In no case is the number of days actually worked given. Assuming this to be the same in all the years, the figures present a remarkable improvement in efficiency, which is probably due to the better fare placed within reach of the workmen by an improved rate of pay. Besides this fact, it will be observed that in times of high wages the workmen's exertions were increased.

The largest quantity of coal raised in these French collieries, per individual employed per annum, was 179 tons in 1872; which, however, only slightly exceeds the half of the average worked per man in the

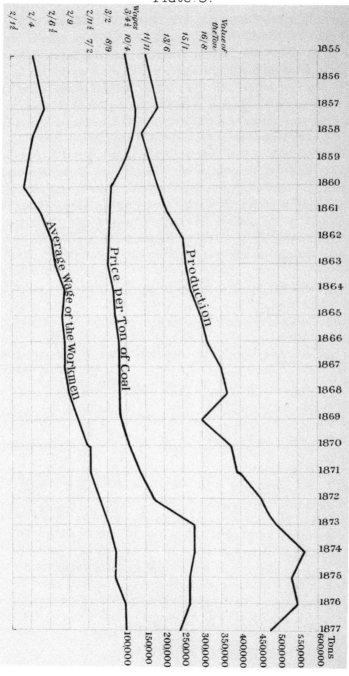

Plate. 9.

SECTION XVI.—THAT OF THE CONTINENT OF EUROPE.

South Durham coal-field during seven years. This latter amounted to 350 tons, according to an official return; and this, it may be observed, was often largely exceeded at some individual collieries, where the work was of a constant character and independent of the shipping trade. The authorities I have consulted mention the fact that at these collieries, between 1844 and 1876, the produce per individual had been increased 70 per cent.; while the wages during the same period had been raised 140 per cent.

At the collieries of Aniche, according to M. Veuillemin, previously quoted, the daily average wage of 2,172 people in the year 1869 was 2s. 2d. per day, which had increased in 1877 to 2s. $9\frac{1}{4}$d.—equal to a rise of 28 per cent.

From an official return from these Aniche mines, it would appear that the earnings of the coal hewers at that concern in 1827 were only 1s. $5\frac{1}{2}$d. per day. In 1869 their wages were 3s. $1\frac{3}{4}$d.—equal to a rise of 113 per cent. In the "wild" years of 1873 and 1874 this class of men in the same locality received 3s. $9\frac{1}{2}$d., which fell in 1877 to 3s. $2\frac{3}{4}$d. These rates were earned on 275 days of the year, and the rise appears to have been as follows:—

1827 to 1845, *i.e.* 18 years, at the rate of about 2 per cent. per annum.
1845 „ 1865 „ 20   „   „   2   „   „
1865 „ 1875 „ 10   „   „ $6\frac{1}{2}$ „   „

At the collieries of Montrambert and de la Béraudière, in Central France, as stated in a document prepared for the Paris Exhibition of 1878, 2,348 persons in 1860 earned on an average 2s. $4\frac{1}{4}$d. In 1877 their pay had risen to 3s. 4d.—equal to a rise of 41 per cent.

A diagram is annexed, Plate IX., on which is shown the gradual increase in the powers of production of the French collieries of Montrambert and de la Béraudière, in the value of the coal, together with the continuous rise in the men's wages, since 1860.

Lest it may be thought that the lower wage in France is due to the employment of young children being more frequent than in Great Britain, I give from the same source the ages of those at work in the Béraudière mines:—

Age—Men and Boys.	Belowground.			Aboveground.			Total.		
	Average Wages per Day.	No.	Per Cent.	Average Wages per Day.	No.	Per Cent.	Average Wages per Day.	No.	Per Cent.
	s. d.			s. d.			s. d.		
12 to 15 years...	1 10¾	22	1·10	1 5¾	12	2·58	1 9	34	1·45
15 „ 20 „ ...	2 7	206	10·94	2 1½	40	8·60	2 6	246	10·48
20 „ 25 „ ...	3 5½	312	16·57	2 7¾	32	6·86	3 4½	344	14·65
25 „ 42 „ ...	3 10¾	1,194	63·44	2 9¾	166	35·62	3 9¼	1,360	57·92
42 „ 55 „ ...	4 0	124	6·58	2 7½	52	11·16	3 7	176	7·49
Above 55 „ ...	3 7¾	24	1·37	2 5½	30	6·43	2 11¾	54	2·30
Women... ...	......	...	...	1 6¾	134	28·75	1 6¾	134	5·71
	3 8¼	1,882	100·00	2 3½	466	100·00	3 4¾	2,348	100·00

The women employed in these last two French collieries were occupied on the surface in picking out stones from the coal. The mode of classifying the ages of the males does not permit a very accurate comparison with the ages of the young people in Great Britain. Having regard to the pay of those classed as between fifteen and twenty years of age, it seems reasonable to suppose that not one-half thereof are under sixteen. If this be correct, we should have only 6 or 7 per cent. of the whole number below the last mentioned age of those employed underground; while in a representative colliery in the county of Durham out of 937 persons the corresponding percentage is about 18. It would therefore appear that the proportion of young people under sixteen is more with us than with the French.

The ages of those employed at the Durham colliery just referred to were as follows:—

Under 12 years of age	7
12 and under 13	13
13 „ 16	160
16 and above	757
	937

No women employed.

In the entire Durham coal-field, according to the official report in 1879, there were 53,152 men and boys employed. Of these 40,664 were engaged underground and 12,488 aboveground.

SECTION XVI.—THAT OF THE CONTINENT OF EUROPE.

The following is the division according to ages:—

UNDERGROUND—		
Above 18 years—Hewers	21,105	
Putters	3,912	
General Work	10,092	
		35,109
Under 18 years and above 16 years	1,579	
Under 16 years	3,976	
		5,555
		40,664
ABOVEGROUND—		
Above 18 years of age	10,763	
Under 18 and above 16 years	538	
Under 16 years	1,187	
	1,725	
		12,488
		53,152

In a previous page reference has been made in detail to the increased exertions put forth by the miners in France and Belgium during the years of high prices and advanced wages, while in England the conduct of the men was the reverse of this.

The following figures are instructive on the subject of production, showing as they do the entire output of coal one year before and one year after the high priced periods of 1873 and 1874:—

		France. Tons.	Belgium. Tons.	Northumberland and Durham. Tons.
1872.	Year of rising prices	15,802,514	15,658,948	30,395,000
1873.	Do. highest prices	17,485,786	15,778,401	29,640,000
1874.	Do. high prices	16,949,032	14,669,629	30,543,800
1875.	Do. prices declining	17,164,794	15,011,331	32,097.323

If the last named year, 1875, be taken as an index of the capacity of the Northumberland and Durham collieries, it would appear that they furnished in the dear times of 1873 and 1874 nearly 7 per cent. less than they were capable of affording; while in the same years the Belgian pitmen worked 4 per cent. and those in the Pas de Calais in France 6 per cent. more per man than they did in the falling market of 1875.

In some parts of Germany, according to Pechar, the average output of lignite[1] per man exceeds the average of the ordinary coal in any of the three countries already dealt with. In the province of Saxony, in 1875, above six millions of tons of this mineral were mined, each workman employed representing a production of 520 tons. The average of Prussia amounted to 450 tons per man, and that of the entire empire, with a total of nearly ten and a half million tons, was 410 tons per man engaged.

The total weight of ordinary coal raised in the German empire during the same year, 1875, was as follows:—

	Tons.		Men.		Tons per Man.
Silesia ...	10,444,364	with	43,506		240
Westphalia ...	10,749,025	,,	54,027	=	199
Rhineland ...	11,645,014	,,	57,258	=	203
Saxony ...	3,061,275	,,	17,272	=	177
Sundries ...	1,391,025	,,	10,602	=	131
	37,290,703	,,	182,665	=	204

The same authority from whom the above figures are taken gives the following information respecting the average weekly earnings of the work people, including women and children:—

	1864.	1871.	1873.	1875.	1877.
	£ s. d.	£ s. d.	£ s. d.	£ s. d.	£ s. d.
Price of coal in Prussia per ton ...	0 5 4	0 6 11	0 10 9	0 7 6	0 5 7
Earnings per Week.					
Ruhr District ...	0 14 5	0 17 1	1 1 11	0 17 1	0 16 4
Saar ,, ...	0 13 7	0 16 5	1 0 10	0 16 8	0 16 3
Upper Silesia ...	0 12 7	0 16 8	0 18 5	0 15 9	0 13 1
Average ...	0 13 6	0 16 8½	1 0 4½	0 16 6	0 15 3

The subjoined table contains the average weekly earnings of German and Belgian coal miners, and also those of the Aniche Colliery in France, as compared with the wages actually paid at a colliery in England, in the years for which I have the necessary data:—

---

[1] Probably mostly brown coal.

SECTION XVI.—THAT OF THE CONTINENT OF EUROPE.

	England, Durham.		Germany.		Belgium.		France.	
	s.	d.	s.	d.	s.	d.	s.	d.
1871	18	4½	16	8½	12	11	14	3
1873	36	9	20	4½	21	5	15	3
1875	29	9	16	6	17	11¼	15	3
1877	27	0	15	3	12	8	15	3
Average	27	11½	17	2½	16	2¾	15	0
Tons worked per man	351		204		147		164	

The proportions, taking England at 31s. to cover house rent and firing as unity, are as follows:—

	England.	Germany.	Belgium.	France.
Money	100	55·37	52·15	48·39
Work performed	100	58·12	41·88	46·72

The actual relative costs of labour, as measured by results, deduced from these figures would be:—

England.	Germany.	Belgium.	France.
100	95	124	103

Thus, partly owing to natural advantages and partly to the superior strength of the English miner, it appears that in spite of the higher wages paid in this country, the actual cost for labour on a ton of coal is usually less than that paid on the Continent of Europe.

The average wages of all the men engaged in the coal mines, employed for an entensive iron work in Westphalia, have experienced the following fluctuations from 1869 up to 1878:—

Year.	Wages per Day.		Price of Scotch Pig.			Increase on 1869. Per Cent.
	s.	d.	£	s.	d.	
1869	2	4·92	2	13	3	—
1870	2	6·72	2	14	4	6·2
1871	2	8·16	2	18	11	11·2
1872	2	11·74	5	2	0	23·6
1873	3	3·39	5	17	3	36·2
1874	3	3·72	4	7	6	37·3
1875	3	0	3	5	9	24·0
1876	2	8·77	2	18	6	13·3
1877	2	5·16	2	14	4	0·1
1878	2	3·36	2	8	5	5·4 decrease.

In this last year the very great depression in the iron trade led to an unusual reduction in the price of all labour connected with the manufacture of the metal. As the quotations at Glasgow influence the value of the metal generally, they are inserted in the above list.

In order to compare the movement in colliers' wages in Great Britain with Westphalia I am able by the courtesy of a friend, largely engaged in the Scotch iron trade, to insert the net earnings of the coal hewers engaged at his establishment during the twenty years ending 1878:—

Year.	Wages per Day. s. d.	Scotch Pig, Average. £ s. d.	Increase on 1859. Per Cent.
1859	3 1	2 11 10	—
1860	3 6	2 13 6	13·5
1861	3 2	2 9 3	2·7
1862	3 1½	2 12 10	1·3
1863	3 7½	2 15 9	17·0
1864	4 0	2 17 3	29·7
1865	4 1	2 14 9	32·4
1866	4 6	3 0 6	45·9
1867	4 1¼	2 13 6	33·1
1868	3 7¼	2 12 9	16·9
1869	3 6¾	2 13 3	15·5
1870	3 9	2 14 4	21·6
1871	4 6	2 18 11	45·9
1872	7 0½	5 2 0	128·3
1873	9 11	5 17 3	221·6
1874	7 2	4 7 6	132·4
1875	5 4	3 5 9	72·9
1876	4 8	2 18 6	51·3
1877	4 1¼	2 14 4	33·1
1878	3 2	2 8 5	2·7

Taking these last-mentioned figures from 1869, and comparing the advance on that year with that of the Westphalian work, we have the following results:—

	1870.	1871.	1872.	1873.	1874.	1875.	1876.	1877.	1878. Decrease.
Westphalia, increase %	6·2	11·2	23·6	36·2	37·3	24·0	13·3	0·1	5·4
Scotland ,,	5·2	26·3	97·0	178·3	101·4	49·7	30·9	15·2	11·1

The following figures relating to Belgium have been calculated in tables given by Pechar. These assume that the year comprises 290 working days, and exhibit the daily earnings of all hands engaged at a colliery:—

Year	1868. s. d.	1869. s. d.	1870. s. d.	1871. s. d.	1872. s. d.	1873. s. d.	1874. s. d.	1875. s. d.	1876. s. d.
Wages	2 2·5	2 3·4	2 5	2 4·5	2 10·2	3 8·7	3 3·1	3 2·4	2 10
Increase on 1869			Per Cent. 5·8	Per Cent. 4·0	Per Cent. 24·8	Per Cent. 63·1	Per Cent. 43·4	Per Cent. 40·1	Per Cent. 24·0

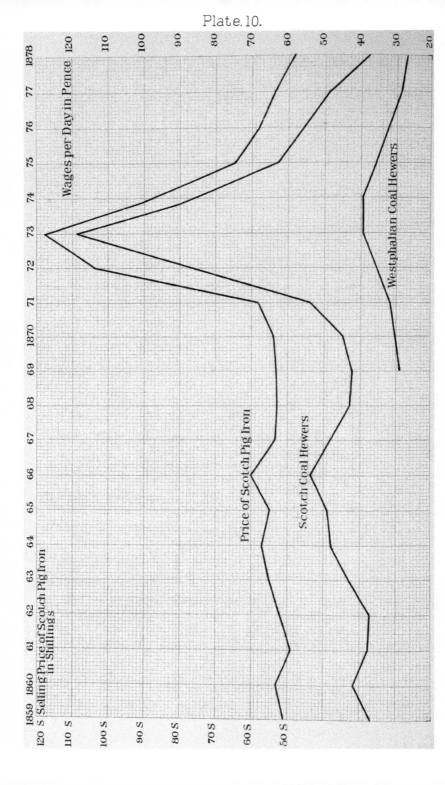
Plate. 10.

SECTION XVI.—THAT OF THE CONTINENT OF EUROPE.

Before leaving these comparative alterations in wages, I would call attention to the accompanying diagram, Plate X., in which the fluctuations given above of Scotch pig iron are laid down in shillings, and those of wages in pence. In it will be seen at once that the Scotch, and I may add the English, miner reaped a larger advantage from the improved position of the iron trade, than fell to the lot of his brethren in Westphalia referred to above.

IRONSTONE.

If ironstone mining does not require a long apprenticeship, neither can it be regarded as a trade which involves no great amount of experienced skill. The drilling and blasting of rock, the risks from the use of explosives, as well as from falls from the roof of the mine, demand on the part of the miner much care and forethought. Besides this, some judgment is necessary in drilling the "shot holes" in such a direction as to bring down the largest possible quantity of material. Notwithstanding these considerations, and notwithstanding the fact that the miner's vocation is usually carried on at a distance from the surface, and shut off from the light of the sun, the wages he earns in most foreign countries are very moderate in amount. This has been seen a few pages back in the case of the colliers, and the miners of ironstone do not appear an exception to the rule.

The miner in Cleveland is eight hours underground and about seven hours at work. His net earnings after paying for powder and oil vary with the state of trade, as will be seen from the following statement:—

PRICE OF PIG IRON No. 3 AND MINERS' WAGES.

	1873.	1874.	1875.	1876.	1877.	1878.	1879.	1880.	1881.	1882.
	s. d.	s. d.	s. d.	s. d.	s. d.	s. d.	s. d.	s. d.	s. d.	s. d.
Price of iron	103 9	74 6	54 0	49 0	43 1½	37 6	43 0	50 6	39 1	43 4
Earnings ...	7 0	6 7¾	5 7½	5 4	4 10½	4 5	3 11	4 5¾	4 6	5 0

The men who work the ironstone in Northamptonshire in the open quarries were only receiving 3s. 2d. in 1878 and 4s. in a mine near the city of Lincoln. The Northamptonshire wages, it will be seen, are lower than those paid in Cleveland. This is partly due to the ironstone requiring less skilful labour for quarry work and partly to the smaller extent of the trade, the requirements of which are easily supplied by the agricultural population.

A year ago I visited a mining district in the South-east of France, where men working ironstone, after paying for oil and powder, only received 2s. 8¼d. per day. For this they were 11 hours in the mine, and had 10 hours of actual work. The day labourers were paid 1s. 9d. to 2s. per day. The workmen here rarely partook of animal food, their nourishment consisting chiefly of maize, which cost 1s. 6d. per stone of 14 lbs.

Near Irun in Spain, in the year 1872, the miners in the iron mines were earning 1s. 11¾d. per day, but near Malaga the pay received by this class was only 1s. 3d. per day.

In Luxemburg the well-known thick bed of oolitic ironstone is worked partly openwork and partly in galleries. Upon the occasion of my visit the miners in the former were earning 2s. 7½d. per day, working from daylight to dark in short days and from 5 a.m. to 7 p.m. in summer with intervals of rest. In the galleries they were 12 hours in the mine, working 10½ hours, for which they were paid 3s. 2d. to 3s. 6d. after paying for powder and candles. They work regularly six days in every week. As nearly as I could make out, the total cost of labour on this ore was 30 per cent. less than that in the Cleveland district, and it possesses the further advantage of not having to be calcined, an operation always practised at the Middlesbrough furnaces.

Upon the occasion of my visit to Luxemburg the necessaries of life were costing:—

Beef	7·12d. per lb.
Bread	1½d. to 2d. per lb.
Potatoes	6d. per stone.
Eggs	½d. each.
Butter	11½d. per lb.
Milk	2d. per litre.
Coals	20s. per ton.
House rent, two rooms	2s. per week.
Shoes	12s. per pair.

A few years ago the earnings in French mines, producing the same quality of ironstone as that at Luxemburg, were 3s. 5d. per day for 10 hours actual work.

At the same time the Cleveland miners were receiving 5s. 5d. for 7 hours actual work (8 hours in the mine) in which time they got 103 cwts.

SECTION XVI.—THAT OF THE CONTINENT OF EUROPE.

The Luxemburg and French mines do not always afford a very fair basis of comparison with those in Cleveland. In each case the ore lies in a seam of 8 to 10 feet in thickness, but while the Cleveland miner generally sends all he excavates to the surface and is paid thereon, the foreign workman has to throw back a considerable quantity of sterile matter.

At one mine I visited in the West of Germany, the quantity of useless mineral was so great, that a miner could only get about 4 cwts. per hour, say 2 tons in a day of ten hours actual labour. For this he received 3s. 4¾d. to 3s. 5½d. per day. Notwithstanding the natural disadvantages referred to, the cost of labour per ton of ore, at the period of my visit, was 7½ per cent. below that paid in Cleveland. In another case in the East of France, although the miners only were making 3s. 3¾d. per day, the cost of labour per ton, on account of the sterile matter, was fully as high as that in Cleveland.

Two men work together in the mine and their wages averaged 3s. 6d. per day. They who raise the ironstone by quarry work only make 2s. 7½d. per day. At Longwy the miners received about 3s. 7½d. per day. The time spent at the mines is 12 hours with intervals of rest.

I am not able to compare the annual output per man and boy employed in France and Germany, with our own, working the same kind of stone. The Report of the British Iron Trade Association (1882, page 76) gives the German and Luxemburg production as follows :—

	Production of Iron Ore in Germany and Luxemburg. Tons.	Average Annual Output per Man. Tons.
1872	5,895	149¼
1873	6,177	156
1874	5,137	161½
1875	4,730	168
1876	4,711	179½
1877	4,980	194½
1878	5,462	196¼
1879	5,859	194
1880	7,238	202
1881	7.573	205

It is inferred from these figures that the men working in the mines have improved in a corresponding degree in efficiency. I have no doubt that they have improved, but it is questionable whether this has been done to the extent set forth in the table. The output per head is so far below that of the Cleveland district, that the only way of accounting for the difference is that the German output includes a quantity of a very different kind of ore from that of Luxemburg, etc. In recent years the produce of the mines in Luxemburg and the West of Germany—Alsace-Lorraine, etc.—has been increasing, and this alone would add to the average annual output of the whole; because a workman can raise more of this ore than of any other.

For the wages received by the Cleveland miners they formerly worked $5\frac{3}{4}$ tons of ironstone; but they adopted the same line of conduct as that pursued by the colliers, viz. a restriction of output. In 1873 the average weight worked per man was 5·25 tons per day, for which he received 7s. This year, 1883, the daily produce has again risen to 5·75 tons and the average earnings have been about 5s. In each case this is after paying for powder and oil.

BLAST FURNACES.

Forty years ago 80 or 100 tons a week would have been considered a fair make for a blast furnace smelting ironstone of the Cleveland character. At that time, generally speaking, the works were deficient in blowing power; there was a prejudice against heating the air beyond 600° F.; and the furnaces did not exceed a capacity of 5,000 or 6,000 cubic feet. In the present day the blast engines are much larger, the heating stoves are able to command a temperature of 1,000° F., and when of fire brick as high as 1,400° F., and the furnaces themselves have usually four times the contents of their predecessors at the time referred to. The consequence is that 450 tons per week is not an unusual make, while occasionally 500 to 550 tons is run from a well-appointed plant at Middlesbrough when smelting Cleveland stone.

In the year 1844 blast furnace keepers were working for 3s. 7d. to 4s. a day, and their slagmen for 3s. Ten years afterwards the discovery of the Whitby beds of ironstone near Middlesbrough had secured for the banks of the Tees a prominent position as a centre of the iron trade. A demand sprung up for suitable men to attend upon the blast furnaces; and labour rose accordingly, until keepers in 1858 reached 7s. and slagmen 3s. 6d. a day, other branches of the work participating in the change of circumstances. At the same time it must be mentioned that the improved plant, introduced about 1851 into an entirely new district, had been accompanied by an increase in the make amounting to 50 per cent. or even more. The great enlargement in the capacity of the furnaces which took place in 1864 and 1865, was followed by a further improvement in the earnings of those engaged in working them; so that all classes received higher wages, keepers and slagmen in 1871 being paid 8s. to 8s. 6d. and 4s. 3d. to 4s. 6d. per day respectively.

The increased production has of course added to the exertions required at the hands of the men, as there is more space to be moulded for receiving the larger quantity of iron; but so far as the actual work at the furnace itself is concerned, the most experienced admit that the labour is less severe than it was when a furnace made only one sixth of the quantity commonly run at present from the Cleveland establishments. This change is entirely due to the improved character of the plant and appliances, as referred to in the previous page.

The extraordinary inflation of the iron trade in the years 1872 and 1873 produced a corresponding effect on labour; so that at one time some sections of a blast furnace staff were receiving fully three times as much as was paid in the year 1850. Subsequent reverses of a very serious character have of course modified this state of things, but a good man can even now, at the worst of times, earn twice as much as his father did at the same occupation.

Although it has not been stated in explicit terms, the ironmaster has been no loser by paying the higher wages as compared with former years. This favourable result has been partly secured by the larger amount of work performed but still more by the introduction of well devised plans for saving labour.

On the Continent the practice differs from that in Cleveland. In cases where the make formerly did not usually exceed two to three hundred tons in the week three men were usually engaged, whereas at Middlesbrough two did the work; and even in the case of a furnace running 400 or 500 tons, two and a half men suffice, *i.e.*, three men on the day turn and two at night. Occasionally in Luxemburg when the make is large, say 100 tons per 24 hours, three men are employed at keepers' work at a cost of about 11s.

To show the differences between the pay of the head furnace keepers in England and on the Continent I have, for the same years, extracted the following from my note books:—

			Cleveland.
1867.—	France, South	3s. 11½d., 4s.	6s. 8½d.
	,, East	2s. 2d., 2s. 8½d.	
	Westphalia	3s., 3s. 6d.. 4s., 4s. 6d.	
	Belgium	2s. 4¾d., 2s. 9½d.	
1868.—	France, East	3s. 6½d.	6s. 2½d.
1872.—	Spain. South	2s.	9s. 0½d.
	,, North	4s.	
1873.—	France, East	3s. 1¼d., 3s. 8d.	10s. 6¼d.
	Luxemburg	4s. 9d.	
1875.—	France, East	2s. 9d.	9s. 1⅓d.
1878.—	,, ,, ...	3s. 6¾d., 3s. 7¼d., 3s. 11½d.	8s. 1¾d.
	Luxemburg	4s.	
	Germany, West	3s. 3½d., 3s. 5d.	
	Westphalia	3s., 3s. 3½d., 4s.	
	Hanover	3s. 7½d.	
	Belgium	3s. 2d., 3s. 0½d., 3s. 5¾d.	
	Germany, West	3s. 11½d.	
	Sweden	2s. 11d	
1881.—	France, East	3s. 3½d., 4s. 4¼d.	7s. 9¼d.
1882.—	Austria	3s. 4d., 4s. 2d.	8s. 1d.

The pay of Continental keepers varies from 3s. to 3s. 9d., and in one case reaches 4s., and in another as much as 5s. per day. The assistants run from 2s. 5d. to 3s. 6d. At almost all the foreign furnaces included in this list the iron was being run in iron moulds; which entails somewhat less labour than in the case of Cleveland, where the metal is received in sand.

According to the foregoing table of wages, the English furnace-keepers receive at least double the pay of those on the Continent. Owing however to differences in the division and nature of the work, it would be very difficult to speak precisely as to the relative amount of duty performed by each. Speaking approximately it may be taken that the British keepers do one-half more work than any of those mentioned in the list.

SECTION XVI.—THAT OF THE CONTINENT OF EUROPE.

With the above-mentioned classes of men on the Continent there has been an advance during the last thirteen years, in cases where the wages were very low. Thus the keepers, who in 1867 received less than 3s. a day, now (1882) get 3s. 6d. to 4s.; while those in Westphalia are only earning about the same wages now as they did in 1867.

The observations which have been made in these pages respecting the gradual rise in wages at the collieries on the Continent, are equally applicable to men generally engaged in the iron works. In the year 1879, a friend living on the banks of the Rhine, and largely interested in the iron trade, gave me the following information respecting the changes in the value of labour in that district during the previous ten years. The establishment consisted of blast furnaces, foundry, and engineering shops, the average yearly and daily earnings per man for the whole establishment being:—

Year	Per Annum. £ s. d.
1869	31 7 0
1870	34 1 0
1871	38 0 0
1872	43 11 0
1873	52 16 0
1874	45 17 0
1875	48 11 0
1876	44 17 0
1877	44 2 0
1878	40 12 0

At the blast furnaces alone the average wages per man per day were:—

1875. s. d.	1876. s. d.	1877. s. d.	1878. s. d.
2 8½	2 9	2 8½	2 7½

Few men would however earn the full amount mentioned above, as from the gross earnings had to be made deductions for sickness and other causes of absence.

It often happens that circumstances connected with the reception and storing of minerals, and a general want of suitable arrangements for economizing labour abroad, give the works of this country a considerable advantage over those of France, Germany, and Belgium. Thus, partly from the causes just named, and to some extent from the coke fillers having to collect the coke from the ovens, which are at the furnaces, often double the number of persons is employed, as compared with the furnaces in the Cleveland district. On the other hand sometimes the conditions of work greatly resemble the Cleveland mode of procedure, but, while employing men at much lower wages (60 per cent. of the Cleveland rates) more hands are required for the same amount of labour.

Having in my possession the numbers of men engaged at several blast furnaces, and the rate of wages they receive, I have been enabled to compare the earnings at each furnace with those of Cleveland. For each sovereign paid in the furnaces on the Tees I estimate the foreign staff to receive as follows:—

		s.	d.
1.—Cleveland		20	0
2.—Westphalian iron work		13	5
3.— ,, ,, ,,		12	0
4.—West of Germany		12	0
5.— ,, ,,		12	0

The larger number of persons required to perform a given amount of work was remarked upon by the authors of the "Voyage Métallurgique," in 1837. They give as examples a Welsh ironwork making 200 tons per week, say 10,000 tons a year, and a French establishment on the English plan making half this quantity, say 5,000 tons a year. The numbers are as follows:—

	Welsh Work.	French Work.
Extraction and transport of coal	317	541
,, ,, ore	521	89
Transport of minerals separately paid in France	—	30
	838	660
Blast furnaces, calcining, etc., etc.	309	245
Forge and mill	205	255
Directors and agents	31	14
	1,383	1,174

Since, they rightly observe, differences in the coal and ore may entail more labour in France than in England, they also compare two forges; and the result of the examination is that the English workmen make twice as much iron as the French.

Notwithstanding the comparative cheapness of labour on the Continent, when contrasted with our own country, the subject of its

SECTION XVI.—THAT OF THE CONTINENT OF EUROPE.

economy is receiving constant attention at the hands of foreign iron-masters. As an example I may quote the case of my friends Messieurs de Wendel, at Hayange in Alsace, where the latest improvements in blast furnace practice, as put in force first at Middlesbrough, have been introduced; without any notable change however having been made in the scale of wages.

In another case, in Rhenish Prussia, there were employed at the blast furnaces in 1875 151 men, who received in that year collectively £7,427, equal therefore per man to £49 3s. 8d. per annum, or about 19s. per week. The same furnaces were worked in 1878 with 117 men, who were paid £5,581, equal to £47 14s. per annum, or 18s. 3d. per week. At the same time the output of the furnaces was increased 30 per cent.

It would appear that the men engaged in the blast furnaces generally are greatly improving in efficiency in Germany and Luxemburg, for according to official figures the average annual output per man for the different years named has been as follows:[1]—

1872.	1873.	1874.	1875.	1876.	1877.	1878.	1879.	1880.	1881.
Tons.	Tons.	Tons.	Tons.	Tons.	Tons.	Tons.	Tons.	Tons.	Tons.
76	79½	78	89	99½	106	132¼	128	129	136

These figures show an increase of 79 per cent. in 1881 as compared with 1872. None of these figures however are any approach to what is done by the workmen at the Cleveland furnaces, and illustrates what has been already observed in these pages, that well paid and well fed men are not always more expensive to the employer than badly paid labour. As a matter of fact, I have rarely found the wages on a ton of the furnace produce to amount to less than what I have found it to be in Cleveland.

MALLEABLE IRON.

In 1840, when I first became occupied with the manufacture of malleable iron, there was still a large quantity of metal used in the puddling furnace, which had previously passed through the refinery. Practically, in Great Britain at least, the only recent improvement in rolling mill work has been the suppression of this preliminary operation, effected by greater attention to the condition of the bottom of the puddling furnace, and by the copious use of "fettling." Upon different occasions

[1] Report of British Iron Trade Association, 1880, p. 77

attempts have been made to replace a portion of the puddler's severe labour by mechanical agency; but little encouragement was given to the ironmaster, in laying out his money in this direction, by the men themselves, who required the same pay as when performing the operation exclusively by hand. This unwillingness on the part of the workmen to assist their employers in improving the manufacture of malleable iron is to be lamented on account of both masters and men. No nation has been more conspicuous in introducing ameliorations into its various branches, than the English. Nevertheless when these are most wanted, and when we find ourselves undersold in foreign markets and in our own, that renown in the march of progress seems to have forsaken us. Some ten years ago, Mr. Danks claimed to have succeeded in dispensing entirely with the severe manual labour required in the actual process itself; but unfortunately his invention came too late; for it was recommended to public notice just about the time when the Bessemer process was apparently in a fair way to extinguish the puddler's occupation altogether. Although this expected event has not so far been realised, there seems to be sufficient uncertainty connected with the future of the iron trade to prevent any general and vigorous attempt being made to economise the labour, and lessen the severity, of the puddler's toil. As a matter of fact, his work is carried on pretty much as it was performed forty years ago—his pay fluctuates with the greater or less prosperity of the iron trade; but his position, one of the most laborious in the entire manufacture, remains unchanged by any such improvements in his appliances, as we have seen introduced in so marked a degree in the case of the men at the blast furnace, and which are perhaps still more conspicuous in the other processes of making malleable iron, which follow that of puddling.

In 1840 the price paid for puddling pig iron in Staffordshire and on the Tyne was 9s. 6d. per ton long weight. In 1842 it fell to 7s. 3d., and I think at one time was as low as 6s. 6d. A ton and a quarter of puddled iron was a good day's work. This at 6s. 6d. per ton amounts to 8s. 1½d., out of which the puddler had to pay probably not less than 2s. 6d. to his underhand, leaving him a trifle above 5s. 6d. for himself.

The following table sets forth the fluctuations to which this kind of work has been subjected in later years:—

Year.	South Wales. s. d.	Cleveland. s. d.	Staffordshire. s. d.	Belgium. s. d.
1866	6 3½ L. W. of 2,400 lbs.	10 0 L. W. of 2,400 lbs.	9 6 L. W. of 2,400 lbs.	—
1868	5 1 ,,	8 0 ,,	8 6 ,,	—
1869	5 1 ,,	8 0 ,,	7 9 ,,	—
1870	5 8 ,,	9 6 ,,	8 10½ ,,	—
1871	5 6½ ,,	9 4 ,,	9 3 ,,	5 9 S. W. of 2,240 lbs.
1872	7 0½ ,,	11 3 ,,	11 6 ,,	6 0¼ ,,
1873	8 7 ,,	12 6 ,,	13 1 ,,	6 1½ ,,
1874	7 0 ,,	11 7½ ,,	11 7½ ,,	6 0¼ ,,
1875	6 0 ,,	9 7½ ,,	9 10½ ,,	5 10¾ ,,
1876	5 4½ ,,	8 3 ,,	9 6 ,,	5 8¼ ,,
1877	4 8 ,,	8 3 ,,	9 6 ,,	5 0¾ ,,
1878	4 4 ,,	7 9 ,,	7 10½ ,,	4 6¼ ,,
1879	3 10 S. W. of 2,240 lbs.	7 0 S. W. of 2,240 lbs.	7 0 S. W. of 2,240 lbs.	4 6¼ ,,
1880	4 5 ,,	8 0 ,,	8 6 ,,	4 9 ,,
1883	4 5 ,,	7 0 ,,	—	—

It will be perceived that the Welsh rates are much below those of England. Including a bonus of 10d. per shift for full time, the head puddler could earn 5s., and his underhand 2s. 6d. per day when the price paid was 4s. 5d. per ton S.W., the produce being 30 cwts.

In Cleveland 27½ cwts. is considered a fair day's work. This quantity at 7s. per ton, with a bonus of 9½d. per ton for full time, enabled the chief man to get 7s. 2d., and his underhand 3s. 6d. per shift.

It is not always an easy task to compare the work of an English puddler with that of a foreigner. In many places abroad it is the practice to have three men, or two men and a strong boy, at the furnace, which of itself enables a little more work to be done. The chief advantage, however, in respect to labour lies in differences in the quality of the iron; which abroad frequently contains only one-fourth of the silicon usually found in Cleveland iron; and, as is well known, this metalloid calls for special exertion on the part of the workman.

In the year 1867, during a visit to the south of France, I found 36½ to 40 cwts. represented the twelve hours' produce of a puddling furnace; the price paid varying with the quality turned out by the puddler, and a premium being often reserved for excellence of produce. The prices paid fluctuated from 5s. 10d. to 7s. 2½d., and the average weight produced was 38½ cwts. The total earnings per shift therefore amounted to about 12s. 4d., divided approximately in the following manner: the first man 6s. 9d., second 3s. 3d., third 2s. 4d. At Middlesbrough the earnings at that period of the head man would be 8s. per day.

On the Rhine, in 1867, 32½ to 36 cwts. seemed to constitute the ordinary twelve hours' work, for which 5s. 6d. to 5s. 9d. per ton of puddled bar was paid. This amounted to about 9s. 3d., to be divided by giving probably 5s. 3d. to the first man, 2s. 6d. to the second, and 1s. 6d. to the boy. In this case the chief puddler earned about 60 per cent. of that of the head man in England; the same weight of iron costing for puddling about one-half more in England than in the German works referred to.

In 1878 I found that in this part of Germany the price averaged for different qualities about 6s. 2d.; which on the weight obtained, $31\frac{1}{2}$ cwts., would give 9s. 10d. Two men only appear to have been working the furnace, something after the fashion of what is known as "level-hand" in England, for the money was divided by giving 5s. 6d. to the best of the two workmen and 4s. 4d. to his mate. The price of puddling in that year at Middlesbrough was 7s. 9d. long weight, or say 7s. 3d. short weight; being about 18 per cent. higher than the German rate.

In Western Germany, in 1878, I found two men doing eleven heats of pig per turn, giving 48 cwts. of puddled iron: the pig only contained ·25 per cent. of silicon. The price paid was 2s. 11d. to the upper and 1s. $9\frac{1}{4}$d. to the underhand, or 4s. $8\frac{1}{4}$d. together. At these prices the former earned 7s., and the other 4s. 1d., per day. The pig iron was preheated in a chamber known as a "dandy." The English price for puddling was therefore fully one-half more than that of the German; but, owing to the large make, the chief puddler earned nearly as much as is paid in England.

In the same portion of the German Empire, a mechanical *rabble* is used, and by its means 65 cwts. of puddled iron are produced from each furnace. The men were paid according to quality, the iron being all sorted. The prices per ton were 3s. $8\frac{3}{4}$d., 4s. $5\frac{1}{2}$d., and 5s. 2d., the average actually paid being 4s. $10\frac{1}{4}$d.; and the individual daily earnings were 5s. to 6s. to the first man, 5s. to the second, 3s. $9\frac{1}{4}$d. to the third, and 1s. $6\frac{1}{4}$d to a boy. This was in 1875, when the price paid in the North-east of England was 9s. 7d., or almost exactly double the price paid by the German ironmaster. It is to be remarked that the relief afforded to the men by the mechanical appliances is such that they are able to work for 12 hours instead of 10 to $10\frac{1}{2}$ hours as at Middlesbrough. An economy, about 50 per cent., in the consumption of coal results from the use of this improved furnace which would represent a saving of about 4s. per ton of puddled iron.

SECTION XVI.—THAT OF THE CONTINENT OF EUROPE.

In Spain, in the year 1872, 36 cwts. of puddled iron were received in twelve hours, for which 6s. was paid ; the cost on the Tees for the same quantity being 11s. 3d., or say 10s. 6d. short weight.

The output in Belgium is usually about 35 cwts. for twelve hours; for which, in 1883, 4s. 9d. was paid per ton, the Cleveland rate being 7s., with an allowance for full time. The staff of three men at the Belgian furnace would therefore earn 8s. 4d., against 9s. 8d. for the two men at the English. The earnings however of the upper-hand in the former case would be probably 5s. 4d., against 6s. 2d., in the latter; so that the Middlesbrough puddler receives 16 per cent. more than the Belgian, and the work costs the English ironmaster 47 per cent. more than his foreign competitor.

Some confusion arises in comparing the entire costs of labour in the puddling forge in different works, owing to the differences of the modes in which the accounts are kept; some charging men's time for repairs under a general head, which includes materials. Besides this source of derangement there are differences arising from the nature of the work performed, and in the general convenience of the appliances. I believe, however, I shall not be far from the truth in laying down the following data for puddling ordinary quality in 1878:—

Average per Ton.
England—Cleveland, with a bonus for full time ... ... 7s. 2d. short wt.
Rhenish Provinces, 4s. 7d., 5s. 2d., 5s. 9d., 6s. 6d., av. 4 works = 5s. 6d ,,
Western Germany ... ... 4s. 10¾d. ,,
Belgium, 4s. 2¼d., 4s. 10½d., 5s. 2¼d., 5s. 2¾d., 5s. 11d., av. 5 works = 5s. 1d. ,,
France—North-East ... ... 5s. 9½d., 5s. 10¼d. = 5s. 9¾d. ,,

On comparing the total wages paid in the puddling forge in the North of England with the payments on the Continent I find that the cost in England is often fully 50 per cent. higher than the average of works I have visited in Western Germany and the East of France. The effect of this is that the expense of making puddled bar, irrespective of the price of pig iron, is some shillings more in England than in those localities with which I have compared it. It may be added that the average earnings of all the staff in one of the foreign establishments were 3s. 6d., whereas in England it is close on 4s. In Germany the puddled bar rollers and shinglers were earning 5s. 3d. per day, whereas in the county of Durham the former were paid in 1878, 13s. to 15s. per day, and the latter 10s. to 18s. per day.

The metal delivered to the foreign puddling furnaces is frequently hard and white, and, to use the workman's language, "soon comes to nature." A material, somewhat of the same character as that in question, is often used in South Wales, for the working of which about the same price is paid as that current in Germany and elsewhere. I believe that the Welsh practice has not been followed, at all events to any extent, elsewhere in the United Kingdom; and it may be observed that wages in the Welsh iron works have always been notably lower than in England or Scotland.

Pig iron intended for forge purposes is distinguished in Belgium under five denominations. These are:—

	Per Cent. Carbon.	Per Cent. Phosphate.
1.—*Metis*—Very common, containing	2·00	2·25
2.—*Ordinaire*—Common ,,	2·50	1·75
3.—*Fort*—Strong ,,	3·00	1·25
4.— ,, —Extra ,,	3·50	0·75
5.—*Aciéreux*—Steely ,,	4·00	0·25

At one work which I was permitted to visit puddling was performed in a double furnace worked from both sides, but without any mechanical aid. No *Metis* was used. Of Nos. 2 and 3 eight charges of pig were worked in every shift weighing about 72 cwts.; but of Nos. 3 and 4, 54 cwts. per shift only were puddled. For Nos. 2, 3, and 4 the prices paid in 1878 were 4s. 2¼d., 4s. 6d., and 4s. 11¼d. per *tonne* of 1,000 kilogrammes.[1] At these rates the head puddler earned on common iron 5s. 2¼d., and his underhand 3s. 5½d. per shift, the quality of the product being very similar to that made at Stockton and Middlesbrough.

When the work of the finishing mills is examined, it will be found in certain instances that the difference between Great Britain and Continental countries far exceeds that just named as obtaining in the puddling forge.

In Great Britain for boiler-plate or ship-plate rolling I find that between 1841 and 1843, the prices paid were 10s. in the former year and 8s. 9d. in the latter per ton for heating, rolling, and shearing. The same work cost in 1875 about 9s. 3d., so that practically there has been no change in the actual tonnage prices paid to the work-

---

[1] On the Continent the tonne is 1,000 kilogrammes = 2,204 lbs.

SECTION XVI.—THAT OF THE CONTINENT OF EUROPE. 531

men in the last forty years. In one respect, however, there is a material change in the conditions of plate-rolling between the two periods. In 1840, 50 tons per week was probably a fair make for a mill, now, in consequence of the use of more powerful machinery, and the extensive character of the orders so largely given out for shipbuilding, the production has risen to 400 and sometimes reaches 600 tons or more in a week.

By the favour of a friend, the manager of a large plate mill in England, I have an account of the earnings of fourteen head plate-rollers. The lowest amount received was 17s. 5d. per shift, the highest £2 0s. 11d., and the average of the whole number was £1 7s. 8d. The wages of the head shearman ranged from 18s. 10d. to £1 16s. 1d. per day, the average being £1 8s. 3½d. The heaters earned from 11s. 4d. to 13s. 8d., the average of fourteen men being 12s. 8d. each per day. These wages were received after all deductions for assistance had been met.

I have in my possession a statement representing the actual payments made in 1879 by a Continental firm for carrying on a work turning out about 300 tons of puddled iron per week, with a plate mill producing 200 tons. I propose comparing the wages, exclusive of those for repairs, paid in the various departments, with those of an English work, turning out also 300 tons of puddled iron, along with a weekly make of 300 tons of plates. In the latter case the deficiency of puddled iron is supplied by purchases from other quarters.

I would first point out the great dissimilarity between the daily pay of most of the workmen in the two instances selected for comparison.

	German Work. s. d.		English Work. £ s. d.
Head puddler	5 8	...	0 6 1
Second ,,	5 3½	...	0 3 6
Third ,,	4 0	...	none.
Boy ,,	1 7½	...	,,
Puddle roller	5 1¾	...	0 15 1
Shingler	5 3	...	1 2 9
Furnace men (plate mill)	6 6¾	...	0 16 1
Plate rollers, average of 5 men	4 6¼	Head roller	2 1 1
Head shear man, average of 8 men	5 3¼	,, shearer	1 14 9
Average	5 2½	...	0 19 11

The enormous differences between most of the figures in the two columns would prepare us for supposing that the cost for labour must necessarily be correspondingly higher in the English than in the German mill. This is more or less true as regards the puddling process, for although the earnings of the puddlers do not differ greatly in the two cases, the actual cost of puddling is above 66 per cent. more in the English than at the German work with which it is compared, viz. 4s. 9¼d. against 7s. 11d. per ton S.W., inclusive of prize money at the latter for full time. This arises partly from differences in the nature of the two kinds of pig iron employed, but more especially from the German workmen having taken kindly to the use of mechanical help, by which, instead of about 30 cwts., each furnace puddles 70 cwts. of pig iron per shift. According to the figures given to me, I have calculated that for each £100 paid in wages in the German puddling mill, made up of the items placed in its column, the cost for the same amount and kind of labour, paid by the English manufacturer is that shown in the adjoining amounts:—

	German Work.	English Work.
Delivering materials	4·42	2·45
Engine and fire men	3·30	8·30
Puddlers	63·25	108·50
Shinglers	4·50	11·10
Bogey men, rollers, draggers, &c.	22·64	12·10
Sundry labour	1·89	2·55
	£100	£145

It is very remarkable that when the actual cost for labour per ton of plates, in the finishing mill, at the two places is compared, there is, in spite of the very much greater individual earnings of the men, no such difference as that just shown as obtaining in the puddling department. In the English works, situate in the County of Durham, it is a common practice to contract with one man who pays for the necessary help at such rates that his own earnings are often greatly increased from this cause. There is no doubt also, that the amount of duty performed per man is much greater in England than it is abroad, as is proved by the larger number of persons employed in the Continental works. As examples, while one man and a boy usually work a mill furnace in England, the men from all the furnaces assist-

SECTION XVI.—THAT OF THE CONTINENT OF EUROPE. 533

ing each other in charging, at this foreign work in question there are two furnace men, one receiving 6s. 6¾d. and the other 4s. per day, besides a boy at 1s. 2¼d. The German plate rollers are, as has been shown, paid an exceedingly small wage as compared with those in the North of England; but then at the former there are no less than 22 men engaged, a much larger number than is to be seen doing the same work in this country.

Adopting the same mode of comparison as that pursued in the puddling department, we have the following approximate results, also exclusive of smiths, fitters, &c.:—

	German Mill.	English Mill.
Delivering materials...	2·13	5·34
Engine men and fire men ...	4·80	5·00
Furnace men ...	16·40	30·66
Filing iron, roller, and shearmen ...	74·03	60·50
Sunday labour	2·64	8·50
	£100	£110

Any one conversant with the subject will recognise the difficulty of making a perfectly correct comparison between any two works, owing to the differences in which the accounts are kept. At the puddling works as well as in the finishing mill, the item of "sundry labour" is much higher in the English than in the German establishment. This is probably due to the greater amount of help afforded to the piece workers in the former. It must also be remembered that the make in the English plate mill being one-half more than in the German, an advantage in point of economy in labour is the natural consequence. Had the German mill been conducted on the same footing as the English it is not unlikely that the difference in its favour would have a second 10 per cent., making the entire excess against the English manufacturer 20 per cent. instead of 10 per cent.

It may be mentioned in conclusion that the average pay of the entire staff in the German mill is 3s. 6½d. per shift in the puddling mill, and 3s. 6d. in the plate mill, which is considerably below that prevailing in England.

I have endeavoured to pursue the inquiry as to the comparative cost of labour in the manufacture of malleable iron in England and Germany. This I have done by dissecting the information I possess so as to obtain the number of men employed along with the wages they receive. The English mill it should be observed enjoys the advantage of being on a much larger scale than the German one with which it is compared, inasmuch as it produces exactly three times the weight of plates made by the other in a given time.

The money paid on each shift at the two places was as follows:—

	ENGLISH MILL.					GERMAN MILL.				
	No. of Men.	Wages.			Average per Day.		No. of Men.	Wages.			Average per Day.
		£	s.	d.	s.	d.		£	s.	d.	s.	d.
Enginemen 2, and 2 boys...	4	0	13	0	3	3	... 2	0	4	9	2	$4\frac{1}{2}$
Boilermen	2	0	9	0	4	6	... 2	0	5	$6\frac{1}{2}$	2	$8\frac{1}{4}$
Wheelers of coal	4	0	18	0	4	6	... 2	0	4	7	2	3
Cutting puddled bars, wheeling, and piling	9	2	9	6	3	$3\frac{1}{2}$	... 9	1	4	$6\frac{1}{4}$	2	$8\frac{1}{2}$
Chargers, 7 men, 5 boys ...	12	2	0	6	3	$4\frac{1}{2}$	... 6	0	15	$4\frac{1}{2}$	2	$6\frac{1}{4}$
Furnacemen	11	6	1	0	10	1	... 9	1	15	$3\frac{1}{4}$	3	11
Men at rolls, etc.	16	8	11	0	10	$8\frac{1}{4}$	... 22	4	2	$1\frac{1}{4}$	3	9
Shearing finished iron ...	10	6	5	0	12	6	... 11	2	12	$1\frac{1}{4}$	4	$8\frac{1}{4}$
Sundry labour	4	0	12	0	3	0	... 3	0	5	$8\frac{1}{4}$	1	$10\frac{3}{4}$
	72	27	19	0	7	$9\frac{1}{4}$	66	11	9	$11\frac{1}{4}$	3	6

In the examples before us notwithstanding that the average earnings of the English workmen are somewhat more than double those of the German, yet owing to the greater power of the English mill and its consequent larger production already mentioned, the total cost of labour for the items enumerated is actually 25 per cent. higher in the German work than in the other. The moment, however, that a foreign mill is established on a similar footing as its English competitor it is most likely that the greater economy of production will lie with the former.

I have dwelt at some length on circumstances which tend to increase the cost of manufactured iron in the United Kingdom; because undoubtedly it is by reason of the cheap production afforded by cheaper labour that Belgium is able to send girders and other forms of malleable iron into our country. It will be a more serious matter if our trade in ship-plates should be interfered with by the same cause; for then, with more economical labour in the shipyards abroad, it might easily happen that we might lose that supremacy in the construction of iron vessels which we at present enjoy. Are there indeed no signs of this state of things visible at the present moment? Some foreign countries are progressing in building ships from iron of their own manufacture, and we hear of others importing plates of English make for shipyards where iron is dearer but labour is cheaper than the present rates ruling in the United Kingdom.

BESSEMER STEEL.

The manufacture of Bessemer steel is, comparatively speaking, so recent an invention that almost every year brings some improvement in the appliances used in its production. The extent to which these have been introduced of course materially affects the cost of labour on the ton of steel.

In respect to the individual earnings of the workmen, although the rates paid in the United Kingdom are much higher than those of the Continent of Europe, the very great differences noticed as existing in certain items between British and foreign malleable iron works, are often not to be found in the manufacture of Bessemer steel.

As examples of this I append the following extracts from memoranda taken in the years 1879 and 1880:—

	Great Britain.				Germany.		
	s. d.		s. d.		s. d.		s. d.
Converter men ...	5 10		—	...	2 9	to 3	7½
Men in ingot pit ...	4 7¾	to	5 1	...	2 7	„ 3	2
Ladle men	4 1	„	4 8	...	2 11	„ 3	2½
Furnace men in rail mill	6 2	„	6 6	...	4 2		—
Rollers	9 0	„	10 10	...	5 8	„ 8	7
Roughers	7 5	„	7 7	...	3 10	„ 7	8
Cold straighteners ...	5 4½	„	—	...	4 0		—

Speaking approximately, there seems to be a difference of 50 per cent. against the British manufacturer in the items just given.

I have had communicated to me, in great detail, the particulars of the wages paid in two steel-rail works. The one is in England, making about double the quantity of ingots of the other, which is in Germany. The weight of rails per shift was about the same in each case. The various items have been divided under the heads given below, in order to compare the average daily earnings at the two places. At both the pig iron was remelted for the converters.[1]

[1] In both cases the cost of delivering the materials is omitted.

536   SECTION XVI.—BRITISH LABOUR COMPARED WITH

	England. Average Wage.	Germany. Average Wage.
	s. d.	s. d.
Converters:—		
Cupola men	6 2½	2 8
Men at converters and ladles	7 0½	2 11½
„ ingot moulds	5 8½	2 10
Engine men	5 7	2 7
Repairs, etc.	4 10¾	2 6¼
Sundry labour	4 6¾	2 0¼
Average of staff at converters	6 2¼	2 8
Rail Mill:—		
Men and boys at furnaces	5 10½	3 10
Head roller	23 2	8 7
Charging furnaces, bogeying, and rolling	7 7¾	3 7
Hot rail sawyers	6 8½	2 6
Cold straightening	3 10¾	2 10½
Engine men	5 6	2 10
Repairs	6 1½	2 8
Sundry labour	3 2½	2 8½
Average of staff at mill	5 3¾	3 3

The information from which these calculations are compiled was given me in detail and in writing. I have thus been enabled to compute the relative cost at the two places referred to.

At the converters the relative amount of work performed per man engaged and its cost are as follows:—

			England.	Germany.
Converters:—				
Work performed per man, England taken as unity			100	81·17
Cost of ingots for the items named	„	„	100	63·24
Average daily earnings	„	„	100	43·24
Rail Mill:—				
Work performed per man	„	„	100	100·
Cost of rails for the items named	„	„	100	86·11
Average daily earnings	„	„	100	61·18

As a proof how the wages of the men are affected by the vicissitudes of trade in Germany, there is given below the daily earnings of various branches of labour in a German steel rail mill, beginning with the year of high prices—1873 :—

	1873.	1874.	1875.	1876.	1877.	1878.	1879.	1880.
First roller ...	10/8¾	10/10	5/8½	6/5½	5/8½	5/8	4/8	4/11¾
Second „ ...	9/8	10/10	5/5¾	6/5	5/8	5/1¾	4/5½	4/4¾
First heater...	6/3¾	6/8	5/6¾	4/4¼	5/3¾	5/1¾	4/3¾	3/5¾
Second „ ...	6/1	6/3½	4/4½	4/2¾	3/11½	4/3	3/5½	3/5
Man at saw ...	5/3¼	6/2¾	3/4	3/1	3/3½	3/1	2/11	3/0¾
Dresser of rails	4/9½	4/5	3/6	3/1½	2/8¾	2/6¾	2/11¾	2/10½
The entire staff at the rail mill	6/1	5/2	5/0	3/6¾	3/2¼	4/1½	3/6¾	3/7½

In further illustration of the fact just referred to I annex the average daily earnings, over a period of ten years, of another establishment in Germany. These embrace the men engaged at the converters and rolling mill:—

1869.	1870.	1871.	1872.	1873.	1874.	1875.	1876.	1877.	1878.
2/11½	2/3¾	2/10¾	3/7	4/0¼	3/8½	3/3	3/1½	3/1	3/0½

I have given these German rates of wages at some length, because it is from that quarter that the British rail-maker meets his most formidable competitors.

In Wales the average earnings of the higher classed men at the Bessemer converters will run from 30s. to 42s. per week. In Germany they are very little more than the half of this, the highest being 20s. and others 18s. The average of all the men employed in Wales at and about the Bessemer pits is about 3s. 7d. per day, as against 2s. 6d. in Germany, or a difference of 33 per cent. against the former. When however the actual money paid for labour in each comes to be compared, the number of men is so much greater abroad than in England, being nearly double, that the actual cost for workmanship, per ton of product, in the Welsh works, is about 22½ per cent. only above that in Germany.

Since the foregoing comparisons were made, very considerable improvements have been introduced into the recently constructed rail mills in England. One of the latest consists of three powerful direct-acting engines, each driving one pair of rolls only. The hot ingot is brought to the cogging mill, from which, by means of a series of rollers driven by steam power, it is conveyed direct to the roughing rolls, and from these in like manner to the finishing mill. After the operation of rolling is completed, a third set of rollers passes the rail on to the saws, from which it is removed by mechanical agency, in which however there is nothing very novel. In this way an ingot, a ton weight or more, is reduced to 100 feet of finished railway bar and cut to length in $4\frac{1}{2}$ minutes. The work is performed with such rapidity that the mere act of compression on the steel is a source of considerable heat so that the rail when completed is almost as hot as the ingot was when it first entered the cogging rolls.

SECTION XVII.

ON LABOUR IN THE UNITED STATES OF AMERICA.

In the preceding section, devoted to a comparison of the conditions and cost of labour in Great Britain and in certain foreign nations, little reference has been made to the United States of America. They were purposely omitted, notwithstanding the fact that in the annual quantity of coal and iron produced the Union stands second only to Great Britain. It has been thought expedient to adopt an arrangement which assigns a separate section to the North American Continent, owing to the important differences in the circumstances which surround the question in the opposite hemispheres.

In Section XIV. it has been assumed that the price at which labour can be supplied is more or less dependent on the terms at which the necessaries of life can be procured; and certainly, so far as human agency is required in the production of such necessaries themselves, some connection between their price and the rate of wages appears to have obtained, at all events in former times, in the old world.

In the Western States of the Union, the extent of land is so immense, as compared with the population, that the element of rent in the cost of growing agricultural produce may be regarded as almost *nil*. It is widely different in Europe, where denser populations have exercised a corresponding influence on the value of land, by the greater demand there exists for its products.

After the elimination of the item of rent it is extremely difficult to form any very satisfactory estimate of the comparative cost of food, as it is produced in the United States and in Europe. This difficulty arises from the totally different nature of the conditions under which the cultivation of the soil is pursued in the two continents. In the Western States, prairie lands are broken up and continue to grow wheat year after year, the produce after a few years gradually diminishing in quantity until the average per acre is only about one-half that in the United Kingdom. According to information given me in 1882 by Mr. Welch, the President of Iowa College of Agriculture at Ames, six or seven men are employed on a farm of 1,000 acres of wheat land. They are retained from April to December, during which time they receive 3s. 2½d. per day, which, with board and lodging, is equal to about 4s. 2d. About half of this staff remain the year round to look after the horses and other work on the place. No manure whatever is purchased, so that any laid on the land is confined to what is produced by the horses employed on the farm or by a cow or two. Upon land so worked, with a produce of about 20 bushels per acre, wheat can be put into the granary on the farm at 1s. 0½d. to 1s. 3d. per bushel. Another estimate brought the cost to 1s. 4½d. including taxes and interest on purchase of land.

Upon large farms the machinery employed is of the most improved type, and in ordinary seasons the pursuit of agriculture is a very lucrative one. The profits of one consisting of 6,000 acres, growing different kinds of grain were spoken of as reaching £9,375 per annum.

To this information I would add the results of my own enquiries when in the United States.

A banker of St. Petersburg in Illinois gave me the following particulars of his own farming in 1874. He is the proprietor of 10,000 acres of land, which he purchased at the rate of £11 per acre, and of it he retains 1,000 in his own hands. He pays his labourers, chiefly immigrants recently arrived, 17s. 6d. per week and their board, equal altogether probably to 30s. Female domestics have 11s. 3d. per week and board. He spoke in disparaging terms of native American labour, and even the fresh importations from Europe only continue to satisfy him in point of efficiency for about three years. His average produce for thirty-five years has been 71 bushels of Indian corn per acre, weighing 56 lbs. per bushel; of wheat he does not exceed 15 bushels.

A gentleman from West Minnesota informed me that the farmers paid about 5s. 9d. per day in harvest time, and that the land can be bought there at 3s. 6d. per acre, which for some years yields from 20

SECTION XVII.—COMPARED WITH THAT OF EUROPE.

to 30 bushels per acre, weighing 65 lbs. per bushel, after which it falls off considerably. In spite of the high price of labour, the actual cost for this item is probably even less than with us, although in England it is little more than half the price paid in Minnesota. On a farm growing nothing but wheat, a force of men at ploughing and seed time, with a second force in harvest, constitute the chief cost for wages. The heads of corn, when ripe, are pulled off, and the straw is burnt; after which the land is ploughed and sown as long as it will bear a crop worth the expense of cultivation. With a price of 30s. per quarter, free on the railroad, they express themselves abundantly satisfied.

I was informed that in Colorado good land could be purchased for 5s. 6d. to 15s. per acre, according to the distance from the railway. Farm labourers get £5 12s. 6d. per month, and their board. The soil immediately after being broken up produces 25 bushels, but this gradually falls off, so that in twelve or fifteen years it has sunk to 12 bushels; after which the location is deserted and other land is purchased.

Upon the occasion of my first visit to Cleveland City, Ohio (1874), a leading banker of that place had just returned from the Red River; and he stated that prairie land in that country could be purchased for prices varying from 11s. to 34s. per acre, which for some years after being ploughed up yielded 46 bushels of wheat per acre. The Red River is about 150 miles from Duluth on Lake Superior, where the grain is shipped direct for Buffalo on Lake Erie, and forwarded thence by rail to New York City.

Mr. Clare Sewell Read, and Mr. Pell, who recently visited the wheat-growing districts of America, as members of a Royal Commission, report to our government that the total cost of labour in some districts, including board, was only 3s. 7d. per day; from which they estimated that wheat could be delivered at the local depôts at 28s. per quarter, and this estimate is based on the yield being only 13 bushels per acre.

According to the authority just quoted, the cost of sending a quarter of wheat by the lakes to Chicago, thence by rail to New York, and by steamer thence to Liverpool, is 19s. $9\frac{1}{2}$d.; which, added to the cost price, brings up the total charge to about 47s. $9\frac{1}{2}$d.

Although the cost of transport from Chicago to Liverpool is about 10s., or 13s. including charges, per quarter, Messrs. Read and Pell mention the fact that, owing to speculation and other causes, wheat is sometimes as cheap at the latter city as at Chicago. This however, I imagine, can only be under exceptional circumstances and possibly of rare occurrence.

According to information received at West Hartlepool, the cost of bringing wheat from the Western States to the seaboard varies from 6s. to 8s. per quarter, and the cost thence to West Hartlepool is 6s. From another source of information I learn that ships have recently been constructed which can bring it to England at a profit at 4s. 3d. Cattle can be conveyed in summer at 50s. and sheep at 4s. per head; but, owing to increased insurance, the freight in the winter months on these descriptions of live stock is 110s. and 8s. respectively.

It is however impossible with us to apply, the factor of rent or indeed any other factors which enter into the cost of farming, to the exclusive production of any particular description of crop. This arises from the circumstances that the soil and climate of the United Kingdom do not admit of any one kind of grain being grown continuously upon the same breadth of ground. The ability therefore of the farmer to sell wheat at any particular price depends to some extent not only on the money he obtains for wheat but also that commanded by oats, barley, etc.

In order to ascertain the proportion which rent bears generally to the cost of raising farm produce in the United Kingdom, my friend Mr. Thomas Gow, a great authority on farming science and practice in Northumberland, kindly prepared for me the following statements. He assumed that £22,500 had been paid for a farm of 500 acres the rent of which was 30s. per acre. This return, viz., £750 per annum, he calculates will only leave, after paying outgoings, $2\frac{1}{2}$ per cent. on the capital invested by the landowner, but such has been the desire to possess a portion of the limited territory in Great Britain on the part of the almost unlimited number of wealthy persons in the country, that many are to be found willing to charge the food of its inhabitants with this very moderate amount of interest on the money invested in its soil.

SECTION XVII.—COMPARED WITH THAT OF EUROPE.

Taking wheat at 44s., barley at 32s., oats at 26s., and beans at 42s. per quarter, with cattle at their then value (1881), viz., about 9s. 6d. per stone, it is doutbtful, everything taken into the account, whether the the farmer would derive more than bare interest on his capital invested in the undertaking. It may be further added, that the income is based on the supposition that the land yields of wheat 32, of barley 40, of oats 42, and of beans 32 bushels per imperial acre; and valuing the grain at the prices already named, we may roughly take the income to consist of one-fourth from each of the descriptions named. It is true, the estimate setting forth the results of an exclusively corn-growing system of tillage appears, under the assumed circumstances, to be the more profitable mode of cultivation: but Mr. Gow doubts its remaining so, owing to the gradual exhaustion of the soil, consequent upon the nature of the farming operations it involves.

Upon such a farm as that just referred to when oats sell at 26s. per quarter, beans at 42s., barley at 32s., and wheat at 44s. will allow about 5 per cent. to the farmer for his capital employed. This estimate is founded on the farm of 500 acres being almost exclusively devoted to growing grain. If instead of this, 150 acres of its area is engaged for grazing purposes with grain at the same values as these just given and meat at 9s. 6d. per stone the profit may at first be a little less than in growing grain alone.

According to an analysis of the cost of tillage farming just mentioned, with which Mr. Gow favoured me, the expenses may be thus divided:—

	All Tillage.
Rent	20·53
Rates and taxes	2·05
Wages	25·57
Wages for repairs including materials	1·78
Manures, etc.	19·03
Seeds	10·26
Horse keep	11·83
Depreciation of implements and horses	4·16
Interest on farmer's capital	4·79
	100·

Although wages constitute 75 per cent. of the cost of growing wheat in America, the dearer labour there does not interfere with cheap production to anything like the extent it would do in England, owing to the very different circumstances attending its cultivation. Nevertheless, as agricultural labour has in the previous section been adopted as a basis of comparison for other kinds of wages I was desirous of having ample information on this point, and hence I applied in 1879 to my friend Mr. J. S. Weeks, the editor of the "Iron Age" published in New York, who kindly obtained for me the current rates payable to farm servants in the vicinity of the chief iron-making centres of the States. The following are some of the results of his elaborate enquiries instituted in the year 1881[1]:—

State.	District.	Prevailing Industry.	Wages. Per Day.		Wages. Per Week. Yearly Service.	
			s.	d.	s.	d.
New York	Lake Champlain	Iron mining and smelting	3	7½ and board	12	6 and board
,,	Salisbury region	Charcoal iron	3 4	1½ with board 2 without board	15 26	7¾ with board 0½ without board
Pennsylvania	Schuylkill Valley	Pig iron	3	1½ and board	9	4½ to 13/6 and board
,,	Lehigh Valley	,,	3	9 and board	17	4 to 26/0½ and board
,,	Pittsburg	Coal and iron	3	1½ to 4/2 and board	10	5 and board
,,	Shenango Valley	,,	2	1 to 3/1½ and board	10	5 and board
,,	Susquehanna Valley	Iron	2	1 to 3/1½ and board		
Ohio	Mahoning Valley	Coal and iron	3	7½ and board	10	5 and board
,,	Stentonville	,,	3	1½ and board	13	6½ and board
,,	Hangingrock	,,	3	1½ and board		
Kentucky	,,	,,	3	1½ and board		s. d.
Maine	Portland and Katahden	Iron	4	2 and board	10	5 to 12 6 and board
Massachusetts		,,	5	2½ to 7 3½ and board	12	6 to 18 9 and board
Virginia		Pig iron	3	1½ to 4 2 and board	8	4 to 10 5 and board
West Virginia	Quinniment	Coal and iron	2 3	1 with board 1½ without board	8	4 to 10 5 and board
,,	Wheeling	Iron	2	1½ and board	12	6 to 21 0 and board
Tennessee	Chattanooga	,,	2	1½ and board	8 12	4 to 10 5 with board 6 to 15 7½ without brd.
Indiana	Greencastle	,,	3	1½ to 4/2 and board	16	8 to 20 10 and board
Illinois	Opposite St. Louis	,,	4	2 and board	10	5 to 12 6 and board
Wisconsin	Millwaukee and Chicago	,,	4	2 and board	12	6 and board
Michigan	Lake Superior	,,	4	2 to 5 2½ and board	27	0½ without board

In the Pittsburg and Susquehanna Valley districts, Pennsylvania, 4s. 2d. per day and board is paid in harvest time; in Stentonville, Ohio, 4s. 2d. and board; in Chattanooga, Tennessee, 4s. 2d. and board; and in Illinois, opposite St. Louis, 10s. 5d. and board.

[1] Board in all these cases including lodgings.

SECTION XVII.—COMPARED WITH THAT OF EUROPE.

The cost of board per week is given by my informant as follows:—

	s.	d.		s.	d.
Lake Champlain	12	6	to	16	8
Salisbury Region	10	5	,,	14	7
Schuylkill Valley	12	6	,,	16	8
Lehigh Valley	12	6	,,	19	9¾
Pittsburg	14	7			
Shenango Valley	10	5	,,	14	7
Mahoning Valley	8	4	,,	14	7
Stentonville	8	4	,,	12	6
Hangingrock	5	2½	,,	9	0¼
,,	5	2½	,,	9	0½
Susquehanna	12	6	,,	18	9
Portland, etc.	10	5	,,	12	6
Massachusetts	10	5	,,	16	8
Virginia	7	3½	,,	14	7
West Virginia	4	2	,,	5	0
,,	8	4	,,	10	0
Chattanooga	4	2	,,	6	3
Greencastle	6	3	,,	8	4
Illinois	8	4	,,	16	8
Millwaukee	8	4			
Lake Superior	7	3½	,,	8	4

I have extracted from Dr. Young's work on American labour the following wages, paid to farm servants in the United States in 1860 and 1874:—

	1860. Dollars.	1874. Dollars.		1860. Dollars.	1874. Dollars.		1860. Dollars.	1874. Dollars.
Massachusetts	1·06	1·50	Ohio	·89	1·13	Michigan	·93	1·25
New York	·89	1·48	Indiana	·96	1·13	Wisconsin	·81	1·50
Pennsylvania	·84	1·13	Illinois	1·02	1·33	Virginia	·60	·64
West Virginia	·77	1·03						

The average for 1864 is ·87 dollars = 3s. 7½d., and that for 1874 1·21 dollars or 5s. 0½d., to which has to be added the cost of board.

These figures exceed considerably those given me by Mr. Weeks, the average of which is only about 3s. 6d., board being found in that case also by the employer. If Dr. Young's figures for the two years are correct, it would appear that in fourteen years the increase in cost of farm labour, irrespective of differences of value in the currency, amounted to 38 per cent.

It is no doubt quite true that in districts remote from manufacturing industry, and where transport is expensive owing to the absence of railways and rivers, agricultural wages in the United States may be greatly under the rates already quoted. Thus on Big Rock Creek among the Smoky mountains in Carolina I found 30s. to 37s. 6d. and board per month, the ordinary pay of a farm servant, which in reality is inferior to what is given in some parts of England.

Dr. Young, in his list of the retail prices of provisions in 1872, quotes the following rates for the articles selected in the last section, as the means of comparing English prices with those of the Continent of Europe:—

	Massachusetts.	New York.	Pennsylvania.	West Virginia.	Average, New England States.	Average, Middle States.	Great Britain, 1874.
	s. d.	s. d.	s. d.	s. d.	s. d.	s. d.	s. d.
Flour, per barrel ...	36 0	31 3	31 8	25 6	30 11	28 8	33 0
Beef, per lb. ...	0 11	0 8	0 8½	0 6	0 9	0 7	0 9¼
Mutton ,, ...	0 9	0 7¼	0 8	0 6	0 8¾	0 7¾	0 8
Lard ,, ...	0 9	0 8¾	0 8½	0 7¾	0 8½	0 8¼	0 8
Cheese ,, ...	0 10	0 9¼	0 9½	1 0	0 9½	0 10¼	0 9
Tea ,, ...	3 11	4 1½	3 9	4 2½	4 2½	3 11	2 6
Roasted coffee, per lb.	1 7	1 4½	1 4¼	1 4	1 4	1 4½	1 4
Brown sugar ,,	0 4¾	0 4¾	0 4¾	0 6	0 5¼	0 5¼	0 4
Potatoes, per bushel	3 5¾	3 2½	4 0½	4 0½	4 0½	3 10	—
,, per stone..	0 7¾	0 7¼	0 9	0 9	0 9	0 8¾	—

It has to be observed that the quotations for England are for 1874, when wheat was selling for 55s. 8d. per quarter, whereas the American prices are for 1872 when wheat however was only 1s. 4d. dearer. With this correction, and with the other prices before us, it is not too much to say that although the United Kingdom draws so large an amount of her grain from the United States, the farmer, miller, and retailer there, manage among them to obtain almost as high prices from the consumers in their own country as they obtain delivered on our shores: in other words they appear to divide the cost of transport as profit among themselves.

Dr. Young's quotations for England (Huddersfield) are slightly different from mine, and are as follows:—

Wheat Flour. Per barrel.	Beef. Per lb.	Mutton. Per lb.	Lard. Per lb.	Cheese. Per lb.	Tea. Per lb.	Roasted Coffee. Per lb.	Brown Sugar. Per lb.
s. d.	d.	d.	d. d.	d. d.	s. d. s. d.	s. d.	d.
35 9	10	10	6 to 9	6 to 10	2 8 to 3 4	1 3	4

SECTION XVII.—COMPARED WITH THAT OF EUROPE. 547

In reality, over the whole of the articles referred to, the difference collectively is not a considerable one, and the results exhibit a much smaller margin than might have been expected in favour of the labouring classes in America, a country upon which British workmen are so largely dependent for a great part of the enumerated items.

Before leaving Dr. Young's figures, I select from examples he gives some of the lowest cases of the expenditures of working men's families.

	Connecticut. 2 Adults—3 Children.			So. Bethlehem, Pa. 2 Adults—3 Children.		
	£	s.	d.	£	s.	d.
Flour and bread	0	4	0	0	3	6½
Butter and cheese	0	2	7½	0	1	10½
Meat, bacon, and lard	0	2	4½	0	2	1
Fish and eggs	0	2	9½	0	1	3½
Vegetables	0	2	0	0	2	1
Milk	0	1	5½	0	1	0
Groceries, including soap, etc. ...	0	2	11¼	0	6	10
Clothing	0	8	0¼	0	3	11
Tobacco, beer, etc.	0	0	2		—	
School, religion, etc.	0	0	5	0	4	2
House rent	0	8	10	0	10	5
Fire and lighting	0	6	1	0	5	0
Taxes	0	0	3½	0	0	2
	£2	2	0½	£2	2	4½
Weekly earnings	£2	3	9	£2	10	0

It will be recollected that the expense in Germany, for families larger in point of numbers than those described by Dr. Young, varied from 12s. 6d. to about 14s. 6d., and in England from 13s. to 18s. 9d. for provisions alone. The only two resembling these which I have been able to select from Dr. Young's numerous list, are the two just mentioned; in which not only are the expenses generally heavier than those of families in Europe, which probably proceeds from a more liberal table being kept, but items which might have been expected lower in America are fully as high or higher than in those countries which are largely supplied from the western hemisphere.

I shall now give some of the results of my own enquiries, during my visits to the United States in 1874 and 1876, on the general questions of cost of living and price of labour.

In 1874 the prices of iron in Great Britain were considerably lower than they had been in the years 1872 and 1873. Scotch pig iron had fallen from an average of 117s. 3d. to 87s. 6d., and Cleveland from 103s. 9d. to 74s. 6d.

The following list contains the prices of different kinds of labour connected with iron-making at the two different periods:—

	1873. s. d.	1874. s. d.	1876. s. d.
Scotch coal hewers, per day	9 11	7 2	4 8
Durham ,, ,, with house and coal	7 6	6 10	5 9
Cleveland ironstone miners	7 0	6 6	5 5
Puddling—South Wales, per ton of 2,400 lbs.	8 7	7 0	5 4½
,, Staffordshire ,, ,,	13 0	12 0	9 0
,, Cleveland ,, ,,	12 10	11 7	8 3
Blast furnace keepers	12 0	11 0	9 6
,, slagmen	7 4½	6 9	5 9

In America prices had also declined after 1873, but the men were still earning high wages, although there had been a general reduction in the prices of labour. The scale paid in different localities was however so dissimilar, that the simplest plan in dealing with the question will be to take each district separately.

Upon the occasion of my visit in 1874 to Lake Champlain, I learnt, as might have been expected, that the high wages paid in the mines and ironworks had also affected the hire of agricultural labour. The farmers were paying 18s. 10¼d. per week, which, added to the cost of board, brought up the total to £1 10s. 2d. per week, or 5s. per day. A miner who, out of a pay of 7s. 6d. to 8s. 6d. per day, had saved £800, and then commenced farming, assured me, that a man could then (1874) dress himself respectably for 2s. 9d. to 3s. 7d. per week; and this was confirmed by a draper of Burlington. This, with board, would come to 17s. 4d. per week, so that for personal expenses, including amusements, etc., 25s. were ample.

Provisions at that time were to be had at the following rates on Lake Champlain:—

	s. d.	
Butcher meat, best	0 8¾	per lb.
,, ,, inferior	4½d. to 0 6½	,,
Butter	1 3¾	,,
Bacon	0 6½	,,
Flour (workmen's)	1 7	per stone.

SECTION XVII.—COMPARED WITH THAT OF EUROPE.

In Cleveland City, Ohio, the following prices were given me:—

		s.	d.	
Butcher meat, best	... ...	0	8¾	per lb.
Flour (workmen's)	... ...	1	11¼	„ stone.
Clothes (working)	... ...	75	6	„ suit.
Shoes	6s. 7d. and	9	5	„ pair.
House rent	37s. 9d. to	75	5	„ month.
House coal	17s. to	18	10	„ ton.

In the Shenango Valley, house rents varied from 11s. 6d. to 19s. per month at the ironworks, and house coal from 11s. 3d. to 12s. 3d. per ton. Butcher meat ranged in 1874 from 3½d. to 8d., that used by ironworkers being 5½d.; pork was 5½d., and flour 1s. 10d. per stone.

Twenty-five years previously the prices in these localities were almost exactly half of the above. At Young's Town, Ohio, a workman, who in the North of England was earning 70s. per week, assured me he could live better on that income at home than on 84s. in America. With the exception of clothing, house rent, and firing, I saw nothing however to justify the complaint; for he admitted that he was buying meat at 6d. per lb., and flour at 1s. 9d. per stone. The price of potatoes was given me at 2s. 3d. per bushel or 4½d. per stone; apples were 2s. 3d. per bushel, peaches 3s. 9d., oats 2s. 1d., Indian corn 3s.

On the other hand, a shingler working at Pittsburg, and earning 9s. 5d. per day, considered that he was decidedly better off in America than in England, so far as cost of living was concerned.

In West Virginia, in 1874, miners were paying 8¾d. for the meat they consumed, 2s. 3½d. per stone for flour, and 3s. 9d. per week for house rent.

In the same State the owners supplied their men, in case the latter chose so to deal, with the necessaries of life, at the following prices:—

	s.	d.	
Bacon	0	7	per lb.
„ shoulders	0	6¼	„
Coffee	1	1½	„
Flour	2	4	per stone.
Sugar	0	7½	per lb.
Molasses	2	3	per gallon.
Lard	0	8¾	per lb.
Clothes (Sunday)	37	9	per suit.
„ (working)	32	0	„
Shoes	11	4	per pair.

The cheapest provisions, etc., I met with were at Charlestown, Ohio, where

	s.	d.	
Butcher meat was selling at	0	3	per lb.
Flour	1	7½	per stone.
Bacon	0	4½	per lb.
Clothes (including shoes)	52	6	per suit.
Board for single men	11	4	per week.

At Ironton, in the same State, flour sold at 1s. 7d. per stone, butcher meat 4½d. to 6½d. per lb., and house rent was 22s. 7½d. to 45s. 3d. per month, according to accommodation, the former giving two and the latter four rooms.

In Missouri, at the Iron Mountain Mines, I obtained the subjoined quotations:—

		s.	d.	
Butcher meat	2½d. to	0	4½	per lb.
Butter		0	11¼	,,
Best bacon		0	7½	,,
Flour		2	2	per stone.

In Indiana, the following rates were given, as showing the almost incredible rise which had taken place in the value of articles of domestic consumption within twenty-four years:—

	1850.		1874.	
	s.	d.	s.	d.
Wheat, per bushel of 60 lbs.	1	2	4	1
Indian corn ,, 56 ,,	0	5	2	2
Beef, per lb.	0	1	5d. to 0	6¼

The board and fare in some of the Southern States are very cheap and very poor. Thus in Tennessee I was told a week's rations for a workman consisted of 1 peck of meal 1s., 10 lbs. of bacon at 6½d. = in all 6s. 5d., supplemented with some vegetables and molasses.

Board for a bachelor is charged 10s. 6d., but in Alabama and Georgia 7s. per week is considered to suffice for keeping a man. Flour was selling at 1s. 10¼d. per stone, bacon at 5½d. per lb., Indian corn 3s. 9d. per bushel, suit of working clothes 38s. to 44s. 3d., and holiday suit 56s. 7d. Upon another occasion the following prices were given me for the same neighbourhood :—

		s.	d.	
Butcher meat		0	3½	per lb.
Flour		1	7	,, stone.
House rent, 2 rooms and kitchen		18	11	,, month.
Clothes	49s. to	52	7	,, per suit.

SECTION XVII.—COMPARED WITH THAT OF EUROPE. 551

In the city of Baltimore as much as 18s. 10½d. per week was paid by a workman for board and lodging, but there day labourers are paid 34s. per week.

At Marquette on Lake Superior in 1876, the following were the retail prices of provisions:—

	s.	d.	
Flour, ordinary	1	8	per stone.
„ finest	2	8	„
Beef	0	7	per lb.
Bacon	0	6¼	„
Butter	1	3¾	„

My informant, an old Clarence ironworks man, stated that he paid 18s. 10d. per week for his board and lodging, which was not superior in any way to what he had in Middlesbrough for 10s.

In the iron mining region around Marquette the following were the prices:—

		s.	d.		
Flour		1	7½	per stone.	
Bacon		0	6	„ lb.	
Beef	7d. to	0	7½	„	„
Butter		0	11¼	„	„
Pork		0	4½	„	„
Eggs		0	5¼	„	doz.

Around the copper mines of Lake Superior:—

	s.	d.		
Flour, ordinary	2	2	per stone.	
Board and lodging, about	16	0	„ week.	
Butcher meat	0	6¾	„ lb.	
Pork	0	6¼	„	„

In the year 1874, wheat averaged in the North of England 55s. 9d. per quarter, and in 1876 46s. 2d.; flour being 2s. 1½d. and 1s. 9d. per stone respectively for these two years.

In 1878 wheat stood in the United Kingdom at 46s. 5d. and flour at 1s. 10d. to 1s. 11d. per stone, so that provisions generally may be considered as occupying an intermediate position between the years 1874 and 1876. On my return from a foreign journey, in November, 1878, I made a note of the retail prices of provisions, etc., at Middlesbrough, from which the following are extracted:—

			s.	d.	
Flour	...	...	1	11	per stone.
Fresh meat	...	6d. to	0	10	per lb.
Bacon	...	5½d. „	0	6	„
Lard	...	...	0	7	„
Cheese	...	...	0	7	„
Butter	...	...	1	3	„
Eggs	...	...	0	9	per doz.
Tea	...	...	3	6	per lb.
Sugar	...	3½d. to	0	4½	„
Potatoes	...	...	0	8	per stone.
House rent	...	3s. to	4	0	per week.
Coals 9s. 11d., at the usual rate of consumption costing	...	...	1	8	„

A comparison of these last quoted figures with those which represent the prices in America, will satisfy any one that, so far as the cost of living is concerned, the ironworkers in Pittsburg and the Lehigh Valley enjoy little if any advantage over those of Cleveland or Scotland.

It may appear a little strange, that while the inhabitants of the United Kingdom derive such large supplies of human food from the United States, the residents of Pittsburg and other large centres of population in the Union should be paying almost as dearly for their provisions as do the people in Liverpool or Birmingham. It must be recollected however that the corn and other agricultural produce, consumed in American centres of industry are saddled, although to a lesser extent than we are, with considerable expenses for transport from the Western States.

The large imports of manufactured articles into the United States are a proof that enough is not produced at home to meet the requirements of the large population. Every object so imported has to bear the charge of transport, together with a heavy duty. As a natural consequence not only is it sold at a much higher price than the same object commands in the exporting country, but it is this artificially raised value, which regulates the selling price of what is produced in the Union itself.

In this way everything of domestic consumption and public use is, generally speaking, dear in the United States. I imagine it is this which exercises a certain influence on the expenses of the retail trade

SECTION XVII.—COMPARED WITH THAT OF EUROPE.

and thus raises the price to the consumer of home grown produce, although it may be obtained on such favourable terms, as to bear heavy transport expenses to Europe, and still be able to meet the farmers of the United Kingdom and the Continent of Europe on their own ground.

Agricultural labour is, as has been shown, unquestionably higher in the United States than in Great Britain; but so far as this element is a factor in the cost of food, it is amply met by the general low price of the land. Taking 4s. per day as the price of farm labour in the United States, and 3s. as that in the neighbourhood of Middlesbrough, we have the former figuring as 33 per cent. higher and sometimes even more, than the latter.

The fact that the United States can and does send annually to the United Kingdom about $2\frac{1}{4}$ millions of tons of different kinds of grain and flour, at a cost of about $12\frac{1}{2}$ per cent. of its value for land carriage, and a similar amount for ocean freight, is an admission that these articles can be had in America at a cheaper rate than that at which they can be obtained in Europe; the actual price varying of course according to the relative situation of the points of consumption and production.

Without pretending that wages and the cost of food have any necessary correspondence, I would draw attention to the fact that in the United States the pay of the very class engaged in the production of food for Europe has been shown to be higher by 20 per cent. or even more than with us. This ceases to be an anomaly, when it is remembered that the United States can only meet the great strain on its industrial activity by largely recruiting its labouring population from the old world. To encourage immigration, larger wages are offered than are current on this side of the Atlantic; and competition for workmen appears to be further stimulated by a wish on their part to acquire land, and work for themselves instead of for others.

Accepting, however, the wages paid to farm servants as running say 33 per cent. above those paid for similar labour in the iron districts of the North of England, it will be found in the sequel that the earnings of those engaged in mines, ironworks, and their cognate branches, in the United States, as compared with those of farm labourers generally show a much greater difference than exists in the old world between these two classes of workmen.

The price of labour in the States themselves, as has already been said, varies very considerably according to locality; the cheapest being in the Southern States, where slavery prevailed until the Civil War. The miserable diet upon which the slaves were fed seems to have left its mark even to the present day. I was informed (in 1874) that the usual slave rations for a week consisted of one peck of meal worth 4½d., and 4 lbs. of bacon costing 1s. 1½d., equal therefore in all to 2½d. per day. Some more indulgent masters allowed their people a little molasses and tobacco, which might bring up the expense to 1s. 10d. or 1s. 11d. per week. They received two suits of clothing in the year, costing under £2; which, with medical attendance, etc., brought up the total annual cost of a slave to his owner to about £9 or £10. The houses or cabins were then, and in many cases are still, most wretched shelters, if buildings can be called shelters which are full of crevices admitting the light, and, fortunately perhaps, the air. Some allowance of course must be made, in considering the amount of food and clothing required, on account of the fineness of the climate, in latitudes 32·5 to 35, therefore some 1,200 miles nearer the equator than the chief mining districts of Great Britain.

Notwithstanding the recent date of the abolition of slavery, common labourers in Alabama and Georgia were being paid in 1874 a paper dollar per day, equal to 3s. 9¼d.; and now of course there is a considerable improvement in the mode of life of the coloured population of these States, who constitute the chief source of labour. Over the Central and Northern States of the American Union a common rate of pay to ordinary labourers in that year was 1½ dollar or 5s. 7¾d. per day. In one case I found the rate as low as 4s. 8½d., viz. in Cleveland City. During the previous year, 1873, from 7s. 6d. to as high as 11s. had been paid in the anthracite coal-fields, during the excited periods of the coal trade; and at Marquette on Lake Superior 8s. 5¾d. was earned by the ordinary labourer.

These quotations have reference to the iron works and their associated mines; for which class of labour at that time (1874) the usual pay in Great Britain was a trifle above 4s. per day. It would therefore appear that in the North of England there was only a difference of 1s. per day between the pay of farm servants and ordinary labourers in the mines of that district. On the other hand, according to Mr.

Weeks, agricultural labour including board was paid at the rate of
of 5s. 6d per day in the iron districts, while ordinary labourers in the
mines in the same vicinity were receiving from 7s. 6d to 11s.

The anthracite colliers' wages are regulated, or were so up to 1879,
by a sliding scale. Probably in the whole history of industry there is
no more remarkable instance of the danger of engrafting artificial
restrictions on a trade, than has accompanied the development of
anthracite mining in the United States. In spite of what turned out
to be the enormous extent of the region containing this valuable
mineral, an attempt was made by some half-dozen corporations to
create a species of monopoly for supplying the produce of the mines
to the consumer. The power possessed by these bodies for controlling
the market was of an unusual character; for they were chiefly the
railway companies, whose lines connected the coal-fields with the shipping ports, with the iron-works, and with other important centres of
consumption. When private enterprise, outside these powerful companies, attempted to supply the market on lower terms than satisfied
the boards of railway directors, the railway dues were raised: on one
occasion, I was informed, they were quadrupled. In a communication
to the Iron and Steel Institute, made in 1875, after my return from
the United States, I ventured to predict the same fate for this American
combination, as that which befel a similar one in the Great Northern
coal-field in England. That prediction has been realised, to the
great embarrassment of at least one great railway company, which had
largely overspeculated in coal lands. Under the monopoly created by
rival companies of railroad owners, immense profits were realised by
the sale of anthracite coal; fresh capital was attracted, and additional
collieries were opened out. This created, of course, a great demand
for men, who naturally enough set an increased value upon their
labour, as the demand for it increased. In this way wages were run
up, until a working collier has been known to realise nearly two pounds
for his day's work. If the railway companies were right in imposing
prohibitive rates of carriage on coal, the men were at least equally
justified in creating, as they did, an artificial scarcity, by a general
abstention from work of thirty days. The fact however is that both
were wrong; and the trade is now, I believe, left to take care of itself;
for the moment the interests of the railway companies began to conflict *inter se*, as they were certain to do in an overstocked market, the

monopoly fell to pieces. It is needless to say that many strikes, one lasting six months, accompanied these excessive fluctuations in the markets; the owners resisting what they regarded as extravagant demands while prices rose, and the men refusing to accept such terms as their employers offered when prices fell. These conflicts between the coal owners and their men were unfortunately the precursors of a much more terrible state of things. A wide spread conspiracy was organised in the anthracite coal districts, which aimed at nothing short of murdering any man who made himself conspicuous in resisting the demands of the workmen. Not only were those who offended in this respect marked by the heads of this foul combination, but no man's life was safe who ventured to give evidence on any trial; so that by false swearing on one side, and fear to tell the truth on the other, conviction for some years became impossible. According to the statement of the counsel, Mr. F. B. Gowen, through whose perseverance the organisation was finally broken up, "county commissioners, high constables, chiefs of police, candidates for associate judgships, were actually guilty of murdering those whom they were from their position bound to protect."[1]

The extent of the undue development of the anthracite coal-fields, fostered as it was by very high prices, may be judged of by the transactions of 1880; for during this year the collieries were actually idle about one-third of their time.

To avoid interruption to work, always fraught with disastrous consequences to both sides, a sliding scale was adoped by one large firm, which for a week's work of 60 hours afforded the following rates of daily pay to the various branches of labour employed:—

	1875. s. d.	1876. s. d.	1877. s. d.	1878. s. d.	1879. s. d.	1880. s. d.
Miners	8 9	8 9	7 5½	8 2	8 9	8 1¼
Labourers belowground	7 11	7 6	6 9½	6 9	7 11	6 10
Driver men	7 11	7 6	6 9½	6 9	7 11	6 10
Engine men	11 2	10 3	10 3	9 4	9 4	
Labourers—average aboveground	7 3	6 3	6 3	5 0	4 2	
Carpenters	10/5½	7/1–10/1½	7/1–10/1½	6/3–8/6	6 8	
Machinists	11/8	7/4–11/8	7/4–11/8	7 4	7 4	

---

[1] Any one wishing for more information respecting the conspiracy entered into by these anthracite coal miners, under the designation of "Molly Macguires," is referred to the most able and intrepid speech in the court of assize, of Mr. Gowen, through whose sole instrumentality the culprits were convicted.

SECTION XVII.—COMPARED WITH THAT OF EUROPE. 557

The scale adopted was to pay the miners 50 cents. (1s. 10½d.) when anthracite was selling at 5 paper dollars (about 18s. 2d.) per ton at New York, with a rise of 12½ cents. (about 5¾d.) per ton for each dollar advance on the price. The pay of all other branches of labour at the pits was regulated by the same course of procedure, and this was popular so long as the market was a rising one, under which, wages of artisans, etc., ranged from 6s. 0¼d. to 7s. 6½d. per day. They were afterwards in consequence of their refusal to abide by the sliding scale, struck out from its operation and their wages were left to be regulated by the rates paid for similar labour in other branches of industry. Under this new code their pay would have amounted to from 5s. 7¾d. to 6s. 0¼d. per day; against which they held out for more than six months and then gave way.

The miners' earnings just given, notwithstanding that they are nearly double those of men of the same class in Great Britain, do not prevent the anthracite coal owner from being able to load his produce into trucks at a cheaper cost than is possible at any colliery I am acquainted with in the North of England. It has further to be remarked that the wages just mentioned do not in point of fact convey a correct idea of the actual earnings; for they are calculated on an assumed weight of coal worked, whereas in reality one-half more is very commonly obtained by a good man. When in an anthracite mine, in 1876, I was informed that the miners own earnings varied from 9s. 5d. to 13s. 2d. per day, out of which he had to pay for powder and oil.

In 1874, the wages given me as earned by a coal hewer in the Schuylkill region (anthracite) varied from 8s. 4d. to 11s. 3d. per day of 10 hours; and in the bituminous coal of Alabama they were from 6s. 7d. to 7s. 6½d.

In the latter State, however, the value of labour is affected by the condition of things antecedent to the abolition of slavery, as well as by the employment of convicts in the mines. These unfortunate people are farmed out by the prison authorities to a contractor, who bargains with the coal owner for the produce of their exertions. At one colliery visited, I was informed that the contractor paid about £2 per annum per convict, which, with watching to prevent escape, feeding, clothing, and idle time, brought up his cost to 3s. per working day. With this treatment they were doing as much work as miners from Wales who were earning 9s. 5d. per day.

In the Connelsville coking-coal pits the hewers in 1874 were receiving above 10s. per day, and at Charleston on the Ohio 9s. 5d. for eight or nine hours' work. In Kentucky State at the same date the miners earned 8s. 1d. to 9s. per shift. In 1876 I heard that in Tuscarawas the gross earnings of the hewers were 6s. 9¼d., which after paying for powder and oil left 5s. 10d. net. In other parts of Ohio, viz., in the Pittsburg district, I was informed that the wages of miners had been reduced 38 per cent. in the two years since 1874, which left their earnings at that time 5s. 8d. to 6s. 7d. per day.

At one large ironwork in Pennsylvania, where the men were paid on a sliding scale determined by the price of rails, the hewers could earn about 7s. 6d. in ten hours after paying for powder and candles; but owing to slack trade their actual pay did not much exceed two-thirds of this sum.

In the years 1874, 1875, and 1876 the rate paid to the coal-hewers in the County of Durham was such that they could earn on an average of the three years 6s. per day, besides free house and firing, equal therefore to about 6s. 8d. In the United States the amount received by this class of men varied from 7s. to 13s. At the then price of farm labour, the collier in the County of Durham could earn double the pay of the agriculturalist; while in the United States the hewer in the anthracite pits could easily make four times as much as the farm servant.[1]

At the same time while the hours of pit work in America were nearly the same as those spent at the plough, the Durham miner remained underground little more than half the time of him who was working in the fields above his head.

The relative amount of work done in the two countries for the same money is difficult of comparison, owing to the varying facility with which different seams of coal are worked. At one ironwork in Eastern Pennsylvania, five tons were given me as the quantity which could be worked by one collier, which is about 10 per cent. better than

[1] *Vide* page 551.

the usual average of a Durham hewer, and nearly 30 per cent. above the amount really worked at the time of my visit, when the latter were restricting their output.

The wages for mining iron ore in America in 1874 varied between 4s. and 8s. 6d. per day of twelve hours. In West Virginia the miners were earning 4s. or a little less, and in Eastern Pennsylvania 4s. 4d.; while in the magnetite mines of New Jersey the rates varied from 5s. 7¾d. to 7s. 6d. In the Lake Superior district they reached as high as 8s. 6d.

In the Cleveland mines the average earnings of the men in the same year were about 6s. 6d.; so that, with the exception of the richer ores of New Jersey and Lake Superior, the English ore miners were as well paid as the same class of men in some parts of the United States.

Having regard to the figures which represent the iron-ore miners' wages, it cannot be pretended that, when compared with English earnings, they are on the whole much out of keeping with the price of agricultural labour in the two countries.

In the year 1874 the following wages were being paid to the different classes of mechanics in one of the largest engine shops in the North of England. Following those are the prices per day paid at the same time in different parts of the United States:—

NORTH OF ENGLAND.

		Fitters and Lathemen.	Blacksmiths.	Strikers to Smiths.	Joiners and Carpenters.	Bricklayers.
		s. d.	s. d.	s. d.	s. d.	s. d.
England	Lowest ...	3 0	4 2	2 10	4 4	3 8
	Average...	5 1¼	5 5¼	3 4¼	5 0	5 0

FITTERS AND LATHEMEN.

	s. d.		s. d.
Ohio	9 5	to	11 3
Pennsylvania	6 9¼	to	11 3
,, ...	6 1	to	10 2
Wisconsin	6 7	to	8 3
Average	8 8½		

UNITED STATES.

Blacksmiths.

Fitters and Lathemen. Black-smiths. Strikers to Smiths. Joiners and Carpenters.

	s.	d.		s.	d.
New Jersey	7	6	to	9	5
New York State				8	5
Ohio				7	6½
Wisconsin				8	3¾
Ohio, Cleveland City	9	5	to	11	3
„ Shenango Valley	7	6½	to	11	0
New York State				8	5¾
Alabama				9	5
Missouri	7	6	to	8	5¼
Average		8	9½		

Strikers to Smiths.

	s.	d.
Ohio, Cleveland City	5	6
„ Shenango Valley	5	7¾
Alabama	5	7¾
Missouri	5	3¼
Average	5	6

Joiners and Carpenters.

	s.	d.		s.	d.
New Jersey				8	5¾
New York				8	5¾
Pennsylvania				9	5
Ohio, Cleveland City	8	5¾	to	9	5
„ Shenango Valley				9	5
Alabama				5	7¾
Missouri	7	6	to	8	5¾
Average		8	4		

Bricklayers.

	s.	d.		s.	d.
New York	8	1	to	8	6
Pennsylvania	13	2	to	15	3
Wisconsin 11s. 3½d.,	13	2	to	15	1
Alabama	7	6	to	8	5
Missouri	7	6	to	8	5¼
Average		10	7		

SECTION XVII.—COMPARED WITH THAT OF EUROPE.

From these figures it appears that the wages paid in the United States are higher than in the North of England iron district, in about the following ratio:—

Fitters and Lathemen.	Smiths.	Strikers.	Joiners and Carpenters.	Masons and Bricklayers.
Per Cent.	Per Cent.	Per Cent.	Per Cent.	Per Cent.
71	60	64	66	110

On referring to the masons and bricklayers, it will be seen that occasionally these classes earn fully three times the wages usually paid in Great Britain. This exceptionally high rate is paid to men engaged exclusively in out-door work, in which the severity of the climate in winter presents many interruptions to steady employment. On the other hand in ironworks, repairs to furnaces, etc. can be carried on at all times, and for such work 8s. to 8s. 6d. may be considered a usual wage. On the whole, therefore, it will not be wide of the truth if it be assumed that in 1874 the pay of mechanics in the United States was nearly 70 per cent. higher than in the English provinces.

The wages paid to the mechanics at ironworks are of a very miscellaneous character, as a good many young people or imperfectly trained men are employed, where the work is not of a very highly finished nature. The subjoined rates, current in an English ironwork, will sufficiently explain this:—

	Fitters and Lathemen.	Blacksmiths.	Strikers.	Joiners.	Bricklayers.
	s. d.	s. d.	s. d.	s. d.	s. d.
Highest	5 10	6 0	4 1	5 0	5 6
Lowest	1 2	4 10	1 6	2 0	3 6
Average—actual	4 2	5 3	3 6	4 6	4 4

In a return sent to me from a large Pennsylvanian work, in 1878, are to be found the following quotations:—

	Fitters and Lathemen.	Blacksmiths.	Strikers.	Joiners.	Bricklayers.
	s. d.	s. d.	s. d.	s. d.	s. d.
Highest	10 0	11 2½	4 3½	11 0	10 2½
Lowest	4 3½	5 4¾	4 2	5 4¼	6 1½

In the North of England, at that time, the returns were as follows:—

	Fitters and Lathemen.	Blacksmiths.	Strikers.	Joiners.	Bricklayers.
	s. d.	s. d.	s. d.	s. d.	s. d.
Highest	5 8	5 8	4 0	5 4	5 4
Lowest	3 0	4 8	3 2	4 6	4 0

BLAST FURNACES.

On comparing American with English blast-furnace work, I have found in some cases that the wages of the best men were a little higher than are current on this side of the Atlantic; but the earnings of the whole staff, and the consequent cost on the ton of iron, are considerably in excess of what is paid with us. This arises partly from the somewhat higher rates paid, and partly from the fact that more hands are engaged per furnace than is the case in the best English and Scotch works. I would also remark that, in spite of high wages in America, neither the rolling stock of the railways which conveys the minerals to the works, nor the mode of dealing with such materials on their arrival, is characterised by the same endeavours to save labour, which are so conspicuous in our best arranged ironworks and rolling stock, particularly in the Cleveland district, as served by the North-Eastern Railway.

A word or two as to the actual earnings at the furnaces. In 1874 the men in the Cleveland district were being paid as follows:—

Keepers. s. d.		Assistant Keepers. s. d.		Slagmen. s. d.		Fillers. s. d.		Chargers. s. d.
10 9½	...	4 3½	...	6 7½	...	5 4¼	...	8 2½

In the United States the conditions and nature of furnace work vary so much that there are wide differences in the wages of the men. There are still, in remote districts, many small furnaces engaged in smelting iron with charcoal, where the make is often very small, and at best rarely reaches a couple of hundred tons per week. In some of these localities the keepers are only paid about 4s. 8½d. per day. In others, particularly where the make is great, as much as 6s. is paid for this kind of work at a charcoal furnace, and 4s. 8½d. to the slagmen, fillers, and chargers. In the Lake Superior district as much as 7s. 6d. to 8s. 6d. was earned by the keepers at the best appointed charcoal furnaces.

The current rates, in 1874, at the larger furnaces using mineral fuel will be seen by reference to the figures below:—[1]

	Keepers. s. d. s. d.	Assistant Keepers. s. d. s. d.	Slagmen. s. d. s. d.	Fillers. s. d. s. d.	Chargers. s. d. s. d.
State of New York ..	8 5 to 9 5	6 7 to 8 0	6 7	6 7 to 7 6	6 7 to 7 6
,, Ohio	7 6½ to 9 3	6 7 to 7 6½	5 7¾ to 7 1¼	6 7 to 7 6½	6 2¾ to 6 11
,, Pennsylvania	7 0 to 9 5	6 2 to 7 6½	5 7¾ to 7 1¼	5 7¾ to 6 7	5 7 to 7 1¾
,, West Virginia	9 5	6 4½	—	5 6	6 4½
,, Indiana ..	7 6½ to 9 5	7 6	8 5¾	6 0½	7 1¾ to 7 6½
,, Missouri ..	9 0½	6 9	—	4 7¾	4 7¾
,, Tennessee ..	7 6	6 7	—	6 7	5 7¾
Averages taken from 18 Works..	8 2½	7 2½	6 5	7 5	6 8½

[1] The dollar is reckoned at 3s. 9½d.

From the figures given we may take the average of the keepers and their assistants put together to be 7s. 6½d. in Cleveland and 7s. 8½d. in America.

Either from the occasional heat in summer, or from an unwillingness to exert themselves as our men do, it is not an uncommon thing in America to have fully twice as many men, to do the same amount of work in keeping and attending to the slag, as we have in the North of England. This, added to inferior arrangements, generally results in the labour on a ton of metal in America amounting to nearly double, and often even more than double, its cost in England.

In certain cases, where the make in the United States is very large, say 700 tons per week, the earnings per man of the *entire staff* in 1879 were fully one-third above those in the Cleveland district. To some extent this is necessitated by the larger make, but not entirely so; for instead of having to receive and handle 3¼ tons of ironstone, as in Cleveland, for each ton of iron, about 1½ tons of ore only are required.

At one work, where the make was less than one-half of that of the Cleveland furnaces, 22 per cent. more men were employed, and the average earnings were 22 per cent. higher; and in another, where the number of men was something like that in Cleveland, notwithstanding that the produce was only about one-half, the average earnings of the chief men were 57 per cent. above the highest in this country.

At two furnaces, making 400 tons of iron per week each, the number of men was 85 per cent. more than is to be found at Middlesbrough, while the average daily earnings per man (5s. 11d.) were 31 per cent. higher.

Speaking generally, I should say that the *entire* staff of men engaged at blast furnaces in the chief seats of the iron trade in the United States receive on an average from 25 to 30 per cent. higher wages than are paid in the best establishments in Great Britain, while at the same time, as formerly observed, the number of men is considerably greater. As already stated however, a good deal of this unnecessary expenditure of labour is to some extent due to the absence of those labour-saving appliances for which the best-equipped works with us are distinguished.

The opinions just given embody the general impressions produced on my mind by visits to many smelting works in different parts of the United States. More recently I have been favoured by American friends with tabular statements of the actual number of men employed at their respective furnaces along with the earnings of the entire staff. By comparing these with what may be considered the average practice in the best arranged of our modern English furnaces, the comparison points to the conclusion that in those of the United States one-sixth more men for each furnace is required for producing less than half the quantity of iron. Besides this difference, the present average earnings of the staff (1883) being higher in America than in England, the cost of labour on a ton of pig iron in the particular cases compared, is in the former more than double of what it is in the latter. On the other hand the wages paid at the English works just referred to average fully one-half more than the earnings of each man of the staff at the German, French, Belgian, and Luxemburg furnaces. As has been stated in the previous section, owing to the larger number of hands employed in these countries, the actual price of the labour in the first three on the ton of product is usually 25 per cent. above that paid in the North of England. In Luxemburg, owing to the larger make of some of the furnaces, this difference against Great Britain is less than one-half of that just named.

In some of the cases which came under my notice on the Continent of Europe, a prominent cause of the large number of men required at each furnace arises from the exceedingly defective arrangements for receiving the materials. In several instances I observed the coke being unloaded from barges by means of baskets carried on men's backs. Of course it may be urged that where labour is so much cheaper than with us, it will scarcely pay to incur the outlay of capital needed for providing the appliances in general use in the vicinity of Middlesbrough

The differences in the arrangement of the works are not however the sole cause of the larger numerical strength of the staff engaged at the foreign furnaces in question. Much must be put down to the greater physical strength of the British workmen. An esteemed friend in Germany wrote to me, in reply to my observations on this question, as follows:—

"I am much amused that you are unable to understand why we require so large a number of men at our blast furnaces; but when I visited England I was also greatly surprised at the few men you employ. The number given in my list is really for one and not for two furnaces. There is no avoiding it. We have often the same technical appliances as you in England; for anything an engineer sees he can imitate and construct. But what we cannot imitate is to work, with our cheaply fed men, with the same vigour that your English workmen labour who enjoy their good meat," etc., etc.

I have also received from a most trustworthy quarter the average weekly earnings over a series of years of the chief workmen at a Pennsylvanian work. These are as follows:—

	Founders. Dollars.	Keepers. Dollars.	1st Helper. Dollars.	2nd Helper. Dollars.	Ore Fillers. Dollars.	Coal Fillers. Dollars.
1870	18·95	17·40	15·55	14·72	13·90	9·90
1871	19·00	18·50	16·65	15·82	15·00	9·30
1872	23·17	22·61	20·39	19·32	18·54	10·50
1873	25·48	24·85	22·40	21·21	20·16	10·50
1874	16·80	15·75	14·35	13·30	13·30	10·15
1875	15·12	14·00	12·95	12·25	12·25	9·10
1876	13·65	12·60	11·76	11·06	11·06	8·40
1877	12·25	11·34	10·57	9·94	9·94	7·70
1878	12·25	11·34	10·57	9·94	9·94	7·70
1879	11·34	11·34	10·57	9·94	9·94	7·70
1880	15·00	14·00	12·95	12·25	12·25	8·40

The make of the furnaces to which the above-mentioned rates apply amounted to 260 tons per week, the fuel used being anthracite. The founder answers to the head keeper in England, and the three other men assist the founder and attend to the slag. Each furnace has four ore fillers and one coal filler on each shift, who discharge the materials into the furnace. This, for the work referred to, makes a total of nine men, against one and a half keepers, one slag man, two and a half fillers, and one top charger—in all six men—at an English furnace making 460 instead of 260 tons per week as in the American.

The following, showing the number of men engaged on each shift at keeping, slagging, filling and charging a furnace, is taken from returns in my possession:—

Locality.	Make of Iron per Shift. Tons.	Number of Men.	Average Daily Pay. s. d.	Cost per Ton of Iron, England taken as unity.
Cleveland	32	6	5 3	100
Germany	50	14	2 10	81
Do.	43	14·5	2 8½	91
France	36	14	2 9½	110
Luxemburg	35	11	2 6	78
United States	25	10	6 8	271

It will be observed that the relative costs in these examples do not correspond with those previously mentioned. This chiefly arises from the remaining branches of the labour differing in the numbers employed as well as in the rate of wages.

Cleveland, which has been taken as the standard of comparison, is no doubt favourably situated for economical cost of labour. Almost day by day the exact quantity of raw material is received at the furnaces for the day's consumption. This enables the furnace manager to deposit his supplies of ore, fuel and flux at the most convenient points for *immediate* consumption. On the other hand on the Continent in some cases, and in the United States in most cases, long distances intervene between the coal and ore, so that considerable stocks of one or other are lying at the furnaces—a condition which interferes with dealing economically with the vast weights of matter involved in the smelting of iron.

MALLEABLE IRON.

Passing on now to the manufacture of malleable iron, we are often met by a much greater dissimilarity between the rates current on the two sides of the Atlantic ocean, than is the case at the blast furnaces. Restricting the comparison at present to prices paid for puddling, in the United Kingdom and in the United States, it may be recollected that the price per ton in Cleveland and Staffordshire in 1874 was 11s. 7½d. In that year I find the following American quotations in my diary:—

AVERAGE FROM VARIOUS WORKS.

	Per Ton.	
	s.	d.
In Susquehanna Valley	18	10
„ Illinois	19	11
„ New York State	19	0
„ Pennsylvania (20s. 8¾d. and 21s. 9d)	21	3
„ Ohio	22	7½
„ Tennessee	24	6

Thus it would appear that 20s. 6d. may be assumed as having been the average cost of hand-puddling in the United States in 1874, or nearly double that paid in Great Britain.

In 1876 the price paid in America had declined to 18s. or 19s., against 9s. in Cleveland including bounty, and 9s. 6d. in Staffordshire;[1] so that England at that time paid less than one-half of the American prices. In 1878 I have one American quotation of 13s. 3¾d., the price with us being then about 7s. 9d. The actual amount of money earned of course depends on the quantity of work done. The Englishman's day's work, for ordinary iron, may consist of 24 to 27 cwts., and the earnings at 7s. 9d. per ton (1878) may be taken at 9s. 9d.; which is divided probably into 6s. 6d. for the chief puddler, and 3s. 6d. for the underhand. In America the quantity worked in a shift is from 22 to 24 cwts, leaving (1878) about 15s. for the head man and 7s. 6d. for his assistant.

In the Lehigh Valley, in 1880, the price paid for puddling was 18s. 9d. per ton short weight. The daily produce is 25 cwts. of puddled bar, so that the daily earnings for the puddler were 15s. 7½d., and for his underhand 7s. 9½d.

In Staffordshire the average for the same year was 8s. per ton short weight. Allowing the produce to be 27½ cwts. the total earnings come to 11s.; giving about 7s. to the chief and 4s. to the underhand.

The Middlesbrough rates in 1880 were 9s., including bounty, per ton short weight; at which the head puddler would earn 7s. 9d., and the helper 3s. 6d. per day.

If we assume that the best paid agricultural labourers in the North of England earn 3s. per day, a head puddler receives a trifle above double the highest pay of the farm servant. In like manner, taking the average countryman in the United States to earn 4s., the chief puddler is paid nearly four times as much as is given for farm labour.

In the United States therefore puddling costs more than double what it does with us, and the receipts of each man are fully twice as much as they are in England.

Excessive, comparatively speaking, as are the rates just referred to, they are considerably less than has been paid in inflated periods of the iron trade. During these, as much as 30s. 2d. per ton has been known

---

[1] The English ton before 1880 is one of 2,400 lbs., whereas I believe in the American works the ton of 2,240 lbs. is always used for puddlers' wages.

to be paid in America for puddling, and I believe the first attempt at a reduction was met by a prolonged strike. In the United Kingdom the highest rate paid for puddling was about 14s., viz. in 1873 and 1874.

I have dwelt at some length on the prices paid for puddling, because there is no better standard of comparison in the manufacture of malleable iron than the price paid for this description of work. Its performance is usually independent of any general convenience of arrangement; and the measure of the amount of labour required for manipulating different kinds of pig is, as a rule, pretty accurately determined by the weight produced in a given time.

I have endeavoured to compare, with proper attention to accuracy, the information I obtained in the United States (in 1874 and 1876) with what I knew to be the daily earnings per man of the entire staff in the puddling forges, as they are paid in the United Kingdom. The result is that I feel pretty confident that the cost to the American iron-master was and is something like 50 per cent. above that which is paid by those of Great Britain.

The same disparity between the rates paid in Great Britain and in the United States exists in all other branches of the malleable iron works, with perhaps one exception, viz. plate rolling; in which I believe the British scale is higher than that adopted in any part of the world.

The following figures show the daily earnings of different classes of workmen engaged in the malleable iron works of Great Britain, and in some of those on the Continent, placed alongside rates paid about the year 1878 in the United States :—

PUDDLING FORGE.

	Britain.		France.	Belgium.	United States.		
Puddlers	7/2	8/1	5/3¼	5/4	9/7	15/6	—
Do. Underhand ...	3/6	3/6	—	3/6	4/9½	7/9	—
Shinglers	22/9	19/3	5/3	—	14/6	24/2	—
Puddle bar rollers	15/0	18/5	5/2	—	11/4	13/2	9/4
Roughers	6/0	6/1	—	—	—	22/1	—
Hookers	5/6	3/9	—	—	—	8/4	—
Draggers out	4/6	4/8	2/8	—	7/7½	—	—

SECTION XVII.—COMPARED WITH THAT OF EUROPE.

FINISHING MILL.

	Britain.	France.	Belgium.	United States.			
Heaters in plate mill ...	16/1	8/5½	7/1½	15/0	13/6	—	—
Helpers to heaters ...	—	5/4½	—	9/2½	8/3	—	—
Heaters for iron rails or bars	—	—	—	18/9	16/3	14/7	—
Helpers at do.	—	—	—	9/4½	7/0½	7/3½	—
First Rollers in plate mill	41/1	7/5½	5/9½	37/1	18/1½	25/0	35/0
Second do.	—	5/11¼	—	—	16/9½	—	—
Shearmen in plate mill ...	34/9	—	2/6	4/7	—	—	—
Do. for rails and heavy bars	20/5	—	—	21/10½	23/1½	15/0	—
Roughing	17/6	—	—	15/7½	16/3	10/0	—
Catchers	—	—	—	9/4½	11/5½	—	—

It will be perceived in these figures that occasionally there are considerable differences in the earnings of the same class of men. This probably arises from differences in the amount of work done or from wages being generally lower in one locality than in others.

For the purpose of comparing the rates paid in the United States over a series of years with those given in the table at page 527, there is inserted below information respecting the prices paid for puddling, together with the daily earnings of the puddlers, at a malleable iron work in Pennsylvania.

PRICE OF PUDDLING AND DAILY EARNINGS OF PUDDLERS AND UNDERHANDS.

Year.	Price per Ton of 2,240 Lbs. Average make 26¾ Cwts. Dollars.[1]	Average Earnings. Headman. Dollars.	Average Earnings. Underhand. Dollars.
1870	6·00	5·00	2·50
1871	5·40	4·50	2·25
1872	6·25	5·21	2·60
1873	7·00	5·83	2·92
1874	5·75	4·80	2·39
1875	5·25	4·38	2·18
1876	4·25	3·54	1·77
1877	4·00	3·34	1·66
1878	3·60	3·00	1·50
1879	3·60	3·00	1·50
1880	4·50	3·75	1·87
1881	4·00	3·34	1·67

[1] The present value of the American dollar may be taken at 4s. 2d. In 1874 and 1876, when I was in the United States, the paper dollar, then the usual currency, did not stand above 3s. 9½d.

The highest price given was in September, 1872, when the rate reached 8 dollars per ton; and the lowest in 1879, viz., 3·40 dollars per ton.

The following table exhibits the fluctuations of wages in the finishing mills:—

AVERAGE DAILY EARNINGS OF HEATERS, THEIR HELPERS, AND CHIEF ROLLERS, IN A MALLEABLE IRON WORK, ROLLING BARS.

Year.	Heaters. Dollars.	Helpers at Furnaces. Dollars.	Head Rollers. Dollars.
1870	6·54	3·09	7·74
1871	6·54	3·09	7·74
1872	6·54	3·09	7·74
1873	7·32	3·45	8·70
1874	6·00	2·82	6·96
1875	5·52	2·61	6·39
1876	3·60	1·80	4·20
1877	3·24	1·62	3·84
1878	3·92	1·92	4·64
1879	3·92	1·92	4·64
1880	5·04	2·52	6·04
1881	4·32	2·16	5·12

An average day's work until January, 1878, for a heater and his helper, was 6 tons (of 2,240 lbs.), after which they turned out 8 tons. The head roller earned his money by rolling the work of two furnaces, say 12 tons per shift up to 1878, and 16 tons subsequently.

Owing to a want of exact information, I am unable to quote precise figures representing the total difference between the cost of labour in the manufacture of malleable iron in European countries and in America. So far however as my information enables me to form an opinion, I think, as compared with England, that the workmen in the German, French and Belgian forges and mills together, may receive about 10s. less on every ton produced than is paid with us. In America, on the other hand, I am of opinion that the wages in the puddling and finishing mills may be 10s. to 15s. more per ton than is paid in Great Britain.

BESSEMER STEEL.

From the manufacture of malleable iron I would now pass on to that of steel, in which the United States have earned a position of great distinction. The converters used in the Bessemer process are usually set in pairs, and, looking at the work they perform, occupy but a very small space of ground. The intense heat which accompanies the operation leads to rapid destruction of the linings of the vessels, particularly at the tuyeres. Unless therefore the arrangements for removing the ingots are very well disposed, and unless the repairs can also be effected with the utmost expedition, serious interruptions may ensue. No one, I believe, has given these important matters more ample consideration than my friend, the late Mr. A. L. Holley of New York, and no one has been more successful in the attainment of the objects he kept so constantly in view. The reward of his intelligence and industry has been the turn out of a larger quantity of ingots, per pair of converters, than that produced at any Bessemer work in the United Kingdom. So far as my information goes, few makers in England have approached within 20 to 30 per cent. of the quantity of work turned out by a pair of converters as designed by Mr. Holley. The number of men engaged in a given amount of work does not appear usually to exceed that in England, but the cost per ton for labour in America at the converters is double that incurred in this country. This difference arises simply from the circumstance that in America the earnings per man per day are twice as much as those paid in England.

According to a paragraph which appeared in the *Scranton Republican* of 10th November, 1883, the Steel Company of that locality averaged from one pair of converters fifty-two blows per turn for a whole week; whereas in England half this number is considered good work. Between 5 A.M. and 4·43 P.M., on one of the turns, as many as sixty blows were made, and from one of the bottoms 42 heats were obtained. The authority quoted from mentions the facts given above as never having been equalled since Bessemer steel was first introduced.

The same publication, in its issue of December, 1883, records the production of 288 tons (2,240 lbs.) of rails in the Lackawanna Company's mill in one turn, and then goes on to state that in the same period (twelve hours) the Scranton Company, with eleven men at the rolls, turned out 318 tons.

If the week is taken as consisting of eleven turns, we have 624 blows of 5 tons each, or 2,860 tons from the converters and 3,498 tons of rails from the mill.

I am not aware of any such quantity of steel as that named having ever been made from one pair of converters, but at Eston above 3,700 tons of rails have been rolled from one mill for several consecutive weeks. As regards the number of men required to work a heavy rail mill, the machinery referred to at the close of the last section, recently erected at Barrow, and equal in power to that at Eston, is served by eleven men.

The United States with their 120,000 miles of railway, against about one-sixth of this in the United Kingdom, are of course much larger consumers of steel rails than we are. Including those required for exportation, there were manufactured in Great Britain in 1881 1,023,740 tons, and in 1882 1,235,785 tons of steel rails. In the United States there were made in the year 1881 1,187,770 gross tons of 2,240 lbs., and 1,284,067 gross tons in 1882. This large production has afforded the American steel-rail makers ample opportunity of improving the necessary appliances used in the manufacture; and, as has been already admitted, they have not failed, so far as a large output is concerned, to profit by the opportunities placed at their disposal. I am unable to refer to any data containing the cost of labour involved in the production of such large makes as those above referred to. According to my friend Mr. E. Windsor Richards, who has himself visited the United States, the great amount of work performed in the American mills is accompanied with a considerable increase in the cost of labour both at the blast furnaces and in the steel works.

The only information I possess of the number of men engaged in rolling steel rails in America, refers, on the whole, to works of somewhat inferior capacity to our English mills, but exceeding in production those of an average German establishment. In the converting department I quote one American mill making more and the other

SECTION XVII.—COMPARED WITH THAT OF EUROPE.

less than the English work with which they are compared. Again adopting England as the standard of comparison, I have arrived at the following results:—

CONVERTER WORK.

	England as Unity.	Germany.	Two Works in United States.	
			A.	B.
Scale of production	100	48	117	65
Relative work performed by each man	100	67	61	91
Average wages per man of whole staff	5/11¾	2/8	8/2	7/10

RAIL MILL.

	England as Unity.	Germany.	A.	B.
Scale of production	100	53	86	56
Relative work performed by each man	100	49	59	77
Average earnings per man of whole staff	5/0¾	3/3¾	7/0	6/5½

As nearly as I am able to calculate from the various figures in my notes the relative costs of labour on a ton of the produce is as follows:—

	England taken as Unity.	Germany.	United States.	
Converting department	100	68	224	144
Rail mill	100	133	234	166

The individual earnings of different classes of workmen in these three cases, viz., England for 1883 and the other two for 1879,[1] are as follows:—

	England.		Germany.		United States.	
	s.	d.	s.	d.	s.	d.
Cupola—1st man	6	0	3	1½	15	7½
„ —2nd „	5	0	2	11	11	7
„ —3rd „	4	4	2	10	10	5
„ —1st charger	5	0	2	10	11	7
„ —2nd „	4	4	2	7½	10	5
„ —3rd „	4	2			6	11½
Converters—1st man	6	8			15	3½
„ —2nd „	5	0	3	7½	11	7
„ —3rd „	4	4	3	1½	9	10
„ —4th „	4	0	2	9	6	11¼
Heaters—1st man	10	0	4	3	13	4
„ —2nd „	7	6	3	8	8	4
„ --3rd „	6	0	2	10	6	5
Roller—head	20	0	8	7½	15	3½
Rougher „	15	0	5	6½	13	0
Catcher	8	4	4	6	9	11

[1] There was little or no difference in rates between the two years.

In each case the daily earnings are what was being paid as nearly at the same period as my information permits, and they all belong to works in which the iron was remelted in cupolas.

Owing to the rails being of different sections in the different countries, and owing to the different manner of keeping the accounts, I can only give what must be regarded as a rough idea only of the relative cost for labour in their manufacture. The average earnings of the entire staff are fully 50 per cent. higher in America than it is in the latest built works of this country; and so far as my data enable me to judge, this difference expresses pretty nearly the higher charge for workmanship on a ton of rails.

Having examined the numerous examples recorded in my memoranda, more or less imperfect in different portions of their details, and having compared them with those which are complete in all respects, I have arrived at the conclusion that we are within the mark in saying that the entire range of iron making, as a whole, costs the American community for labour something like double what it costs the British nation. In some instances this arises from the price per ton or the daily wage actually paid being in accordance with this difference; but it is equally certain that there is another way in which a man may improve his position, as he will consider it, viz., by insisting on additional help, and by himself doing less work. The effect is the same, the employer first, and the community subsequently, have to pay for it.

In a new country like the United States, where the labour market is so largely recruited by immigration, we must expect wages to be higher than they are in those countries from which the emigrants are drawn. Land is so cheap on the other side of the Atlantic that the farmer there can afford to offer a considerable inducement to agricultural labourers to quit the land of their birth. This inducement takes the form of wages varying from 20 to 30 per cent. above the best rates paid in the United Kingdom. But when the grower of corn consumes iron, and has to pay for the labour required in mining, smelting and working the metal, he does so at double and sometimes far more than double the price it commands among European nations.

Since my own experience of the cost of labour in the United States, given in the present section, was printed, I have been enabled to consider the opinions of an American authority on the relative rates of wages paid in his own country at the present time and those current in Great Britain. Mr. J. D. Weeks, a name respectfully quoted more than once in these pages, visited the Cleveland district in the autumn of 1883, and a portion of the results of his enquiries have appeared in the columns of the *Philadelphia Press*. Unfortunately only two of what is intended to be a series of articles, have reached this country, but enough has already been published to show that Mr. Weeks confirms what I have said on the subject.

I select from the lists given by Mr. Weeks a few of the rates, showing in each case the net earnings in dollars of the men at iron ore mines in Cleveland and in the United States.

	Cleveland.	Lake Superior.		State of New York.	
Deputies	1·084 to 1·148	Pit Boss	2·66	Pit Boss	2·23
Miners	1·168	...	2·16	...	1·75
Enginemen	1·041 to 1·213	...	2·02	...	2·16
Blacksmiths	·994 to 1·234	...	2·66	...	2·50
„ Strikers	·780	...	1·58	...	1·40
Joiners	1·148	Carpenter	2·20	...	1·73

The hours of labour are seven in Cleveland of actual work at the face for the miners, and eight to ten down the pit or at bank for other branches.

At the mines of Lake Superior all men work ten hours per day. The period of labour is not given at those of New York State.

Comparing the mines of Cleveland and Lake Superior Mr. Weeks gives the average earnings of all the staff in the former at 90 cents against 2·10 dollars in the latter. I believe that Mr. Weeks understates the average of the English mines by about 10 per cent. in his tables, and it has also to be observed that the American miners work nearly one-half longer hours than those in Cleveland. Notwithstanding these differences, it is quite clear, from the average earnings of all the men employed, that, so far as the mines of Lake Superior are concerned, the wages may be taken to be double those paid in Cleveland. If Mr. Weeks should afterwards maintain that wages generally, certainly in the Northern States were higher than in Great Britain, it is nothing more than has been already admitted in this work.

Making proper allowance for the difference of hours of labour,

Mr. Weeks considers that the Lake Superior miners earn one-half more money than do the men in the Cleveland mines. He further remarks that the shorter hours in the latter are "not the result of a desire to promote the physical or intellectual well-being of the workmen who in most cases would only be too happy to earn the extra 5 or 10 per cent. and work the additional half-hour or hour."

I believe however that the period of labour has been exclusively settled by the Cleveland miners themselves and that "the physical and intellectual well-being" of the men was not more considered by the employers of the Lake Superior mines, in fixing ten hours as a day's work, than it was by the Cleveland mine-owners in consenting that seven hours' labour should be so regarded.

Mr. Weeks mentions the fact that when labour can be obtained in the United States at lower rates than those current in new and remote districts like that of Lake Superior very much lower prices are paid than those he quotes. Mr. Weeks illustrates this by a reference to the earnings of the mining class in the Southern States, of which mention will be found in these pages. He explains this by stating that their work does not require the skill beyond that possessed by an ordinary labourer. My own enquiries in the Southern States as well as in some of the Northern led me to conclude that in many cases the *same kind of labour* was very differently paid in different places, cheaply where it could be so produced, dearer when where it could not.

The second of the two articles referred to concludes as follows:—
"As we go eastward in this country" (United States) "and competition, or the danger of competition with foreign products, becomes sharper and more imminent, wages for the same or similar work decrease. . . . . As a rule, when an article is produced in this country at the same cost as in England the rates of wages paid the workmen producing it are the same or nearly the same as the English wages, and as the American cost approximates the English so the American wages approximate to the English wages."

I am, from what follows, justified in supposing that this language is intended as an argument for the maintainance of protective duties in the United States. If this be true, we cannot in this country complain of the willingness or even the wish of the consumer there to pay the manufacturer an enhanced rate for his products. I think however that the Cleveland miner will repudiate the doctrine laid down in a short leader of the issue of the *Philadelphia Press* containing one of his communications and to which apparently special attention is directed. Speaking of Cleveland the editor observes:—"It is the most wonderful iron and coal district in the world. Nature has endowed it with untold riches, yet Free Trade has forced wages down to a point where the workman is compelled to labour for the common necessaries of life. The truth is that under Free Trade the masses in England have steadily grown poorer and the rich richer."

As a matter of fact the Cleveland miners know perfectly well that men of their class can earn nearly double the money received by their predecessors forty or fifty years ago in the times of protective duties in this country, while at the same time every shilling worked for to-day can purchase more than a shilling could do before the repeal of the Corn Laws. But whether this be true or not, our miners have no wish, as the language of the newspaper article would suggest, to have the necessaries of life enhanced by heavy import duties. Indeed their opinions on such a change would probably not be found to differ materially from those entertained by the *agricultural* population on Mr. Weeks' side of the Atlantic. Not even the competition offered to the Cleveland miner by his Belgian or German rival will induce him to alter his views on this question. With the knowledge that some 300,000 tons of foreign iron are brought into this country, he remembers, that in one shape or another, more than one-half the pig iron made in Great Britain is exported, where it has to meet the competition of those countries in which the principles of Free Trade have not been adopted, and where wages, contrary to the doctrine laid down by the editor of the *Philadelphia Press*, are very much lower than they are in Great Britain, notwithstanding its Free Trade policy.

I refrain from quoting the circumstance upon which the opinion of the Philadelphia newspaper is grounded, that "a few monopolists" (in Cleveland) "live in princely splendour where the workman is compelled to labour for the common necessaries of life." It is notorious at the present moment that the margin of profit throughout Great Britain is a very small one, and it is equally true that in former years of prosperity the ironstone miners of Cleveland earned as high a wage as is paid in the mine in the State of New York quoted by Mr. Weeks.

SECTION XVIII.

THE CHIEF IRON-PRODUCING COUNTRIES COMPARED.

In the two sections immediately preceding, it has been attempted to shew that while our immediate neighbours on the Continent of Europe obtain the labour in their mines and iron works at considerably cheaper prices than those current in Great Britain, the iron-masters in the United States pay their workmen at much higher rates than those prevailing among ourselves. The mere ascertainment however of wages in various countries is only one step in comparing the resources of any one nation with those of others. To do this satisfactorily in the iron trade, we have, in addition to the cost of labour, to consider the nature of the natural deposits from which the minerals are drawn, the facility and consequent economy with which these raw materials are extracted, and finally the position of the markets in relation to the centres of production.

Although Sweden, Russia, and Austria are not without a certain degree of importance in the manufacture of iron, our attention will be confined, among European nations, to Great Britain, Germany, Belgium, and France, for they alone occupy conspicuous positions as exporters of the metal. Compared with its vast mineral resources the United States have up to this time made but little progress in dealing with markets outside their own limits. Having regard however to the fact, that North America has risen with unprecedented rapidity to the rank of being the second iron-making nation in the world, these resources will demand careful consideration; for upon their geographical position and character will depend the ability of the United States to meet European nations in the markets of the world.

Reference has already been made to the competition offered by the Continent of Europe to the iron trade of Great Britain. This is not

SECTION XVIII.—CHIEF IRON-PRODUCING COUNTRIES COMPARED. 579

only encountered in neutral markets, but something like 312,000 tons of manufactured iron and steel were actually imported into the United Kingdom itself in the year 1882, chiefly from Germany and Belgium.

In order to exhibit the nature of the development of Germany as an exporting nation the following figures are given for the seven years ending with 1882:—

Exports from the Zoll Verein	1876. Tons.	1877. Tons.	1878. Tons.	1879. Tons.	1880. Tons.	1881. Tons.	1882. Tons.
Pig and scrap iron, puddled bars, and steel ingots	301,675	360,015	410,231	426,715	314,166	348,027	275,084
Metal castings, malleable iron and steel	315,729	507,046	559,275	552.553	664,012	770,874	775,395
Machinery	40,862	48,871	75,113	63,637	62,137	65,617	83,624
Totals	658,266	915,932	1,044,619	1,042,905	1,040,315	1,184,518	1,134,103

The exports from Great Britain, Belgium, and France, for the last of these years, 1882, were as follows:—

	Great Britain. Tons.	Belgium. Tons.	France. Tons.
Pig iron, puddled bars, and steel ingots...	1,942,582	24,708	68,229
Iron and steel	2,407,715	440,129	36,604
Machinery, roughly estimated from values[1]	478,506	64,627	32,908
Totals...	4,828,803	529,464	137,741
Value of the machinery exported	£11,962,660	£1,615,675	£977,715

I have endeavoured to draw up a statement[2] of the weight of iron exported from the United States, but this from the way in which the accounts are kept must only be regarded as a very rough approximation. In it I have, as in the case of Great Britain, Belgium, and France, assumed £25 per ton as the value of the exported machinery.

	1876.	1877.	1878.	1879.	1880.	1881.
Iron and steel of all kinds, weights given in net tons of 2,000 lbs.	15,416	23,559	20,406	9,867	8,416	13,219
Machinery including agricultural implements, estimated weight in tons of 2,000 lbs....	121,506	156,489	137,374	130,018	132,721	160,605
Net tons ...	136,922	180,048	157,780	139,885	141,137	173,824

[1] The estimates are based on the assumption of the average value of the machinery being £25 per ton.

[2] Compiled from figures furnished by Mr. J. S. Jeans, Secretary of Iron Trade Association.

Notwithstanding the mere approximate character of this estimate it seems quite certain that in spite of the higher wages paid in the United States, the exports of that class of goods in which labour most largely enters occupies a very prominent position; the amount of machinery being apparently larger than that of the united quantities sent from Germany, France, and Belgium. This speaks very favourably either for the labour saving machines employed in the construction of the machinery, or for the superior excellence of the articles exported. Possibly both these advantages may play a part in securing for the American manufacturer so large a share of the world's custom in mechanical appliances. The value of the articles exported annually during the six years given above ranges from £3,000,000 to £4,000,000 against perhaps something like £2,500,000 from Germany, France, and Belgium.

The vast extent of the exportation from Germany might reasonably lead to the inference that in the Zoll Verein we had a formidable rival in economy of production. I am not prepared to say that in periods of moderate prosperity a considerable trade may not be carried on by the German manufacturers, and this in some cases with a fair amount of profit. When however the article to be supplied is steel rails taken in large quantities, such as those required by the Italian railways, at the low prices which often prevail, it was formerly impossible to avoid a considerable loss falling to the share of the Westphalian and Belgian houses who compete with the English makers. About three years ago the quotations for a particular order were, from a Westphalian firm £5 2s. 6d., from an English house £5 2s., and from a Belgian company £5 1s. 6d., delivered at an Italian port. At that period I believe the native pig iron used in the manufacture of ingots by the three competitors was largely smelted from Spanish ore. Practically the sea transport was the only charge between Bilbao and the works of the firm who sent the English quotation referred to. So far as ocean freight is concerned, England stands in a somewhat more favourable position than Rotterdam or Antwerp, at which ports the ore is received for the Westphalian and Belgian works. Without claiming however any advantage in respect to this item, the conveyance of a ton of ore from Rotterdam to Westphalia costs about 5s., equal to about 13s. on every ton of rails. To this has to be added 5s. 6d. for the transport

SECTION XVIII.—CHIEF IRON-PRODUCING COUNTRIES COMPARED. 581

of the rails for shipment at the loading port, so that the German maker is thus placed at a disadvantage of nearly 20s. per ton as compared with his rival in England. Now the total cost for labour on a ton of rails, from the ore to the finished article, is not such as to admit, with the cheaper labour referred to at page 573, of more than a mere fraction of this 20s. of disadvantage being recovered.

Of course when the steel rails produced in Germany have to be employed for the convenience of the German people, the German rail-maker enjoys considerable advantages as compared with his British competitor. It will cost the latter about 12s. per ton to deliver his rails at the door of a Westphalian rail mill, in addition to which an import duty of 25s. per ton has to be paid, raising thus the additional burden to be borne by the English producer to 37s. per ton.

The unprofitable nature of the export trade carried on by Germany in steel rails in times of low prices may be inferred from the action of her own manufacturers. About the time at which the transaction referred to occurred, their government, moved by the representations of their inability to compete with English houses, imposed heavy duties on all those classes of raw and manufactured iron which for some years past had been admitted duty free. The result of this has been that the German consumer pays 40s. or 50s. per ton above the price obtained by the German rail-makers in their export trade. Thus the home consumer was not only called upon to pay the manufacturer a profit upon what is required for domestic use, but he has to make good a loss incurred in rails or other articles for foreign countries. The duties levied on iron and steel entering the German Zoll Verein are as follows:—

Pig Iron.	Boiler Plates.	Bar Iron, Iron Rails, &c.	Steel Ingots.	Steel Rails.
10s.	30s.	25s.	15s.	25s.

Coal and iron ore are imported duty-free.

The charges of conveying ore to a Belgian ironwork, and of taking the rails back to Antwerp, are less than to Westphalia; but on the other hand coal costs more in Belgium than in Germany, so that, from this circumstance, the positions of the two are probably nearly equalised.

In respect to the importations of iron into Great Britain from Germany and Belgium, to which reference has been made in the present section, I am inclined to think that the cost of manufacture at the works is actually something under that at which English houses with their more highly paid labour can produce the articles in question. Admitting any difference arising from this cause to be considerably less than the cost of sending the articles usually dealt with to a British port, it must yet be recollected that the British manufacturers do not pay much less than the German or Belgian makers for transport from their works to the port of reception, which is usually London.

So far as skill in the conduct of the various metallurgic operations themselves is concerned, I have never heard it pretended, and I have failed to discover, any superiority in the practice of foreign manufacturers over ourselves. Mention has been made in these pages however of the greater readiness displayed by foreign workmen in assisting to economise production by the adoption of certain labour-saving machines, as compared with their English rivals. Mechanical puddling was given as an instance of this. The mechanical drills for working ironstone may serve as another. I would not have it inferred that there has been an actual refusal on the part of the English workmen to use the means which mechanical inventions have offered to lighten their labours, much less that there was any want of skill in their management or application. There has however been a want of that ready co-operation on the part of the men which tends so much to facilitate the development of new processes. Certainly in the two inventions just referred to this has been the case. It was evinced either by an unwillingness on their part to perform the work at lower wages than those paid for the less expeditious and more laborious system of hand labour, or, by not availing themselves of the full advantages in point of amount of work, which the machines are capable of affording.

The observations made respecting the exportation from Germany of steel rails in the year 1879 are no longer fully applicable to the transactions of the present day. The works in the Rhenish provinces, although deficient in ores sufficiently free from phosphorus for Bessemer steel, can be supplied on very moderate terms with Luxemburg ironstone or with Luxemburg pig iron. The dearer ores, foreign and domestic, required for Bessemer iron raised the cost of the latter about 30s. per ton above the less pure iron obtained from ores like

SECTION XVIII.—CHIEF IRON-PRODUCING COUNTRIES COMPARED. 583

that of Luxemburg. On the other hand on the banks of the Tees the difference of cost between the two kinds of iron is not much, if indeed it is anything, above one-third of the sum just mentioned. When therefore Mr. E. Windsor Richards had, by his intelligence and perseverance, demonstrated at the works of Messrs. Bolckow, Vaughan, and Co., Limited, the practicability of freeing pig iron from its phosphorus by the Basic process, he placed the Westphalian makers in a better position to compete with the works of his own company at Eston by something like 20s. per ton, than they were prior to the introduction of this recent invention. It is of course this great difference between the cost of Bessemer and phosphoric pig that has stimulated the more general adoption of the basic process in Germany as compared with Great Britain; for it is natural to suppose that a difference of 30s. per ton between the two kinds of pig iron affords a much better opportunity of meeting the expense attending the basic process than does the 10s. alluded to as constituting the margin between hematite and phosphoric pig in the Cleveland district.

In illustration of the different position now occupied by the Rhenish provinces, as compared with former times, I copy a published return of the quotations for 9,000 tons of steel rails contracted for towards the close of 1883 to be delivered at Civita Vecchia in Italy. The names of the firms who sent in the tenders are given, but it is needless to reproduce them here :—

			£	s.	d.
French Company near St. Etienne			6	18	6½
Belgian Company			5	18	5½
,,	,,		5	16	2
,,	,,		5	5	5
German	,,		6	1	6¼
,,	,,		5	19	2¼
,,	,,		5	17	10½
,,	,,		5	15	7
,,	,,		5	10	5¼
,,	,,		5	1	1½
,,	,, received the order		4	19	4¼
English	,,		5	15	9
,,	,,		5	10	10
,,	,,		5	8	0¾
,,	,,		5	5	8¼

According to the return furnished to me, the net price after paying railway dues, sea freight, insurance, commission, and loss in exchange, leaves £4 11s. 2½d. at the works to the English house which quoted £5 8s. 0¾d., whereas, according to a newspaper report, the successful competitor would obtain net at the mill £4 3s. 3½d. This is lower by 7s. 11d. than the figure quoted by the English firm, and I doubt whether in point of fact the actual difference is not more than this, probably nearer 12s.

Having dealt in this somewhat general way with the question before us, it is now proposed to consider the relative costs of the minerals employed in the manufacture of iron, the facilities with which they are brought together, and how the produce has to be delivered to the consumer.

There are two ways of looking at the price of a mineral as it enters into the cost of the metal. It may either be calculated according to the actual expense incurred in raising and delivering it at the iron works, or it may be reckoned according to the market price. The latter is perhaps the more convenient way; firstly, because many manufacturers, not being mine owners, purchase the coal and ore they require, and secondly, because the market prices being matters of public notoriety, use can be made, in any estimates, of such information, without disclosing communications of a private nature with which I have been confidentially entrusted. At the same time it will be my endeavour, during the course of the enquiry, to show the relative positions of different localities by stating them in the comparative form already adopted in these pages, and to give such further information, when I am at liberty to do so, as will assist me in the object in view.

It should further be remarked that, the manufacture of iron being carried on under such a variety of conditions in the same country, the cost of producing the pig may vary by 10s. to 20s. or even more per ton. These differences must of course always constitute a difficulty in settling the amount of those protective measures which are intended to shut out all foreign competition. As an example we may take the present condition of the American iron trade. Stimulated by high prices and large profits the powers of production are now far beyond the necessities of the country. In consequence many furnaces—above

SECTION XVIII.—CHIEF IRON-PRODUCING COUNTRIES COMPARED. 585

one-half of the whole—which are unable to compete with those more favourably situated have been extinguished.[1] Has the import duty to be arranged for the protection of the weaker portion? If so, it is more than sufficient for the stronger; if security is only to be afforded to the latter then the former must go to the wall.

To show more exactly the extent of the difficulty referred to, I have lately examined an account issued by an iron-making firm in the Southern States, from which it would appear that their pig is costing 45s. 10d. per ton at the works with an immediate prospect of seeing it reduced to about 42s. 4d. In the Northern seats of the iron trade a blast furnace must be pretty well situated to be able to deliver its produce at something like 30s. a ton above the last-mentioned figure.

On my return from the United States in 1874, in a paper prepared for the Iron and Steel Institute,[2] I mentioned that the then "comparatively undeveloped resources of Tennessee, Georgia and Alabama will prove a match for any part of the world in the production of cheap iron." The language of the report just referred to would indicate that some considerable advance has already been made in the direction in question. Recently my attention has been drawn to a confirmatory opinion expressed by no less an authority than my friend the Hon. Abram S. Hewitt, of New York. Speaking of Alabama he says:—"It is in fact the only place upon the American Continent where it is possible to make iron in competition with the cheap iron of England, measured, not by the wages paid, but by the number of days' labour which enter into its production.[3] In Alabama the coal and the ore are in many places within half a mile of each other, and the cost of the iron is only about ten days labour to the ton, or not far from the labour cost in Cleveland. Throwing aside all questions of tariffs for protection, here is a possibility upon the American Continent of producing iron at as low a cost in labour as in

[1] FURNACES IN AND OUT OF BLAST IN THE UNITED STATES.

	In Blast.	Out.	Total.	Percentage out of blast.
1881	455	261	716	36·4
1882	417	270	687	39·3
First 6 months, 1883	334	354	688	51·4

—Report of Directors of Iron and Steel Works Association of Virginia.

[2] Transactions, 1875, No. 1.

[3] It remains, I apprehend, to be proved that Tennessee and Georgia may not be able to compete with Alabama in cheapness of production.—I. L. B.

the most favoured region in the world, and allowing for the expense of transportation to compete with them, paying a higher average rate of wages than is paid in Great Britain."[1]

If the statistician, fifty or sixty years ago, had proposed to himself the solution of the problem contained in the heading of the present division of this work, he would have had to deal with factors of very different values to those involved in its consideration at the present day.

Selecting a railway bar, as the largest single object of manufacture, for the purpose of examination, we have an article one ton of which in Scotland, anterior to the introduction of hot blast, would have consumed, for its production, 10 to 12 tons of coal in the blast furnaces, forges, and mills.[2] Besides this wasteful expenditure of fuel, something like one-third of the metal contained in the ore at that time never made its appearance in the final product, but was tipped over the spoil heap in the forge and mill cinders.

In place of this extravagant use of raw material, railway bars of steel are now turned out with a total consumption of about $2\frac{1}{2}$ tons of coal, and practically all the metal in the ore is delivered to the consumer in the rail he buys.

From what has been advanced in these opening remarks, it is obvious that the relative position between two iron-producing countries has been entirely altered by the changes referred to. In the case of a nation having cheaply worked and favourably situated coal, the iron manufacturer there would have formerly enjoyed an advantage over a less favourably placed rival measured by the difference of cost on each ton of fuel, multiplied by 10 or 12 tons instead of by $2\frac{1}{2}$ tons. In like manner a nation possessing cheaper ore than another, having regard to the content of iron in each, suffers somewhat by any economy in the waste of the metal it contains. Thus if 4 tons of ore at 10s. per ton were required to produce 1 ton of iron rails, we have 40s. as the value of the ore consumed. Against this let us assume the case of an ore of the same richness, but costing only 6s. per ton, and this on 4 tons gives 24s. only, or an advantage of 16s. on the ton of rails. But

---

[1] "The Hill Country of Alabama," 1878, p. 37.

[2] I think no iron rails were at that time made in Scotland. The estimate therefore is based on the consumption of coal in the blast furnaces to which an addition is made for puddling and mill work.

SECTION XVIII.—CHIEF IRON-PRODUCING COUNTRIES COMPARED. 587

if the progress which has been made in the manufacture diminishes the consumption of ore to 3 tons per ton of rails, this difference of 16s. will be reduced to one of 12s. per ton on the final product.

This more favourable position in respect to the raw materials employed in the manufacture of iron, as described in the last paragraph, may as a whole perhaps be claimed by Great Britain; but as a set off, the British iron masters labour under the disadvantage of dearer labour as compared with their continental competitors. The alterations just spoken of in the mode of producing rails have on the other hand operated in favour of the maker in our own country; because the amount of labour to produce a given quantity of the object in question is not one-half what it was half a century ago. In consequence of this change any economy effected by cheaper labour on the Continent has been correspondingly reduced. It must thus be admitted that the puddling furnace of Cort, Neilson's discovery of the hot blast, the Middlesbrough enlargement of the blast furnace, Bessemer's pneumatic method for making steel, and the application of the Basic process to the Bessemer converter—all British inventions—have tended to diminish the distance which separated this country from Continental Europe in cheapness of production. On the other hand it is equally certain that the economy effected by means of the discoveries just enumerated has extended the use of iron in its various forms to a point which, under the mode of manufacture pursued less than 100 years ago, would have been perfectly impossible.

Whatever historical interest a comparison of the past and present state of the iron trade may have, the question is less important than that which connects itself with the future. The extent to which the consumption of fuel and the waste of metal has been reduced leaves a greatly diminished margin for further economy: indeed, as regards waste, the loss in the manufacture of a steel rail may be regarded as perfectly insignificant. Small as is the weight of fuel which enters into the production of steel rails, it may be well to consider shortly the prospect there is of any notable reduction in the present quantity employed.

Beginning with the blast furnace I showed, in a paper read before the Iron and Steel Institute,[1] that out of 22·32 cwts. of coke required in smelting Cleveland ironstone a quantity of heat was carried off

[1] Transactions, 1883, No. 1. p. 130.

in the escaping gases equivalent to 2·21 cwts. per ton of pig iron made, or equal to 9·9 per cent. of the whole. This observation was made in connection with a furnace containing 11,500 cubic feet; but I found in the same furnace that the loss upon another occasion from this cause was reduced to 6·21 per cent. By the use of a furnace containing 35,013 cubic feet at the Ormesby works of Messrs. Cochrane I ascertained this loss in the escaping gases to be 7 per cent. of the entire heat evolved.[1] So far then as loss from this particular cause is concerned, it may be inferred that there seems little hope of amendment: indeed in Section V., pp. 74 *et seq*. I gave my reasons at some length for concluding that heat is evolved in the upper portion of the furnace, that is in the reducing zone, by the generation of carbonic acid, and that this takes place too near the escaping point of the gases to permit its complete absorption by the descending solid contents of the furnace. It is quite true that upon some occasions the elevation of temperature due to the cause just referred to appears to be absent; but this, I apprehend, only happens when an excess of moisture is present in the materials, and then an addition to the fuel consumed is required for its evaporation.

As I have pointed out, in Section VI., all the expectations which were at one time entertained of reducing the consumption of fuel by a great addition to the temperature of the blast are not likely to be realized. These expectations were based on the idea that some approach to the saving which had been effected by the first addition of 600° F. to the air entering the furnace would be realised by a further like increase of temperature.

Very little consideration of the subject must convince us of the fallacy involved in such an opinion. If we assume the requirements of a well appointed modern furnace, smelting Cleveland or other analogous mineral, to be 86,000, or any other number of, Centigrade heat-units, it is easy to calculate what temperature the air must have to supply this quantity of heat when burning any given quantity of carbon.

The amount of heat capable of being afforded by the combustion of the coke in a blast furnace with air at 0° C. (32° F.), or indeed at any other temperature, depends of course on the proportion which can be burnt to the state of carbonic acid. In practice however, as soon

[1] Transactions of Mechanical Engineers, August, 1882, No. 3, p. 306.

as out of 3·22 units of carbon burnt, we have 1 unit in the state of carbonic acid, the reduction of the oxide of iron, as it exists in Cleveland calcined ore, is suspended. Basing my calculations on this view, I estimated the temperature the blast must have to supply the difference to make up the 86,000 calories given as the total quantity required. At pp. 89 *et seq.* the reasons are given for believing that when burning 16 cwts. of carbon in the manner just described (^{1}1 of C to CO_2 and 2·22 to CO) the blast will require to have a temperature of 900° C. (1,652° F.). This quantity of carbon, with ash, moisture and carbon absorbed by the iron means theoretically[1] fully 18 cwts. of coke per ton of iron produced from an ore of the Cleveland type. Messrs. Cochrane claim to have made a ton of iron with 18·7 cwts. of coke, and with the blast at 1,406° F. in a furnace of 35,013 cubic feet; but when I was permitted to examine the process, the consumption amounted to 19·69 cwts., the air being heated to 1,507° F.[2]

Notwithstanding the low rates of consumption disclosed, by this last-mentioned instance, I am not very sanguine of their being maintained under all circumstances over a long period of time. Mr. Hawdon, in a paper read before the Iron and Steel Institute,[3] gives the particulars of a furnace under his management at Newport, near Middlesbrough, containing 29,410 cubic feet. The average temperature of the blast over seven weeks was 1,328° F., and the coke consumed was reduced to 22·3 cwts. per ton of foundry iron.

The best return I have from a furnace using hematite ore shows with a blast of 1,300° F. a consumption of about 20 cwts. of coke per ton of pig, which for a rich material and a consequent low production of slag is not an extraordinary rate of working when compared with the practice in Cleveland.

Neglecting therefore exceptional cases, and bearing in mind the difficulty of maintaining the air at the higher temperatures, I am not prepared to give hopes that we shall see the coke used in smelting Cleveland iron permanently reduced below 21 cwts., nor in hematite iron below 19 cwts. per ton. Should this surmise turn out to be correct, there is evidently nothing to render it probable that the

[1] Paper by Mr. Charles Cochrane, Tran. Inst. Mech. Engineers, August, 1882, No. 3, p. 279.
[2] Discussion on a paper by Mr. Charles Cochrane, Trans. Inst. Mech. Engineers, August, 1882, No. 3, pp. 307 and 308.
[3] Trans. Iron and Steel Institute, 1883, No. 1, p. 101.

present relations of different nations can be materially affected by any future changes in the manufacture of pig iron, so far as the consumption of fuel is concerned.

In Section XI. (p. 315 *et seq.*) the use of raw coal in the blast furnace is considered as a question of heat development. It may be useful in a section devoted more particularly to the economic aspect of the subject to say a few words on this form of combustible.

The inconveniences attached to employing the fuel in the form of coke, as produced in the ordinary bee-hive oven, are as follows:—

a.—The loss of a portion of the fixed carbon, amounting to about 10 per cent. of the whole.

b.—The combustion of this carbon, together with that of the combustible gases of the coal, of which a small portion only is utilised in the distillation of the volatile portions of the fuel.

c.—The loss of certain volatile products, viz., condensable hydrocarbons and compounds of ammonia.

d.—The loss of interest on the capital invested in coke ovens and the wages and other expenses connected with the process of coking.

The fact that whenever the quality permitted, the coal is used in the raw state in the blast furnace, may be accepted as a proof of the greater economy of this mode of procedure as compared with using it in the form of coke.

According to experimental research, a given weight of coal and coke has approximately the same value in a heat producing point of view, when both are completely oxidised. I have had this verified by the duty performed by the two kinds of fuel in a locomotive engine. The experiments were extended over fourteen days, using the same engine, and carrying the same loads over the same country.

It has been shown in Section XI. however that matters are entirely changed when raw coal is used in the blast furnace; for it appears (page 318) that when smelting with coal an ore somewhat richer than the Cleveland, but using the same quantity of limestone, 22·65 units of fixed carbon in the coal, equal to about 24·5 of good Durham coke, are required to smelt 20 units of iron.

SECTION XVIII.—CHIEF IRON-PRODUCING COUNTRIES COMPARED. 591

No doubt the loss of fixed carbon, as mentioned under *a* when speaking of the bee-hive ovens, is here avoided; but it will be observed (see page 319), that out of about 110,000 calories 24,560 calories are absorbed in expelling the gases. Against this loss must be set the value of the volatile portion of the coal which is applicable for raising steam and for many other purposes.

If however everything is taken into account, there is no doubt, omitting the value of the tar or other hydro-carbons and the ammoniacal compounds in both cases, that the use of raw coal is more economical than that of coke, obtained from coal charged at the same price.

In recent years however, as was stated at page 314, Messrs. Baird of the Gartsherrie works have succeeded in condensing the ammonia and hydro-carbons. According to a notice which appears in the Transactions of the Iron and Steel Institute,[1] the net value of these compounds, at present prices, is equal to 3s. 2d. per ton of coal employed, or about 6s. on the iron produced. If this be true, the Scottish ironmaster may be said, so far as net cost is concerned, to be able to make his iron without any outlay for fuel, *i.e.*, so long as the bye-products in question can command, in face of a greatly increased production, their present prices.

So far, the comparison, economically speaking, between coke and raw coal has been on the supposition that the former was produced in the ordinary bee-hive ovens. In this not only are the bye-products wasted but a large amount of heat escapes unutilised into the air. This form of oven however is capable of some amelioration. The loss of fixed carbon (*a*) by careful attention may be somewhat reduced and a portion of the volatile constituents, valued at about one shilling on each ton of coal employed in the Brancepeth district, may be partially condensed according to an invention of Mr. Jameson.[2]

A simple mode of utilising a portion of the heat wasted (*b*) in the ordinary oven consists in employing it for raising steam. This has

---

[1] 1884, No. 2, page 698.

[2] Mr. Jameson's process consists in aspirating through suitable condensers a certain quantity of the volatile constituents of the coal. In other respects the operation of coking remains unaltered, so that a considerable proportion of the gases, being burnt as heretofore in the oven, yield neither tar nor ammonia. Mr. Jameson's invention however possesses the great recommendation of being, with little change, applicable to existing ovens. With certain coals Mr. Jameson values the products at 3s. 9d. per ton of coal (v. p. 328 *ante*).

been done with perfect success at certain collieries in the County of Durham, among others at those belonging to my own firm. Including the saving of wages formerly paid to firemen and the value of the coal employed in raising steam the saving amounts to about 3d. per ton on the whole output of the colliery. Mr. Jameson's system is in the act of being tried at one of the pits referred to and, if the expectations of it are realised, there will result a total saving as compared with the original form of the bee-hive oven of 1s. 3d. per ton of coal raised and coked. This would represent about 2s. on the ton of iron.

Many years ago my firm erected several ovens in which the coke was distilled as in a retort. The gases were cooled in suitable appliances and the ammonia, tar, etc., were thus condensed. The difficulty of maintaining the structure sufficiently tight and the then value of the products led to the abandonment of the system. Since then some improvements in the form of the oven and the increased value of tar and ammonia have induced my friends Sir J. W. Pease & Co. to erect an experimental bench according to the plans of Messrs. Carvés & Simon, and they are about to construct a second lot.

According to public rumour the condensed substances obtained from the so-called Simon-Carvés oven are worth about 3s. 6d. on each ton of coal treated. The extra yield of coke—75 per cent. instead of 65 per cent.—may be taken at 1s.; making in all 4s. 6d. from which something must be deducted for expenses of working the condensers, etc. Calling the net value 4s. per ton, we have a gain of about 7s. on each ton of pig iron manufactured, with coke so obtained.

It has to be observed however that many trials have been made in the County of Durham with ovens having for their object a better yield of coke from the coal employed. As a rule, one after another these trials have ended in failure, and the old bee-hive form has been adopted in their room. The cause of this has been the inferior value of the product, due as I have supposed (*vide* p. 100) to the less perfect mode of carrying on the operation of coking. The effect of this is that the coke is softer in texture, and more susceptible of being acted on by carbonic acid in the upper region of the furnace. Whether the same defect will be perceptible in the coke obtained from English coal made in the Simon-Carvés oven remains to be seen, but in France I heard no complaint on this head of the coke obtained from French coal.

SECTION XVIII.—CHIEF IRON-PRODUCING COUNTRIES COMPARED.

The observations just made on the fuel employed in smelting iron are more or less applicable to all districts. The extent of the saving however, to be effected on each ton of iron will depend on the nature of the coal employed—as a rule the greater the amount of volatile matter the fuel contains, the greater will be the quantity of tar and ammonia condensed. All this saving may, it is true, not fall to the share of the iron-masters; but may be divided between them and the consumers, by a decline in the market values of coal-tar and sulphate of ammonia.

After what has been said in Section XI. on the action of hydrogen in the blast furnace, and on the use of the so-called water gas for smelting iron (*vide* pp. 330 *et seq.*), I need add nothing to prove my belief that no help is to be expected from that quarter.

When we come to the conversion of pig iron into ingots, and of ingots into rails, there is still less probability of there being any disturbing cause in the direction referred to in connection with the manufacture of pig iron. There are many mills in which a ton of rails is produced by the use of 15 cwts. of coal, reckoned from the pig to the finished bar. The molten pig iron brings its own heat from the blast furnace; and this, added to that evolved by the combustion of the associated metalloids, suffices for its conversion. We have then merely to provide the motive power and coal to reheat the ingots. Since my short allusion to the soaking pits of Mr. Gjers (Sec. XIV., p. 440) this gentleman has furnished me with *data* respecting the actual working of this system of preparing the ingots for the rolls, without the use of any fuel in the mill. Not short of 200,000 tons of ingots have been so rolled down to blooms but by far the greater part has then been reheated before going to the finishing mill. In one case, however, 3,000 tons of heavy double-headed rails have been rolled without any heating after leaving the ingot moulds; and those who have adopted Mr. Gjers' invention express great hope of being able, with sufficiently strong machinery, to make every description of rail without burning any coal for heating the ingots. If so, then 50 cwts. of coal, the estimated quantity supposed to be required to produce a ton of rails from ore, will probably be reduced to 45 cwts. or less. With either quantity it seems very unlikely that any future improvements can greatly alter the relations of iron producing countries as they at present exist.

The quantity of limestone used for a ton of iron, and its low price, afford but a small margin for economy so far as the mere value of the mineral affects the question. In smelting Cleveland ironstone $10\frac{1}{2}$ to 12 cwts., worth from 1s. 9d. to 2s., suffice for the ton of pig. The presence of carbonate of lime in the blast furnace is however attended with other consequences, which are of almost as serious import as the price itself. At page 198 it was stated to have been proved by direct experiment that at the temperature at which carbonic acid is expelled from limestone, the coke is attacked by the acid with formation of carbonic oxide. Coke so oxidised of course never reaches the tuyeres, and is so much loss; a loss equal, when using $10\frac{1}{2}$ cwts. of limestone per ton, to 1·26 cwts. of carbon per ton of iron. Again, at page 126 it was held that the coke consumed for melting the slag produced per ton of pig, amounted to 5·794 cwts.; and, taking 26 cwts. of slag to be the quantity to be fused, we have a consumption of ·22 cwt. of coke per cwt. of slag. The lime in $10\frac{1}{2}$ cwts. of limestone is 5·88 cwts., requiring for its fusion in the form of slag 1·29 cwt. of coke. The use therefore of $10\frac{1}{2}$ cwts. of flux involves the expenditure of nearly 2·55 cwts. of coke, now worth at Middlesbrough about 1s. 6d.

In reference to any possible diminution in the quantity of flux, it has to be observed that fusion of the earthy portions of the ore is only one of the objects sought to be attained by the addition of limestone. On consulting Berthier's experiments,[1] I was led to infer that the earths, in the proportions in which they exist in Cleveland stone, constituted a very fusible compound. A mixture was therefore made of silica, alumina, lime, and magnesia, analagous in composition to the earthy constituents of the ore referred to. This at a very moderate temperature ran down into a perfectly vitreous slag.

The attempt, referred to at page 169, was then made to discontinue the use of limestone; but it was found, as indeed was expected, that blast furnace cinder, in order to free the iron from sulphur, must have a distinctly basic composition; and to obtain this a further addition of lime was indispensable.

The figures mentioned in reference to the cost attending the use of limestone show that it is necessary to avoid employing a flux containing any large quantity of carbonate of magnesia; for this earth

[1] Essais par la Voie seche.

SECTION XVIII.—CHIEF IRON-PRODUCING COUNTRIES COMPARED. 595

appears powerless in decomposing sulphur compounds. In the Pennsylvanian works a limestone is used so rich in magnesia carbonate that, although the mixture of ores smelted contains more than 50 per cent. of iron, 25 cwts. of flux are consumed for each ton of iron, or fully 15 cwts. more than would be needed according to the Middlesbrough standard. This at Middlesbrough prices means an addition to the cost of a ton of pig of 2s. 3d. for extra fuel, added to about 2s. 6d. for the extra weight of limestone used.

RAILWAY DUES.

When the raw material weighs five or six times as much as the marketable product, and where the quantities dealt with are so large as in the case of iron in its various forms, the item of transport occupies a very important place in the cost of production. Cleveland pig iron has been sold at as low a price as 33s., and out of this it is calculated that about 7s. 6d. to 8s., or 23 per cent. of the actual value, was paid to the North-Eastern Railway Company.

Speaking broadly, I take it we may assume that the mere act of transport over a great distance can be more cheaply done by water than by any other mode of conveyance. A remarkable case in point was given me in Pittsburg. Coal was being conveyed from the neighbourhood of that city to New Orleans (a distance, including sinuosities of the river, of probably 1,500 miles) for 1s. per ton. This was done by loading barges sufficient in number to contain 20,000 tons of coal, which, under the guidance of a steamer, were towed down the Ohio and Mississipi to their destination. This is a state of things, however, which finds no parallel in any manufacturing operation. The quantities dealt with in the latter are much smaller in amount than that named, and have to be brought to different centres from various quarters. Canals could rarely be constructed with economy to convey coal or ironstone over the uneven country lying between the mines and the furnaces; and even with a river like the Rhine it is generally found more economical to transport the ore brought from Spain, or the produce of the works on its way for shipment, by land rather than by water. Such being the case, the rates levied by the railway companies in different countries on iron and on iron making materials form not an unimportant subject of enquiry.

Fortunately for Great Britain, the distances which separate the different materials used in her iron works are, necessarily, from the size of the country, much shorter than those which intervene in the case of most foreign nations; while the works themselves are at no great distance from the sea-board. In the United States iron ore is brought in considerable quantities from the mines of the Lake Superior district to Pittsburg—a distance of about 800 miles. Pittsburg in its turn is about 440 miles from New York. Railways have been constructed to connect the mines with the shipping ports of Marquette and Escanaba, at distances which may be approximately taken at 25 and 50 miles respectively. These railways have been made exclusively for the accommodation of the mines; and looking at the somewhat uncertain and changeable nature of the traffic, are perhaps justified in demanding the high rates charged, which when I visited the country appeared to be about 1·3d. per ton per mile. This is however an exceptionally high rate; and there are others which, owing to the peculiar circumstances of the trade, may be considered as exceptionally low. Thus in Tennessee (in 1874) the cost of bringing pig iron from the furnaces to the forge was given as being 1½ dollars per ton equal at the then rates of exchange to ·68d. per ton per mile, or ·75d. per ton per mile in gold. The distance between Chattanooga and Chicago is about 550 miles, and I was informed that on merchandise travelling south the charge for distances above 100 miles was 2 cents per ton per mile, say 1d.; while in the opposite direction it was only 1 cent. It was further hinted that, in order to promote the establishment of iron works, the railway company would prefer conveying pig metal at ½ a cent per ton per mile rather than take the wagons back without any load.

I have from various sources compiled tables of the rates charged by railway companies, in different iron making centres which I have visited in the course of my travels.

Railway Rates in Cleveland.

So far as the rolling stock and general railway accommodation are concerned, I have seen no district either in Europe or America better served than that of Cleveland. The wagons are all supplied by the North-Eastern Railway Company and are of the best descrip-

SECTION XVIII.—CHIEF IRON-PRODUCING COUNTRIES COMPARED. 597

tion, furnished with "bottom boards," so that the cost of unloading by the iron manufacturers is effected with great expedition and economy. The generally understood charge for wagons is ·125d. per ton per mile, but in the rates quoted below this item is included.

For the distances given the rates on minerals and pig iron are given underneath per ton per mile:—

Miles	10.	12.	15.	18.	20.	22.	30.	40.
	d.	d.	d.	d.	d.	d.	d.	d.
Rate on minerals	1·10	·97	·95	·88	·85	·83	·78	·74
,, pig iron	1·80	1·60	1·53	1·44	1·40	1·36	1·23	1·00

These are the charges when No. 3 pig iron is at 45s. to 46s. per ton. For each 1s. of fall in price, an allowance is made on the mineral rates of 1 per cent. till the price reaches 36s., by which time the rebate amounts to 10 per cent. On the other hand a similar addition is made to the rates until the price rises to 55s., when it ceases.

Reckoned upon the average distances along which the minerals are carried, the dues received by the railway company for such service will amount to about 7s. 6d. to 8s. per ton of pig iron, according to the value of the latter.

RAILWAY RATES IN SOUTH WALES.

The close proximity of the collieries to the mountain limestone and to the works enables the smelter to obtain his flux and fuel at a small expense so far as cost of transport is concerned. This and the moderate distance in some cases of the blast furnaces from the ports at which iron ore is discharged, confer great advantages on South Wales in the manufacture of pig iron. The Taff Vale Railway Company, as a rule I believe, do not supply wagons for the mineral traffic on their line. To cover this therefore ·125d. per ton per mile is added to the rates charged which are those paid by a firm about midway between Merthyr and the sea.

Miles	7.	12.
	d.	d.
Rate on minerals	1·318	·928
,, pig iron	—	·928

The dues on the minerals for one ton of pig to furnaces situated as above described including dock dues amount to 3s. 3d. to 4s.

598 SECTION XVIII.—CHIEF IRON-PRODUCING COUNTRIES COMPARED.

RAILWAY RATES IN THE WEST OF SCOTLAND.

So far as the information I have been able to obtain enables me to judge, the dues levied by the Scotch railway companies are 15 to as much as 30 per cent. higher than those charged by the North-Eastern Railway Company to the Cleveland iron-masters. This however presses much less heavily on the Scotch smelters than it would on the English, owing to the much shorter distances over which the minerals are conveyed to the furnaces in Scotland. While however in Cleveland many of the works are placed on a navigable river, involving only a few pence (3d. to 6d.) for shipping, it costs most of the West of Scotland furnaces about 2s. 6d. per ton to reach the Clyde. The extent to which ironmaking materials are conveyed by railway varies so much that I have preferred inserting the estimated cost for carriage per ton of iron in actual cases of which I possess the particulars. To this the charge for shipping at Glasgow is added:—

	s. d.	s. d.	s. d.	s. d.	s. d.
Dues on minerals	4 2	4 0	5 3	3 9	5 0
Carriage on pig to Glasgow	2 6	2 6	2 6	2 6	2 6
	6 8	6 6	7 9	6 3	7 6

RAILWAY RATES IN THE MIDLAND COUNTIES OF ENGLAND.

The Midland Railway Company does not at present always find wagons for coke and ironstone. The blast furnaces generally speaking are situate either at or near the collieries or the ironstone mines; and the distances to be traversed being not inconsiderable (54 to 84 miles) the rates are low, sometimes only $\frac{1}{2}$d. per ton per mile, or including wagons say ·625d. per ton per mile. On the Derbyshire coal-field, according to my information, the actual cost for carriage including that on the short distance for coal, will amount to from 7s. 9d. to 9s. 3d. per ton of pig iron. At or near Northampton the charge for carriage is from 10s. to 12s. on the ton of metal. The coal here and in Derbyshire being generally used raw.

RAILWAY RATES IN LINCOLNSHIRE.

For smelting the Lincolnshire ironstone South Durham coke was largely used. Coke obtained from the washed small coal of the South Yorkshire coal-field is also employed to some extent.

SECTION XVIII.—CHIEF IRON-PRODUCING COUNTRIES COMPARED. 599

The distance of the South Durham pits from Frodingham is 112 miles, the cost of carriage being about 5s. 9d. The South Yorkshire coal-field and the Lincolnshire iron ore district are only about one-third of this distance apart. Allowing a trifle for getting the ironstone to the furnaces, 6s. to 7s. 6d. per ton of iron probably represents the dues paid for conveyance of minerals.

RAILWAY RATES IN CUMBERLAND AND LANCASHIRE.

An ordinary rate for conveying iron ore and limestone to the furnaces is 1s. 6d. per ton from the mines and quarries, while that paid on the coke varies from 5s. 6d. to 7s. 6d.—the distances ranging from 60 to nearly 100 miles. At these prices, the carriage on the minerals amounts to an average of about 10s. 9d. on the ton of pig iron.

RAILWAY RATES IN SOUTH STAFFORDSHIRE.

In former years the expense of bringing the raw materials together at the South Staffordshire blast furnaces was very trifling, both fuel and ironstone being obtained from the coal measures adjoining the furnaces. Subsequently hematite was imported from Lancashire or Cumberland, and black band from North Staffordshire, so that the actual sum paid for carriage on the iron would depend upon the relative quantities of these ores used in its manufacture. Now the South Staffordshire furnaces receive considerable supplies of ironstone from Northamptonshire, distant about 60 miles. The ore raised in the district itself will not furnish above one-third of the pig iron made there, so that probably 7s. to 12s. 6d. represents the cost of transport on each ton of metal produced.

RAILWAY RATES IN NORTH STAFFORDSHIRE.

The position of North Staffordshire as an iron-making district is one of recent creation. The pits supplying the ironstone and the fuel, which is used uncoked, are in consequence still quite near the furnaces. The coal basin is one of limited area, and the furnaces I was permitted to examine are so situated that the calcined ironstone is conveyed over distances varying from 6 to 9 miles. Assuming four tons to be required for each ton of pig iron, and to cost 9d. to 1s. for carriage, we have 3s. to 4s. as the expense of transport.

Two new furnaces have recently been erected at Great Fenton, where the cost of bringing the material together will, I apprehend, be less than the above-mentioned sum.

FOREIGN RAILWAY RATES.

Official returns have been published containing the railway charges paid in France, Belgium, Germany, and the United States. From these I have copied the rates and distances given below.

FRANCE.

The rates per ton per mile are thus given on different lines:—

Coal and Coke.		Ore from Mines.		Ore from Ships.		Pig Iron.	
Miles.	d.	Miles.	d.	Miles.	d.	Miles.	d.
18	1·52	108	·97	108	·64	66	·90
27	1·52	170	·70	109	·68	84	·90
62	·90	174	·59	109	·70	207	·70
74	·70	182	·59	127	·78	212	·70
85	·70	262	·59	150	·56	221	·70
114	·90	443	·54	209	·50	—	
125	·90	—		—		—	
141	·70	—		221	·62	—	
147	·59	—		228	·50	—	
149	·90	—		326	·44	—	
206	·59	—		364	·46	—	
244	·58	—		394	·46	—	

BELGIUM.

Coal and Coke.		Ore.		Pig Iron.		
				Home Consumption.		Exportation.
Miles.	d.	Miles.	d.	Miles.	d.	d.
19	1·1	28	·66	68	·93	·68
31	·90	—		70	·88	·62
71	·62	40	·55	72	·86	·61
74	·60	59	·48	74	·86	·62
87	·43	—		—		—

SECTION XVIII.—CHIEF IRON-PRODUCING COUNTRIES COMPARED.

GERMANY.

Coal and Coke.		Ore.		Pig Iron.	
Miles.	d.	Miles.	d.	Miles.	d.
6	1·5	4	2·09	8	1·5
7	1·7	18	1·00	41	·78
11	1·2	31	·87	88	·59
—		45	·60	206	·50
—		46	·60	247	·51
—		85	·52	318	·43
—		86	·58	339	·50
—		97	·51	569	·45
—		143	·45	—	

UNITED STATES.

Coal and Coke.		Ore.		Pig Iron.	
Miles.	d.	Miles.	d.	Miles.	d.
208	·75	66	·75	360	·55
235	·49	110	·575	452	·49
345	·42	151	·66	—	
—		280	·46	—	
—		385	·36	—	

Taken as a whole the French rates are higher than those charged in the North of England. In Belgium the charges for coke for short distances are somewhat higher, while on long distances they are lower than with us. There is evidently a wish to promote an export trade on the part of the Belgian Railway authorities, as is evidenced by a difference in the rates: thus the average charge on pig iron for home use is ·88d. per ton per mile, against ·63d. per ton per mile for export. The same charges are levied on malleable as for pig iron for home consumption and for exportation. In the United States the rates suffered an average reduction of 53 per cent. in 1878 as compared with 1868.

The charges per ton per mile for conveyance by railway in the four foreign countries are thus summarized in the return now quoted:—

	Coal and Coke.			Iron Ore.		
	Highest.	Lowest.	Average.	Highest.	Lowest.	Average.[1]
France ...	1·52	·58	·85	·97	·54	·66
Belgium ...	1·1	·43	·73	·66	·48	·56
Germany ...	1·7	1·20	1·46	·60	·45	·51
Do. Luxemburg	—	—	—	·49	·42	·46
United States ...	·75	·42	·55	·75	·36	·56

[1] The averages are not those of the highest and lowest rates, but of the whole of the rates for each particular product in their respective quantities.

Iron for home use is thus charged:—

	Highest.	Lowest.	Average.
France	·90	·70	·78
Belgium	·93	·86	·88
Germany	1·50	·43	·66
United States	·55	·49	·70 (*sic*)

ROYALTY DUES.

The item of railway charges, frequently amounting, as we have seen, to a large item per ton of pig iron, is, when levied by a public company, a source of expense beyond the control of the manufacturer. The same may be said of the royalty dues, as they are called in England, paid to the land-owner for permission to work the minerals.

The sums received by the mineral owners vary considerably in the same localities, according to the situation and character of the mines and the periods at which the leases were granted. For hematite and calcined black band, 2s. 6d. per ton is about the usual royalty paid to the land-owner; but in some cases, when prices of iron and therefore of ore were very high, the sum paid by speculators, forgetful of former trade vicissitudes, was fully three times the amount just mentioned. Cleveland, Lincolnshire and Northamptonshire ironstone being much poorer, and the metal from these varieties of ore commanding a much lower price than that from hematite or Scotch calcined black band, the sums paid for permission to work the former are much lower. At the present time (February 1884) the royalty dues for these descriptions amount to about 9 per cent. of the selling price of Cleveland, and about 12 per cent. of that of Hematite and Scotch pig iron.

In the United States, on the other hand, vast tracts of mineral lands are sold, including the surface, at from 3s. to 30s. per acre. Practically therefore the actual purchase price for royalty of a ton of coal, or of ironstone, is merely nominal. For reasons, however, which will be spoken of hereafter, it is only under exceptional circumstances that the American pig iron maker reaps any advantage from the absence of a charge which often presses pretty heavily on the British iron-master.

At the iron works of the Northern States, ore is so scarce that large quantities are conveyed from Lake Champlain and Lake Superior;

SECTION XVIII.—CHIEF IRON-PRODUCING COUNTRIES COMPARED. 603

the carriage from the former to the furnaces near Philadelphia costing about 14s. per ton, and that from Lake Superior to Pittsburg 20s. per ton. Under such circumstances the mineral owners near the furnaces have no difficulty in realising a good return on such ore as they may let to the furnace owners. Thus, for brown hematite containing under 35 per cent. of iron, and situate near the Lehigh furnaces, 50 to 75 cents royalty was being paid in 1874; and on magnetite containing 66 per cent. of metal 75 cents were paid, under an old lease, at a mine near the town of Dover in New Jersey.

At Lake Champlain I heard of a very productive mine, situate within a short distance of the Lake, which it was said had been purchased about the year 1834 for 4,500 dollars; but as the produce, worked from a solid vein of ore, was being sold at prices from $5\frac{1}{2}$ to $7\frac{1}{2}$ dollars, according to differences in quality, it is easy to suppose how different a value would be placed on it, in the event of its now changing hands. In other words the present owners were, and probably are receiving, a very high royalty in the shape of very handsome profits as mine owners.

At Marquette a certain mine was described to me as having been opened out on an area of 40 acres by three men in comparatively humble circumstances. The royalty originally agreed to be paid to the then owner of the soil was stated to be 50 cents. per ton, his predecessor having probably obtained that soil for a dollar an acre or less. After spending 15,000 to 20,000 dollars, the concern was sold to an iron company, already possessing furnaces and mills, for 375,000 dollars.

The iron mountain at Cornwall in Pennsylvania is a mass of ore rising above the plain in which it stands. The mineral is so easily worked that one man can, it is said, fill 10 tons for a day's work. A railway company, I was informed, offered one million pounds sterling for the property, which was declined.

All this goes to prove that whatever the original terms may be, the value of the mineral itself is determined subsequently by the ordinary rules of supply and demand, coupled with any peculiarities of a local character.

From what has just been advanced, it would appear that the selling price of ore in the United States, at all events in the northern division of the Union, for some years past has been sufficiently high to embrace within its amount, as compared with this country, a very high profit to the mine owners, as well as a pretty high royalty rent—a fact which will be more particularly seen hereafter. This state of things has been brought about by the condition of the American iron-trade, which it may be convenient to consider at the present moment.

Some twenty years ago very large quantities of iron were imported into the United States, almost exclusively from Great Britain. At this time the total value of the imports and exports, exclusive of the precious metals, from and to the Union nearly balanced each other in point of value; for in 1865 the former were £21,624,125 against £25,170,787. From the year 1865 to the end of 1869 the average annual freight on pig iron from England to New York varied from 20s. 4d. to 24s. 2d., of itself a very effective protective duty to the American iron-masters. In the year 1870 the balance of trade had already shown signs of considerable change, and with it the protection afforded by high freights began to diminish. The following table, compiled from the figures given in the "Financial Almanac," from the Reports of the American Iron and Steel Association, and from information kindly supplied to me by the Inman Steam Navigation Company, is instructive on the question before us:—

Value of Imports and Exports into and from the United Kingdom from and to the United States exclusive of Bullion and Specie, together with Rates of Freight on Pig Iron from England, and other particulars.

	1870.	1872.	1873.	1874.	1875.	1876.	1877.	1878.	1879.	1880.	1881.
Imports from U.S. in thousands of pounds	£	£ 49,804	£ 54,663	£ 71,471	£ 73,897	£ 69,590	£ 75,899	£ 77,825	£ 89,146	£ 91,818	£ 107,081	£ 103,267
Exports into U.S. in thousands		31,306	45,907	36,698	32,238	25,062	20,062	19,885	17,531	25,818	37,954	36,783
Freight average on pig		19/	16/	15/9	6/5	4/3	1/9	3/	2/6	4/6	10/	7/
Imports into U.S. of iron and steel in thousands of net tons		946	1,325	717	337	268	228	236	236	862	2,112	1,322

It will thus be perceived that in the year 1876 the weight of iron and steel imported into the United States was almost exactly one-sixth of that in 1872. At the same time the value of all exports sent out of the United States to the United Kingdom had increased about 40 per cent., while that of merchandise going in the other directions had fallen off to the extent of 44 per cent. This of course meant a

SECTION XVIII.—CHIEF IRON-PRODUCING COUNTRIES COMPARED. 605

greatly diminished weight of material to be transported to America, against a great increase in that coming from America to this country. Pig iron under such a state of things was considered as ballast; and the British iron-masters, instead of paying 24s. 2d. freight as they did in 1867, were only charged 1s. 9d. in 1876. In other words the American iron-masters lost 22s. 5d. per ton of the natural protection afforded to them by the cost of transporting British pig iron to New York.

A remarkable change took place all over the world in the value of iron of every description in the years 1872 and 1873, influenced in a great measure by the altered selling prices of these commodities in Great Britain.[1] The following figures will best illustrate the approximate position of the fluctuations in question:—

	1870.	1871.	1872.	1873.	1874.
Cleveland pig iron, No. 3, sold at Middlesbrough	50/3	54/6	92/-	103/9	74/6
Scotch pig iron, mixed numbers, at Glasgow	55/-	60/-	98/-	110/-	90/-
Iron rails, f.o.b.	130/-	135/-	240/-	260/-	200/-
Steel rails, f.o.b.	205/-	220/-	280/-	307/-	233/-
Coke at Middlesbrough	11/-	12/6	18/-	36/-	28/-
Locomotive coal at pits	6/6	6/9	9/6	14/-	14/-

The United States were unprepared for the sudden demand upon their resources, particularly in the matter of iron ore, which require time for their development; and in consequence very high prices ruled throughout the Union.

The producing powers in America, it should be observed, had been strained for some time previous to the period we are considering; for in the year 1865 the make of pig iron was less than one million tons. By 1871 the production had risen to nearly two million tons and in the following year it was close on three million tons. The vast mining district known as that of Lake Superior, was only in its infancy at that time, so that great difficulty was experienced in supplying the rapidly increasing number of furnaces with the ore they required. The same observation applied to coal and to labour; and as a natural consequence every item in the cost of making pig, as well as other descriptions of iron, increased enormously. The following

[1] The high price of iron and steel in Great Britain arose partly no doubt from the great demand which had sprung up in the United States.

figures will best illustrate the nature of the change. They set forth the costs of pig iron, as copied by myself from the books of a firm who kindly placed them at my disposal:—

Table showing the make of pig iron in the United States and the increased cost of production per ton in one particular locality.[1]

	1860.	1871.	1874.
Cost of ore per ton of iron	2·95	6·84	9·78
„ coal do.	6·46	8·72	11·07
„ limestone do.	·10	·29	·36
Wages per ton	1·72	2·60	2·75
Repairs, general charges, &c.	1·47	2·27	1·86
	$12·70	$20·72	$25·22

The great increase in the price paid by the smelters for the minerals they required led to a corresponding rise in the value of labour at the mines, some particulars of which will be reserved until a later period. Of course, so long as the selling price of pig iron remained at the point which justified the increase in the rates paid for the raw materials, nobody complained; but unfortunately, when the demand slackened, or perhaps more properly when the make was increased, prices fell, and yet the demand on the collieries and mines being still very active, coal, ore and labour were paid for at rates which left good profits to the mine owner and high wages to the labourer, but an actual loss to the smelter.

Everything which tends to inflated prices has always been, and probably always will be attended with the same results, viz., the undue attraction of fresh capital, succeeded by ruinous competition. Such has been our experience in Great Britain, and the same has fallen to the lot of the American iron-masters; but with the latter the evil was actually intensified by the impediments to natural and artificial foreign competition offered by high transport charges and an import duty of nearly 30s. per ton on pig iron. It cannot be a matter of astonishment that a competition, far more dangerous than that possible for any European nation to offer, was entered into by domestic speculators, when the enormous dividends paid by certain companies were a matter of sufficient publicity.

[1] The increase of the cost of pig iron is one of which I have the particulars, but the same change in point of character took place in all other districts.

SECTION XVIII.—CHIEF IRON-PRODUCING COUNTRIES COMPARED.

To pursue the history of the development of the pig iron trade in America, the make fell from 2,854,558 net tons in 1872 to 2,093,236 net tons in 1876, or by nearly 29 per cent.

It was not until 1879 that the production in the United States regained the position it occupied in 1872, the make in the former year having reached 3,070,875 tons. By this time there had arisen an extraordinary demand in the Union for iron required in the construction of new railways, so that not only did the home make rise between 1878 and 1882 from 2,577,361 net tons to 5,178,122 gross tons, but the imports from Great Britain were increased from 159,017 tons in the former year to 1,175,259 tons in 1881 and 1,211,710 tons in 1882. Freights at once felt the influence brought about by this change of trade: as high a rate as 15s. was paid in 1880, 10s. in 1881, and 11s. 6d. in 1882, for cargoes from Liverpool, while 17s. 6d. in 1880 and 15s. in 1882 were the average charges from Glasgow to New York. That these quotations did not reach a still higher point was, no doubt, due to the enormous quantities of merchandise brought from the United States to British ports. This for each of the three years ending 1881 was more than double what it was in 1870, viz. above 100 millions sterling instead of about $49\frac{3}{4}$ millions.

Now it is needless to say that such rates of freight as were paid on pig iron in 1877 and 1878, viz. 3s. and 2s. 6d., involve, so far as the actual expenses of navigation from the British Isles to New York are concerned, a very great loss to the shipowner, which loss is reimbursed by an increased charge for transport on the return cargo.

To show how this operates, I have compared two years, viz. 1877, when the freight on pig iron averaged 3s. from Liverpool and Glasgow, with 1881, when it was 11s. 6d. from the former port, and 9s. from the latter :—

Average Freight from New York to Liverpool for the Years 1877 and 1881.

	Grain. Per Bush., 60 Lbs.	Cotton. Per Lb.	Cheese. Per Ton.	Bacon. Per Ton.
1877	6·37d.	·26d.	40s. 8d.	30s. 8d.
1881	4·28d.	·19d.	23s. 9d.	19s. 7d.
Decrease of 1881 compared with 1877	Per Cent. $48\frac{1}{4}$	Per Cent. 40	Per Cent. $71\frac{1}{2}$	Per Cent 56

These increased costs of transport of course come out of the pockets of the agricultural portion of the population of the United States. This statement however does not necessarily involve the supposition that under any system of international commerce between the American Union and the United Kingdom, a higher charge for transport would not fall to the lot of the agriculturist, because with our enormous demand for food, *so long as that is supplied from the United States*, the actual weight brought to our shores will, in all probability, greatly exceed that leaving Great Britain for American ports. Burthened, however, as the American farmer is by heavy transport charges, rendered still heavier by legislative enactment, he may surely be expected to demur to having to pay an import duty of 70s. 10d. per ton on steel rails supplied by his British customers, and £5 12s. per ton on any wire fencing he may require on his farm.

Returning now to the question of royalty, a payment of some kind may in certain cases have to be borne by the consumer, whether received by the landowner or not. The value of the mineral may be raised by the supply being insufficient as compared with the demand: or in the case of a district like Cleveland, whose mineral resources are practically unlimited, mere geographical position in relation to the fuel confers a value upon certain districts not enjoyed by those more remote from the coal-field. If two mine owners pay the same royalty, the difference of cost of carriage from one furthest from the point of consumption may probably mean so much more gain to the better placed mine.

The same observations which a few pages back have been made on the rise in value of lands containing iron ore apply also to those containing coal. As already stated, thousands of acres in comparatively undeveloped States of North America can still be purchased for what may be regarded as a nominal sum. In other localities, like the anthracite region of Pennsylvania, as much as 60 cents (2s. 6d.) is paid as royalty on each ton of this valuable fuel. This however may be considered as a very high rate for coal, and is confined to anthracite; for I was informed in Pittsburg that the rent paid by my informant there on bituminous coal was only 6d. per ton.

SECTION XVIII.—CHIEF IRON-PRODUCING COUNTRIES COMPARED. 609

Notwithstanding the difficulty of assigning a correct position to the royalty rents, in a general estimate of the cost of iron, it may be interesting, having partly dealt with the question in reference to Great Britain and the United States, to record in the present place the payments under this head in the three iron-producing countries on the Continent of Europe.

ROYALTIES PAID IN FRANCE.

All ironstone, save that belonging to alluvial deposits, which latter are of insignificant importance, and all coal, became under the law of 1810 the property of the State.

Concessions to applicants are granted and worked on payment of a fixed rent of 10 francs—say 8s.—per square kilometre, about 245 acres. In addition to this small payment an impost is paid, which does not exceed 5 per cent. on the *profits*. If then the latter amounted to as much as 2s. per ton, the entire sum paid by the mine owner would be below 1¼d. per ton of coal and ore.

ROYALTIES IN BELGIUM.

In the case of iron ores an opposite course to that pursued in France seems to have been adopted in Belgium; for in 1830 property in such ores was transferred from the State to the owners of the soil. By the proprietor leases are granted, on payment of royalties varying from 4¾d. to 2s. (·50 to 2·50 francs) per ton.

Coal remains the property of the State. Leases are granted on a yearly payment of 10 centimes (less than 1d.) per hectare (2 acres 1 rood 37 poles) together with 2½ per cent. on the net annual profit. If this amount to 2s. per ton, the royalty would be less than ¾d. per ton.

ROYALTIES IN GERMANY.

Coal and iron ore are vested in the State. The royalty paid on the former by the colliery owner is 2 per cent. on the profits. Hence if the selling price is 6s. per ton the royalty would probably be less than 1½d. Iron ore pays no royalty.

According to the information obtained in the three countries just named, damage done to the surface in working the minerals is made good by a payment to the landowner.

M M

610 SECTION XVIII.—CHIEF IRON-PRODUCING COUNTRIES COMPARED.

The charge upon one ton of pig iron for royalty on ore and coal, in the four European nations just described, is roughly as follows:—

Great Britain			Germany.	France.	Belgium.
Cleveland.	Scotland.	Cumberland.			
3s. 3d.	6s.	6s. 3d.	6d.	8d.	1s. 3d. to 4s.

Having thus dealt at some length with questions of a somewhat general character, it is now intended to treat the subject of the present section in its more direct application to the four countries under consideration. The seats of the iron trade of most nations are too numerous to permit their all being examined: indeed for the object I have in view this is not necessary. In some districts the manufacture of the metal, though expensive, may be carried on profitably for purposes of local consumption, because the transport charges from places of cheaper production may more than cover the difference of cost. A description of works embraced in this latter division would greatly exceed the limits laid down for my guidance; and I shall therefore content myself with references to those localities whose facilities for the manufacture of iron are most typical of the four countries, or where the advantages they possess entitle them to be considered more or less from an international point of view.

In pursuance of this object, it is proposed to give some figures showing the comparative prices and values of the minerals used in the manufacture of pig iron. The crude metal itself is selected as the means of illustration, because upon its cheap production the position of a district or nation is mainly dependent.

COAL.
GREAT BRITAIN.

Middlesbrough.—The coal which is employed to smelt the ironstone got from the Cleveland hills is conveyed over distances of 20 to 30 miles, chiefly from the pits of South Durham. Much of it is obtained from shafts having a depth of 100 fathoms or less, and the measures are not, as a rule, much troubled with serious dislocations or with water. Two seams, the "Brockwell" and the "Busty," are practically all from which fully $5\frac{1}{4}$ millions tons of coal are thus annually wrought for the use of the Cleveland smelters; and as the former is known to be the lowest of the beds in this North-Eastern district, no future discovery can add to their number. From information I have received,

SECTION XVIII.—CHIEF IRON-PRODUCING COUNTRIES COMPARED.

I conclude that, with the exception of Scotland and some few instances in South Yorkshire and Lancashire, the present average cost of raising a ton of coal in the County of Durham is as low as it is in any part of the United Kingdom. In the important question of purity the Brockwell seam will, as a whole, bear comparison with the coal of any part of the world. On the other hand the Busty seam, at all events in the Southern division, not only contains more sulphur, but its earthy constituents are occasionally so large in quantity as to require the coal to be crushed and washed.

The following analyses exhibit the composition of these two beds, as worked in the Brancepeth district of South Durham:—

	Brockwell. 1.	Brockwell. 2.	Busty. 1.	Busty. 2.
Carbon	81·47	80·45	81·03	79·11
Hydrogen	4·57	4·29	4·59	4·49
Oxygen and nitrogen	6·45	9·55	7·46	8·67
Sulphur	1·22	1·22	1·30	1·29
Water	1·23	1·10	·70	1·01
Ash	5·51	3·84	5·40	5·91
	100·45	100·45	100·48	100·48
Of the carbon the quantity in the fixed state is	72·89	71·90	70·17	66·32
Volatile above 212° F.	20·84	23·40	23·73	26·76
„ below 212° F.	1·23	1·10	·70	1·01

By computation based on the foregoing analyses, the coke made from the Brockwell ought to contain somewhat less ash and sulphur than that obtained from the Busty seam. As a matter of fact this is so, but the inferiority of quality of the coke so obtained from the two seams in the Brancepeth district, is greater than the difference of ash and sulphur would appear to indicate. This I conceive to be due to the less compact nature of the coke made from the Busty, permitting its reaction on the carbonic acid present in the furnace in the manner already referred to in these pages. Generally speaking, the produce of the two beds is mixed for the coke ovens; and the average composition of two samples so made, as received at the Clarence works during one month, was as follows:—

	No. 1.	No. 2.
Ash	6·48	6·72
Sulphur	·95	·97
Moisture	·30	·28

The fuel used in the Cleveland furnaces is invariably in the form of coke, and the average cost of conveying it from the pits to the works is a little above 2s. per ton.

The following are the average selling prices of coke delivered at Middlesbrough, as furnished me by an eminent house in the trade:—

1870.	1871.	1872.	1873.	1874.	1875.	1876.
s. d.	s. d.	s. d.	s. d.	s. d.	s. d.	s. d.
11 0	12 6	18 0	36 0	28 0	17 6	14 0
1877.	1878.	1879.	1880.	1881.	1882.	1883.
s. d.	s. d.	s. d.	s. d.	s. d.	s. d.	s. d.
13 2	12 7	11 6	12 7	11 8	12 3	12 4

The selling prices of best screened steam coal at the pits in Northumberland were as follows:—

1870.	1871.	1872.	1873.	1874.	1875.	1876.
s. d.	s. d.	s. d.	s. d.	s. d.	s. d.	s. d.
6 3	6 6	9 6	14 $1\frac{1}{2}$	14 $1\frac{1}{2}$	11 0	9 9
1877.	1878.	1879.	1880.	1881.	1882.	1883.
s. d.	s. d.	s. d.	s. d.	s. d.	s. d.	s. d.
8 $10\frac{1}{2}$	7 $7\frac{1}{2}$	6 $1\frac{1}{2}$	6 9	6 9	7 $1\frac{1}{2}$	7 3

Coal in Scotland.—The ironworks in the neighbourhood of Glasgow are perhaps exceptionally favoured in the cost and quality of the celebrated "splint" of the coal-field in which they are located. The pits are of a moderate depth, the water is not excessive in quantity, their produce is cheaply wrought, and the quality is all that could be desired for the blast furnace. The following analysis shows the composition of an average sample of this splint, brought from one of the leading works in Lanarkshire, as determined in the Clarence laboratory:—

Carbon	66·00
Hydrogen	4·34
Oxygen and nitrogen	12·03
Sulphur	·59
Water	11·62
Ash	5·42
	100·00
Fixed carbon	53·41
Volatile above 212° F.	28·90
„ below 212° F.	11·62

SECTION XVIII.—CHIEF IRON-PRODUCING COUNTRIES COMPARED. 613

The following analysis made in Glasgow of coal used in another Scotch ironwork shows the fixed carbon as under:—

Volatile.—Gas, tar, &c.	34·17	
Sulphur	·32	
Water expelled at 212° F.	6·82	
		41·31
Non-volatile.—Fixed carbon	55·14	
Sulphur left in coke	·38	
Ash ,, ...	3·17	
		58·69
		100·00

As a source of heat development this splint coal is greatly inferior to the Brancepeth coals, as described page 611. This will appear from the following estimate, containing the respective values of the two coals in centigrade calories:—[1]

	Brockwell.		Busty.		Scotch Splint.
	No. 1.	No. 2.	No. 1.	No. 2.	
C to CO_2	6,518	6,436	6,482	6,329	5,280
H to H_2O	1,554	1,459	1,560	1,527	1,476
Total developed..	8,072	7,895	8,042	7,856	6,756
Heat absorbed—					
Expulsion of H_2O	6	6	4	5	63
,, hydro-carbons, &c. ...	417	490	475	535	578
	423	496	479	540	641
Actual development..	7,649	7,399	7,563	7,316	6,115

Taking the mean of the four Durham coals, viz. 7,481 calories, it would appear that they possess a higher heating power by 22 per cent. than the Scotch splint.

So far as the mere composition of the above-mentioned specimens of coal is concerned, it might be inferred that the four from Durham would be preferred to the Scotch splint for use as raw coal in the blast furnace. The average content of fixed carbon in the former is 70·32

---

[1] The factors used in these calculations are:—

Heat evolved by carbon burnt to carbonic acid ...	8,000 cal.
Do. by hydrogen burnt to water	34,000 ,,
Heat absorbed by expulsion of hydro-carbons, &c., including deficiency in heat development of C and H through combined O	2,000 ,,

per cent., against 53·41 in the Scotch, so that the Durham coals contain close on 32 per cent. more than that from the Ayrshire pit. However, owing to its bituminous nature the English coal would make the contents of the furnace cohere together; the materials would thus be caused to descend by slips, and so prevent the uniform passage of the gases through the materials. On the other hand lumps of the Scotch splint coal ignite and burn apparently without any adhesion of one piece to its neighbour. Dry burning as this coal seems to be, when it is pounded and exposed to heat in a crucible, the mass unites and forms a tolerably compact piece of coke. To some extent this process of agglomeration happens in the blast furnace, for in Scotland there is more hanging of the materials than where coke is employed. To such an extent does this occur that the lofty furnaces in use at Middlesbrough have been found impracticable, particularly with those mixed qualities of coal which iron-masters, for want of the best splint, are now obliged to employ. Where large furnaces are still in blast, they are, as was stated page 316, not kept filled to the top. Nor does it indeed appear from later experience, that much is to be gained from any great addition beyond a height of 51 feet; for it will be observed that the average temperature of the escaping gases (*vide* p. 316) is, in both the cases given, under 400° F. This cooling effect on the gases is doubtless due to the gasification of the hydro-carbons and water contained in the raw coal. It is true that the low ratios of carbon as carbonic acid to that as carbonic oxide (1 to 5·28 and 1 to 5·32, as stated in page 317, instead of 1 to 2·32 or thereabouts, as in a coke furnace), would suggest that the reducing gases are not kept long enough in contact with the oxide of iron. It must moreover be borne in mind that the low temperature of these gases is inconsistent with rapid action on the ore.

The difference in the richness of the ores used in Cleveland and Scotland is such that the slag from the furnaces of the former requires, from its larger weight, one-half more heat for its fusion than that from those of the latter (15,356 cal. against 10,054 cal.—*v.* pp. 126 and 319). Roughly this difference may be regarded as equivalent to 1 cwt. of carbon per ton of metal produced. Notwithstanding this difference in favour of the Scotch furnace, and notwithstanding that ·54 cwt. of hydrogen appears to be oxidised in its gases (*v.* p. 320),

SECTION XVIII.—CHIEF IRON-PRODUCING COUNTRIES COMPARED.

yet, in the case given p. 318, 22·65 cwts. of fixed carbon accompanied with 5·34 cwts. of volatile carbon were consumed per ton of No. 1 iron. If we assume 23 cwts. of coke as being required for this quality of Cleveland pig, we have about 21·40 cwts. of carbon to produce one ton of the metal.

In the matter of economy of money, as will hereafter appear, the Scotch smelters enjoy a marked advantage over those of Cleveland. In the meantime it may be observed that, owing to the waste of fixed carbon in the coking process, a larger quantity of this constituent is consumed per ton of iron in Cleveland, than in a Scotch furnace using splint coal of the composition formerly given.

	Cwts.
23 cwts. of coke is assumed as the weight required to produce one ton of Cleveland No. 1 pig iron. Raw coal reckoned to yield only 65 per cent. of coke, although, including ash, it contains more than 75 per cent., gives ... ... =	35·38
Suppose the average percentage of fixed carbon in South Durham coal to be 71, then 35·38 cwts. will contain of fixed carbon...	25·11
The weight of raw coal used in the Scotch furnace, burthened to produce one ton of No. 1 pig, was 42·31 cwts., containing 53·41 per cent. of fixed carbon, equal therefore to fixed carbon	22·60
Excess of fixed carbon consumed in the Cleveland furnace ...	2·51

It has to be remembered however, that in smelting Cleveland iron there is a larger weight of slag to be melted than in the Scotch furnace above referred to. In one case the additional heat absorbed for this cause alone was 6,648 calories, viz., the difference between 16,702 calories and 10,054 calories.

Now at page 244 it was shown that with the blast heated to 1,045° F. each unit of coke evolved, in a furnace 80 feet high, 3,785 calories, or say 4,000 for each unit of pure carbon. From this we have $\frac{6,648}{4,000}$ or 1·662 cwts. of fixed carbon for fusing the additional slag generated in the furnace smelting Cleveland stone. This reduces the apparent difference to (2·51 − 1·662) or ·848 cwts.

616 SECTION XVIII.—CHIEF IRON-PRODUCING COUNTRIES COMPARED.

In the estimate just given the effect of two very different qualities of coal has been compared. It may interest some to compare the same kind of coal, as used in its raw state and in the form of coke.

For this purpose we will assume a coal of the average composition of the four Brancepeth coals given above, viz.:—

Fixed carbon	70·32
Water	1·00
Volatile above 212° F.	23·68
Ash and sulphur	5·00
	100·00

That portion of its fixed carbon which is really effective for iron smelting is arrived at, if from the 70·32 per cent. we subtract so much as is needed to gasify the volatile constituents; and here we must assign to the unit of C a calorific power of only 4,000[1] as it is burnt under the disadvantageous conditions of the blast furnace.

	Calories
The expulsion of 1·00 H_2O requires	540
,, 23·68 other volatile matter requires 23·68 × 2,000 =	47,360
Total	47,900

The carbon needed to supply this heat is $\frac{47,900}{4,000} = 11\cdot97$ units; so that the portion of the fixed carbon really effective for iron making is only $(70\cdot32 - 11\cdot97) = 58\cdot35$ per cent. of the coal.

Assuming now that 23 cwts. coke, containing 92 per cent. or 21·16 cwts. carbon, are needed to make one ton of pig, its equivalent of coal will be so much as contains the same weight of effective carbon, viz., $\frac{21\cdot16 \times 100}{58\cdot25} = 36\cdot32$ cwts.

---

[1] The assumption of 4,000 calories by the combustion of each additional unit of carbon is founded on the supposition that each such unit is oxidised to the same extent as that which preceded it. This is however by no means certain, or even probable, but any difference arising from this cause is scarcely worth consideration. It may also be observed in this place that it is assumed in the calculations, that we have to deal with a coal capable of being used either as coke or as raw coal in the furnace. This, as is well known, is very far from being the case.

SECTION XVIII.—CHIEF IRON-PRODUCING COUNTRIES COMPARED. 617

The constituents of the coal are, in the two cases, disposed of as follows:—

	In Coke Furnace.	In Coal Furnace.	Difference in favour of Coke Furnace.
Volatile matters expelled—			
In coke oven	8·66	...	...
In blast furnace	...	8·88	...
Ash and sulphur melted in blast furnace	1·84	1.90	...
Fixed Carbon—			
Expelling volatile matters in coke ovens	3·72	...	...
Do. in blast furnace	...	4·38	...
Smelting ore in blast furnace	21·16	21·16	...
Total fixed carbon	24·88	25·54	·66
Total coal required	35·38	36·32	·94

According to this mode of comparison it seems practically a matter of indifference, so far as the furnace is concerned, whether the coal be coked or used raw; for we have 35·38 cwts. of coal required to give the 23 cwts. of coke, and 36·32 cwts. of coal calculated as the weight necessary to smelt the iron and to gasify the volatile matter, when employed raw in the furnace itself.

There is, of course, a much greater loss from radiation, &c., in the coke oven than in the blast furnace, but this appears to be compensated by the fixed carbon burnt, as well as by the more complete oxidation of both carbon and hydrogen burnt during the process of coking.

So far the question has been regarded as one of heat evolution only, but the practical smelter has to deal with its financial aspects, rather than with the mere calorific power of the two kinds of fuel. For coking the coal a large and expensive plant is required, representing not less probably than £3,600 for each blast furnace—so that with interest, wear and tear, and wages, each ton of pig iron may involve a cost of 1s. 6d. for converting the coal into coke. From this charge a very variable allowance has to be made by the opportunity afforded of separating the larger coal and selling it at a higher price.

The use of raw coal in smelting iron permits a far more complete utilisation where a use can be found for them, of the volatile parts, than can be effected in the coke oven. In the latter case I have estimated that the heat obtained from each ton of coal coked represents about 6d. on the ton of iron made, whereas the 8 cwts. or so of gases distilled from the coal in the furnace may be safely taken at 2s.

There is a further important matter for consideration, viz., the relative facilities possessed by the two systems of collecting the bye products, viz., the ammonia and tar.

Messrs. Bairds at Gartsherrie have demonstrated the practicability of condensing these substances from the waste gases of their furnaces using raw coal. The same object, as already has been mentioned, has been attained at the coke ovens, and in these undoubtedly the process ought to be more easily accomplished than in the furnace; because from a furnace using raw coal the volume of gas will be about thirteen times as great as that from ovens making coke to produce a similar quantity of iron.

An important point in connection with this branch of the subject is the cost of plant for the condensation of the ammoniacal gas and tarry matters. A sum of £16,000 has been named as being required to separate these substances from the escaping gases of blast furnaces capable of making 500 tons a week. So far as my own experience enables me to judge, one-third of this sum will represent the extra cost of the coke ovens required for effecting the same object.

The natural conditions under which coal is found in Scotland enable it to be wrought at a very low cost: indeed I heard of it as being sold for locomotive purposes at 3s. 9d. to 4s. per ton.

These natural advantages are further supplemented by the cheaper rates of labour which prevail in the iron making districts of Scotland. The average daily earnings of the entire staff at one colliery were given me in 1879 at 2s. $10\frac{1}{2}$d. In the Ell coal a hewer can work 4 tons in the whole coal, and as much as 7 tons in pillars; in the main coal he can get 3 tons and 4 tons respectively, and in the splint $2\frac{3}{4}$ and $3\frac{1}{2}$ tons respectively. The average earnings of the hewers at the period in question were 3s. $1\frac{1}{4}$d. per day in the Ell coal and 2s. $10\frac{3}{4}$d. per day in the hard coal (main and splint), the men paying for house rent. At another ironwork the colliers earned from 3s. 1·51d. to 3s. 5·56d. per day, and the drawers 3s. to 4s., working $5\frac{1}{4}$ days in the week; and notice had been given of a reduction of 10 per cent.

SECTION XVIII.—CHIEF IRON-PRODUCING COUNTRIES COMPARED. 619

Cumberland and Lancashire.—Coal is found in both these counties, in the latter in very large quantities. That obtained from the Cumberland pits, although in immediate proximity to some of the blast furnaces, is only sparingly used for smelting purposes. The coke made from the coal, as drawn from the pit, would contain as much as 12 to 14 per cent. of ash; hence it requires washing, but even then the coke is often associated with $1\frac{1}{4}$ per cent. of sulphur. From whatever cause, the blast furnaces in both counties use almost exclusively coke produced from the coals of Durham and Northumberland, which is brought to the ironworks on the West Coast at a cost for carriage varying from 5s. 5d. to 7s. 6d. per ton.

Coal in South Staffordshire.—The quality of the coal obtained in South Staffordshire adapts it admirably for the purposes of the iron-smelter. The large demands however levied upon its resources for many years past have materially increased its present value in the market. The general practice forty years ago was to coke it for blast furnace use, but now it is largely used in its raw state. The operation was performed in the open air, with no building beyond a temporary central erection of loose brick work, to act as a chimney. Of course a large waste of fuel attended such a rude course of procedure, and the money loss was further increased by the exclusive use of large coal, which in the year 1871 I found was selling at 10s. to 12s. per ton when used for other purposes than that of making pig iron. Its present market value I am informed is 9s. 6d. For many years past raw coal has been almost exclusively used in the blast furnaces of South Staffordshire, and when coke is employed it is generally brought from other coal-fields such as those of South Wales.

The furnaces in this old seat of the iron trade are of moderate dimensions, nevertheless I have heard of as small a quantity as 31 cwts. of raw coal having sufficed to produce a ton of grey iron. It would however not surprise me, iron and coal being measured by the same standard, if 40 cwts. were nearer the mark. This indeed, according to the Government returns, which will be referred to at a later period, appears to be the actual rate of consumption, which includes the coal used for calcining and other purposes at the furnaces.

Coal in North Staffordshire.—The beds of coal in this division of the county vary from 3 to $7\frac{1}{2}$ feet in thickness. A hewer, I was informed, cuts from 3 to 4 tons for his day's work, and the daily production for the whole staff is about 22 cwts. per individual employed—equal therefore to something over 350 tons per annum. The royalty paid in one case was 10d. per ton, and the total expense in delivering the coal at the pit's mouth is pretty nearly the same as that in the County of Durham. The average selling price in the latter, according to the last ascertainment, was 4s. 7½d. per ton—a figure stated by the coal-owners as being unremunerative.

The furnaces in North Staffordshire are usually greatly inferior in dimensions to those in use at Middlesbrough; the height does not usually exceed 50 or 60 feet. In one case I found one of 70 feet; but the increased size did not appear to be accompanied by any economy in the consumption of coal.

The greater part of the fuel is used raw, when 36 to 38 cwts. are consumed to make a ton of iron. A portion of the small coal is coked, in which event the weight required may be taken at 29 cwts. of raw coal and 7 cwts. of coke per ton of forge iron. In such cases 40 cwts. of a mixture of calcined black and clay band ironstone and 6½ cwts. of limestone go to produce one ton of pig.

Coal in South Wales.—According to the evidence of the late Mr. William Menelaus, given before a committee of the House of Commons about the year 1866, 25 cwts. of the Dowlais four-feet upper vein were capable of smelting a ton of No. 3 foundry iron from an ironstone yielding only 38 per cent. of pig iron. Indeed Mr. Menelaus stated that occasionally 20 cwts. of this exceptionally fine coal sufficed for the purpose. I had an opportunity afforded me of examining the Dowlais furnace books, and there it was stated that for 13 consecutive weeks the average consumption had been 19 cwts. 3 qrs. 13 lbs.

The composition of this seam was given as follows:—

Carbon fixed and volatile	89·33
Hydrogen	4·43
Nitrogen	1·24
Oxygen	3·20
Sulphur	0·55
Ash	1·25
	100·00

SECTION XVIII.—CHIEF IRON-PRODUCING COUNTRIES COMPARED.

The carbon in the fixed state was stated to be 80 per cent. of the whole weight; and at Cyfarthfa, a neighbouring work, the non-volatile carbon amounted to 77 per cent.

I am without any precise data upon which to found an exact estimate of the quantity of heat required to smelt such an ironstone as that in use at Dowlais at this time. Applying however the same rules as those applied with success to the treatment of other ores, I doubt whether it would be safe to adopt a less weight of coal than the larger one mentioned by Mr. Menelaus, viz., 25 cwts., and even then this must be regarded as very good work.

At the period referred to by Mr. Menelaus the valuable vein of coal just described must have been far from sufficient for the Dowlais requirements, because for the production of a ton of forge iron $32\frac{1}{2}$ to 35 cwts. of raw coal were being used, although the mixture of ore yielded 41·75 per cent. of pig iron instead of 38 per cent. as in the former case. Indeed Mr. Menelaus himself gave 38 cwts. of coal as the usual rate of consumption in the South Wales blast furnaces.

Bessemer pig iron is now the chief object of production in this division of the Principality. The fuel usually employed is coke, made from a more bituminous coal than that of the four-feet vein. A common rate of consumption is 22 to 23 cwts. per ton of pig, although there are instances of 20 cwts. sufficing for the purpose.

The practice of giving the consumption of fuel in the production of pig iron, reckoned in the form of raw coal, was commenced by my friend Robert Hunt, F.R.S., in the Mining Records. The plan has also been adopted in the continuation of his excellent work, prepared by Her Majesty's Inspector of Mines. From these figures the Table at page 622 has been prepared. The results will not be found always to correspond with those already given in this work. This partly arises from the latter representing exclusively the fuel used *in* the blast furnaces, whereas the official returns included any coal used in calcining and at the boilers, &c. The Government report also deals with the average of the whole, while mine gives the consumption of particular works.

SUMMARY OF THE PRODUCTION OF PIG IRON IN GREAT BRITAIN IN THE YEAR 1882, AND OF THE COAL USED (INCLUDING COAL CONVERTED INTO COKE) IN ITS MANUFACTURE.

ENGLAND AND WALES—	Number of Furnaces. In Blast.	Number of Furnaces. Out of Blast.	Pig Iron Made. Tons.	Coal used. Tons.	Yearly Make per Furnace. Tons.	Coal per Ton of Pig. Cwts.
Cumberland	45	10	1,001,181	1,826,393	22.249	36·48
Denbighshire and Flintshire	6	4	53,138	154,888	8,856	58·30
Derbyshire	40	13	372 650	930,360	9,316	49·93
Durham	34	36	815.671	1,634,584	23.990	40·08
Glamorganshire	27	64	404,350	853,642	14,976	42·22
Gloucestershire, Hampshire, Somersetshire, and Wiltshire	4½	8½	52,991	108.721	11,775	41·03
Lancashire	35	14	790,999	1,345,863	22,600	34·03
Lincolnshire	17	4	201,561	464,600	11,856	46·10
Monmouthshire	35	13	530,084	991,969	15,145	37·42
Northamptonshire	15	11	192,115	445,300	12,807	46·35
Northumberland	4	3	93,422	143.237	23,355	30·66
Nottinghamshire	5	—	73,085	182,206	14,617	49·86
Shropshire	10	14	80,475	240,000	8,047	59·64
Staffordshire, North	25	12	275,577	542,572	11.023	39 38
Staffordshire, South	31¼	61½	247,667	558,966	7,862	45·14
Worcestershire	16	24	150.776	399,416	9,423	52·98
Yorkshire, North	82	9	1,803,508	3,694,857	21,994	40·97
Yorkshire, West	28¼	18¾	321,430	726,482	11,377	45·20
TOTAL, ENGLAND AND WALES	460¼	319¾	7,460,680	15,244,056	16,210	40·86
SCOTLAND—						
Ayrshire	28½	14½	350,423	731,004	12,295	41·72
Linlithgowshire, Lanarkshire, and Stirlingshire	81½	24½	775,577	1,821,241	9,516	46·96
	110	39	1,126.000	2,552,245	10.236	45·33
TOTAL PRODUCTION OF GREAT BRITAIN	570¼	358¾	8,586,680	17,796,301	15,058	41·45

Using the information contained in this Government report I would now compare the actual work performed by the fuel in some of the chief centres of the iron manufacture in Great Britain and elsewhere. In coming to any conclusion upon the subject a mere statement of the coal consumed per ton of pig iron produced does not convey all that is required. The richness of the ore and the quantity of limestone needed as a flux must also be taken into the account.

SECTION XVIII.—CHIEF IRON-PRODUCING COUNTRIES COMPARED.

In the short Table which follows will be found the combined weights of ore and flux, together with the official return of the coal used. From these two factors the quantity of these two ingredients smelted by one ton of the British raw coal has been estimated, and to the results thus obtained others, from private sources of information, have been added to exhibit the duty performed by foreign fuel:—

HEMATITE PIG IRON, CHIEFLY BESSEMER.

	Cwts. of Ore and Limestone per Ton of Metal.	Cwts. of Raw Coal per Ton of Metal.	Cwts. of Materials Smelted per Ton of Coal.
Northumberland	45·50	36·00	25·27
Cumberland	46·00	36·48	25·22
South Wales	48·00	39·50	24·30

PIG IRON FROM OOLITIC IRONSTONE, AVERAGE ABOUT No. 3·50

Cleveland[1]— foundry	59·00	44·00	26·82
Lincolnshire ,,	62·22	46·08	27·00
Northamptonshire ,,	55·43	46·35	23·92

PIG IRON FROM CLAY IRONSTONE.

North Staffordshire—forge	48·50	36·00	26·94
Scotland—chiefly No. 1	48·39	42·39	22·83

FOREIGN COAL.

Westphalia—Bessemer	53·26	37·28	28·54
Do. do.	54·00	45·47	23·75
Belgium do.	56·25	27·30	41·20
France do.	44·90	34·76	25·54

In the case of the Belgium furnace the yield of coke from the coal was 74 per cent.

COAL IN GERMANY.

Coal in the Rhenish Provinces.—The weight of coal given me as hewn by each miner in eight hours actual work in Rhenish Prussia was 43 cwts. For this, in 1878, the hewer earned 2s. 9d., without house or firing, and the average weight extracted by the entire staff of the mine was about 17½ cwts. for a day's work, the average earnings for the year in question being 2s. 4d. The selling price at the pit was stated to be 5s. 2½d. to 5s. 5d.

---

[1] The Government returns include hematite pig. For the purpose of the estimate 23 cwts. of coke has been assumed as being the average consumption for Cleveland iron alone, to which 3·75 cwts. of coal has been added for calcining and other purposes.

per ton, which frequently left only a moderate profit. The average ash in the coal of these districts is considered to be 10 per cent.; or about twice the amount contained in the coking coal of Durham. At the market value, say 5s. 3d. per ton, the cost of coke from unwashed coal was mentioned as being 9s. 3d. at the furnaces, where the ovens are usually situated. From washed coal it would of course be something more. The older furnaces are of very moderate dimensions— about 50 to 54 feet in height; and at these $24\frac{1}{2}$ to $29\frac{1}{2}$ cwts. of coke are used per ton of white sparry (strahliges) iron, 28 cwts. for common white pig, and often above 30 cwts. for Bessemer iron. The sparry iron contains 2·6 of carbon, ·13 of silicon, and ·8 of phosphorus. With furnaces of 71 to 72 feet in height, blown with air heated in fire-brick stoves to a temperature of low redness in the dark, the coke consumed was 24 to 26 cwts. per ton of Bessemer pig, the make being at the rate of 560 tons per week. The mixture of ore was such that $41\frac{1}{4}$ cwts. of it and $11\frac{1}{2}$ cwts. of limestone were required to produce one ton of metal.

In 1882 the cost of coal near Dortmund was given at 5s. 2d. to 5s. 5d. per ton. The hewers were earning 4s. 6d. to 5s. per shift of eight hours' actual work. The average output of about 700 men, all told, was given in one pit at 19 cwts.; in a second, 18·9 cwts.; and in a third, 20 cwts. per diem. The average earnings of the whole staff were £40 per annum, probably, therefore, about 2s. 9d. per shift.

In 1883 the selling price, in the Dortmund neighbourhood, of coal suitable for the manufacture of coke, and yielding 70 per cent. of coke, was given at 7s.

What has been said goes to prove that the coke of the Rhenish Provinces is decidedly inferior to that employed in the northern counties of England, the difference being about 25 per cent. in favour of the latter. The pits being usually nearer to these German furnaces than the Durham collieries are to Middlesbrough, an average saving of probably 1s. per ton in carriage is enjoyed by the former; against which something must be allowed for inferiority of quality.

Coal at Saarbrück, Prussia.—This district is an important seat of the German coal trade, its annual produce being four to five million tons. The mines are, I believe, chiefly worked by the Government, and in 1878 the selling price of small coal fit for coking was quoted as having been 4s. $6\frac{1}{2}$d. at the pit; but the coal as worked, large and small, commanded double this price.

SECTION XVIII.—CHIEF IRON-PRODUCING COUNTRIES COMPARED.

The average annual output over the five years ending 1877, per man employed at the mines, was stated as having been 196 tons per annum. The daily earnings during the period in question, according to a printed statement to which I had access, ranged between 2s. 10½d. and 3s. 4½d. per day; the average of the five years being 3s. 0½d. The wages per ton amounted to 4s. 5½d., and the selling price of the entire produce varied from 8s. 4d. in 1877 to 16s. 8d. in 1873, the average for the five years being 12s. 1d.

From another quarter the following information was received:—

	1873. Tons.	1874. Tons.	1875. Tons.	1876. Tons.	1877. Tons.
Production per man employed	204	194	200	196	187
	s. d.	s. d.	s. d.	s. d.	s. d.
Wages per ton raised	4 9¾	5 2	4 7¼	4 5¼	4 2½
Ditto (another quotation)	4 6¼	4 10	4 4	4 5¼	4 2½
Weekly earnings per man employed	18 10¾	19 4¼	17 9½	16 6	16 0
Daily do. (another quotation)	3 4½	3 5¾	3 2¼	2 11½	2 10¼
Selling price of coal, per ton	16 8	14 10½	11 0¼	9 7¾	8 4

Speaking generally, it may be assumed I think that in point of quality all the coke produced in the western part of Germany is inferior to that obtained from the North-Eastern coal-field of England.

The small coal of the Saarbrück district is impure, containing as it sometimes does 20 per cent. of ash, so that much of it has to be washed, when it affords a coke containing from 8 to 12 per cent. of ash. The loss of weight in washing amounts to nearly one fourth, and the yield of coke is usually about 68 per cent. of the coal; but it is stated that as much as 80 per cent. is obtained in the Appolt oven. In some cases a mixture of washed and unwashed coal is coked, giving a product containing 15 per cent. of ash.

With small coal bought at 4s. 6½d. and washed, the coke costs about 12s. at the pits, which, as compared with the County of Durham, is rather a high price, particularly when we bear in mind that the German coke contains 11 to 12 per cent. of ash. Notwithstanding this high rate of impurity, in smelting Luxemburg ore in a furnace of 48 feet by 17 feet, 20·80 cwts. suffice for the production of a ton of white iron, the blast being only heated to 350° C. (662° F.) The power of getting such good duty from so small a furnace is greatly due to the easily reducible character of the Luxemburg ore.

I imagine we may assume that the cost of 12s. per ton is as low a price as coke can be made at, in any large quantity, in the Saarbrück coal-field, even if the firms worked their own collieries; because for any increase in the production of coke, the large as well as the small coal would have to be used, and it would manifestly be greatly to the iron smelter's disadvantage if he had to make coke from large coal, which probably before it was crushed and washed would cost at its market value, even in the cheaper years, 10s. to 11s. per ton.

It is so long since I visited the collieries of Prussian Silesia, that I am unable to speak positively as to the quality of the coal of that part of the German Empire. My own impressions, however, confirmed by recent enquiry, do not lead me to think that it is much if indeed at all superior to the coal of Western Germany just described.

Coal in France.

Of the five nations we are considering, none is so poorly provided in the matter of coal, having regard to its size, as France. At pp. 448 and 449 the output is given at $20\frac{3}{4}$ million tons, while the weight imported exceeds that exported by about $9\frac{3}{4}$ million tons. On referring to some notes taken in the year 1865, the average cost of working coal in 13 French collieries was about 30 per cent. higher than that in the United Kingdom at that time; since which the cost of working has increased, probably to a greater extent than with ourselves. The exact amount of the relative increase is perhaps not very material; because what the French iron-makers have to look at, is not the cost alone, but the value of coal when used for other purposes, for that is the nature of the competition they have to encounter. Now, according to returns in my possession, the selling price of coal in the Pas de Calais varied from 11s. $3\frac{1}{2}$d. in 1877 to 15s. 6d. in 1874, the average during the five years ending in the former year being 14s. 2d. So late as the year 1878, in one of the chief iron-making centres, situate on a coal-field which I visited, the best coal was selling at 12s. 8d. to 13s. 6d., and commoner coal at 9s. 6d. per ton; coal in England only commanding about half this price at the same period.

Pas de Calais.—From a printed statement furnished to me as a juror of the Paris Exhibition in 1878, I extract the following figures respecting this now important coal-producing locality:—

	1873.	1874.	1875.	1876.	1877.
Production in thousands of tons...	2,982	2,973	3,257	3,324	3,435
Tons worked per annum per man employed	166	152	151	148	148
	s. d.	s. d.	s. d.	s. d.	s. d.
Earnings per day (year taken at 290 days)	2 10¼	2 9½	2 9½	2 9½	2 4½
Cost of labour per ton of coal ...	5 3½	5 7¾	5 8	5 9¾	4 11
Selling prices of coal	15 5½	15 6½	14 5¾	14 0¾	11 3½

According to this information the working expenses of this coal-field must be regarded as high.

According to the same printed statement, given me by a French coal-owner, the average earnings for the year 1876 of the men in the County of Durham was 3s. 9¾d. per day,[1] and the wages per ton about 2s. less than in the Pas de Calais. The total cost of coal for that year was quoted as being 3s. 8d. per ton dearer in these French pits than in those of Durham.

In a paper read by M. Laporte before the members of the Northern Institute of Mining Engineers, at their meeting in Paris in 1878, it is mentioned that the wages of the hewers have increased in the northern coal-field of France 161 per cent. since 1825. Their daily earnings were as under:—

	1825.		1835.		1865.		1875.
	s. d.		s. d.		s. d.		s. d.
Per day	1 6	...	2 0	...	2 6	...	3 10

According to M. Laporte the wages per ton of coal in 1875 amounted to about 6s. 3d.

St. Etienne Coal-field.—When I visited this district in 1867, the selling price of unwashed coking coal delivered at the ovens was 9s. 1d., which included a charge of about 1s. for carriage.

At one mine, where the hewers were only earning 2s. 5¾d. per day of ten hours, the cost of extraction at the pit's mouth was stated to be as low as 3s. 0½d. per ton. The items making up this sum were also given, and were as follows:—

Wages.	Sinking Fund.	Management.	Sales.	Rent.	Repairs and General Charges.	Total.
2s. 2¼d.	4¾d.	1¼d.	1½d.	½d.	2d.	3s. 0½d.

[1] This quotation, in my opinion is lower than the wages have been for many years.—I. L. B.

Under any circumstances this figure must be taken as an exceptionally low one, for when the hewers' earnings amounted to sums varying from 2s. 9d. to 3s. 7¼d. per day, the cost, at other pits, of working a ton of coal was stated as being 6s. 7¾d., of which 2s. 10¾d. was for men's wages.

The expenses of working coal in the neighbourhood of St. Etienne ranged, in 1867, from 5s. to 7s., exclusive of royalty which in some cases was high.

The yield of coke from the coal was 60 per cent., and the cost of labour at the ovens amounted to 1s. 3½d.

The selling prices at that time of the output of one mine visited were as follows:—

	Per Cent.				Per Ton. s. d.
Small for coking	... 60	of saleable produce	...	7s. 5d. to	8 0
Lumps	... 33	,,	,,	...	12 9¼
Large blocks ...	... 7	,,	,,	...	18 3
	100		Average about	...	10 2

For Bessemer pig smelted in 1878 from ores, chiefly foreign, yielding 56½ per cent. pig, smelted in old and very small furnaces blown with air at 350° C. (662° F.), 30 cwts. and upwards of coke were used and 14½ cwts. of limestone. With fire-brick stoves delivering the blast at 600° C. (1,112° F.) 20·86 cwts. of the coke sufficed. In this case only 9½ cwts. of limestone were required. The make was 245 tons per week from the furnaces with pipe stoves, and 350 tons from those with fire-brick stoves.

The iron works near the town of St. Etienne are surrounded by collieries of the character just referred to. Many iron smelters purchase washed coal for their ovens, at a cost (in 1878) of 13s. 10½d. per ton, at which price they can produce coke at 20s. In one case the coke so made, having to be conveyed for twenty-two miles, involved a charge of 2s. 6d. per ton, equal therefore to 1·36d. per ton per mile. It contains 12 to 14 per cent. of ash.

Coal in Central France.—Commentry.—The iron works in the vicinity of Commentry were supplied, in 1867, when I visited them, with fuel from the adjacent collieries. At that time the coke ovens were charged 8s. to 9s. per ton for small coal, of so impure a quality that the coke made from it contained 18 per cent. of ash. By washing, this was reduced to 12 to 14 per cent., the product costing 16s. to 18s. per ton, and being sold at 20s.

In furnaces of 48 feet in height and using an ore yielding 37 per cent. of pig, consumption of coke for white iron was only $21\frac{1}{2}$ cwts., while 30 cwts. were required for foundry and 26 cwts. for grey forge pig. The ore, in the form of small gravel from Bourges, costing then 11s. 9d. delivered at the furnaces, required as a flux 14 cwts. of limestone per ton of metal. The temperature of the blast was 350° to 400° (652° to 752° F.)

Coal of the Department of the Gard.—In 1867 I visited a colliery belonging to the field of the Grande Combe, in which the coal rarely exceeded $3\frac{1}{2}$ feet in thickness. The cost of working at that time was stated as varying from 5s. to 5s. 3d. per ton; no royalty was paid, the lands being the property of the company. The entire output of the district, at the period of my examination, was 1,300,000 tons, and the selling price of small coal in 1867 ranged from 6s. $4\frac{3}{4}$d. to 7s. 2d. per ton, and of the large 14s. 3d. The produce of the pit being half large and half small, the average market price was above 10s. A hewer was cutting $4\frac{1}{2}$ tons in ten hours, earning thereby 3s. $2\frac{1}{4}$d. to 4s. per day without house or fire. The coal was being washed for the iron works, and then sold for 8s. $4\frac{3}{4}$d. per ton: it yielded 57 per cent. of coke; the selling price of the latter was 17s. $7\frac{1}{2}$d. per ton. The ore used was partly got in the neighbourhood, and the remainder was brought from La Voulte on the Rhone, and from Africa.

COAL OF BELGIUM.

In comparison with its area Belgium furnishes a greater weight of coal than any other nation. The annual quantity raised is about $17\frac{1}{2}$ million tons (*v.* p. 448), and the coal exported exceeded the imports in 1882 by about $4\frac{3}{4}$ million tons. Moreover this prominent position, in a coal producing point of view, is attained in spite of mining difficulties greatly beyond the average of most other localities. The mines are often of great depth, sometimes above 1,000 yards, and the strata are often so contorted that the shaft passes through the same bed of coal several times. Besides these disadvantages the seams are often very thin; and taking the quality throughout, the produce of the Belgian mines is, like those of France and Germany, inferior to most of the coal in Great Britain. Under such circumstances as those referred to, the cost of extraction is necessarily high, and in consequence the selling prices are above those of the United Kingdom. The quotations of one large company in the year 1875 per ton at the pit were as follows :—

Gas Coal.	Coking Coal.	For Brick and Lime Burning.	Coke from Washed Coal.
9s. 6d. to 10s. 4d.	7s. 9d. to 9s.	4s. 9d. to 6s.	14s. to 15s.

According to a Government return, the published value of all the coal raised in Belgium is thus stated :—

1870.	1871.	1872.	1873.	1874.	1876.	1878.	1879.
8s. 8d.	8s. 11½d.	10s. 7¾d.	17s. 9½d.	13s. 1½d.	10s. 9½d.	7s. 11d.	7s. 6d.

At page 509 the average output per individual engaged per annum is given for the province of Hainaut alone. According to returns given to me the following figures represent the annual production for men and boys over the whole of Belgium. As a means of comparison, some of the corresponding figures are added for South Durham :—

	1870.	1871.	1872.	1873.	1874.	1876.	1878.	1879.	1880.	1881.
Belgium ... tons	149	146	159	146	134	132	150	158	—	—
South Durham ,,	—	—	—	335	330	341	356	350	401	398

During a visit I paid to Belgium in 1873, the year of highest prices, a large coal-owner named 8s. 8½d. as the cost of raising coal near Liége, to which it had risen from prices varying from 5s. 7d. to 6s. 5d. At the date of my inquiry the selling price however was 22s. 4¾d., so that the profit per ton was nearly 14s. From another quarter 8s. per ton was given as a common cost, while 5s. 7d. had been the figure at this particular mine before the great rise in the value of coal.

By a very trustworthy authority the following information was given me in the year 1873 as to the increase in the cost of raising a ton of coal in his own district :—

1869.	1870.	1871.	1872.	1873.
3s. 9½d. ...	4s. 3d. ...	5s. 0d. ...	5s. 4¾d. ...	6s. 7¾d.

SECTION XVIII.—CHIEF IRON-PRODUCING COUNTRIES COMPARED. 631

The details of the last year—1873—were as follows :—

	s.	d.
Labour	2	10¾
Stores	1	5½
Redemption	0	2¼
Loading	0	1¼
General expenses	1	7¾
Extraordinary work	0	4¼
	6	7¾

In many cases however the cost in 1873 would be as high as 9s. 7¼d. per ton. In all these there is an allowance for redemption of capital and royalty.

Unwashed coal afforded in one case 68 per cent. of coke, containing 10 to 12 per cent. of ash, and of it 22·40 cwts. were used to give a ton of white iron in furnaces having a height of about 63 feet. The blast was heated in fire-brick stoves, and the ore used was chiefly brought from Luxemburg.

In 1878 the small coal sold to iron-works for making coke was charged at prices varying from 7s. 1½d. to 10s. 3½d. per ton. It contains only from 15 to 20 per cent. of volatile matter, so that in an Appolt oven the yield was 80 per cent. of coke, and in an ordinary (Coppée) oven it was 74. Taking coal at a mean price of 8s., the coke would therefore probably cost from 12s. 6d. to 13s. per ton. In 1878, at one work I visited, 8s. 3d. to 8s. 8½d. was being paid for unwashed coking coal, from which the coke, made in Appolt ovens, was said to cost 11s. 1d. per ton, the yield from the coal being 80 per cent. This coke was soft and poor looking, and contained 12 to 14 per cent. of ash. The ore used, chiefly that of Luxemburg, yielded 38 to 40 per cent. of pig, and the consumption of coke per ton of white pig was 23 cwts.

In 1878 the cost of working coal in the Liége district was stated to me to vary from 6s. 4d. to 7s. 1½d. per ton. The selling prices were as follows :—[1]

	s.	d.		s.	d.
Best coal in very large pieces	14	3	to	15	10
Large coal free from small	11	1	to	11	10½
Rough small	9	6	to	10	3½
Dead small	—			7	1½

[1] Paper read by M. Laporte at meeting of members of North of England Institute of Mining Engineers held at Paris, 1878. *Vide* Transactions for that year.

The coal in the Charleroi district is rarely washed, and contains on an average 12 per cent. of ash; washing when adopted reducing the ash to 7 or 8 per cent. Coke is sometimes used in the furnace containing as much as 22 per cent. of ash. At an establishment also examined in 1878, unwashed coal was delivered at 6s. to 8s. 8½d. per ton. The coke obtained from it made a ton of white pig with 21 to 23 cwts. For grey foundry pig 27 cwts. were consumed. The furnaces were 52 feet high, blown with air at 450° to 500° C. (842° to 932° F.)

The exportation of so large a proportion of the coal raised in Belgium, as may be supposed, is not due either to superior excellence or to greater cheapness of extraction as compared with Great Britain, but to the readier and cheaper access to the iron-making districts of France and Luxemburg.

Coal in the United States.

Fifty years ago or less the known areas of coal deposits on the Continent of Europe were such as to have led to the belief that Great Britain possessed almost a monopoly of the carboniferous formation. Subsequent experience has done much to dispel this illusion, and if we, in this country, are entitled to claim, and this even to-day with some show of reason, to be exceptionally favoured in our coal beds, more recent discoveries have abundantly proved how far from correct was the common opinion of former times, for many extensive tracts of coal have been discovered and opened out within the fifty years referred to.

In a country like the United States, where half-a-century may be said to embrace the whole of its industrial life, this remark as to its coal resources pre-eminently applies; for at the present moment the known coal-fields extend over an area fully eight times as large as those of the United Kingdom, and to those already discovered others are frequently being added. Notwithstanding this wealth, possessed as it were by one nation, it seems scarcely probable, with our actual knowledge of the geology of North America, that for the same area of territory, the United States is equal in coal resources to the parent country.

Anthracite Regions of Pennsylvania.—Whatever advantages Great Britain, area for area, may possess over the United States in a coal-producing point of view, those connected with the anthracite deposits of the latter, so far as iron-making is concerned, may be said to be without a rival. It is true that in South Wales this form of coal is found; but it is in limited quantity and, for the blast furnace, appears inferior in quality to the almost inexhaustible beds of North America.

In the mining district surrounding Hazelton I was informed that there are three seams of anthracite varying in thickness from 13 to 18 feet. They lie at a very steep slope, 45° to 60°, and after reaching the maximum depth very often run in an opposite direction to the surface. To win the coal thus situated a large timber erection is constructed, one side of which is an inclined plane, being a continuation of the inclined road which is formed in the coal-seam itself. The use of the building is to separate the largest coal for blast furnace or other purposes, or to break and sort it along with the smaller into pieces of uniform size, for other manufacturing and for domestic fires. During the extraction and breaking of the mineral a considerable quantity of dust is produced, and although coming into partial use for artificial fuel, enormous mounds of rejected small coal were to be found at all the older pits I examined. The cost of opening out an anthracite colliery capable of furnishing 250,000 tons a year, including the "Breaker," was stated to be 150,000 dollars. There is a considerable loss, partly in the dust just referred to, partly in stony matter, and partly in the amount of coal to be left unworked, in consequence of the difficulty of supporting the roof in a seam lying at such a steep angle. The yield given me was 15,000 tons per acre from a seam of 14 feet, which is probably less than half the actual weight contained in the bed.

In some cases six or seven tons of water have to be pumped for every ton of coal drawn.

Although the thicknesses already mentioned represent a fair average of the whole, I descended one mine where the seam was 32 feet in thickness; and in another place, after taking off 40 feet of superincumbent rock, an anticlinal axis was bared, from which in a face averaging 50 feet in thickness the coal was quarried.

At page 557 some rates of labour are given; but at the periods of highest prices the hewers earned from 8 to 10 dollars per day and ordinary labourers in the mines 3 dollars. At the period of my visit, in 1874, the former class was getting 4 to 5 dollars and the latter 2 dollars. The period of labour for men working on contract was eight hours, for daymen working at the surface ten hours, and for those underground nine hours.

In 1876 I visited the Schuylkill anthracite region, when I was informed that the coal there could be worked for 78 cents. per ton (2s. 11d. at the then rate of exchange), of which about one-third was for expenses of breaking and sorting. At other collieries a cost of 1·30 dollars per ton was given (4s. 10¾d.) The vein of coal in one colliery I visited was named as having a thickness of 43 feet. The mines are about 110 miles from Philadelphia, and the average dues are about ·85d. per ton per mile, which includes cost of putting on board.

The prices given me in 1874, as being paid at the pits for furnace coal, was 11s. 7d.;[1] which with carriage of 40 miles brought up the cost, depressed as the iron trade was, to 16s. 7d. delivered at the furnaces. At another work the price given was 16s., and at a third distant 34 miles from the pit the price at the furnaces was 16s. 11½d. At this work, a few years previously, the same coal had been laid down at the furnaces at 7s. 6d.

In some cases this coal is conveyed for great distances for iron making, involving heavy charges for carriage. Thus at Lake Champlain it cost, in 1874, 24s. to 26s. laid down at the furnaces.

As a mere question of suitability for furnace work, there is no doubt that anthracite is inferior to coke. This does not arise from any tendency of this variety of coal to cake, as the splint coal of Scotland does to a certain degree. On the contrary I picked up pieces of anthracite twice the size of a man's fist, and found on breaking them that internally they had all the characteristic lustre of the original coal. The defect of this kind of fuel is its tendency to decrepitate, and thus impede the passage of the blast. To overcome this obstruction, blast is commonly used at a pressure of 7 lbs. per square inch; and I have found instances of 12 lbs. being used. In consequence of the inconvenience just referred to, the chargers do not fill any small which may be produced by handling, in loading or unloading the wagons which bring the coal from the mines. This precaution involves a rejection of 3 to 13 per cent. of the whole, which has to be used for other purposes.

[1] Dollar taken at 3s. 9½d.

SECTION XVIII.—CHIEF IRON-PRODUCING COUNTRIES COMPARED.

On looking over the notes taken upon the occasion of several visits to smelting establishments using anthracite coal in the United States, I have arrived at the conclusion that in furnaces up to 60 feet or thereabouts in height, blown with air at 750° to 800° F., 35 to 37½ cwts. of coal is used for a ton of foundry grey iron, the average make being under 300 tons per week. The yield of the ore was equal to 50 per cent. of pig, but, owing to the flux containing a considerable quantity of carbonate of magnesia, 20 to 25 cwts. of limestone were used.

By raising the furnaces to a height of 70 to 75 feet, with a capacity of something over 8,000 cubic feet, and by using well selected coal burnt with air at 1,100° to 1,200° F., the consumption of anthracite is reduced to 23 to 25 cwts. per ton of iron; but it does not appear that the weekly produce exceeds that of the smaller furnaces.

My friend Mr. Witherbee read a paper on the use of anthracite, reported in the "American Mining Journal." The results were very favourable, but in this case the blast reached nearly 1,400° F. Mr. Witherbee's figures are as follows:—

1875.	Weekly Make.	Av. Quality. No.	Yield of Ore. Per Cent.	Coal used. Cwts.	Temp. of Blast. Deg. F.
September	239¾	1·61	54·4	24·1	1,336
October	294⅞	1·60	59·2	23·1	1,370
November	291$\frac{4}{10}$	1·95	55·6	23·1	1,397
December	282$\frac{9}{10}$	1·96	54·8	22·7	1,398
1876. January	291$\frac{4}{10}$	2·03	53·5	22·6	1,384

These figures point to the conclusion that in high furnaces, blown with air from firebrick stoves, it is possible to obtain a ton of iron with a consumption of about 24 cwts. of anthracite coal; which is about 15 per cent. more than is required of best Durham coke. It must be recollected, however, that although this variety of fuel is very poor in volatile matter, it is by no means to be regarded as always consisting exclusively of fixed carbon and ash. This is apparent from the following analyses:—[1]

[1] Transactions American Institute Mining Engineers. Paper by Professor Frazer, jun. Vol. VI., p. 430.

636   SECTION XVIII.—CHIEF IRON-PRODUCING COUNTRIES COMPARED.

Name of Coal.	Fixed Carbon.	Volatile Matter.	Ash, Water, &c.
1.—Rhode Island	77·00	3·00	20·00
2.—Nesquehoning	86·60	6·40	7·00
3.—Summit Mines, Lehigh County	88·50	7·50	4·00
4.— Do. do.	87·70	6·60	5·70
5.—Tamaqua	92·07	5·03	2·90
6.— Do.	89·20	4·54	6·26
7.— Do.	87·45	7·55	5·00
8.—Beaver Meadow	88·20	7·50	4·30
9.—Schenoweth Bed	90·20	2·52	7·28
10.—Third Coal, Pottsville	94·10	1·40	4·50
11.—Forest Improvement	89·20	5·40	5·40
12.—Sharp Mountain	90·70	3·07	6·23

Among these it will be perceived that some, such as Nos. 5 and 10, containing 92·07 and 94·10 per cent. of fixed carbon, are almost equal in point of chemical composition to the best Durham coke; while others have a composition which easily explains the reason of the larger quantity of anthracite used in smelting a ton of pig iron. In some cases, where the fixed carbon in this natural coke, so to speak, approaches that found in the best English coke, the duty of the former appears inferior to the latter by about 10 to 15 per cent.

Coal of Pittsburg.—The valleys of the Youghiogheny and Monongahela, in the neighbourhood of Pittsburg, contain immense resources of bituminous coal, fit for the iron-smelter's wants. The district known as Connellsville occupies the foremost place as a coke-producing centre. I was informed by a large firm engaged in the trade, that the coking seam there has a thickness of 11 feet, and is frequently obtained from levels which pierce the sides of the rising ground. The workings are neither troubled with gas nor water. The hewers, who load the trains, work about $3\frac{3}{4}$ tons in 8 hours' actual labour. For this amount, in 1874, they were receiving 3 dollars. At the then value of paper currency (3s. $9\frac{1}{2}$d.) I estimated the cost of a ton of coal at the pit to be 4s. 6d., which included 6d. per ton for royalty.

I cannot say that the coke made from the Connellsville coal, high as its reputation is in the American ironworks, always impressed me strongly in its favour. This may be accounted for by the amount of foreign matter it sometimes contains; for, according to the Geological Survey, it is stated to contain 12 per cent. of ash.

SECTION XVIII.—CHIEF IRON-PRODUCING COUNTRIES COMPARED.

The following analyses of the coal used in the Connellsville ovens are copied from the Geological Survey of Pennsylvania, 1876-8, p. 11 et seq. :—

	Volatile.	Fixed Carbon.	Sulphur.	Ash.	Water.	
Roof coal ...	38·490	45·895	2·905	11·690	1·020	= 100
,, ...	36·770	51·467	2·098	8·890	·775	= 100
Upper bench ...	40·510	41·324	7·566	9·090	1·510	= 100
,, ...	37·375	54·561	1·499	4·475	1·730	= 100
,, ...	40·350	50·311	2·594	5·665	1·080	= 100
,, ...	35·830	58·154	·761	4·075	1·180	= 100
Main bench ...	36·490	59·051	·819	2·610	1·030	= 100
,, ...	37·225	56·608	·982	4·145	1·140	= 100
,, ...	35·580	54·185	1·290	5·095	·850	= 100
Lower bench ...	38·720	40·253	3·722	16·175	1·130	= 100
,, ...	35·275	58·167	·758	4·660	1·140	= 100
,, ...	36·880	56·829	·796	4·070	1·425	= 100
,, ...	34·655	60·414	·766	3·045	1·120	= 100
,, ...	34·125	57·979	·586	6·020	1·290	= 100

The above are apparently from different parts of the seam, taken from different localities. Those which follow are not so distinguished, and are, it is to be supposed, the average of the bed.

Origin	Volatile.	Fixed Carbon.	Sulphur.	Ash.	Water.	
Near Monongahela City... ...	35·075	55·030	1·910	7·335	·650	= 100
,, ...	35·315	57·332	·648	5·595	1·110	= 100
,, ...	35·350	55·010	·895	7·745	1·000	= 100
,, ...	35·420	60·537	·658	2·165	1·220	= 100
,, ...	36·810	55·312	·643	6·345	·890	= 100
Westmoreland County (slack) ...	29·330	57·399	1·308	11·063	·810	= 100
,, (small coal)	28·145	58·511	1·019	11·115	1·210	= 100
,, ...	27·365	49·651	·859	21·195	·930	= 100
Indiana County	27·385	49·748	3·017	19·000	·850	= 100

I find the average of the 23 samples is 54·04 of fixed carbon and 7·87 of ash, together 62·9 per cent. of carbon and ash. According to the laboratory returns, the average yield of coke of the same samples works out to 63·57 per cent., which would leave ·66 for sulphur. It happens that these examples, taken at random from many others, would indicate a content of 12·38 per cent. of ash in the coke, which corresponds closely with the information obtained in the country.

The average sulphur in the 23 samples of coal is 1·65 per cent. If one half of this, supposing it to exist as Fe S_2, be expelled, we have ·825 left in the coal; and this, concentrated in the coke, would represent 1·29 per cent. in the latter.

In smelting ore, chiefly from Lake Superior, giving 64 per cent. of iron, with air heated to 950° F., 25 cwt. of Connellsville coke was being used, upon the occasion of my visit, for grey forge iron. The furnace was 75 feet high, with 18-feet boshes, and the limestone used was 15, cwts. per ton of pig. This is an excessive consumption of coke, but it is only right to state that the furnaces in the neighbourhood of Pittsburg are frequently driven at an unusual speed, 600 to 610 tons per week being often produced. In later years the make, I learn, has sometimes reached 1,000 tons per week. In 1874 the price of Connellsville coke was 10s. 6d. delivered at the furnaces—the distance of the ovens being 60 miles. From another source I was informed that the price of Connellsville coke was only about 6s. at the ovens.

Since the foregoing on the Connellsville coke was written, I am indebted to my friend Mr. Jos. D. Weeks, of the "Iron Age," for an interesting article on the same subject. He first gives the average annual values of Durham coke at the ovens, which are as follows:—

1870.	1871.	1872.	1873.	1874.	1875.	1876.	1877.	1878.	1879.	1880.	1881.	1882.
10/9	12/6	25/6	40/6	21/-	15/-	12/6	11/6	10/-	11/-	12/6	11/-	11/6.

These prices, it will be found, do not exactly correspond with the quotations given at page 612.

Mr. Weeks admits the inferiority of Connellsville coke as compared with that of Durham, but as he justly observes the market price of the former (1884) is also much lower, viz., 1·12 dollars or 4s. 8d. at the ovens; a rate never heard of in Great Britain. He gives the field which yields the coal as having an area of 150 to 180 square miles, situate at a distance of 60 miles from Pittsburg. Twenty-five tons have been dug and loaded in ten hours by a man and a boy, the chief labour being loading.[1] The seam is 8 to 10 feet in thickness, from which the miners average 6 tons per diem, and for this they receive 30 cents (1s. 3d.) per ton, say 1·80 dollars or 7s. 6d. per day, including the loading.[2]

[1] This greatly exceeds anything I heard of.—I.L.B.
[2] The currency being now in gold the dollar is reckoned at 4s. 2d.

SECTION XVIII.—CHIEF IRON-PRODUCING COUNTRIES COMPARED. 639

Mr. Weeks then proceeds to contrast the earnings of the Connellsville workmen with those of Durham, as follows:—

	Connellsville.			County of Durham.		
Class of Labour.	Hours.	Average Daily Earnings. Dollars.		Class of Labour.	Hours.	Average Daily Earnings. Dollars.
Miners...	7 to 10	1·84	...	Hewers ...	7½	1·15
Deputies	10	2·00	...	Under-boss	8	1·88
Drivers	10	1·50	...	Drivers ...	10	·33
Firemen	10 to 12	1·75	...	Firemen...	11½	·85
Stablemen	12	1·60	...	Horsekeepers	11	·73
Dampers	8 to 10	1·62	...	Banksmen	10	·97
Trackmen	8 „ 10	1·88	...	Rolleywaymen	11	·97
Onsetters	8 „ 10	1·75	...	Onsetters	10	·95
Haulers	8 „ 10	1·88	...	Putters ...	10	·79

According to these figures the men in the American mine average exactly double the earnings of those in the County of Durham. The same difference appears to hold good with the men engaged at the coke ovens.

A large quantity of coal being raised in the vicinity of Pittsburg, a considerable weight of small is thrown into the market. This is disposed of at a mere nominal price. It is then washed and coked, the coke costing about 4s. to 5s. per ton. With blast at 950° F. and in a furnace 75 feet by 20 feet, 26 cwts. of this coke from washed coal was used for hard iron, and 30 cwts. for Bessemer pig.

Ohio Block Coal.—I examined some mines near Youngstown producing a laminated splint coal, which is used raw in the blast furnaces of the Chenango and Mahoning Valleys. The furnaces in use when I was in that country were small, and although they were using very rich ore—yield 66 per cent.—and only 10 cwts. of limestone, 2 to 2½ tons of coal were being consumed per ton of iron. The cost delivered at the furnaces was 5s. 7¾d. in 1874, reckoning the paper dollar at 3s. 9½d.

West Virginian Coal.—A considerable quantity of coal is worked by levels which run into the mountains which form the valley of the New River. The yield of coke is very like that of South Durham, viz. 65 per cent. in beehive ovens. The coal is so cheaply worked that the coke delivered at the Quinnimont furnace only cost about 9s. 6d. per ton in 1875. Using an ore containing 43 per cent. of iron, and as much as 26 cwts. of limestone per ton of iron, 30 cwts. of the coke sufficed to make a ton of grey No. 2 iron. The coal contains above 75 per cent. of fixed carbon, and the coke has the following composition :—

	Carbon.	Ash.	Sulphur.	
Quinnimont coke, dry ...	93·83 ...	5·85 ...	·30 =	100·08
Do. do.	93·11 ...	5·94 ...	·82 =	99·87

The furnace is 60 by 15 feet, blown with air at 800° F.

Coal in Kentucky.—The coal I had an opportunity of examining was that obtained from the eastern division of this State, which supplies the Hanging Rock iron furnaces with fuel. This coal-field has been estimated to contain nearly 9,000 square miles. From the analyses I have seen, it does not appear to be very rich in fixed carbon, which varies from 53 to 59 per cent. The beds vary from 3 to 5 feet in thickness, and the cost at the pit was given me in 1874 at 4s. 3d. per ton: a hewer being able to cut $3\frac{1}{2}$ to 4 tons of large coal for a day's work, by which he earned 8s. to 9s. Nearly half a ton of small is produced, but this being worthless the workman receives no pay for it. It is used raw in the blast furnace, but is not a strong coal, and is apt to fall into small pieces. For forge iron 40 cwts. were given as the smallest rate of consumption, and for foundry as much as 50 to 60 cwts. per ton of pig. The furnaces, generally speaking, are of moderate dimensions, although one, at the Etna works, was then in course of construction $87\frac{1}{2}$ feet high, with boshes of 18 feet.

Coal in Alabama.—In 1874 very little had been done in opening up coal-fields in this State for iron-making purposes. Two furnaces had been erected by the Red Mountain Iron Co., but they were in liquidation at the period of my visit. Since then they have passed into other hands, and are, I am informed, doing well. I went into one heading, in the Cahawba coal-field, but the seam was there only a little above 3 feet thick. I also examined a mine in the so-called Warrior coal-field. Access to it, as in the case of the previous one, was by means of a drift. The coal was also 3 feet thick, and the men were cutting about $2\frac{1}{2}$ tons per shift, for which they were receiving 2 to $2\frac{1}{2}$ dollars (7s. $6\frac{1}{2}$d. to 9s. 5d. at the then value of the paper dollar). The coal appeared soft. At another colliery the coal was dirty, and required washing before it was coked. This concern was largely worked with convicts, farmed from the State for 75 cents per day, which, with food, clothing, and watching, brought the cost up to about one dollar.

SECTION XVIII.—CHIEF IRON-PRODUCING COUNTRIES COMPARED. 641

The best coal from the Warrior field had been tried by a railway company, and found quite equal to that of Pittsburg and Kentucky for locomotive purposes.

The Alabama coal, so far as I could learn, is highly bituminous, and like that of Eastern Kentucky is low in fixed carbon. The following analyses are copied from "Macfarlane's Coal Regions of America:"—

	Cahawba Level Bed.	Cahawba, Mulberry Creek.	Cahawba.	Warrior, Tuscaloosa.
Volatile combustible	35·51	36·68	34·49	40·60
Fixed carbon	57·42	57·23	60·09	54·07
Ash	6·31	5·30	4·32	3·09
Sulphur	trace	trace	·17	1·06
Moisture	·76	·79	·93	1·18
	100·	100·	100·	100·

According to an analysis of the coal of one of the collieries I visited, viz. that of the Newcastle Coal and Iron Company in the Warrior coal-field, the content of fixed carbon considerably exceeds that given above. It is as follows:[1]—

Volatile Combustible.	Fixed Carbon.	Ash.	Sulphur.	Moisture.	
26·11	71·64	2·03	·10	·12	= 100

There is however some inconsistency in the statements which follow, for the author mentions 43,800 lbs. of coal having only given 24,090 lbs. of coke in a gas work, which is a yield of only 55 per cent.[2] The gas obtained was at the rate of 10,976 cubic feet per ton of coal, having an illuminating power of 14·5 candles.

It would have been premature to speak of the cost of working the coals in Alabama at the period of my visit, because the output at any one mine did not exceed 300 tons per day. On the other hand, the mines being recently opened out, the expenses attending the working coal close to the drift mouth were much less than they would be at a later period. Very low prices were given as sufficing to cover all expenses: in one case it was less than 3s. exclusive of royalty, for which there was nothing to pay.

In the work already quoted—"Hill Country of Alabama"—published in 1878, it is stated that the cost on a large output should not

[1] "Hill Country of Alabama," p. 46.
[2] Possibly no allowance is made for the coal consumed in firing the retorts.—I. L. B.

exceed 1·75 dollars per ton put into railway waggons. Other authorities express a belief that the coal can be worked for 3s., but think it safer to reckon upon 6s. All these speculations are so entirely dependent on the cost of labour that it is useless to pursue the inquiry. All that need be said at present is, that with six seams of coal from $3\frac{1}{2}$ to 4 feet in thickness, I saw nothing to induce me to think that they, with labour at the same price, could not be worked as cheaply, or, as long as the produce was obtained by drifts, perhaps even more cheaply than coal was got in most parts of the world.

In the event of the Alabama coals proving suitable for iron making, of which there is little doubt, their importance, in an industrial point of view, will be immense, on account of the inexhaustible beds of ore which are to be found in close proximity to the coal-bearing strata.

Coal in Georgia.—I only had an opportunity of visiting one coal mine in this State. I measured the seam at the face, where it varied from 5 to 7 feet in thickness, but it often was as low as 3 feet. The average was considered to be $4\frac{1}{2}$ feet. One hundred and fifty convicts were employed here, whose cost was mentioned as being 80 cents per day. They worked as well as men earning 2 to $2\frac{1}{2}$ dollars. The price paid for free labour, hewing and loading into trains, in 1874, was 2s. $7\frac{1}{2}$d. (70 cents paper) per ton of large and small. Forty coke ovens were at work, the labour at which I calculated to cost 1s. 1d. per ton. The coal is so soft that, as the proprietor informed me, a man could hew 7 tons per day, and $5\frac{1}{4}$ tons would be an easy day's work. As a fact, much less than either of these quantities was got.

Coal in Tennessee.—This State possesses very large resources in coal. One working I went into was approached by a level. The coal had a thickness of $3\frac{1}{2}$ to 4 feet. In 1874 I made out the cost of hewing and loading to be 2s. $7\frac{1}{2}$d. per ton, and that the total cost was 5s. 3d. Each hewer worked 70 cwts. for his day's work. To my eye the coal from this colliery was dirty, and the coke made from it is described in my notes as rough, brittle, and slaty. Nevertheless, from a work on the resources of Tennessee, the analysis of the coal, as regards ash, seems highly favourable :—

Volatile Combustible.	Fixed Carbon.	Ash.	Moisture.		Sulphur in Coke. Per Cent.
38·87	56·44	3·65	1·04	= 100	·59

SECTION XVIII.—CHIEF IRON-PRODUCING COUNTRIES COMPARED.

The compositions of other samples from this State are given as follows:—

Volatile Combustible.	Fixed Carbon.	Ash.	Moisture.			Sulphur in Coke. Per Cent.
38·82	... 57·52	... 2·67	... ·99	=	100	... ·13
27·70	... 63·10	... 7·70	... 1·50	=	100	... ·45
30·10	... 65·80	... 2·80	... 1·30	=	100	... ·52
29·90	... 63·50	... 6·60	... —	=	100	... —

I had an opportunity of witnessing the duty performed by coke in the blast furnace near Chattanooga; and there certainly, neither as to appearance nor duty, was I favourably impressed with its quality. In using an ore yielding, in a furnace 63 feet in height, 48·74 per cent. of pig, the coke used for foundry iron was 38·34 cwts. per ton. At this establishment it is only right to mention that the heating stoves were deficient in power.

At another work, with an ore yielding 53·7 per cent. of pig iron, and with a consumption of 9½ cwts. of limestone per ton of metal, the consumption of fuel was 18·23 cwts. of coke, used with 12·57 cwts. of raw coal for forge pig.

With regard to the cost of this Tennessee coal, my information leads me to set it down at 5s. to 5s. 6d. per ton.

With the exception of the anthracite beds, the whole of the American coal-fields mentioned in the preceding pages belong to one huge deposit. In shape it is an irregular isosceles triangle, with a base of about 150 miles, commencing about fifteen miles from Lake Erie. The perpendicular runs in a south-westerly direction over a distance of above 500 miles, constituting probably the largest known coal-bearing area in the world. At its north-east angle it is about 150 miles from the harbour of New York. Its eastern flank is little more than 100 miles from the port of Baltimore; but further south towards Tennessee it recedes from the Atlantic to a distance of 250 to 300 miles. The nearest points of the anthracite beds to the ocean are New York and Philadelphia, both cities being separated from these coal-fields by a distance of 50 or 60 miles only.

Coal in Illinois and Indiana.—The information respecting American coal mines referred to in the present section is all derived from personal visits. I had an opportunity of inspecting certain iron works in

Illinois and Indiana, and, although I did not visit the collieries situate on the immense coal-field which covers a large area of these states, I obtained many particulars of their produce.

The furnaces using it were 62 feet high, with boshes of only 11 feet, blown with air at 800° to 900° F. The ore used yielded 60 to 65 per cent. of pig, for every ton of which $12\frac{1}{4}$ to $13\frac{3}{4}$ cwts. of limestone were consumed. The fuel required, taken over periods of four weeks, varied from 30 to 32 cwts., which consisted of five-sixths of raw coal and one-sixth of coke. The pig iron averaged two-thirds foundry and one-third forge, each furnace running about 250 tons per week.

The fuel used for smelting iron is chiefly block-coal, of which the following is the composition:—

Volatile Combustible.	Fixed Carbon.	Ash.	Water.	
34·70	57·30	3·50	4·50	= 100

If this analysis represents an average content of carbon, the weight of this element consumed per ton of pig is under 22 cwts., which, looking at the quantity of the fuel used in its raw state, may be considered as very good work.

Charcoal.—The abundance of iron ores in certain localities in the United States, and the great distances which separate the mines from the colliery districts, preserve the manufacture of iron by means of charcoal from becoming an obsolete industry. The enormous tracts of forest, and the excellence of the quality of the metal produced from the fuel they supply, still maintain for "charcoal iron" a very important position in the make of the American blast-furnaces.

According to the returns furnished by my friend Mr. James M. Swank, in his excellent reports of the American Iron and Steel Association, the following shows the extent to which charcoal pig iron is still manufactured in the United States:—

	1879.	1880.	1881.	1882.
Bituminous coal	1,438.978	1,950,205	2,268,264	2,438,078
Anthracite	1,273,024	1,807,651	1,734,462	2,042,138
Charcoal	358,873	537,558	638,838	697,906
	3,070,875	4,295,414	4,641,564	5,178,122

SECTION XVIII.—CHIEF IRON-PRODUCING COUNTRIES COMPARED.

The make of the three kinds of iron have risen almost proportionately with the general increase in the production. That of 1882 was 70 per cent. above that of 1879, and for the four years just given consisted of:—

	Per Cent.	Per Cent.	Per Cent.	Per Cent.
Bituminous coal	47	45	48	$47\frac{1}{2}$
Anthracite	41	42	38	39
Charcoal	12	13	14	$13\frac{1}{4}$
	100	100	100	100

Thus the make of charcoal iron in 1882 was nearly double that of 1879 and constituted in the last mentioned year about $13\frac{1}{2}$ per cent. of the entire production. Cheap however as timber is in the United States, the labour and cost of conveyance to the furnaces are such, that the price of charcoal at the furnaces varies all the way from 20s. to sometimes as high as 40s. per ton. Important therefore as the manufacture of this variety of pig iron is in a domestic point of view, the high price of the fuel employed, united to the great distance of the furnaces using it from the sea coast, will probably prevent this branch of American industry ever becoming an important item in the exports of that nation.

In Great Britain Messrs. Harrison Ainslie & Co. are the only manufacturers of charcoal iron, and according to information kindly communicated, their make does not reach 2,000 tons per annum. Their cost of charcoal varies from 45s. to as much as 75s. per ton.

Sweden and Austria are the largest producers of charcoal iron on the Continent of Europe. In France I imagine about 5 per cent. only of the pig iron made is smelted by means of charcoal, and in the Zoll Verein the proportion is still less.

The cheapest charcoal I have heard of was obtained in Norway, where it was chiefly obtained from refuse timber from the saw mills. It was valued at 20s. per ton, but this by no means represents the average cost; which may be taken to range from 30s. to 40s. per ton, laid down at the furnaces. A large portion of the expense is often incurred by having to convey the charcoal for considerable distances over the ordinary country roads.

IRON ORES.
GREAT BRITAIN.

In an industrial point of view Great Britain founded its greatness as an iron making centre, such as it was in the first three decades of the present century, on the least valuable of its ores. The furnaces of Wales and Staffordshire for many years were supplied almost exclusively from nodules, and from thin bands of clay ironstones, obtained from the shales of the coal measures.

In the year 1880 (*v.* p. 455) only 20·59 per cent. of the entire make of the United Kingdom was made from carboniferous clay ironstones, and of this probably nearly one-half was obtained from the black band of Scotland and North Staffordshire, a variety of ore discovered by David Mushet in the year 1801. In 1830 the total make of pig iron in Great Britain was 678,417 tons, a very insignificant quantity compared with our present modern ideas, but quite large enough then to place this country at that time far above every other in this branch of industry. Soon after 1830, aided by the introduction of the hot blast, black band commenced to be largely used in Scotland, and hematites from Cumberland or Lancashire were also being taken in small quantities to South Wales and South Staffordshire as an admixture with the poorer clay ironstones. In 1880[1] 21·04 per cent. of the entire make of the kingdom, close on 8,000,000 tons, was estimated to be smelted from native hematites, and 19·85 per cent. from imported ores of this class—in all 40·89 per cent. thus coming from an ore but sparingly used before 1830. Excluding hematites and carbonates, the remainder of the make, viz. 38·52 per cent. of the whole, is made up of the produce of the lias measures, a mineral scarcely known to the iron trade until after 1850.

Middlesbrough.—The furnaces of this district are, as is well known, supplied chiefly from the famous bed of ironstone of the Cleveland hills. I say chiefly, because, owing to the demand for hematite pig for the steel works and a general depression in the iron trade, a certain number of the furnaces in the north-eastern corner of England is

---

[1] The quantities of ores raised in the United Kingdom are given up to 1882, *v.* p. 454, but I have preferred quoting from an analysis of the consumption given at p. 455.

SECTION XVIII.—CHIEF IRON-PRODUCING COUNTRIES COMPARED. 647

now smelting ore brought from Bilbao. In 1881 however no less than 6,538,471 tons, and in the following year 6,326,314 tons, were worked from the hills adjacent to Middlesbrough.

Few words need be said respecting the bed referred to. It varies from 8 to 10 feet as an average thickness, and varies in yield of pig iron from 32 per cent. at the mines on the northern edge of the bed to 25 per cent. at its more southern extremity near Whitby.

The net earnings of a miner have already been given at page 517. The average weight worked per man over a series of years is as follows :[1]—

1868.	1869.	1870.	1871.	1872.	1873.	1874.	1875.	1876.	1877.	1878.	1879.	1880.
Tons.	Tons.	Tons.	Tons.	Tons.	Tons.	Tons.	Tons.	Tons.	Tons.	Tons.	Tons.	Tons.
5·40	5·55	5·00	5·00	5·10	5·05	5·20	5·25	5·25	5·25	5·25	5·35	5·25	5·35

The actual rate for mining was from 10d. per ton, in 1868, to prices varying from 1s. 4½d. to 1s. 7d. in 1873 and 1874; being an increase of 60 per cent. The price paid in 1883 to the miners was above 10½d. per ton. In all cases oil and powder are found by the men. In 1868 the average price of No. 3 Cleveland pig was about 44s.; but in 1883 the price was only 36s. 10d., so that, having regard to the value of iron, which now regulates the wages, the cost of hewing is considerably increased since the former year.

The average earnings of the entire complement of men above and below ground, during 1883, were about 4s. 4d. per shift. The output per man engaged in some Cleveland mines reaches 900 tons per annum. So far as my data enable me to judge, the men in the Nancy and Longwy mines, working a similar kind of ironstone, do not turn out above 500 tons per head.

According to the Government returns the average numbers of tons worked in the Cleveland district per man per annum has been as as follows :—

| 1873. | 1874. | 1875. | 1876. | 1877. | 1878. | 1879. | 1880. | 1881. | 1882. | 1883. |
| 581 | 557. | 623 | 667 | 735 | 784 | 659 | 809 | 798 | 791 | 822 |

The average selling prices of ironstone per ton of 20 cwts. delivered at Middlesbrough as a central point, during the years in question, are given me by the largest vendors in the trade as under:—

| 1870. | 1871. | 1872. | 1873. | 1874. | 1875. | 1876. | 1877. | 1878. | 1879. | 1880. | 1881. | 1882. | 1883. |
| 5s. | 5s. | 6s. | 7s. | 7s. | 6s. 6d. | 5s. 6d. | 5s. | 4s. 8d. | 4s. 4d. | 5s. | 5s. | 5s. | 5s. |

[1] These are considered as the average of the district. The figures given p. 520 were for a few mines only.

These quotations and costs of labour may serve as a means of comparison with other districts, some of which latter I have not visited for many years.

The average cost of transport of ironstone from the mines to the Middlesbrough furnaces, may be regarded as amounting to 1s. 3d. per ton.

The composition of the ironstone of Cleveland has already been given in these pages. In practice 65 to 68 cwts. of the mineral, as it is received from the mines, are required to produce a ton of pig iron.

Lincolnshire.—The geological measures found in the Cleveland hills pass in a south-westerly direction through Lincolnshire and Northamptonshire on their way to the English Channel, where they reach the sea in the county of Wilts. It is in the first two alone, however, that the ironstone they contain has been wrought to any extent.

The mode of working the ore in Lincolnshire is very simple. A bed of sand, 2 or 3 feet in thickness, is removed, and in many places immediately below it is a seam of ironstone, measuring 10 to 18 feet from top to bottom. It is so loose that for 1s. per ton it can be put into wagons. At another quarry 1s. to 1s. 6d. was given as the total cost of labour for working. In consequence of its very cheap extraction the landowners charge 1s. per ton as a royalty. It was sold, delivered at the works which are close to the ironstone workings, at 3s. 6d. in 1874, and yielded at this place 34 per cent. of pig iron.

The iron it contains has no doubt been in the form of carbonate, but exposure to atmospheric influence has converted it almost entirely into a hydrated peroxide. The mineral has a very varied composition, changing at every foot of depth in a remarkable degree. An analysis was made of each foot, where the bed had a thickness of 20 feet. The composition of the first 8 feet is given as examples of what has just been said:—

	Iron.	Silica.	Lime and Carb. Acid.	Water.
1st foot	21·09	5·71	36·86	12·13
2nd ,,	26·92	5·59	25·46	13·98
3rd ,,	23·16	4·41	39·14	8·11
4th ,,	37·13	5·06	18·47	7·17
5th ,,	20·09	3·92	46·43	5·44
6th ,,	37·05	4·35	4·54	17·86
7th ,,	36·09	4·28	10·11	16·06
8th ,,	28·25	7·99	17·18	14·12

SECTION XVIII.—CHIEF IRON-PRODUCING COUNTRIES COMPARED. 649

The average of these 8 feet is 28·72 per cent. of iron as it is worked, and 32·84 when dried at 212° F. Some contains only 11 to 15 per cent. of metal, and, of the whole 20 feet, above one-third, viz., 7 feet contains less than 20 per cent. of iron. The average of the entire bed at this place is 22·4 per cent. as worked, and 25·31 per cent. when in the dried state. For the furnace the poorer portions are thrown back, so as to give a mineral yielding about 34 per cent. of pig iron. It is worked raw, and requires no lime; indeed an admixture of a less calcareous stone is needed for working.

Northamptonshire.—Near Bletchley, the ore is worked opencast from a bed 18 feet in thickness, the covering of clay being 6 or 10 feet. It is like that of Lincolnshire, in the form of a hydrated peroxide. It is so easily worked that it only costs 5d. per ton to put it in wagons. The royalty varies from 6d. to 1s. per ton. Instead of being calcareous, as the Lincolnshire, the Northamptonshire stone is highly siliceous. Labour is cheap here: the miners were not receiving at the time of my visit above 3s. 2d. per day. The mineral, giving 38 per cent. of pig iron, costs 2s. 8d. to 3s. 9d. or 4s. delivered at the furnaces, according to the distance from the quarries.

Scotland.—No place presents greater uncertainty in arriving at any conclusion respecting the cost, delivered at the furnaces, of the ore employed. The black band in former times was chiefly obtained near the furnaces; but now when this mineral is smelted it has to be brought from various distances, and being scarce it is worked under circumstances which have greatly increased its cost. The quantity deficient is partly made up by working clay band ironstone, which also varies considerably in cost of extraction and in the distances from which it is brought. To some extent also the deficiency of ore is made good by the importation of the mineral from Spain. From a conversation I had with the late Mr. William Baird, the senior partner at Gartsherrie, I believe I am within the mark in stating that at the present moment, with cheaper labour than is paid in North Yorkshire and South Durham, pig iron is costing 60 per cent. more to produce than it did in former times, as described to me by Mr. Baird. There is moreover an important difference between the two periods in question, viz., that when the iron was being cheaply manufactured it was sold at a high price, whereas now (in 1883) that the expenses of production are greatly increased, the present market value of the result is generally stated by the makers themselves to be below cost price.

In 1879 I heard of calcined black band being sold at 14s. 9d. at the pits, to which 1s. 6d. had to be added for conveyance to the furnaces. This price would leave a profit to the mineral owner, for in some cases I heard of it being raised at 10s. to 11s. at the pits. The yield would be about 63 per cent. of pig.

At another work I visited calcined black band was being delivered at the furnaces at 13s. At the same time I heard of calcined clay band costing to buy 10s. to 13s. per ton.

A hewer in the ironstone pits of Scotland cuts about $1\frac{1}{2}$ tons of raw black band or 1 ton of clay band for his days work. The average earnings of the staff engaged in an ironstone pit was, at the period of my enquiry, only 2s. 8d. per day. Nothing is allowed for house rent.

At one work the ironstone miners averaged from 3s. 1·51d. to 3s. 9d. per day, and the drawers from 3s. to 4s., working $5\frac{1}{4}$ days in the week. At that time the men were under notice of a reduction of 10 per cent.

South Staffordshire.—The native ironstone lies in thin bands or in nodules embedded in shale of the coal formation. To get the stone the workman has to excavate so much shale that he only gets about 25 cwts. for his day's work. The actual cost of working probably varies considerably. Its market value is about 15s. per ton, yielding 35 per cent. of pig iron. This price—equal to 43s. for ore alone on every ton of iron made from it—easily explains the decadence of ironstone mining in this old and celebrated seat of the iron trade. Considerable quantities of Northamptonshire stone are now smelted in South Staffordshire, costing about 7s. delivered, so that the cost for ore per ton of pig will be about one half that obtained from native ironstone.

North Staffordshire.—The ironstone worked in this district consists largely of black band. In one pit I descended, the property of His Grace the Duke of Sutherland, the bed had a thickness of about 4 feet. At one ironwork I was permitted to visit, three seams of ore

SECTION XVIII.—CHIEF IRON-PRODUCING COUNTRIES COMPARED.

are found :—(1) "Black band," 14 inches thick, lying on the top of 18 inches of poor coal; (2) "Red slag ironstone," 16 inches thick, lying above 2 feet of poor coal; (3) the "Red Mine stone," 20 inches thick, with 18 inches of coal. Besides these there is a bed of clay ironstone $3\frac{1}{2}$ feet in thickness.

These valuable beds of ironstone lie in the carboniferous measures, where the seams of coal vary from 3 feet to 7 feet in thickness. The yield of pig iron from the calcined minerals used was about 50 per cent. The coal was chiefly used raw in the furnace, and something under two tons was required per ton of pig, a very high temperature of blast being maintained.

This mineral field is not a large one, but as far as it extends it is, from what has been stated, eminently valuable.

Cumberland and Lancashire—Hematites.—For many years the chief importance of the valuable ores of these two counties arose from their use as an admixture with the poorer ironstones of Staffordshire and Wales. This was particularly the case in the Principality, where the Welsh ironstone did not greatly exceed 25 per cent. of metal, and where the employment of forge cinders, rich in phosphorus, required correction by the presence of an ore containing but a small percentage of this metalloid. After the invention of the Bessemer process, in which a very pure pig iron was indispensable, the demand for metal made exclusively from hematites increased so greatly that ultimately large importations of foreign ores into Great Britain have been required to supplement the produce of our native mines.

In Cumberland and Lancashire the hematite ore lies in pockets, as they are termed, or in veins. Both are so irregular in point of size and in general conditions that it would be impossible to say what constitutes the average output per man. This varies, as might be expected, not only in different mines but in different places of the same mine.

In two of four mines known to me the average output was a trifle under 15 cwts per day for all the hands employed above and below ground. In the third it was about $33\frac{1}{3}$ cwts., and in the fourth 36 cwts. per diem.

652   SECTION XVIII.—CHIEF IRON-PRODUCING COUNTRIES COMPARED.

The following analyses of samples taken from the depôts where the ore is stored ready for delivery to the furnaces give a fair idea of the composition of the produce of the Lancashire mines:—

									Average.
Peroxide of iron...	72·21	74·24	70·17	80·90	75·34	73·21	76·73	78·07	75·11
Moisture ...	6·91	7·20	6·61	7·32	5·14	6·47	6·91	7·06	6·70
Insoluble in acid, chiefly silica ...	16·18	12·14	19·36	5·90	17·69	12·47	9·68	9·99	12·93
Soluble do., chiefly lime	4·70	6·42	3·86	5·88	1·83	7·85	6·68	4·88	5·26
	100·00	100·00	100·00	100·00	100·00	100·00	100·00	100·00	100·00
Metallic iron ...	50·55	51·97	49·12	56·63	52·74	51·25	53·71	54·65	52·58

The average wages of all engaged above and below ground may be taken at 4s. to 4s. 3d. per day.

In the present depressed state of the trade the price of ore, containing from 51 to 52 per cent. of iron, is about 10s. per ton, to which may be added 1s. 6d. for delivery at the furnaces.

ORE IN GERMAN ZOLL VEREIN.

Rhenish Provinces.—These are an old seat of the iron trade. In the Lahn valley red and brown hematites, as well as spathose ore, have been wrought for many years. The average yield of the hematites in iron was given me at 52 per cent., and the cost a little above 10s. per ton.

In the coal district of the Ruhr clay ironstone is obtained. There is besides a certain quantity of black band found in Westphalia, which costs 10s. to 12s. per calcined ton, yielding 50 to 55 per cent. of pig iron. The Nassau red ores yield from 45 to 48 per cent., and were quoted at 13s. to 15s. in 1878, delivered at the furnaces. The cost of roasted spathose ore, containing 48 per cent. was given in 1882 as high as 22s. per ton; it is very valuable for the steel makers, as it contains 8 per cent. of manganese. In 1878 this valuable ore was quoted at 20s.

At the market rates of ore and their yields of pig iron I estimated that in 1878 the cost for ore per ton of metal was as follows:—

Common Pig for Puddling.	Best Puddling Pig.	Foundry Iron.	Bessemer Pig.
22s.	28s. 6d.	27s.	38s. 6d.

SECTION XVIII.—CHIEF IRON-PRODUCING COUNTRIES COMPARED.

At the present time large quantities of ore are brought in from Bilbao, which in 1882 cost 23s. per ton delivered at the works. Since that time freights as well as the ore itself have receded considerably in price. Ships bring it to the United Kingdom and take their chance of finding a market for it on their arrival. In this way it has been delivered occasionally at as low a price as 10s. 6d. at Cardiff, so that it is possible moderate quantities may now find their way to Westphalia at 18s. to 20s. per ton.

Ore was being used near Dortmund, obtained from Sauerland, which only cost 8s. to 9s. per ton and yielded 40 to 48 per cent. of iron.

At the period of my visit very sanguine expectations were entertained of obtaining supplies from a bed of oolitic ore in Hanover. It was said to contain 40 per cent. of iron, and to cost 6s. 6d. delivered at the furnaces, the actual cost of extraction not exceeding 1s. 8d. per ton: but I have since been informed that the results have not proved as satisfactory as were anticipated.

OOLITIC IRONSTONE OF WESTERN GERMANY AND EASTERN FRANCE.

The great bed of this ore (corresponding to a great extent with the geological position of the Cleveland and Northamptonshire ironstone) in thickness and point of quality is at about its best at Metz and Luxemburg. It commences near the town of Luxemburg, and extends in great perfection to the south of Nancy, a distance of about 70 miles. It is worked considerably to the south of the latter town by the Creusot Iron Company, but there the bed is thinner and not so rich in iron as it is further north.

Department of the Moselle.—When I first visited this district, in 1867, the miners near Metz were being paid 1s. 7½d. per ton for working the bed of ironstone mentioned above. This included working the stone, loading it into wagons, and pushing them to the main roads, often a distance of about 100 yards. Two men worked together as in Cleveland, and in a shift of 10 hours they got 4 tons of good stone. The total cost of the stone at that time did not exceed 2s. 6d. per ton. The seam is about 13 feet thick, and is approached by a level. Interspersed in its substance are thin irregular masses of calcareous matter, which contain only 15 per cent. of iron, and are rejected as ore, but are sometimes used as flux. The stone is chiefly a hydrated peroxide, losing only 15 per cent. on ignition, and hence is used uncalcined in the furnace, in which state it yields 37 per cent. of iron. Owing to the calcareous nature of the mineral little limestone is required, and only for foundry pig. The stone is easily reduced, so that in a furnace 49 feet high a ton of white iron in 1875 was made with 22·80 cwts. of coke, partly brought from Saarbrück and partly from Belgium, although the blast had only a temperature of 350° C. (662° F.) By raising the furnaces to 66 feet, the coke consumed was reduced to 20·84 cwts. For grey foundry iron however something over 28 cwts. of coke was formerly consumed. Subsequently the temperature of the blast was raised to 440° C. (824° F.), and the consumption of coke was reduced in a 50-feet furnace to 21·40 cwts. for white iron. Firebrick stoves were applied to a furnace of 60 feet, but I believe that no economy was obtained by raising the blast above 500° C. (932° F.), at which temperature the coke for white iron was reduced to 18·74 cwts.

In the production of one ton of foundry iron, in 1878, 26 cwts. of coke were consumed in a furnace 50 feet high, blown with air at 957° F.; while in another furnace, 58 feet high, blown with air at 1,337° F., the consumption for the same quality was only 23 cwts. per ton of metal.

In 1878 the cost of the coke imported from Saarbrück was 15s.; from Westphalia, 17s. 6d.; and from Belgium, 20s. per ton delivered at the furnaces. In each case the coal used at the coke ovens was small, and sold at a lower price than the cost price of the entire output of the mine. The average cost of conveying the coke from the mines in the proportions used was about 4s. to 5s. per ton, making the average cost about 18s. or 19s. at the iron works.

A valuable deposit of the same ore as just described is worked near Longwy in the same department. It is so siliceous that the limestone added reduces the yield of the mixture to 25 per cent.

The following analyses indicate the composition of this ore:—

$Si_2 O_3$	...	14·76	...	25·29	...	25·78	...	14·30
$Fe_2 O_3$	...	55·58	...	41·18	...	47·98	...	58·66
$Fe O$	...	—	...	3·44	...	3·69	...	—
$Al_2 O_3$	...	4·51	...	1·80	...	·54	...	6·20
$Ca O$	...	8·13	...	8·91	...	5·38	...	5·30
$Mg O$	...	·94	...	·51	...	·61	...	1·97
$S O_3$	...	·10	...	trace	...	trace	...	·093
$P_2 O_5$	...	2·21	...	2·68	...	2·23	...	2·33
$C O_2$	...	4·91	...	7·62	...	4·92	...	4·07
$H_2 O$	...	8·86	...	8·60	...	8·85	...	7·41

SECTION XVIII.—CHIEF IRON-PRODUCING COUNTRIES COMPARED.

Luxemburg.—There are two beds of ironstone in the locality I visited—the red and the grey; the one 8 to 9¾ feet in thickness and the other about 11 feet. They crop out along a hill side, and are won in open work as long as the baring is under 30 feet in thickness, after which they are mined in the ordinary way. Two men in 12 hours, resting for meals, load a railway wagon holding 10 tons, for which they received (in 1878) 9s. 7d., they finding their own powder and oil. This is equal to 11½d. per ton, and left 3s. 2d. net for each man.

Some of the iron exists as silicate and none as carbonate. The metal is chiefly in the form of a hydrated oxide, the water of hydration amounting to 10 per cent. of the whole. The following analyses show some of the constituents of the two seams:—

	Red Seam.	Grey Seam.	Other Samples.		
Silica	9·16	8·10	—	—	—
Iron	37·63	34·40	42·45	35·50	31·40
Alumina	5·42	4·50	—	—	—
Lime	12·60	16·50	—	—	—
Magnesia	·49	·70	—	—	—
Phosphorus as phos. acid	·778	·83	·855	·72	·68

The following are complete analyses of various specimens of the produce of the Luxemburg mines:—

	Minerai Gris I.			Minette Rouge II.		Minette Rouge III.	
SiO_2	7·96	8·10	7·50	9·16	9·61	25·43	27·40
Fe_2O_3	46·19	49·15	47·21	53·75	60·06	52·02	42·47
FeO	·45	—	·62	—	—	—	—
Al_2O_3	2·05	4·50	3·50	5·42	6·30	5·07	5·93
CaO	18·86	16·35	16·82	11·60	6·23	4·08	9·17
MgO	·58	·70	1·53	·49	·39	·41	·50
MnO_2	trace	trace	trace	·10	trace	—	—
ZnO	—	—	trace	trace	trace	—	—
P_2O_5	1·92	1·91	1·75	1·78	1·98	1·89	1·61
SO_3	·12	·09	·10	·093	·17	·05	·12
CO_2	13·42	12·71	13·24	8·72	4·88	3·29	7·15
H_2O	8·34	6·53	7·15	8·93	10·45	7·72	6·12
Metallic iron	31·02	34·30	33·21	37·62	42·04	36·41	29·72
Thickness of bed	6½ to 13 feet.			6½ to 16¼ feet.		6½ to 9¾ feet.	

We have thus a total thickness of 19½ to 39 feet of mineral, averaging about 35 per cent. of metallic iron. No. I. is the lowest, No. II. lies 32½ to 40 feet above No. I., and No. III. 32½ to 36 feet above No. II., so that there is approximately an average of 29 feet of ironstone beds comprised in a space of 160 feet of strata. It would be very difficult to find a place richer in this ore or where the mineral is more easily won than it is in this part of the duchy.

It will be seen that both seams are rich in phosphorus, as indeed is proved by the composition of the pig iron, as below:—

	Analyses of Luxemburg Iron.	
	White Forge Iron.	Grey Foundry Iron.
Silicon	·460	2·012
Phosphorus	1·892	1·902
Sulphur	·431	·060

Including a royalty of 6d. to $9\frac{1}{2}$d., the cost of the Luxemburg ore varies from 2s. 6d. to 3s. per ton at the mines, which are sometimes close to the furnaces.

The coke is brought from Westphalia and Belgium. The former cost 17s. $2\frac{1}{2}$d. in 1878 delivered at one set of furnaces, the price at the ovens being 8s. 4d. to 9s. $1\frac{1}{4}$d. That from Belgium was charged 11s. $10\frac{1}{2}$d. at the ovens. The cost of transport varied from 4s. $9\frac{1}{2}$d. to 5s. $11\frac{3}{4}$d., or an average of 5s. 3d., so that the average cost at the furnaces was 17s. $1\frac{1}{2}$d. The Westphalian coke contained $7\frac{1}{2}$ to 10 per cent. of ash and the Belgian 11 to $13\frac{1}{2}$ per cent.

This Luxemburg ore is smelted in furnaces $58\frac{1}{2}$ feet high, with a content of about 12,000 cubic feet. The blast is heated to 350 to 400° C. (672 to 752° F.) They were running 700 tons of white iron per week at the period of my visit (1872), but this greatly exceeds the average hitherto attained. In 1872 the make of each furnace worked out to 19,230 tons (of 1,000 kilogrammes), and in 1874 to 18,461 tons, chiefly white iron. The coke consumption for this quality was 22·60 cwts.; but for foundry iron it was very much higher. It is not usual to employ any limestone. As many as 100 men are required to man each furnace, which is more than double the number employed in Cleveland.

Department of the Meurthe.—The ironstone which commences in Luxemburg is worked very extensively in this department between Pont à Mousson and in the Nancy district. The beds which furnish it are much thinner than in the duchy, 1 to 2 metres ($3\frac{1}{4}$ to $6\frac{1}{2}$ feet) being the usual thickness. It was sold (1878) at prices varying from 2s. $4\frac{1}{2}$d. to 3s. $11\frac{1}{2}$d. delivered into boats on the canal or into railway wagons. The average cost of extraction was given at about 2s. 6d.

SECTION XVIII.—CHIEF IRON-PRODUCING COUNTRIES COMPARED. 657

The richest ore I heard of contains 45·91 per cent. of iron, but the average is not above 37 per cent. The ore generally contains 10 to 16 per cent. of silica, and 5 to 6 per cent. of lime. The phosphorus varies from ·5 to nearly ·9 per cent.

At one place the furnaces were 55¼ feet in height, and contained a trifle under 6,000 cubic feet. The blast was heated to 400° C. (752° F.) The ore cost 3s. 0½d. delivered at the mines, with 11½d. transport to the furnaces. The make was about 315 tons of grey foundry iron per week, for which, 28·60 cwts. of coke were used; whereas 20 cwts. per ton sufficed for making white pig. The coke used was brought from Anzin, Westphalia and Saarbrück, and cost 20s. 10d. to 21s. 8d. delivered alongside the works.

At other works the coke consumed had the following composition:—

	Saarbrück.	Anzin.	Ruhr.
Ash	11·12	13·10	9·50
Water	5·70	2·00	4·10
Carbon	83·18	84·90	86·40
	100·00	100·00	100·00
Cost at ovens	11s. 6d.	13s. 5½d.	8s. 8½d.

The cost of transport from the Ruhr was about 12s. 9d., duty and expenses 1s. 3d., total 14s. The duty and carriage from Saarbrück was 9s. if by rail and 4s. 6d. by canal, and from Anzin 7s. 8d. by boat. The cost of coke at the Nancy group of furnaces, in 1878, would be 20s. 6d. to 21s. 1½d. There is a considerable loss from breakage on coke brought from so great a distance, Saarbrück being 62 miles, Anzin 277 miles, and Ruhrort 215 miles from the furnaces. With firebrick stoves I heard of 21 cwts. of coke being consumed per ton of grey iron.

At another mine the cost of working the stone was named at 2s. 5d., which with carriage brought the price to 3s. to 3s. 2½d. delivered at the furnaces. The coke contained 14 per cent. of ash, and sometimes as much as 10 per cent. of water. That brought from Saarbrück cost 18s. in 1878 by boat, and from Anzin 22s. also by canal. For white iron 21·60 to 23·60 cwts. of coke were consumed per ton of pig, with blast at 375° C. Upon certain occasions no limestone is used, upon others flux is required, according as the ore is calcareous or the reverse.

P P

St. Dizier.—It is many years (1867) since I visited this neighbourhood. It is an old seat of the French iron trade, where until recently charcoal was the only fuel used. A canal in process of construction enabled Prussian coke to be brought at a cost of 26s. 4¾d. per ton. Upon its completion it was expected to reduce this price to 23s. 2½d.

The ore employed is a brown oxide, found in pockets, and was conveyed to the works along the ordinary roads of the country. The items of cost were 1s. 7d. for working, 4¾d. for royalty, and 2s. 0¼d. for carting, a total of 4s. It contains 32 per cent. of iron. A more distant mine afforded an ore which yielded 42 per cent., but the cost was 8s. per ton. The proportions used were three parts of the former and one of the latter. The furnaces at that time were 42 to 45 feet high, making from 56 to 115 tons per week. The coke required for a ton of mottled and white iron ranged from 22 to 24 cwts., but for a whole year at the works of my informant, the work was done for 20 cwts. The native ore by itself was said to be unfit for making foundry iron.

Central France.—The works at Fourchambault, etc., in 1867 were supplied with ore from irregularly-shaped pockets found in the limestone rock. Pits were sunk through the superincumbent strata (about 100 feet in thickness) to the ore, which occurs in the form of gravel, and being mixed with earthy matter requires washing. It is an expensive ore, costing about 10s. per ton, and only yielding 35 to 40 per cent. In smelting it, half its weight of limestone was being used. The coke was then brought from the Commentry coal-field, at a cost of about 16s. 6d. per ton. In furnaces of 40 feet the consumption of coke was 32 to 35 cwts. per ton of grey iron. At Commentry itself white iron was being made from ore similar in nature to that used near Fourchambault, etc., with 21·46 cwts. of coke and 14 cwts. of limestone.

It would appear from what has preceded that the works of Central France are by no means favourably situated in respect to the ores of iron. The same may be said with regard to those of St. Etienne. In 1867 the furnaces there were being supplied with ore from La Voulte on the Rhone yielding 45 per cent. at 12s. 9½d., with Mokta ore yielding 65 per cent. at 28s., and with Barcelona ore, containing 38 per

SECTION XVIII.—CHIEF IRON-PRODUCING COUNTRIES COMPARED. 659

cent. of iron besides some manganese, at 25s. 7d. per ton. A certain quantity of an ore, apparently oolitic, containing 30 per cent. of iron, was used for common purposes.

In furnaces of very moderate dimensions, with blast at 280° C. (only 536° F.), they were making Bessemer iron with something under 26 cwts. of coke, and using about $9\frac{1}{2}$ cwts. of limestone.

Pas de Calais.—Here in the green-sand formation, as I was told, and immediately below the clay, a bed of brown hematite, 2 to 4 feet in thickness is worked, which would appear to correspond with the position of the mineral worked for centuries in Sussex, and only discontinued when charcoal iron could no longer compete with that obtained by the aid of fossil coal. The baring is nearly 10 feet in thickness, and the ore is mixed with so much sand and clay that it loses half its volume in the process of washing. The royalty, $10\frac{1}{2}$d. per ton, and the value of the land which is damaged makes it somewhat expensive. It has to be carted to the works over a distance of a mile or two, so that laid down at the furnaces the cost is about 9s. 6d. per ton for a material yielding 35 per cent. of metal.

It was being smelted with expensive coke, made from coal brought from the county of Durham, and therefore costing 23s. to 24s. per ton. Of this 22 cwts. were used with 10 cwts. of limestone per ton of white iron.

Iron Ore in Belgium.

At one time the pig iron made in Belgium was wholly from native ores. But in the year 1882 the total quantity of pig iron smelted was 717,000 tons[1], and the native ores used amounted to 250,000 tons, while the imports reached 1,206,717 tons: thus the native ore was only about one-sixth of the entire consumption. In 1867 a dozen varieties of native ores were being used in the Liége group of furnaces. At that period Luxemburg ore was being imported at a cost of 5s. 6d. to 6s. 9d. for carriage alone, the yield in pig iron being then given at 33 to 35 per cent. In furnaces about 50 feet in height forge iron was being made at that time with $20\frac{1}{2}$ cwts. of coke, the limestone being 14·80 cwts. per ton of pig. For foundry iron 30 cwts. of coke or more was the rate of consumption.

[1] Iron and Steel Institute Foreign Reports 1883, p. 398.

In 1878 I found that at one work near Liége the use of native ores had been entirely abandoned, being replaced by that of Luxemburg, for pig of ordinary quality. The cost of transport had been reduced to rates varying from 4s. 9d. to 5s. 6½d. per ton. The price at the mines was only about 2s. 6d. per ton. The yield in the furnace was 33 to 34 per cent., and the white pig iron thus made was got by a consumption of 21 to 23 cwts. of coke per ton of whitish mottled metal, the limestone used weighing 7·20 cwts.

At that time (1878) Spanish and African ores were being used for Bessemer iron, the inland cost of transport being 2s. 4½d. by boat and 3s. 6¾d. by rail. The sea freight was then 14s. 3d. from Bilbao to Antwerp. The reduced rates of freight will enable this ore to be delivered at the present day (1884) much below the price just named, for, as already mentioned, it is being supplied in limited quantities at Cardiff for 10s. 6d. per ton. The cost price of coke, taking the selling prices of small coal to be 7s. 1½d., would be nearly 12s. 6d. per ton unwashed.

The coke consumed at one work was given as follows :—

	Cwts.	
For commonest quality of forge iron (*métis*)	21	Blast heated
„ common forge pig (*fer ordinaire*)	22	in
„ strong forge pig (*fer fort*)	23	metal stoves.
„ Bessemer iron from rich ores, furnace 60 feet in height	20	Fire-brick stoves.

At another work 23 cwts. of coke were mentioned as being consumed per ton of forge iron, chiefly smelted from Luxemburg ore. It contained 12 to 14 per cent. of ash. It was made from coal paid for at the rate of 8s. 8½d. per ton, and yielded 80 per cent. of coke at a cost of 11s. 6d. per ton.

The blast was heated to 450° to 500° C. (842° to 932° F.) at one establishment; and the coke consumed was 21 to 22 cwts. per ton of white and 27 cwts. per ton of foundry iron. Coking coal was bought at the rate of 7s. 11d. to 8s. 8½d. per ton.

Iron Ore in the United States.

The production of pig iron in recent years in the United States has been as follows :—

1879.	1880.	1881.	1882.
2,741,853	3,835,191	4,144,254	4,623,323 gross tons.

SECTION XVIII.—CHIEF IRON-PRODUCING COUNTRIES COMPARED. 661

The make of 1882 was 5,178,122 tons (of 2,000 lbs.) and was thus distributed:—

States.	Net Tons.	States.	Net Tons.
Pennsylvania	2,449,256	Connecticut	24,342
Ohio	698,900	Colorado	23,718
New York	416,156	Massachusetts	10,335
Illinois	360,407	Indiana	10,000
Michigan	210,195	Minnesota	8,126
New Jersey	176,805	Oregon	6,750
Tennessee	137,602	Maine	4,100
Missouri	113,644	Texas	1,321
Alabama	112,765	Vermont	1,210
Virginia	87,731	North Carolina	1,150
Wisconsin	85,859	California	987
West Virginia	73,220	Utah Territory	57
Kentucky	66,522		
Maryland	54,524		5,178,122 net tons.
Georgia	42,440	equal to	4,623,323 gross tons.

The division according to the fuel used stands thus in net tons:—

Anthracite Coal.	Bituminous Coal.	Charcoal.	Total Net Tons.
2,042,138	2,438,078	697,906	5,178,122

The ore obtained from the American mines, so far as they are given by Mr. Swank in the report of the American Iron and Steel Association, is as follows in gross tons:—

	1879.	1880.	1881.	1882.
Lake Superior, including Menominee Range	1,414,182	1,987,598	2,321,315	2,943,314
New Jersey	488,028	745,000	737,052	900,000
Lake Champlain	—	—	637,000	675,000
Cornwall Bank, Pennsylvania	—	—	249,050	309,680
			3,944,417	4,827,994
Imported	—	493,408	782,887	589,655
			4,727,304	5,417,649

The statistics of the iron ores of the United States are in a very incomplete condition, as may be inferred from the fact, that the figures given above embrace all the information on the subject which the industry of Mr. Swank has been able to collect.

The only other data I have been able to lay my hands on is a return from Ohio for the year 1872[1] and another from Missouri for 1875, from a work by Pechar.[2] The former produced 336,758 tons and the latter 370,000.

According to the information received at the various works during my two visits to the United States, I have estimated the average yields of the ores already enumerated in gross tons, to be as follows :—

	1882. Tons Ore.	Per Cent.		Yield of Pig. Tons.
Lake Superior	2,943,314	63	=	1,854,287
New Jersey	900,000	60	=	540,000
Lake Champlain	675,000	63	=	425,250
Cornwall Bank	309,680	52·5	=	162,582
Imports	589,655	52	=	306,620
Gross tons of 2,240 lbs.	5,417,649			3,288,739
Weight of iron in net tons				3,683,387

Supposing the furnaces in the following States to have been chiefly supplied from the sources above-mentioned, we know that the pig iron made in them was as below :—

	1882.
Pennsylvania	2,449,256
Ohio	698,900
New York	416,156
Illinois	360,407
Michigan	210,195
New Jersey	176,805
Connecticut	24,342
Massachusetts	10,335
Indiana	10,000
Maine	4,100
Vermont	1,210
	4,361,706
Supposing half the cinder made in the mills and forges to be returned to the furnaces and each ton of malleable iron to afford 2 cwts. of such pig, we have approximately from this source[3]	124,691
Pig from ore, net tons	4,237,015

[1] Ohio Statistics, 1873, p. 269.
[2] Pechar, Coal and Iron, p. 177.
[3] Malleable iron made was 2,493,830 tons.

SECTION XVIII.—CHIEF IRON-PRODUCING COUNTRIES COMPARED. 663

The malleable iron made in America is chiefly produced in the States just given.

It should be observed that in the year we are considering 48,354 tons of blooms were made in Catalan fires, also chiefly in the twelve last-mentioned States. If we regard this quantity as pig iron, which, although not strictly correct, will not materially affect the estimate, we have :—

	1882. Net Tons.
Pig due to ore	4,237,691
Blooms do. considered as pig	48,354
	4,286,045
Pig iron equivalent to ore worked in localities specified ...	3,683,387
Pig iron due to sources not specified	602,658

As a considerable portion of the unspecified sources are clay iron-stones of Ohio, etc., we shall probably be quite safe in assuming their average yield not to exceed 50 per cent.—if so, the weights of ore from these sources ought to weigh for the 602,658 tons of pig iron 1,205,316 net or 1,076,175 gross tons.

The make of pig iron in the Southern and Western States for the two years in question was :—

	1882.
	816,416 net tons.
Equal to about	728,943 gross tons.

The ironstone of the Hanging Rock region being carbonate, and the brown hematites of Tennessee, etc., not being rich, we may roughly take the whole at 53 per cent. of iron, and regard the weight of the ore smelted in 1882 to produce the 728,943 tons of pig iron as amounting to 1,375,330 gross tons.

664 SECTION XVIII.—CHIEF IRON-PRODUCING COUNTRIES COMPARED.

The total quantities of ores dealt with will be as follows :—

	1882. Gross Tons.
Ore worked at Lake Superior, New Jersey, Lake Champlain, Cornwall Bank, and imported	5,417,649
Assumed produce of other localities beyond these just specified	1,076,175
Assumed produce of Southern and Western States	1,375,330
	7,869,154
Imported ore to be deducted	589,655
Ore raised in United States	7,279,499
The total weight of pig and blooms made was	4,671,677
Pig due to imported ores ... 306,620	
Do. from mill cinders... 124,691	
	431,311
Due native ores	4,240,366

Viewed in the manner just described the average yield of the native ores works out to 58·25 per cent. for 1882.

Calculated in the same way the British ores afford, according to the Mine Inspector's returns, the following results :—

	1882.
Raised—West Coast hematites	3,134,171
Other native ores	14,897,786
Total native ores	18,031,957
Imported ores	3,284,946
Total	21,316,903
Pig iron made	8,586,680
Manufactured iron made—	
1882.—2,841,534 tons cinder from which gave of pig say	284,153
1882.—Due to ores native and imported	8,302,527
Due to the imported ores at 52 %	1,708,171
Due to native ores	6,594,356
Average yield of British ores	36·01 %

SECTION XVIII.—CHIEF IRON-PRODUCING COUNTRIES COMPARED.

It will thus be seen that the American ores yield above 60 per cent. more iron than those of Great Britain. Notwithstanding this great advantage enjoyed by the former, it will be seen when we come to examine the other conditions, that the manufacturers of Great Britain are much more favourably placed than the average of those of the United States.

DISTRICTS—LAKE SUPERIOR AND MENOMINEE RANGE.

Within the memory of men I conversed with at Marquette, the the wild Indian held almost undisputed sway over all the country adjacent to this field. Even now descendants of the old possessors of the soil reside in their wigwams in their primeval forests on the edge of the lake, and I saw them engaged in unloading cargoes of ore at the Bay furnaces. Now the country is studded with mines, as may be supposed from the fact that about 3,000,000 tons of ore are obtained from the veins which traverse it in various directions.

Not only does the district abound in iron, but it is one of the richest copper regions in the world. As an example may be mentioned the Cliff mine. In 1848 the Company bought 5,000 acres of land from the government for 11,000 dollars, and in a very short time they began to divide 20,000 dollars per month as profits. The copper is native, and in 1854 a mass 80 feet in length and 20 feet in width was met with.

About a dozen miles from Houghton I descended one of the slopes belonging to the owners of the Heckla and Calumet mines. There the copper, although often occurring in great masses, is usually disseminated through the rock in small grains, which amount to about 5 per cent. of the whole. In 1875 these mines produced 13,165 tons of metallic copper, the total make of the district being for that year 22,941 tons and 22,225 tons for 1874. The original capital was 800,000 dollars. They have accumulations valued at 4,000,000 dollars, and have paid in dividends 9,440,000 dollars.

Upon the occasion of my visit in 1876, I examined more or less attentively ten of the chief iron mines. It would occupy too much space were I to attempt to describe the nature of the pockets or veins of ore. The important matter for consideration is the expense at which the mineral can be obtained and its value in the market.

A very important mining company raised 110,150 tons of ore in the year 1873. For this the miners and men working by the piece averaged 74·5 in number for the whole year. The output was therefore close on 148 tons per man engaged. The men working by the day averaged 86·3, making in all an average of 160·8 hands employed, so that the actual output for the entire staff was 68·38 tons per annum. The daily output for each shift worked was almost exactly 2 tons. The cost for working each ton was as follows :—

	Dollars.		Dollars.		Per Ton. s. d.
Piece-work	·21 days at 3·40	=	·71	=	2 8¼
Day work	·24 days at 2·27	=	·55	=	2 0¾
			Total labour	...	4 9¼

The lands being the property of the Company there is no royalty to pay, so that if to the labour we add 1s. 8¾d. for stores, etc., we have 6s. 6d. as the cost put into wagons on the railway.

In the year 1877 the output of this mine was something above one-half of that raised in 1873. For each ton of ore worked there was ·36 ton of rock brought to the surface. This proportion by no means always represents the total sterile matter which has to be dealt with, a good deal often remaining below.

During two months of the year (1877) 239 men worked in all 12,166 days, each man thus averaging 51 days. The average product per man per day had fallen from 2 to 1·04 tons. The average earnings of the entire staff was 1·80 dollars against 1·88 dollars in 1873. A good deal of preparatory work was in hand, although the actual mining labour was something less, viz., 1·16 dollars against 1·27 dollars in 1873. The cost of the ore loaded into wagons had risen fully 22 per cent. in 1877.

At these rates of labour the piece men were earning about 3·40 dollars per day, and the day men 2·27 dollars; which at the then rate of exchange (3s. 9½d.) was equal to about 12s. 10d. and 8s. 7d. per diem respectively.

The cost just given for raising the ore was considered an exceptionally low one—the more common figures varying from 2½ to 3 dollars, say 9s. 4d. to 11s. 3¾d. per ton.

The cost of conveying the ore to Cleveland City, where there are blast furnaces and which is the terminus of railways to Ohio and Pennsylvania, is very fluctuating in character. The railway charge is

SECTION XVIII.—CHIEF IRON-PRODUCING COUNTRIES COMPARED. 667

high to Escanaba on Green Bay, viz., 1·3d. per ton per mile, and amounts to 6s. 9d. per ton. The lake freight when I was first in the country in 1874 was 8s.; but it varies very much according to the quantity of grain that has to be carried. At one time it was as low as 4s. 6d., and at others as high as 15s. to 19s. If we assume 14s. 9d. as a mean cost for conveyance from the mines to Cleveland City, and 9s. 3d. as an average cost of the ore at the mines, we have 24s. as the total cost delivered at a wharf ready to be forwarded by rail to any part of the States. Now the selling price for 4 or 5 years previously to 1874 had varied from 30s. 2d. to 32s., leaving a profit of about 7s. per ton.

In 1876 the following particulars respecting the cost of raising ore were given me as appertaining to a very extensive mine :—

For every 100 tons of ore drawn to bank there were brought up 40 tons of rock, and nearly 300 tons of water. The expenses of working were as follows :—

	Per Ton.
	s. d.
Labour	5 4
Explosives	0 6¼
Steel	0 0¾
Pumping and drawing...	0 6¾
Tools	0 1
Horse keep	0 3½
Railways	0 1½
General repairs, &c.	0 6¼
Explorations	0 6¾
	8 0¾

The information given me at another mine shewed almost exactly 10 cwts. of sterile matter drawn to bank for each ton of ore. This mine, like many others, was a quarry. Its depth was about 150 feet. The total cost of the ore delivered into wagons was 8s. 5½d. in 1876, and the selling price was 15s. 1d. It yielded 65 per cent. of pig. The rail dues to Marquette were 3s. 1d. per ton, and to Escanaba 4s. 3¾d. The lake freight to Chicago from Escanaba had been sometimes 7s. 6½d., at other times as high as 11s. 3¼d., but at the moment it was only 3s. 9¼d. from Escanaba, and from Marquette 5s. 7¾d.

I obtained the particulars of work done at one of the blast furnaces near the mines. It had a height of 41½ feet with a bosh of 9½ feet in diameter, blown with air varying from 720° to 820° F.

Week.	Weight of Iron (Grey No. 1). Net Tons.	Bushels Charcoal.	Yield of Ore.
1st	216¼	107·92	67·56
2nd	228¼	109·21	67·60
3rd	262	106·28	66·84
4th	261	110·63	66·28
5th	253	107·39	69·12

Taking the bushel of charcoal at 20 lbs., and the consumption at 110 bushels, we have 2,200 lbs. of fuel for 2,000 lbs. of iron, or 24·6 cwts. per ton. They were not at this furnace exceeding 1 cwt. of limestone per ton of pig.

According to the figures already given, about 40 per cent. of all the iron made in the United States is the product of the mines in the district we are considering. The weight of the iron according to the moderate estimate employed—63 per cent.—was 1,854,287 gross tons. In Michigan itself only 187,665 gross tons (210,195 net tons) were smelted in 1877, which means that about 2,600,000 tons of ore were sent to other seats of the iron trade (*vide p.* 622). The furnaces in the Mahoning and Chenango Valleys, at Pittsburg, the Lehigh Valley and Cleveland City are in fact largely dependent on the ores of Lake Superior.

Lake Champlain.—A continuation of the Adirondack range of mountains skirts the western shore of Lake Champlain, and near the summit of one of these, six miles from the lake and 1,200 or 1,300 feet above the level of its waters, lies the most famous deposit of the ore of this district. Embedded in the mineral for which Lake Champlain in recent years has become so famous, are layers of apatite, or natural phosphate of lime. It constituted, I was informed, 2 per cent. of the vein, and for some time a company extracted the apatite for agricultural purposes, handing over the iron ore it was necessary to move to the owners of the property. Although the existence of the vein has been known for about 60 years, it is only about 20 years since its value was rendered available to any extent. At the period of my visit in 1874 to this extraordinary deposit, the output was about 400,000 tons per annum. In 1882, according to the report of Mr. Jas. M. Swank, the production of the entire Lake Champlain district was 675,000 tons. At one point in the vein above referred to, the ore reaches the surface or is covered with a little soil: afterwards rock overlies the vein. The thickness of the latter was given me as varying between 150 to 200 feet. In working the ore a depth of 125 feet had been penetrated, forming a large chasm open to the day, but beneath the floor of the mine they had bored 80 feet at one place, 125 feet at another, and 147 feet at a third, before reaching the bottom of the vein. At that time the ore was being extracted under the covering of rock, the roof being supported by means of five huge columns of ore said to contain 70,000 to 80,000 tons of this valuable mineral. The miners had been earning 2·60 dollars per day, but this had been reduced to 1·75 per day in paper currency or 6s. 7d. for ten hours actual work. There is very little aboveground labour, beyond that for removing the covering of rock; but this occupies as many men or thereabouts as are engaged in the mine. I understood for every man working at the bottom of the mine 2 tons of ore per day were delivered at the surface. If so it is not likely, everything included, that the cost could exceed 6s. or 7s. per ton for labour.

The selling prices in 1874 were quoted as follows, in paper currency:—

	Dollars.		Per Ton. s. d.
Furnace ore	5½	=	20 8¾
Screened ore, small, for Catalan fires ...	6	=	22 7½
General run as worked	6½	=	24 6
Ore for puddling furnaces—lumps ...	7½	=	28 3¼

These figures included delivery on board craft on Lake Champlain, involving carriage on six miles of railway.

It was reported that the present owners paid only a few thousand dollars for about half the ground occupied by the vein. This and the figures just given justify what has already been said that the margin on selling such ore even in bad times of the iron trade such as 1874 suffices for a high royalty rent and handsome profit to those who work the ore.

Some portions of the produce of this mine contain a considerable portion of phosphoric acid; whereas others are sufficiently free from this ingredient to enable it to be used for Bessemer steel. The ore is chiefly magnetic. The following analyses by Prof. Chandler show its constituents :—

Magnetic oxide of iron	95·99 = Fe 69·51	Peroxide of iron	59·84	} = Fe 62·61	
		Protoxide of iron	26·69		
Alumina	2·00	Alumina	1·87		
Lime	·52	Lime	6·02		
Magnesia	·60	Magnesia	—		
Oxide of manganese ...	·10	Oxide of manganese	·55		
Silica	·64	Silica	3·45		
Sulphur	·10	Sulphur	·20		
Phosphoric acid ...	·10 = P. ·04	Phosphoric acid ...	1·94	= P. ·76	
	100·05		100·56		

The freight to Philadelphia by water was 13s. 11¼d. (3·75 dollars). The anthracite used in the furnaces is brought from a distance of 400 miles. It was usually received at New York, and cost as follows :—

				Per Ton.	
	Dollars.		s. d.		s. d.
Price at New York	4·70	=	17 8½		
Carriage from New York ...	1·37 to 2	=	5 2½	to	7 6
Making the cost delivered			22 11	to	25 2¼

From another quarter 23s. 9d. was given as the cost for smelting coal delivered at the lake side, or 26s. 4¾d. at the furnaces.

In 1872 and 1873 foundry iron—not fit for Bessemer steel—had been sold in this district for £9 8s. 6d. (50 dollars) per ton, and for some small lots as much as 70 dollars had been paid. These fluctuations in the iron trade were accompanied by reductions in wages. This led to resistance on the part of the men, who immediately before my visit had wrecked a house, and conducted themselves with such violence, that bills were displayed on the walls offering 1,000 dollars for the discovery of the offenders.

From data received the yield in the furnace appears to vary from 60·6 to 66 per cent. Six to seven cwts. of limestone are consumed per ton of metal. In 1860 the coal required—anthracite—was 30 cwts. but this has now been reduced to about 24 cwts. per ton of grey foundry iron.

New Jersey.—The iron mines of the state of New Jersey enjoy great advantages in point of geographical position as compared with

SECTION XVIII.—CHIEF IRON-PRODUCING COUNTRIES COMPARED. 671

those of Lake Superior. Instead of being 400 to 500 miles, as Marquette is, from the nearest coal-field, the produce of the New Jersey ore deposits are but 80 miles from the great Pennsylvanian anthracite coal basins.

The ore is of the hematite class, yielding from 55 to 66 per cent. of pig iron. So far as quantity and facility of extraction are concerned, the mines are greatly inferior to those in the states of Michigan and Wisconsin. In 1870 the output of New Jersey was 362,636 tons. In 1874, when I visited this mining field, the produce was about 500,000 tons, a quantity which afterwards was increased to 900,000 tons in 1882. The cost of bringing the coal and ore to a point midway between the two, such as to the furnaces in the Lehigh Valley, is about 3s. 4d. per ton. If $3\frac{1}{2}$ tons have to be moved, 11s. 8d., or inclusive of limestone say 12s. 6d., will represent the cost of transport charges on each ton of pig iron.

I visited two of the chief mines of the district. The vein at one varied from 12 to 40 feet in thickness, but there was a good deal of sterile matter interspersed with the ore. The staff of men engaged consisted of the following:—

Miners	82
Labourers underground...	16
Men at Surface	15
Sundry ,, ...	8
	121

When working full time 60 to 70 tons per day were considered good work; which is equivalent to 11 cwts. per man per day all told, or say 165 tons per annum for 300 days, making a total output of 21,000 tons. Good miners were earning $1\frac{1}{2}$ to 2 dollars per shift, the pay being at the rate of about 2 dollars per ton. The total cost, in 1874, was stated to be:—

	Dollars. Paper Currency.		s.	d.
Cost of working	5·0	=	18	$10\frac{1}{4}$
Royalty	·75	=	2	$9\frac{3}{4}$
Haulage to railway	·30	=	1	$1\frac{1}{2}$
Dues to furnaces...	1·10	=	4	$2\frac{1}{2}$
	7·15		27	0 at furnaces.

The yield of the ore is 66 per cent. of pig iron.

The other mine visited was producing 48,000 tons per annum, and cost as follows :—

	Dollars. Paper.		s.	d.
Cost delivered at the mine ...	3·75	=	14	1½
Royalty ... ... ·45 to ·75 say	·60	=	1	9
Carriage to furnaces	1·30	=	4	10¾
	5·65		20	9¼

The yield was given me as being only 55 per cent.

Below are sundry assays taken from the "Geological Survey" of New Jersey for 1879 :—

Metallic iron...	53·75	57·68	63·399	56·13	56·97	45·22	63·12	66·98
Sulphur ...	3·33	2·66	·068	7·59	·088	·028	trace	trace
Phosphorus ...	·036	·025	·010	·29	·367	·09	·017	·052

From these figures it is apparent that the mines vary much in per centage of iron, as well as in respect to sulphur and phosphorus, the former occasionally being very high.

Pennsylvania.—Cornwall Ore Bank presents an extraordinary deposit of ore. The mineral constitutes a somewhat irregularly shaped mass about three-quarters of a mile in length, with a mean width of about 500 feet. This ridge of ore is intersected by two small valleys dividing it into eminences known as the Big, Middle and Grassy hills. Bore-holes had been put down in one of these valleys, and had gone through the ore at depths varying from 50 to 134 and 179 feet. A spiral railway runs up Big Hill, and the ore is excavated by open work and thrown direct into the wagons. At the base of the hill the ore forms a cliff 350 feet in height. It is so soft that one man can blast and load into trucks 10 tons of ore for his day's work, at a cost, in 1874, of less than 6d. per ton, the entire cost of raising it being about 1s. per ton. The ore is very sulphureous, and contains from 50 to 55 per cent.—sometimes. indeed under 50 per cent.—of iron. The market price in 1874 was stated to be 4¾ dollars paper, or 17s. 10d. per ton. It is the property of one family who work it. At the year of my first visit the yearly output was 220,000 tons. In 1881, according to Mr. Swank, this was increased to 249,050 tons and to 309,680 tons in 1882.

The undertaking is connected with the public railway by a private line 7 miles long. There were in 1874 eleven furnaces, which are

SECTION XVIII.—CHIEF IRON-PRODUCING COUNTRIES COMPARED. 673

supplied with coal from the Schuylkill region, distant about 87 miles, for 1·80 dollars paper or 6s. 9¼d. freight, equal to ·80d. per ton per mile. The coal costs at the pit 2·63 dollars, making 14s. 5d. delivered at Labanon where the furnaces chiefly are, and perhaps 16s. at Cornwall Bank. In a furnace 55 feet high they were running 150 tons per week, using 37½ cwts. of anthracite and 12¾ cwts. of limestone per ton of iron. The keepers were only paid 1·65 dollars per shift, and assistant keepers and slag men 1·35 dollars.

Black Band of Tuscarawas Valley, Ohio.—This is the only locality in which I heard of this kind of ironstone being smelted. I had previously visited an opening into what was called Black Band on Davis Creek in Kentucky, a few miles to the south of Charleston on the Ohio. The distance from the Tuscarawas district is about 200 miles, and what I saw was the reverse of promising. On my return to America I visited Tuscarawas to see the Black Band, about which I had heard much in its favour, far more indeed than it deserved.

Instead of being situate in what may be termed the sub-carboniferous strata, like that in Scotland, the Black Band of this Ohio field is above the coal, and so high up that it forms in some places the tops of the hills and in others has no great thickness of rock above it. Under these circumstances it rarely happens that any particular working extends to above 20 or 30 acres, and they are frequently less than this—sometimes only a fraction of an acre. It lies thick in the ground, in one drift I entered it was 8 feet, but the stone is poor looking and when calcined only yields 50 per cent. of pig, and sometimes not so much. There is a pretty extensive coal-field in the neighbourhood, which is cheaply worked, 3s. 9d. per ton, including 1s. for royalty, being given me as the cost of extraction. I measured the coal at 4 feet 8 inches. It was being worked by level and had been recently opened, which may account for the low cost. The coal near the ironworks however is unfit for furnace purposes, so that the supplies are brought from the neighbourhood of Massillon, a distance of 18 or 20 miles, and cost about 7s. per ton at the ironworks.

A very unfortunate adventure was made at Port Washington on this field of ore. Two excellent furnaces were built 70 feet high. The coal turned out so bad that they brought coke from Connellsville, a distance of 150 miles. The black band seams were speedily exhausted

Q Q

and so poor that mixed with one-sixth of the rich ore of Lake Superior the yield was only 50 per cent. of metal. The limestone used was enormous in point of quantity—viz. 20 to 30 cwts. per ton of iron. The make was about 250 tons per week of No. 1 pig.

The cost of pig iron at another work in this neighbourhood was 91s. 8d., of which 30s. 7d. was for ironstone alone. Wages were low, in spite of which the sum for "labour, &c.," was set down at 15s 1½d. per ton.

My friend Prof. J. S. Newberry, in a report contained in the Statistics of Ohio (year 1873, p. 283) gives a complete analysis of some specimens of this black band. From these the following figures are selected:—

	No. 1.	No. 2.	No. 3.	No. 4.	No. 5.	No. 6.
Metallic iron ...	32·46	24·25	48·95	34·74	22·96	24·06
Phosphoric acid...	·41	·31	1·08	·51	·63	·49
Sulphur	·16	·28	·14	·35	·21	·11

Hanging Rock Region of Ohio and Kentucky.—This district may be regarded as a counterpart of that of South Staffordshire in England. The ironstone lies in three layers, or in nodules embedded in fire-clay or shale. It consists of several varieties as follows :—

Block or Limestone Ore.—Worked partly opencast and partly in mines. In the former, if there is much baring, a man will only get half a ton of ironstone per day. In underground work he will average 15 cwts. The earnings of the miners in 1874 were 2 dollars paper (7s. 6½d.), but in 1873 they were getting 3 dollars (11s. 3¾d.) The most industrious class is to be found among the Germans and Swedes. As a rule the miners do not save, the exceptions being those of the nationalities just named. Twelve to fourteen inches are considered a fair thickness for the layers which constitute this deposit, but occasionally they have been found 3 to 4 feet thick.

Block Ore somewhat resembles the last, but lying embedded in the shale above it are nodules.

Kidney Ores.—This is a local name for the nodular masses also found embedded in shale. The bed in which they occur varies in thickness from 3 to 6 feet. This variety constitutes one-half to two-thirds of the ore smelted in this district.

SECTION XVIII.—CHIEF IRON-PRODUCING COUNTRIES COMPARED. 675

Brown Hematite or Limonite.—A certain quantity of ore so designated is used, but it is merely one or other of the three kinds of clay ironstone "weathered" at the outcrop of the seams, the carbonate of iron being converted by atmospheric influence into hydrated peroxide.

The following figures are taken from the "Geological Survey of Kentucky," Vol. 1, New Series :—

Siderite or Carbonate of Iron. Unaltered Limestone Ore,—

Metallic iron ...	32·57	35·54	40·46	31·59	24·59	36·62
Sulphur	—	·53	1·85	·04	·10	·19
Phosphorus ...	·13	·11	·31	·21	·16	·18

Limestone Ores as Brown Hematite—

Metallic iron ...	49·99	47·50	39·02	51·07	34·73	56·14
Sulphur	—	—	—	·07	—	—
Phosphorus ...	·27	·06	·16	·21	·23	·05

Lower Block Ores unaltered—

Metallic iron ...	33·34	....	29·68	...	26·07	...	29·85
Sulphur	·08	...	·06	...	·10	...	·10
Phosphorus ...	·29	...	·05	...	·08	...	·19

Lower Block Ore as Brown Hematite—

Metallic iron ...	38·17	35·00	28·97	29·81	44·73	41·96	25·38
Sulphur	·13	·14	·06	·07	·06	·33	·04
Phosphorus ...	·42	·33	·25	·01	·14	·36	·16

Upper Block unaltered—

Metallic iron ...	38·14	...	36·10	...	32·46	...	30·97
Sulphur	·52	...	·41	...	·35	...	·48
Phosphorus ...	·22	...	·20	...	·30	...	·18

Upper Block as Brown Hematite—

Metallic iron ...	39·96	45·95	46·86	44·89	45·20	23·59
Sulphur	·15	·23	·27	·10	07	·07
Phosphorus ...	·16	·39	·05	·26	·07	·22

Kidney Ores unaltered—

Metallic iron ...	42·94	...	43·47	...	41·27	...	37·83	...	46·34
Sulphur	·01	...	·03	...	·08	...	·03	...	·07
Phosphorus ...	·34	...	·22	...	·16	...	·03	...	·05

The ironstone from any of these beds is evidently expensive from the quantity of sterile matter to be handled, and being situate from 3 to 13 miles from the works cost, in 1874, from 3 to 4 dollars paper—11s. 3¾d. to 15s. 1d.—per ton laid down at the furnaces.

At one establishment I visited they were smelting a mixture consisting of $\frac{2}{5}$ calcined native stone, $\frac{2}{5}$ Missouri ore costing 43s. 4d. delivered, and $\frac{1}{5}$ mill cinder. The cost of the ore per ton of pig iron, using this mixture, was about the same as if all native ironstone had been used. The furnaces are 60 to 65 feet in height, making 250 to 280 tons per week.

The coal, which has to be brought from distances of 5 or 6 miles, is used raw, and cost laid down at the furnaces about 8s. 6d. As much as 2 to $2\frac{1}{2}$ tons of coal are used per ton of pig iron, and 20 cwts. of limestone.

The iron made of this ironstone is much esteemed, but it is clear the Hanging Rock region can never, with such dear ironstone, be a cheap iron-making district.

Wages paid at the furnaces are not high—keepers when I was there getting 7s. $6\frac{1}{2}$d., helpers 6s. 1d. to 6s. 9d., fillers 6s. 7d.

Missouri.—The so-called Iron Mountain secures for the State of Missouri a high position in an iron-producing point of view. The produce of the deposit, distinguished by this somewhat high-sounding name, is a rich and therefore tolerably pure peroxide, slightly magnetic, and has, like some of the purer ores of the north, an extensive sale as a fettling for the puddling furnaces. In the matter of quantity however Missouri occupies a very inferior position to most of the great American iron-mining centres. I am without any recent data as to the exact production; but from such as I do possess it seems doubtful whether it has ever reached half a million of tons per annum. The surface of the ground consists of gravel of decomposed porphyry and clay, intermixed with a considerable proportion of rounded masses of the ore itself. This is washed to separate the impurities, when the mineral is ready for the market. The cost of separating the ore from the foreign matter is about 2s. per ton.

The main sources of supply however are derived from veins which run through the rock, chiefly porphyry. The upper portion of the latter, varying from 20 to 50 feet in depth, is decomposed and of a buff colour. It is so soft that by directing a stream of water, having a pressure of 70 or 80 lbs. on the square inch, against the face, the porphyry is removed, leaving veins varying from 1 inch to 3 or 4 feet in thickness. In other places however the iron ore forms very powerful veins. At one place it had a thickness of 30 to 60 feet, the total depth ascertained by boring being about 200 feet. The ore where I saw it was hard and solid, without any admixture of rock, with a face 50 feet wide and 70 feet high. In other places, where rock was embedded in the ore, the lines of separation were quite distinct, so that there was no difficulty in separating the two.

The ore being hard and there being a good deal of sterile matter to deal with, the output I understood did not exceed, reckoned on the entire staff, one ton per man per day. If 290 days are regarded as the year's work, that number represents the tons worked annually by each man. The mineral however being very rich—66 to 67 per cent.—the labour for working the necessary quantity to furnish a ton of pig iron is correspondingly diminished. I was informed that the earnings of the miners varied from 4s. 6d. to 5s. 4d. per day. This was in 1874 when the paper dollar was current. It did not seem to me, everything included, that this very valuable ore cost above 5s. 9d. put into wagon. Labour here did not generally appear high. Thus at the charcoal furnaces the following rates per day were given me in 1874:—

Keepers.		Fillers.		Carpenters.		Masons.		Smiths.		Smiths' Strikers.	
s.	d.	s.	d.	s.	d.	s.	d.	s.	d.	s.	d.
6	4¾	5	3¼	7	6½ to 8	5¾	7	6½ to 8	5¾	7	6½ to 8	5¾	5	3¼

There is indeed no reason why wages should be dear, for provisions were very moderate in price. The quotations given were:—

Butcher Meat. Per lb.		Butter. Per lb.		Best Bacon. Per lb.		Best Flour. Per barrel.
2½d. to 4½d.	...	11½d.	...	7½d.	...	30s. 2d.

Single men pay 13s. per week for board and lodging. The men are allowed house-rent free, and liberty to cut what wood they require for firing. Cash payments were then made only once in three months, orders on "the shop" being given in the mean time. This was therefore a very objectionable case of the truck system: at the same time the above figures do not look as if the prices charged were otherwise than reasonable.

678 SECTION XVIII.—CHIEF IRON-PRODUCING COUNTRIES COMPARED.

The following analyses, taken from the Geological Survey of Missouri, show the composition of the ore of the Iron Mountain:—

	Vein Ore.	Surface Ore.
Insoluble, chiefly silica	4·71	1·88
Peroxide of iron	91·45	95·04
Protoxide of iron	2·34	2·57
Alumina	·93	·75
Lime	·45	·15
Magnesia	·19	·12
Phosphoric acid	·252	·071
Sulphur	—	·005
	100·322	100·586
Metallic iron	65·78	68·53
Phosphorus	·110	·031

This mine is 80 miles from St. Louis. The transport is effected by a railway at a cost of 2¼ dollars per ton, equal therefore to about 1¼d. per ton per mile. The cost of the ore put into wagons was given me by a furnace owner in the Carondalet, near St. Louis, as not exceeding 5s. 9d., or say with carriage to St. Louis 13s. 6d. per ton. The selling price at the time was about 26s. 6d., leaving 13s. per ton profit at a period when the iron masters could barely obtain the cost price of the pig iron made from this ore. As much as 10 dollars per ton had been charged for it when iron was selling at a higher price.

Although the coal-fields of Kentucky and other Southern States, and those of Illinois and Indiana, are nearer to St. Louis than Pittsburg, it would appear that the Connellsville coke was preferred here although the distance is about 600 miles. The cost delivered at the furnaces was close on 30s. per ton.

Pilot Knob was the other mine I examined in Missouri. The ore seems to be a regular seam lying in porphyry, with a pretty regular thickness of 30 feet. In some places it is extremely hard and difficult to work. The seam is very variable in the iron it contains, being interstratified with porphyry. The best quality I understood to give 57 per cent. of pig iron, although according to the "Geological Survey

SECTION XVIII.—CHIEF IRON-PRODUCING COUNTRIES COMPARED. 679

of Missouri" it yields in the furnace 60 per cent., but a large portion of the seam is much poorer. The vein is divided by a "slate-seam," and according to the authority just quoted the iron contained in the ore is as follows :—

	Ores in the Ore-bed below the Slate Seam.					Ores above the Slate Seam.		
	1.	2.	3.	4.	5.	6.	7.	8.
Iron ...	61·03	58·29	59·15	64·91	47·16	53·91	44·01	36·52
Phosphorus ...			·015	·031	·041		·044	

The cost of raising the best ore was mentioned as being 9s. 6d. per ton. Some of it was being smelted on the spot by means of charcoal which cost 45s. per ton, bringing up the cost of the pig to something like 95s. per ton.

Southern States—Alabama, Georgia, Tennessee, North Carolina, and Virginia.—These five States are grouped together, because the two beds which constitute the chief sources of ore supply are common to them all. These are limonite or brown hematite, and a red hematite known as the Fossiliferous or Dyestone Belt. Besides these there are clay ironstones and black band, but hitherto they have been of no account. In the Smoky Mountains of North Carolina I visited a small vein of magnetic oxide, a variety of ore which is pretty extensively worked at the Cranberry mine and probably elsewhere. That of Cranberry is much esteemed, owing I believe to its very small content of phosphorus.

With the exception of the Hanging Rock region and the unimportant district of the Tuscarawas Valley black band in Ohio, the distance between the ore and fossil coal, as described in these pages, has, looking at the main portion of the requirements, been considerable. But in a work on the "Resources of Tennessee" it is stated that "the Dyestone Belt lies at the very base of the coal measures. Here, then, we have sandwiched, coal, iron ore, limestone and sandstone, the latter suitable for hearths." A letter from my friend General J. T. Wilder, inserted in the book just quoted, mentions the fact that for a distance of 160 miles the coal-beds are never more than half a mile from the iron ore.

This is confirmed by what I saw in Alabama. On one side of Jones' Valley a bed of fossiliferous ore, 30 feet thick, crops out. Of this 8 feet only were being worked, and of this portion the following are analyses from two different localities :—

Silica	...	...	...	...	17·09	...	17·22
Peroxide of iron	...	...	...		77·32	...	78·67
Lime	...	...	...	...	2·10	...	1·87
Magnesia	...	...	...	...	·43	...	·76
Alumina	...	...	...	...	1·75	...	1·59
Phosphoric acid	...	...	...		·51	...	·49
					99·20	...	100·60
Metallic iron	...	...	...		54·40	...	54·90
Phosphorus	...	...	...		·51	...	·49

The valley itself was limestone, and the opposite hills consisted of the coal measures.

I was shown the books of the Red Mountain Iron Company, and from them it appeared that 2·06 tons of this ore sufficed to make one ton of pig, equal therefore to a yield of 50 per cent.

The bed of ore, where it was being worked, had a covering of 2 to 4 feet of soil. The cost of tearing up the mineral was under 2s. per ton.

Labour, as is generally the case in the Southern States, is much cheaper than in Pennsylvania, as may be seen from the following rates :—

	Keepers.	Assistant Keepers.	Fillers.	Chargers.	Coke Fillers.
	s. d.	s. d.	s. d.	s. d.	s. d.
Per day ...	6 0¼	5 3¼	4 5¾	4 5¾	3 9¼

From some cause or another however the wages on a ton of pig iron at the Red Mountain furnace amounted to 13s. 3d. This must have been the result of faulty arrangements, which probably accounted for the concern being then in difficulties.

The brown hematite forms masses very irregular in point of outline and of dimensions. They constitute a superficial deposit and are worked quarry fashion. I visited several localities where the ore evidently exists in large quantities, and I inspected one quarry out of which the contractor excavated the mineral and having delivered it at the furnace received 8s. 6d. per ton of iron made. The yield of pig iron was given me as being 56 per cent.

The furnace using it, blown with air at 500° F., was making iron with 105 bushels of charcoal, weighing it was said 1,890 lbs. The limestone required was 8 cwts. per ton of pig.

In justification of the favourable views just expressed in reference to the iron-making facilities of the district thus briefly described, I would again quote the experience of a Southern States firm (*v.* p. 585) who mention their ability to make iron at about 42s. per ton. This undoubtedly is a higher cost than would attend the development of a similar mineral field in Great Britain. The Southern States however are in their infancy as iron-makers. Give them a little time, and I see nothing to prevent their producing pig, as well as other forms of iron, as cheaply, perhaps more cheaply, than it can be manufactured in most other known quarters of the globe.

According to a hand sketch made for me, showing a section from east to west,—beginning at the eastern limit—we have first the valley containing the Cahawba Coal-field. To the west of it is a ridge containing the red fossiliferous ore, with some brown hematite in the valley. Jones' Valley already mentioned follows, with another bed of the red ore in the hill beyond, 6 to 9 feet in thickness, and immediately behind it is the Black Warrior Coal-field, already referred to in the present section. To the east of the Cahawba Valley are other deposits of brown ore, but these I was unable to visit, there being no railway accommodation; besides I had seen enough to satisfy me that want of raw materials could never be a barrier to the production of cheap iron in this part of the world.

LIMESTONE.

It would be useless labour to devote many words to the conditions and expense attending the working of limestone. It usually occurs in such vast masses in nature and so near the surface that it is obtained from quarries at a cost, including royalty, of about 1s. 6d. per ton. With dear labour as in the United States the price of quarrying is of course proportionately increased. The chief outlay therefore, in very many instances, is that attending its transport to the furnaces. In some cases, it is true, limestone is got by actual mining, and under such circumstances considerable expense is incurred in winning it. Practically however the question is the cost at the ironworks, and hence I will content myself with naming the prices given me at the point of consumption.

Middlesbrough	3/6 to 4/- per ton.		Westphalia, Germany	3/- to 4/-	per ton.
Scotland	4/3 to 5/-	„	Saarbruck „	2/1	„
South Wales	1/6 to 3/-	„	Hayange „	2/3	„
South Staffordshire	4/6	„	United States—		
North Do.	4/3 to 5/-	„	New York States	5/- to 6/3	„
Shropshire	4/9	„	Andover	3/3	„
Cumberland	2/9 to 3/-	„	Youngstown	3/-	„
Northamptonshire	3/-	„	Pittsburg	6/9	„
Derbyshire	3/2	„	Virginia	3/-	„
Lancashire	1/6 to 3/6	„	Indiana	3/9	„
Longwy, France	2/3	„	Alabama	2/6	„
St. Dizier „	1/3	„	Turcarawas	4/8	„
Nancy „	1/3	„	Chicago	4/6	„
Belgium	1/2 to 2/6	„	Kentucky	6/-	„

In many cases, the information given in the present section respecting the prices and other particulars of the minerals used in the production of pig iron, affords the necessary data for estimating their cost per ton of the product. There is however a want of entire uniformity in the data to which the information in question applies, as well as in the manner in which one or more of the minerals are valued. At one time, where the furnace owner raises his own coal and ore, the cost prices are quoted, at others, where the minerals are purchased, the market quotations form the basis of calculation. In order to secure as far as practicable a uniform standard of comparison, a moderate sum has been added to cover the miner's profit; and a period of low prices has been selected, when, in the majority of cases, this profit would be small, and when, it may be assumed, the makers were endeavouring to work under the most economical conditions. At the time in question Cleveland No. 3 was selling at 37s. 6d., mixed numbers of Scotch pig at 48s. 6d. at Glasgow, and for hematite iron in Cumberland and Lancashire the ore is taken at 13s. per ton.

It is not proposed to carry the comparison beyond the cost of the minerals required to produce a ton of metal, because the smelting charges vary considerably, even in the same locality, according to the more or less perfect character of the furnace plant. In other cases,

SECTION XVIII.—CHIEF IRON-PRODUCING COUNTRIES COMPARED. 683

where no difference arises from this cause, but where the cost is dependent on the price and efficiency of labour, differences of the latter class will in most cases be equally applicable to the wages paid at the furnaces as at the mines.

TABLE showing approximately the Comparative Costs of Coke (or Raw Coal), Ore, and Limestone required to produce One Ton of Pig Iron in various Localities, calculated upon the Selling Prices of the Materials used—given in percentages of No. 3 Cleveland Pig.

MIDDLESBRO'—

Cleveland No. 3 100 Forge iron ... 98 Bessemer ... 123

In the beginning of 1884 the cost of Spanish ore has fallen to an abnormally low price. In the estimate of Bessemer iron the ore has been taken at 13s. per ton in the Tees. On the Cleveland iron the cost of carriage, per ton of iron, is about 7s. 9d., and on the Bessemer iron about 17s. The royalties on the minerals, for the Cleveland iron, may be taken at 3s. 9d. per ton of pig.

NORTHUMBERLAND—

Bessemer ... 120

The carriage on fuel from the collieries to the furnaces in Northumberland costs something less than it does in the Middlesbrough district; and the Tyne being a better loading port than the Tees, the freight from Bilbao is frequently rather less to the former. These advantages combined may sometimes reduce the cost in the Northumberland furnaces to something below the figure given.

LINCOLNSHIRE—

Foundry
No. 3 ... 99

In this case, using South Durham coke, the conveyance of the minerals per ton of iron is probably 7s. 9d. to 8s. 6d., and the royalties amount to about 4s.

NORTHAMPTONSHIRE—

Foundry
No. 3 ... 93

The carriage on the minerals to the furnace is believed to amount to about 10s. per ton of metal, and the royalties to 3s. 9d.

DERBYSHIRE—
Foundry
No. 3 ... 104

The railway dues on the minerals are assumed to be 10s. 6d., and the royalties 3s. 9d. on the ton of iron.

NORTH STAFFORDSHIRE—
Forge iron ... 97

The railway carriage on the minerals does not probably exceed 3s. 6d. to 4s. per ton of metal, and the royalties are about 4s.

SOUTH STAFFORDSHIRE—
Foundry
No. 3 ... 120

This price is based on data upon which, as I have been informed, a cheap foundry iron can be made chiefly from Northampton ore, with a portion of mill cinders.

CUMBERLAND AND LANCASHIRE—
Bessemer ... 125

The ore has recently fallen considerably in price. In the estimate it is taken at 13s. at the furnaces. The dues on the minerals will come to about 10s., and the royalties to 6s. 6d. per ton of iron. The figure given above represents the ratio of cost for Cumberland, but the average of that of Lancashire is about the same.

SOUTH WALES—
Forge ... ... 120 Bessemer ... 125

The forge iron is entirely from Northamptonshire stone, and the Bessemer from Spanish ore, taken at the same price, at Bilbao, as that smelted in Middlesbrough. The cost of conveying the minerals for the forge iron will probably be 13s. and the royalties 3s. 9d. The cost of conveying the minerals for the Bessemer iron may be taken at 20s. 6d. on the ton of iron, including freight from Spain.

SCOTLAND—
Mixed Nos. 112

The railway dues on the minerals and conveyance of the pig iron to the place of shipment amount to 6s. 3d. to 7s. 6d. and the royalties (assuming the ironstone to be half calcined black band and half clay ironstone) may be reckoned at 4s. 4½d. per ton of iron.

SECTION XVIII.—CHIEF IRON-PRODUCING COUNTRIES COMPARED. 685

The localities just given comprise the chief iron-making centres in the United Kingdom. Of the number Middlesbrough, Northumberland, Cumberland, Lancashire, Scotland, and South Wales alone can be regarded as occupying a position which enables them to export pig iron. South Wales indeed can scarcely be regarded in this light, at all events in periods of depressed trade. In the matter of cost at the furnaces it enjoys no advantages over Middlesbrough and Northumberland, while the cost of conveying the iron to Cardiff or Newport will average at least 2s. 6d. per ton. Cheaper labour no doubt greatly assists South Wales, but this is too insignificant in amount, in the manufacture of pig iron, to be of much service: when however the pig iron has to be converted into steel rails, cheap labour and cheap fuel enable this part of the principality to compete successfully with the rail mills at the sea-ports in the North of England, where wages are higher and the coal more expensive, because more distant from the steel works.

Staffordshire is certainly not without some export trade; but this is, I apprehend, principally confined to the higher qualities of malleable iron, such as the produce of the works of Lord Granville, Lord Dudley, and a few others.

The calculations which follow exhibit the position of some of the chief iron-making centres on the Continent of Europe and in the United States.

GERMAN ZOLL VEREIN.

WESTERN GERMANY—
Foundry ... 100   Forge iron ... 85

The above figures represent the cost, as compared with Cleveland No. 3, of pig iron produced from the oolitic ironstone of Western Germany, Luxemburg and France, under the most favourable circumstances which came under my observation. At the particular work referred to, situate close to the mine, the charge for conveyance of minerals was almost exclusively that incurred in bringing the coke from Belgium, Westphalia and Saarbrück. The cost of carriage therefore per ton of iron would not exceed 5s. on forge and 6s. 6d. to 7s. on foundry iron; and the royalties were under 1s. on the ton of pig.

At another work, situate in the Saarbrück coal-field, the cost was higher, say for forge iron 103. The carriage of the ore from Luxemburg, a distance of about 80 miles, costs 4s. 4d. per ton. Dues on minerals about 14s. per ton and royalties about 1s. on the ton of metal.

LUXEMBURG—
    Forge iron ... 90

This cost is that of one of the most favourably situated and best appointed works in the Duchy. The iron made here and at other works is chiefly white forge, and as the coal-fields, Belgian and Westphalian are distant, little beyond the manufacture of pig iron is attempted. The average dues on coke are about 7s. per ton. The total cost of conveying the minerals will be about 9s. and that of royalties about 1s. per ton of iron made.

ILSEDE—
    Forge iron ... 80

This place, although about 140 miles from the Ruhr coal, is able to make very cheap pig iron, owing to the very inexpensive ore got in the neighbourhood. This mineral contains a good deal of phosphorus, which fits the iron produced admirably well for the Bessemer basic process. A small addition has been made for supposed profit on the ironstone. The fuel is partly coked at the furnaces.

The probable cost of conveying minerals I estimate at 12s., and royalties at about 1s. per ton of iron made.

WESTPHALIA—

Foundry ... 140 Forge (Common) 103 Bessemer ... 140
      Do. (Best) 135

The ores employed at the work affording the above results are obtained from such a variety of different places that I am unable to give any precise information respecting the cost of conveying the minerals for the foundry and forge iron. Although for Bessemer iron a mixture of ores is used, nothing at the present moment—1884—is cheaper than that of Bilbao. Inclusive of sea freight, the carriage on the ore and coal will cost about 23s. on the ton of iron and the royalties 1s.

It is very difficult to give a correct average of the Westphalian works owing to the great differences in the character of the blast furnaces.

At the period when the above information was collected I was informed that the white pig of Luxemburg was being delivered at the Rhenish works at 45s. per ton.

FRANCE.

DEPARTMENT OF MEURTHE—

Foundry iron, 137   Forge iron ... 110

The coke used for making the above-mentioned iron is brought from the Ruhr, a distance of 215 miles; from Anzin, 277 miles; and from Saarbrück, 61 miles. The cost of conveying the materials to the furnace will probably be from 8s. to 10s. per ton of iron, according to the proportions of German or French coke which is used. The royalties will not amount to 1s. It is perhaps worthy of note that grey pig iron was selling in the neighbourhood of this work for 30s. more than the price at Middlesbrough.

NEIGHBOURHOOD OF NANCY—

At one work Forge iron...103
At a second work 90

The ore was from the oolitic bed, and the coke partly from Anzin and partly from Saarbrück. In both cases the conveyance was by canal, costing 7s. 2½d. in the former and 2s. 9½d. in the latter case. The cost of conveying the ore was only 7d. per ton, by rail. Approximately therefore we may take the charge of bringing the minerals to the furnace to be 7s. 9d. on the ton of iron, and the royalties under 1s.

From a work further south, and not so favourably situated, using oolitic stone for forge iron, a mixture of ores for foundry purposes, and chiefly African ore for Bessemer, I have the following estimates in comparison with Middlesbrough costs:—

Foundry iron, 129   Forge iron ... 123   Bessemer ... 214

NEIGHBOURHOOD OF ST. ETIENNE—

Bessemer iron 221
A second work Bessemer iron... 230
Do.   in 1867 Bessemer iron... 212

Although St. Etienne is plentifully supplied with coal, the great demand upon its resources keeps up the price. In illustration of this I have inserted the cost of Bessemer iron in the year 1867, when coke was valued at 22 francs, whereas nearly 20 years afterwards, with the iron trade generally speaking in a depressed state, it was selling at 25 francs, and the washed coal from which it was made at 13s. 6d. at the furnaces. Neither can the district be said to be well situate as regards ore, being largely dependent on that from Mokta in Africa. The price of this ore at that time was 12s. 8d. per ton f.o.b., the sea freight and railway carriage 19s. 4d. Elba and other ores were used, but they were equally dear. The mere cost of carriage on the minerals from the mines must be more than 30s. on the ton of iron.

CENTRAL FRANCE (Four examples)—
Forge iron ... 158
Do. ... 175
Do. ... 187
Do. ... 156

It is many years since I visited the works of the district in the department of Allier. I have no reason for thinking fuel is any cheaper now than it was when I examined the works, and the ore was, when I was there expensive to obtain. Since then the ironstone of Eastern France, between Metz and Nancy, has been developed, and, as will be seen on comparing the districts, is infinitely more favourably situate than that of any other part of the French territory.

BELGIUM.

As already intimated, page 659, the ironworks of Belgium are almost entirely dependent on imported ores. For common purposes the supplies are brought from Luxemburg, and for Bessemer iron the mines of Bilbao furnish the largest quantity. Regarding these two varieties of ore as those upon which Belgium trade rests we have:—

Foundry iron, 120   Forge iron ... 102   Bessemer ... 140

The Luxemburg ore is taken at 7s. per ton, of which 4s. is carriage; and the Bilbao ore at 17s. of which 3s. is carriage from Antwerp partly by boat and partly by rail. The ore is valued at 14s. *ex* ship at Antwerp; freight from Bilbao 8s. Approximately the conveyance of the minerals per ton of iron made from Luxemburg ore is 12s. 6d. and 13s. 6d., and the total conveyance on the Bessemer iron will probably stand at 22s. to 24s. on the ton of pig.

SECTION XVIII.—CHIEF IRON-PRODUCING COUNTRIES COMPARED. 689

UNITED STATES.

Mention has already been made of the excessive fluctuations in costs of production which have marked the progress of the American iron trade. According to a statement which has already appeared in these pages (p. 606) the cost of minerals in a ton of iron made on Lake Champlain rose from 15·85 dollars in 1871 to 21·21 dollars in 1874, or from 100 to 128 per cent. The starting point in Cleveland was of course considerably below 15·85, but the fluctuation of cost of minerals was such that what might be taken as 100 in 1871 rose to 174 in the year 1874, and fell to 110 in 1876.

It has not been my good fortune to visit America since 1876, and I have not sufficient data to enable me to compare the present costs in England with those of the United States. I have therefore found it impossible to construct a comparative table of costs of pig iron, which, instead of being useful, would not have been calculated to mislead. Not only upon the two occasions of my visits to America were the prices paid for labour in connection with mining and iron making, in many cases, far above those current on our side of the Atlantic, but they were disproportionate to those received by the workmen in other branches of American industry. Besides having the costs unduly raised from this cause, the iron masters complained of the rates charged for carriage on the railways, and of the price of coal forced up by a combination among the railway companies.

Under such circumstances no actual estimate of the cost of pig iron based on the prices of 1874 or 1876 would have had any value in estimating the position of the United States to compete with the iron-making centres in Europe in an export trade. I have upon one or two occasions inserted information obtained since my visits to the iron works in America, and this, together with the summary which follows hereafter of the costs of bringing the minerals to the furnaces, will give a general idea of the relative cost of manufacture.

As a rule we may take it that the manufacture of pig iron is carried on in the Northern States under circumstances of considerable difficulty. Practically this means, at present, the entire make of the country, because in 1882 the quantity made in the Southern States did not exceed 15 per cent. of the whole production. A chief and at the same time an insurmountable impediment to the manufacture of cheap pig

R R

iron in the North is the distance generally speaking between the ore and the fuel. I propose shortly to point out the extent to which the question of transport affects the cost of pig iron.

At pp. 662 and 664 it was estimated that in 1882 out of 7,869,154 gross tons of ore dealt with 5,417,649 were either the produce of four of the Northern States or were imported. The different sources of origin of this last named quantity were as follows:—

Lake Superior including Menominee Range	2,943,314
New Jersey	900,000
Lake Champlain	675,000
Cornwall Bank	309,680
Imported	589,655
Total	5,417,649

The chief centres of consumption of the ore obtained from these sources will be consecutively examined with a view to ascertain the cost of bringing together the minerals required to produce one ton of pig iron.

Cost of conveying minerals for a ton of iron made from Lake Superior ore.—A certain quantity of the produce of the mines of this district is smelted on the spot by means of charcoal. In this case the carriage of the ore will probably not exceed 1s. or 2s. per ton. The cost of conveying the charcoal varies exceedingly, but supposing it to average 5s. the total expense of bringing the raw materials together will work out to about 8s. on the ton of pig iron. Taking Cleveland City as the nearest point of consumption of pig, the carriage on the metal including lake freight, will be about 12s., bringing up the total to 20s. per ton of iron. Delivered at Pittsburg the total cost for carriage and freight would come to about 25s. per ton of pig. To Chicago, where there are Bessemer works, the cost of transport would be something less—say about 23s.

A certain quantity of the Lake Superior ore is smelted at Chicago, and the cost of conveying it from the mines to the works will probably average 12s. per ton: the fuel used in the blast furnaces is coke from Connellsville, the carriage of which in 1876 was about 20s. to 22s. From this it is estimated that the cost of transport on the minerals, per ton of Bessemer iron was not less than 46s.

The carriage of a ton of ore from the Lake Superior Mines to Cleveland City in 1874 was 14s. 9d. The fuel is chiefly raw coal, of which 2¼ tons are used per ton of grey iron, the railway dues being 4s. 8½d. From these data it is estimated that the transport on the materials for a ton of pig will be about 32s.

A considerable quantity of Lake Superior ore goes to the Pittsburg furnaces. The carriage of the ore from the mines was about 23s., and that of coke from Connellsville about 4s., and of limestone about 1s. 6d. per ton. This will bring the cost of carriage on the minerals, for a ton of iron, to about 40s. 6d.

Cost of conveying minerals per ton of iron made from New Jersey ore.—This ore is chiefly conveyed to the furnaces situate to the west of the mines at Andover, Glendon, &c. The dues by rail are about 5s. on coal and ore. Three and a half tons are used of the two together costing for transport 17s. 6d.; which with that on 12 cwt. of limestone brings the total charge to 18s. per ton of metal.

Cost of conveying minerals per ton of iron made from ores of Lake Champlain.—The transport charges on the materials per ton of iron smelted on the Lake are thus made up. In 1874 the cost of conveying the coal, partly by rail and partly by water, was 14s. 9d., and that of the ore about 2s. At these figures the conveyance of the minerals was about 26s. per ton of pig. But to get the iron to market costs 6s. more, making the total cost 32s.

If the Lake Champlain ore is smelted on the Anthracite coal-field, the charges for transport will be about the same as at the furnaces on the Lake itself, the quantity of material to be carried either way not differing greatly.

Cost of conveying the minerals per ton of iron made from Cornwall ore.—The expense of bringing coal to Lebanon, where the furnaces are chiefly situated, in 1874 was 5s. 6d. per ton; that incurred in carrying the ore to Lebanon is 1s. 2d. The average consumption is about 2 tons of ore and the same of coal, and ¾ ton of limestone; which brings the cost per ton of iron for carriage of materials to about 13s.

Cost of conveying minerals per ton of iron made from imported ores.—I saw at a large work in the Lehigh Valley a considerable quantity of ore imported from Europe. The chief source of supply is from Bilbao. Ships which have brought American produce to the United Kingdom &c. occasionally go round by the Bay of Biscay and load ore. The freight varies considerably, but for the purpose of this estimate we will assume the rate to be 12s. 6d. Including unloading at New York and railway dues to the furnaces, the cost of carriage will not fall short of 17s. 6d. The coal is 34 miles distant, and the charge made by the railway company was 3s. per ton. The expense therefore on the ton of iron for conveying the minerals will not fall short of 40s.

It must not be supposed that the sums just mentioned, as representing the cost of bringing the minerals to the blast furnaces, always give the amount actually entailed in the manufacture of every ton of iron made at any particular work. In some cases no doubt the whole of the materials have to be conveyed over the full distances upon which the estimates just given have been calculated. In others however more or less of the ore which is smelted may be obtained from mines more favourably situated than those of Lake Superior or Lake Champlain. Looking at the question however as a whole, we have something like $3\frac{1}{4}$ million tons of pig iron produced in the Northern States under the unfavourable conditions as to carriage just described.

Cost of conveying minerals per ton of iron in Kentucky.—The distances of the mines from the furnaces vary very much. The ironstone has sometimes to be brought by rail 3 and at other times 13 miles. In my notes I find as much as 3s. per ton of ore was paid for dues. There are collieries on the river itself; and from one of these the cost of bringing the fuel to one furnace I visited was only about 6d. per ton. Under such circumstances as those just referred to it is not probable that the charge for carrying the minerals for a ton of iron can exceed that at Middlesbrough, say 7s. 9d.

Cost of conveying the minerals per ton of iron made from the Iron Mountain and Pilot Knob of Missouri.—The cost of bringing the produce of these mines to the Carondalet furnaces near St. Louis was about 7s. 9d. in 1874. The fuel preferred was Connellsville coke brought from a distance of 600 miles, at a cost for conveyance of about 23s. per ton. The fuel employed however consists of 60 per cent. of coke and 40 per cent. of raw coal, brought from the Big Muddy River coal-field. This latter was being delivered at

SECTION XVIII.—CHIEF IRON-PRODUCING COUNTRIES COMPARED. 693

St. Louis for 11s. 9d. by the Mississippi. Including the river freight, which is 2s. 9¾d., I suppose 4s. 6d. would be the outside cost for carriage. These figures would bring the average rate of carriage on the fuel to 15s. 6d. The weight consumed was nearly 30 cwt. per ton of iron, so that the total expense of conveying the raw materials to the furnaces will amount to about 35s. per ton of the metal produced.

In some of the above-mentioned instances of costs of transport of iron-making minerals in the United States there are rates, such as those on the Marquette and on the Iron Mountain and St. Louis Railways, which may be susceptible of reduction; but in many cases, and on long distances, the charge does not appear excessive.

Before considering the position of Tennessee, Alabama, and Georgia I would refer to the difficulties the American iron smelters in the Northern States have had to encounter in the high prices they have had to pay for their fuel and ore.

As regards the fuel, the observation is limited to the furnaces using anthracite; for when I visited the ironworks in 1874 the price of Connellsville coke was only about 5s. at the ovens, and since then, according to Mr. Swank's returns, it has been sold at 90 cents, and generally in other localities the price of coal was moderate.

As regards anthracite however, although in 1874 the price of iron was such that the iron masters were making no profit, the cost of this fuel was 11s. 6d. per ton at the pit, leaving, as I estimate, from 5s. to 6s. profit to the mine owner. This excessive charge, it must be remembered, did not arise from any difficulty in supplying the demand, but from a combination among the railway companies, who, being also coal-owners, not only controlled the prices of their own coal, but by raising the dues for conveying the produce of independent mine owners, exercised for a time all the privileges of a very dangerous monopoly.

For the excessive price of ore the iron trade had itself to blame. The rapid expansion of the production of pig iron, already spoken of in these pages, was more than the mines were able to meet, and in consequence, owing to the competition offered by the furnace owners themselves, the mine owners had no difficulty in maintaining a very high price for their ore.

According to my own observations, and guided by the information I obtained, I estimate that on every ton of iron manufactured from the following ores, the owners of the best mines received in the form of profits the amount attached to each :—

Lake Superior	15s.
Lake Champlain	18s. to 20s.
Cornwall Bank	28s.

If to these figures 9s. to 10s. 6d. of profit to the coal-owners has to be added for each ton of iron smelted by means of anthracite it must be admitted that an unprofitable trade was being made to carry a very heavy burthen on its shoulders.

In 1876 the price of coal delivered at some of the furnaces was reduced to 12s. 6d. per ton, viz., 10s. 3d. at Mauch Chunck, with 2s. 3d. for railway dues. The values given me for the various ores when at the mines near Marquette in 1876, were as under :—

		Per Ton.
		s. d.
Ore of	Jackson Mine	16 11½
Do.	Cleveland Mine	15 1
Do.	Lake Superior Mine	15 1
Do.	Michigamme Mine	14 4¼

Practically therefore the ore was sold at the same rates in 1876 as it was in 1874; pig iron at Middlesbrough in the meantime having fallen 23s. and Scotch pig at Glasgow 29s. per ton.

Cost of conveying minerals to make a ton of iron in the Southern States.—For the purpose of examining this branch of the subject I prefer taking Alabama, as being the state in which, in all probability, the cheapest iron will hereafter be made in all the American Union. The close proximity of the coal and ore has already been described, but the introduction of the manufacture of pig iron by means of fossil fuel, is too recent to have afforded opportunities of opening up the country by sufficient railway accommodation. I cannot doubt however, from what I saw and heard, that many locations will present themselves where the distance of transport of one of the

SECTION XVIII.—CHIEF IRON-PRODUCING COUNTRIES COMPARED. 695

materials to furnaces situated on the others will not exceed 3 or 4 miles. If I am correct in this, then probably something under 2s. 6d. per ton of pig iron will cover the cost of carriage; but should it from any circumstances unknown to myself amount to 5s., still it would seem, from this and from what has preceded, that there are few if any countries, where pig iron can be produced on cheaper terms than it will be manufactured in Alabama.

With the enormous resources possessed by the United States in ore and coal, the question of the capability of America to become a great exporter of iron is one which greatly interests the British producer in the future. Of the ability of the United States to occupy this position some public writers there as well as private individuals entertain somewhat sanguine expectations. In 1876 Mr. W. E. S. Baker, of Philadelphia, Secretary to the Eastern Ironmasters' Association, compiled tables showing the cost of pig and bar iron at the works, beginning in 1850 and ending at the end of March, 1876. Their contents, coming as they do from an official authority, are interesting at the point we have now reached, and I therefore insert the tables on page 696 compiled by this officer of the Eastern iron trade.

The *Iron Age* newspaper of New York offers the following observations upon Mr. Baker's statements :—

"Mr. Baker's figures are compiled with great care, and fairly represent the cost of pig and bar iron at Pennsylvania favourably situated for economy. Some furnace and mill owners can probably make iron for less than the totals given in Mr. Baker's table; a great many others think they can, and the results of mistakes of this kind as to the cost of iron are found in the failure of upwards of 40 per cent. of our iron works, east and west, since 1873."

"Comparing the totals of costs for a series of years, we find that pig iron can be made cheaper to-day than at any time since 1863; while the cost of converting pig iron into bars is less than at any time since 1862. Examining the items entering into the cost of pig iron, we find that labour has been reduced to ante-war rates. In some places of which we know, labour employed in mines is down to 80 cents per day, and in others men can be had for the rough work about blast

Average Cost of Pig Iron per Ton at Furnace Bank.

	1850.	1851.	1852.	1853.	1854.	1855.	1856.	1857.	1858.	1859.	1860.	1861.	1862.	1863.	1864.	1865.	1866.	1867.	1868.	1869.	1870.	1871.	1872.	1873.	1874.	1875.	1876.
	Dolls.	Dolls.	Dolls.	Dolls.	Dolls.	Dolls.	Dolls.	Dolls.	Dolls.	Dolls.	Dolls.	Dolls.	Dolls.	Dolls.	Dolls.	Dolls.	Dolls.	Dolls.	Dolls.	Dolls.	Dolls.	Dolls.	Dolls.	Dolls.	Dolls.	Dolls.	Dolls.
Ore	5·75	5·44	5·55	5·97	6·65	7·51	7·50	7·75	7·66	7·08	7·45	7·35	7·08	7·49	9·12	13·13	12·19	11·71	10·92	11·86	12·96	12·67	13·64	14·27	12·94	11·95	9·54
Coal	3·70	3·36	3·65	3·23	3·53	4·63	3·90	3·89	4·06	3·26	3·49	3·26	3·68	3·42	5·41	9·66	7·55	7·44	7·11	7·41	7·08	8·59	7·28	7·45	7·94	8·01	6·79
Limestone	·93	·96	1·09	1·06	1·38	1·26	1·16	1·14	1·18	1·15	1·21	1·17	1·11	1·20	1·93	2·85	2·65	2·76	2·51	2·14	2·44	2·08	2·04	1·98	1·84	1·14	1·01
Labour	2·22	1·61	2·02	2·00	2·45	2·85	2·58	2·30	2·10	1·82	1·87	1·97	1·57	2·07	2·85	4·56	3·46	3·99	3·86	3·46	3·89	3·54	4·69	5·11	3·65	2·97	2·54
Sundries	1·65	1·93	2·03	2·62	1·99	2·62	2·91	2·16	2·73	2·83	2·83	2·86	2·67	2·35	1·66	2·01	2·03	1·98	1·90	1·96	3·67	2·77	2·93	3·00	2·26	2·10	1·73
	14·25	13·30	14·34	14·88	16·00	18·87	18·05	17·24	17·73	16·14	16·85	16·61	16·11	16·53	20·97	32·21	27·88	27·88	26·30	26·83	30·04	29·65	30·58	32·41	28·63	26·17	21·61
Interest	1·05	1·05	1·15	1·22	1·37	1·29	1·21	1·47	1·22	1·28	1·36	1·57	1·67	1·40	1·59	1·61	1·64	1·80	1·63	1·71	1·85	1·82	1·75	2·08	1·90	1·70	1·59
	15·30	14·35	15·49	16·10	17·37	20·16	19·26	18·71	18·95	17·42	18·21	18·18	17·68	17·93	22·56	33·82	29·52	29·68	27·93	28·54	31·89	31·47	32·33	34·49	30·53	27·87	23·20

Average Cost of Bar Iron per Ton.

	1850.	1851.	1852.	1853.	1854.	1855.	1856.	1857.	1858.	1859.	1860.	1861.	1862.	1863.	1864.	1865.	1866.	1867.	1868.	1869.	1870.	1871.	1872.	1873.	1874.	1875.	1876.
Pig used	25·65	24·90	25·71	25·25	42·17	42·64	32·84	33·34	30·61	26·54	25·61	25·35	24·36	27·90	41·40	68·60	50·77	50·64	44·53	43·29	43·63	40·52	49·11	43·24	31·69	29·12	25·19
Coal	5·70	5·61	5·61	5·81	6·00	8·28	6·59	6·00	5·49	5·17	5·27	5·39	6·19	7·66	8·44	13·03	10·92	9·13	8·64	8·33	8·55	7·55	8·13	8·55	6·98	8·73	6·85
Labour	10·43	10·17	10·37	11·06	15·12	14·70	12·85	13·06	11·77	10·68	10·90	11·12	11·78	15·14	18·94	27·45	20·61	22·02	19·87	20·65	18·57	17·70	21·55	20·37	14·37	16·87	15·74
Sundries	4·64	4·83	4·88	7·05	10·39	10·78	8·88	10·38	10·84	7·91	8·78	8·71	10·03	7·66	9·15	8·03	9·50	9·44	7·70	7·75	7·03	7·85	5·74	5·83	6·15	4·79	4·73
	46·42	45·51	46·57	49·17	73·68	76·40	61·16	62·78	58·71	50·30	50·56	50·57	52·36	58·36	77·93	117·11	91·80	91·23	80·74	80·02	77·78	73·62	84·83	77·99	59·19	59·51	52·51
Interest	1·56	1·49	1·64	1·50	1·80	1·63	1·59	1·89	1·65	1·60	1·71	1·90	1·75	1·77	1·80	2·80	2·00	2·05	1·96	2·09	2·15	2·20	2·22	2·25	1·74	1·86	1·70
	47·98	47·00	48·11	50·67	75·48	78·03	62·75	64·67	60·36	51·90	52·27	52·47	54·11	60·13	79·73	119·91	93·81	93·28	82·70	82·11	79·93	75·82	87·05	80·24	60·93	61·37	54·21
Price of Scotch Pig	..	..	45/5	61/5	79/9	71/-	..	69/2	54/5	..	53/6	..	52/10	55/9	57/3	54/9	60/6	53 6	52 9	..	54/4	58/11	102/-	117/3	87/6	65/9	58/6

In one or two cases errors in the figures in these tables have been approximately corrected.—I, L, B, furnaces for 90 cents. There is no room for contraction here, nor can rates of wages be maintained on the present starvation basis, unless we are prepared to see our working men reduced to a level scarcely above that of the agricultural labourer in England. With meat and provisions at present prices, working men cannot support even small families in comfort on 80 or 90 cents a-day. The labour employed in rolling mills has not, as will be seen from the lower table, declined in the same proportion, as blast furnace labour; but it is down very close to starvation rates in the east, especially as the men seldom have work more than half the time. Examining the cost of ore, we find that the furnaces are paying more by 2 dollars per ton than in 1863. This is chiefly because of its scarcity near furnaces, and partly because of high rates of freight by rail. Looking at the cost of anthracite coal per ton of pig iron, we find that, notwithstanding a decline from the extortionate prices charged from 1865 to 1875, it is still nearly double the price of the monopoly which now controls the anthracite mines and their outlets. We shall have a few words to say on this subject further on. The price of bituminous coal has declined about in proportion to the decline in ore, for the reason that the sources of supply are so extensive and widely separated that no combination among producers is possible. The cost of pig iron per ton of bars has not been so low since 1851, and, but for the high freights, would cost the mill less than before the war."

"The tables, and the conclusions we have taken the liberty of drawing from them, furnish material for much profitable thought."

"All that stands between us now and the attainment of cheapness of production, which would place our iron in the world's market in competition with the cheapest foreign irons of marketable quality, is the rapacity of the coal operators and the greed of the railroads. It is well-known that the anthracite monopolists fix their prices two months a-head—a thing unheard-of in any other branch of trade—and by combinations clearly illegal, control the amount of production, the price, and the cost of transportation and handling. The railroads arrange their tariffs so that they can pay ten per cent. on watered capital; and the policy of their management in a majority of instances, is so to operate the wires that one end shall reach the Stock Exchange and the other terminate in the pockets of the directors."

"With our present capacity for producing iron, it is obvious that its utilisation depends upon the building-up of an export trade; but, before we can do this with the certainty of holding such advantage as we may gain, we must be able to reduce the cost of production even below the present lowest average. To secure a further reduction in the cost of materials we must break up the coal ring and the railroad ring. This could be done, as we shall show in a future issue, by legal means; whether it will or not depends upon whether the iron markets of the East have the courage to grapple with the evils they have allowed to grow during the years of their prosperity."

Mr. Baker mentions that the ore used in the blast furnaces consists of a mixture of Juinata and Montour hematites, with a little Cornwall, and that the average consumption over 20 years was 2·17 tons per ton of pig—equal therefore to 46·08 per cent. The coal was anthracite, chiefly from the Wyoming and Lehigh Valleys: 40 cwts. were used per ton, and 20·08 cwts. of limestone.

The coal used in the mills was Broadtop and Cumberland. Mr. Baker gives 20·32 cwts. of coal to have been consumed per ton of finished bars, and 20·10 cwts. of a mixture of white pig iron. Both these quantities however are, I cannot help thinking, erroneous.

Now I cannot resist the impression that some very important matters have been overlooked in commenting upon these figures, which are calculated, according to the *Iron Age,* to convey a lesson to the iron manufacturers of America as to the means to be pursued to obtain an export trade.

The railway companies are held partly responsible by the journal in question, for the high price of ores in the later years given in Mr. Baker's tables; but in reality the other cause referred to is far more responsible for the change than the railways viz.: the scarcity of ore near the furnaces. It was some 15 years after the periods of cheap production, in 1850 and the following years, before Lake Superior and Lake Champlain did much to rescue the American iron trade from the impending ore famine which threatened to overtake it. When supplies commenced to reach the blast furnaces from these distant mines, they came loaded with a weight of something like 60s. on every ton of iron they furnished, in the shape of carriage and profit to the mine owners, charged at famine prices, as commodities usually are in times of scarcity.

SECTION XVIII.—CHIEF IRON-PRODUCING COUNTRIES COMPARED. 699

I have already pointed out the circumstances which enabled the railway companies and coal owners to inflict "the extortionate prices" complained of in the *Iron Age*, upon a trade in which there had been "failures of 40 per cent. of the ironworks since the year 1873." The combination met, as I foretold in 1874 it would meet, its own reward ; but after all the line is perhaps not a very strong one which divides the policy under which railroads and coal owners, independently of State assistance, raise the price of coal, from that under which iron masters, demand and secure that State co-operation which enables them to tread in the footsteps of the very monopoly so distasteful to the editor of the journal just quoted.

Before noticing what is advanced by the *Iron Age* on the labour question, I would call attention to the unexplained causes which have raised the cost on bar iron for "sundries" (which must mean stores, repairs, rates and taxes, management &c.) from 4·64 dollars in 1850 to above 10 dollars in subsequent years; and the apparently wasteful method in which the pig iron has been treated in the mills, sometimes as much as 33 cwts. (and in one case 40 cwts.) having apparently been consumed to produce a ton of bars.

With regard to the subject of labour, the *Iron Age*, writing in 1876, mentions that the rates of wages were then "on a starvation basis," although in a ton of pig iron they amount to 10s. 4d. and of bar iron to 59s. 8d. ; and these excessive prices appear in the list to have been only one-half what they were in some of the previous years. The writer goes on to say that, unless, in America, they are prepared to see their working men reduced to a level scarcely superior to that of the agricultural labourer in England, wages must be increased. The fact however is that the wages of ironworkers at that time (1876) were quite as high or higher than they were at Middlesbrough, when Mr. Weeks, the present editor of the *Iron Age*, visited that district in 1882. The results of his enquiries have already been referred to in these pages, and need not be further enlarged upon. One remark may however be permitted in reference to such charges for labour as 10s. 4d. on pig iron and 59s. 8d. on bars (12s. 5d. and 66s. 9½d. per imperial ton), namely that they can only have been the result of most defective arrangements in the plant itself, or of a deficiency in the amount of work performed by the men, such as did not entitle them to higher wages than those of the most unskilled "agricultural labourers in England" or elsewhere. It is only due to the iron masters and to the men to say that no such incapacity on either side came under my notice during very extensive examinations of what was doing in 1874 and 1876. Passing by the difficulty of reconciling the necessity of guarding against foreign importations (already burthened with cost of freight) by a protective duty, with the expectation of a future export trade, which in its turn must be saddled with transport charges, it seems to me that so long as the Northern States are dependent upon these present mining resources as regards ore, it is futile to hope for any export trade from that division of the Union. On the contrary the iron masters of the North must prepare themselves for importations, not from Europe but from a quarter against which the present legislative constitution of the States will afford no protection.

The quarter alluded to is of course the Southern States. Very trifling extensions of the present railways will place the whole of Tennessee, Alabama and Georgia in direct communication with the Tennessee river. I understand one impediment only exists, which impedes free navigation. This removed, the Mississippi and Ohio will become accessible from those States by steam navigation. The distance from a central point, say Chattanooga, to Pittsburg by river is probably 1,000 miles, for which the freight will probably not exceed that from Great Britain. In these Southern States coal can be worked nearly as cheaply as at Connellsville, while the labour on the whole of the ore entering into the manufacture of a ton of iron is not more than that expended on the extraction of a single ton of ore near Marquette. Besides this there is the fact that the bringing of the minerals together in the Northern States often costs 30s. to 40s. per ton of iron made. With these elements of cost, it seems impossible to deny that, in the absence of fresh ore discoveries in the North, time alone is required to produce a considerable change in the seats of the American iron trade.

This state of things naturally suggests the enquiry as to the ability of the South to enter the markets of the world in competition with Great Britain. It cannot be disputed that up to this time pig iron has never been produced in Alabama or in its vicinity, within some shillings per ton, of the price at which it can be made from Cleveland ironstone in England. The removal of difficulties which always beset the introduction of new industries may partly equalise these differences, but by that time labour probably will no longer be to be had in the Southern States upon so much lower terms than it commands in the North. Be this however as it may, there remains the insurmountable difficulty of the cost of transport to the chief iron-consuming populations in the world, viz. those of Europe. The nearest point of the Alabama mineral field cannot be short of 150 miles from the seaboard. Admitting the carriage from the works to be done for $\frac{1}{2}$d. per ton per mile, this added to the Atlantic freight, would, probably entail a cost of 20s. per ton of iron delivered on the shores of Great Britain or of Northern Europe, above that paid by ourselves or by Germany. This extra charge for freight no doubt which would be reduced when competing with us for the custom of the Mediterranean ports or those of Asia, South Africa, Australia, &c.

In contemplating the future of such a trade as that of iron, we naturally are led to consider the possible sources of further economy. Little need be said in respect to improvements in the process itself; firstly because we have arrived at a point at which any great relief from this quarter can scarcely be hoped for, and secondly because, were it otherwise these same improvements would speedily be everywhere adopted.

Accepting Cleveland, as has already been done, as a standard of comparison, the expenses of manufacturing pig iron included about 7s. 9d. for railway charges and 3s. 9d. for royalty dues. These together amount to 11s. 6d., or nearly 33 per cent. of its entire cost, whereas in the Southern States these two items are not half the sum just named. Can the English railway companies abate their charges, and will the English landowners be satisfied with more moderate royalties? The first of these probabilities depends on the ability of the great carrying interests in England to afford any relief in the direction suggested; the second on the willingness of the landowners to be content with such an amount as will enable the British manufacturers, in case of need, to compete with foreign nations.

It is almost unnecessary to mention that the observations just made in reference to the relative position of Great Britain and the

Southern States of America may be materially altered, not immediately, but within a few generations. The capability of the Durham coal-field to furnish cheap fuel to the ironworks will be gradually curtailed; and as this takes place coke and coal will have to be brought from greater distances. At present the States in question may be regarded as virgin ground, in which iron ore and coal, as I understand, exist in sufficient abundance to endure long after the north-eastern coal-fields of England are exhausted.

The question which has just been raised respecting the United States brings us as a natural consequence to consider the prospects of the three European nations more prominently referred to in these pages, as regards competition with us in the exportation of iron and steel.

The quarter, indeed it may be said the only quarter, from which Great Britain is threatened, at present at all events, with successful competition is from those districts which are either situate upon or are within easy reach of the great iron ore deposit stretching from Western Germany through Luxemburg into Eastern France. From some cause or another, there seems occasionally a greater difficulty in producing foundry iron from this variety of mineral than obtains in the use of the Cleveland ironstone. It is however in the subsequent products of malleable iron and steel that we have most to fear the rivalry of our continental neighbours, rather than in pig iron. The reason of this is that labour being cheaper with them than it is with ourselves, it is only when wages enter more largely into the cost of the product than they do in the case of pig iron, that the continental advantages from this source operate seriously to the prejudice of the British manufacturers. At the same time, cheap pig iron is an all-important factor in the economical production of iron and steel, and so far as the raw materials for the blast furnaces are concerned, it would appear, from what has been said a few pages back, that Western Germany, Luxemburg and Eastern France enjoy facilities, in some cases, even superior to those possessed by the district in England most favourably situated for exportation. Belgium on the other hand, having to bring its ore from Luxemburg at a cost of 4s. 8d. per ton for carriage, although it has coal on the spot, is unable to obtain the raw materials for its pig iron quite as cheaply as they are delivered to the Cleveland furnaces.

SECTION XVIII.—CHIEF IRON-PRODUCING COUNTRIES COMPARED. 703

Immediately after completing my work at the French Exhibition in 1878, I visited the districts from which girders and other forms of malleable iron, including considerable quantities of wire rods, were being sent to London. This was at a time when the iron trade was by no means in a flourishing condition, either there or with ourselves. Pig iron No. 3 averaged at Middlesbrough 38s. 2d., and forge iron probably about 37s.

Girders were quoted at one work in the West of Germany at £6 up to 8 inches in depth and £7 10s. above this size. This establishment, although it had to bring the Luxemburg ore from a considerable distance, was able to produce the pig iron used in the mills upon terms quite as low as that named as the selling price at Middlesbrough. According to the costs furnished to me, the price of £6 for girders left, considering the state of the trade at the time, a fair and reasonable profit. The cost of conveying these girders to London was 11s. 10½d. to Antwerp plus 4s. 6½d. from Antwerp to London, or together 16s. 7d. per ton. The freight from Middlesbrough to London may be taken at 5s.; so that with the cheaper wages in the mill and a trifling sacrifice of the profit, there is no doubt the German manufacturer is quite able to compete with anyone in Middlesbrough in the article in question. There are however mills in Western Germany at which pig iron can be made on more advantageous terms than at the work just referred to, and where girders will cost a proportionately lower price, while the carriage to London is no higher. From such establishments wire rods of excellent quality were being delivered in London at £7 5s. per ton, a price at which one of our makers, using however a superior class of pig iron, informed me at the time he could not compete.

Belgium undoubtedly is not so well situated as the last mentioned works, having, as was just mentioned, to bring its ore from Luxemburg at a cost of 4s. to 4s. 6d. per ton for carriage. Pig iron can however be produced in Belgium for making girders as cheaply as at the work in West Germany first spoken of, and it enjoys an advantage over both in a cheaper cost of transport to London, viz. 8s. 1d. instead of 16s. 7d.

As an element of cheap production in the works just referred to as engaged in the manufacture of girders (and also of iron sleepers), may be mentioned the suitability of their pig iron for the purposes in question. The puddled iron it affords is somewhat cold-short in its character, which enables the manufacturers to use a smaller quantity of roughed-down iron than would, I think, be found practicable in England. This however, in our present state of knowledge, is a matter within the power of the English smelters and mill owner. Pig iron for Bessemer purposes can be made at the will of the smelter, and, what is more germane to the present argument, pig iron is being produced of the exact composition, as regards silicon and phosphorus, which will meet the requirements of the basic steel maker.

What has been advanced in regard to certain forms of malleable iron is no doubt equally applicable to the subject generally. An exception however must be made to the manufacture of Bessemer steel by the old, or the so-called acid process. This arises, as has already been explained, from the greater or less dependence of England and the continent of Europe on the mines of Bilbao for the ore required in this process. So long as the necessary supplies of this mineral are brought to a seaport, so long will furnaces situate like those on the Tyne or Tees or indeed in South Wales, enjoy an advantage over Belgium and Westphalia. On referring back to the comparative costs of pig iron made from Bilbao ore, the cost taken at 13s. delivered *ex* ship, the figures, taking Cleveland No. 3 as unity, stood thus :—

The Tyne	120
Tees	123
South Wales	125
Belgium	140
Westphalia	140

The entire position however is altered by the adoption of the basic process, in which we start with a quality of pig iron not much if any dearer than forge iron.

The figures representing the relative costs of the materials for forge iron were as follows :—

Tees	100
Western Germany	75
Ilsede	81
Belgium	102
Westphalia	103

The result of the discovery of the basic process is, that places like

Western Germany and Ilsede—without any ores suitable for the acid process and so far distant from a seaport that the carriage of the ore from the ship would forbid its use—are no longer shut out from the manufacture of steel.

If, as it has been pretended, the cost of "basic material" for pig iron in extra waste and additional labour, is covered by 10s. per ton of steel or even per ton of pig iron used, and adding 11s. in the case of Western Germany and Ilsede for dues to a seaport, 4s. in that of Belgium and 6s. of Westphalia, the figures will be approximately thus altered:—

Tees	100
Western Germany	118
Ilsede	114
Belgium	114
Westphalia	121

Great Britain therefore, although it still possesses some advantage over other European nations, in this new process in relation to its export trade, has had this advantage materially lessened by the introduction of the basic process. So much so indeed that it is, in some instances, questionable whether the cheaper labour in the steel processes themselves, may not place the two in a position of equality, when they meet each other in neutral markets. Such certainly appears to have been the case even when the higher price of the pig iron required in the acid process was included; for certain it is that as regards the dearer kinds of steel, such as that for springs, railway axles, &c., Belgium and Westphalia have been sending considerable quantities for the use of English railways.

The figures given, which I believe will be found fairly correct, go to prove that the abrogation of all import duties in Belgium or Germany would not place Great Britain in a better position, as regards the manufacture of forge iron, than that occupied by the makers, even in the least favoured of the districts just named. The force of this observation is still stronger when applied to iron in the more advanced stages of malleable iron and of steel.

The stretch of country between Luxemburg and Nancy would be somewhat less favourably placed for the manufacture of steel rails than those localities above described, because of the greater distances from the coal; but any importations required for Eastern France would, under the conditions of the cost of manufacture in Western Germany, be obtained from works there which are much nearer the point of consumption than those of Great Britain.

As regards Central and Southern France, their geographical position and the price of fuel forbid the cultivation of a large foreign trade. I have been informed, that it has been in contemplation to remove one of the rail mills in the neighbourhood of St. Etienne to the coast, in order to render the Bilbao ore available for export orders. In such a case the coal would have to be conveyed over a considerable distance; but whether, looking at the resources of Bilbao in respect to ore, already placed under very heavy contributions, and also at the expense of fuel such a course would be a prudent one, may be open to question.

Since writing what has appeared in the present work on the construction of iron vessels, I have received particulars of an establishment in Norway, where something like 800 men are engaged in building steamers of iron. These are so instructive that I think it desirable to quote them upon the present occasion together with the opinions of Mr. Raylton Dixon the well-known shipbuilder of Middlesbrough.

On comparing the wages paid at the yard in question Mr. Dixon estimates that upon a vessel of a given size the labour will cost 25 per cent. less than that expended on a similar vessel built in England. From the advantage accruing from this source must be deducted the freight on the materials used in its construction, which at the present time, with the exception of timber, may be said to be taken from England.

The net result of these two factors on the cost is such that, on a vessel of 1,500 tons dead-weight capacity, the Norwegian builder can turn out his work for £525 less than his English competitor is able to do it.

We still however, in the competition which may arise between English and Norwegian shipbuilders, enjoy the privilege of furnishing the iron and coal. As soon however as the German mills are placed upon the footing of those in our country which are laid out for rolling large quantities of ship-plates, we shall, I fear, behold this trade slip away from us, unless our plate-rollers, etc., are willing to accept something approaching to the same prices as those paid in Germany and elsewhere.

SECTION XVIII.—CHIEF IRON-PRODUCING COUNTRIES COMPARED.

I have been at some trouble in collecting and arranging the data necessary for comparing the position of the chief iron-producing countries in the world; being impressed with its importance to those who are deeply interested in the iron trade of the United Kingdom. The following figures will give an idea of the relative extent of our home and foreign transactions in the metal.

In the year 1882 the quantity of pig iron smelted in the United Kingdom was 8,493,287 tons. At the same time the following quantities of iron of different kinds were exported, and to these I have appended the approximate weight of pig iron consumed in their manufacture :—[1]

	Exported.	Equivalent in Pig.
Pig iron	1,758,152	1,758,152
Old iron for re-manufacture	131,393	
Bar, angle, bolt and rod iron	313,645	
Railroad iron of various sorts	933,123	
Wire	86,686	
Hoops, sheets and boiler plates	343,287	3,240,181
Tin plates	265,021	
Iron, cast or wrought and manufactures	329,399	
Steel unwrought	171,653	
Manufactures of steel or steel and iron	17,938	
	4,350,297	4,998,333

Thus it would appear that in one form or another close on 59 per cent. of all the pig iron produced in Great Britain has been sent out of the country. During the same year there was imported of different kinds 312,676 tons, equal roughly in the form of pig to 390,845 tons.

If we consult the list of our exports of iron of all kinds, over the last ten years, there has been, between the first and last years of the period, a considerable expansion of our foreign trade; but it has to be remarked that until 1880 there was no increase but the reverse. The last three years exhibit a marked improvement, chiefly I apprehend due to a sudden demand for iron in America and elsewhere. Besides the

[1] Compiled from Board of Trade Reports by E. Schmitz & Co. Report of Iron Trade Association, 1883, p. 139.

weight of iron exported the imports of the bar iron and other forms of manufactured iron and steel are given. The actual weights are as follows:—

	Exports.	Imports. Bar Iron.	Iron and Steel Manufactured.	Total Imports.
1873	2,957,813	74,666	30,767	105,433
1874	2,487,522	73,469	52,700	126,169
1875	2,457,306	34,228	58,002	92,230
1876	2,224,470	37,068	69,310	106,378
1877	2,346,370	42,254	83,981	126,235
1878	2,296,860	102.742	105,719	208,461
1879	2,883,484	95,458	112,335	207,793
1880	3,792,993	120,114	155,293	275,407
1881	3,820,225	111,700	175,648	287,348
1882	4,350,297	139,652	173,027	312,679

It may be urged that the actual quantity of iron brought into this country is insignificant; but the steady increase is not unimportant, and I would observe that the motive formerly mentioned in treating the export of steel rails does not apply here. In transactions of this latter description, the object in view was the maintenance of large foreign establishments at full work; in these now under consideration the individual orders are small, and cannot, I apprehend, be regarded otherwise than as an indication of the ability of the exporting nations to compete with British manufacturers on their own ground. It is further worthy of notice that the increased facilities in economical production offered by the basic process, as already described, cannot as yet have had time greatly to affect the subject before us.

Should the views I have laid down be correct we naturally look to the future for any possible change in circumstances calculated to operate in our favour.

I attach little, indeed no importance to any improvements in the processes connected with the manufacture itself; if for no other reason than that of their immediate adoption by competing nations. But further I am not very sanguine of the possibility of the iron-masters being able greatly to reduce, by further change, the cost of production. In the blast furnaces not only is there practically no waste of iron contained in the ore, but scientific considerations seem to justify the inference that if any economy of fuel is possible, its extent is so small as scarcely to be worth consideration. I do not say that in the malleable ironworks some amelioration is not possible, but the uncertainty of the future of this branch of the trade is so great, as to offer little inducement to a manufacturer to incur the necessary outlay and risk connected with mechanical puddling and other cognate improvements. When we come to Bessemer steel, the labour and expenses connected with the act of conversion have been brought down to a point which leaves but little margin for any further economy. In the rail mill, such has been the perfection to which the rolling machinery has been brought that the work in connection with this department can be done for less than 1s. per ton. In waste of metal and in fuel some small improvement is not impossible, but they cannot, I think, lessen very materially the cost of production.

In connection with the question before us I cannot refrain from calling the very serious attention of our workmen to their position in connection with an industry, in the welfare of which they and their employers have a common interest. Free Trade has placed within their reach the necessaries of life upon terms practically as favourable as those enjoyed by any European nation. An exception in their favour, by imposing a duty on iron so as to shut out foreign competition from our own shores, cannot, as they know quite well, be entertained. Besides, were the idea of a return to protective measures to extend itself beyond the very limited circle of the so-called fair traders, where is the promised relief to come from? Not certainly in the inevitable consequence of an increased price of the provisions consumed by the working classes of this country. The iron trade of the United Kingdom is greatly dependent on its exports, amounting as we have seen to close on 60 per cent. of its entire production, and no domestic legislation worthy of the name has, so far as I know, been suggested which can improve our position in the development of its foreign commerce.

No one who, like myself, has had nearly five and forty years experience with the workmen in the iron trade of this country, can fail to have noticed in later years an immense improvement in their capacity to appreciate the fundamental truths connected with political economy which regulate the affairs of their calling. The boards of conciliation and arbitration have afforded an opportunity to both employers and men, of attaining a better understanding of the difficulties common to the situation of both; but I cannot help feeling, and therefore expressing my opinion, that upon some occasions the immediate future is too exclusively regarded in the settlement of questions of difference which arise between the two parties.

The difficulties in question are of most frequent occurrence when the state of the iron trade compels the iron masters to seek for assistance from their men, in order to keep the works employed at the low prices which prevail at the time. At all events it is upon such occasions that the future of our iron industry is most threatened by foreign competition. The question in dispute is then usually referred to arbitration; but upon these occasions, I have never yet heard any emphasis laid upon the effect foreign labour has upon the iron industry of this country. The loss and financial inconvenience incurred by a stoppage of work, render it difficult, often indeed impossible, for the employer to risk having recourse to so extreme a measure, while the men are naturally anxious to secure the best terms they can for themselves. Both sides, in point of fact, behave in the struggle as if its adjustment were to be arrived at, not upon the actual exigencies of the case, but upon what may be termed a question of immediate expediency. If circumstances are of a character which leave the foreign maker a profit, when none falls to the share of the English makers, our trade must shrink in its dimensions, while that of other nations will expand. By a successful resistance to the demands of the employers the men no doubt reap a present gain; but what follows on a state of things, the effect of which must necessarily be a reduced demand for their labour? The answer is found in the distress which to-day has overtaken many deserving men, who are dependent upon making or working iron for their livelihood. In this connection, the importance cannot be exaggerated of the workmen being disposed to lend loyal and ready help in promoting the adoption of every contrivance having for its object the reduction of the cost of production.

On the other hand it is only reasonable and just that in times of prosperity the men should participate in the advantages offered, as they have been called upon to bear their share of the privations entailed upon the trade in periods of adversity. That the employers

SECTION XVIII.—CHIEF IRON-PRODUCING COUNTRIES COMPARED.

assent to this view of their relations with their workmen may be assumed by their ready adoption of a sliding scale for the regulation of wages. By this system an increase in the rates paid follows an increase in the value of the commodity they manufacture and *vice versa*. It will be generally admitted that labour for some time past has been more highly paid in comparison with the value of iron than it was in former years, and this without the iron masters having, up to a certain point, greatly suffered by the change. This ability to pay, without loss to themselves, higher rates is due to the great improvements which have been introduced into the various departments of the process within the last twenty years, by means of which the cost of production has been much reduced to the great advantage of masters and men.

With regard to the present prospects of the iron trade, in a cosmopolitan point of view, it is difficult if not impossible to speak with any degree of certainty. The commerce of the world is dependent on such a vast variety of circumstances that it is impossible to predict what may be its fate even in the immediate future. A season of general prosperity restores commercial confidence; new public works such as railways &c. are undertaken, activity once more prevails in our mines and forges and prices rise, often more in consequence of an anticipated scarcity than from an actual difficulty on the part of the ironworks in meeting the demands made on their resources.

This difficulty in most cases does not arise from there being any want of the appliances for the manufacture of iron. In illustration of this, we may quote the operations of 1879 as an example. In that year there arose a sudden demand for iron in the United States. On 31st December there were 388 furnaces in blast, instead of 265 as at the close of the previous year. The quantity of pig iron produced rose from 2,577,361 net tons in 1878 to 3,070,875 tons in 1879. In addition to this the stocks of pig iron in America fell from 574,565 tons to 141,674 tons in the latter year. Such however was the demand for iron that the imports of various kinds of the metal increased from 236,434 net tons in 1878 to 862,382 tons in 1879. Notwithstanding all this activity, out of the 697 furnaces in existence at the time 309 furnaces or 44 per cent. were out of blast.

SECTION XVIII.—CHIEF IRON-PRODUCING COUNTRIES COMPARED.

At the end of 1882 the furnaces built in Great Britain were 926 in number, of which only 565 were in blast. In the United States, out of 697 furnaces 417 were blowing. Thus taking the United States and Great Britain alone, 40 per cent. of the furnaces in existence were idle.[1]

With a dormant power like this at our backs, any great improvement in the price of iron must be the result of an inability to divert at once the necessary amount of labour to its production, and not to the want of the necessary plant for its manufacture.

The present position of the steel rail trade is one of some difficulty and seems to have been brought about by a combination of circumstances. When the Bessemer process was first introduced, and the steel made by its means was, to a considerable extent, confined to purposes for which a higher price could be afforded than could be given for rails, Sheffield, the old and chief seat of the steel trade, was a sufficiently appropriate locality for this new industry. At first, owing to their greater price, rails of steel were sparingly used, until their excellence and reduced cost of manufacture extended their application. The pig iron used in the process was, at that time, almost exclusively the produce of the Cumberland and Lancashire mines, and as re-melting the pig iron for the converters was deemed indispensable for success, Sheffield was perhaps as convenient a centre for supplying the United Kingdom as any other locality. When however it was discovered that the iron could be run direct from the blast furnace there was an evident advantage in placing the steel works where the pig iron was made. This led to the establishment of rail mills in mining districts near the coast, which were thus also more favourably situated than Sheffield for sending off the manufactured article by sea. In the meantime Germany and France, favoured by cheaper labour, and a much higher import duty levied on steel rails than on pig iron, began to erect converters and to roll steel rails in mills hitherto used for rolling those of iron, although they had to contend with bringing a considerable quantity of ore from foreign countries. Belgium and France also devoted attention to this new process, although in the latter case expensive ore had to be brought from Africa and carried inland for considerable distances. The cheaply wrought hematite of

---

[1] It by no means follows that the proportion of furnaces out of blast implies a corresponding deficiency of make, arising from their being idle; because it is pretty certain that those not at work are below the average capacity of the whole.

SECTION XVIII.—CHIEF IRON-PRODUCING COUNTRIES COMPARED. 713

Bilbao has in its turn greatly assisted in the development of the production of steel rails, which in later years has received a further impetus by the course of events in South Wales.

As is well known, Glamorganshire and Monmouthshire were the chief seats of the iron rail trade. Labour, almost as cheap as in any place on the Continent, and cheap fuel had enabled the manufacturers there to produce rails probably more economically than was done in any other district in the world, although two-thirds of the pig iron used was obtained from imported ore; for the produce of the Welsh mines was not only expensive to get but was deficient in quantity. The gradual extinction of the iron rail trade compelled the Welsh iron masters to seek for other employment for their establishments. Steel rails offered a means of escape from what would have been all but an entire sacrifice of the capital invested in the works of South Wales, and to this escape the rapid development of the Bilbao mines has afforded most valuable aid. In the meantime the Basic process, as we have already seen, relieved the German steel manufacturers from the disadvantages of their position—not only so, but two large establishments have been erected, one at Hayange in Lorraine and the other on French territory, the former by Messrs. De Wendel and the other by the combined efforts of Messrs. De Wendel and M. Henri Schneider, to supply rails from the phosphoric ore already frequently mentioned in these pages.

Thus in the last few years a great extension in the steel rail making powers has taken place in Europe, not always with reference to any extended demand for their produce, but because certain changes, such as those just mentioned, prompted the erection of additional converters and mills. Besides these additions in Europe a similar line of conduct has been pursued in the United States, which now in the production of steel rails stands at the head of the list.

According to a statement prepared for me by Mr. Jeans, the following quantities were made in 1883:—

United States	1,243,925
United Kingdom	1,097,174
Germany	505,133
France	381,178
Russia	230,000
Belgium	173,000
Austria and Hungary	130,000
	3,760,410

Mention has already been made of the importation of steel into Great Britain, chiefly from Germany. Since writing what was said on the subject, my attention has been called by an article in the *British Trade Journal* to the *continued* importation of steel tram-rails for English use. In this last case the rails—a heavy section—were delivered at Birmingham—the very centre of England—at 10s. per ton cheaper than the lowest quotation from the British makers. The Editor explains the transaction by the wages in Germany being cheaper than in England. There is expended on the article in question much more labour than on an ordinary railway bar, so that I am quite prepared to believe that the ability of the German manufacturer to undersell his English rival is dependent on the cause assigned in the *British Trade Journal*. This view of the case indeed only confirms what has already been advanced on the subject in these pages, and must be regarded as an indication of what must be the conduct of all interested in the iron trade of this country, viz., that we must be prepared to accept foreign competition as an important factor in the iron trade of the world.

There is one matter in connection with the future requirements for steel rails which deserves consideration, viz., the manner in which their substitution for iron rails for the maintenance of the permanent way will ultimately affect the demand and consequently the price.

The rail trade, even when iron was the material employed, occupied a somewhat unique position when comparing the powers of production with the sources of consumption. Previous to the introduction of railways, iron once applied to the construction of gas and water pipes, of steam engines, or to the various minor offices to which its properties suited it, remained, generally speaking, undisturbed for many years. The excessive wear to which rails were exposed, in cases of large traffic, speedily rendered them unfit for service, for in many parts of the North-Eastern system the average life of iron rails is about seven years while those of steel are expected to last double the time. Let us assume ten and twenty years to represent the average duration of all the iron and steel rails in the world. According to the returns published by the Iron Trade Association at the end of 1882 the total length of railways amounted to 247,529 miles. If to cover sidings and double road we assume 325,000 miles to represent the extent in single way, there will be nearly thirty-five million tons of rails in use at the present time. If this were all iron, with a life of ten years, 3,500,000 tons of rails per annum would be needed for relaying, while in steel only half the quantity or 1,750,000 tons would be required.

The construction of railways at first was so gradual and so continuous that this change in the supply of old material, out of which new had to be manufactured, was not seriously felt. At the present day the cost of steel rails has been reduced below that at which iron rails can be made, so that all the renewals now made are of the former material. So long therefore as railway companies were taking out iron and replacing it with steel the larger annual demand represented by the life of iron rails would be required. When however all the lines in the world are laid with steel, instead of $3\frac{1}{2}$ million tons being annually required for repairs $1\frac{3}{4}$ million tons only will be needed.

If Mr. Jeans' figures of the weight of steel rails actually rolled in 1883 are correct we are within the mark in setting down the present producing powers at 4 million tons, so that over and above the quantity required for relaying ($1\frac{3}{4}$ million tons) we shall, when all the iron rails are removed, have an annual surplus of $2\frac{1}{4}$ million tons. To find a market for this quantity it will be necessary to lay 21,000 single miles of new railroad, an amount of work of which there is perhaps not an immediate prospect.

The figures contained in this estimate must only be regarded as very approximate; but they serve to show my meaning, which is that the annual quantity of rails required for maintenance of the road will be very much less when steel instead of iron is the material exclusively used. It is even now open for consideration whether a diminishing demand for renewals is not already being felt, and is the cause which has led to a reduction of make among British and Continental steel manufacturers.

In conclusion it is satisfactory to remember that Nature has been sufficiently bountiful to Great Britain, and that its position as an iron-making community is, all things considered, not inferior to that of any other nation. The mineral wealth underlying its soil, taken as a whole, is so accessible in a commercial point of view, that our miners, per man, can deliver a larger quantity at the surface than is the general practice elsewhere. If their exertions, and those of our workmen at the furnaces, forges and mills, are paid for at a higher rate than is found to prevail in some other countries, it is only fair to say, that some part at least of this difference is, as I have previously shown, honestly earned by greater efficiency.

With such natural resources as we possess, directed by skill not inferior to that found among our competitors, I cannot doubt, when the necessity arises, that a community of interests will suggest a course of action on the part of all concerned, which will still secure for the iron-masters and iron-workers of Great Britain a foremost place in the development and pursuit of an industry to which they have rendered such signal service.

CORRECTIONS AND ADDITIONS.

Page 6.—9th line from foot. The position of Great Britain and the Continent in reference to imports and exports is somewhat modified since the short expression of opinion given here was written. The subject is referred to at page 443, as well as elsewhere, and again under the present heading.

Page 10.—4th line from foot, and page 31, 7th line from top, for "*trombe*" read "*trompe*."

Page 12.—12th line. In the profusely illustrated work by Agricola referred to there is no figure of a blast furnace. Dr. Percy, however, in his treatise on *Iron and Steel*, quotes another writing by Agricola—*Metallurgische Schriften*—in which it is stated that "iron smelted from ironstone is easily fusible, and can be tapped off."

Page 57.—8th line, for "15" read "25."

Page 62.—Last line, for "$CO_2 \times C = 2CO$" read "$CO_2 + C = 2CO$."

Page 80.—6th line from foot, omit words "burnt to carbonic oxide."

Page 85.—Last paragraph. Mr. Horton recently informs me that I have somewhat overstated the economy of fuel effected by enlarging the cold blast furnace, but the error does not effect the nature of the argument made use of.

Pages 92 and 146.—It may be well to observe that the $20\frac{1}{2}$ cwts. of coke mentioned as sufficing to produce one ton of pig iron must be considered as an exceptionally low rate. The subject is further referred to at page 106.

Page 103.—2nd line, for "two hours" read "seven hours."

CORRECTIONS AND ADDITIONS.

Page 105.—18th line. Mr. Parry's analysis is given by volume, which reduces the carbon fully oxidised to about two-thirds, instead of three-fourths, as stated. This will slightly alter the figures in the subsequent paragraph.

Page 117.—2nd line, for "3,137" read "3,107."

Page 138.—2nd line from foot, and page 144, 5th line. In page 138 the excess of air in the steam boiler fireplace is given at 100 per cent. the authority being *Rankine on the Steam Engine*, Fourth Edition, page 281. At page 144, 5th line, the excess of air (20 per cent.) was the result of several analyses at the Clarence boilers, blast furnace gas being the fuel used, instead of coal, as in Rankine's computation.

Page 191.—Table at top of page, first column, for "·018" read "·108"; second column, for "1·118" read "1·188."

Page 214.—8th line from foot, for "xNaOKO" read "x[NaK]O;" 6th line from foot, for "·72N" read "·42N;" 2nd line from foot, for "NaOKO" read "[NaK]O."

Page 218.—Table of Cyanides.—1st line, for "15·06" and "3·77" read "49·06" and "4·73;" 2nd line, for "29·11" and "9·07" read "112·70" and "5·19."

Page 230.—Table of earths in furnace fume. In the escaping gases the composition, instead of the figures given, should be

Silica 70 ... Lime 4 ... Alumina 26 ... Magnesia 0 = 100.

Page 241.—4th line from foot, after words "requiring heat which," add words "latter will."

Page 260.—7th line, for the words "can exceed" read "can, unit for unit, exceed."

Page 261.—In table at foot of page, where word "cwt." occurs, read "unit."

Page 266.—20th line, for "981" read "925."

Page 283.—5th line, for "fully 9 per cent. more than coke, viz., 4,376 calories," read "fully 17 per cent. more than coke, viz., 4,084 calories." On referring (page 282) to the heat generated by dry coke and dry charcoal, each burnt with hot air, the difference between the two kinds of fuel is under 5 per cent., viz., 4,168 calories for coke, and 4,376 calories for charcoal. This arises from the quantity of heat in the blast being much larger in the case of the former than in the latter. The percentage of carbonic acid in the charcoal furnace gases being as great as it can be maintained, it is possible that a higher temperature in the blast has been found useless.

Page 289.—15th line, omit word "not."

Page 314.—5th line from foot. The oxygen in Durham coal is overstated; the content given being that of Scotch coal. *Vide* page 315. The following is the correct analysis of the coal from a South Durham colliery:—

Carbon	82·27 per cent.
Hydrogen	4·68 ,,
Oxygen	5·66 ,,
Nitrogen	·91 ,,
Water	1·00 ,,
Sulphur	1·22 ,,
Ash	4·26 ,,
	100·00 ,,

Page 328.—4th line from foot, omit words "at 3d."

Page 337.—1st line, omit word "blast."

Page 339.—11th line, for "15·3" read "15·37"; 10th line from foot, for "35·0" read "35·2."

Page 468.—2nd line. For the words "nearly the whole" read "the greater part." In later years the exports, as will be observed in XV. and in future sections, have largely increased from other countries as well as from Great Britain.

Pages 469 and 470.—The weight of iron consumed in the United Kingdom as computed by Mr. Jeans is 4,618,932 tons, whereas I make it only 3,495,000 tons. I am ignorant of the basis upon which Mr. Jeans' estimate is framed. The figures made use of by myself were obtained by deducting the exports from the total make of pig iron. *Vide* page 444.

Page 576.—15th line from foot, for "produced" read "procured."

Page 583.—13th line, for "30s." read "20s."

The 20s. here referred to represents the difference of cost between pig iron for the acid and for the basic processes. It is quoted on the authority of Ritter von Tunner, who reported on the relative merits of the two methods to the Iron Masters Union, of Styria and Carinthia.[1] The cost of hematite metal suitable for the acid process is given at 64s. 8d., while that employed for the basic mode of treatment is only 45s.

I may add that both these quotations, particularly the first, are considerably higher than those adopted by myself. This want of agreement is probably due to the fact that I have been guided, not by the actual cost at any particular work, but by what I conceived to be the capabilities of the country generally. Thus, in furnaces of antiquated description, as much as 10 cwts. more coke were being consumed to make a ton of Bessemer pig iron than was employed in establishments of a more modern type.

Under any circumstances the comparison of the position of different countries, in an iron-making point of view, is a very difficult problem. This arises from the varying values of the raw materials and from the improvements which have been made in the smelting of the ore and in the conversion of the metal into the finished article.

As far as steel rails are concerned, the result of the consideration I have given to the subject is, that as between the best situated works here and similar works on the continent of Europe there is an advantage of at least 10s. per ton in favour of the British maker in the cost of the article put on board a vessel at a sea port.

[1] *Vide* Transactions of Iron and Steel Institute, 1880, pp. 296 and 297.

Important as a sum of 10s. is in an article not worth more than about 90s. per ton, it is obviously not sufficient to enable the manufacturers of Great Britain, except under very exceptional circumstances, to export rails for consumption in countries where the difference in cost does not exceed the 10s.

In an article like an ordinary steel rail the amount paid in wages is too small to permit the foreign maker, by the cheaper labour at his command, to overcome the other disadvantages of his position. When however 40s. or 50s. is added to the cost price of an object it seems very probable that we must prepare ourselves to have to meet the competition of German and Belgian manufacturers even in our home markets.

Page 606.—7th line from foot, for "natural and artificial foreign" read "native and foreign."

Page 668.—18th line from foot, for words "*vide* page 622" read "*vide* page 661."

Page 704.—Foot of page, for "Tees 100" read "Tees 98"; and for "Western Germany 75" read "Western Germany 85."

Page 706.—Since the reference made to iron shipbuilding in Norway Mr. Raylton Dixon has kindly furnished me with some further particulars. The saving of £525 in the cost of a ship built in Norway for labour was effected in the wages paid to the iron workers alone. To these my informant includes that connected with the carpenters, joiners, painters, &c., in the following terms:—

"The net result of the two factors on the cost is such that on a vessel of say 1,250 tons gross register and 1,500 tons dead weight capacity, the Norwegian builder can turn out his work for at least £850 less than his English competitor."

Reference has been made in these pages to the occasional want of readiness on the part of English workmen to co-operate with their employers in the adoption of improvements for the saving of labour. The mechanical drill for mining ironstone was instanced as a case in point. It is only right now to mention that a great advance has since been made in the manipulation, at the Cleveland mines, of the machine in question.

INDEX.

Acid and basic processes, cost of, compared, 424; heat required in, compared, 421. (See also *basic process* and *Bessemer*, various heads of.)
Acid, carbonic (see *carbonic acid*).
Acid process, sources of heat in, 422.
Advantages, comparative, of different iron-making countries, 453.
Agricola and Dud Dudley, writings of, 12.
Agricultural and ironworks labour in Great Britain and America compared, 554.
Agricultural labour, cost of, in England and other countries compared, 496; cost of, in France, 496; introduction of new industry, effect on, 476; price of in America and North of England, 553; price of, on Continent of Europe and in Great Britain, 496.
Airdrie splint coal, analysis of, 120.
Air, atmospheric effect of temperature and conditions of, on combustion of coke, 238; atmospheric, means of heating, by escaping gases, 251; atmospheric, quantity of required for combustion of coke, 264 atmospheric, superheated, effect of application of, to worn-out furnaces, 259. (See also *blast*.)
Åkerman, comparison of coke and charcoal with anthracite by, 290; on working of charcoal furnaces in Sweden by, 276.
Alabama, convict and slave labour in, 557; iron manufacture of, 700.
Alkalies, supposed deoxidation of, in blast furnaces, 214.
Alkaline matter, conditions of increase of, in blast furnace, 224.
Alkaline compounds, possible cause of differences in quantities of oxygen and carbon in blast furnace gases, 218.
Alumina and lime least affected by sublimating influences in blast furnaces, 232.
Alumina, effect of presence of, in limestone, 57.
Aluminium, experiments to obtain, by means of cyanide of potassium, 230; occasional presence of, in pig iron, 167.
America, cost of a ton of charcoal in, 54; freight on pig iron to, 604; high transport charges on ore and coal in, 606; import duty and transport charges on pig iron in, 606; imports and exports into values of 1870 to 1881, 604; increase of cost of pig iron in, 606; increased cost of ore and coal in, 606; iron and stee imports into, 604; iron trade of, 604; ocean freights to and from, 607; sudden demand upon iron resources of, in 1872 and 1873, 605. (See also *United States*.)
American charcoal furnaces compared with Styrian and Swedish, 300.
American experience of relative qualities of hot and cold blast iron, 152.
Ammonia and tar, collection of, from blast furnace, 314, 326; condensation of, at coke ovens, by Jameson, 591; condensation of, at Gartsherrie furnaces, 591. (See also *coke* and *coking process*.)
Ammonia, presence of, in blast furnace, 228; sulphate of, produced in Scotch furnaces, 327.
Ammoniacal compounds, recovery of, in coking process, 52.
Ammoniacal salts, collection of, from Clarence furnace, 326.
Analyses, ammonia salts in blast furnace gases, 228; Bessemer and blast furnace slags, 392; Bessemer and Cleveland pig iron, 388; Bessemer gases in blowing, 390, 392, 418, 419; Bessemer slag, 418, 420; Bessemer steel, 408, 414, 417, 421; Bessemer steel, basic, 408; Black Band, Tuscarawas, 674; boiler-plate made in different kinds of puddling furnaces, 373; Bowling iron before and after refining, 359; Cast iron, 147, 148, 150, 153, 155, 157, 158, 168, 172, 316, 345, 346, 354, 355, 360, 367, 388, 398, 404, 416, 417, 419, 421, 438, 656; after exposure to heat, 159; cast iron, Cleveland, and products therefrom, 398; cast iron during conversion in Bessemer process, 391; cast iron, glazed, 162; cast iron, hæmatite and rails therefrom, 430; cast iron heated in contact with wrought iron, 160; Cinder in Bessemer converter, 392, 394, 418, 420; cinder in boiler-plates, 369; cinder, blast furnace, 168, 170, 392, 394; cinder, blast furnace gases from, 173; cinder, Lancashire hearth, 346; cinder, blast furnace, making ferro-manganese, 166; cinder, blast furnace without lime, 169; cinder, mill, 369; cinder, puddling, 361, 395; cinder, puddled steel, 438; cinders produced at refineries, 359; cinder, purified iron, 404; cinder, refinery, 354, 355, 358, 359, 361, 394, 396; Cleveland

Analyses— *Continued.*
iron blown at Eston, 409; Cleveland pig and refined iron from, 354; coal, 120, 127, 511, 612, 616, 620, 641, 643, 644; of coal, 120, 644, 645; anthracite and bituminous, 127; Lanarkshire, 315; splint and bituminous, 120; coke, 104, 657; coke, by Muck, 105; coke, by Parry, 105; fume, blast furnace, 225, 231, 232; fume, from cinder, 174; gases, Bessemer converter, 390, 392, 418, 419; gases, blast furnace, 68, 73, 102, 107, 108, 110, 114, 142, 196, 200, 204, 208, 210, 214, 227, 292, 293, 308, 309, 317, 318; iron ore, 648, 652, 654, 655, 670, 672, 674, 675, 678, 679, 680; iron rails, 365, 428; malleable iron, 345, 373; materials used in basic process, 419; metal at different stages in acid process, 391; metal at different stages of basic process, 408; metal before and after purification, by Bell, 398; metal before and after purification, by Krupp, 404; open hearth steel, 434, 435; open hearth steel with injected steam, 433; phosphorus and sulphur in flux, fuel, and ore, 357; phosphorus and sulphur in pig iron and in slag, 165, 357; phosphorus, excessive quantity in steel rails, 428; pig iron, Swedish and English, 346; puddled steel, 438; puddling slags, 395; purified iron, 398, 404; rail steel from Cleveland pig, 427; refined metal, 354, 358, 359, 360; rivet iron, 367; slag from basic blow, 409, 420; slag from Siemen's steel furnaces, 431; slags from basic Bessemer converter, 409, 420; Spiegel iron, 419, 421; steel basic blow, 408; steel rails, acid, 414; steel rails, basic, 414; steel rails with excess of phosphorus, 428; sulphur and phosphorus in flux, fuel, and ore, 165, 357; sulphur and phosphorus in pig and slag, 357; West Yorkshire malleable iron, 434; white iron before and after blowing in basic converter 417 (see *pig iron, &c.*)

Angles of boshes, effect of, 268.

Anthracite coal, combinations to keep up prices of, 555; comparison of, with coke and charcoal, 290; composition of, 127; consumption of, compared with charcoal and coke, 130; consumption of, in American blast furnaces, 127; cost of, 557; price of, 693; earnings of men at mines of, 557; profit on, 693; sliding scale of wages at mines of, 556; strikes in mines of, 555; suitability of, for iron smelting, 46, 126; tendency to splinter requires correction by strong blast, 127; wages in mines of, 557.

Armstrong, Sir W. G., application of malleable iron for ordnance, 27.

Ash, effect of, in blast furnace, 237. (See also under heads of *analyses, coal and coke.*)

Atmosphere in Bessemer converter, 392.

Austria, cost of a ton of charcoal in, 54; wages paid at ironworks of, 477.

Austrian furnaces, low consumption of charcoal in, 274.

Azote in coke, 105. (See also *analyses; coal; coke.*)

Baker, W. E. S., on costs of iron in Eastern States, 695; tables of costs of iron, 696; remarks thereon, 698.

Bar iron, cost of producing, in 1727, 350; low hearths first used for manufacture of, 344; waste of pig in manufacture of, 350. (See also heads *malleable iron, boiler plates, puddled bar, puddling and puddled iron.*)

Basic and acid Bessemer processes, comparison of, 407.

Basic blow, heat carried off in gases of, 424; heat evolved by different bodies in, 423; loss of iron in, 421; quantity of heat evolved in, 421; temperature of, 424; weight of substances expelled during, 420.

Basic process, analysis of pig iron and steel of, 417; consumption of materials in, 421; cost of, compared with acid process, 425; effect of silicon in, 416; effect on relative position of certain European countries as steel makers, 705; expulsion of metalloids in, 420; heat required in, 412; inconvenience of, 411; inexhaustible supply of ores for, 451; in Western Germany and Eastern France, 705; order of removal of metalloids in, 408; overblowing in the, 413; oxidation of iron in, 423; quantity of iron lost in, 411; refractory cinder in, 411; separation of phosphorus by, 4, 413; time required for blowing in, 418; use of white or hard grey iron in, 416. (See also *acid process, Bessemer*, various heads of *steel.*)

Basic slag, analysis of, 409, 420; oxides of iron in, 418.

Basic steel rails, analyses of, 415.

Beer and butcher meat, cost of, in 1685, 478.

Beer, consumption of, in different countries, 493.

Belgium, coal of, 629; coal working in, 509; competition of, with England, 443; cost of iron in. 688; exports of iron from, 466; import duties of, 465; iron ore of, 659; position of, in reference to iron ore

INDEX. 725

Belgium—*Continued.*
supplies, 452; progress of iron trade in, 464; transition in iron trade of, 464; weekly expenditure of a family in, 486.
Bessemer acid process, weight of substances expelled during, 421.
Bessemer and blast furnace slags, analyses of, 394.
Bessemer and open hearth steel compared, 435.
Bessemer blow, carbonic oxide, &c., in, 390; gases in (see *gases*); gases taken at various periods of, 390; oxidising and reducing gases in, 390.
Bessemer Cleveland iron, margin of economy in use of, compared with Hematite, 406.
Bessemer converter, and the puddling furnace compared, 382; atmosphere in, 392.
Bessemer iron, fuel consumed in manufacture of, 259.
Bessemer pig, impurities contained in, 388.
Bessemer process, acid and basic compared, 407; advantages of, over puddling, 384; difficulty of expelling phosphorus by, 383; intense temperature of, 383; use of spiegel and ferro-manganese in, 383; use of Swedish iron for, 383.
Bessemer, Sir Henry, invention of pneumatic process for making steel, by, 19.
Bessemer steel, acid process, relative costs of pig iron in Europe for, 704; basic process, relative costs of pig iron in Europe for, 704; cost of production of, compared with open hearth, 432; development of manufacture of, 384; earnings of men at, 573; earnings of workmen at, in Great Britain and Germany compared, 535; economic conditions of manufacture of, 384; number of men required for manufacture of, in different countries, 573; ores used for manufacture of, 386; production of, in different countries, 453; quality of Bessemer steel, compared with open hearth, 434; statistics of manufacture of, 432; superiority in strength over wrought iron, 385; works in United States, excellent character of, 571. (See also *acid process* and *basic process.*)
Birkinbine's comparison of charcoal, coke, and anthracite, 290.
Black Band, Tuscarawas Valley, 673.
Blair, T. S., direct process of, 34.
Blauofen, 11.
Blast, greater weight of, consumed in smaller furnace. 204; heating of, by escaping gases, 251; heating of, by firebrick stoves, 234; heat of, as commonly used, 88; heat unit of, may be equivalent to heat unit from fuel, 255; hot and cold compared, 86; increased pressure of, in furnaces using anthracite, 92; low temperature of, in Swedish furnaces, 132, 283; number of calories required in, as supplementary to those afforded by coke, 252; saving of fuel by successive additions of heat to, 265; superheated, value of, in worn-out furnaces, 259; temperature of, according to units of coke employed, 252; temperature of, in Vordernberg furnace, 283; temperature of, limit to which it can be raised, 266; units of coke burnt by, and by carbonic acid in limestone, 107. (See also *air, atmospheric.*)
Blast furnace, action of, as affected by size, 72; advantages possessed by large, 203; chemical action of, 65; Cleveland and Continent of Europe, wages, &c., at, compared, 521; completeness of duty performed by, 138; condition of working of, indicated by colour of cinder, 171; conditions under which fuel are burnt in, 140; date of invention of, 12; defective examples of, 263; derangements in, 100; difficulty of proving hydrogen in gases of, 306; dimensions of, in Cleveland, 22; disturbance in composition of gases of, caused by cyanides, 216; driving, effect of varying rates of, 203; earnings in 1870 to 1880, 565; effect of derangement of, on cinder, 172; effect of overburdening of, 172; effect of structural errors in, 202; experiments on, by Ebelmen, Bunsen, and Playfair, 308; extent of reduction performed by H in, 315 fume, description and occurrence of, 174 gases, analysis of, 68; gases in, above the tuyeres, 177; gases, quantities of hydrogen in, 306; gases, waste of heat by escape of, from throat of furnace, 73; height of, no uniform law for, when smelting ore, 145; hematite, use of raw coal in, 323; imperfect combustion of carbon in, 67; improvements for saving labour at, 525; increased production and its influence on wages at, 521; increase of dimensions of, 23; instant decomposition of water in, at hearth, 322; limit of useful extension of size in, 74; men at, cost of living in North of England, 487; mode of action in, 67; nature of process of combustion in, 63; no uniform law for height of, 145; number of men at, different countries compared, 524; reactions in, 186; results obtained by use of raw coal in, 315; Scotch, particulars of working of, 316; sequence of chemical action in, 176; sizes of, required for best results, 202;

Blast furnace—*Continued.*
small and large compared, 203; slag, absence of lime in, 169; slags, description of, 169; starting-point of iron-making operations, 46; summary of improvements in, 24; superiority of over direct process, 61; three functions of, 241; twofold duty of, 65; use of water-gas in the, 337; utilisation of waste gases from, 23; wages at, 520; work, disturbing influences of, 259; working average of, 268; work of, 61. (See also *atmospheric air, ammonia, carbonic acid, carbonic oxide, charcoal, coal, coke, cyanides, cyanogen, fuel, furnace blast, heat, hot blast, hydrogen compounds, iron stone and iron ores, labour, lime and limestone, metalloids, ore, oxygen, phosphorus, pig iron, silicon, slag and slags, sulphur, reducing zone, temperature, tuyeres, water gas, white iron, zinc.*)
Blow. basic process (see *basic*).
Board and lodging, cost of, on Continent and in Great Britain, 489. (See also *food, provisions.*)
Board in Southern States, cost of, 550. (See also *food; provisions.*)
Boiler-plates, analyses of different varieties of iron for, 373; cinder in, 369; unsound welding of, 369.
Boshes, angle of, in blast furnace, 268.
Bowling cinders from refinery, analyses of, 359. (See also *refinery; refined iron, analyses.*)
British Iron Trade Association, publications of, 474.
British made iron, exports to iron-making countries, 469; protective duties levied on, 467.
Bunsen, Ebelmen, and Playfair, experiments on blast furnaces by. 308.
Butcher meat, cost of, in 1685, 478; cost of, in 1866, 480; in France and Germany in 1847, 480. (See also *provisions.*)
Bunsen and Playfair's experiments on blast furnace gases, 308.
Calcination, imperfect, of ironstone, 240; iron ores, change effected by, 56; limestone effect of, 58; limestone, want of economy in large furnaces. 60; ores containing carbonic acid, before being used in blast furnace, 56.
Calcining kilns used in Cleveland, 56.
Calcium found in pig iron, 167.
Capacity of furnace, advantage derived from increase in, 247.
Capital attracted to most remunerative trades, 463.
Carbon, action of, on oxide of iron, 197.
Carbon and carbonic oxide, temperatures at which they begin to act on peroxide of iron, 70.
Carbon and oxygen, alteration in quantities of, in furnace gases, 214; at different levels of furnace, 212; in furnace, irregularities in quantities of, 208; in gases, alteration of, 215; possible cause of irregularity in quantities of, caused by alkaline compounds, etc., 218. (See also *oxygen.*)
Carbon and silicon, oxidation of, in Bessemer "blow," 391.
Carbon as an agent for work performed in the blast furnace, 62; as carbonic acid, diminution of, in blast furnace, 214; as carbonic acid, extent of existence of, in furnace gases, 107; ascertainment of, in gases at different levels of the furnace, 208; calories provided in blast furnace by oxidation of, 88; changes caused in condition of, by refinery, 355; character of change in, by long exposure to high temperature, 158; combination of, in iron in cementation process, 160; combination of, in pig iron, 155; combination of iron with, in atmosphere of carbonic oxide, 160; combined, found most largely in white iron, 158; combustion of, 62; combustion of iron in low fires, 63; condition of, in gases, 270; condition of, in pig iron affected by temperature, 158; consumption of, compared in iron smelting with charcoal and coke, 129; deposited, use of, in protecting masonry of blast furnace, 222; deposition, conditions favourable for, 189; deposition due to action of escaping gases on oxide of iron, 202; deposition of, without formation of metallic iron, 189; derived from coke, fermula for estimation of, 273; derived from fuel, calculated from composition of gases, 269; effect of lowering 6·58 of as CO_2, 89; escaping from blast furnace as carbonic acid and loss of heat from diminution in quantity of, 83; exceptionally low content of, in iron, 157; fixed in coal, duty obtained from, in furnaces using coke and raw coal, 325; fixed in coal, heating power of, 235; fixed in coal, heat obtained from, 236; fixed in coal, loss of in coke-making, extent of, 51; fixed in coal, value of fuel dependent on, 234; gaseous condition not necessary for combining with iron, 160; heat evolved by, in basic process, 418; heat evolved by, limits of, 69; heat evolved by one unit of, in blast furnace, 96; heat evolved by 20 units of, in blast furnace, 195; highest record of, in pig iron at Clarence works, 157; in coke and char-

INDEX.   727

Carbon—*Continued.*
coal practically performs the same duty, 132; loss of heat in reducing iron by, 197; loss of, in ordinary coking process, 234; oxidation of, causes which affect, 247; oxidation of, in coke heated *in vacuo*, 105; oxides of, laws governing mixtures of in furnace gases, 69; oxides of, point of equilibrium in furnace gases, 292; power of, to split up carbonic acid, 194; proportion of, found in gases as carbonic acid, 252; quantity and conditions of, in gases escaping from blast furnace, 68; quantity of, as carbonic acid, in gases from Cleveland furnace, 195; quantity of, consumed in coke and raw coal furnaces compared, 325; quantity of required in blast furnace, 93; quantity or quality of, does not necessarily affect grade of iron, 158; quick and slow currents of CO in reducing iron and depositing, 190; rationale of formula for determining quantity used in blast furnace, 270; reduction of oxide of iron by, and deposition of, compared, 192; reduction of quantity of, in gases, and effect of on CO_2 present, 89; required to generate water-gas, 331; separation of, in graphitic state interfered with by sulphur, 103; solid, tendency to uniformity of composition of it and iron, 161; state of oxidation in which it leaves the furnace, 238; substitution of in blast furnace by heat in the blast, 89; used per 20 units of hot and cold blast iron, 149; vegetable, compared with mineral, 304; waste of, in coking. 51.

Carbonic acid, action of carbon on, in blast furnace, 214; action on spongy iron, 185; and carbonic oxide, reducing powers of mixtures of, 313; antagonistic influence in preventing reduction, 184; at different depths of coke and charcoal furnaces, 292; carbon in, from furnaces using coke and raw coal, 322; cause of disappearance of, from furnace gases, 112; compared with quantity of carbonic oxide in charcoal furnaces, 282; decomposition of, by hydrogen, 312; disappearance of, in blast furnace, 70; disappearance of. in gases of furnace using raw coal, 324; effect of, on being liberated from flux in blast furnace, 198; effect of, on charcoal and coke. 287; effect of, on fuel in furnace, 295; effect of, on soft coke, 290; extent of removal of, by calcination of limestone, 60; greater disappearance of, from gases, the more heat must be contained in blast, 88; in products of combustion of different fuels, 128; less quantity in gases, furnaces using fire-brick stoves, 267; of limestone split up by coke, 198; loss of, in being split up by carbon, 194; presence of, in limestone, and effect of in blast furnace, 59; ready dissociation of, by reduced iron, 65; reducing powers of mixture with CO, intensified by hydrogen, 313; small quantity of, at tuyeres, 199; split up by carbon, 194; tendency of, to react on carbon, 188; unexpelled, in Cleveland ore, 240. (See also *blast furnace, carbon; and corbonic oxide.*)

Carbonic oxide, abstraction of, from blast furnace, 99; acidification of, by oxide of iron, 118; action of, on other substances than iron, 193; and carbonic acid, reducing powers of mixtures of, affected by hydrogen, 313; and unreduced ore, action between, in blast furnace, 78; as a reducing agent, 305; decomposition of, by hydrogen, 312; direct generation of, by carbon with air, 199; dissociation of, by peroxide of iron, 188; effect of mixtures of, and CO_2, on ores and fuel, 287; effect of, on Spathose and Cleveland ores compared, 285; effect of, on spongy iron, 185; excess of, in escaping gases. 106; excess of, required for reduction of oxides of iron, 186; function of, in blast furnace, 65; incapable of completely reducing oxide of iron, 291; in the Bessemer "blow," 390; point of saturation by oxygen of, in blast furnace, 98; power of oxide of iron in different states to dissociate, 189; power of, to take up oxygen from ore, 78; ratio which it bears to CO_2 in blast furnace, 294; reoxidation of iron by. 220; susceptibility of strongly roasted ore to impair action of, 323. (See also *carbon, carbonic acid, dissociation of carbonic oxide, oxide carbonic.*)

Caron on dissociation of carbonic oxide, 189.

Carondalet, iron works of, 678.

Carvés, description of a coke oven by, 327.

Cast iron, decomposition of phosphate of lime by. 410; importance of, in arts, 343. (See also *pig iron.*)

Catalan, and other low fires, 10; furnace, waste of metal in, 344; hearth and refinery, process of combustion in, 63.

Cenentation process, 436; carbon in blister steel produced by, 160.

Chalk, dried. content of carbonate of lime in, 57. (See also *limestone.*)

Charcoal, Akerman's list of furnaces using, 276; American furnaces compared with coke and coal. 129; and Cleveland furnaces, heat equivalents of, compared, 275; and coke compared with anthracite, 290; and coke, difference of action

Charcoal—*Continued.*
between, 297; and coke, effect of high temperature on, 289; and coke furnaces, carbonic acid (CO_2) at different depths of, 292; and coke furnaces, increase of temperature towards hearths of, 295; average consumption of, in making grey iron, 278; Birkinbine's comparison with anthracite and coke, 290; consumed in Carinthian furnaces, compared with coke, 128; consumed in Swedish furnaces, 276; cost of, in America, 54; consumption of, compared with coke in making white iron, 285; consumption of, in Austrian and Swedish furnaces, 274; contains less ash than coke, 134; cost of, 54, 135; diminution of heat caused by inferior oxidation of, 203; foreign matter in, 235; furnaces, American, compared with Styrian and Swedish, 300; furnaces, carbonic acid per 20 units of iron less than in coke furnaces, 291; furnaces compared with coke furnaces, 277; furnaces, duty performed by, in Sweden, 278; furnaces in Sweden, working of, 130; furnaces making white iron, 284; furnaces, ore not reduced $\tfrac{4}{5}$ height of, 298; furnaces, ratio of CO_2 to CO in, more than in those using coke, 289; furnaces, tables of duties of, 279; furnaces, weight of burden in, 277; furnaces, withdrawal of O from ore in, 296; hard and soft coke compared with, 281; heat, equivalent of, compared with coke, 281; heat evolved per unit of, 281; iron, cold blast American preferred for railway wheels, 152; iron, cost of in Austria, 54; iron, extent of manufacture of, in United Kingdom, 55; low consumption of, in smelting iron, 274; manufacture of, in close kilns on Lake Superior, 54; moisture in, 282; percentage yield of, from wood, 53, 135; remarkably low consumption of, in blast furnaces, 274; super-heated air applied to, 298; usually made where wood is grown, 54; yield of, from acre of ground in United States, 53. (See also *blast furnace, carbon, carbonic oxide, and carbonic acid.*)

Chemical action in blast furnaces, 176; sequence of, in blast furnace, 176.

Chemicals, use of, in puddling, 375.

Chemical works, cost of labour at, 504.

Chief iron-making countries, importance of iron exportation from, 707.

Chill, deeper, taken by cold blast iron than by hot, 153.

Chilled railway wheels, cold blast charcoal iron preferred for, 152.

Chlorides, supposed decomposition of, by sodium, in blast furnace, 167.

Chlorine, presence of, in blast furnace, 167.

Cinder, colour of, indicates condition of furnace, 171; conditions requisite for separation of, from metal, 169; effect on quality of pig-iron produced from, 358; furnace (blast) seldom without traces of iron, 170; emission of white smoke from, 174; from purifying iron, analysis of, 404; in boiler-plate, composition of, 369; occlusion or absorption of gases by, 173; produced at Bowling, analyses of, 359; quantity of, produced in making pig iron, 168; refactory, in basic process, 411; scouring, causes of production of, 161. (See *blast furnace, slag.*)

Clarence furnace compared with two blown with superheated air (Ormesby), 107; summary of working of two, 97. (See *blast furnace.*)

Clay ironstones, extent of foreign matter in, 56; modes of calcination of, 56; of coal measures, changes in use of, 458; position of, among ores of iron, 449; superiority as an iron-making nation of Great Britain, referred to, 454. (See *iron ores. Ores.*)

Cleveland, and charcoal furnace compared, 275; and Luxemburg ores, earthy constituents of, compared, 171; calcined stone, loss of oxygen in, by exposure to carbonic oxide, 184; calcined stone, reduction of, by CO, more rapid and at lower temperature than by coke, 198; district, work of modern type of plant in, 261; furnaces, dimensions of, 22; hot and cold blast, heat absorbed in furnaces using, 242; iron, care required in forge with, 364; iron, coke consumed in smelting 20 units of, 126; iron, heat units required for smelting of, 96; iron, lowest rate of consumption of coke required in smelting, 96; iron (No. 3), theoretical weight of coke required to make a ton of, 105; iron, purification of, by oxide of iron, 401; iron rails, contents of metalloids in, 365; iron, refining of, 354; iron, slags from, in basic process, 409; ironstone, character of deposits of, 119; ironstone, reduction of by H, 310; ironstone, value of superheated air in connection therewith, 119; ironstone, reduction of, by hydrogen, 310; ore, effect of CO on, compared with foreign ores, 285; pig iron, extent of impurities in, 69; pig, means of expelling phosphorus from, 387; seams of ironstone, differences in quality of, 119.

Clothing, quality of, compared, 489.

Coal, Belgian, 629; bituminous, &c., analyses of, 120, 127, 511, 612, 616, 620, 641, 643, 644; bituminous, consumption of, in Scotch blast furnaces, 120; bituminous,

INDEX. 729

Coal—*Continued.*
reducing properties of, when used raw, 120; coking of, to fit it for use in blast furnace, 47; condensation of water from, in gas works, 314; consumption of, per ton of puddled bars, 363; exports and imports by five nations of, 449; expulsion of moisture of, in furnace, 324; French, 626; Gartsherrie, average content of oxygen and moisture in, 314; German, 623; Great Britain, 610; importance of, to iron-exporting countries, 448; in Belgium, cost of labour at working in, 509; increased cost of labour in working of, 508; inferior, now used in coke manufacture, 92; in France, cost of labour at, 510; kinds of, capable of being used in raw state for iron smelting, 46; labour cost in working, in Great Britain of, 505; labour cost of, in America and Durham, 558; miners, average earnings of, in different countries, 514; miners, fluctuations in earnings of, 517; of County of Durham unfit for use in raw state in the blast furnace, 120; of Durham, volatile substances in, 49; of North Staffordshire compared with coke as used in Cleveland furnaces, 123; output of, in Germany and cost of labour in, 514; output of, in times of high and low prices, 513; production of, in chief producing countries in 1882, 448; production of, per head, 506; raw and coke, comparison of, in blast furnace, 325; raw, use of, in blast furnace, 315; raw, use of, in furnace smelting Spanish ore, 323; raw, use of, in smelting oolitic ores, 324; raw, water gas obtained from, 335; splint, of Airdrie, 120; tar, recovery of, in coking process, 52; United States of, 632; working, ages of persons engaged in, 512; yield per acre of, 136. (See *coke, Durham coal, fuel.*)

Cochrane, Chas., on Ormesby furnace, 264.

Coke, action of, on CO_2 (carbonic acid), 101; air required for combustion of, 264; analysis of, from Belgium, 104; analysis of, used at Clarence furnaces, 104; and charcoal compared with anthracite, 290; and charcoal, consumption of, compared with anthracite and coke, 130; and charcoal, consumption in making white iron, compared, 285; and charcoal furnaces, CO_2 at different depths of, 292; and charcoal furnaces, increase of temperature towards hearth in, 295; and charcoal, possible effect of high temperature on, 289; average consumption of, in hematite furnaces, 255; average consumption of, in making Bessemer pig, 255; azote in, 105; black ends of, 104; causes of oxidation of, 247; compared with raw coal for smelting, 325; consumed in smelting Cleveland iron, 126; consumption of, as effected by increasing temperature of blast, 249; continental, large quantity of earthy matter in, 238; cooling capacity of, on escaping gases, 179; difference in consumption of, as between rich and poor ores, 116; economy resulting from reduction of waste of, 52; effect of foreign matter in, 235; effect of impurities in, 100; effect of introduction of, on carbonic acid, 179; effect of oxygen in, on production of heat, 240; effect of temperature and condition of air in burning, 238; estimated saving of, due to preliminary calcination of limestone. 59; estimated weight of required to smelt Cleveland iron, 87; formula for estimating carbon derived from, in furnace gases, 273; furnaces compared with charcoal, 277; furnaces, weight of burden in, 277; gases given off by, heated *in vacuo*, 105; heat equivalent of, 260; heat equivalent of, compared with charcoal, 281; heating power of, in blast furnaces, 104; importance of quality of, 100; inferior, use of, 92; larger rateable weight of, burnt in smaller than in larger furnace, 204; limit of economy of, by increase of temperature in blast, 88; limit of economy in, by raising blast from 1,000° to 1,700° F., 92; mechanical texture of, 101; mode of checking consumption of, 269; ordinary yield of, in bee-hive oven, 235; ovens, attempts to improve the yield from, 592; ovens, Simon Carvés', 328, 592; ovens, utilisation of waste heat at, 591; oxygen and azote in, 104; oxygen contained in, 240; percentage of earthy and volatile matters in, 104; quality of, affected by mode of burning, 51; quality of, from Carvés' and bee-hive ovens, 328; quantity of oxygen contained in, 237; quantity of, required to make pig iron, 261; quantity of, required to smelt Cleveland iron, 106; soft, aversion of furnace managers to, 101; soft, effect of CO_2 on, 290; soft, effect of, in splitting up CO_2, 196; substitution of, by heat in hot air, action of, 88; substitution of heat in blast for, possible extent of, 91; temperature at which CO_2 begins to act on, 102; theoretical weight of, required to make the ton of No. 3 Cleveland iron, 106; useful effect of, diminished by presence of oxygen in, 105; waste of different qualities of, in reducing zone, 102. (See *ammonia and tar, coal, coke, coking process, Durham coal, fuel, ovens.*)

U U

730   INDEX.

Coking process, extent and nature of waste in, 49; generally rude and unscientific, 47; in open air, wasteful of coal, 53; Knab's, description of, 52; less wasteful in France and Belgium than in England, 49; loss of fixed carbon in, 51; loss of heat in Durham county, 50. (See also *coke.*)

Cold blast furnace, elements of heat absorption in, 242; evolution of heat in, 243.

Cold blast iron, and hot blast compared, 148; and hot blast iron of America compared, 152; composition of, and hot blast iron, compared, 150.

Cold blast, substitution of, for hot, effect of, on composition of iron, 150.

Coltness furnace slags, analysis of, 233.

Comparison of, acreage yield of charcoal and coke, 132; blast furnace with other furnaces, 140; charcoal and coke for blast furnaces, 133; consumption of charcoal, coke, and anthracite, smelting the same ore, 130.

Connelsville coke, price of, 693; wages at collieries of, 558.

Consumption of iron by different countries, 470.

Conveyance by water in United States, cheap instance of, 595.

Convict and slave labour in Alabama, 557.

Corn, averages of, 481.

Cornwall ore, cost of conveying minerals for iron made from, 691.

Cort, Henry, first attempt to puddle on sand bottoms by, 351; inventions of, 14.

Cost, of living, in North of England, 487; of living not materially different in different countries, 490; of making iron in Lancashire fires, 347; of making pig iron, in 1727, at Stour works, 349; of raw materials used in steel manufacture, 387; of transport of iron ores in United States and United Kingdom compared, 473. (See also *board and lodging, butcher meat, food, provisions.*)

Creusot, working of revolving puddling furnaces at, 371.

Cyanides, behaviour of, in blast furnaces, 223; decomposition of, in blast furnace, 327; decrease of, during ascent in furnace, 218; supposed disturbance in composition of gases caused in blast furnace by, 216.

Cyanogen, compounds of, conversion into carbonates, 224; reducing power of, on oxide of iron, 223.

Danks furnace, dephosphorizing agency of, 395; revolving furnace, 365; revolving furnace, conditions for success of, 366. (See also *puddling.*)

Darby, Abraham, inventions of, 13.

Dawes, trial of water gas by, in Staffordshire, 329.

Delhi, iron column at, 10.

Deoxidation, of alkalies, supposed possibility of, in blast furnaces, 214; of iron regarded as a preliminary process in the blast furnace, 207. (See also *alkaline matter,* &c.)

Deoxidising action, in charcoal furnaces making white iron, 201; power of furnace, addition to, by use of raw coal, 121.

Deposition of carbon, quick and slow currents of CO in promoting, 190.

Diameter of blast furnaces, useful limits of, 124.

Direct process, advantages of, 32; and blast furnace compared, 40; as carried on at Lake Champlain, 43; as means of removing phosphorus compared with puddling process, 374; by means of the carbon and silicon in pig iron, 35; compared with blast furnace and puddling process combined, 44; comparison of, with blast furnace, 61; cost of, compared with that of Bessemer steel, 34; description of, 31; in closed vessels and in reverberatory furnaces, 32; in low fires or in reverberatory furnaces, reduction of ore by. 66; iron from, for steel-making, 35; of Clay and of Chenot, 33; of M. Dupuy, 37; of making malleable iron, 30; of Siemens, 37 of T. S. Blair, 34; prospect of revival of, 5; reasons for following it in United States, 45; Ritter v. Tunner on Siemens' plan, 39; Y. Barra, Messrs., as practisedby, 33.

Dissociation, of carbonic oxide by different oxides, 189; of carbonic oxide, effect of presence of carbonic acid, 190; of carbonic oxide, presence of metallic iron not indispensable for, 193; of carbonic oxide, varying power of different samples of peroxide of iron for, 190; of products of combustion, heat absorbed by dissociation of, 48. (See also *blast furnace, carbonic oxide.*)

Dixon, Raylton, on cost of labour, 706, 721.

Dudley, Dud, inventions of, 13.

Dufrenoy, examination of results of hot blast by, 81.

Dunlop & Co.'s ironworks, extent of saving at, by hot blast, 81.

Durham coal-field, future exhaustion of, 702.

Durham coal, quantity of, annually converted into coke, 47.

Durham coke, average yield of, in county of Durham, 49; manufacture of, in flued ovens, 51; waste in manufacture of, 49.

Duties, import, extension of the iron trade caused by, 474; probable effect of repeal of, 447; protective, effect of, 446.
Earnings, agricultural, 472, 496 544; at blast furnaces, 562; at blast furnaces in Europe, 521; at Bessemer steel works, 535; at coal mines in Europe, 505; at iron ship building, 501; at ironstone mines, in Cleveland, 517; France, 518; Luxemburg, 518; Northamptonshire, 517; Spain, 518; Bessemer steel, 571; coal mines, 554; iron ore mines, 575; malleable iron, 566.
Earnings at blast furnaces in U.S.A. from 1870 to 1880, 565.
Ebelman and Tunner on increase of oxygen at tuyeres, 212.
Ebelman's experiments on blast furnace gases, 308.
Economy, margin of, in Bessemer and Cleveland iron, 406.
Economy of fuel effected by alteration in mode of charging furnaces, 100.
England, imports of iron from Belgium and Germany to, 466.
English workmen, cost of living of, 486.
Essen, production of water-gas at, 331.
Eudiometers, errors of, 306.
European countries as competitors in manufacture of iron, 702.
Exhibitions, international, value of, 441.
Exports of iron, from chief iron-making countries, 706; of all kinds from Great Britain, 444; to iron-making countries from Great Britain, 469.
Exports of iron manufactures from Belgium, 579; France, 579; Great Britain, 579; United States, 579.
Fairbairn, Sir W., on comparative strength of hot and cold blast iron, 150; on effect of application of heat to solid iron, 148.
Farm and ironworks labour in Great Britain and America, differences in, compared, 554.
Farmer, continental, relations with British consumers, 480.
Farming, in America, 540; England, 542.
Farm labour, cost of, in United States, 544; in America, Dr. Young on, 545.
Ferrie's furnace, and its results, 122.
Ferro-manganese, use of, in Bessemer process, 383.
Fettling employed, in Dank's furnace, 395; for puddling furnaces, 362.
Fire-brick stoves, and metal-pipe stoves, comparison of, 107; application of, to heating blast, 234; theoretical evolution of heat in burning coke with air heated in, 264; value of, for poor minerals, 119. (See also *hot blast*.)

Fireplace of boiler, analysis of deposit gathered in, 232.
Flints and blast furnace slag, experiments with, to determine evolution of heat in reducing zone, 76.
Flour, improvements in grinding of, 488.
Food, cost of, in United States, 539; cost of, relatively to wages, 477; machinery, use of, in U.S.A., for production of, 540; rent as an element in cost of, 542; transport of, in U.S.A., 541; weight of, required per family, 479. (See also *board and lodging, provisions*.)
Foreign duties, supposed effect of repeal of, 461.
Foreign, iron trade, alarm in respect to, 6; iron trade, compared with British, 7; iron trade, extension of, 5.
Foreign matter, different modes of freeing pig iron from, 376; matter, effect of, in blast furnace fuel, 235.
Foreign workmen, cost of living of 484; mode of living of, 484.
Fossil coal, use of, in blast furnace, 13.
France, Central and Southern, as exporter of iron, 706; coal in, 449; coal working in, 510; importations into and make of iron, 1866 to 1880, 467; imports of iron, 466; iron ore raised in, 467; iron trade between, and Great Britain, 466; North of, cost of provisions in, 482; position of, in the iron manufacture, 442.
Free trade, benefits conferred by, 488; effect of, in Great Britain and on the Continent of Europe, 495; Right Hon. John Bright's remarks on, 490.
Freight on pig iron to America, 604.
Friderici on use of charcoal in blast furnaces, 284.
Fuel, combustion of, in blast furnaces, 140; consumed in Lancashire fires, 347; diminution of full heating power of, 99; economy of, by alteration in construction of furnace, 73; economy of, by altering method of charging, 100; economy of, effected by hot blast in different localities, 82; effect of foreign matter in, 235; expulsion of gaseous matter in blast furnace from, 234; in furnace, effect of CO_2 (carbonic acid) on, 295; loss of, in smelting 20 units of iron, 237; minimum of, required in smelting Cleveland iron, 99; possibilities of further reductions in consumption of. 145; preliminary treatment of, 46; small consumption of, in certain Cleveland furnaces, 117; used in heating blast, cost of and money saving effected by, 91; value of, in blast furnaces dependent on fixed carbon, 234; waste of, by deficient production of carbonic acid,

Fuel—*Continued.*
107; waste of, by irregular distribution of reducing gases, 111; waste of, in blast furnace where flux contains carbonate of magnesia, 58; waste of, in furnaces of small capacity, 123. (See also *coal, coke, charcoal, blast furnace.*

Fume, analysis of, 224; blast furnace, occurrence of, 174; composition of, at different levels of furnaces, 230; of blast furnace, substances found in, 229.

Furnace, blast, capacity of and temperature of blast convertible terms, 247; blown with superheated air, working of, 107; charcoal, capable of turning out large quantities of iron, 133; charcoal, in Sweden, working of, 130; decreased make at, amount of loss entailed by, 254; importance of shape of, 268; diameter at tuyeres of, 124; effect of enlargement of, and its analogy with the action of hot blast, 84; effect of increased dimensions compared with that of heating the blast, 86; enlargement of diameter of, 124; importance of shape of, 124; increase of height and increase of diameter considered, 124; large make in small, in Scotland, 121; of different dimensions, duty performed in, 75; of different dimensions, results obtained by, 85; of 48 feet and 80 feet in height compared, 112; point at which further enlargement is necessary in, 123; size and temperature of blast of, 124; smelting Bilbao ores, working of, 113. (See also *blast furnaces.*)

Fusible slags, production of, in blast furnace, 57.

Gartsherrie, coals, average quantity of oxygen in, 314; furnaces, production of sulphate of ammonia in, 327; furnaces, recovery of tar and ammonia from gases at, 314.

Gaseous fuel, probable future use of, 329.

Gases, action at different levels of blast furnace, 222; action of, on spongy iron, 221; alteration in quantities of carbon and oxygen in, 215; bulk of, passing over ore in furnace using raw coal, 323; Bunsen and Playfair's experiments on, 308); carbon from fuel calculated by average composition of, 269; carbonic acid, disappearance at Ormesby furnace in, 248; carbonic acid in, at Ormesby, 252; carbon in, at different levels of furnace, 208; changes in composition in, as indication of changes in ore, 207; changes of composition of, as they rise in blast furnace, 201; composition of, at different levels of furnace, 210; composition of, in furnaces of 48 feet and 80 feet, 73; conditions of carbon in the, 270; emitted in smelting iron with raw coal, 121; escaping, carbon deposited by, 202; escaping, composition and temperature of, 177; escaping, effect of charging on temperature of, 179; escaping, estimate of oxygen, in, 207; escaping, exhaustion of reducing power of, 201; escaping from blast furnace, analysis of, 68; escaping from blast furnace, carbon contained in, 68; escaping, from Scotch furnaces, analysis of, 317; escaping, loss consequent on weight of, 111; escaping oxidation of carbon and hydrogen in, compared, 319; escaping, reducing by, as affected by scaffolding, 253; escaping, use of, in heating blast, 251; escaping, utilisation of, 23; escaping, volumetric composition of, 178; escaping, weights of, 318; furnace, adaptability of, for steam boilers, etc., 327; given off by coke heated *in vacuo*, 105; heat escaping unutilised in, 240; heat lost in, at Ormesby, 248; in Bessemer "blow," alteration in 391; in Bessemer "blow," reducing power of, 392; in converter, nature of, 389; increase of CO_2 at blast fuanace, 178; of basic blow, heat carried off in, 424; of blast furnace, hydrogen present in, 306; loss of heat in, 73; oxygen in, 270; prolonged action of, on solids by diminution of volume of, 85; recovery of tar and ammonia from, 314; reducing energy of, intensified by hydrogen, 313; relation of nitrogen to carbon and oxygen in, 270; saturation of, by oxygen, 98; unable to reduce iron when one-third of carbon is in form of carbonic acid, 239; waste of heat caused by escape of, from blast furnaces, 73; weight and composition of, from large furnaces, 204; weight and composition of, from small furnaces, 206. (See also *blast furnaces.*)

German export trade, nature of, 474.

Germany, coal resources, 448; coal working in, 514; cost of living at different towns of, 485; cost of maintenance at ironworks in, 484; cost of provisions in, 483; effect of protective duties in, 461; exports of iron from, 474; exports of pig iron, &c., from, 445.

Girders, &c., export of, from Germany and Belgium to Great Britain, 703.

Gjers, J., "soaking pits" of, 440.

Glazed iron, analyses of, 162; mode of correcting production of, 163.

Gow, T., farming in Great Britain. 542.

Grade of iron not necessarily affected by quantity or quality of carbon, 158.

Graff, Bennett & Company's experience of revolving furnaces, 370.
Granville's (Lord) cold blast furnaces, economy of fuel by increase of size of, 85.
Graphite, analyses of, separated from pig iron, 156.
Gratz, manufacture of Siemens-Martin steel rails at, 426.
Great Britain as an iron-producing country, 715; change in character of minerals used in, 457; coal resources of, 448; comparative cost of living in, 495; cost of agricultural labour in, 496; exports of iron from 1871 to 1882, 444; iron ore resources of, 449; labour, increased cost of, in, 447; lias measures of, 457; limestone resources of, 461; make of pig iron in, in 1740, 443; make of pig iron in, 1830 to 1882, 454; malleable iron, production of, in, 460; position for manufacture of Bessemer pig in, 451; progress of invention in, 443; quantities of pig iron made in, 443; relative position of mining resources of, 587; sources of iron ore production of, 453; supremacy of iron manufactures of, threatened with abatement, 446.
Grey iron, average consumption of charcoal in making, 278.
Greyness of iron as affected by temperature, 149.
Grüner, Prof., on influence of Silica in preventing removal of phosphorus from iron, 397.
Hanging Rock, ironstone of, 674.
Harrison, T. E., on weight of rails laid down in United Kingdom, 430.
Hearth of blast furnace, oxygen separated at, 213.
Hearth of cold blast furnace, heat of. 149.
Hearths, low, first used for making bar iron, 344.
Heat absorbed by dissociation of products of combustion, 48; absorption of, in basic process at Hörde, 422; amount of, for gasification of volatile constituents of coal, 319; amount of, provided by hot blast and by oxidation of carbon, 88; application of in evaporation of water, 138; avoidance of waste of, 247; calculations, basis of, 47; carried off in gases of basic blow, 424; computation of, obtained from raw coal in blast furnace, 320; desirability of knowing quantity of, required in smelting, 95; developed by oxidation of carbon, 205; developed in iron smelting per unit of carbon, 96; developed in larger and smaller furnaces compared, 206; development in cold blast furnaces at tuyeres, 85; development in hearth of blast furnace, 64; different quantities of, afforded by different bodies, 48; due to oxidation of iron in basic process, 423; economy effected by successive additions of, to blast, 265; effect of increase of, on elements contained in iron, 149; effect of repeated application of, to solid iron, 149; emitted by combustion of different varieties of coal. 50; equivalent of charcoal of, in the blast furnace, 274; equivalent of Cleveland furnaces using coke, and charcoal furnaces compared, 275; equivalent of coke and charcoal compared, 281; equivalent of, obtained from coke, 241, 260; equivalent per unit of charcoal in Swedish furnaces, 303; evolved and absorbed in blast furnaces, comparison of, 241; evolved by combustion with air in charcoal furnaces, 303; evolved in basic blow, 420; evolved in furnaces of different dimensions, 74; evolved per 20 units of carbon, 195; evolved per unit of carbon, 69; evolved per unit of charcoal, 281; evolved per unit of coke, 117; evolution of, by increase of carbonic acid in gases, 178; evolution of, in reducing zone, 76; expenditure of, on limestone used in blast furnace, 594; function of, in upper part of blast furnace, 46; given off by fixed carbon in blast furnaces, 235; high coefficient of, at Ormesby blast furnaces, 254; in blast may be equivalent to heat from fuel, 255; in blast, substitution of, for heat in coke, 253; in escaping gases, 248; in escaping gases leaving the blast furnace, 143; in hearth, black slag proof of insufficiency of, 291; interception of, by materials in blast furnace, 64; leaving blast furnace, utilisation of, 143; loss of, from chimneys, 143; loss of, in coke-making, 49; loss of, in escaping furnace gases, 74; loss of, in escaping gases, 142; loss of, in escaping gases, incapable of reduction below certain limits, 76; loss of, in process of cokemaking, 50; of hearth of cold blown furnace, 149; only portion of, not usefully appropriated in blast furnace, 142; possible extent of substitution of in blast for coke, 91; previous exposure of fuel to, effect of, in enabling it to resist action of ore, 288; produced in blast furnace, sources of, 80; proportion of evolved in blast furnace due to oxidation of carbonic oxide, 127; prospective economy of, in blast furnace, 145; quantity and intensity of, from combustion of water gas, 336; quantity of, as distinguished from in-

Heat—Continued.

tensity, 49; quantity of, distinguished from intensity, 49; quantity of, evolved in coke-making by bee-hive oven, 50; quantity of, in blast, per unit of coke burnt, 264; quantity of, lost annually in Durham in coke-making, 50; quantity of, which escapes unutilised in gases, 240; required in basic and acid processes, comparison of, 421; requirements of Austrian and Cleveland furnaces compared. 129; requirements for smelting two different varieties of ironstone, 116; sources of, in acid and basic processes, 419; sources of loss of, in blast furnace, 80; specific, 48; theoretical evolutions by coke of, using fire-brick stoves. 264; units of, lost in combustion of raw coal, 139; units of, required to smelt 20 units of Cleveland iron, 106; units or calories required to smelt Cleveland iron, 95; units required per 20 kilos of pig iron smelted from Cleveland stone, 98; used and wasted in blast furnace, summary of, 144; useful, in blast furnace, afforded by coke, 47; values for different bodies in basic blow, 423; varying measures of, required to smelt similar quantities of iron, 95; varying quantities of, afforded by different degrees of oxidation of the fuel, 48; waste, application in raising steam by, 139; waste, at coke ovens applied at collieries, 50; water gas, heat evolved by. (See also *blast furnace*.)

Heated air, figures illustrating the saving of fuel by, 86.

Heath, R., experience of, with revolving furnaces, 370.

Heath's, I. M., addition of manganese to crucible steel, 383.

Heath's, I. M., process of steel making, 426.

Height of blast furnaces, no uniform law for determining, 145.

Hematite furnaces, smelting Bilbao ores, work of, 113.

Hematite ores, resources of, in United Kingdom, 450.

Hewitt, A., on cost of pig iron in Southern States of America, 585.

High Level Bridge (Newcastle) quality of iron selected for, 150.

Hindoo process of steel making. 436.

Hot blast and cold blast iron, differences of quality, 152; Fairbairn's experiments on, 150.

Hot blast, Dufrenoy's examination of, 81; Dunlop & Company's saving by, 81; E. A. Cowper, improvements by, 24; effect of, compared with that of enlargement of furnaces, 84; effect of, on quality of iron, 148; examination and theory of action of, 81; examination of, by Dufrenoy, 81; explanation of action of, 82; fuel used for heating, 91; gain supposed possible by, at 1954° F., 90; general use of, in Sweden, 154; invention of, 16; and cold blast iron, 148; saving by, 241; saving effected by application of, to iron smelting, 81; temperature of, in early days of, 22. (See also *blast furnace*.)

Hours of labour, British and foreign compared, 500.

House rent and board, cost of, 489; observations on, 489.

Hydro-carbons, value of, in blast furnace, 235.

Hydrogen and compounds of, in blast furnace, 305; and other gases at tuyeres. 307; and oxygen in escaping gases of Scotch furnaces, 320; as a deoxidising agent, 322; compared with CO as reducing agent, 310; content of, in Gartsherrie coal, 314; decomposition of CO_2 and CO by, 312, 324; difficulty of proving exact quantity of, in blast furnace gases, 306; doubtful utility of, in coke furnaces, 309; enquiry into value of, for smelting iron, 329; generally present in the blast furnace, 305; in hearth of furnace, caused by water, 305; mixture of, with carbonic oxide, in reducing Cleveland ore, 310; not proved to be a reducing agent in coke furnaces, 309; oxidation in blast furnace, 320; oxidation of, in blast furnace by oxygen in coal, 321; presence of, due to imperfectly charred wood or coal, 305; proposed use by Mickle in smelting iron, 329; quantity of, ascertained by its conversion into water, 306; quantities of, in blast furnace gases, 306; reducing energy of blast furnace gases intensified by, 313; reducing power of, in blast furnace, 315; reducing powers of mixtures of CO_2 and CO intensified by, 313; superior affinity of, for O compared with CO, 310. (See also *blast furnace*.)

Import duties, difficulties attending levying of, 584; in Germany, 581; with transport charges on pig iron, America, 606.

Imports, of iron by non-producing countries, 467; of iron into Great Britain, 581; of ores and exports, 452; of ores from Spain, 450.

Impurities present in limestone, 57.

Ingot iron v. welded iron for boilers, 376.

Ingots, steel, Alex. Wilson's plan of avoiding reheating of, 439.

Intensity of heat distinguished from quantity, 49.

International Exhibitions, value of, 441.

INDEX. 735

Intoxicating drinks, Dr. Young on, 490; use of, in Great Britain and in foreign countries, 491.

Iodine, action of, on metallic iron, 193.

Iron Age, remarks of, on Baker's costs of iron, 695.

Iron, and carbon, temperatures at which they are acted on by carbonic acid, 71; and carbon, tendency to combine, 160; and steel, future improvements in manufacture of, 708; and steel, probability of future improvements in manufacture of, 708; and steel, quality of, determined by composition, 405; changes in manufacture of, 2; charcoal, expensive manufacture of, in United Kingdom, 55; Cleveland No. 3, theoretical weight of coke required to make a ton of, 105; consumption of, per head in different countries, 470; cost of minerals for manufacture of, 584; costs of, by W. E. S. Baker, in Eastern States of America, 695; difference in make of, when using charcoal, coke and anthracite respectively, 132; disappearance of, in ancient monuments, 8; extended use of, 17; foundries, early mention of, 12; from blast furnace containing no carbon, specimen of, 161; grey, refining of, 356; hardening of, by excess of sulphur in coke, 103; imports into Great Britain from Germany and Belgium, 581; improvements in manufacture of, effect of, 586; increase of foreign manufacture of, 2; indication of, in ores, 9; large mass of, in column at Delhi, 10; making countries, comparative advantages of, 403; malleable, future of, 28; manufacture, effect of improvements in, on British iron trade, 6; manufacture, future of, 28; manufacture, historical notice of, 8; manufacture, position of workmen connected therewith, 708; manufacture of, progress of, in 18th century, 15; manufacture, prospect of further improvements in, 587; manufacture, skill of different nations in, 582; manufactures, exports from Belgium, 579; manufactures, exports from France, 579; manufactures, exports from Great Britain, 579; manufactures, exports from United States, 579; manufactures in Germany, development of exports of, 579; manufactures of Great Britain, competition with, 578; meteoric, 8; native, 8; older modes of manufacture of, compared with modern, 27. pig and bar, table of costs of, in Eastern States of America, 696; possible early formation of tools of, 10; present prospects of trade in, 711; primitive modes of manufacture of, 9; producing countries compared, 578; prospect of future improvements in manufacture of, 708; reduced cost of, in consequence of improvements in the manufacture of, 29; refined, analyses of, at Tudhoe and Bowling, 355; refined and pig, analyses of, 354; reoxidation of, by carbonic oxide, 193; spongy, effect of carbonic acid on, 185; trade, future of, 587; trade of America, 604; workmen engaged in manufacture of, 709. (See also *blast furnace*.)

Iron ores, calcination of, 56; cause of high price of, 693; carriage on minerals for making iron in United States, 690; classification of, 455, 499; classification of, as delivered to blast furnaces, 55; division into those fit for steel-making and those unfit for steel, 449; effect of using uncalcined in blast furnace, 56; extent of lias measures of, 457; fuel consumed in calcination of, 56; hematite, resources of, in Great Britain, 450; imports and exports of, in different countries, 452; imports into U.S.A. of, 692; imports of, from Spain, 450; inexhaustible supply of, for basic steel manufacture, 451; labour in mining at, 517; of Belgium. 659; of France. 656; of German Zoll Verein, 652; of Great Britain, 646; of U.S.A., 660; of Western Germany and Eastern France, 653; preliminary treatment of, 55; production of, in different countries, 451; resources of, in United Kingdom, 449; resources of United States, 472; restricted output per man, of 520; sources of production of, in United Kingdom, 449, 453; spathose, position, of, in Great Britain, 450; transport of, in United States, 473; wages in U.S.A. at, 559. (See also *blast furnace, ironstone*.)

Ironstone, action of carbon on, 197; cooling capacity on escaping gases of, 179; earnings in Cleveland at, 517; earnings in France, 518; earnings in Luxemburg, 518; earnings in Northamptonshire at, 517; earnings in Spain, 518; effect of imperfect calcination on, 240; temperature required for deoxidation of, 191; wages and production at working, comparison of, 518. (See also *blast furnace, iron ore*.)

Iron shipbuilding in Norway, 706, 721.

Iron trade, depression of, 1; present prospects of, 711.

Jameson's coking process, oil and tar from, 328.

Kentucky ore, 692; cost of conveying minerals for iron made from, 692.
Kish given off hot metal, 155.
Knab's process of coke-making described, 52; results of trials of, 53.
Knights, Messrs., manufacture of iron in 1727, 348.
Krupp, on removal of phosphorus from pig iron, 404.
Labour, agricultural, in England and other countries, 475; and prices of iron, and prices of provisions in America, 548; and prices of iron in Great Britain compared, 548; and provisions, connection between, cost of, 539, 553; at Middlesbrough, Mr. Weeks thereon, 699; convict and slave, in Alabama, 557; cost in chemical works in England, 504; cost of, in Northern and Southern States of America, 554; cost of, in working coal in Great Britain, 505; cost on Rhine and in Great Britain, 500; difficulties in, 710; economy of, at blast furnaces, 525; farming, in America and North of England compared, 553; in America, J. D. Weeks on, 574; in America, higher cost of, 574; in Great Britain and Continent of Europe, 495; in United States of America, 539; Irish, character of, 494; price dependent on cost of provisions, 539; proportion of cost of pig iron, 475; severity of, in puddling, 365; on building of iron ships, 501; on building of iron ships, English and Foreign compared, 502, 706; slave, in Alabama, 557.
Labouring classes, improved condition of, 478.
Lake Champlain ore, cost of conveying minerals for iron made from, 691.
Lake Superior ore, cost of conveying minerals for iron made from, 690.
Lancashire and West Cumberland, iron ores of, 458.
Lancashire fires, analysis of slag from, 346; continued use of, in Sweden, 154; fuel consumed in, 347; method of working of, 347; quality of iron produced by, 344; quality of iron used in, 276; still used in Sweden and Russia, 345.
Lead, presence of, in blast furnaces, 228.
Life of rails, duration of, 378.
Lignites and peat, composition and qualities of, 136; inferiority of, for smelting iron, 137.
Lime, affinity of, for sulphur at high temperatures, 58; associated with carbonate of magnesia in United States, 58; behaviour of, with carbonic acid at high temperatures, 60; effect of sublimating influences on, in blast furnace, 232; excess of, required to free iron from sulphur,
58; lining, use of, for Bessemer converters, 406; not indispensable in blast furnace slags, 169; retention of carbonic acid by, 198. (See also *blast furnace*.)
Limestone, 681; alumina in. 57; calcination of, before being used in blast furnace, 58; calcined, use of, in blast furnace, 209; carbon carried off by carbonic acid of, 216; carbon oxidised by carbonic acid in, 107; dispensing with use of, 594; expense attending its use, 594; geological deposits of, 57; greater quantity required of, when using mineral fuel, 130; in different countries, 681; magnesian, content of magnesia in, 58; mountain, impurities present in, 57; preliminary treatment of, 57; quarrymen, cost of living in North of England, 487; raw, objection to use of, in blast furnace, 59; resources of, in United Kingdom, 461; silica and alumina in, 57. (See also *blast furnace, carbonic acid*.)
Limits of useful size of furnaces and temperature of blast, 124.
Liquidity of iron necessary for removal of phosphorus, 398.
Lithia, presence of, in blast furnace flame, 229.
Living, cost of, in different countries, 490; cost of, in the North of England, 487.
Locomotive and marine boilers, iron suitable for, 373.
Locomotive engines, loss of heating power of fuel in, 139.
Low fire, result of throwing ore on to, 66.
Luxemburg, cost of food in, 518; ore, composition of, 654.
Machinery, exports of from Great Britain, Belgium, and France, 579; exports from United States, 580.
Macaulay on cost of living in England, 478.
Magnesia, compound of, with silica and lime, 58; effect of presence of, in limestone, 58.
Magnesium found in pig iron, 167.
Malleable iron, brands of pig preferred for, 153; competition from Belgium in girders, 534; cost of labour in different countries, comparison of, 527; cost of labour at, in Great Britain and on the Continent compared, 529; cost of making in 1727 at Stour Works, 349; direct processes for making, 30; direct process in America, &c., 31; direct process in Asia and Africa, 30; effect of steel trade on production of, 459; English and Swedish, analysis of, 345; fluctuations in cost of labour in manufacture of, 526; from cold blast Clarence pig not superior

INDEX. 737

Malleable iron—*Continued.*
to hot blast, 150; labour at forges and mills, 530; labour at manufacture of, 525; make in Great Britain of, 460; make of, in the North of England, 459; manufacture of, in low hearths, 344; manufacture of, in United States, 566; mechanical puddling in manufacture of, 528; production of, in United Kingdom, 460; puddling, cost of, in United States and England, 567; quantity of hearth-refined, produced in Russia and Sweden, 348; superiority in strength of Bessemer steel over, 385; wages in mills, United States, from 1870 to 1881, 570. (See also *boiler plates*, 369, 373.)

Manganese, as a constituent of iron ores, 166; effect of, in iron and steel, 166; Heath's addition of, to crucible steel, 383; requires high temperature for reduction, 165; use of, in Bessemer process, 20.

Manufactures, effect of introduction of, 475.

Markets open to British manufacturer, 489.

Martin, E. P., on consumption of coke with fire-brick stoves, 257.

Masonry, protection of, by deposited carbon in blast furnaces, 222.

Materials, irregular descent of, in blast furnaces, 92.

Mechanical puddling, trials with, 365.

Metallic iron, effect of, in inducing carbon deposition, 182.

Metalloids, behaviour of pig iron in relation to. 154; behaviour of, while charge is being refined, 359; contents of, in steel and iron, 435; expulsion of, in Bessemer and other processes, 388; influence of, on quality of pig iron, 155; reaction of, in Siemens process, 430; removal of, as affected by high temperature, 400; removal of, by various processes, 398; removal of, from pig iron as affected by temperature, 400; removal of, from pig iron in Styria, 397; removal of, in Bessemer process, 391; removal of, in refining process, 355; separation of, in puddling furnace, 360; transference of, from cast to wrought iron at high temperature, 160. (See also *blast furnace.*)

Mickles's proposal to use hydrogen in blast furnace, 329.

Mineral carbon compared with vegetable, 304.

Minerals used, connection with quality of iron, 151.

Miners, cost of living in the North of England of, 487.

Missouri, iron mountain of, 676; ore, cost of conveying minerals for iron made from, 692; wages of, 677.

Mixtures of carbonic oxide and carbonic acid, reducing powers of, intensified by hydrogen, 313; of carbonic oxide and carbonic acid, reducing power of, 313. (See also *blast furnace.*)

Moisture contained in coke and charcoal, 236; in air blown into furnace, 241.

Monkbridge, economy of revolving puddling furnace at, 374.

Monopolists in Great Britain, Philadelphia press on, 577.

Mushets connection with Bessemer process, 383; discovery of value of manganese in Bessemer process, 20.

Nail rods, cost of, in 1727, 350.

Nancy ore, earthy constituents of, compared with Cleveland, 171.

Nasmyth, hammer of, 26.

Neilson, J. B., application of hot blast by, 81; hot blast, invention by, 16.

New Jersey, flux required in smelting magnetic ores of, 58; ore, cost of conveying minerals for iron made from, 691.

Nitrogen, estimate of O and C in furnace gases therefrom, 271; in coal dissipated by coking, 326; in furnace gases, 270; proportion of to carbon in blast furnace gases, 68; quantity of, in Scotch coal, 327.

North America, coal resources of, 448.

North of England, cost of living in, 487; production of malleable iron in, 459.

North Staffordshire smelters, use of raw coal by, 122.

Northern States of America, difficulty attending manufacture of iron in, 689; weight of iron ore produced in, 690.

Norway, iron shipbuilding in, 706, 721.

Oolitic ores, smelting of, with Derbyshire raw coal, 324.

Open-hearth, and Bessemer steel compared, 432, 435; steel boiler plates, &c., analyses of, 434; steel, certain costs of manufacture of, 432; steel, quality of, compared with that of Bessemer steel, 434; steel, statistics of production of, 432.

Ore, advantages of their free exposure to reducing gases, 133; and coal, profit on, per ton of Iron in U.S.A., 694; and fuel, mixture of, exposed to mixtures of carbonic oxide and carbonic acid, 286; changes in, during descent in furnace, 219; changes in, estimated by changes in gases, 208; cost of, for steel making, 387; differences in reduction of, 82; effects of difference in, in blast furnace, 94; imported, cost of conveying minerals for

Ore—*Continued.*
iron made in U.S.A. from, 691; in charcoal furnaces not reduced in four-fifths of height of furnace, 298; in charcoal furnace, withdrawal of oxygen from, 296; of iron, importations of, into Great Britain, cause of, 3; output of, in Cleveland, 520; output of, in Luxemburg, 519; pig iron obtained in United Kingdom from, 386; process, a form of direct process, 35; process for making open-hearth steel, 430; quantity of carbon required for reduction of iron in, 33; quantity used in United Kingdom, 386; reduction of, in blast furnace by carbonic oxide, 82; re-oxidation of reduced iron in by watery vapour, 321; suitable for manufacture of steel, 386; temperature required for reduction of iron in, 33. (See also *blast furnace.*)

Ormesby furnaces, consumption of coke at, 269; formula for estimating carbon from coke used in, 273; heat coefficient at, 254; loss of heat in escaping gases at, 248.

Osmund furnace, 11.

Ovens, flued, use of, in Durham for coke-making, 51.

Overblow in Bessemer blow, dense brown fumes caused by, 418.

Oxidation, of carbon, degrees of, 251; different quantities of heat afforded by, 48; of carbon, heat developed by, 205; of carbon, heat produced by, 88; of charcoal when burnt with superheated air, 303; of iron by carbonic acid, 184; of iron in basic process, heat from, 423.

Oxide, carbonic (see *carbonic oxide*); of iron, amount of, used in puddling Cleveland iron, 365; of iron, action of carbonic oxide on, 70; of iron, action of carbon on, 197; of iron, artificially prepared experiments on, 83; of iron, bottoms for puddling furnace, suggested by S. B. Rogers, 351; of iron, change of grey to white iron by contact with, 159; of iron, disassociation of carbonic oxide by, 189; of iron, effect of increased use of, in puddling, 362; of iron, experiments with, for dephosphorizing iron, 396; of iron, in basic slag at different periods of blow, 418; of iron, incapable of being completely reduced by carbonic oxide, 291; of iron, reduction of, by carbonic oxide, 186; of iron, reduction of, near to tuyeres, 173; of iron, use in converter, 406.

Oxides of carbon, antagonistic forces of, in blast furnace, 184; effect of, on behaviour of materials in furnace, 182. (See *blast furnace.*)

Oxygen, absorption of, at different stages of Bessemer blow, 391; absorption of, by silicon, &c., in Bessemer blow, 390; amount of, required to oxidise iron and metalloids in refinery, 352; and carbon, alteration in quantities of, in furnace gases, 213, 214; and carbon, at different levels of furnace, 212; at tuyeres, Ebelmen and Turner thereon, 212; and carbon in furnace, irregularities in quantities of, 208; comparative amount of, in iron and basic steel, 413; contributed by calcined limestone, 211; estimation of, in gases, 207; in coke, 104; in gases, estimation of, 270; in minerals and in air, mode of calculating, 271; in oxide of iron, acidification of carbonic oxide by, 118; loss of, by oxide of iron by exposure to reducing gases, 121; of gases of blast furnace, limits of mean saturation by, 79; quantity of, in coke, 237; quantities of, in gases at different levels, 208; required for acidification in Bessemer process, 410; saturation of gases of blast furnaces by, 77; separated at hearth of blast furnace, 213; Tucker's mode of estimating quantity in malleable iron, 373; withdrawal of, from ore in charcoal furnace, 296. (See also *blast furnace, carbon and carbonic oxide and carbonic acid.*)

Peat, use of, in Vordenburg furnaces, 136.

Pecten seam of ironstone, Cleveland, 119.

Percy, Dr., work on iron by, 25.

Pernot furnace, 433; use in purifying iron by, 433.

Petroleum, experimental use of, in blast furnace, 137; injected at tuyeres, iron smelted by, 342.

Phosphate of lime, decomposition of, by iron, 410.

Phosphoric acid volatile at high temperatures, 397.

Phosphorus, almost invariable presence of, in iron-making materials, 165; and sulphur, quantity of, in pig and slag, 357; behaviour of, in Bessemer converter, 361; behaviour of, under different conditions, 389; conditions affecting separation of, from pig iron, 396; content of, affects quality of iron, 345; content of, in iron rails, 428; expulsion of, in different processes, 389; in basic process, replacement of silicon by, 416; minute proportions only of, admissible in steel, 165; removal from white iron by basic process, 417; removal of, by oxide of iron, 401; removal in puddling of, by oxide of iron bottoms, 352; removal of, by puddling process and by direct process, 374; separ-

INDEX. 739

Phosphorus—*Continued.*
ation by basic process, 4; separation from pig iron of, 3; taken up by pig iron, 165. (See also *blast furnace and pig iron*).

Pig iron, Abram Hewitt on cost in Southern States of America, 585; behaviour of carbon in, 155; behaviour of, in relation to metalloids, 154; behaviour of silicon in, 162; Bessemer and Cleveland, analyses of, 388; Cleveland and Bessemer, metalloids in, 388; comparative costs in different lolalities of, 682; comparative strength of hot and cold blast, 150; composition of Cleveland, 147; composition of, from Nancy ore, 172; composition of, suitable for mill purposes, 153; cost of, in Southern States of America, 585; cost of making in 1727, 348; estimation of coke required for smelting, 261; exports of, and other iron, 1871 to 1882, 444; exports of, from Great Britain and other countries, 1871 to 1880, 445; extended use of, 25; fluidity of, promoted by silicon, 163; glazed, analyses of, 162; grade of, not necessarily affected by quality and quantity of carbon, 158; greyness of, affected by temperature, 149; highest recorded carbon in, at Clarence Works, 157; increase in production of, in principal countries, 445; increase of production of, 16; made from ores consumed in United Kingdom, 455; make of, in districts where no ore is raised, 455; make of, in Great Britain, in 1740, 443; manufacture, economy of labour in, 61; manufacture, recent history of, 607; maximum carbon in, 157; occurrence of sulphur in, 163; partly grey and partly white from Clarence furnaces, analyses of, 154; presence of calcium, magnesium, and aluminium in, 167; prices of, from 1782 to 1826, 16; production of cinder in making, 168; production of, up to 1788, 14; proportion of phosphorus and sulphur contained in minerals which passes into pig ir5n, 357; purer kinds, analyses of 148; quality of, as affected by minerals, used, 151; quantities of, made in Great Britain and in all other countries, 444; railway dues in manufacture of, 595; separation of phosphorus from, 3, 396; Scotch, analysis of, 316; statistics of make of, in United Kingdom for years 1860-70-80, 456; Swedish and English, analyses of, 346; tendency of iron, to combine with carbon in, 160; waste of, in making bar iron in 1727, 350; white, analyses of 150. (See also *blast furnace and cast-iron.*)

Pilot Knob, ore of, 678.
Port Washington, iron works of, 673.

Potassium, existence of, in the blast furnace, 167.
Pressure of blast, great, used in United States, 92.
Price, J., furnace of, 21; removal of phosphorus from pig iron, by, 403.
Protective duties, against British-made iron, 467; effect of, 446; effect of variation of, in Germany, 461.
Provisions and labour, connection between prices of, 553.
Provisions, American imports of, into Great Britain, 480; connection between labour and cost of, 553; cost in America, 548; cost of, America and Great Britain compared, 546; at Lake Superior, 551; expense of, in America, 552; fluctuations in prices of, 487; in America, prices of, in 1874, 549; increase in prices of, at collieries in France, 482; increased cost in Europe, Dr. Young on, 483; in France, Dr. Young on, 482; in Germany and Great Britain, 483; in France, &c., 482; in Germany and England per family, 547; in Germany and Great Britain compared, 483; in Great Britain, 551; in Switzerland, 482; occasional high prices of, 487; recent alterations in value of, 477; recent increase of prices of, in America, 549; Southern States of, 558; supplies of, to Great Britain from America, 552; transport charges on, 483. (See also *board and lodging, butcher meat, cost of living, food.*)
Prussia, development of iron trade in, 463.
Public houses, effect of increase of, 493.
Puddlers and underhands, wages of, from 1870 to 1881 in United States, 569.
Puddled bar, coal used in manufacture of, 363; quantity of pig used per ton of, 363.
Puddling, cinder at Bowling, composition and quality of, 361; different modes of, compared, 372; forge, prices of labour in different countries, 586; furnace and Bessemer converter compared, 382; furnaces and Lancashire fires compared as regards quality of products, 154; furnaces, cost of different systems of, 372; furnace, Danks', 365, 366, 395; furnaces, decrease in number of, in United Kingdom, 460; furnace, fettling for, 362; improvements in, by S. B. Rogers, 15; loss of iron in, 361; mechanical and hand, compared, 371; mechanical, of Creusot, 371; number of furnaces at work in 1873 and 1882, 377; rationale of, 360; separation of metalloids in, 360; services rendered by, 377; slags from hand and revolving furnaces, 395.

INDEX.

Puddled iron, future of, 370; great irregularity in wear of, 378; production of, in 1873 and 1882, 377. (See also *bar iron, boiler plates, puddled bar, puddling, and rails.*)
Puddling, invention of, 14; iron on sand bottoms in, 351; pig iron, labour at, 363; process, evils of, 368.
Puddled steel, 437.
Purified metal, analyses of, 404; application to manufacture of iron and steel, 405.
Purifying process, results obtained by, 401.
Rails, average life of iron and of steel, 379; basic steel, durability of, 425; content of phosphorus in samples of, 365; displacement of iron by steel in manufacture of, 459; duration of life of iron, on N.E. Railway, 378; iron and steel compared, 380; iron and steel, economical results of use of, compared, 429; iron, content of phosphorus, &c., in, 428; life of, when made by refinery process, 364; Losh's patent fish-bellied, life of, 378; manufacture of, by Siemens-Martin process, 430; old iron, use of, in making steel, 427; steel, power of, to resist abrasion, 380; weight of, laid down in United Kingdom, 430. (See also *puddling.*)
Railway bars, economy in use of coal for manufacture of, 586.
Railway rates, British and foreign, compared, 601; foreign, 600; in Great Britain, 596; in United States, low rates of, 596; on minerals used in manufacture of pig iron, 595.
Railways, in United States, extent of, 572; use of iron in, 18.
Ramsbottom, reversing engine of, 26.
Rapidity of current of reducing gas, effect on deoxidation of ore, 190.
Raw coal, condensation of ammonia and tar from, 591; use in blast furnace of, 315, 590; use of, in smelting hematite, 323. (See also *blast furnace.*)
Raw materials, choice of, for Bessemer process, 385.
Reaumur's process for making steel, 426.
Redshortness of Bessemer iron removed by Speigeleisen, 383; in puddled iron made in revolving furnace, 373.
Reducing zone, equilibrium of, established by composition of gases in, 78; shown to be of a heat-producing character, 77. (See also *blast furnace.*)
Reduction of ore, causes of acceleration of, 201; languid at low temperatures, 184.
Refined iron, analyses of, 354.
Refined metal, advantages of use of, in puddling furnace, 356; composition of, 405; puddling of, 19; puddling of, combination of principle of low hearth and Cort's furnace, 362.
Refinery, action of the, 352; change in condition of carbon on pig iron, by, 355; cinders, analyses of, 359; cost of making iron in the, 362; effect of, on quality of iron, 363; employment of, in blowing cinder pig, 358; oxygen required in, to burn coke to carbonic acid, 353; phenomena of, 19; quantity of air delivered to, 352.
Refining and Bessemer process compared, 382.
Refining fire, production of steel in a, 436.
Rent as an element in the cost of food, 539.
Re-oxidation of iron by carbonic oxide, 193.
Repeal of foreign duties, supposed effect of, 447.
Results obtained from fuel in blast furnace working compared with those of other furnaces, 141.
Reverberatory furnaces, cause of loss of heat in, 139; waste of heat in, 139.
Revolving furnace, at Creusot, working of, 371; at Monkbridge, economy of, 372; economy of, 370; quality of iron produced by, 368; slag in iron produced by, 368; Williams, E., and Tunner on, 375.
Richards, E. W., association of, with basic process, 407.
Rivet iron, by refining and mechanical puddling, 367; strength of, 367.
Rocholl's experiments on blast furnace gases, 209.
Rogers, S. B., suggested use of oxide of iron bottoms by, 351.
Rolling mills, extended application of, 25.
Royalty dues, incidence of, 608; on minerals in different countries, 602.
Running-out fires, analyses of slags from, 394; refining crude iron in, 352.
Russia, protective duties imposed in, 467.
Sand bottom for puddling furnaces, 351.
Saving by successive additions of heat to blast, 265.
Savings' banks, deposits in, 494; increase of deposits in, 494.
Scaffolding, effect of, on reducing gases, 253.
Schinz on dissociation of carbonic oxide, 189.
Schwechat furnaces, use of coke in, 298.
Scotch, bituminous coal, analysis of, 120; blast furnaces, use of raw coal in, 315; pig iron, analysis of, 316.
Scotland, use of raw coal in blast furnaces of, 120. (See also *blast furnace.*)
Scouring cinder, causes of, 161.

INDEX. 741

Shipbuilding, consumption of iron in, 472; iron, development of, 471; iron, in Norway, 706; iron, labour at, 501.
Siemens' furnace, extent of power of fuel utilised by, 139; for steel making, 21; principle of, described, 140; the probable future of, 429; value of, 425.
Siemens-Martin, process applied to rail manufacture, 430; process, oxidation of metalloids in, when steam is blown into bath, 432; process, use of in Styria, 426; steel, quality of, compared with Bessemer, 432.
Siemens, reverberatory gas furnace of, in steel-making, 426. (See also *steel*.)
Silica, almost invariably found in ironmaking materials, 161; and other earths in furnace fume possibly due to reoxidation of sodium, 168; condition of, in fume, 230; effect of presence of, in limestones, 57; expulsion of acids by, 393; extent of, in slags permitted in basic process, 408; in slags prevents removal of phosphorus, 400; power of, to neutralise oxide of iron, 397; presence of, affecting the removal of phosphorus in purifying process, 403; presence of, in blast furnace, 356. (See also *blast furnace*.)
Silicic acid, decomposition of, in blast furnace, 161.
Silicon and carbon, oxidation of, in Bessemer " blow," 391.
Silicon, behaviour of, in Bessemer converter, 359; brittleness of iron containing high percentage of, 162; difference caused in, by substitution of cold for hot blast, 150; occurrence of, in pig iron accounted for, 161; presence of, in Cleveland iron, 162; prevention of removal of phosphorus by, 399; removal of, in puddling, 402; replacement of, by phosphorus in pig for basic process, 416; separation of, in old fires, 352; useful in increasing fluidity of iron, 163; use of, in Bessemer converter, 163. (See also *blast furnace*.)
Slag, additional fuel required to melt larger quantity of, 208; analyses of, from Lancashire hearth, 346; Bessemer and blast furnace, analyses of, 392; black, produced by imperfect reduction, 200; black, proof of insufficient heat in hearth, 291; composition of, in Rhenish Prussia, 169; from basic blow, composition of, 420; from Clarence furnace, character of, 154; from Siemens' steel furnace, 431; heat units required for fusion of, 118; in iron made by revolving furnace, 368; produced by furnace making ferromanganese, 166; quantity of, in smelting a 42% ore, 118. (See also *blast furnace*.)

Slags, analyses of blast furnace and Bessemer, 394; basic matter in, 394; Bessemer acid slags, silica found in, 393; composition of, at different periods of Bessemer blow, 392; composition of basic, at end of blow, 419; from puddling furnace, 395; from running-out fires, 394; oxides of iron in, during basic blow, 418; production of, in making malleable iron, 356; weight of, compared with weight of basic steel from converter, 419. (See also *blast-furnace slags*.)
Slave labour in U.S.A., 554.
Smoke, absence of, in process of utilising waste heat of coke ovens, 51; white, from Cleveland furnaces, 147.
Snelus, G. J., on the Danks furnace, 366; use of lime in converter by, 406.
Soaking pits of John Gjers, 440, 593.
Soda waste, production of sulphide iron from, 164.
Sodium, supposed decomposition of chlorides by, 167.
Southern States of America, as competitors with Northern States in manufacture of iron, 700; as exporters of iron to foreign countries, 700; cost of living in, 554; Hewitt on, 585; ironmaking in, 681; ore and coal of, 679 694; ore, cost of conveying minerals for iron made from, 683.
Spain, cost of living in, 477; wages paid at ironworks of, 476.
Spanish hematite ores, imports of, into United Kingdom, 450.
Spathose ore, effect of carbonic oxide on, 285; resources of United Kingdom in, 450.
Specific heat, 48.
Spiegel iron, application of, to Bessemer process, 383.
Spirituous liquors, consumption of, 490.
Splint coal of Airdrie, 120.
Splitting up carbonic acid, by carbon, loss of heat by, 194; soft coke by, 196.
Static equilibrium in composition of blast furnace gases, 266.
Steam and raw coal, water-gas obtained from, 335.
Steam, effect of, in Siemens-Martin process, 433; raising of, at collieries by waste heat of coke ovens, 51.
Steel, Bessemer process for making of, 19; cast, extended use of. 26; importation of, into Great Britain, 702, 714; manufacture of, from phosphorus pig, 407; manufacture of, in low hearths, 439; necessity for, 18; ore required for manufacture of, 2; probability of, superseding iron, 451; progress of manufacture of, threatened by want of raw material, 387; Siemens'

Steel—*Continued.*
furnace in manufacture of, 20; use of, for making tin-plate bars, 348.
Steel ingots and steel rails, consumption of coal for, 593.
Steel rails, basic process applied to manufacture of in Germany, 582; consumption in Germany of, 581; cost in Germany and England compared, 580; future requirements for, 714; quotations of, 583; unprofitable nature of export trade in Germany of, 581.
Steel rail trade, present extent of, 713; present position of, 712.
Stour works, make of raw iron at, in 1727, 348.
Stoves of fire-brick, errors in estimating value of, 113.
Strength, comparative, of hot and cold blast iron, 150; of cast iron, effect of strain and temperature on, 152.
Strikes in anthracite mines, 556.
Stückofen, 11.
Styria and Carinthia, occurrence of carbonic acid in making white iron in, 201.
Styrian charcoal furnaces compared with Swedish, 302.
Sulphur, affinity of iron and lime for, 164; and phosphorus, quantity of, in pig and slag, 357; content of, in Durham coke, 104; counteracting of evils of, by lime, 103; effect of, on pig iron, 163; effect of presence of, in coke, 103; found more largely in white than in grey iron, 164; occurrence of, in pig iron, 163; prevents separation of graphitic carbon in pig iron, 103. (See also *blast furnace.*)
Superheated air, advantage of, 90, 92, 118; and charcoal, 298; examples of working with, 107.
Superiority of blast furnace to any other used for iron-making, 140.
Sweden, low temperature of blast in, 283; performances of 27 charcoal furnaces in, 130; wages paid at ironworks of, 477.
Swedish charcoal, furnaces duty performed by, 278; furnaces, consumption of charcoal in, 274; furnaces, low temperature of blast at, 132, 283; iron, analyses of, 345; iron, early use of, in Bessemer process, 383; iron, high price of, 347; method of making malleable iron by, advantages of, 346; works making Bessemer iron, particulars of, 277; works making grey forge iron, results at, 276.
Switzerland, cost of provisions in, 482.
Tar and ammonia, recovery of, from furnace gases, 314, 326.
Temperature, and condition of air, effect of, on combustion of coke, 238; at which carbon and carbonic oxide act on peroxide of iron, 71; at which carbonic acid begins to act on iron and carbon, 71; effect of differences in, on quality of metal, 158; effect of, in blast, on certain substances found in pig iron, 150; effect of increase of, on behaviour of sulphur in pig iron, 149; effect of, in removal of metalloids, 400, 402; effect of, in removing phosphorus, 397; effect of, on phosphorus in pig iron, 149; effect of, on strength of cast iron, 152; high, effect of, on coke and charcoal, 289; high, effect of exposure, cast iron to, on condition of carbon, 159; high, effect of, on reduction of silicon, 162; high, in Bessemer converter, maintained by silicon, 163; in basic process, 422; increased rate of, towards hearth in charcoal furnaces, 295; influence of, in removal of phosphorus in puddling, 403; low, in blast furnace, effect of, on sulphur, 164; low, of blast, in Sweden, 283; mode of ascertaining, in charcoal furnace, 295; most favourable for carbon deposition, 190; of basic blow, 425; of blast and capacity of furnace, tables showing results effected by, 244; of blast, each successive addition to, not attended with same economy, 87; of blast in Vordernberg furnace, 283; of blast, limit to which it can be raised, 266; of blast, limit of profitable increase of, 86; of blast, limits of useful elevation of, 124; of blast, value of successive additions to, 249; of different levels of blast furnaces, 203; of escaping gases, changes in, before and after charging furnace, 180; of escaping gases, effect of by a round of materials on, 180; of escaping gases, in determining action of mixtures of carbonic oxide and carbonic acid in reduction of ore, 185; of gases determined by chemical action in upper zone of furnace, 99; of gases, increase of, while discontinuing charging, 178; of materials filling two furnaces, diagram of, 203; of water-gas as generated from coke, 332; required for deoxidation of Cleveland ore, 191; required for oxidation of iron by carbonic acid, 184. (See also *blast furnace.*)
Thickness of the Cleveland ironstone, 119.
Thomas and Gilchrist, use of lime lining by, 406.
Timber as a source of heat in the blast furnace, 134.
Tin-plate bars, iron used for, in Great Britain, 348.
Titanium, presence of, in iron, 167.
Tooth's revolving furnace, 365.

Transport charges, America, to, on pig iron, 604; cheap instance of, by water in the United States, 595; coal and ore, on, 693; increase of, and effect on American agriculture, 608; in Southern States of America, 694; of iron ore in United States, 473; on iron from imported ores, 691; on iron from Iron Mountain of Missouri, 692; on iron from Lake Champlain ore, 691; on iron from Lake Superior ore, 690; on iron from New Jersey, 691; on railways, British and Foreign, compared, 601; United Kingdom and United States compared, 473; wheat, on, from United States to Liverpool, 541. (See also *railway rates*.)

Tudhoe refined iron, analyses of, 355.

Tunner, examination of Wrbna furnace by, 293.

Tuyeres, blowing-in fuel for smelting iron at, 137; carbon burnt at, in blast furnace, 68; composition of gases at, 177; decomposition of water at, 172; disappearance of carbonic acid at, 200; effect of derangement at, 168; effect of increase of temperature at, 104; heat evolved at, 162; increase of oxygen and carbon at, 211; injecting gas at the, 137; quantity of oxygen which should be found at, 214; small quantity of carbonic acid at the, 200. (See also *blast furnace*.)

United Kingdom, consumption of drink in, 492; exports of iron from, and imports into, 707.

United States, as an iron exporting centre, 472, 695; Bessemer works in, 571; blast furnace labour in, 562; cheap conveyance by water in, 595; colliery labour in, 558; costs of iron in, 689; cost of iron ores in, 472; cost of labour and cost of food, connection between, 553; earnings in, 565; expense of provisions in, 552; export of food from, 553; farming in. 540; farming in, Clare Sewell Read thereon, 541; finishing mills, wages at, 509; increased production of iron in, 446; iron ore, resources of, 472; labour in, 539; living and labour, cost of, 547; malleable iron labour at. 566; mechanics in, compared with England, 559; ore mines in. compared with England, 557; position of iron trade of, 468; provisions from, 480; puddling, cost of, in, 566; railways in, 572; Weeks, J. D., on labour in, 575; wheat, cost of growing in, 544; wheat freight into England from, 542; wheat in, Read and Pell on, 541. (See also *America*.)

Utilisation of waste heat of coke ovens in raising steam, 51.

Value of products recovered from coking coal, 328.

Variations in the conditions of making pig iron, 262.

Vaughan, John, improvements in blast furnaces, by, 23.

Volatile constituents should be expelled in upper part of blast furnace, 46.

Vordernberg furnaces, temperature of blast at, 283.

Wages, agricultural, 475; agricultural, determine to some extent wages in manufactures, 477; agricultural, on Mediterranean, 476; blast furnaces, Cleveland and Continent of Europe compared, 521; blast furnaces, England and United States compared, 562; effect of gold discoveries on, 497; effect of high, on iron trade, 464; effect of improvement in, 497; high, connection of, with drinking habits, 491; in anthracite mine of America, 555; in coal work, America and Durham, 558; in Connellsville coal region, etc., 558; in mining ore, in England and America, 559; mechanics', in England, 559; mechanics', in United States, 560; mechanics', England and United States compared, 561; of mechanics, British and foreign compared, 499; of mechanics, increase of, 497; paid at Indian tea plantations and blast furnaces, 476; paid at Spanish ironworks, 476; paid at works in Austria and Sweden, 477; purchasing power of, 479.

Wales, South, experience of fire-brick stoves in, 257; waste of iron in puddling in, 363.

Walloon fires, 345.

Waste in basic and acid Bessemer processes compared, 412; in combustion of raw coal, percentage loss of heat evolved by, 139; of heat, avoidance of, 247.

Water, amount of, in furnaces using raw coal, 324; decomposition of, in hearth of blast furnace, 322; effect of, in blast furnace, 172; evaporation of, in best marine engine, 138; quantity condensed at gas works per ton of coal, 314; uncombined effect of presence of, in blast furnace, 236.

Water-gas, analysis of. 333; as obtained from steam and raw coal, 335; carbon required for generation of, 331; Dawes trials of, in blast furnace, 329; generated from coke, temperature of, 332; heat from, compared with that from raw coal, 335; most recent mode of manufacture of, 330; product obtained from coal, 334; production of at Essen, 329; proposed as a fuel for smelting iron, 329; proposed use in blast furnace. 593; pure, heat from

Water-gas—*Continued.*
 combustion of, 334; reducing power and use of, in blast furnace, 337; temperature of combustion of, 336; trial of, in Staffordshire, 329; unsuited for reducing iron in blast furnace, 341; use of, in blast furnace with coke, 340. (See also *blast furnace.*)
Water-jacket for puddling furnaces, 371.
Weeks, J. D., at Middlesbrough, 699; English and American labour compared by, 575; on cost of farm labour in U.S.A., 544.
Weight of coke required per ton of pig, 262.
Welding properties of puddled iron, 381.
West coast of England, experience attending use of superheated air on, 256.
West Cumberland and Lancashire, iron ores of, 458.
West Yorkshire iron, analyses of, 434.
Wheat, cost of carriage to Liverpool from United States, 541; cost of, in reign of Charles II., 478; decennial average prices of, 481; importations of, 481; Read and Pell, cost of growing of, in U.S.A., 541.
White cinder, oxide of iron in, 291.
White iron, cause of production of, 161; combined carbon usually most largely found in, 158; contains more sulphur than grey, 164; due to imperfect furnace working, 161; make of, in charcoal furnace, 284; make of, in Cleveland furnaces, 285; occurrence of carbonic acid in production of, 201; presence of sulphur in, 417; produced by contact of oxide of iron with grey cast iron, 159; use of, in basic process,417. (See also *blast furnace.*)
White smoke, emission of, from furnace cinder, 174; from slag, composition of, 174.
Williams, E., on importance of shape of furnace, 268; on make of malleable iron in North of England, 459; on revolving furnace, 539.

Williams, Price, on average life of rails, 379.
Wilson, Alexander, plan of avoiding in heating of ingots, 439.
Wood, annual production of, from acre of ground, 135; dry, composition of, 135; preliminary treatment of, 53; use of, in blast furnaces, 134; yield of charcoal obtained from, 53.
Woods of United States, yield of timber from, 53.
Wootz steel, 436.
Work at different furnaces, 250.
Working of Scotch blast furnaces, particulars of, 315.
Working of two furnaces blown with superheated air, 107.
Workmen engaged in iron trade, position of, 709; English and German, compared, 564; Foreign and British, compared, 484; number of, at blast furnaces in different countries, 565.
Wrbna furnace, examination of, by Tunner, 293.
Wylam furnaces, phenomena observed at, 157.
Y-Barra, direct process practised by, 33.
Yorkshire West, iron, analyses of, 345; high quality of. 153.
Young, Dr., on cost of living in Germany, 485; on drinking habits of workmen, 490; on labour and cost of living in Europe and U.S.A. compared, 482; on labour at farming in U.S.A., 545; on workmen's expenditure in Germany, 547.
Zinc metallic, behaviour of, in contact with carbonic acid, 227; collection of, in blast furnace, 228; in furnace fume, 226. (See also *blast furnace.*)
Zoll Verein, export of steel rails from, 580.
Zones of reduction, of heat interception and of fusion, 72.